Human Biology

Human Biology

Third Edition

Sylvia S. Mader

WCB Wm. C. Brown Publishers

Book Team

Editor *Kevin Kane*
Developmental Editor *Carol Mills*
Production Editor *Renée A. Menne*
Designer *David C. Lansdon*
Art Editor *Miriam J. Hoffman*
Photo Editor *Carrie Burger*
Permissions Editor *Vicki Krug*
Visuals Processor *Andréa Lopez-Meyer*
Visuals/Design Consultant *Marilyn Phelps*

 Wm. C. Brown Publishers

President *G. Franklin Lewis*
Vice President, Publisher *George Wm. Bergquist*
Vice President, Operations and Production *Beverly Kolz*
National Sales Manager *Virginia S. Moffat*
Group Sales Manager *Vincent R. Di Blasi*
Vice President, Editor in Chief *Edward G. Jaffe*
Marketing Manager *Paul Ducham*
Advertising Manager *Amy Schmitz*
Managing Editor, Production *Colleen A. Yonda*
Manager of Visuals and Design *Faye M. Schilling*
Production Editorial Manager *Julie A. Kennedy*
Production Editorial Manager *Ann Fuerste*
Publishing Services Manager *Karen J. Slaght*

WCB Group

President and Chief Executive Officer *Mark C. Falb*
Chairman of the Board *Wm. C. Brown*

Cover photos: (Embryo) © Omikron/Photo Researchers, Inc.; (Mother and Child) © Smith & Garner/The Stock Market

Copyedited by Barbara R. Day

The credits section for this book begins on page 490, and is considered an extension of the copyright page.

Library of Congress Catalog Card Number: 90-85165

ISBN 0-697-12333-2 (paper)
ISBN 0-697-13837-2 (case)

Printed in the United States of America by Wm. C. Brown Publishers, 2460 Kerper Boulevard, Dubuque, IA 52001

10   9   8   7   6   5

For my family

Brief Contents

Contents

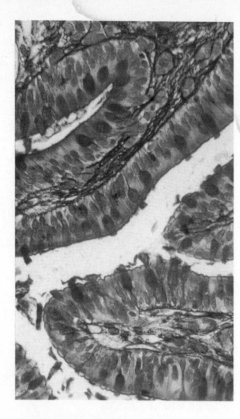

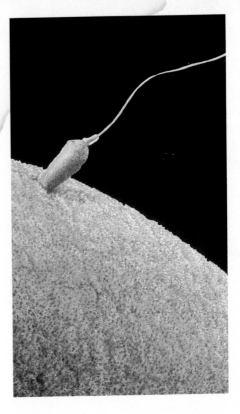

Preface

Human Biology is suitable for use in one-semester biology courses that emphasize human physiology and the role that humans play in the biosphere. All students should leave college with a firm grasp of how their bodies normally function and how the human population can become more fully integrated into the biosphere. This knowledge can be applied daily and helps assure our continued survival as individuals and as a species.

The application of biological principles to practical human concerns is a relatively new approach for instructional courses in biology. It has gained even wider acceptance because it fulfills a great need. Human beings are frequently called upon to make decisions about their bodies and their environment. Wise decisions require adequate knowledge.

The third edition of *Human Biology* has the same style and organization as the previous edition. Each chapter presents the topic clearly, simply, and distinctly so that the student will feel capable of achieving an adult level of understanding. Detailed, high-level scientific data and terminology are not included because I believe that true knowledge consists of working concepts rather than technical facility.

Students and instructors alike will find this text stimulating and a pleasure to read and study especially since the illustration program has been completely revamped and there are many new four-color illustrations. As before, however, the pedagogy is extremely strong and instructors will be pleased to know that there are now critical thinking case studies called What's Your Diagnosis? at the end of some chapters. Students are asked to use the knowledge they have acquired from the chapter to diagnose an illness. Their diagnosis must be substantiated by logical thinking and presentation of their ideas. Also, there are critical thinking questions at the end of

each chapter. These ask students to apply their knowledge to new and different situations.

Organization of the Text

An introductory chapter precedes the 23 chapters of *Human Biology* which are grouped in six parts.

Introduction: A Human Perspective

The introductory chapter lays a foundation and provides a rationale for the text as a whole. It discusses the biological characteristics of humans, including our cultural heritage which separates us from other living things and makes it difficult for us to understand that we are part of the biosphere. This chapter also defines and explains the scientific method.

Part One: Human Organization

Part One presents principles needed for the parts that follow. It includes chemistry, cell structure and function, and body organization. Homeostasis, which is introduced in this part, is a recurring theme in the chapters that follow.

Part Two: Processing and Transporting in Humans

Part Two covers those systems of the human body that could be described as vegetative, and tells how they contribute to homeostasis.

There is a new chapter in this section called "The Lymphatic System and Immunity." Both the general and specific defenses of the body are discussed in this newly written chapter.

The chapter on excretion was completely rewritten to increase student understanding. A new illustration that explains nephron anatomy now precedes the discussion of urine formation.

Part Three: Integration and Coordination in Humans

The control of homeostasis by the nervous and hormonal systems is found in Part Three. It also includes the musculo-skeletal system and sense organs.

There are several new illustrations in this part, all of which will make it easier for instructors to teach and for students to learn.

Part Four: Human Reproduction

Part Four deals with human reproduction and development. A new reading on circumcision addresses the recent concerns about this procedure. Updated and expanded information about AIDS, a new supplement for the text, appears in this part.

The development chapter now contains the most recent findings regarding differentiation and morphogenesis, still within the context of human development.

Part Five: Human Genetics

The transference of traits from one generation to the next and associated topics are considered in Part Five.

This part has been reorganized. The chapter "Genes and Medical Genetics" has been revised to give the most recent information about human genetic disorders. The chapter "DNA and Biotechnology" has also been updated and includes a section on the human genome project. There is now a chapter about cancer, and this replaces the cancer supplement which appeared in previous editions.

Part Six: Human Evolution and Ecology

The first chapter in this part, which concerns evolution, has an extended section on human evolution. The last two chapters are about ecology.

The ecology chapters have been reorganized. The first ecology chapter explains the principles of chemical cycling and energy flow through an ecosystem. The second ecology chapter considers population growth and modern ecological concerns such as global warming, tropical rain forest destruction, and the growing ozone depletion.

To the Student

Human Biology includes a number of aids that will help you study biology successfully and enjoyably.

Text Introduction

The introduction discusses the characteristics of humans and presents an overview of the book. It outlines the biological principles that will be important to understanding the text and also shows you the possibility of using these principles to make bioethical decisions.

Part Introductions

An introduction to each part highlights the central ideas of that part and specifically tells you how the topics within each part contribute to biological knowledge.

Chapter Concepts

Each chapter begins with a list of concepts stressed in the chapter. This list introduces you to the chapter by organizing its content into a few meaningful sentences. The concepts provide a framework for the content of each chapter.

Boldfaced Words

Terms that are pertinent to the topic being discussed appear in boldfaced print. The first time these terms appear, they are defined in context. All boldfaced terms are defined in the glossary. Each entry in the glossary is accompanied by its phonetic pronunciation.

Tables and Illustrations

Numerous tables and illustrations appear in each chapter and are placed near their related textual discussion. The tables clarify complex ideas and summarize sections of the narrative. Once you have achieved an understanding of the subject matter by examining the chapter concepts and the text, these tables can be used as an important review tool. The photographs and drawings have been carefully chosen and designed to help you visualize structures and processes.

Boxed Readings

Two types of boxed readings are included in the text. Readings chosen from popular magazines illustrate the applications of concepts to modern concerns. These spark interest by illustrating that biology is an important part of everyday life. The second type of reading is designed to expand, in an interesting way, on the core information presented in each chapter. Topics pertaining to human concerns are addressed in these readings.

In-Chapter Summaries

Summary statements are placed strategically within the body of each chapter. These give you periodic reinforcement of the information being presented. Such statements are highlighted for easy identification.

Human Issue Boxes

Human Issue Boxes are included in every chapter throughout the text. These general discussion boxes are designed to stimulate interest and thought, especially about how the chapter topics can be applied to human concerns. New topics have been included such as the advisability of fetal surgery and research; the rights of the individual to his or her own tissues; and animal rights.

What's Your Diagnosis?

New to this edition are critical thinking case studies called What's Your Diagnosis? that appear at the end of some chapters. You are asked to use the information in the chapter to diagnose an illness. You are then required to use critical thinking to substantiate and present your diagnosis.

Chapter Summaries

Chapter summaries offer a concise review of material in each chapter. You may read them before beginning the chapter to preview the topics of importance, and you may also use them to refresh your memories after you have a firm grasp of the concepts presented in each chapter.

Chapter Questions

Study questions, objective questions, and critical thinking questions are at the close of each chapter. The study questions allow you to test your understanding of the information in the chapter. The objective questions allow you to quiz yourself with short fill-in-the-blank questions. New to this edition is the frequent use of labeling exercises in the objective questions section. Critical thinking questions require you to use the information presented in the chapter just studied and in previous chapters in a creative manner.

Chapter Glossaries

Selected key terms are listed with their phonetic pronunciations, definitions, and page references in the key-term glossary at the end of each chapter. All boldfaced terms are still listed alphabetically with their pronunciations, definitions, and page references in the text glossary at the end of the book.

AIDS Supplement

All material about AIDS in the previous edition has been consolidated. The new AIDS Supplement also includes expanded and current information on this topic. It appears in Part Four: Human Reproduction.

Further Readings

For those of you who would like more information about a particular topic or who are seeking references for a research paper, each part ends with a listing of articles and books to help you get started. Usually the entries are *Scientific American* articles and specialty books that expand on the topics covered in the chapter.

Appendixes and Glossary

The appendixes contain optional information for your referral. Appendix A is an expanded Periodic Table of the Elements; Appendix B discusses the metric system.

The text glossary defines the terms most necessary for making the study of biology successful. By using this tool, you can review the definitions of frequently used terms.

Additional Aids

Instructor's Manual/Test Item File

The Instructor's Manual/Test Item File revised by Andy Anderson is designed to assist instructors as they plan and prepare for classes using Human Biology. Possible course organizations for both semester and quarter systems are suggested, along with alternate sequencing of chapters. An extended lecture outline is provided for each chapter as well as comments on the Human Issue boxes. Answers to the Critical Thinking Questions

and What's Your Diagnosis? case studies are also included.

Approximately 50 objective test questions and several essay questions are provided for each chapter. A list of suggested audiovisuals for the various topics and a list of suppliers are included at the end of the Instructor's Manual.

Student Study Guide

To ensure close coordination with the text, the author has written the *Student Study Guide* that accompanies this text. Each text chapter has a corresponding study guide chapter that includes a listing of behavioral objectives, study exercises, and a chapter test. Answers to study guide questions are provided to give students immediate feedback.

Laboratory Manual

The author has also written the *Laboratory Manual* that accompanies *Human Biology*. With few exceptions, each chapter in the text has an accompanying laboratory exercise in the manual (some chapters have more than one accompanying exercise). In this way, instructors will be better able to emphasize particular portions of the curriculum if they wish. The twenty laboratory sessions in the manual are designed to further help students appreciate the scientific method and to learn the fundamental concepts of biology and the specific content of each chapter. All exercises have been tested for student interest, preparation time, and feasibility.

Laboratory Resource Guide

More extensive information regarding preparation is found in the *Laboratory Resource Guide*. The guide includes suggested sources for materials and supplies, directions for making up solutions and otherwise setting up the laboratory, expected results for the exercises, and suggested answers to all questions in the *Laboratory Manual*. It is available for free to all adopters of the *Laboratory Manual*.

Transparencies and Lecture Enrichment Kit

A set of 125 transparency acetates also accompanies the text. Most feature key illustrations from the text and are in full color. They are accompanied by a Lecture Enrichment Kit, which provides additional high-interest information not presented in the text.

WCB TestPak with Enhanced QuizPak and GradePak

WCB TestPak, a computerized testing service, provides instructions with either a mail-in/call-in testing program or the complete test item file on diskette for use with the IBM PC, Apple, or Macintosh computer. WCB TestPak requires no programming experience.

WCB QuizPak, a part of TestPak, provides students with true/false, multiple choice, and matching questions for each chapter in the text. Using this portion of the program will help students to prepare for examinations. Also included with the WCB QuizPak is an on-line testing option to allow professors to prepare tests for students to take on the computer. The computer will automatically grade the test and update a gradebook file.

WCB GradePak, also a part of TestPak, is a computerized grade management system for instructors. This program tracks student performance on examinations and assignments. It will compute each student's percentage and corresponding letter grade, as well as the class average. Printouts can be made utilizing both text and graphics.

The Gundy-Weber Knowledge Map of the Human Body by G. Craig Gundy, Weber State University

This thirteen disk, Mac-Hypercard program is for use by instructors and students alike. It features computer graphics, animations, labeling exercises, self-tests, and practice questions to help students examine the systems of the human body. The knowledge maps 1–5 will be available by August, 1991. The entire set will be available by August, 1992. Contact your local Wm. C. Brown sales representative or call 1–800–351–7671.

The Knowledge Map Diagrams

1 Introduction, Tissues, Integument System
   (0–697–13255–2)

2 Viruses, Bacteria, Eukaryotic Cells
   (0–697–13257–9)

3 Skeletal System
   (0–697–13258–7)

4 Muscle System
   (0–697–13259–5)

5 Nervous System
   (0–697–13260–9)

6 Special Senses
   (0–697–13261–7)

7 Endocrine System
   (0–697–13262–5)

8 Blood and the Lymphatic System
   (0–697–13263–3)

9 Cardiovascular System
   (0–697–13264–1)

10 Respiratory System
    (0–697–13265–X)

11 Digestive System
    (0–697–13266–8)

12 Urinary System
    (0–697–13267–6)

13 Reproductive System
    (0–697–13268–4)

Demo–(0–697–13256–0)

Complete Package–
(0–697–13269–2)

Acknowledgments

The personnel at Wm. C. Brown Publishers are due many thanks for their help in developing and producing this edition of *Human Biology*. Kevin Kane, biology editor, not only oversaw the development of the manuscript, but also made important production decisions. He was assisted by Carol Mills, developmental editor, who helped on a more daily basis. Renee Menne was the production editor who coordinated the efforts of many and who smoothly guided the book through all stages of production.

A number of people contributed to the fine illustration program. Carlyn Iverson did the many new rendered drawings; the beauty of her work gives a new dimension to the text. Other drawings were provided by Cris Creek and Kathleen Hagelston. Miriam Hoffman was the art editor who cheerfully oversaw the preparation of the drawings and Carrie Burger was the photo researcher who found just the right photographs. Vicki Krug was the permissions editor who patiently made sure the credits were correct. Each illustration was processed by Andréa Lopez-Meyer.

David Lansdon designed the book with skill. He put all the pieces—illustrations, tables, readings, text—together into an attractive whole. It was David who gave us a book several pages shorter than the previous edition!

Many instructors have contributed to this revision of *Human Biology*. I especially want to thank Jay Templin of Widener University who wrote the "What's Your Diagnosis" boxes; he made them not only instructional but also fun. Other instructors commented on the entire or a portion of the text. I am extremely thankful to each one, for we have all worked diligently to remain true to our calling and to provide a product that will be the most useful to our students. In particular, it seems proper to acknowledge the help of the following individuals:

Donald Mykles
Colorado State University

Kathleen S. Morgan
Rutgers University

Roberta Williams
University of Nevada, Las Vegas

D. Andy Anderson
Utah State University

D. J. Burks
Wilmington College

Vic Chow
City College of San Francisco

Sheldon R. Gordon
Oakland University

Laszlo Hanzely
Northern Illinois University

Keith Knutson
St. Cloud State University

Cran Lucas
LSU–Shreveport

Joyce B. Maxwell
California State University– Northridge

John McCue
St. Cloud State University

Sally Sapp Olson
Winthrop College

Raymond R. White
City College of San Francisco

Finally, I wish to express appreciation to my family for their constant support. My children, Karen and Eric, and my sister, Rhetta, were always ready to offer advice and encouragement.

Sylvia S. Mader

Introduction

A Human

Perspective

Introductory Concepts

1 Human beings have characteristics in common with all living things.

2 The scientific method is the process by which scientists gather information about the material world.

3 All persons have to be prepared to use scientific information to make value judgments.

Human beings share the environment with other living things.

Figure I.1   Human beings, regardless of race, share certain characteristics, many of which are also typical of living things in general.

This book has two primary functions. The first is to explore human anatomy and physiology so that you will know how the body functions (fig. I.1). The second is to take a look at human evolution and ecology in order to understand the place of humans in nature. Both the human body and the environment are self-regulating systems that can be thrown out of kilter by misuse and mismanagement. An appreciation of the delicate balance present in both systems provides the perspective by which future decisions can be made.

Characteristics of Human Beings

You are about to launch on a study of human biology. Before you begin, it is appropriate to discuss the biological characteristics of human beings. The following have been singled out as being especially pertinent.

Human beings are a product of the evolutionary process. Life has a history that began with the evolution of the first cell(s) about 3.5 billion years ago. Thereafter, living things became increasingly complex. Figure I.2 only shows the evolutionary history of vertebrate animals. *Human beings are vertebrate animals* because they have a dorsal hollow nerve cord protected by vertebrae. The repeating units of the vertebrae indicate that we are segmented. Segmentation leads to specialization of parts as is well exemplified in humans (fig. I.2).

Today most biologists classify living things into the five kingdoms or groups noted in figure I.3. Because *human beings are related to other living things,* they contain the same type of chemicals, like DNA, the substance of genes, and ATP, the

energy currency of cells. It's even possible to do research with bacteria (kingdom Monera) and have the results apply to humans. For example, it is common practice today to test food additives first by applying the chemicals to bacterial culture plates. If the chemical causes mutations in bacteria, it is considered unsafe for human consumption.

Human beings reproduce. Reproduction is that part of the human life cycle that assures continuance of the human species. It also permits evolution to occur. It is possible to trace human ancestry through a series of prehistoric ancestors until modern-day humans finally evolved. Human beings are distinguishable from their closest relatives, the apes, by their highly developed brains, completely upright stance, and the power of creative language.

When human beings reproduce, they pass on their organization to their offspring. The sperm and egg contain chromosomes contributed by each parent, and these chromosomes carry the blueprint of life that is essential to the offspring. While evidence is strong that genes on chromosomes control our physical traits, we are less certain and less willing to entertain the belief that genes also control our behavior.

Human beings are highly organized. The organs depicted in figure I.2 are composed of tissues that contain cells of a particular type. In order to maintain this organization, living things must take chemicals and energy from the environment. Only plants and plantlike organisms are capable of utilizing inorganic chemicals and energy of the sun in the process of making their food. Other organisms, such as humans, must take in preformed food as a source of chemicals and energy.

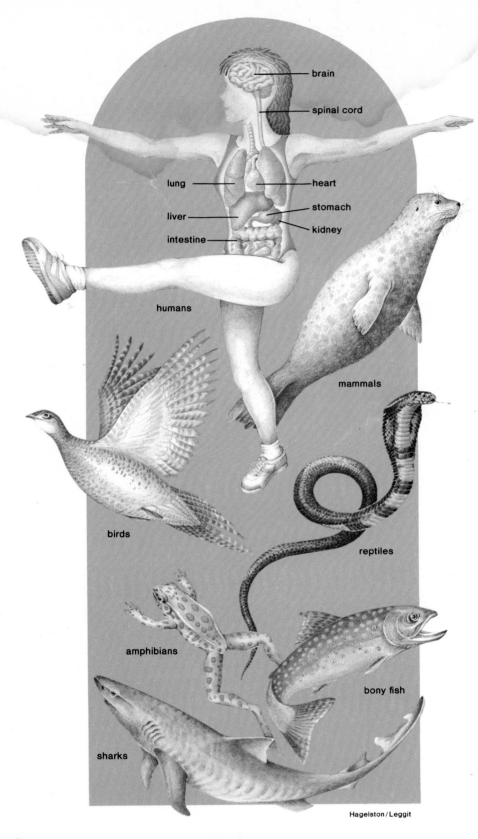

brain

spinal cord

lung

heart

stomach

liver

kidney

intestine

humans

mammals

birds

reptiles

amphibians

bony fish

sharks

Hagelston / Leggit

Figure I.2   Human beings, like all living things, have an evolutionary history. In the case of humans, this means that their organ systems are similar to those of other vertebrates. The evolutionary tree of life has many branches; the vertebrate line of descent is just one of many.

Kingdom	Representative Organisms	Description
Monera (monerans)	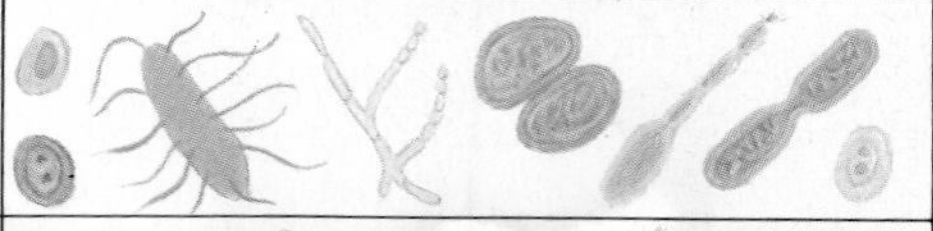	bacteria, including cyanobacteria
Protista (protists)	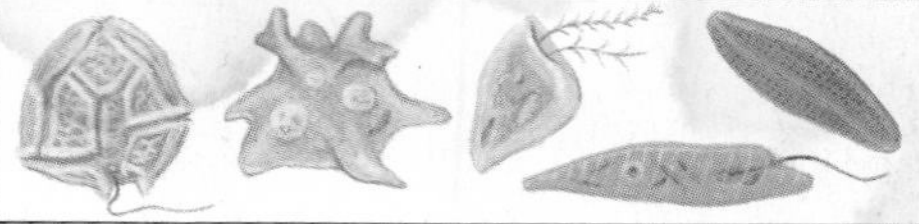	protozoans, all algae, and slime molds
Fungi (fungi)		molds and mushrooms
Plantae (plants)		mosses, ferns, various trees, and flowering plants
Animalia (animals)	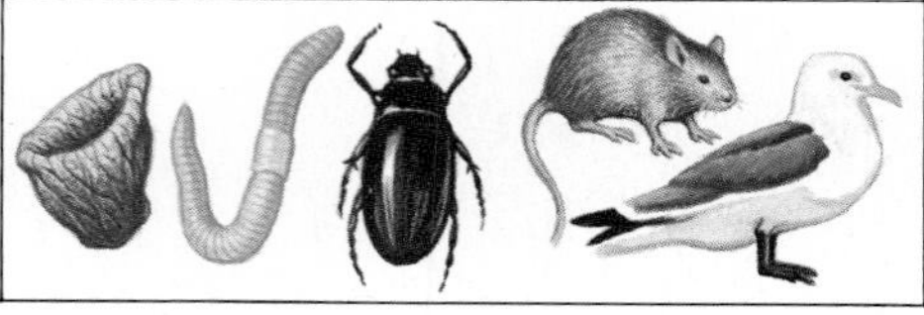	sponges, worms, insects, fishes, amphibians, reptiles, birds, and mammals

Figure I.3   All living things are placed in one of these five kingdoms, the major categories of classification. Humans are mammals in the animal kingdom. All mammals have hair and mammary glands.

Human beings need an internal environment that remains fairly constant. The body's organ systems work together to maintain a homeostasis in which bodily activities fluctuate minimally. Certain of the systems are directly concerned with maintenance (digestive, circulatory, immune, respiratory, and excretory), while others are concerned with coordination of these systems and interaction with the environment (nervous, sensory, muscular and skeletal, and hormonal). The ability to respond to the environment is a unique ability of living things.

Humans are also a product of a cultural heritage. We are born without knowledge of civilized ways of behavior, and we gradually acquire these by adult instruction and imitation of role models. Unfortunately, it is our cultural inheritance that separates us from nature and makes it difficult for us to see our dependence on the natural world. On the contrary, our industrialized society has an extreme ability to exploit and alter the environment (fig. I.4) for its own purposes. The end result is often a degradation of the environment that is harmful to all living things, including human beings. In recent years human beings have become aware of their destructive influence and are seeking ways to work with nature rather than against nature. The desire to exploit nature is being replaced by a concern to be wise managers of nature.

Like all living things, humans are a product of the evolutionary process. They reproduce, are highly organized, and maintain a dynamic constancy of the internal environment. Unlike other living things, humans are also a product of a cultural inheritance.

Human Issue

Scientists frequently hold differing opinions about the same social issue. For example, there are some scientists who believe that the world is not overpopulated and population control is not necessary, while there are other scientists who believe that the world is overpopulated and we should do all we can to control population growth. How would you decide which group of scientists is correct? Are value judgments made by scientists any more valid than those made by nonscientists? Who should be involved in making social value judgments—only scientists, only nonscientists, or both?

a.

b.

Figure I.4   Natural area versus area developed by humans. Humans
need both types of places. *a.* They often benefit psychologically from
visiting natural areas that provide a home for many plants and animals.
b. Humans carry on most of their activities in developed areas.

. . . Us and the Apes

Most agree that the anatomical differences between us and the apes stem from our habit of walking upright all the time. Apes tend to knuckle-walk, with hands bent, while on the ground. In keeping with walking on all fours, the ape pelvis is long and narrow and the hands and feet tend to resemble each other. In humans the shorter pelvis is designed to bear the weight of the trunk, and it also better serves as a place of attachment for the muscles of the longer and more muscular legs. The human foot is markedly different from the hand; it supports the body on a broad heel and a thick-skinned ball cushioned by fat. Still, the foot has a springy arch needed for our striding gait.

The upright stance of humans frees the hands so that they are ready to do manipulative work. Although both apes and humans have a thumb that is "opposable" in the sense that it can reach over and touch the fingers, the human thumb is longer and rotates. It is this remarkable thumb that allows us to be such great tool users. Tool use and the human brain are believed to have evolved together, each one promoting an increase in the other:

tool use ↔ intelligence

Apes use tools but they don't make highly specialized tools for specific purposes. They also don't carry tools about with them for when they might be needed.

The human brain really sets us apart from the apes. If an ape were the size of a human, its brain would still be only ⅓ as large. Not surprising is the fact that the lobes of the brain needed for thinking and language are proportionately larger in humans also. Apes make faces and sounds that serve to communicate with others, but only humans are capable of communicating many different types of ideas. It's possible that human intelligence and language evolved as a means to understand and function within a complex social system.

Because the large brain of humans develops after birth, infants are born in an immature state and are dependent upon their mothers for quite some time. Perhaps this caused the women to "stay home" while the men went out to hunt for food that they later brought home to share with the other members of their group. Again, apes form social groups but an organized system of food sharing is not characteristic of the group.

Learning how to think and use tools within a social group is the mainstay of human culture. Today we humans are enveloped by our culture; most of us live in cities, existing on food and using materials grown or made far away. We no longer have a sense of our evolutionary past nor of our place in nature. We don't realize that even though we can control our environment, to a large extent we are still dependent upon natural ecological systems. Consider, for example, our dependence on the rains to bring us fresh water, on plants to supply us with food and to purify the air, and on the past remains of plants and animals to give us a supply of energy. Only when we fully appreciate our place in nature will it be possible for us to wisely guide our future destiny. If we greatly interfere with the cycles of nature, as when we overdam rivers, cut down forests, or burn fossil fuels to excess, it can only be to our detriment.

Was it the human hand that led to evolution of culture? *a.* Apes use their hands primarily to climb trees. *b.* They live in social groups where offspring are cared for. *c.* Humans use their hands primarily to manipulate tools. *d.* They live in a society where children learn the culture of that society.

Hagelston/Leggitt

a.

c.

b.

d.

Science

Scientific Method

Science is a process resulting in a body of knowledge that allows us to understand and manipulate the material world. Scientists often employ a methodological approach such as is described in figure I.5. For example, suppose physiologists wanted to determine if sweetener S was a safe additive for foods. First they would study any *previous data,* or objective information, collected on the subject. Then on the basis of this information, they might *hypothesize* that sweetener S would be safe at a low concentration but unsafe at a high concentration. Next, the physiologists would design an experiment in order to collect new data. They might decide to feed sweetener S to groups of mice at ever greater concentrations.

Group 1: diet contains no sweetener S
Group 2: 5% of diet contains sweetener S
Group 3: 10% of diet contains sweetener S
$\downarrow$
Group 11: 50% of diet contains sweetener S

No doubt the physiologists would keep the different groups of mice in separate cages in the laboratory. The laboratory setting is much preferred for scientific experiments because it is here that all aspects of the experiment can be manipulated. For example, all the groups of mice would be kept under the same environmental conditions and fed the same diet except for the amounts of sweetener S. Notice that one group is fed no sweetener S at all. This is the *control group,* the sample that goes through all the steps of the experiment except the one being tested. Use of a control group, in which no effect is expected, gives greater validity to the results of the experiment (fig I.6).

The hypothetical results of the sweetener S study are given in figure I.7. On the basis of this new data, the physiologists might *conclude* that mice fed an even greater amount of sweetener S above 20% are more likely to develop bladder cancer at an ever-increasing rate. Scientists often prefer such mathematical data because it is objective and easily evaluated.

Any experiment must be repeatable. Other scientists using the same design and carrying out the experiment under the same conditions are expected to get the same results. If they do not, the original experiment is called into question.

In order to make the experimental design and results available to the scientific community, scientists often publish their results in scientific journals, where they may be read and studied by all those interested. It's also possible that scientists might go on to design and do other experiments with sweetener S.

Human Issue

You are constantly exposed to all kinds of information. Information comes to you from a variety of sources, such as television, newspapers, magazines, the *Farmer's Almanac,* and horoscopes. Even as you read this textbook, you are encountering more information! In your opinion, is there anything special and unique about scientific information based on scientific data compared to conclusions that are not based on such data?

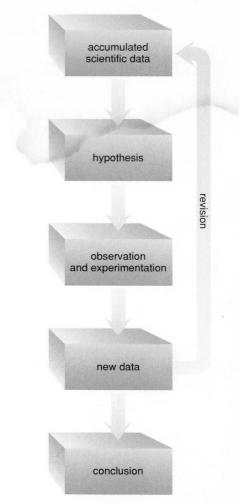

Figure I.5 The scientific method often includes these steps.

The ultimate goal of science is to understand the natural world in terms of concepts, interpretations that take into account the results of many experiments and observations. These concepts are stated as theories. In a movie, a detective might claim to have a "theory" about the crime, or you might say that you have a "theory" about the win-loss record of your favorite baseball team. But in science, the word *theory* is applied to those hypotheses that have been supported by a large number of observations and are considered valid by an overwhelming majority of scientists.

Social Responsibility

There are many ways in which science has improved our lives. The most obvious examples are in the field of medicine. The discovery of antibiotics such as penicillin and the vaccines for polio, measles, and mumps have increased our life spans by decades. Cell biology research is helping us understand the mechanisms that cause cancer. Genetic research has produced new strains of agricultural plants that have eased the burden of feeding our burgeoning world population.

Science has also produced conclusions that we find disturbing and has fostered technologies that have proven to be

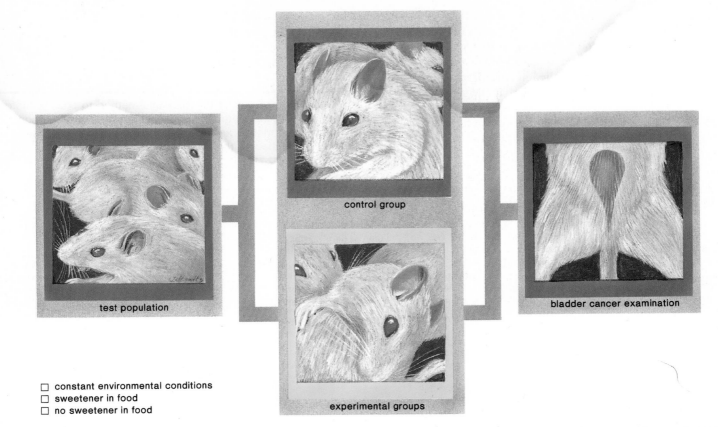

□ constant environmental conditions
□ sweetener in food
□ no sweetener in food

test population

control group

experimental groups

bladder cancer examination

Figure I.6   Design of a controlled experiment. From *left* to *right:* genetically identical mice are divided randomly into the control group and the experimental groups. All groups are exposed to the same environmental conditions, such as housing, temperature, and water supply. Only the experimental groups are subjected to the test. At the end of the experiment, all mice are examined for bladder cancer.

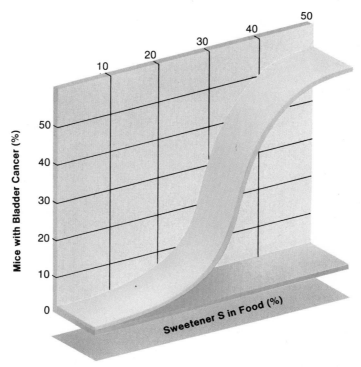

Figure I.7   Presenting the data. Scientists often acquire mathematical data that they report in the form of tables or graphs. Mathematical data are more decisive and objective than visual observations. The data in this instance suggest that there is an ever-greater chance of bladder cancer if the food is more than 10 percent sweetener S. Similar experiments will be repeated many times to test these results, and statistical analyses will be done to see if the results are significant or due to chance alone.

ecologically disastrous if not controlled properly. Too often we blame science for this and think that scientists are duty bound to pursue only those avenues of research that will not conflict with our system of values and/or result in environmental degradation, particularly when applied in an irresponsible manner. Yet science, by its very nature, is impartial and simply attempts to study natural phenomena. Science does not make ethical or moral decisions. Instead, all men and women have a responsibility to decide how best to use scientific knowledge so that it benefits the human species and all living things. Therefore, this text includes suggestions for the discussion of various human issues so that you will have an opportunity to think about the application of scientific, especially biological, information to everyday human concerns.

Scientists make use of the scientific method to discover information about the material world. It is the task of all persons to use this information as they make value judgments about their own lives and about the environment.

Summary

Human beings, just like other organisms, are a product of the evolutionary process. They are members of the animal kingdom and are most closely related to vertebrates and specifically to the apes. Like other living things, human beings reproduce, are highly organized, and maintain a fairly constant internal environment. Unlike other living things, they are also the product of a cultural inheritance.

The scientific method often includes studying previous data, making hypotheses, doing experiments, and coming to a conclusion based on the results (data) from these experiments. It is the responsibility of all to make ethical and moral decisions about how best to make use of the results of scientific investigations.

Part One

Human Organization

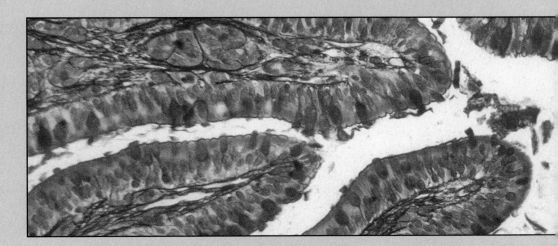

Columnar epithelium of small intestine

The human body is composed of cells, the smallest units of life. An understanding of cell structure, physiology, and biochemistry serves as a foundation for understanding how the human body functions.

Principles of inorganic and organic chemistry are discussed before a study of human cell structure is undertaken. The human cell is bounded by a membrane and contains organelles, many of which are also membranous. It is membrane that regulates the entrance and exit of molecules and determines how cellular organelles carry out their functions.

The many cells of the body are specialized into tissues that are found within the organs of the various systems of the body. All body systems help maintain a dynamic constancy of the internal environment so that the proper physical conditions exist for each cell.

Chapter One

Chemistry of Life

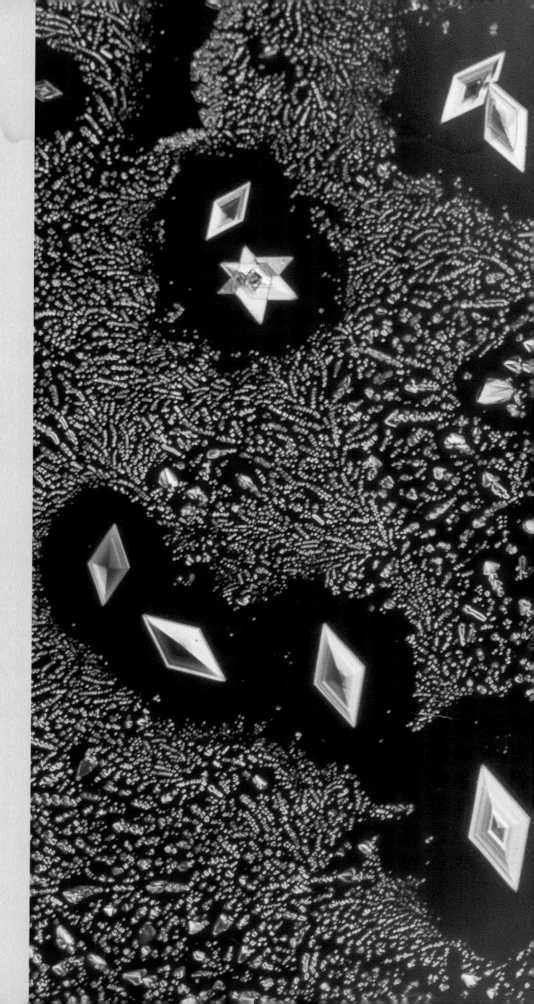

Chapter Concepts

1 Atoms, the smallest unit of an element, combine with one another to form compounds and molecules.

2 Reactions can be ionic as when a salt results or covalent as when water forms.

3 The pH scale indicates the acidity and basicity of a solution.

4 Proteins, carbohydrates, fats, and nucleic acids are the unique molecules found in living things. Each of these is composed of smaller molecules joined together.

Outside the body, asparagine, one of the amino acids found in your muscles, forms crystals of various sizes. Asparagine was first isolated from asparagus. Since human cells can also make asparagine, it is nonessential in the diet.

Table 1.1 Common Elements in Living Things*

Element	Symbol	Atomic number	Atomic weight†	Comment
hydrogen	H	1	1	These elements make up most biological molecules.
carbon	C	6	12	
nitrogen	N	7	14	
oxygen	O	8	16	
phosphorus	P	15	31	
sulfur	S	16	32	
sodium	Na	11	23	These elements occur mainly as dissolved salts.
magnesium	Mg	12	24	
chlorine	Cl	17	35	
potassium	K	19	39	
calcium	Ca	20	40	

*A periodic table of the elements is given in appendix A.
†Average of most common isotopes.
The atomic number gives the number of protons (and electrons in electrically neutral atoms). The number of neutrons is equal to the atomic weight minus the atomic number.

Table 1.2 Subatomic Particles

Name	Charge	Weight
electron	one negative unit	almost no weight
proton	one positive unit	one atomic unit
neutron	no charge	one atomic unit

Inorganic Chemistry

Although inorganic chemistry pertains to nonliving matter, inorganic chemicals are important constituents of the body. Also, some knowledge of inorganic chemistry is necessary for understanding the unique molecules of life, the biomolecules, discussed later.

Elements and Atoms

All living and nonliving things are composed of **elements,** basic substances that cannot be broken down further into simpler substances. Considering the variety of the world about us, it's quite remarkable that there are only 92 naturally occurring elements. It is even more surprising that over 90% of the human body is composed of just three elements: carbon, oxygen, and hydrogen. Every element has a name and a symbol; for example, the symbol for carbon is C (table 1.1).

An **atom** is the smallest unit of an element that still retains the chemical and physical properties of an element. While it is possible to split an atom by physical means, an atom is the smallest unit to enter into chemical reactions. For our purposes, it is satisfactory to think of each atom as having a central nucleus, where subatomic particles called **protons** and **neutrons** are located, and shells, where **electrons** orbit about the nucleus. Two important features of protons, neutrons, and electrons are their weight and charge, which are indicated in table 1.2. The structure of the carbon atom is given in figure 1.2.

Figure 1.1   Human beings, like all living and nonliving things, are composed only of chemicals.

It is not always easy to understand that human beings, and indeed all living and nonliving things, are composed of chemicals. After all, it is not possible to see the chemicals that make up the body (fig. 1.1). However, a few minutes' reflection regarding the dietary needs of the body usually convinces us that humans, indeed, are made of chemicals. For example, calcium is needed to maintain the bones, iron is necessary to prevent anemia, and adequate amino acid intake is required to build muscles.

Human Issue

Can—and should—life be described in terms of molecules? Are human beings chemical and physical machines? For some, such explanations detract from the beauty and wonder of life. For others, it increases the possibility of understanding life and ultimately of promoting the health of humans.

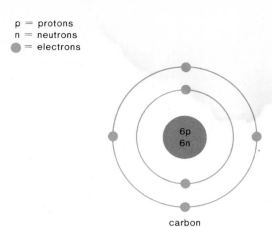

carbon
$^{12}_{6}C$

Figure 1.2   All the atoms of a particular element have the same atomic symbol and the same atomic number. The symbol for carbon is C and the atomic number is 6. The atomic number tells the number of protons (positive charges), which is equal to the number of electrons (negative charges) in an electrically neutral atom. Protons and neutrons are located in the nucleus of an atom. They account for the weight of an atom: The usual weight of carbon is 12. Electrons are located in the shells. The first shell of any atom can contain up to 2 electrons: Thereafter, each shell of the atoms noted in table 1.1 can contain up to 8 electrons. Carbon has 4 electrons in its outer shell because its atomic number is 6.

Isotopes

Isotopes are atoms of a particular element that differ only by the number of neutrons they contain. For example, all carbon atoms have 6 protons and 6 electrons; most carbon atoms have 6 neutrons but others have more or less than that number. Certain isotopes called *radioactive isotopes* are unstable and as they decay, they emit radiation, which may be detected by a Geiger counter. Carbon fourteen, which contains 6 protons and 8 electrons, is an isotope of carbon that is radioactive.

Radioactive isotopes are widely used in biological research because it is possible to trace their presence in chemical substances and tissues. For example, because the thyroid gland uses iodine (I), it is possible to administer radioactive iodine and then scan the gland to determine the location of the iodine isotope (fig. 1.3).

Compounds and Molecules

When two or more atoms of different elements combine, a **compound** results. Each compound has a particular chemical formula which tells how many of each type of atom are in the compound:

$CaCl_2$

In the compound calcium chloride, each molecule has 1 calcium atom and 2 chlorine atoms.

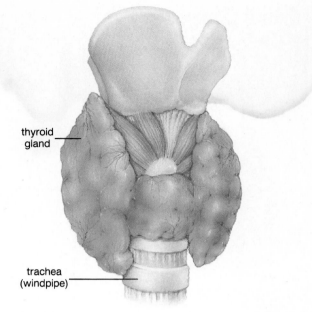

thyroid gland

trachea (windpipe)

a.

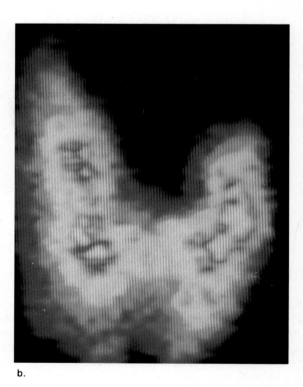

b.

Figure 1.3   Use of radioactive iodine. *a.* Drawing of the anatomical shape and location of the thyroid gland. *b.* A scan of the thyroid gland 24 hours after the patient was administered radioactive iodine.

A **molecule** is the smallest part of a compound that still has the properties of that compound. Molecules can also form when 2 or more atoms of the same element react with one another. For example, Cl_2 is a molecule that contains 2 atoms of chlorine.

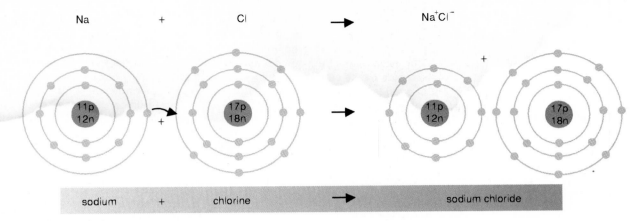

Na + Cl → Na⁺Cl⁻

+

sodium + chlorine → sodium chloride

Figure 1.4   Formation of the salt sodium chloride. During this ionic reaction, an electron is transferred from the sodium atom to the chlorine atom. Each resulting ion carries a charge as shown. Most people use the term *salt* to refer only to sodium chloride, but chemists use the term to refer to similar combinations of positive and negative ions.

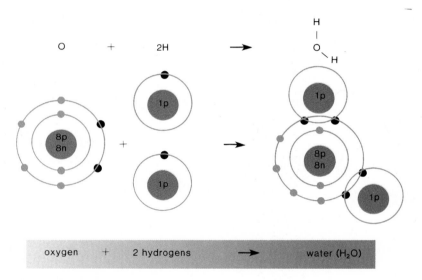

O + 2H →

oxygen + 2 hydrogens → water (H₂O)

Figure 1.5   Formation of water. Following a covalent reaction, oxygen is sharing electrons with two hydrogen atoms.

Reactions between Atoms

Compounds and/or molecules result when atoms react with one another. Usually, reactions between atoms involve the electrons in their outer shell. The *octet rule,* based on chemical findings states that atoms react with one another in order to achieve 8 electrons in their outer shell. Hydrogen (H) is an exception to this rule. In hydrogen, the first shell, which is complete with 2 electrons only, is the outer shell.

In one type of reaction, there is a transfer of an electron(s) from one atom to another in order to form a molecule. Such atoms are thereafter called **ions,** and the reaction is called an *ionic reaction.* For example, figure 1.4 depicts a reaction between sodium (Na) and chlorine (Cl) in which chlorine takes an electron from sodium. Now the sodium ion (Na⁺) carries a

positive charge, and the chlorine ion (Cl⁻) carries a negative charge. Notice that a negative charge indicates that the ion has more electrons (−) than protons (+) and a positive charge indicates that the ion has more protons (+) than electrons (−). Oppositely charged ions are attracted to one another, and this attraction is called an **ionic bond.**

In another type of reaction, atoms form a molecule by sharing electrons. The bond that forms between these atoms is called a **covalent bond.** For example, when oxygen reacts with two hydrogen atoms, water (H₂O) is formed (fig. 1.5).

Sometimes the atoms in a covalently bonded molecule share electrons evenly, but in water the electrons spend more time encircling the larger oxygen than the smaller hydrogens. Therefore, there is a slight positive charge on the hydrogen atoms

Acid Rain and Colon Cancer Are Linked

Acid rain could be responsible for elevated death rates from cancer of the colon and other organs in the northeastern United States according to two California researchers. They say carbon dioxide gas, a major component of acid rain, absorbs ultraviolet light that normally would fuel the body's production of vitamin D. Without vitamin D, people cannot absorb enough calcium to protect tissues from becoming cancerous, said Frank Garland, chief of occupational medicine at the Navy Health Research Center near San Diego.

Scientists have long suspected that acid rain contributes to lung cancer, but Garland said it also could be responsible for deaths from cancer of the colon, which is dramatically higher in the Northeast than in other areas of the nation. "If we know sulfur dioxide levels, we can predict with 60% accuracy the rate of colon cancer in that area," Garland said. Sulfur dioxide, released mainly from plants burning coal and oil, also could cause cancer of the breast and ovaries, he said.

"That is a very imaginative and exciting hypothesis," said Dr. Richard Rivlin of the Memorial Sloan-Kettering Cancer Center in New York. He added that lack of vitamin D could be one of several dietary factors responsible for colon cancer and that further research is needed on air pollution's role in blocking production of the vitamin. Dr. Kurt Isselvacher, director of Massachusetts General Hospital's cancer center, called the theory "surprising" but "worthy of examination and review." And Senator John Kerry, a leader in pushing for acid rain controls, said the new research would "add to the impetus" for a strong cleanup program, "although this is a tragic way to have to add to the impetus."

. . . Garland said he first suspected something was wrong when, 12 years ago, he saw the National Cancer Institute's earliest maps showing cancer deaths across the country. Red dots indicated the greatest numbers of colon cancer deaths, he explained, and almost every single red spot was in the northeastern quadrant of the United States. Diet has long been suspected as a cause of colon cancer, but nothing seemed to distinguish diets in this part of the country, said Garland, who conducted the research with his brother Cedric, a senior cancer researcher at the University of California in San Diego.

The pair noticed, however, that the map on skin cancers was "exactly opposite" from the one on colon cancers. Sunlight causes skin cancer, and the pair suspected that certain rays of the sun were blocked in the Northeast, but not in other regions with comparable climates, increasing colon cancer cases and decreasing the number of skin cancer cases. Further studies showed that sulfur dioxide absorbs a limited spectrum of ultraviolet light—precisely the same waves that fuel production of vitamin D in the skin. While ozone and other pollutants also block sunlight, he said, they do not affect the rays that produce vitamin D.

. . . The research, if it holds up, could have major implications for the heated debate over acid rain, which so far has focused mainly on damage to lakes, trees, and buildings. But Garland said his work was not designed with political motivations. "Sulfur dioxide did not become acid rain to us," he said, "until someone mentioned, 'That's the same stuff that makes acid rain. Do you realize the implications of that?'"

By Larry Tye. *The Boston Globe* February 10, 1988. Reprinted courtesy of *The Boston Globe*.

Air pollutants such as sulfur dioxide contribute to the acidity of rain and may also be detrimental to our health.

and a slight negative charge on the oxygen atom. For this reason, water is called a *polar molecule,* and hydrogen bonding occurs between water molecules (fig. 1.6). **A hydrogen bond** occurs whenever a partially positive hydrogen is attracted to a partially negative atom in another molecule. The hydrogen bond is represented by a dotted line in figure 1.6 because it is a weak bond that is easily broken.

Atoms react with one another to form molecules. In one type of reaction, positively and negatively charged ions are formed when an electron(s) is (are) transferred from one atom to another. In another type of reaction, the atoms share electrons in a molecule where the atoms are covalently bonded to one another.

Oxidation—Reduction Reactions   When one atom gives electrons to another, it has been oxidized. This terminology recognizes that an oxygen atom can act as a receiver of electrons. In the figure 1.4, then, sodium (Na) has been oxidized when it reacts with chlorine (Cl). Correspondingly, when an atom receives electrons, it has been reduced. This terminology recognizes that during reduction, an atom receives negative charges. Therefore in figure 1.4, chlorine has been reduced.

The terms **oxidation** and **reduction** also are applied to certain covalent reactions. In this case, however, oxidation is the loss of hydrogen (H) atoms and reduction is the gain of hydrogen atoms. A hydrogen atom contains one proton and one electron; therefore, when a molecule loses a hydrogen atom, it has lost an electron, and when a molecule gains a hydrogen atom, it has gained an electron. For example, in figure 1.5, oxygen has been reduced when it becomes water.

pH Scale

The **pH** scale ranges from zero to 14 (fig. 1.7); the numbers indicate the hydrogen ion concentration $[H^+]$ and also the hydroxide ion concentration $[OH^-]$. The pH of water (pH 7) is neutral pH because pure water breaks up or dissociates in this manner:

$$H - O - H \rightarrow H^+ + OH^-$$

In any quantity of water, most of the water molecules are intact but some have dissociated to give an equal number of H^+ and OH^- ions. The fraction of water molecules that dissociate is 10^{-7} (or 0.0000001) which is called a pH of 7. In other words, the pH scale was devised to simplify discussion of the hydrogen ion concentration $[H^+]$.

Acids and Bases

Acids are substances that increase the $[H^+]$ of a solution. For example, hydrochloric acid (HCl) dissociates in this manner:

$$HCl \rightarrow H^+ + Cl^-$$

In the pH scale, the numbers 0 to 7 indicate an acidic pH; each lower unit has 10 times the number of hydrogen ions as the number above.

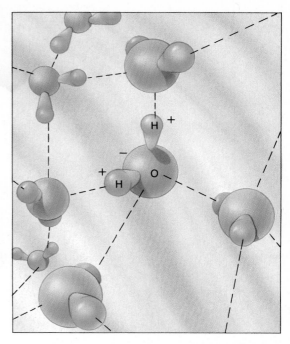

Figure 1.6   Hydrogen bonding between water molecules. Water molecules are polar; each hydrogen (H) carries a partial positive charge and each oxygen (O) carries a partial negative charge. The polarity of the water molecules brings about hydrogen bonding between the molecules in the manner shown. The dotted lines represent hydrogen bonds.

Polarity causes water to be an excellent solvent; it dissolves various chemical substances, particularly other polar molecules. Hydrogen bonding allows water to absorb a great deal of heat and still remain liquid; therefore, water is liquid at body temperature. Eventually, however, water will vaporize; when humans sweat, body heat is used to vaporize water.

Bases are substances that increase the $[OH^-]$ of a solution. For example, sodium hydroxide (NaOH) dissociates in this manner:

$$NaOH \rightarrow Na^+ + OH^-$$

In the pH scale, the numbers 7 to 14 indicate a basic pH; each higher unit has 10 times the number of hydroxide ions as the number below.

Buffers

All living things need to maintain the hydrogen ion concentration, or pH, at a constant level. For example, the pH of the blood is held constant at about 7.4, or we become ill. The presence of buffers helps keep the pH constant. **A buffer** is a chemical or a combination of chemicals that can take up excess hydrogen ions or excess hydroxide ions. When an acid is added to a buffered solution, a buffer takes up excess hydrogen ions, and when a base is added to a buffered solution, a buffer takes up excess hydroxide ions. Therefore, the pH changes minimally whenever a solution is buffered.

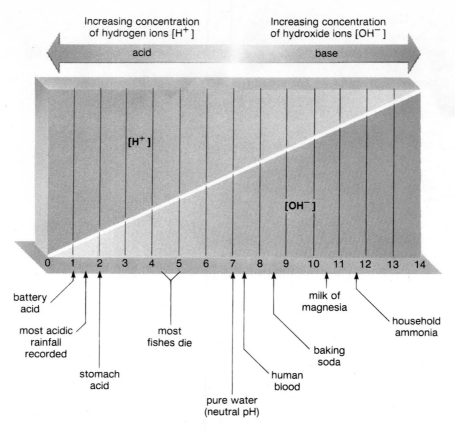

Figure 1.7   The pH scale. The proportionate amount of hydrogen ions (H⁺) to hydroxide ions (OH⁻) is indicated by the diagonal line. Any pH above 7 is basic, while any pH below 7 is acidic. The pH of some common substances is given.

Table 1.3   Inorganic versus Organic Chemistry

Inorganic compounds	Organic compounds
usually contain positive and negative ions	always contain carbon and hydrogen
usually ionic bonding	always covalent bonding
always contain a small number of atoms	may be quite large with many atoms
often associated with nonliving elements	associated with living organisms

Acids have a pH that is less than 7, and bases have a pH that is greater than 7. The presence of buffers helps keep the pH of body fluids constant at about neutral, or pH 7, because a buffer can absorb both hydrogen and hydroxide ions.

Organic Chemistry

Table 1.3 contrasts inorganic compounds with organic compounds. Both types of compounds are necessary to the proper functioning of the human body.

Unit Molecules

The chemistry of carbon accounts for the formation of the very large number of organic compounds we associate with living organisms. Carbon shares electrons with as many as four other atoms. Many times, carbon atoms share electrons with each other to form rings or chains of carbon atoms. These act as skeletons for the unit molecules found in the biomolecules—carbohydrates, fats, and nucleic acids. Therefore, the properties of carbon are essential to life as we know it.

Human Issue

A number of organic and inorganic compounds are used as drugs for medical and nonmedical purposes. As you know, the federal government maintains a close watch over the production and marketing of drugs. Some believe that taking virtually any drug should be a matter of personal choice; the government should merely be required to inform us of the biological effects of the drugs and then we can make our own choice. Others believe that it should be illegal to produce and sell certain types of drugs to the public. Which approach do you think is better?

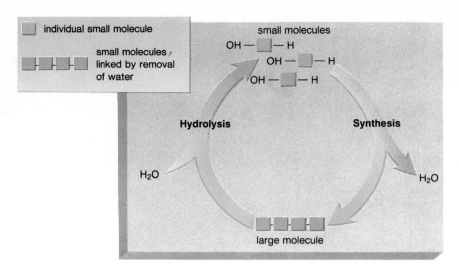

Figure 1.8   Synthesis and hydrolysis of an organic polymer. When unit molecules, or monomers, join to form a polymer (synthesis), water is released; when a polymer is broken down (hydrolysis), water is added.

Synthesis and Hydrolysis

Biomolecules are **macromolecules,** large molecules built up by the joining together of unit molecules. These are the macromolecules and unit molecules we will be studying:

Macromolecules	Unit Molecule(s)
polysaccharide	monosaccharide
lipid	glycerol and fatty acid (for fat)
protein	amino acid
nucleic acid	nucleotide

A body that joins two unit molecules together is created after the removal of H^+ from one molecule and OH^- from the next molecule. As water forms, dehydration **synthesis** occurs. Macromolecules, called *polymers* when they are chains of unit molecules, can be broken down in a manner opposite to synthesis: the addition of water leads to the disruption of the bonds linking the unit molecules together. During this process, called **hydrolysis,** one molecule takes on H^+ and the next takes on OH^- (fig. 1.8).

Proteins

Functions

Proteins are large polymers that sometimes have mainly a structural function. For example, in humans, keratin is a protein that makes up hair and nails, and collagen is a protein found in all types of connective tissue, including ligaments, cartilage, bone, and tendons. The muscles (fig. 1.9) contain proteins (actin and myosin) that account for their ability to contract.

Figure 1.9   The well-developed muscle cells of an athlete contain many protein molecules. Today, a major concern is the use of steroids to promote the buildup of muscles. This practice can be detrimental to health.

Some proteins function as **enzymes,** necessary contributors to the chemical workings of the cell and therefore of the body. Enzymes are organic catalysts that speed up chemical reactions. They work so quickly that a reaction that might normally take several hours or days takes only a fraction of a second when an enzyme is present.

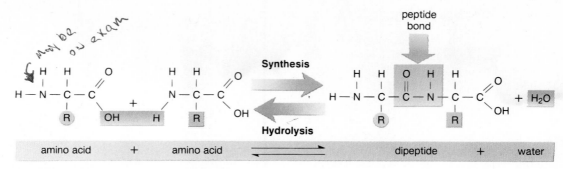

Figure 1.10   Formation of a peptide. On the left-hand side of the equation, there are 2 different amino acids, as signified by the difference in the R group notations. During synthesis, as the peptide bond forms, water is given off—the water molecule on the right-hand side of the equation is derived from components removed from the amino acids on the left-hand side. During hydrolysis, water is added as the peptide bond is broken.

Structure

The unit molecules found in proteins are called **amino acids.** The name amino acid refers to the fact that the molecule has two functional groups, an *amino group* and an *acid group.*

R = remainder of molecule

Amino acids differ from one another by their *R groups,* which vary from being a single hydrogen atom to a complicated ring. Because there are 20 different amino acids commonly found in the proteins of living things, there are also 20 different types of R groups.

The bond that joins two amino acids together is called a **peptide bond.** As you can see in figure 1.10, when synthesis occurs, the acid group of one amino acid reacts with the amino group of another amino acid and water is given off. A dipeptide contains only two amino acids, but when 10 or 20 amino acids have joined together, the resulting chain is called a **polypeptide.** Proteins are made of one or more very long polypeptide chains.

Levels of Structure   Proteins are said to have at least three levels of structure: primary, secondary, and tertiary. The *primary structure* is simply the sequence, or order, of the different amino acids. Any number of the 20 different amino acids may be joined in various sequences, and each type of protein has its own particular sequence. The resulting chain is like a necklace comprising up to 20 different types of beads that reoccur and are linked in a set way. The *secondary structure* is the usual orientation of the amino acid chain. One common arrangement of the chain is the alpha helix, a right-handed spiral, called an α helix, held in place by hydrogen bonding between members of the various peptide bonds. The *tertiary structure* of a protein refers to its final three-dimensional shape. In a structural pro-

tein like collagen, the helical chains lie parallel to one another, but in enzymes the helix bends and twists in different ways. The final shape of a protein is maintained by various types of bonding between the R groups. Covalent, ionic, and hydrogen bonding are all seen. Figure 1.11 illustrates the main features of protein chemistry.

If a protein has more than one polypeptide chain, each chain has its own primary, secondary, and tertiary structures. Within the protein, these separate chains are arranged to give a fourth level of structure termed the *quaternary structure.* Hemoglobin is a complex protein having a quaternary structure.

Amino acids are the unit molecule for peptides and proteins. Proteins have both structural and metabolic[1] functions in the human body.

Carbohydrates

Carbohydrates are characterized by the presence of $H-C-OH$ groupings in which the ratio of hydrogen atoms to oxygen atoms is approximately 2:1. Since this ratio is the same as the ratio in water, the meaning of this compound's name—hydrates of carbon—is very appropriate. If the number of carbon atoms in the compound is low (from about three to seven), the carbohydrate is a monosaccharide. Larger carbohydrates are created by joining together monosaccharides in the manner described in figures 1.8 and 1.12.

Monosaccharides and Disaccharides

As their name implies, **monosaccharides** are sugars having only one unit. These compounds are often designated by the number of carbons they contain; for example, pentose sugars have five carbons, and hexose sugars have six carbons. **Glucose** is a six-carbon sugar, with the structural formula shown in figure 1.12. Although there are other monosaccharides with the molecular

[1]Metabolism is all the chemical reactions that occur in a cell.

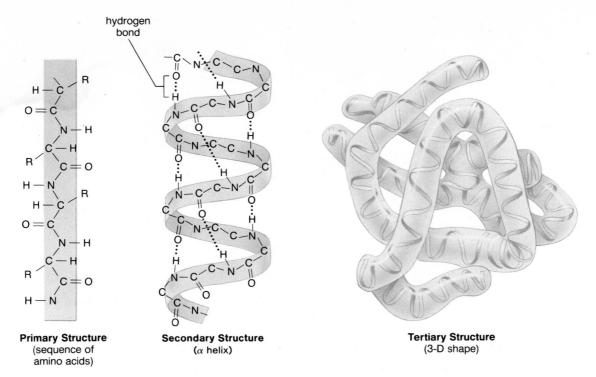

Primary Structure
(sequence of
amino acids)

Secondary Structure
(α helix)

Tertiary Structure
(3-D shape)

Figure 1.11   Levels of organization in the structure of a protein. Primary structure of a protein is the order of the amino acids; secondary structure is often an alpha (α) helix in which hydrogen bonding occurs along the length of a polypeptide, as indicated by the dotted lines; and in globular proteins, the tertiary structure is the twisting and the turning of the helix that takes place because of bonding between the R groups. Enzymes are globular proteins.

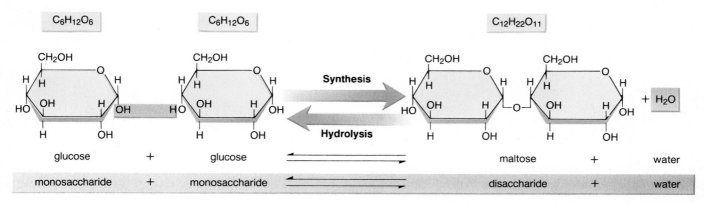

Figure 1.12   Synthesis and hydrolysis of maltose, a disaccharide containing 2 glucose units. During synthesis, a bond forms between the 2 glucose molecules as the components of water are removed. During hydrolysis, the components of water are added as the bond is broken.

formula $C_6H_{12}O_6$, in this text we will use the molecular formula $C_6H_{12}O_6$ to mean glucose, since glucose is the most common six-carbon monosaccharide found in cells. Cells use glucose as an immediate energy source.

The term **disaccharide** tells us that there are two monosaccharide units joined together in the compound. When two glucose molecules join together, maltose is formed (fig. 1.12). You may also be interested to know that when glucose and another monosaccharide, fructose, are joined together, the disaccharide called sucrose is formed. *Sucrose* is derived from plants and is commonly used at the table to sweeten foods.

Polysaccharides

A **polysaccharide** is a carbohydrate that contains a large number of monosaccharide molecules. Three polysaccharides are common in animals and plants: glycogen, starch, and cellulose. All of these are polymers of glucose. Even though all three chains contain only glucose, they are distinguishable from one another.

Glycogen (fig. 1.13), the storage form of glucose in humans, is a molecule with many side branches. After eating, the liver stores glucose as glycogen; in between eating, the liver releases glucose so that the blood concentration of glucose is always near 0.1%.

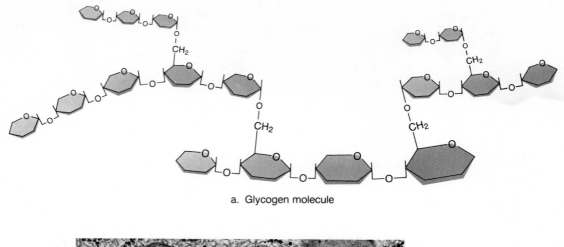

a. Glycogen molecule

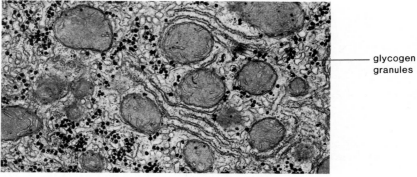

glycogen granules

b. Glycogen granules in liver cell

Figure 1.13 Glycogen structure and function. *a.* Glycogen is a highly branched polymer of glucose molecules. The branching allows breakdown to proceed simultaneously at several points. *b.* Electron micrograph showing glycogen granules in liver cells. Glycogen is the storage form of glucose in humans.

Starch and *cellulose* are found in plants. Plants store glucose as starch, a polymer similar in structure to glycogen except that it has few side branches. Starch is an important source of glucose energy in our diet because it can be hydrolyzed to glucose by digestive enzymes. In cellulose, a common structural compound in plants, the glucose units are joined by a slightly different type of linkage compared to that of glycogen and starch. For this reason we are unable to digest cellulose, and it passes through our digestive tract as roughage (fiber). Recently it has been suggested that the presence of roughage in the diet is necessary to good health and prevention of colon cancer.

In this text, $C_6H_{12}O_6$ stands for glucose, the unit molecule in glycogen, our storage form of glucose, and starch, an important energy source for humans.

Lipids

Many **lipids** are nonpolar and therefore are insoluble in water. This is true of fats, the most familiar lipids, such as lard, butter, and oil, which are used in cooking or at the table. In the body, fats serve as long-term energy sources. Adipose tissue is composed of cells that contain many molecules of neutral fat.

Neutral Fats (Triglycerides)

A neutral (nonpolar) fat contains two types of unit molecules: **glycerol** and **fatty acids.** Each fatty acid has a long chain of carbon atoms with hydrogens attached, ending in an acid group (fig. 1.14). Fatty acids are either *saturated* or *unsaturated.* Saturated fatty acids have no double bonds between the carbon atoms. The carbon chain is saturated, so to speak, with all the hydrogens that can be held. Unsaturated fatty acids have double bonds in the carbon chain wherever the number of hydrogens is less than two per carbon atom. Unsaturated fatty acids are most often found in vegetable oils and account for the liquid nature of these oils. Vegetable oils are hydrogenated to make margarine. Polyunsaturated margarine still contains a large number of unsaturated, or double, bonds.

Glycerol is a compound with three $H-C-OH$ groups attached by way of the carbon atoms. When fat is formed, by dehydration synthesis, the $-OH$ groups react with the acid

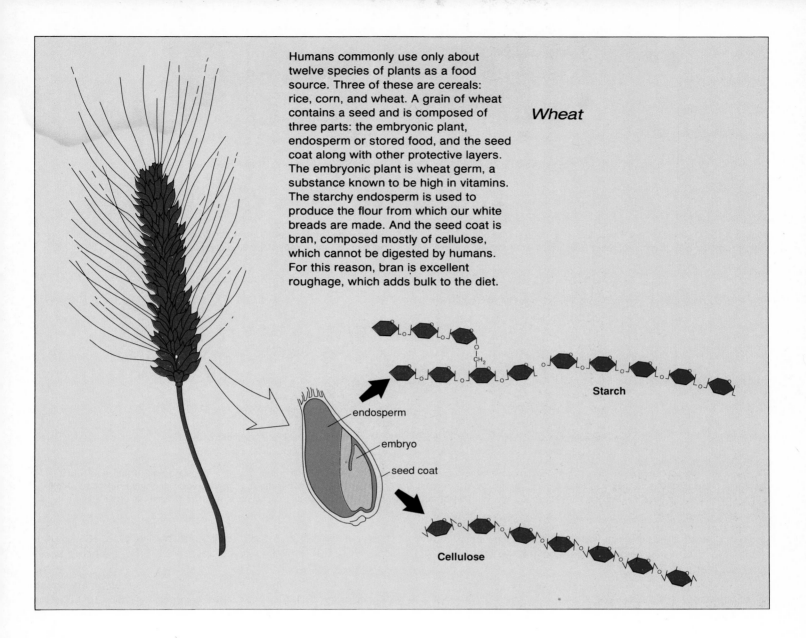

Humans commonly use only about twelve species of plants as a food source. Three of these are cereals: rice, corn, and wheat. A grain of wheat contains a seed and is composed of three parts: the embryonic plant, endosperm or stored food, and the seed coat along with other protective layers. The embryonic plant is wheat germ, a substance known to be high in vitamins. The starchy endosperm is used to produce the flour from which our white breads are made. And the seed coat is bran, composed mostly of cellulose, which cannot be digested by humans. For this reason, bran is excellent roughage, which adds bulk to the diet.

Wheat

endosperm

embryo

seed coat

Starch

Cellulose

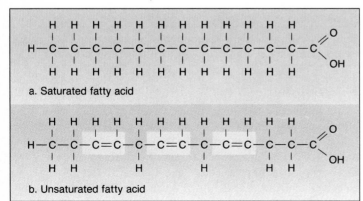

a. Saturated fatty acid

b. Unsaturated fatty acid

Figure 1.14   **Saturated versus unsaturated fatty acids.** Fatty acids are either saturated (have no double bonds) or unsaturated (have double bonds). *a.* In a saturated fatty acid, the carbons carry all the hydrogen atoms possible. *b.* In this unsaturated fatty acid, there is a double bond at the third from last carbon and at other carbons.

portions of three fatty acids so that three molecules of water are formed. The reverse of this reaction represents hydrolysis of the fat molecule into its separate components (fig. 1.15).

Phospholipids

Phospholipids, as their name implies, contain a phosphate group:

$$^-O - \overset{\overset{\displaystyle O}{\displaystyle \|}}{\underset{\underset{\displaystyle ^-O}{\displaystyle |}}{P}} - {}^-O$$

Essentially, phospholipids are constructed like neutral fats; in place of fatty acid, there is a phosphate group or a grouping that contains both phosphate and nitrogen (fig. 1.16). These molecules are not electrically neutral as are the fats because the phosphate group can ionize. Notice, then, that the phospholipids have a nonpolar region that is not soluble in water and a

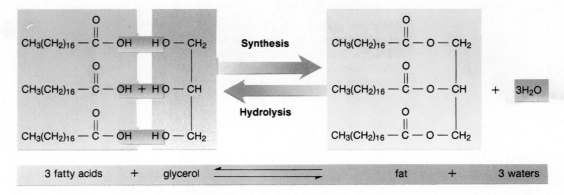

3 fatty acids   +   glycerol   ⟶   fat   +   3 waters

Figure 1.15   Synthesis and hydrolysis of a neutral (nonpolar) fat. Three fatty acids plus glycerol react to produce a fat molecule and three water molecules. A fat molecule plus three water molecules react to produce three fatty acids and glycerol.

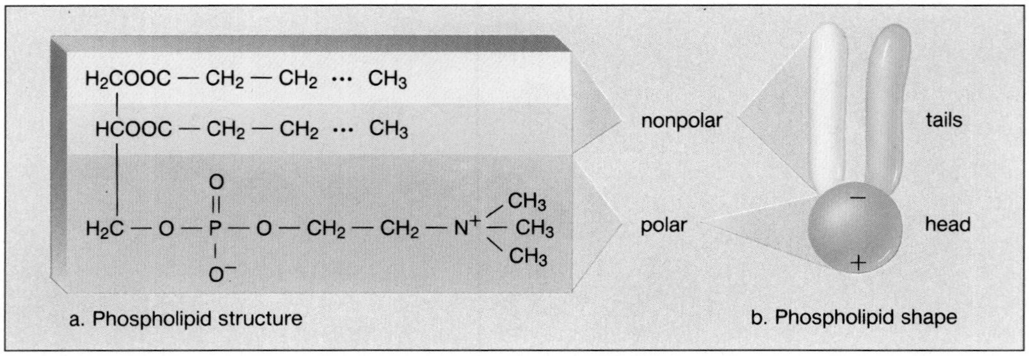

| a. Phospholipid structure | b. Phospholipid shape |

nonpolar   tails

polar   head

Figure 1.16   Phospholipid structure. Phospholipids are constructed similarly to fats except that they contain a phosphate group. a. Lecithin, shown here, has a side chain that contains both a phosphate group and a nitrogen-containing group. b. The polar portion of the molecule is soluble in water, whereas the 2 hydrocarbon chains are not soluble in water. This causes the molecule to arrange itself as shown.

polar region that is soluble in water. This latter property makes them very useful compounds in the body, as we will see in the next chapter.

Cholesterol and Steroids

Cholesterol and **steroids** have similar structures. They are constructed of four fused rings of carbon atoms to which is usually attached a carbon chain of varying length (fig. 1.17). Today there is a great deal of interest in cholesterol because a high cholesterol blood level is associated with development of coronary heart disease as discussed on page 108. Steroids, however, are very necessary compounds in the body; for example, the sex hormones are steroids.

Lipids include nonpolar fats, long-term energy-storage molecules that form from glycerol and 3 fatty acids, and the related phospholipids, which have a charged group. The steroids have an entirely different structure, similar to that of cholesterol.

Nucleic Acids

Nucleic acids are huge, macromolecular compounds with very specific functions in cells; for example, the genes are composed of a nucleic acid called **DNA** (deoxyribonucleic acid). DNA has the ability to replicate, or make a copy of itself. It also controls protein synthesis. Another important nucleic acid, **RNA** (ribonucleic acid), works in conjunction with DNA to bring about protein synthesis.

Both DNA and RNA are polymers of nucleotides and therefore are chains of nucleotides joined together. Just like the other synthetic reactions we have studied in this section, when these units are joined together to form nucleic acids, water molecules are removed.

Both DNA and RNA are composed of nucleotides. DNA makes up the genes and along with RNA controls protein synthesis.

Nucleotides

Every **nucleotide** is a molecular complex of three types of unit molecules: phosphoric acid (phosphate), a pentose (five-carbon) sugar, and a nitrogen base. In DNA the sugar is deoxyribose and in RNA the sugar is ribose, and this difference accounts for their respective names. There are four different types of nucleotides in DNA and RNA. Figure 1.18 shows that a DNA base can be one of the **purines,** adenine or guanine, which have a double ring, or one of the **pyrimidines,** thymine or cytosine, which have a single ring. These structures are called bases because they have basic characteristics that raise the pH of a solution. RNA differs from DNA in that the base uracil is used in place of the base thymine.

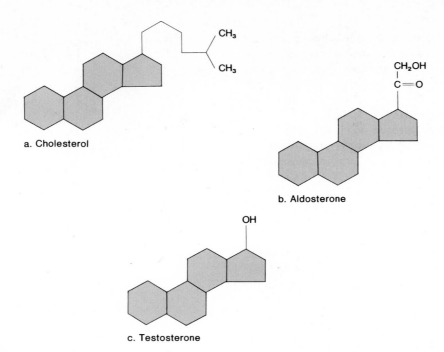

a. Cholesterol

CH₂OH
C=O

b. Aldosterone

OH

c. Testosterone

Figure 1.17   Like cholesterol, *a.,* steroid molecules have four adjacent rings, but their effects on the body largely depend on the type of chain attached at the location indicated. The chain in *b.* is found in aldosterone, which is involved in the regulation of sodium and potassium blood content. The chain in *c.* is found in testosterone, the male sex hormone.

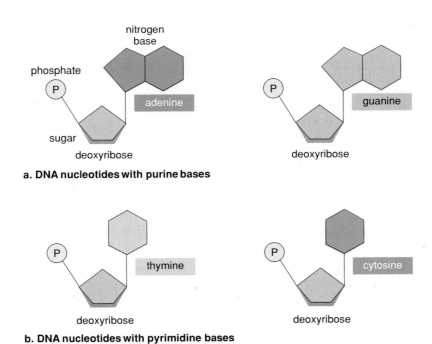

nitrogen base

phosphate

P

adenine

sugar

deoxyribose

a. DNA nucleotides with purine bases

P

guanine

deoxyribose

P

thymine

deoxyribose

P

cytosine

deoxyribose

b. DNA nucleotides with pyrimidine bases

Figure 1.18   Nucleotides in DNA. Each nucleotide is composed of phosphate, the sugar deoxyribose, and a base. *a.* The purine bases are adenine and guanine. *b.* The pyrimidine bases are thymine and cytosine.

Strands

When nucleotides join together, they form a linear polymer, called a strand, in which the so-called backbone is made up of phosphate-sugar-phosphate-sugar. The bases project to one side of the backbone. RNA is single stranded (fig. 1.19), but DNA is double stranded. The two strands of DNA twist about one another in the form of a **double helix** (fig. 1.20a and b). The two strands are held together by hydrogen bonds between purine and pyrimidine bases. Thymine (T) is always paired with adenine (A), and guanine (G) is always paired with cytosine (C). This is called complementary base pairing. If we unwind the DNA helix, it resembles a ladder (fig. 1.20c). The sides of the ladder are made entirely of phosphate and sugar molecules, and the rungs of the ladder are made only of the *complementary paired bases*. The bases can be in any order, but A is always paired with T, and G is always paired with C, and vice versa. Therefore, no matter what the order or the quantity of any particular base pair, the number of purine bases always equals the number of pyrimidine bases.

DNA has a structure like a twisted ladder: sugar-phosphate backbones make up the sides of the ladder, and hydrogen-bonded bases make up the rungs of the ladder. The base A is always paired with the base T, and the base C is always paired with the base G. RNA differs from DNA in several respects (table 1.4).

ATP

ATP, adenosine triphosphate (fig. 1.21), is a very special type of nucleotide. It is composed of the base adenine and the sugar ribose (together called adenosine) and three phosphate groups. The wavy lines in the formula for ATP indicate high-energy phosphate bonds. When these bonds are broken, an unusually large amount of energy is released. Because of this property, ATP is the energy currency of cells; when cells "need" something, they "spend" ATP.

ATP is used in body cells for synthetic reactions, active transport, nervous conduction, and muscle contraction. When energy is required for these processes, the end phosphate group is removed from ATP, breaking down the molecule to ADP (adenosine diphosphate) and (P) (phosphate) (fig. 1.21).

ATP Cycle

The reaction shown in figure 1.22 occurs in both directions; not only is ATP broken down, it is also built up when ADP joins with (P). Since ATP breakdown is constantly occurring, there is always a ready supply of ADP and (P) to rebuild ATP again.

Figure 1.22 illustrates the ATP cycle in a diagrammatic way. Notice that when ATP is broken down, energy is released, and when it is built up, energy is required. We shall see that

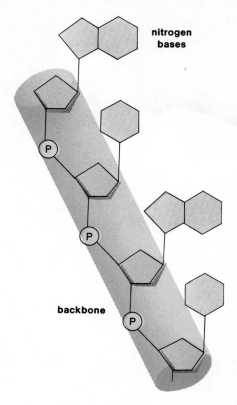

Figure 1.19   Generalized nucleic acid strand. Nucleic acid polymers contain a chain of nucleotides. Each strand has a backbone made of sugar and phosphate molecules. The bases project to the side.

Table 1.4   DNA Structure Compared to RNA Structure

	DNA	RNA
sugar	deoxyribose	ribose
bases	adenine, guanine, thymine, cytosine	adenine, guanine, uracil, cytosine
strands	double stranded with base pairing	single stranded
helix	yes	no

aerobic cellular respiration, a metabolic pathway that takes place largely within mitochondria, produces the energy needed for ATP buildup.

ATP is the energy molecule of cells because it contains high-energy phosphate bonds.

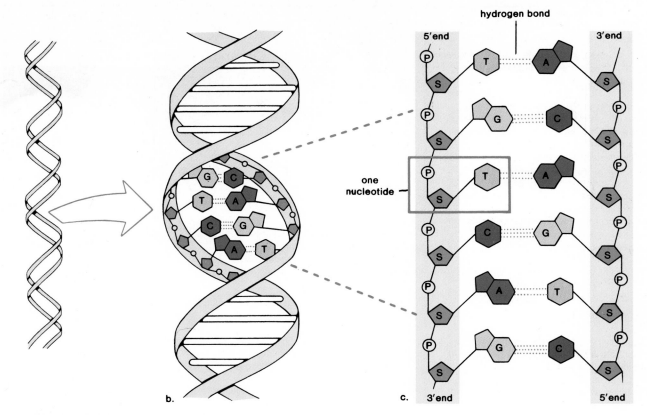

Figure 1.20 Overview of DNA structure. *a.* Double helix. *b.* Complementary base pairing. *c.* Ladder configuration. Notice that the uprights are composed of sugar and phosphate molecules and the rungs are complementary paired bases.

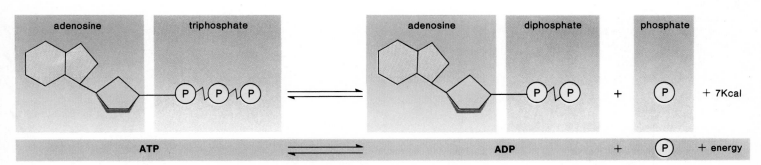

Figure 1.21 ATP reaction. ATP, the energy molecule in cells, has 2 high-energy phosphate bonds (indicated in the figure by wavy lines). When cells require energy, the last phosphate bond is broken, and a phosphate molecule is released.

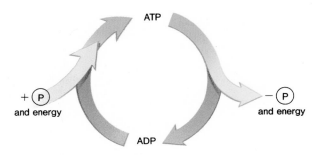

Figure 1.22 The ATP cycle. When ADP joins with Ⓟ, phosphate, energy is required; but when ATP breaks down to ADP and Ⓟ, energy is given off.

Summary

All matter is made up of atoms, each having a weight that is dependent on the number of protons and neutrons in the nucleus and chemical properties that are dependent on the number of electrons in the outermost shell. Atoms react with one another in order to form molecules. In ionic reactions, one atom gives electrons to another and in covalent reactions, atoms share electrons.

Water, acids, and bases are important inorganic compounds. Water has a neutral pH; acids decrease and bases increase the pH of water. The organic molecules of interest are proteins, carbohydrates, lipids, and nucleic acids, each of which has (a) particular unit molecule(s) (table 1.5). Dehydration synthesis joins unit molecules together and hydrolytic degradation releases them. Some proteins are enzymes; carbohydrates serve as immediate energy sources; and fats are a long-term energy source for the individual. Nucleic acids are of two types, DNA and RNA. DNA is the genetic material, and both of these have functions related to protein synthesis, which will be discussed in chapter 19.

Table 1.5 Organic Compounds of Life

Macromolecules	Unit molecule	Usual atoms
protein	amino acid	C, H, O, N, S
carbohydrate, e.g., glycogen	glucose	C, H, O
lipid	glycerol and fatty acids	C, H, O
nucleic acid	nucleotide	C, H, O, N, P

Study Questions

1. Describe the composition of an atom, and give the weight and charge of an atom's components. (p. 13)
2. Give an example of an ionic reaction, and define the term *ion*. (p. 15)
3. Give an example of a covalent reaction, and define the term *covalent bond*. (p. 15)
4. On the pH scale, which numbers indicate a basic solution? an acidic solution? (p. 17)
5. What are buffers, and why are they important to life? (p. 17)
6. Name four general differences between inorganic and organic compounds. (p. 18)
7. Explain synthesis by dehydration and breakdown by hydrolysis of organic compounds. (p. 19)
8. Describe the primary, secondary, and tertiary structure of proteins. What functions do proteins serve in the body? (p. 19)
9. Name a monosaccharide, disaccharide, and polysaccharide, and state appropriate functions. What is the most common unit molecule for these? (p. 20)
10. What type molecules react to form a neutral fat? Explain the difference between a saturated and an unsaturated fatty acid. (p. 22)
11. Name several types of lipids, and state their functions. (pp. 22–24)
12. What are the two types of nucleic acids in cells, and what are their functions? What is the unit molecule of a nucleic acid? (p. 24) Name four differences between DNA and RNA. (pp. 24–26)

Objective Questions

1. _Atoms_ are the smallest units into which matter can be chemically broken.
2. Isotopes differ by the number of _neutrons_ in the nucleus.
3. The two primary types of reactions and bonds are _ionic_ and _covalent_.
4. A type of weak bond, called _hydrogen_ bonding, exists between water molecules.
5. Acidic solutions contain more _hydrogen_ ions than basic solutions, but they have a _lower_ pH.
6. The primary structure of a protein is the sequence of _amino acids_; the secondary structure is very often an alpha _helix_; the tertiary structure is the final _shape_ of the protein.
7. _enzymes_ speed up chemical reactions.
8. Glycogen is a polymer of _glucose_, molecules that serve to give the body immediate _energy_.
9. A neutral fat hydrolyzes to give one _glycerol_ molecule and three _fatty acid_ molecules.
10. The genes are composed of _DNA_ a nucleic acid made up of _nucleotides_ joined together.

Answers to Objective Questions

1. Atoms 2. neutrons 3. ionic, covalent 4. hydrogen 5. hydrogen, lower 6. amino acids, helix (or spiral), shape 7. Enzymes 8. glucose, energy 9. glycerol, fatty acid 10. DNA, nucleotides

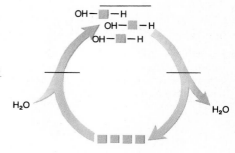

Label this Diagram.
See figure 1.8 (p. 19) in text.

Critical Thinking Questions

1. The human body is composed of organic molecules but also contains inorganic molecules. Argue for the importance of inorganic molecules in the body.

2. Compare a glycogen molecule to a protein molecule. How are the differences you note related to the function of these molecules in the body?

3. What type of cellular reaction forms water in one direction and uses it up in the reverse direction? Explain.

Selected Key Terms

acid (as'id)   a solution in which pH is less than 7; a substance that contributes or liberates hydrogen ions in a solution.   17

amino acid (ah-me'no as'id)   a unit of protein that takes its name from the fact that it contains an amino group (NH_2) and an acid group (COOH).   20

atom (at'om)   smallest unit of matter nondivisible by chemical means.   13

ATP   adenosine triphosphate; a compound containing adenine, ribose, and three phosphates, two of which are high-energy phosphates. It is the "common currency" of energy for most cellular processes.   26

base (bās)   a solution in which pH is more than 7; a substance that contributes or liberates hydroxide ions in a solution; alkaline; opposite of acidic. Also, a term commonly applied to one of the components of a nucleotide.   17

buffer (buf'er)   a substance or compound that prevents large changes in the pH of a solution.   17

DNA   deoxyribonucleic acid; a nucleic acid, the genetic material that replicates and directs protein synthesis in cells.   24

electron (e-lek'tron)   a subatomic particle that has almost no weight and carries a negative charge; travels in an orbital, called a shell, about the nucleus of an atom.   13

enzyme (en'zīm)   a catalyst that speeds up a specific reaction or a specific type of reaction.   19

hydrogen bond (hi'dro-jen bond)   a weak attraction between a hydrogen atom carrying a partial positive charge and another atom carrying a partial negative charge.   17

ion (i'on)   an atom or group of atoms carrying a positive or negative charge.   15

isotopes (i'so-tōps)   atoms with the same number of protons and electrons but differing in the number of neutrons and therefore in weight.   14

lipid (lip'id)   a group of organic compounds that are insoluble in water; notably fats, oils, and steroids.   22

neutron (nu'tron)   a subatomic particle that has a weight of one atomic mass unit, carries no charge, and is found in the nucleus of an atom.   13

nucleic acid (nu-kle'ik as'id)   a large organic molecule made up of nucleotides joined together; for example, DNA and RNA.   24

peptide bond (pep'tīd bond)   the bond that joins two amino acids.   20

pH   a measure of the hydrogen ion concentration; any pH below 7 is acid and any pH above 7 is basic.   17

polysaccharide (pol''e-sak'ah-rīd)   a macromolecule composed of many units of monosaccharide.   21

protein (pro'te-in)   a macromolecule composed of one or several long polypeptides; polypeptides contain many amino acids joined together.   19

proton (pro'ton)   a subatomic particle found in the nucleus of an atom that has a weight of one atomic mass unit and carries a positive charge.   13

RNA   ribonucleic acid; a nucleic acid that assists DNA in the production of proteins within the cell.   24

Chapter Two

Cell Structure
and Function

Chapter Concepts

1 The fundamental unit of life is the cell, which is highly organized and contains organelles that carry out specific functions.

2 The cell membrane regulates the entrance and exit of molecules to and from the cell.

3 The nucleus, a centrally located organelle, controls the metabolic functioning and structural characteristics of the cell.

4 Endoplasmic reticulum, the Golgi apparatus, and lysosomes are all membranous tubules or vesicles concerned with the entrance, production, digestion, excretion, or transportation of molecules.

5 Mitochondria are organelles concerned with the conversion of glucose energy into ATP molecules.

6 Centrioles and related structures are concerned with the shape and/or movement of the cell.

7 The cytoplasm contains metabolic pathways, each a series of reactions controlled by enzymes.

The body contains many types of cells, each with a specific structure and function. The blood contains 5 recognizable types of white cells. This is a section of an eosinophil that has many cytoplasmic granules and a bilobed nucleus. Eosinophils increase in number during parasitic infections such as a hookworm infection.

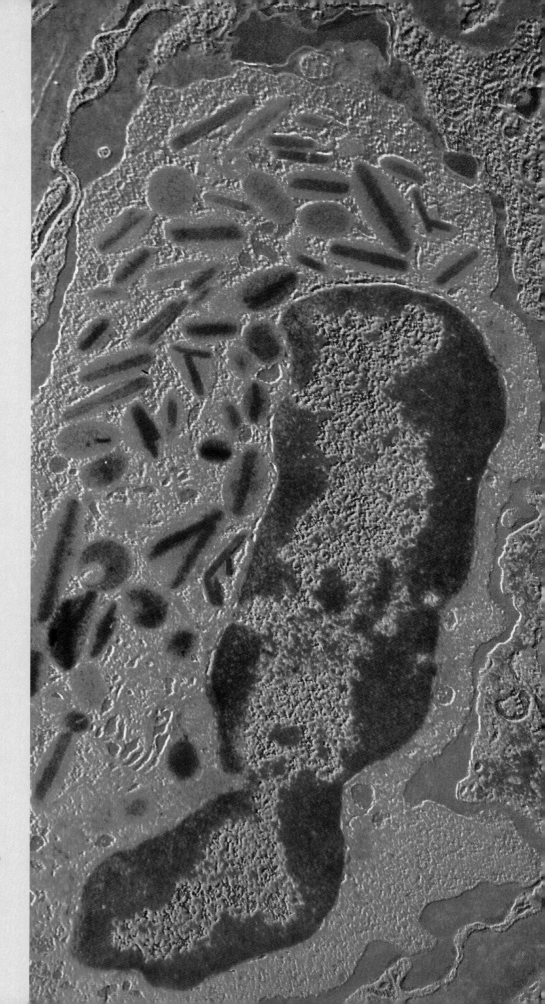

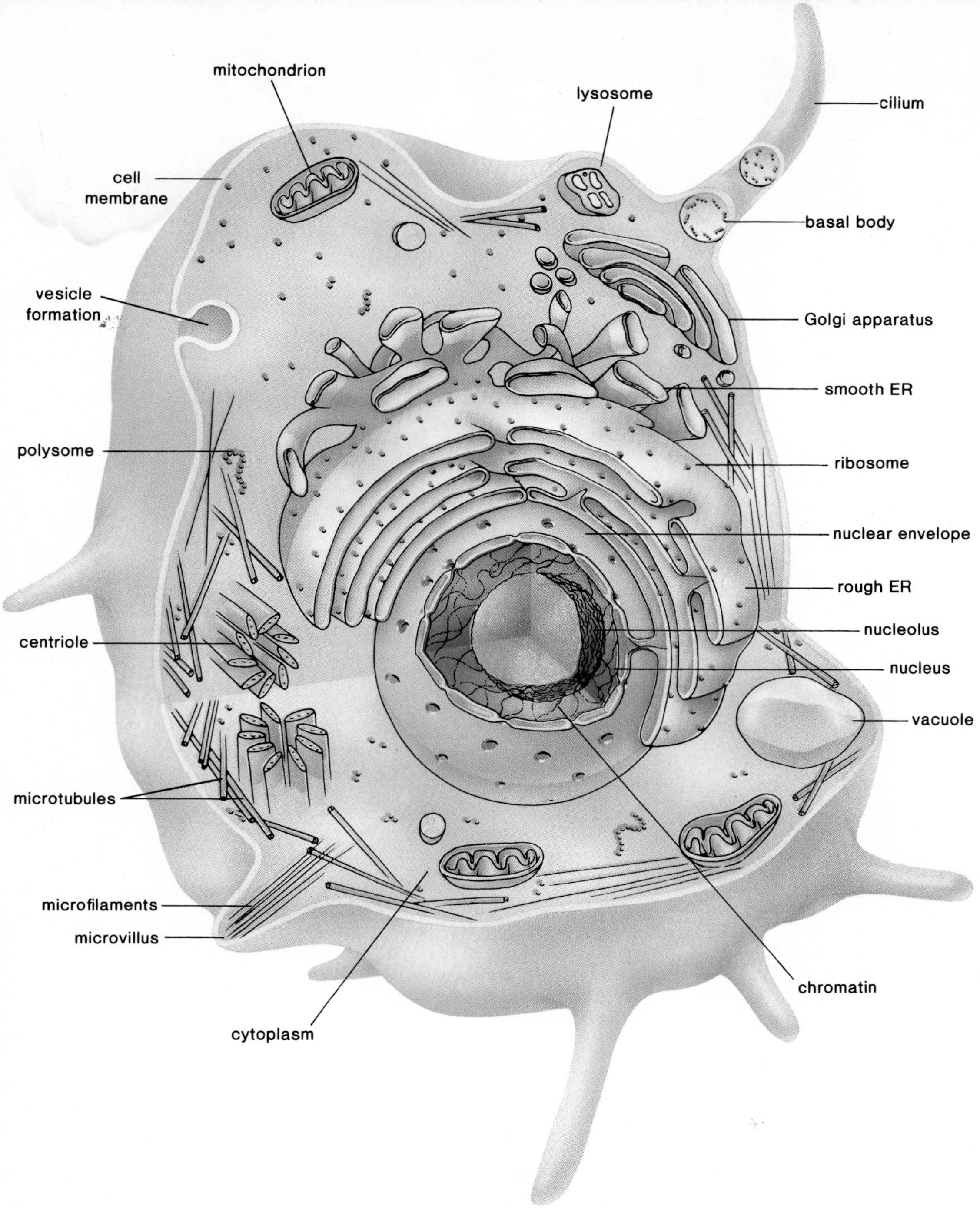

Figure 2.1   Animal cells. These generalized representations are based on electron micrographs.

We are multicellular animals. The cell (fig. 2.1) is the fundamental unit of our bodies, and it is at the cellular level that we must understand health and disease. Because cells are microscopic, it is sometimes hard to imagine that it is not the intestine or heart that is causing the difficulty—it is the cells that make up the intestine or heart. Nowhere is this more evident than when we study the cause of cancer of the uterus, the lungs, the colon, and so forth. In all cases, cancer is characterized by uncontrolled growth of cells due to irregularities of cell structure and function.

Human Issue

Most scientists, on the basis of data, believe that a chemical evolution produced the first cell(s), and thereafter all organisms evolved from this (these) cell(s). Many lay people, on the basis of faith, believe that God created all living things. Advocates of divine creationism argue that biology texts should include creationism because students have the right to be presented with an alternative viewpoint to evolution. The vast majority of scientists argue that only ideas generated by scientific investigation should be presented in biology texts. Should biology texts be required to include the theory of divine creation? If you were a school board member and this issue arose in your school district, what position would you take?

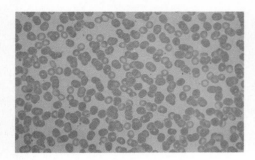
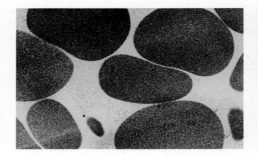
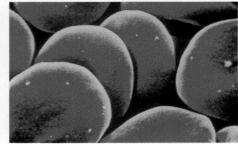

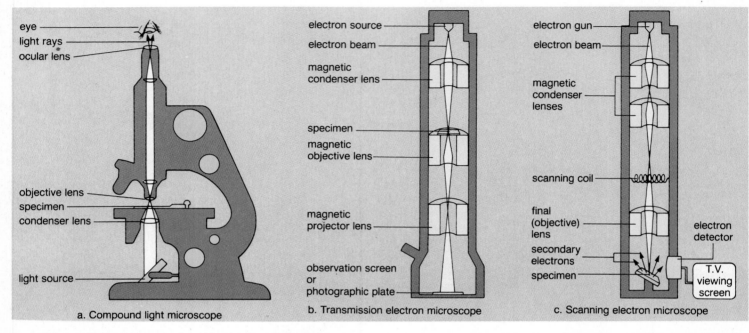

a. Compound light microscope

eye
light rays
ocular lens
objective lens
specimen
condenser lens
light source

b. Transmission electron microscope

electron source
electron beam
magnetic condenser lens
specimen
magnetic objective lens
magnetic projector lens
observation screen or photographic plate

c. Scanning electron microscope

electron gun
electron beam
magnetic condenser lenses
scanning coil
final (objective) lens
secondary electrons
specimen
electron detector
T.V. viewing screen

Figure 2.2   *a.* Light micrograph of red blood cells. A light microscope uses light to view a specimen. The specimen can be living if it is thin enough to allow light to pass through. A light microscope does not magnify or distinguish as much detail as the electron microscope. *b.* Transmission electron micrograph (TEM) of red blood cells. The transmission electron microscope uses electrons to "view" the specimen. The specimen is nonliving and must be thin enough to allow electrons to pass through. The magnification and amount of detail seen is far greater than with the light microscope. *c.* Scanning electron micrograph (SEM) of red blood cells. The scanning microscope scans the surface of the specimen with an electron beam. The secondary electrons given off are collected and the result is a three-dimensional image of the specimen.

Generalized Cell

Your body and all living things are made up of cells. Human cells come in many different shapes and sizes, but no matter what the shape or the size, each one has the same basic organization revealed by microscopy.

Microscopy

Two types of microscopes are commonly used to look at cells. *Light microscopes* utilize light to view the object, and electron microscopes utilize electrons. A *transmission electron microscope* gives us an image of the interior of the object, while a *scanning electron microscope* provides a three-dimensional view of the surface of an object.

The useful magnification of a transmission electron microscope ($100,000\times$) is much greater than the useful magnification of a light microscope ($1,000\times$) because the resolving power is much greater (fig. 2.2). Resolving power is the ca-

pacity to distinguish between two points. If two points are seen as separate, then the image appears more detailed than if the two points are seen as one point. At the very best, a light microscope can distinguish two points separated by 200 nm (nanometer = 1×10^{-6} mm), but the transmission electron microscope can distinguish two points separated by about 0.2 nm.

In the case of scanning electron microscopy, a narrow beam of electrons is scanned over the surface of the specimen, which has been coated with a thin metal layer. The metal gives off secondary electrons that are collected in order to produce a television-type picture of the specimen's surface on a screen.

New types of scanning microscopes, called scanning-probe microscopes, have been invented recently. In one type, laser light is focused on a tiny probe that is pressed against an organic polymer. Reflected light tells the position of the probe as it moves up and down the polymer, and these data allow a computer to give a three-dimensional picture of a single biological molecule. The probe moves up and down in response to electron move-

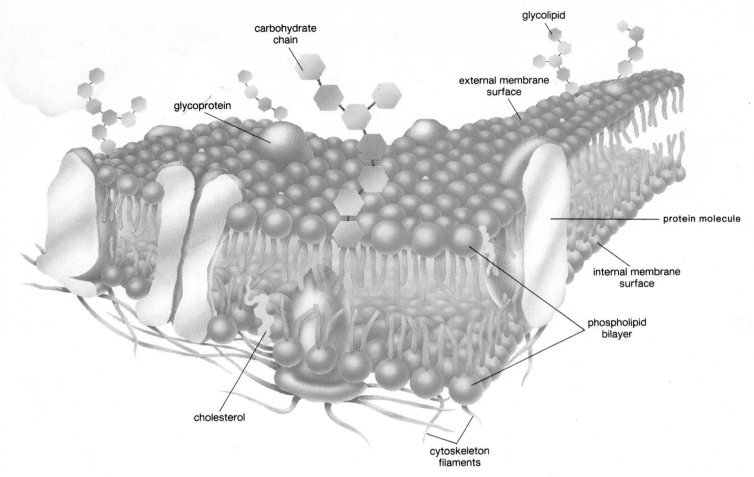

carbohydrate
chain

glycoprotein

glycolipid

external membrane
surface

protein molecule

internal membrane
surface

phospholipid
bilayer

cholesterol

cytoskeleton
filaments

Figure 2.3   Fluid-mosaic model of the cell membrane. The membrane is composed of a phospholipid bilayer with embedded proteins. The polar heads of the phospholipids are at the surfaces of the membrane; the nonpolar tails make up the interior of the membrane. The carbohydrate chains of glycolipids and glycoproteins are involved in cell-to-cell recognition. Proteins also function as receptors for chemical messengers, conductors of molecules through the membrane, and as enzymes in metabolic reactions.

ments and atomic forces that exist between the probe and the material.

A picture obtained by using the light microscope sometimes is called a photomicrograph, and a picture resulting from the use of electron microscopes is called a transmission electron micrograph (TEM) or a scanning electron micrograph (SEM), depending on the type of microscope used.

Electron micrographs have helped biologists develop an understanding of cell structure.

Membrane

Figure 2.1 shows that an animal cell is surrounded by an outer membrane or **cell membrane,** within which is found the **cytoplasm,** the substance of the cell outside the nucleus. Membrane not only surrounds the cell, it also makes up many of the organelles, small bodies found within the cytoplasm that have specific structures and functions.

Membrane Structure

The *fluid-mosaic model* of membrane structures tells us that protein molecules have a changing pattern (form a mosaic) within a bilayer of phospholipid molecules that are fluid, having a consistency of light oil. Notice the manner in which the phospholipid molecules arrange themselves in figure 2.3. Their structure, discussed in chapter 1, causes each molecule to have a polar head and nonpolar tails. Within the phospholipid bilayer, the tails face inward and the heads face outward, where they are likely to encounter a watery environment. Membranes also contain cholesterol, another type of lipid, which is arranged as depicted in figure 2.3.

Short chains of sugars are attached to the outer surface of some protein and lipid molecules (called glycoproteins and glycolipids, respectively). It is believed that these carbohydrate chains specific to each cell mark it as belonging to a particular individual and account for such characteristics as blood type or why a patient's system sometimes rejects an organ transplant. Other glycoproteins have a special configuration that allows

Table 2.1 Passage of Molecules into and out of Cells

Name	Direction	Requirements	Examples
diffusion	toward lesser concentration	concentration gradient	lipid-soluble molecules, water, and gases (oxygen and carbon dioxide)
transport			
facilitated	toward lesser concentration	carrier	sugars and amino acids
active	toward greater concentration	carrier plus energy	sugars, amino acids, and ions
exocytosis	toward greater concentration	vacuole release	secretion of substances
endocytosis	toward greater concentration	vacuole formation	phagocytosis of substances

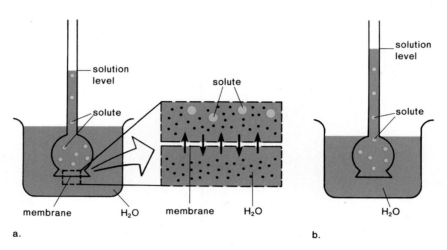

a.   b.

Figure 2.4   Osmosis demonstration. *a.* A thistle tube, covered at the broad end by a membrane, contains a solute (*large circles*) in addition to a solvent (*small circles*). The beaker contains only solvent. The solute is unable to pass through the membrane, but the solvent passes through in both directions. *b.* There is a net movement of solvent toward the inside of the thistle tube. This causes the solution to rise in the thistle tube until a pressure equivalent to osmotic pressure builds.

them to act as a receptor for a chemical messenger like a hormone. Some cell membrane proteins form channels through which certain substances can enter cells or are **carriers** involved in the passage of molecules through the membrane.

Membrane Function

The cell membrane forms a boundary between the outside of the cell and the inside of the cell. It allows only certain molecules to enter and exit the cytoplasm freely; therefore, the cell membrane is said to be **selectively permeable.** Small molecules that are lipid soluble, such as oxygen and carbon dioxide, can pass through the membrane easily. Certain other small molecules, like water, are not lipid soluble but still cross the membrane passively by moving through a protein channel. Still other molecules require the use of a carrier in order to enter a cell.

The cell membrane, composed of phospholipid and protein molecules, is selectively permeable and regulates the entrance and exit of molecules from the cell.

Diffusion and Osmosis

Lipid-soluble, including gases, and water molecules pass through the membrane by **diffusion** (table 2.1), the movement of molecules from the area of greater concentration to the area of lesser concentration until they are equally distributed. To illustrate diffusion, imagine opening a perfume bottle in the corner of a room. The smell of the perfume soon penetrates the room because the molecules that make up the perfume have drifted to all parts of the room. Another example is putting a tablet of dye into water. The water eventually takes on the color of the dye as the tablet dissolves.

Osmosis is the diffusion of water across a cell membrane. It occurs whenever there is an unequal concentration of water on either side of a selectively permeable membrane. For example, figure 2.4 represents a thistle tube covered with a selectively permeable membrane. The tube contains a sugar solution and the beaker contains distilled water. Because of the presence of the sugar (solute), there is a lesser concentration of water (solvent) inside the tube than there is outside the tube. Since the sugar cannot cross the membrane, there will be a net move-

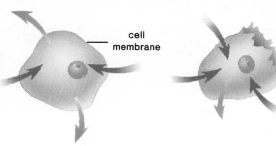

cell membrane

a. Under isotonic conditions, there is no net movement of water.

b. In a hypotonic environment, water enters the cell, which may burst (lysis) due to osmotic pressure.

c. In a hypertonic environment, water leaves the cell, which shrivels (crenation).

Figure 2.5   Osmosis. The arrows indicate the movement of water. *a.* In an isotonic solution, a cell neither gains nor loses water. *b.* In a hypotonic solution, a cell gains water. *c.* In a hypertonic solution, a cell loses water.

ment of water to the inside of the tube. The final height of the solution in the tube indicates the amount of **osmotic pressure** signified by the flow of water into the tube.

Since cytoplasm contains proteins and salts and is surrounded by a selectively permeable membrane, a cell exerts osmotic pressure when it is placed in a **hypotonic** solution, which contains a greater concentration of water (lesser concentration of solute) than does the cell. Under these circumstances the cell will swell or even burst (fig. 2.5). When a cell is placed in a **hypertonic** solution, which contains a lesser concentration of water (greater concentration of solute) than does cytoplasm, the cell loses water and becomes dehydrated. In an **isotonic** solution, the osmotic pressure is similar on both sides of the membrane, and there is no net movement of water.

Carriers

Most solutes do not simply diffuse across a cell membrane; rather, they are transported by means of protein carriers within the membrane. During **facilitated transport,** a molecule (e.g., an amino acid or glucose) is transported across the cell membrane from the side of higher concentration to the side of lower concentration. The cell does not need to expend energy for this type of transport because the molecule is moving in the normal direction.

During **active transport** a molecule is moving contrary to the normal direction; that is, from lower to higher concentration. Iodine collects in the cells of the thyroid gland; sugar is completely absorbed from the gut by cells that line the digestive tract; and sodium (Na^+) is sometimes almost completely withdrawn from urine by cells lining the kidney tubules. Active transport requires a protein carrier and the use of cellular energy obtained from the breakdown of ATP (p. 27). When ATP is broken down, energy is released and in this case the energy is used by a carrier to carry out active transport.

Certain small molecules, like water and gases, diffuse across a cell membrane. The diffusion of water, termed osmosis, can cause a cell to swell or dehydrate depending on the environmental medium. Other molecules must be transported by means of protein carriers found in the membrane.

Cellular Organelles

The cell contains a number of **organelles,** small bodies with specific structure and functions (table 2.2). These help the cell carry out its many activities.

Nucleus

The **nucleus,** the largest organelle found within the cell, is enclosed by a double-layered **nuclear envelope** that is actually continuous with the endoplasmic reticulum discussed in the following paragraphs. As illustrated in figure 2.6, there are pores, or openings, in the envelope through which large molecules pass from the *nucleoplasm,* the fluid portion of the nucleus, to the cytoplasm or vice versa.

The nucleus is of primary importance in the cell because it is the control center that oversees the metabolic functioning of the cell and ultimately determines the cell's characteristics. Within the nucleus there are masses of threads called **chromatin,** so called because they take up stains and become colored. Chromatin is indistinct in the nondividing cell, but it condenses to rodlike structures called **chromosomes** at the time of cell division. Chemical analysis shows that chromatin, and thus chromosomes, contain the chemical DNA (deoxyribonucleic acid) along with certain proteins and some RNA (ribonucleic acid). This is not surprising because we already know

Table 2.2 Cellular Organelles (Simplified)

Name	Structure	Function
cell membrane	bilayer of phospholipid and globular proteins	passage of molecules into and out of cell
nucleus	nuclear envelope surrounds chromatin, nucleolus, and nucleoplasm	control of cell
nucleolus	concentrated area of RNA in the nucleus	ribosome formation
chromatin (chromosomes)	composed of DNA and protein	contains hereditary information
endoplasmic reticulum	folds of membrane forming flattened channels and tubular canals	transport by means of vesicles
rough	studded with ribosomes	protein synthesis
smooth	having no ribosomes	lipid and carbohydrate synthesis
ribosome	RNA and protein in two subunits	protein synthesis
Golgi apparatus	stack of membranous saccules	packaging and secretion
vacuole and vesicle	membranous sacs	containers of material
lysosome	membranous container of hydrolytic enzymes	intracellular digestion
mitochondrion	inner membrane (cristae) within outer membrane	cellular respiration
cytoskeleton	—	cell shape and subcellular movement
microfilament	actin and myosin proteins	same
microtubule	tubulin protein	same
cilium and flagellum	9 + 2 pattern of microtubules	movement of cell
centriole	9 + 0 pattern of microtubules	organization of microtubules; associated with cell division

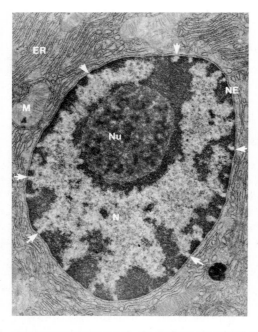

Figure 2.6 An electron micrograph of a nucleus (N) with a clearly defined nucleolus (Nu) and irregular patches of chromatin scattered throughout the nucleoplasm. The nuclear envelope (NE) contains pores indicated by the arrows. This nucleus is surrounded by endoplasmic reticulum (ER), and its size may be compared to the mitochondrion (M) that appears to the left.

that chromosomes contain the genes and that the genes are composed of DNA. DNA, with the help of RNA, controls protein synthesis within the cytoplasm, and it is this function that allows DNA to control the cell.

Nucleoli

One or more **nucleoli** are present in the nucleus. These dark-staining bodies are actually specialized parts of chromatin in which a type of RNA called ribosomal RNA (rRNA) is produced. Ribosomal RNA joins with proteins to form organelles, subunits that become part of the ribosomes.

The nucleus contains chromatin, which condenses into chromosomes just prior to cell division. Chromosomes contain DNA that, with the help of RNA, directs protein synthesis in the cytoplasm. Another type of RNA, rRNA, is made within the nucleolus before migrating to the cytoplasm, where it is incorporated into ribosomes.

Membranous Canals and Vacuoles

The endoplasmic reticulum (ER), the Golgi apparatus, vacuoles, and lysosomes (fig. 2.1) are structurally and functionally related membranous structures. Ribosomes are not composed of membrane, but they are included in this category because they often are associated intimately with the ER.

Endoplasmic Reticulum

The **endoplasmic reticulum** (fig. 2.7) forms a membranous system of tubular canals that is continuous with the nuclear envelope and branches throughout the cytoplasm. Ribosomes can be attached to the ER, and if so, the reticulum is called **rough ER**; if ribosomes are not present, the ER is called **smooth ER** (fig. 2.7b). Rough ER specializes in protein synthesis, but smooth ER has different functions in different cells. Sometimes it spe-

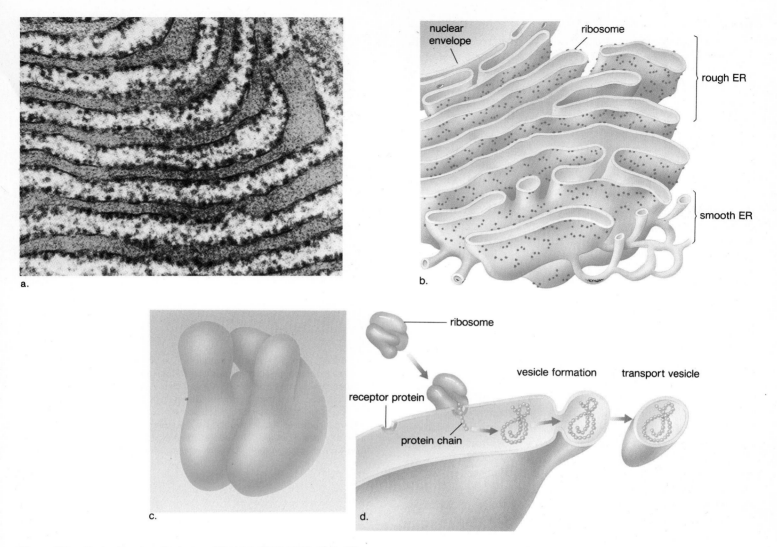

Figure 2.7   Endoplasmic reticulum. *a.* Electron micrograph of rough ER shows a cross section of many flattened vesicles with ribosomes attached to the sides that abut the cytoplasm. *b.* Drawing that shows the 3 dimensions of the ER. *c.* Model of a single ribosome illustrates that each is actually composed of 2 subunits. *d.* Method by which the ER acts as a transport system.

cializes in the production of lipids. For example, smooth ER is abundant in cells of the testes and the adrenal cortex, both of which produce steroid hormones. In muscle cells, smooth ER acts as a storage area for calcium ions that are released when contraction occurs. In the liver, it is involved in the detoxification of drugs, including alcohol. It is quite possible that drugs are detoxified within structures called **peroxisomes,** membrane-bound vacuoles often attached to smooth ER that contain enzymes capable of carrying out oxidation of various substances.

Ribosomes   **Ribosomes** look like small, dense granules in low-power electron micrographs (fig. 2.7*a*), but higher-resolution micrographs show that each contains two subunits (fig. 2.7*c*). Each of these subunits has a particular mix of rRNA and proteins. We already have mentioned that rRNA is joined with proteins within the nucleus, but the two subunits are not assembled into one ribosome until they reach the cytoplasm.

Ribosomes can lie free within the cytoplasm, where they also are involved in protein synthesis. In these instances, several ribosomes, each of which is producing the same type of protein, are arranged in a functional group called a *polysome.* Most likely, these proteins are for use inside the cell.

The ribosomes attached to rough ER are making proteins for export from the cell. The proteins enter the lumen (interior space) of the rough ER (fig. 2.7*d*) and proceed to the lumen of the smooth ER. A small vesicle then pinches off from the smooth ER. Most vesicles formed in this way move through the cytoplasm to the Golgi apparatus, where the proteins are received and are processed further. This is how the ER serves as a transport system.

Golgi Apparatus

The **Golgi apparatus** (fig. 2.8) is named for the person who first discovered its presence in cells. It is composed of a stack of about a half-dozen or more saccules that look like hollow pancakes. In human cells, one side of the stack (the inner face) is directed toward the nucleus, and the other side of the stack (the outer face) is directed toward the cell membrane. Vesicles, small vacuoles, occur at the edges of the saccules.

Biochemical analyses suggest that the Golgi apparatus receives protein-filled vesicles from the ER at its inner face. After this, the proteins move from one saccule to the next via newly formed vesicles. In the meantime, the proteins are modified by the enzymes present within the saccules. Vesicles containing the modified proteins move to different locations in the cell (fig. 2.8). Some of the vesicles formed by the Golgi apparatus move to the cell membrane, where they discharge their protein contents in a process called exocytosis (table 2.1). Because this is *secretion,* it often is said that the Golgi apparatus is involved in processing, packaging, and secretion.

Lysosomes

Lysosomes (fig. 2.8) are membrane-bound vesicles formed by the Golgi apparatus that contain *hydrolytic enzymes.* Hydrolytic enzymes digest macromolecules in the manner described in figure 1.8, p. 19. Macromolecules sometimes are brought into a cell by vesicle formation at the cell membrane, a process called endocytosis (table 2.1). A lysosome can fuse with an endocytic vesicle and can digest its contents into simpler molecules that then enter the cytoplasm. Some white blood cells defend the body by engulfing bacteria, a process that involves vesicle formation at the cell membrane. When lysosomes fuse with these vesicles, the bacteria are digested. It should come as no surprise, then, that even parts of a cell are digested by its own lysosomes (called autodigestion because auto = self). Normal cell rejuvenation most likely takes place in this way, but autodigestion is also important during development. For example, the fingers of a human embryo are at first webbed, but they are freed from one another following lysosomal action.

Occasionally, a child is born with a metabolic disorder involving a missing or inactive lysosomal enzyme. In these cases, the lysosomes fill to capacity with macromolecules that cannot be broken down. The best-known of these lysosomal storage disorders is Tay-Sachs disease, discussed on page 365. Scientists have found a cure for Gaucher's disease, another lysosomal storage disorder (p. 381).

The ER is a membranous system of tubular canals that can be smooth or rough. Proteins synthesized at the rough ER are processed and are packaged in vesicles by the Golgi apparatus. Some vesicles discharge their contents at the cell membrane, and some are lysosomes that digest any material enclosed therein.

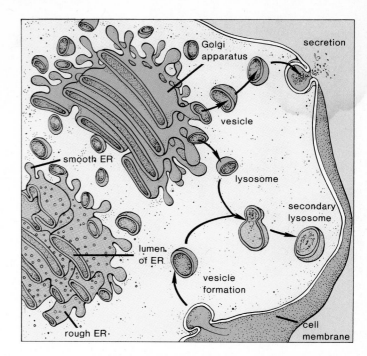

Figure 2.8 Golgi apparatus function. The Golgi apparatus receives vesicles from the smooth ER and thereafter forms at least 2 types of vesicles, lysosomes and secretory vesicles. Lysosomes contain hydrolytic enzymes that can break down large molecules. Vesicles bringing large molecules into a cell sometimes join with lysosomes, forming structures called secondary lysosomes. Thereafter, the molecules are digested. The secretory vesicles formed at the Golgi apparatus discharge their contents at the cell membrane.

Human Issue

Human cells contain mitochondria in which glucose products are broken down to release energy. Plant cells have mitochondria but they also have chloroplasts in which glucose is made using solar energy. Plants make food for themselves and all living things. Since humans are dependent on plants, do you think human biology courses should include more information about plants? Or do you think it is sufficient for us to understand the workings of our own bodies?

Mitochondria

A **mitochondrion** (fig. 2.9) is bounded by a double membrane. The inner membrane is folded to form little shelves, called *cristae,* which project into the *matrix,* an inner space filled with a gel-like fluid.

Mitochondria produce cellular energy in the form of ATP (adenosine triphosphate). Every cell uses a certain amount of ATP energy to synthesize molecules, but many cells use ATP

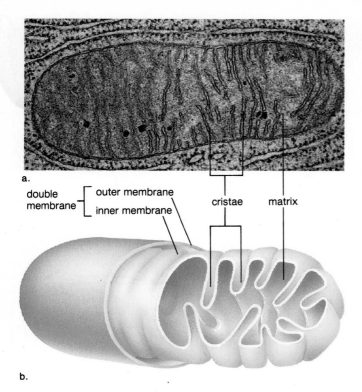

a.

double
membrane { outer membrane
         inner membrane   cristae   matrix

b.

Figure 2.9   Mitochondria structure. *a.* Electron micrograph.
b. Generalized drawing in which the outer membrane and a portion of
the inner membrane have been cut away to reveal the cristae.

to carry out their specialized function. For example, muscle cells use ATP for muscle contraction that produces movement, and nerve cells use it for the conduction of nerve impulses so we are aware of our environment.

Mitochondria often are called the powerhouses of the cell: just as a powerhouse burns fuel to produce electricity, the mitochondria use glucose products to produce ATP molecules. In the process, mitochondria use up oxygen and give off carbon dioxide and water. The oxygen you breathe in enters cells and then the mitochondria; the carbon dioxide you breathe out is released by the mitochondria. Because gas exchange is involved, it is said that mitochondria carry on **aerobic cellular respiration.** A shorthand way to indicate the chemical transformation associated with cellular respiration is

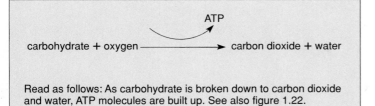

ATP

carbohydrate + oxygen ⟶ carbon dioxide + water

Read as follows: As carbohydrate is broken down to carbon dioxide and water, ATP molecules are built up. See also figure 1.22.

Mitochondria are the sites of cellular respiration, a process that provides ATP energy molecules to the cell.

Cytoskeleton

Several types of filamentous protein structures form a **cyto-skeleton** (fig. 2.10) that helps maintain the cell's shape and either anchors the organelles or assists their movement as appropriate. The cytoskeleton includes microfilaments and microtubules. **Microfilaments** are long, extremely thin fibers that usually occur in bundles or other groupings. Microfilaments have been isolated from a number of cells. When analyzed chemically, their composition is similar to that of actin, one of the two proteins responsible for muscle contraction.

Microtubules are shaped like thin cylinders and are several times larger than microfilaments. Each cylinder contains 13 rows of tubulin, a globular protein, arranged in a helical fashion. Aside from existing independently in the cytoplasm, microtubules are also found in certain organelles, such as cilia, flagella, and centrioles.

Remarkably, both microfilaments and microtubules assemble and disassemble within the cell. When they are assembled, the protein molecules are bonded together, and when they are disassembled, the protein molecules are not attached to one another. When microfilaments and microtubules are assembled, the cell has a particular shape, and when they disassemble, the cell can change shape.

The cytoskeleton contains microfilaments and microtubules. Microfilaments, thin actin strands, and microtubules, thirteen rows of tubulin protein molecules arranged to form a hollow cylinder, maintain the shape of the cell and also direct the movement of cell parts.

Centrioles

Centrioles are short cylinders with a 9 + 0 pattern of microtubules. There are 9 outer microtubule triplets and no center microtubules (fig. 2.11). There is always one pair of centrioles lying at right angles to one another near the nucleus (fig. 2.1). Before a cell divides, the centrioles duplicate, and the members of the new pair are also at right angles to one another.

Centrioles are part of a microtubule organizing region which also includes other proteins and substances. Microtubules begin to assemble in this region, and then they grow outward, extending through the entire cytoplasm. Centrioles give rise to basal bodies that direct the formation of cilia and flagella. Centrioles may also be involved in other cellular processes that use microtubules, such as movement of material throughout

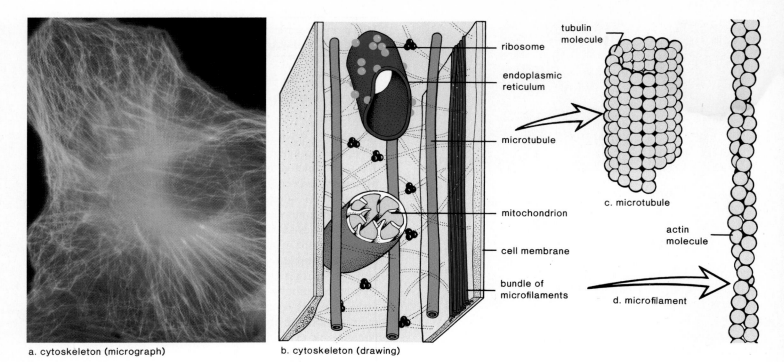

a. cytoskeleton (micrograph)

b. cytoskeleton (drawing)

ribosome

endoplasmic reticulum

microtubule

mitochondrion

cell membrane

bundle of microfilaments

tubulin molecule

c. microtubule

actin molecule

d. microfilament

Figure 2.10   The cytoskeleton. *a.* Light micrograph of a cell treated to reveal microfilaments (red) and microtubules (green). Notice how the microfilaments extend into the processes, and the microtubules seem to emanate from a region near the nucleus (oval area). *b.* Drawing based on electron micrographs shows how the cytoskeleton anchors the organelles. The cytoskeleton also can allow these organelles to move about. *c.* A microtubule is a cylinder composed of tubulin molecules. *d.* A microfilament is composed of actin molecules.

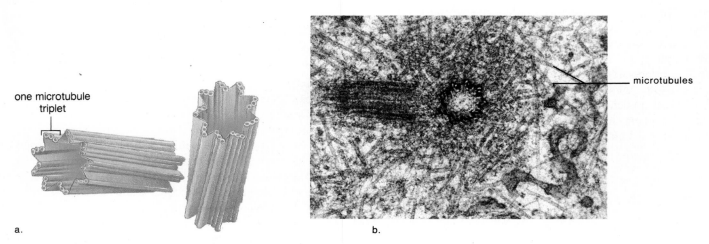

one microtubule triplet

a.

b.

microtubules

Figure 2.11   Centrioles. *a.* Drawing of centrioles showing their 9 + 0 arrangement of microtubule triplets. (The zero in this equation means that there are no microtubules in the center of the organelle.)
b. Electron micrograph of centrioles verifies that they lie at right angles to one another. Notice the large number of microtubules near the centrioles, which lie within the microtubule organizing region.

the cell or the appearance and disappearance of the spindle apparatus (p. 341). Their exact role in these processes is uncertain, however.

Cilia and Flagella

Cilia and **flagella** are hairlike projections of cells that can move either in an undulating fashion, like a whip, or stiffly, like an oar. Cells that have these organelles are capable of movement. For example, sperm cells, carrying genetic material to the egg, move by means of flagella (fig. 2.12). The cells that line our respiratory tract are ciliated (fig. 3.4). These cilia sweep debris trapped within mucus back up the throat, and this action helps keep the lungs clean.

Each cilium and flagellum has a basal body lying in the cytoplasm at its base. **Basal bodies,** like centrioles, have a 9 + 0 pattern of microtubule triplets. They are believed to organize the structure of cilia and flagella even though cilia and flagella have a 9 + 2 pattern of microtubules. In cilia and flagella, there are nine microtubule doublets surrounding two central microtubules. This arrangement is believed to be necessary to their ability to move.

Centrioles have a 9 + 0 pattern of microtubules and give rise to basal bodies that organize the 9 + 2 pattern of microtubules in cilia and flagella. Centrioles may be connected in some way to the origination of microtubules and to the spindle fibers that are seen during cell division.

Cellular Metabolism

Cellular **metabolism** includes all the chemical reactions that occur in a cell. Quite often these reactions are organized into metabolic pathways:

$$1 \quad 2 \quad 3 \quad 4 \quad 5 \quad 6$$
$$A \rightarrow B \rightarrow C \rightarrow D \rightarrow E \rightarrow F \rightarrow G$$

The letters, except A and G, are *products* of the previous reaction and the *reactants* for the next reaction. A represents the beginning reactant(s), and G represents the end product(s). The numbers in the pathway refer to different enzymes. *Every reaction in a cell requires a specific enzyme.* In effect, no reaction occurs in a cell unless its enzyme is present. For example, if enzyme number 2 in the diagram is missing, the pathway cannot function; it will stop at B. Since enzymes are so necessary in cells, their mechanism of action has been studied extensively.

Metabolic pathways contain many enzymes that are arranged to perform their reactions in a sequential order.

Enzymes

When an enzyme speeds up a reaction, the reactant(s) that participates in the reaction is called the enzyme's *substrate(s)*. Enzymes are often named for their substrate(s) (table 2.3).

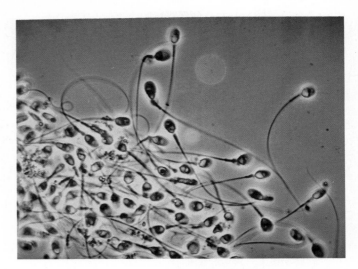

Figure 2.12   Sperm cells use long, whiplike flagella to move about.

Table 2.3   Enzymes Named for Their Substrates

Substrate	Enzyme
lipid	lipase
urea	urease
maltose	maltase
ribonucleic acid	ribonuclease
lactose	lactase

Enzymes have a specific region, called an **active site,** where the substrates are brought together so that they can react. An enzyme's specificity is caused by the shape of the active site, where the enzyme and its substrate(s) fit together in a specific way, much as the pieces of a jigsaw puzzle fit together (fig. 2.13). After one reaction has been completed, the product or products are released, and the enzyme is ready to catalyze another reaction. What we have said can be summarized in the following manner:

$$E + S \rightarrow ES \rightarrow E + P$$

(where E = enzyme, S = substrate, ES = enzyme substrate complex, and P = product).

Environmental conditions such as an incorrect pH or high temperature can cause an enzyme to become denatured. A denatured enzyme no longer has its usual shape and is therefore unable to speed up its reaction.

Coenzymes

Many enzymes have **coenzymes,** nonprotein portions that assist the enzyme and may even accept or contribute atoms to the reaction. It is of interest that vitamins are often components of coenzymes. The vitamin niacin is a part of the coenzyme NAD

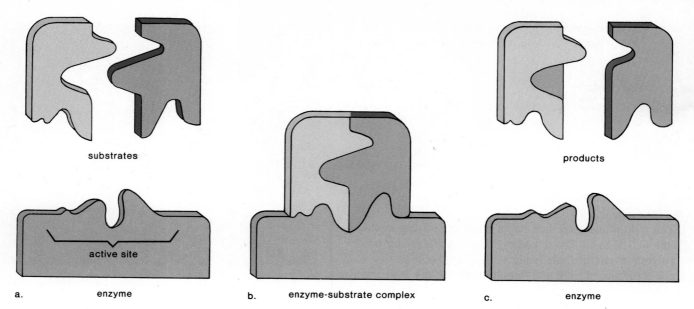

Figure 2.13 Enzymatic action. *a.* An enzyme has an active site where the substrates and enzyme fit together in such a way that the substrates are oriented to react. *b.* Following the reaction, the products are released. *c.* The enzyme assumes its prior shape.

that removes hydrogen (H) atoms from substrates and therefore is called a *dehydrogenase.* NAD that is carrying hydrogen atoms is written as $NADH_2$ because NAD removes two hydrogen atoms at a time.

When molecules are broken down, hydrogen atoms are sometimes removed by NAD. As we shall see below, the removal of hydrogen atoms, a form of *oxidation,* releases energy that can be used for ATP buildup.

Enzymes are specific because they have an active site that accommodates their substrates. Enzymes often have nonprotein helpers called coenzymes. NAD is the usual coenzyme of oxidation and reduction within a cell.

Cellular Respiration

Cellular respiration is an important part of cellular metabolism because it accounts for ATP buildup in cells. Cellular respiration includes *aerobic* (requires oxygen) cellular respiration and fermentation, an *anaerobic* (does not require oxygen) process.

Aerobic Cellular Respiration

During **aerobic cellular respiration,** glucose is broken down to carbon dioxide and water. Even though it is possible to write an overall equation for the process, it does not occur in one step. Glucose breakdown requires three subpathways: *glycolysis,* the *Krebs cycle,* and the *respiratory chain* (fig. 2.14). The location of these subpathways is as follows:

Glycolysis—occurs in the cytoplasm, outside a mitochondrion
Krebs cycle—occurs in the matrix of a mitochondrion
Respiratory chain—occurs on the cristae of a mitochondrion

Each arrow you see in glycolysis and the Krebs cycle (fig. 2.14) represents a different enzyme, and the letters represent the product of the previous reaction and the substrate for the next reaction. Notice how each pathway resembles a conveyor belt in which a beginning substrate continuously enters at the start and, after a series of reactions, end products leave at the termination of the belt. It is important to realize, too, that all three pathways are going on at the same time. They can be compared to the inner workings of a watch, in which all parts are synchronized.

It is possible to relate the reactants and products of the overall equation for aerobic cellular respiration to the subpathways:

1. Glucose, C_6, is to be associated with **glycolysis,** the breakdown of glucose to two molecules of pyruvate (PYR), a C_3 molecule. During glycolysis, energy is released as hydrogen (H) atoms are removed. This energy is used to form two ATP molecules as described in figure 1.21.

2. Carbon dioxide, CO_2, is to be associated with the transition reaction and the Krebs cycle. During the transition reaction, PYR is converted to active acetate (AA), a C_2 molecule, after CO_2 comes off. Because the transition reaction occurs twice per glucose molecule, two molecules of CO_2 are released. Hydrogen (H) atoms are also removed at this time.

AA enters the **Krebs cycle,** a cyclical series of reactions that give off CO_2 molecules and produce one ATP molecule. Since the Krebs cycle occurs twice per glucose molecule, altogether four CO_2 and two ATP are produced per glucose molecule. Hydrogen (H) atoms are also removed as the Krebs cycle occurs.

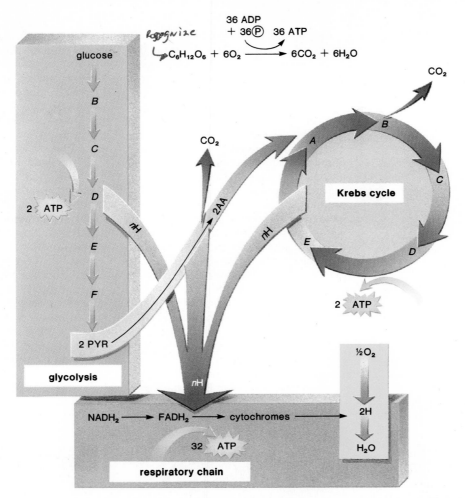

Figure 2.14 Aerobic cellular respiration contains three subpathways; glycolysis, the Krebs cycle, and the respiratory chain. As the reactions occur, a number of hydrogen atoms (nH) and carbon dioxide molecules are removed from the various substrates. Oxygen acts as the final acceptor for the hydrogen atoms.

3. Oxygen, O_2, and water, H_2O, are to be associated with the **respiratory chain.** The respiratory chain begins with $NADH_2$, the coenzyme that carries most of the hydrogen (H) atoms to the chain. The respiratory chain is a series of molecules that pass hydrogen atoms from one to the other until they are finally received by oxygen, which then becomes water. As hydrogen atoms pass down the chain, oxidation occurs and energy is released to allow the buildup of ATP.

4. ATP is to be associated with glycolysis, the Krebs cycle, and the respiratory chain. Most ATP, however, is produced by the respiratory chain. Usually 32 ATP are produced altogether by the respiratory chain.

Table 2.4 summarizes the discussion of aerobic cellular respiration.

Aerobic cellular respiration requires glycolysis, which takes place in the cytoplasm; the Krebs cycle, which is located in the matrix of the mitochondria; and the respiratory chain, which is located on the cristae of the mitochondria.

Table 2.4 Overview of Aerobic Cellular Respiration

Name of pathway	Result
glycolysis	removal of H from substrates produces 2 ATP
transition reaction	removal of H from substrates releases 2 CO_2
Krebs cycle	removal of H from subtrates releases 4 CO_2 produces 2 ATP after 2 turns
respiratory chain	accepts H from other pathways and passes them on to O_2, producing H_2O produces 32 ATP

Fermentation

Fermentation is an anaerobic process. When oxygen is not available to cells, the respiratory chain soon becomes inoperative because oxygen is not present to accept hydrogen atoms. In this case, most cells have a safety valve so that some ATP can still be produced. The glycolytic pathway will run as long as it is

Alcoholic Beverages

Wine, beer, and whiskey production all require yeast fermentation. To produce wine, grape juice is allowed to ferment. After the grapes are picked, they are crushed in order that the juice may be collected. In the old days, wine makers simply relied on spontaneous fermentation by yeasts that were on the grape skins, but now many add specially selected cultures of yeast. Also, it is now common practice to maintain the temperature at about 20° C for white wines and 28° C for red wines. Fermentation ceases after most of the

sugar has been converted to alcohol. Various methods are used to clarify the wine; that is, to remove any suspended materials. Also many fine wines improve when they are allowed to "age" during barrel or bottle storage.

Brewing beer is more complicated than wine production. Usually grains of barley are first *malted;* that is, allowed to germinate for a short time so that amylase enzymes that will break down the starch content of the grain are produced. After the germinated grains have been crushed and mixed with water, the *malt wort* is separated from the spent grains and traditionally boiled with hops (an herb derived from the hop plant) to give flavor to the beer. Next, the *hop wort* is seeded with a strain of yeast that converts the sugars in the wort to alcohol and carbon dioxide. At the end of fermentation, the yeast is separated from the beer, which is then allowed to mature for an appropriate period. After filtration and pasteurization, the beer is packaged.

The production of whiskey (from grains), brandy (from grapes), and rum (from molasses) differs from wine and beer production chiefly in that the alcohol is removed from the fermented substance by distillation. Most often in the United States, corn or rye is used in the production of whiskey. These grains are ground up and mashed to release their starch content. Then amylase enzymes are added to convert the starch to fermentable sugars. Now yeast is added so that fermentation can occur. Following fermentation, the alcohol is concentrated by distillation. A warm temperature causes the alcohol to become gaseous and rise in a column where it condenses to a liquid before entering a collecting vessel. The alcohol content of the collecting vessel is much higher following this distillation process. The distillate is usually stored, quite often in an oak barrel, to improve the aroma and taste of the final product.

a.

b.

c.

Fermentation of *a.* corn, *b.* barley, and *c.* grapes produces whiskey, beer, and wine respectively.

supplied with "free" NAD; that is, NAD that can pick up hydrogen atoms. Normally, $NADH_2$ passes hydrogen to the respiratory chain and thereby becomes "free" of hydrogen atoms. However, if the chain is not working due to lack of oxygen, $NADH_2$ passes its hydrogen atoms to PYR as shown in the following reaction:

$$NADH_2 \longrightarrow NAD$$
$$\text{pyruvic acid} \longrightarrow \text{lactic acid}$$

The Krebs cycle and respiratory chain do not function as part of fermentation, but when oxygen is available again, lactic acid can be converted back to pyruvic acid and metabolism can proceed as usual.

Fermentation is an impractical process for two reasons. First, since glycolysis alone is occurring, it produces only two ATP per glucose molecule. Second, it results in lactic acid buildup. Lactic acid is toxic to cells and causes muscles to cramp and fatigue. If fermentation continues for any length of time, death will follow.

It is of interest to know that fermentation takes its name from yeast fermentation. Yeast fermentation produces alcohol and carbon dioxide (instead of lactic acid). When yeast is used to leaven bread, it is the carbon dioxide that produces the desired effect. When yeast is used to produce alcoholic beverages, it is the alcohol that humans make use of.

Fermentation is an anaerobic process, a process that does not require oxygen, but produces very little ATP per glucose molecule and results in lactic acid buildup.

Summary

The human cell is surrounded by a cell membrane, which regulates the entrance and exit of molecules. Some molecules, such as water and gases, diffuse through the membrane. The direction in which water diffuses is dependent on the tonicity of the cell.

Table 2.2 lists the cell organelles we have studied in the chapter. The nucleus is a large organelle of primary importance because it controls the rest of the cell. Within the nucleus lies the chromatin, which condenses to become chromosomes during cell division.

Proteins are made at the rough endoplasmic reticulum before being packaged at the Golgi apparatus. Golgi-derived lysosomes fuse with incoming vesicles to digest any material enclosed within, and lysosomes also carry out autodigestion of old parts of cells.

Mitochondria are the powerhouses of the cell. During the process of aerobic cellular respiration, mitochondria convert carbohydrate energy to ATP energy.

Microfilaments and microtubules are found within a cytoskeleton that maintains the shape and permits movement of cell parts. Centrioles are associated with the spindle apparatus during cell division, and they also produce basal bodies that give rise to cilia and flagella.

Cellular metabolism uses pathways in which there are a series of reactions that proceed in an orderly step-by-step manner. Each of these reactions requires a specific enzyme. Sometimes enzymes require coenzymes, nonprotein portions that participate in the reaction. NAD is a coenzyme.

Aerobic cellular respiration (the breakdown of glucose to carbon dioxide and water) requires three subpathways: glycolysis, the Krebs cycle, and the respiratory chain. If oxygen is not available in cells, the respiratory chain is inoperative and fermentation (an anaerobic process) occurs. Fermentation makes use of glycolysis only, plus one more reaction in which PYR is reduced to lactic acid.

Study Questions

1. Describe the structure and biochemical makeup of membrane. (p. 33)
2. What are the three mechanisms by which substances enter and exit cells? Define isotonic, hypertonic, and hypotonic solutions. (p. 34)
3. Describe the nucleus and its contents, including the terms DNA and RNA in your description. (p. 35)
4. Describe the structure and function of endoplasmic reticulum. Include the terms rough and smooth ER and ribosomes in your description. (p. 36)
5. Describe the structure and function of the Golgi apparatus. Mention vesicles and lysosomes in your description. (p. 38)
6. Describe the structure of mitochondria, and relate this structure to the pathways of aerobic cellular respiration (pp. 38, 42)
7. Describe the composition of the cytoskeleton. (p. 39)
8. Describe the structure and function of centrioles, cilia, and flagella. (p. 41)
9. Discuss and draw a diagram for a metabolic pathway. Discuss and give a reaction to describe the specificity theory of enzymatic action. Define coenzyme. (p. 41)
10. Name and describe the events within the three subpathways that make up aerobic cellular respiration. Why is fermentation wasteful and potentially harmful to the human body? (p. 42)

Objective Questions

KNOW

For questions 1–5, match the organelles in the key to their functions.

Key:
- a. mitochondria
- b. nucleus
- c. Golgi apparatus
- d. rough ER
- e. centrioles

1. packaging and secretion ___C___
2. cell division ___e___
3. powerhouses of the cell ___A___
4. protein synthesis ___a___
5. control center for cell ___B___
6. Microfilaments and microtubules are a part of the _Cytoskeleton_ the framework of the cell that provides its shape and regulates movement of organelles.
7. Water will enter a cell when it is placed in a _hypotonic_ solution.
8. Substrates react at the _active site_, located on the surface of their enzyme.
9. During aerobic cellular respiration, most of the ATP molecules are produced at the _respiratory chain_, a series of carriers located on the _cristae_ of mitochondria.
10. Fermentation produces only _2_ ATP compared to the _36_ ATP produced by aerobic cellular respiration.

Answers

1. c 2. e 3. a 4. d 5. b 6. cytoskeleton
7. hypotonic 8. active site 9. respiratory chain,
cristae 10. 2, 36

Label this Diagram.
See figure 2.1 (p. 31) in text.

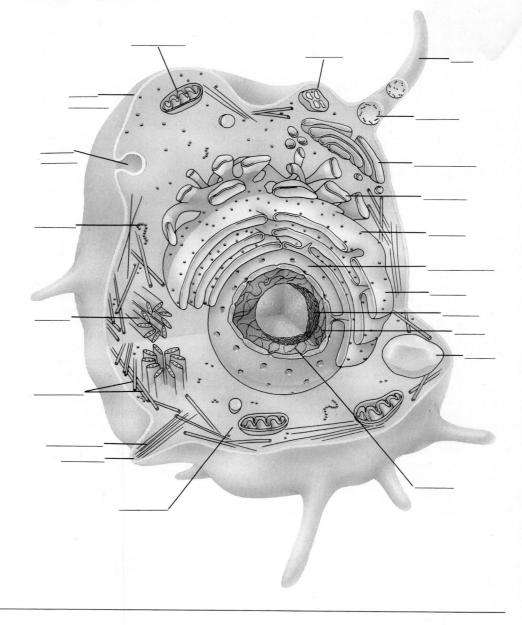

Critical Thinking Questions

1. What is the advantage of organelles in complex cells?

2. Present evidence that the cell is dynamic rather than static as it appears to be in drawings and micrographs.

3. Show that human beings have a cellular basis by describing the function of the oxygen we breathe in and the origination of the carbon dioxide we breathe out.

Selected Key Terms

active site (ak′tiv sīt)   the region on the surface of an enzyme where the substrate binds and where the reaction occurs.   41

centriole (sen′tre-ōl)   a short, cylindrical organelle in animal cells that contains microtubules in a 9 + 0 pattern and is associated with the formation of the spindle apparatus during cell division.   39

chromosomes (kro′mo-sōmz)   rod-shaped bodies in the nucleus, particularly during cell division, that contain the hereditary units or genes.   35

coenzyme (ko-en′zīm)   a nonprotein molecule that aids the action of an enzyme, to which it is loosely bound.   41

cytoplasm (si'to-plazm'') the ground substance of cells located between the nucleus and the cell membrane. 33

cytoskeleton (si''to-skel'ĕ-ton) filamentous protein structures found throughout the cytoplasm that help maintain the shape of the cell. 39

endoplasmic reticulum (en''do-plas'mik rĕ-tik'u-lum) a complex system of tubules, vesicles, and sacs in cells; sometimes having attached ribosomes. 36

fermentation (fer''men-ta'shun) anaerobic breakdown of carbohydrates that results in organic end products such as alcohol and lactic acid. 43

glycolysis (gli-kol'ĭ-sis) the metabolic pathway that converts sugars to simpler compounds and ends with pyruvate. 42

Golgi apparatus (gol'je ap''ah-ra'tus) an organelle that consists of a stack of saccules and functions in the packaging and secretion of cellular products. 38

Krebs cycle (krebz si'kl) a series of reactions found within the matrix of mitochondria that give off carbon dioxide; also called the citric acid cycle because the reactions begin and end with citric acid. 42

lysosome (li'so-sōm) an organelle in which digestion takes place due to the action of hydrolytic enzymes. 38

microfilament (mi''kro-fil'ah-ment) an extremely thin fiber found within the cytoplasm that is involved in the maintenance of cell shape and movement of cell contents. 39

microtubule (mi''kro-tu'būl) an organelle composed of thirteen rows of globular proteins; found in multiple units in several other organelles, such as the centriole, cilia, and flagella. 39

mitochondrion (mi''to-kon'dre-on) an organelle in which aerobic cellular respiration produces the energy molecule ATP. 38

nucleolus (nu-kle'o-lus) an organelle found inside the nucleus; composed largely of RNA for ribosome formation. 36

nucleus (nu'kle-us) a large organelle containing the chromosomes and acting as a control center for the cell. 35

organelles (or''gah-nelz') specialized structures within cells, such as the nucleus, mitochondria, and endoplasmic reticulum. 35

respiratory chain (re-spi'rah-to''re chān) a series of carriers within the inner mitochondrial membrane that pass electrons one to the other from a higher energy level to a lower energy level; the energy released is used to build ATP; also called the electron transport system; the cytochrome system. 43

ribosomes (ri'bo-sōmz) minute particles, found attached to endoplasmic reticulum or loose in the cytoplasm, that are the site of protein synthesis. 37

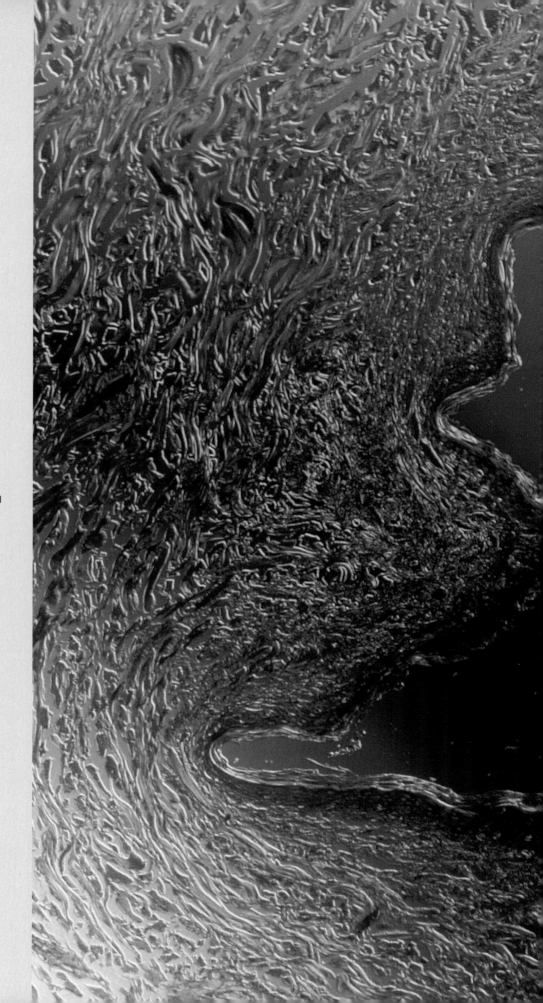

Chapter Three

Human

Organization

Chapter Concepts

1 Human tissues can be categorized into four major types: epithelial, connective, muscular, and nervous tissues.

2 Organs usually contain several types of tissues. For example, although skin is composed primarily of epithelial tissue and connective tissue, it also contains muscle and nerve fibers.

3 Organs are grouped into organ systems, each of which has specialized functions.

4 Humans exhibit a marked ability to maintain a relatively constant internal environment.

5 All organ systems contribute to homeostasis.

Our skin is an outer boundary that serves as a barrier between ourselves and the environment. This cross section shows the hardened cells that make up the ridges of a fingertip above a much larger layer containing cells carrying out various functions.

It's all very well to think of humans as being multicellular animals, but actually the cells are differentiated and have specific structures and functions. Like cells form tissues that are grouped into muscular tissue, nervous tissue, connective tissue, and epithelial tissue (fig. 3.1). We all know that muscular tissue contracts, nervous tissue conducts, and connective tissue binds, but what about epithelial tissue? Epithelial tissue forms a protective covering on the exterior of the body and all its various orifices. Because of epithelial tissue, the body is like a citadel that forbids the entrance of microorganisms.

Types of Tissues

Figure 3.2 shows that the human body has levels of organization. Cells of the same type are joined to form a tissue. Different tissues are found in an organ, and various types of organs are arranged into an organ system. Finally, the organ systems make up the organism.

Epithelial Tissue

Epithelial tissue, also called epithelium, forms a continuous layer, or sheet, over the entire body surface and most of the body's inner cavities. On the external surface, it forms a covering that protects the body from injury and drying out. On internal surfaces, epithelial tissue may be specialized for other functions in addition to protection; for example, it secretes mucus along the digestive tract; it sweeps up impurities from the lungs by means of hairlike extensions called cilia; and it efficiently absorbs molecules from kidney tubules because of fine cellular extensions called microvilli.

There are three types of epithelial tissue. **Squamous epithelium** (fig. 3.3a) is composed of flat cells and is found lining the lungs and the blood vessels. **Cuboidal epithelium** (fig. 3.3b) contains cube-shaped cells and is found lining the kidney tubules. In **columnar epithelium** (fig. 3.3c), the cells resemble pillars or columns, and nuclei usually are located near the bottom of each cell. This epithelium is found lining the digestive tract. An epithelium can have microvilli or cilia as appropriate for its particular function. For example, the oviducts are lined by ciliated columnar cells that beat to propel the egg and the embryo toward the uterus, or the womb.

An epithelium can be simple or stratified. Simple means that the tissue has a single layer of cells, and **stratified** means that the tissue has layers piled one on top of the other. Table 3.1 gives the body locations for squamous and stratified squamous epithelium, for example. One type of epithelium is **pseudostratified**—it appears to be layered, but actually true layers do not exist because each cell touches a baseline. The lining of the windpipe, or trachea, is *pseudostratified ciliated columnar epithelium* (fig. 3.4).

An epithelium sometimes secretes a product, in which case it is described as glandular. A gland can be a single epithelial cell, as in the case of the mucus-secreting goblet cells found within the columnar epithelium lining the digestive tract (fig. 3.3c), or a gland can contain numerous cells. Glands that secrete their product into ducts are called **exocrine glands,** and those that secrete their product directly into the bloodstream are called **endocrine glands.**

Epithelial tissue is classified according to the shape of the cell. There can be one or many layers of cells, and the layer lining a cavity can be ciliated and/or secretory.

Junctions between Cells

Epithelial tissue cells are packed tightly and are joined to one another in one of three ways: spot desmosomes, tight junctions, and gap junctions (fig. 3.5). In a *spot desmosome,* internal cytoplasmic plaques, firmly attached to the cytoskeleton within each cell, are joined by intercellular filaments. The cells are joined more closely in a *tight junction* because adjacent cell membrane (tight junction) proteins actually attach to each other, producing a zipperlike fastening. A *gap junction* is formed when two identical cell membrane channels join. This lends strength, but it also allows substances to pass between the two cells.

An epithelium often is joined to underlying connective tissue by a so-called *basement membrane*. We now know that the basement membrane is glycoprotein reinforced by fibers supplied by the connective tissue.

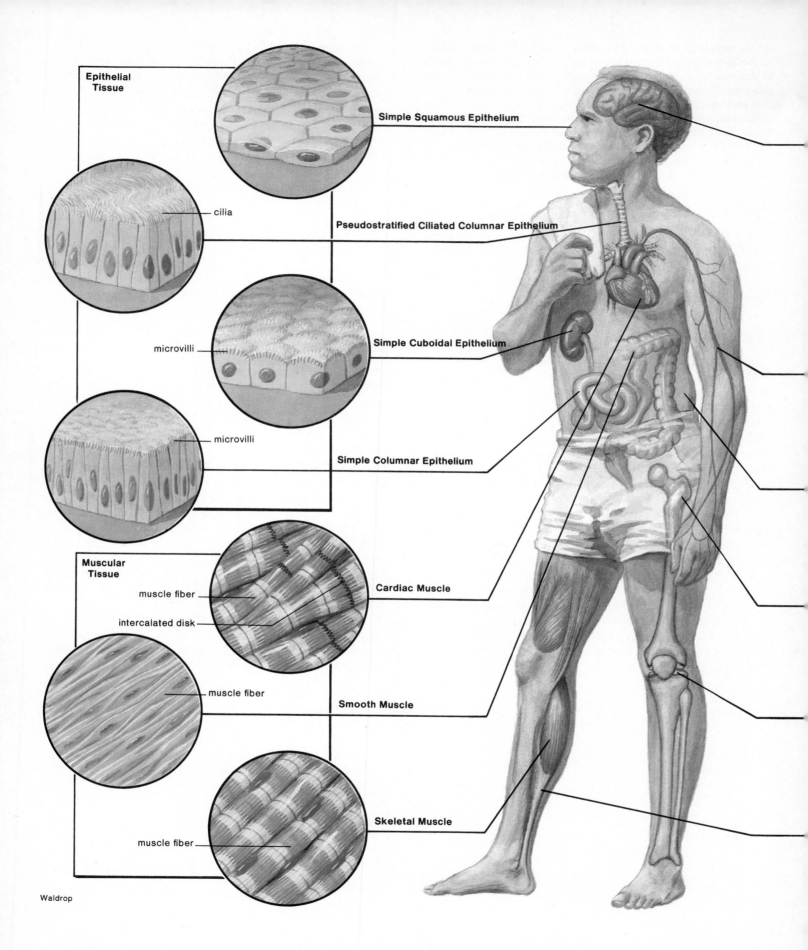

Epithelial Tissue

Simple Squamous Epithelium

cilia

Pseudostratified Ciliated Columnar Epithelium

microvilli

Simple Cuboidal Epithelium

microvilli

Simple Columnar Epithelium

Muscular Tissue

muscle fiber

intercalated disk

Cardiac Muscle

muscle fiber

Smooth Muscle

muscle fiber

Skeletal Muscle

Waldrop

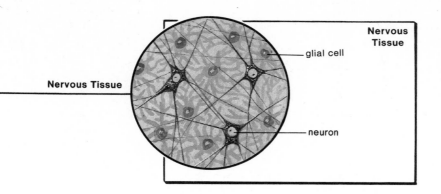

Nervous Tissue

Figure 3.1   The major tissues in the human body. Reading clockwise, observe that the nervous tissue contains specialized cells called neurons. Connective tissue includes blood, adipose tissue, bone cartilage, and fibrous connective tissue. Muscular tissue is of 3 types: skeletal, smooth, and cardiac. Epithelial tissue includes columnar, cuboidal, ciliated columnar, and squamous epithelium. Each type of epithelium can be stratified or ciliated; not all types are shown.

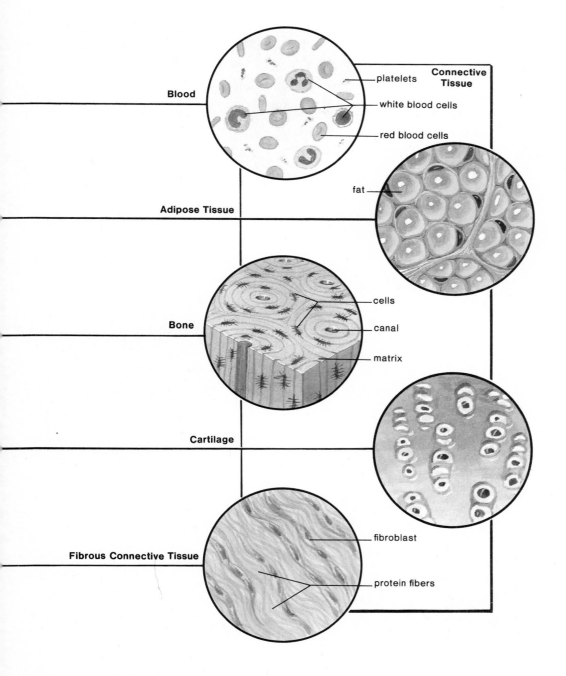

Blood

Adipose Tissue

Bone

Cartilage

Fibrous Connective Tissue

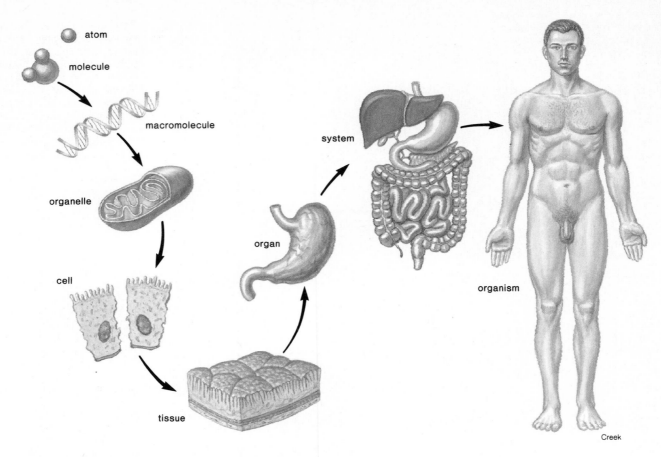

Figure 3.2 Levels of organization in the human body. Cells are composed of molecules, tissues are made up of cells, organs are composed of tissues, and the organism contains organ systems.

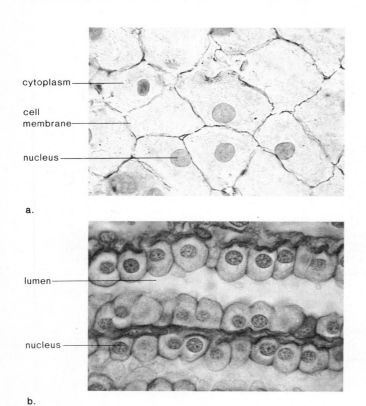

a.

b.

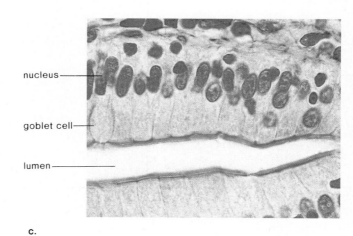

c.

Figure 3.3 Simple epithelial tissue. a. Simple squamous epithelium consists of a single layer of thin cells. b. Simple cuboidal epithelium is composed of cells that look like cubes. c. Simple columnar epithelium contains cells that resemble columns because they are elongated. Epithelial tissue lines cavities (lumen = cavity).

Table 3.1 Epithelial Tissue

Type	Function	Location
simple squamous	filtration, diffusion, osmosis	air sacs of lungs; walls of capillaries; lining of blood vessels
simple cuboidal	secretion, absorption	surface of ovaries; lining of kidney tubules
simple columnar	protection, secretion, absorption	lining of uterus; tubes of the digestive tract
ciliated columnar	movement of sex cells	lining of various tubes of reproductive system
pseudostratified ciliated columnar	protection, secretion, movement of mucus	lining of respiratory passages
stratified squamous	protection	outer layer of skin; lining of mouth cavity, vagina, and anal canal

From John W. Hole, Jr., *Human Anatomy and Physiology,* 5th ed. Copyright © 1990 Wm. C. Brown Publishers, Dubuque, Iowa. All Rights Reserved. Reprinted by permission.

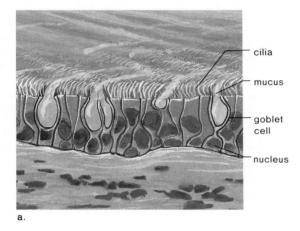

a.

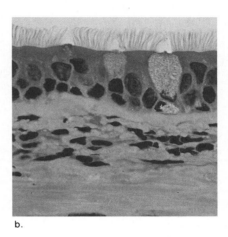

b.

Figure 3.4 Pseudostratified ciliated columnar epithelium forms the lining of the windpipe. *a.* Note the goblet cells and that all cells touch the baseline. When you cough, material trapped in the mucus secreted by goblet cells is moved upward to the throat, where it can be swallowed. *b.* Photomicrograph of pseudostratified ciliated columnar epithelium.

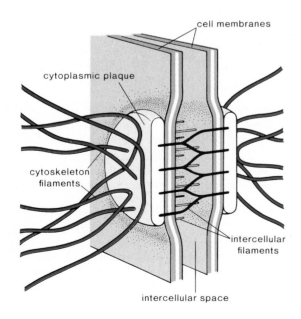

a. spot desmosome

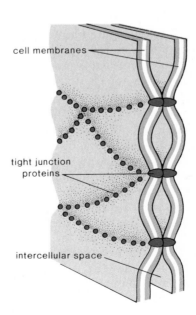

b. tight junction

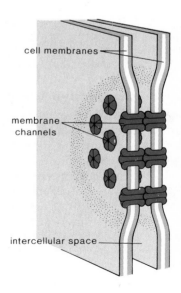

c. gap junction

Figure 3.5 Junction between epithelial cells. Epithelial tissue cells are held tightly together by *a.* spot desmosomes and *b.* tight junctions. *c.* Gap junctions allow materials to pass from cell to cell.

Table 3.2 Connective Tissue

Type	Function	Location
loose connective tissue	binds organs	beneath the skin; beneath most epithelial layers
adipose tissue	insulates; stores fat	beneath the skin; around the kidneys
fibrous connective tissue	binds organs	tendons; ligaments
cartilage		
hyaline cartilage	supports; protects	ends of bones; nose; rings in walls of respiratory passages
elastic cartilage	supports; protects	external ear; part of the larynx
fibrocartilage	supports; protects	between bony parts of backbone and knee
bone	supports; protects	bones of skeleton
blood	transports gases, nutrients, and wastes about body; infection fighting; blood clotting	blood vessels

From John W. Hole, Jr., *Human Anatomy and Physiology*, 5th ed. Copyright © 1990 Wm. C. Brown Publishers, Dubuque, Iowa. All Rights Reserved. Reprinted by permission.

Connective Tissue

Connective tissue (table 3.2) binds structures together, provides support and protection, fills spaces, stores fat, and forms blood cells. As a rule, connective tissue cells are separated widely by a **matrix,** in this instance a noncellular material found between cells. The matrix may have fibers of two types. White fibers contain collagen, a substance that gives them flexibility and strength. Yellow fibers contain elastin, a substance that is not as strong as collagen but is more elastic.

Loose Connective Tissue

Loose connective tissue binds structures (fig. 3.6). The cells of this tissue, which are mainly *fibroblasts,* are located some distance from one another and are separated by a jellylike matrix that contains many white collagen fibers and yellow elastic fibers. The collagen fibers occur in bundles and are strong and flexible. The elastic fibers form networks that when stretched return to their original length. As discussed previously, loose connective tissue commonly lies beneath epithelial layers. In certain instances, epithelium and its underlying connective tissue form body membranes (p. 62). In addition, adipose tissue (fig. 3.7) is a type of loose connective tissue in which the fibroblasts enlarge and store fat and in which the intercellular matrix is reduced.

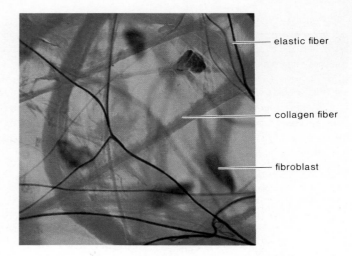

— elastic fiber
— collagen fiber
— fibroblast

Figure 3.6 Loose connective tissue. Loose connective tissue has plenty of space between components. This type of tissue is found surrounding and between the organs.

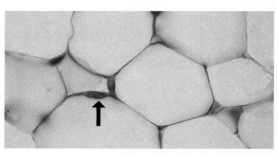

Figure 3.7 Adipose tissue. Adipose cells look like white ghosts because the fat has been washed out during preparation of the tissue. The nucleus of one cell is indicated by the arrow.

Fibrous Connective Tissue

Fibrous connective tissue contains many collagenous fibers that are packed closely together. This type of tissue has more specific functions than loose connective tissue. For example, fibrous connective tissue is found in **tendons,** which connect muscles to bones, and in **ligaments,** which connect bones to other bones at joints. Tendons and ligaments take a long time to heal following an injury because their blood supply is relatively poor.

Loose connective tissue and fibrous connective tissue, which bind body parts, differ according to the type and the abundance of fibers in the matrix.

Cartilage

In **cartilage,** the cells lie in small chambers called **lacunae** (singular, lacuna), separated by a matrix that is solid yet flexible. Unfortunately, because this tissue lacks a direct blood supply, it heals very slowly. There are three types of cartilage, distinguished by the type of fiber in the matrix.

Hyaline cartilage (fig. 3.8), the most common type of cartilage, contains only very fine collagenous fibers. The matrix has a milk-glass appearance. Hyaline cartilage is found in the nose, at the ends of the long bones and the ribs, and in the supporting rings of the windpipe. The fetal skeleton also is made of this type of cartilage. Later, the cartilaginous fetal skeleton is replaced by bone.

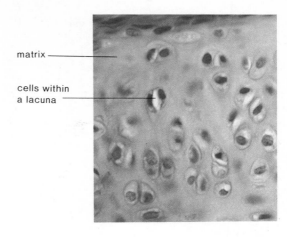

Figure 3.8   Hyaline cartilage. Hyaline cartilage cells, located in lacunae, are separated by a flexible matrix rich in protein and fibers. This type of cartilage forms the embryonic skeleton, which is later replaced by bone.

Human Issue

Human beings put a great deal of emphasis on safety—safety crossing the street, in the workplace, and in a car or plane, for example. Yet, humans will play football, take to the slopes, and do all manner of recreational activities that might injure the body. Every year, thousands of operations are done to remove damaged cartilage from the knee. Are such injuries justifiable?

Elastic cartilage has more elastic fibers than hyaline cartilage. For this reason, it is more flexible and is found, for example, in the framework of the outer ear.

Fibrocartilage has a matrix containing strong collagenous fibers. Fibrocartilage is found in structures that withstand tension and pressure, such as the pads between the vertebrae in the backbone and the wedges found in the knee joint.

Bone

Bone is the most rigid connective tissue. It consists of an extremely hard matrix of calcium salts deposited around protein fibers. The minerals give bone rigidity, and the protein fibers provide elasticity and strength, much as steel rods do in reinforced concrete.

The shaft of a long bone is compact bone (fig. 3.9). In **compact bone,** bone cells (osteocytes) are located in lacunae that are arranged in concentric circles around tiny tubes called Haversian canals. Nerve fibers and blood vessels are in these canals. The latter bring the nutrients that allow bone to renew itself. The nutrients can reach all of the cells because there are minute canals (canaliculi) containing thin processes of the osteocytes that connect them with one another and with the Haversian canals.

The ends of a long bone contain spongy bone (fig. 11.3), which has an entirely different structure. Spongy bone contains numerous bony bars and plates separated by irregular spaces. Although lighter than compact bone, **spongy bone** still is designed for strength. Just as braces are used for support in buildings, the solid portions of spongy bone follow lines of stress.

Cartilage and bone are support tissues. Cartilage is more flexible than bone because the matrix is rich in protein and not calcium salts like that of bone.

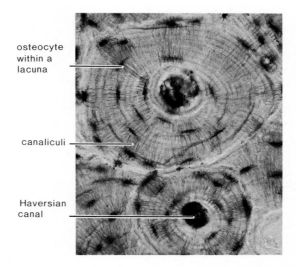

Figure 3.9   Compact bone. Compact bone is highly organized. The cells are arranged in circles about a central (Haversian) canal that contains a nutrient-bearing blood vessel.

Blood

Blood (fig. 3.10) is a connective tissue in which the cells are separated by a liquid called plasma, the contents of which are listed in table 3.3. Blood cells are of two types: red blood cells (**erythrocytes**), which carry oxygen, and white blood cells (**leukocytes**), which aid in fighting infection. Also present in plasma are *platelets,* which are important to the initiation of blood clotting. Platelets are not complete cells; rather, they are fragments of giant cells found in the bone marrow.

Blood is a connective tissue in which the matrix is plasma.

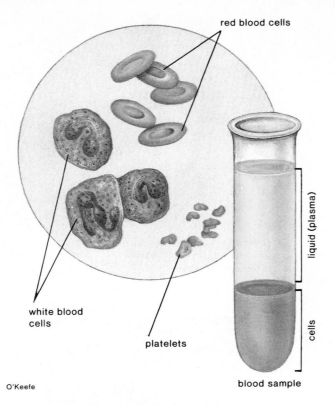

red blood cells

white blood cells

platelets

liquid (plasma)

cells

blood sample

O'Keefe

Figure 3.10 Blood, a liquid tissue. Blood is classified as connective tissue because the cells are separated by a matrix—plasma. Plasma, the liquid portion of blood, contains several types of cells (red blood cells, white blood cells, and platelets).

Table 3.3 Blood Plasma

water	92% of total
inorganic ions (salts)	Na^+, Ca^{++}, K^+, Mg^{++}, Cl^-, HCO_3^-, HPO_4^-, SO_4^-
gases	O_2, CO_2
plasma proteins	albumin, globulins, fibrinogen
organic nutrients	glucose, fats, phospholipids, amino acids, etc.
nitrogenous waste products	urea, ammonia, uric acid
regulatory substances	hormones, enzymes

Blood is unlike other types of connective tissue in that the intercellular matrix (i.e., plasma) is not made by the cells. Plasma (table 3.3) is a mixture of different types of molecules that enter the blood at various locations. Some people do not classify blood as connective tissue; instead, they suggest a separate tissue category for blood called vascular tissue.

Muscular Tissue

Muscular (contractile) tissue is composed of cells that are called *muscle fibers.* Muscle fibers contain actin filaments and myosin filaments, whose interaction accounts for the movements we associated with animals. There are three types of muscles: *skeletal, smooth,* and *cardiac* (table 3.4).

Table 3.4 Muscular Tissue

Type	Fiber appearance	Location	Control
skeletal	striated	attached to skeleton	voluntary
smooth	spindle shaped	internal organs	involuntary
cardiac	striated and branched	heart	involuntary

Skeletal muscle (fig. 3.11*a*) is attached to the bones of the skeleton; it moves body parts. It is under our voluntary control and contracts faster than all the other muscle types. Skeletal muscle cells are cylindrical and quite long—they run the length of the muscle. They arise during development when several cells fuse, giving one multinucleated cell. The nuclei are placed at the periphery of the cell, just inside the cell membrane.

Skeletal muscle cells are **striated.** There are light and dark bands perpendicular to the length of the cell. These bands are due to the placement of actin filaments and myosin filaments in the cell.

Smooth muscle is so named because the cells lack striations. The spindle-shaped cells form layers in which the thick middle portion of one cell is opposite the thin ends of adjacent cells. Consequently, the nuclei form an irregular pattern in the tissue (fig. 3.11*b*). Smooth muscle is not under voluntary control and therefore is said to be involuntary. Smooth muscle, found in walls of viscera (intestine, stomach, and other internal organs) and blood vessels, contracts more slowly than skeletal muscle but can remain contracted for a longer time. When the smooth muscle of the intestine contracts, it moves the food along, and when the smooth muscle of the blood vessels contracts, it constricts the blood vessels, helping to raise the blood pressure.

Cardiac muscle (fig. 3.11*c*), which is found only in the heart, is responsible for the heartbeat. Cardiac muscle seems to combine features of both smooth muscle and skeletal muscle. It has striations like skeletal muscle, but the contraction of the heart is involuntary for the most part. Cardiac muscle cells also differ from skeletal muscle cells in that they have a single, centrally placed nucleus. The cells are branched and seemingly fused one with the other, and the heart appears to be composed of one large interconnecting mass of muscle cells. Actually, cardiac muscle cells are separate and individual, but they are bound end to end at intercalated disks, areas of folded cell membrane between the cells.

All muscular tissue contains actin filaments and myosin filaments; these form a striated pattern in skeletal and cardiac muscle, but not in smooth muscle.

Nervous Tissue

The brain and the nerve cord (also called the spinal cord) contain conducting cells termed neurons. A **neuron** (fig. 3.12) is a specialized cell that has three parts: (1) *dendrites* that conduct

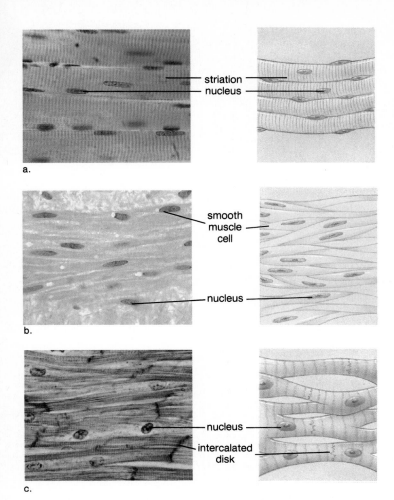

Figure 3.11   Muscular tissue. How do you distinguish a plant from an animal? One way is to detect rapid motion—only animals have contractile fibers that permit movement. *a.* Skeletal muscle is found within the muscles attached to the skeleton. Note the striations and the peripheral location of the nuclei in the multinucleate fibers. *b.* Smooth muscle is found in the walls of internal organs. Note the single nucleus and the lack of striations. *c.* Cardiac muscle pumps the heart. Note the branching of the fibers, the central position of the nuclei, and the presence of intercalated disks, folded cell membranes between adjacent individual cells.

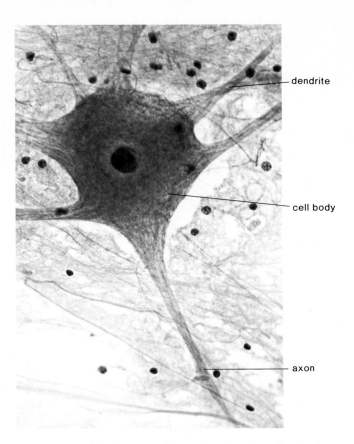

Figure 3.12   Photo of neuron. Conduction of the nerve impulse is dependent on neurons, each of which has the 3 parts indicated. A dendrite takes nerve impulses to the cell body, and an axon takes them away from the cell body.

impulses (send a message) to the cell body; (2) the *cell body* that contains most of the cytoplasm and the nucleus of the neuron; and (3) the *axon* that conducts impulses away from the cell body.

When axons and dendrites are long, they are called *nerve fibers*. Outside the brain and the spinal cord, nerve fibers are bound by connective tissue to form **nerves.** Nerves conduct impulses from sense organs to the spinal cord and the brain, where the phenomenon called sensation occurs. They also conduct nerve impulses away from the spinal cord and the brain to the muscles, causing them to contract.

In addition to neurons, nervous tissue contains **glial cells.** These cells maintain the tissue by supporting and protecting neurons. They also provide nutrients to neurons and help to keep the tissue free of debris.

Organs and Organ Systems

We tend to think that a particular organ contains one type of tissue. For example, we associate muscular tissue with muscles and nervous tissue with the brain. However, these organs also contain other types of tissue; for example, they contain loose connective tissue and blood. An **organ** is a structure that is composed of two or more tissues. An *organ system* contains many different organs.

We are going to consider the skin as an example of an organ. Some people even like to call the skin the *integumentary system,* especially since it cannot be placed in one of the other organ systems; however, the skin does not really have distinct organs.

Skin

The skin (fig. 3.13) covers the body, protecting underlying parts from physical trauma, microbial invasion, and water loss. Skin also helps to regulate body temperature (p. 63), and because it contains sense organs, the skin helps us to be aware of our surroundings and to communicate with others.

Skin has an outer epidermal layer (the epidermis) and an inner dermal layer (the dermis). Beneath the dermis, there is a subcutaneous layer that binds the skin to the underlying organs.

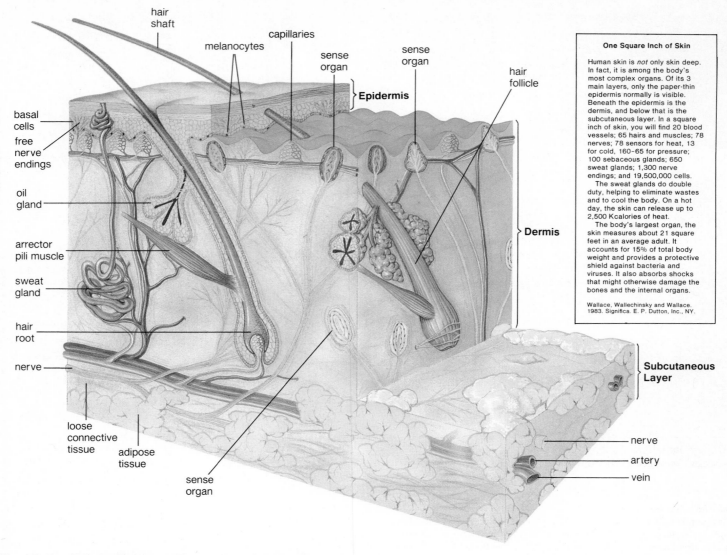

One Square Inch of Skin

Human skin is *not* only skin deep. In fact, it is among the body's most complex organs. Of its 3 main layers, only the paper-thin epidermis normally is visible. Beneath the epidermis is the dermis, and below that is the subcutaneous layer. In a square inch of skin, you will find 20 blood vessels; 65 hairs and muscles; 78 nerves; 78 sensors for heat, 13 for cold, 160–65 for pressure; 100 sebaceous glands; 650 sweat glands; 1,300 nerve endings; and 19,500,000 cells.

The sweat glands do double duty, helping to eliminate wastes and to cool the body. On a hot day, the skin can release up to 2,500 Kcalories of heat.

The body's largest organ, the skin measures about 21 square feet in an average adult. It accounts for 15% of total body weight and provides a protective shield against bacteria and viruses. It also absorbs shocks that might otherwise damage the bones and the internal organs.

Wallace, Wallechinsky and Wallace. 1983. Significa. E. P. Dutton, Inc., NY.

Figure 3.13 Human skin anatomy. Skin contains 3 layers: epidermis, dermis, and the subcutaneous layer.

A burn can affect one or all of these layers. A first-degree burn, which affects only the epidermal layer, is characterized by redness, pain, and swelling. As in a sunburn, the skin usually peels in a few days. A second-degree burn, which affects both the epidermis and dermis, usually causes blistering. A third-degree burn is most serious because it leaves underlying parts with no protection at all. When a person is burned over a large portion of the body, it is sometimes difficult to find enough skin to make autografting (graft from skin remaining) possible. Under these circumstances, physicians can now make use of artificial skin, consisting of two layers. The inner layer is a lattice made from shark cartilage and collagen fibers from cowhide.

The outer layer is a rubberlike silicone plastic. After the artificial skin is sewn in place, the lattice is slowly digested away and replaced by the patient's own cells. At that time, the silicone layer can be safely removed.

Skin Layers

The **epidermis** is the outer, thinner layer of the skin. It is made up of stratified squamous epithelium, which is produced continually by a bottom layer of cells termed basal cells. As newly formed cells are pushed to the surface, they gradually flatten and harden. Eventually, they die and are sloughed off. Hardening is caused by cellular production of a waterproof protein called *keratin*. Over much of the body, keratinization is minimal, but the palm of the hand and the sole of the foot have a particularly thick outer layer of dead keratinized cells arranged in spiral and concentric patterns. We call these patterns fingerprints and footprints.

Specialized cells in the epidermis called *melanocytes* produce melanin, the pigment responsible for skin color in dark-skinned persons. When you sunbathe or visit a tanning salon, the melanocytes become more active, producing melanin in an attempt to protect the skin from the damaging effects of the ultraviolet (UV) radiation from sunlight or tanning machines.

Human Issue

The U.S. government has played an active role in protecting us from drugs, food additives, chemicals, and contagious diseases. For example, the clear role that smoking plays in causing cancer is reflected by the required Surgeon General's warning on the back of cigarette packages. In recent years, the incidence of skin cancer in the United States has dramatically increased. Since there is good evidence that this increase in skin cancer is associated with increased tanning of the skin, should the Surgeon General's office mount a public awareness campaign about the dangers of sunbathing and using tanning machines? How similar should our government's role in reducing skin cancer be to its role in reducing lung cancer?

The epidermis, the outer layer of skin, is made up of stratified squamous epithelial cells. New cells that are produced continually in the innermost layer of the epidermis push outward, become keratinized, die, and are sloughed off.

The **dermis** is a layer of fibrous connective tissue that is deeper and thicker than the epidermis. It contains elastic fibers and collagen fibers. The collagen fibers form bundles that interlace with each other and, for the most part, run parallel to the skin surface. As a person ages and is exposed to the sun, the number of fibers decreases, and those remaining have characteristics that make the skin less supple and cause wrinkling.

There are several types of structures in the dermis. A hair, except for the root, is formed of dead, hardened epidermal cells; the root is alive and resides in a *hair follicle* found in the dermis. Each follicle has one or more *oil (sebaceous) glands* that secrete sebum, an oily substance that lubricates the hair and the skin. Particularly on the nose and the cheeks, the sebaceous glands may fail to discharge, and the secretions collect and form "whiteheads" or "blackheads." The color of blackheads is due to oxidized sebum. If pus-inducing bacteria also are present, a boil or a pimple may result.

A smooth muscle called the *arrector pili* muscle is attached to the hair follicle in such a way that when contracted,

the muscle causes the hair to "stand on end." When you have had a scare or are cold, goose bumps develop due to the contraction of these muscles.

Sweat (sudoriferous) *glands* are quite numerous and are present in all regions of the skin. A sweat gland begins as a coiled tubule within the dermis, but then it straightens out near its opening. Some sweat glands open into hair follicles, and others open onto the surface of the skin.

Small *sense organs* are present in the dermis. There are different sense organs for touch, pressure, pain, and temperature. The fingertips contain the most touch receptors, and these add to our ability to use our fingers for delicate tasks. The dermis also contains nerve fibers and blood vessels. When blood rushes into these vessels, a person blushes, and when blood is reduced in them, a person turns "blue."

The dermis is composed of fibrous connective tissue and lies beneath the epidermis. It contains hair follicles, sebaceous glands, and sweat glands. It also contains sense organs, blood vessels, and nerve fibers.

The **subcutaneous layer,** which lies below the dermis, is composed of loose connective tissue, including adipose tissue. Adipose tissue helps to insulate the body from either gaining heat from the outside or losing heat from the inside. A well-developed subcutaneous layer gives a rounded appearance to the body. Excessive development of this layer accompanies obesity.

Organ Systems

In this text, we are going to study the organ systems depicted in figure 3.14. Each of these systems has a specific location within the body. The central nervous system is located in a *dorsal* (towards the back) cavity, where the brain is protected by the skull, and the spinal cord, which gives off spinal nerves, is protected by the vertebrae (fig. 3.15). The repeating units of vertebrae and spinal nerves show that humans are segmented animals, meaning that body parts recur at regular intervals.

The other internal organs are found within a *ventral* (front) body cavity (fig. 3.15). This cavity is divided by a muscular diaphragm that assists breathing. The heart, a pump for the closed circulatory system, and the lungs are located in the upper *thoracic* (chest) cavity. The major portion of the digestive system and the kidneys are located in the *abdominal* cavity; much of the reproductive system, the urinary bladder, and the terminal portion of the large intestine are located in the *pelvic* cavity.

The musculoskeletal system is another internal organ system. The skeleton provides the surface area for attachment of striated muscles, which are well developed and powerful. The musculoskeletal system makes up most of the body weight and is specialized for locomotion.

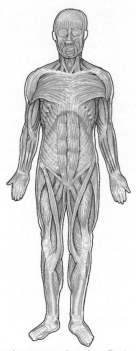

Skeletal system function: Internal support and flexible framework for body movement; production of blood cells

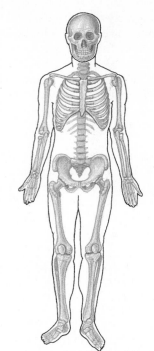

Muscular system function: Body movement; production of body heat

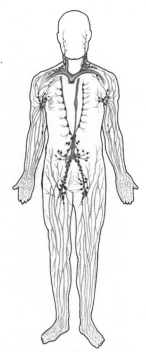

Lymphatic system function: Body immunity; absorption of fats; drainage of tissue fluid

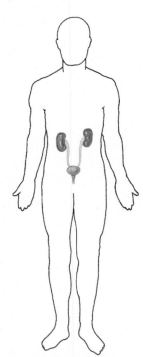

Urinary system function: Filtration of blood; maintenance of volume and chemical composition of the blood

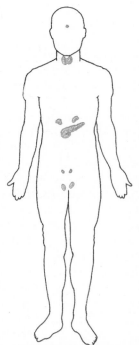

Endocrine system function: Secretion of hormones for chemical regulation

Figure 3.14   Organ systems of the human body.

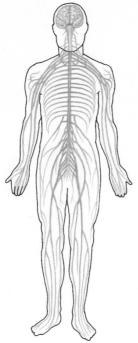

Nervous system function: Regulation of all body activities; learning and memory

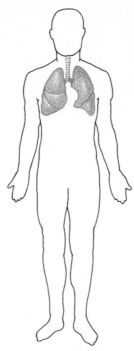

Respiratory system function: Gaseous exchange between external environment and blood

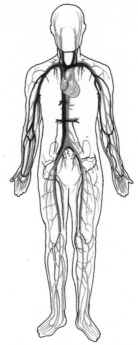

Circulatory system function: Transport of life-sustaining materials to body cells; removal of metabolic wastes from cells

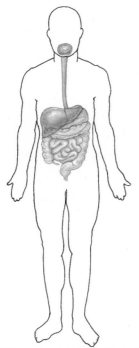

Digestive system function: Breakdown and absorption of food materials

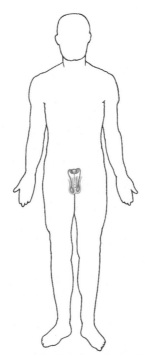

Male reproductive system function: Production of male sex cells (sperm); transfer of sperm to reproductive system of female

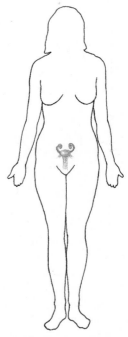

Female reproductive system function: Production of female sex cells (ova); receptacle of sperm from male; site for fertilization of ovum, implantation, and development of embryo and fetus; delivery of fetus

Figure 3.14   Continued

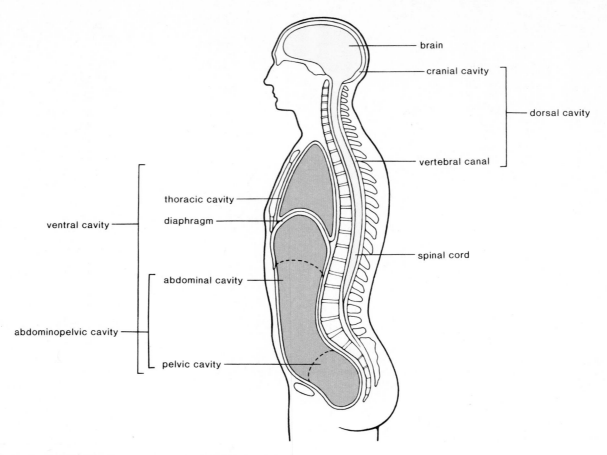

brain
cranial cavity
dorsal cavity
vertebral canal
thoracic cavity
diaphragm
ventral cavity
spinal cord
abdominal cavity
abdominopelvic cavity
pelvic cavity

Figure 3.15   Organs are located in cavities of the human body. Notice that we are obviously vertebrates since we have a backbone composed of vertebrae.

Body Membranes

The term *membrane* on the organ level generally refers to a thin lining or covering that is composed of a layer of epithelial tissue overlying a layer of loose connective tissue. For example, mucous membrane lines the organs of the respiratory and digestive systems. This type of membrane, as its name implies, secretes mucus. Serous membrane lines enclosed cavities and covers the organs that lie within these cavities, such as the heart, lungs, and kidneys. This type of membrane secretes a watery lubricating fluid.

There are numerous cavities within the body. These cavities are usually lined with membrane made of a layer of epithelial tissue and a layer of loose connective tissue.

Homeostasis

Homeostasis means that the internal environment remains relatively constant regardless of the conditions in the external environment. For example:

1. Blood glucose concentration remains at about 0.1%.
2. The pH of the blood is always near 7.4.
3. Blood pressure in the brachial artery remains near 120/80.
4. Blood temperature remains around 37° C (98.6° F).

The ability of the body to keep the internal environment within a certain range allows humans to live in a variety of habitats, such as the arctic regions, deserts, or the tropics.

The internal environment includes a tissue fluid that bathes all of the tissues of the body. The composition of tissue fluid must remain constant if cells are to remain alive and healthy. Tissue fluid is created when water (H_2O), oxygen (O_2), and nutrient molecules leave a capillary (the smallest of the blood vessels). Tissue fluid is purified when water, carbon dioxide (CO_2), and other waste molecules enter a capillary from the fluid (fig. 3.16). Tissue fluid remains constant only as long as blood composition remains constant. Although we are accustomed to using the word *environment* to mean the external environment of the body, it is important to realize that it is the internal environment of tissues that is ultimately responsible for our health and well-being.

The internal environment of the body consists of blood and tissue fluid, which bathes the cells.

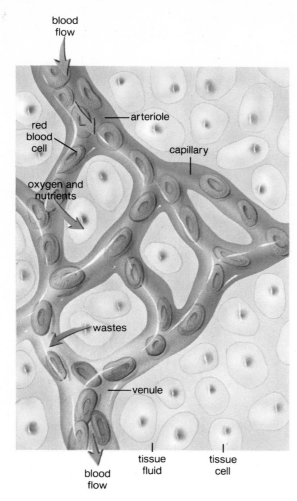

Figure 3.16   Formation of tissue fluid from blood. The internal environment of the body is the blood and the tissue fluid. Tissue cells are surrounded by tissue fluid that is continually refreshed because nutrient molecules constantly exit the bloodstream and waste molecules continually enter the bloodstream.

Most systems of the body contribute to maintenance of a constant internal environment. The digestive system takes in and digests food, providing nutrient molecules that enter the blood and replace the nutrients that are constantly being used up by the body cells. The respiratory system adds oxygen to the blood and removes carbon dioxide. The amount of oxygen taken in and carbon dioxide given off can be increased to meet bodily needs. The chief regulators of composition, however, are the liver and the kidneys. They monitor the chemical composition of plasma (table 3.3) and alter it as required. Immediately after glucose enters the blood, it can be removed by the liver for storage as glycogen. Later glycogen can be broken down to replace the glucose used by the body cells; in this way, the glucose composition of the blood remains constant. (The hormone insulin, secreted by the pancreas, regulates glycogen storage.) The liver also removes toxic chemicals, such as ingested alcohol and drugs, and ammonia given off by the cells. These are converted to waste molecules that can be excreted by the kidneys. The

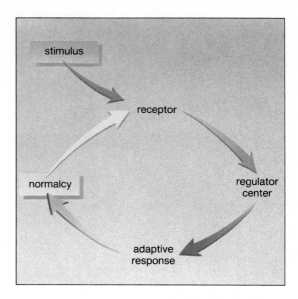

Figure 3.17   Diagram illustrating the principle of negative feedback control. A receptor (sense organ) responds to a stimulus, such as high or low temperature, and signals a regulator center to direct an adaptive response such as sweating. Once normalcy, such as a normal temperature, is achieved, the receptor no longer is stimulated.

kidneys are also under hormonal control as they excrete wastes and salts, substances that can affect the pH level of the blood.

All of the systems of the body contribute to homeostasis; that is, maintenance of the relative constancy of the internal environment.

Although homeostasis is, to a degree, controlled by hormones, it is ultimately controlled by the *nervous system*. The brain contains centers that regulate such factors as temperature and blood pressure. Maintaining proper temperature and blood pressure levels requires receptors (sense organs) that detect unacceptable levels and signal a regulator center. If a correction is required, the center then directs an adaptive response (fig. 3.17). Once normalcy is obtained, the receptor no longer signals the center. This is called control by **negative feedback** because the adaptive response cancels the stimulus that brought about the response. This type of homeostatic regulation results in fluctuation between two levels, as illustrated by temperature control (fig. 3.18). Notice in the following example that feedback control is a self-regulatory mechanism.

Body Temperature Control

The receptor and the regulator center for body temperature are located in the hypothalamus. The receptor is sensitive to the temperature of the blood, and when the temperature falls below normal, the regulator center directs (via nerve impulses) the blood vessels of the skin to constrict. This conserves heat. Also, the arrector pili muscles pull hairs erect, and a layer of insulating air is trapped next to the skin. If body temperature falls

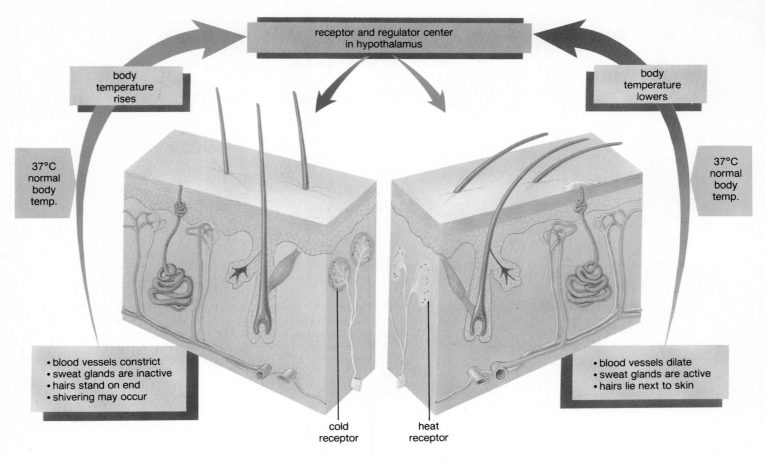

receptor and regulator center
in hypothalamus

body
temperature
rises

body
temperature
lowers

37°C
normal
body
temp.

37°C
normal
body
temp.

• blood vessels constrict
• sweat glands are inactive
• hairs stand on end
• shivering may occur

• blood vessels dilate
• sweat glands are active
• hairs lie next to skin

cold
receptor

heat
receptor

Figure 3.18   Temperature control. When the body temperature rises, the regulator center directs the blood vessels to dilate and the sweat glands to be active. The body temperature lowers. Then, the regulator center directs the blood vessels to constrict, hairs to stand on end, and even shivering to occur if needed. The body temperature rises again. Because the regulator center is activated only by extremes, the body temperature fluctuates above and below normal.

even lower, the regulator center sends nerve impulses to the skeletal muscles, and shivering occurs. Shivering generates heat, and gradually body temperature rises to 37° C and perhaps higher. During the period of time that the body temperature is normal, the receptor and the regulator center are not active, but once body temperature is higher than normal, they are reactivated. When this happens, the regulator center directs the blood vessels of the skin to dilate. This allows more blood to flow near the surface of the body, where heat can be lost to the environment. The regulator center activates the sweat glands and the evaporation of sweat helps to lower body temperature. Gradually, body temperature decreases to 37° C and perhaps lower. Once body temperature is below normal, the cycle begins again.

Homeostasis of internal conditions is a self-regulatory mechanism that results in slight fluctuations above and below a mean. For example, body temperature rises above and drops below a normal temperature of 37° C.

Summary

Human tissues are categorized into four groups. Epithelial tissue covers the body and lines its cavities. Connective tissue often binds body parts. Contraction of muscular tissue permits movement of the body and its parts. Nerve impulses conducted by neurons within nervous tissue help to bring about coordination of body parts.

Different types of tissues are joined to form organs, each one having a specific function. Organs are grouped into organ systems. In humans, the brain and spinal cord are located in a dorsal cavity and the internal organs are located in a ventral cavity that contains the thoracic, abdominal, and pelvic cavities.

All organ systems contribute to the constancy of the internal environment. The nervous and hormonal systems regulate the other systems. Both of these are controlled by a feedback mechanism, which results in fluctuation above and below a mean.

Study Questions

1. State in order the levels of organization of the human body. (p. 49)
2. Name the four major types of tissues. (p. 49)
3. What are the functions of epithelial tissue? Name the different kinds, and give a location for each. (p. 49)
4. What are the functions of connective tissue? Name the different kinds, and give a location for each. (p. 54)
5. What are the functions of muscular tissue? Name the different kinds, and give a location for each. (p. 56)
6. Nervous tissue contains what type of cell? Which organs in the body are made up of nervous tissue? (p. 56)
7. Describe the structure of skin, and state at least two functions of this organ. (p. 57)
8. In general terms, describe the location of the human organ systems. (p. 59)
9. Distinguish between cell membrane and body membrane. (p. 62)
10. What is homeostasis, and how is it achieved in the human body? (p. 62)

Objective Questions

1. Most organs contain several different types of _____ .
2. Kidney tubules are lined by cube-shaped cells called _____ epithelium.
3. Pseudostratified ciliated columnar epithelium contains cells that appear to be _____ , have projections called _____ , and are _____ in shape.
4. Both cartilage and blood are classified as _____ tissue.
5. Cardiac muscle is _____ but involuntary.
6. Nerve cells are called _____ .
7. Skin has three layers: epidermis, _____ , and the subcutaneous layer.
8. Outer skin cells are filled with _____ , a waterproof protein that strengthens them.
9. Mucous membrane contains _____ tissue overlying _____ tissue.
10. Homeostasis is maintenance of the relative _____ of the internal environment, that is, the blood and _____ fluid.

Answers to Objective Questions

1. tissue 2. cuboidal 3. layered (stratified), cilia, columnar (elongated) 4. connective 5. striated 6. neurons 7. dermis 8. keratin 9. epithelial, loose connective 10. constancy, tissue

Critical Thinking Questions

1. In this chapter, the skin is said to be one of the body's organs and is also called the "integumentary system." Explain why the skin should be considered as an organ and which structures should be included in the integumentary system aside from the skin.
2. Classification schemes are arbitrary in nature. Make up a new classification system for the tissues described in this chapter.
3. Explain why it is beneficial for humans to have homeostatic control systems.

Selected Key Terms

bone (bōn) connective tissue having a hard matrix of calcium salts deposited around protein fibers. 55

cartilage (kar'tĭ-lij) a connective tissue in which the cells lie within lacunae embedded in a flexible matrix. 54

compact bone (kom'pakt bōn) hard bone consisting of Haversian systems cemented together. 55

connective tissue (kŏ-nek'tiv tish'u) a type of tissue characterized by cells separated by a matrix that often contains fibers. 54

dermis (der'mis) the thick skin layer that lies beneath the epidermis. 59

epidermis (ep''ĭ-der'mis) the outer skin layer composed of stratified squamous epithelium. 58

epithelial tissue (ep''ĭ-the'le-al tish'u) a type of tissue that lines cavities and covers the external surface of the body. 49

homeostasis (ho''me-o-sta'sis) the relative constancy of conditions, particularly the internal environment, such as temperature, blood pressure, and pH. 62

hyaline cartilage (hi'ah-līn kar'tĭ-lij) cartilage composed of very fine collagenous fibers and a matrix of a milk-glass appearance. 55

lacuna (lah-ku'nah) a small pit or hollow cavity, as in bone or cartilage, where a cell or cells are located. 54

ligament (lig'ah-ment) dense connective tissue that joins bone to bone. 54

muscular tissue (mus'ku-lar tish'u) a type of tissue that contains cells capable of contracting; skeletal muscles are attached to the skeleton, smooth muscle is found within walls of internal organs, and cardiac muscle comprises the heart. 56

negative feedback (neg'ah-tiv fēd'bak) a self-regulatory mechanism that is activated by an imbalance and results in a fluctuation about a mean. 63

neuron (nu'ron) nerve cell that characteristically has three parts: dendrites, cell body, axon. 56

pseudostratified (su''-do strat'ĭ-fīd) the appearance of layering in some epithelial cells when actually each cell touches a baseline and true layers do not exist. 49

spongy bone (spun'je bōn) porous bone found at the ends of long bones. 55

stratified (strat'ĭ-fīd) layered, as in stratified epithelium, which contains several layers of cells. 49

striated (stri'āt-ed) having bands; cardiac and skeletal muscle are striated with bands of light and dark. 56

subcutaneous layer (sub''ku-ta'ne-us la'er) a tissue layer found in vertebrate skin that lies just beneath the dermis and tends to contain adipose tissue. 59

tendon (ten'don) dense connective tissue that joins muscle to bone. 54

Further Readings for Part I

Allen, R. D. February 1987. The microtubule as an intracellular engine. *Scientific American.*

Berns, M. W. 1983. *Cells.* 3d ed. New York: Holt, Rinehart & Winston.

Bretscher, M. S. October 1985. The molecules of the cell membrane. *Scientific American.*

Caplan, A. I. October 1984. Cartilage. *Scientific American.*

Dautry-Varsat, A., and H. F. Lodish. May 1984. How receptors bring proteins and particles into cells. *Scientific American.*

Dustin, P. August 1980. Microtubules. *Scientific American.*

Hole, J. W. 1990. *Human anatomy and physiology.* 5th ed. Dubuque, Iowa: Wm. C. Brown Publishers.

Kessel, R. G., and R. H. Kardon. 1979. *Tissues and organs: a text-atlas of scanning electron microscopy.* San Francisco: W. H. Freeman.

Lack, J. A. August 1981. Ribosome. *Scientific American.*

McIntosh, J. Richard, and Kent L. McDonald. October 1989. The mitotic spindle. *Scientific American.*

Porter, K. R., and J. B. Tucker. March 1981. The ground substance of the living cell. *Scientific American.*

Sloboda, R. D. 1980. The role of microtubules in cell structure and cell division. *American Scientist* 68(3):290.

Swanson, C. P., and P. L. Webster. 1985. *The cell.* 5th ed. Englewood Cliffs, New Jersey: Prentice-Hall.

Todorov, Igor N. December 1990. How cells maintain stability. *Scientific American.*

Weber, K., and M. Osborn. October 1985. The molecules of the cell matrix. *Scientific American.*

Weinberg, R. A. October 1985. The molecules of life. *Scientific American.*

Wickramasinghe, H. Kumar. October 1989. Scanned-probe microscopes. *Scientific American.*

Part Two

Processing and Transporting

in Humans

All of the systems of the body help maintain homeostasis, the relative constancy of the internal environment. The internal environment is the blood and the fluid that surrounds the cells of the tissues. The heart pumps the blood and sends it to the tissues, where exchange occurs with tissue fluid. At the same time, blood is continually being renewed by the digestive, respiratory, and excretory systems. Nutrients enter the blood at the small intestine, external gas exchange occurs in the lungs, and waste products are excreted at the kidneys. The immune system prevents microorganisms from taking over the body and interfering with its proper functioning.

Chapter Four

Digestion

Chapter Concepts

1 Small molecules, such as amino acids, glucose, and fatty acids, that can cross cell membranes are the products of digestion that nourish the body.

2 Regions of the digestive tract are specialized to carry on specific functions; for example, the mouth is specialized to receive and to chew food, and the small intestine is specialized to absorb the products of digestion.

3 Digestion requires enzymes which are specific to their function and have a preferred temperature and pH.

4 Proper nutrition requires that the energy needs of the body be met and that the diet be balanced so that all vitamins, essential amino acids, and fatty acids are included.

Food! One of the challenges of the modern age is to choose foods that will not only nourish but also maintain the body in a healthy state.

Table 4.1 Path of Food

Organ	Function	Special feature	Function
mouth	receives food; digestion of starch	teeth tongue	chewing of food formation of bolus
esophagus	passageway	————	
stomach	storage of food; acidity kills bacteria; digestion of protein	gastric glands	release gastric juices
small intestine	digestion of all foods; absorption of nutrients	intestinal glands villi	release intestinal juices absorb nutrients
large intestine	absorption of water; storage of nondigestible remains	————	————
anus	defecation		

The discovery of digestive enzymes involved the use of a human as the experimental material. William Beaumont was an American doctor who had a French Canadian patient, Alexis St. Martin. St. Martin had been shot in the stomach, and when the wound healed, he was left with a fistula, or opening, that allowed Beaumont to collect gastric (stomach) juices and to look inside the stomach to see what was going on there. Beaumont was able to determine that the muscular walls of the stomach contract vigorously and mix food with juices that are secreted whenever food enters the stomach. He found that gastric juice contains hydrochloric acid (HCl) and a substance (enzyme) active in digestion. Gastric juices are produced separately from protective mucous secretions of the stomach, he said. Beaumont's work, which was very carefully and painstakingly done, pioneered the study of the physiology of digestion.

Digestive System

Digestion takes place within a tube, often called the gut, that begins with the mouth and ends with the anus (table 4.1 and fig. 4.1). Digestion of food in humans is an extracellular process. Digestive enzymes are secreted into the gut by glands that reside in the gut lining or lie nearby. Food is never found within these *accessory glands,* only within the gut itself.

While the term *digestion,* strictly speaking, means the breakdown of food by enzymatic action, in this text, the term is expanded to include both physical and chemical processes that reduce food to small soluble molecules. Only small molecules can cross cell membranes and be absorbed by the gut lining. Too often we are inclined to think that since we eat meat (protein), potatoes (carbohydrate), and butter (fat), these are the substances that nourish our bodies. Instead, it is the amino acids from the protein, the sugars from the carbohydrate, and the glycerol and fatty acids from the fat that actually enter the blood and are transported throughout the body to nourish our cells. Any component of food, such as cellulose, that is incapable of being digested to small molecules leaves the gut as waste material.

Mouth

The mouth receives the food in humans. Most people enjoy eating because of the combined sensations of smelling and tasting food. The olfactory receptors, located in the nose, are responsible for smelling; tasting is a function of the taste buds, located primarily on the tongue. (See chapter 12 for a description of these sense organs.)

The teeth chew the food into pieces convenient to swallow. During the first two years of life, the 20 deciduous or baby teeth appear. These will be replaced eventually by the adult teeth. Normally, adults have 32 teeth (fig. 4.2). One-half of each jaw has teeth of four different types: two chisel-shaped *incisors* for biting; one pointed *canine* for tearing; two fairly flat *premolars* for grinding; and three *molars,* more flattened for crushing. The last molars, called the wisdom teeth, may fail to erupt, or if they do, they are sometimes crooked and useless. Oftentimes, the extraction of the wisdom teeth is recommended.

Human Issue

An adult's teeth are sometimes malpositioned and/or crooked. In these instances, a person can wear braces to straighten the teeth. The dentist may recommend that this be done to improve one's bite, but sometimes it is done for purely aesthetic reasons—the person wishes to look better. Should elective procedures such as this be covered by insurance companies or Medicaid? Some would argue that they should because they improve the self-image and allow persons to function better. Others argue that since the procedure is elective, the individuals alone should bear the cost. What do you think?

Each tooth (fig. 4.3) has a crown and a root. The crown has a layer of enamel, an extremely hard outer covering of calcium compounds; dentin, a thick layer of bonelike material; and an inner pulp that contains the nerves and the blood vessels. Dentin and pulp are also in the root. Tooth decay, or *caries,*

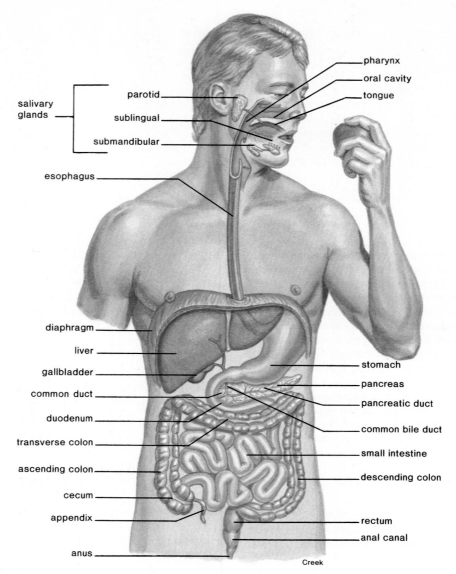

salivary glands
- parotid
- sublingual
- submandibular

pharynx
oral cavity
tongue

esophagus

diaphragm
liver
gallbladder
common duct
duodenum
transverse colon
ascending colon
cecum
appendix
anus

stomach
pancreas
pancreatic duct
common bile duct
small intestine
descending colon

rectum
anal canal

Creek

Figure 4.1   Trace the path of food from the mouth to the anus, and note the placement of the accessory organs of digestion, the liver and the pancreas.

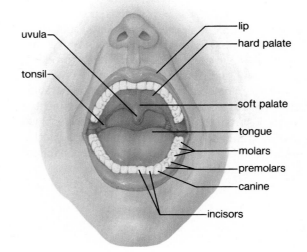

uvula
tonsil

lip
hard palate

soft palate
tongue
molars
premolars
canine
incisors

Figure 4.2   Diagram of the mouth showing the adult teeth. The sizes and shapes of the incisors, canine, premolars, and molars correlate with their function.

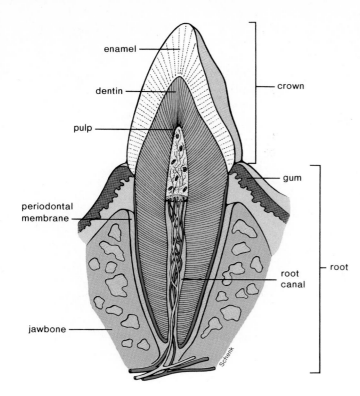

Figure 4.3   Longitudinal section of a canine tooth. Nerves and blood vessels are found within the pulp of a tooth.

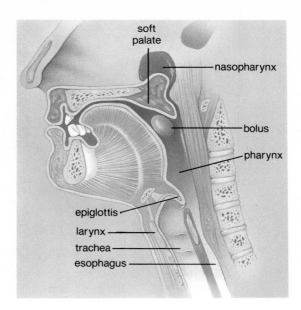

Figure 4.4   Swallowing. When food is swallowed, the soft palate covers the nasopharyngeal openings and the epiglottis covers the glottis so that the bolus must pass down the esophagus. Consequently, you do not breathe when swallowing.

commonly called a cavity, occurs when the bacteria within the mouth metabolize sugar and give off acids that corrode the tooth. Two measures can prevent tooth decay: eating a limited amount of sweets and daily brushing and flossing of teeth. It also has been found that fluoride treatments, particularly in children, can make the enamel stronger and more resistant to decay. Gum disease is more apt to occur with aging. Inflammation of the gums (gingivitis) can spread to the periodontal membrane (fig. 4.3) that lines the tooth socket. The individual then has **periodontitis,** characterized by a loss of bone and loosening of the teeth so that extensive dental work may be required. Stimulation of the gums in a manner advised by dentists is helpful in controlling this condition.

In humans, the roof of the mouth separates the air passages from the mouth cavity. The roof has two parts: an anterior **hard palate** and a posterior **soft palate** (fig. 4.2). The hard palate contains several bones, but the soft palate is only muscular. The soft palate ends in the *uvula,* a suspended process often mistaken by the layperson for the tonsils. In fact, the tonsils are at the sides of the oral cavity (fig. 4.2), at the base of the tongue, and in the nose (called adenoids). The tonsils play a minor role in protecting the body from disease-causing organisms, as is discussed in chapter 7.

There are three pairs of **salivary glands** that send their juices (saliva) by way of ducts to the mouth. The parotid glands lie at the sides of the face immediately below and in front of the ears. They become swollen when a person has the mumps, a viral infection most often seen in children. Each *parotid gland*

has a duct that opens on the inner surface of the cheek at the location of the second upper molar. The *sublingual glands* lie beneath the tongue, and the *submandibular glands* lie beneath the lower jaw. The ducts from these glands open into the mouth under the tongue. You can locate these openings if you use your tongue to feel for small flaps on the inside of your cheek and under your tongue. An enzyme within saliva begins the process of digesting food. Specifically, this enzyme acts on starch.

The tongue, which is composed of striated muscle with an outer layer of mucous membrane, mixes the chewed food with saliva. It then forms this mixture into a mass called a bolus in preparation for swallowing.

The salivary glands send saliva into the mouth, where the teeth chew the food and the tongue forms it into a bolus for swallowing.

Pharynx

Swallowing (fig. 4.4) occurs in the **pharynx,** a region between the mouth and the esophagus, which is a long muscular tube leading to the stomach. Swallowing is a *reflex action,* which means the action usually is performed automatically and does not require conscious thought. Normally, during swallowing, food enters the esophagus because the air passages are blocked. Unfortunately, we have all had the unpleasant experience of having food "go down the wrong way." The wrong way may be either into the nose or into the trachea (windpipe). If it is the latter, coughing usually forces the food up out of the trachea into the pharynx again. Food usually goes into the esophagus because the openings to the nose, called the *nasopharyngeal openings,* are covered when the soft palate moves back. The

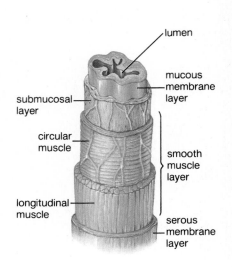

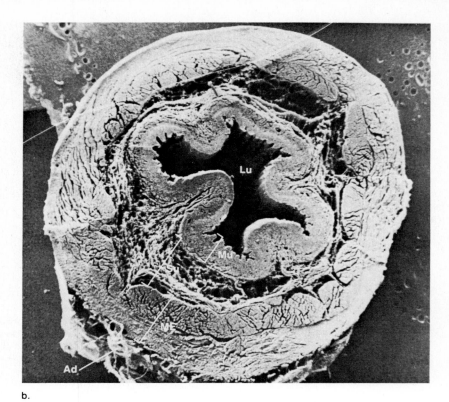

Figure 4.5   Wall of the esophagus. Like the rest of the digestive tract, several different types of tissues are found in the wall of the esophagus. Note the placement of circular muscle inside longitudinal muscle. This arrangement ensures that the action of the circular muscles does not interfere with that of the longitudinal muscle. *a.* Diagrammatic drawing. *b.* Scanning electron micrograph. (Lu = central lumen; Mu = mucous membrane; Su = submucosa; Me = muscular layer; and Ad = adventitia, or serous layer.)

opening to the larynx (voice box) at the top of the trachea, called the **glottis,** is covered when the trachea moves up under a flap of tissue called the **epiglottis.** This is easy to observe in the up-and-down movement of the *Adam's apple,* a part of the larynx, when a person eats. Notice that breathing does not occur during swallowing because air passages are closed off.

The air passage and the food passage cross in the pharynx. When you swallow, the air passage usually is blocked off, and food must enter the esophagus.

Esophagus

After swallowing occurs, the **esophagus** conducts the bolus through the thoracic cavity. The wall (fig. 4.5) of the esophagus is representative of the gut in general. A *mucous membrane layer* lines the **lumen** (space within the tube); this is followed by a *submucosal layer* of connective tissue that contains nerve and blood vessels, a *smooth muscle layer* having both longitudinal and circular muscles, and finally, a *serous membrane layer.*

A rhythmic contraction of the esophageal wall, called **peristalsis** (fig. 4.6), pushes the food along. Occasionally, peristalsis begins even though there is no food in the esophagus. This produces the sensation of a lump in the throat.

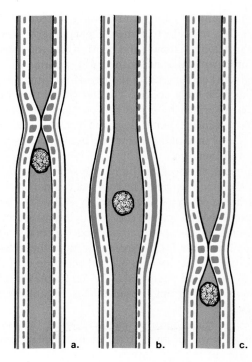

Figure 4.6   Peristalsis in the digestive tract. Rhythmic waves of muscle contraction move material along the digestive tract. The 3 drawings show how a peristaltic wave moves through a single section of gut over time.

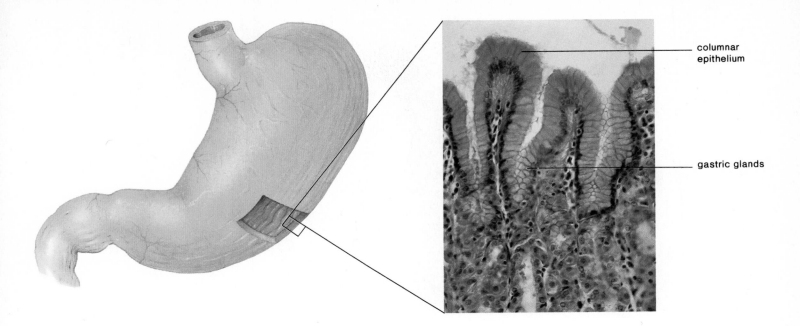

columnar
epithelium

gastric glands

Figure 4.7   Photomicrograph of the mucous membrane layer of the stomach. Gastric glands produce gastric juice rich in pepsin, an enzyme that digests protein to peptides.

The esophagus extends from the back of the pharynx to just below the diaphragm, where it meets the stomach at an angle. The entrance of the esophagus into the stomach is marked by the presence of a constrictor, called the lower esophogeal sphincter, although muscle development is less than in a true sphincter. **Sphincters** are muscles that encircle tubes and act as valves; tubes close when sphincters contract, and they open when sphincters relax. When food is swallowed, the sphincter relaxes, allowing the bolus to pass into the stomach. Normally, this sphincter prevents the acidic contents of the stomach from entering the esophagus. Heartburn, which feels like a burning pain rising up into the throat, occurs when some of the stomach contents escape into the esophagus. When vomiting occurs, a reverse peristaltic wave causes the sphincter to relax, and the contents of the stomach are propelled upward through the esophagus.

Stomach

The **stomach** (fig. 4.7) is a thick-walled, J-shaped organ that lies on the left side of the body beneath the diaphragm. The stomach is continuous with the esophagus above and the duodenum of the small intestine below. The stomach stores food. The wall of the stomach has three layers of muscle and also contains deep folds that disappear as the stomach fills. The muscular wall of the stomach churns, mixing the food with gastric secretions. When food leaves the stomach, it is a pasty material called *acid chyme.*

The columnar epithelium lining the stomach contains millions of microscopic digestive glands called **gastric glands** (the term gastric always refers to the stomach). The gastric glands produce gastric juice. Gastric juice contains hydrochloric acid (HCl) and pepsin, a digestive enzyme that digests protein. The high acidity of the stomach (about pH 2) is beneficial because it kills most bacteria present in food. Although hydrochloric acid does not digest food, it does break down the connective tissue of meat and activates the digestive enzyme in gastric juice. Normally, the wall of the stomach is protected by a thick layer of mucus, but if by chance hydrochloric acid does penetrate this mucus, autodigestion of the wall can begin, and an ulcer results. An *ulcer* is an open sore in the wall caused by the gradual disintegration of tissue. It is believed that the most frequent cause of an ulcer is oversecretion of gastric juice due to too much nervous stimulation; persons under stress tend to have a greater incidence of ulcers. However, there is now evidence that a bacterial (*Campylobacter pyloridis*) infection may impair the ability of cells to produce mucus, which protects the wall from ulcer formation.

Normally, the stomach empties in about two to six hours. Acid chyme leaves the stomach and enters the small intestine by way of the pyloric sphincter. The pyloric sphincter repeatedly opens and closes, allowing acid chyme to enter the small intestine in small squirts only. This means that digestion in the small intestine will proceed at a rate slow enough to ensure thoroughness.

The stomach expands and stores food. While food is in the stomach, it churns, mixing food with acidic gastric juice.

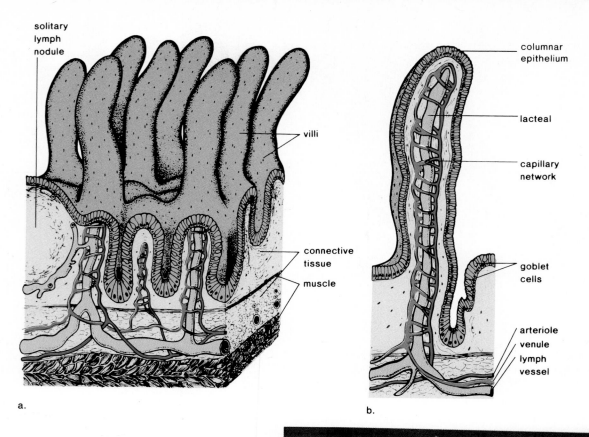

solitary
lymph
nodule

villi

connective
tissue

muscle

columnar
epithelium

lacteal

capillary
network

goblet
cells

arteriole
venule
lymph
vessel

a.

b.

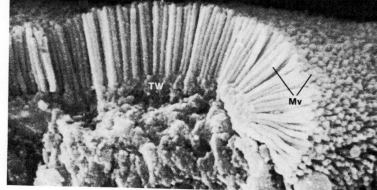

TW

Mv

c.

Figure 4.8   Anatomy of intestinal lining. *a.* The products of digestion
are absorbed by villi, fingerlike projections of the intestinal wall. *b.* Each
villus contains blood vessels and a lacteal. *c.* This scanning electron
micrograph shows that the villi themselves are covered with microvilli
(Mv).

Small Intestine

The **small intestine** gets its name from its small diameter (compared to that of the large intestine), but perhaps it should be
called the long intestine because it averages about 6.0 m (20 ft)
in length compared to the large intestine, which is about 1.5 m
(5 ft) long. The first 25 cm (10 in) of the small intestine is called
the **duodenum.** Duodenal ulcers sometimes occur because gastric juice within acid chyme can digest the intestinal wall in this
region.

Ducts from the gallbladder and the pancreas join to form
the common duct, which enters into the duodenum (see fig. 4.1).
The small intestine receives secretions from these organs by way

of the common duct; chemically and mechanically breaks down
acid chyme; absorbs nutrient molecules; and transports undigested material to the large intestine.

The wall of the small intestine contains fingerlike projections called **villi** (fig. 4.8), and the villi themselves have microvilli that are visible microscopically. Because they are so
numerous, the villi give the intestinal wall a soft, velvety appearance. Each villus has an outer layer of columnar epithelium
and contains blood vessels and a small lymphatic vessel called
a **lacteal.** The lymphatic system is an adjunct to the circulatory
system and returns fluid to the veins. Villi cells produce an intestinal digestive juice that contains enzymes for finishing the

digestion of acid chyme to small molecules that can cross cell membranes. They also greatly increase the surface area of the small intestine for absorption of water and nutrients.

Absorption of nutrient molecules across the wall of each villus continues until all small molecules have been absorbed. Therefore, absorption is an active process involving active transport of molecules across cell membranes and requiring an expenditure of cellular energy. Sugars and amino acids cross the columnar epithelial cells to enter the blood, and the components of fats rejoin before entering the lacteals.

The small intestine is specialized to absorb the products of digestion. It is quite long (6.0 m) and has fingerlike projections called villi, where nutrient molecules are absorbed into the circulatory and lymphatic systems.

Large Intestine

The **large intestine,** which includes the cecum, the colon, the rectum, and the anal canal, is larger in diameter than the small intestine (6.5 cm compared to 2.5 cm). The large intestine absorbs water and salts. It also prepares and stores nondigestible material until it is defecated at the anus.

The *cecum,* which lies below the entrance of the small intestine, has a small projection called the veriform **appendix** (veriform means wormlike) (fig. 4.9). In humans, the appendix, like the tonsils, may play a role in immunity. This organ is subject to inflammation, a condition called appendicitis. It is better to have the appendix removed before the fluid content rises to the point that the appendix bursts because this can lead to peritonitis, a generalized infection of the serous membranes of the abdominal cavity.

The **colon** has three parts: the *ascending colon* goes up the right side of the body to the level of the liver; the *transverse colon* crosses the abdominal cavity just below the liver and the stomach; and the *descending colon* passes down the left side of the body to the rectum, the last 20 cm of the large intestine. The colon is subject to the development of *polyps,* small growths arising from the epithelial lining. Polyps, whether they are benign or cancerous, can be removed individually. If colon cancer is detected while it is still confined to a polyp, the outcome is expected to be a complete cure. The cause of colon cancer is not known yet, but a low-fat, high-fiber (roughage) diet that promotes regularity has been recommended as a protective measure.

The *rectum* ends in the anal canal, which opens at the **anus.** The stimulus to defecate arises from stimuli generated in the colon and the rectum (fig. 4.10). In addition to nondigestible remains, feces also contain certain excretory substances, such as bile pigments and heavy metals, and large quantities of the bacterium *Escherichia coli* and other bacteria.

The large intestine normally contains a large number of normally noninfectious bacteria that live off any substances that were not digested earlier. When they break this material down, they give off odorous molecules that cause the characteristic odor

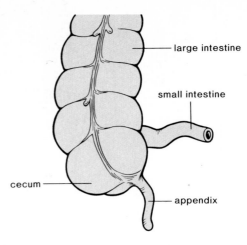

Figure 4.9   The anatomical relationship between the small intestine and the colon. The cecum is the blind end of the ascending colon. The appendix is attached to the cecum.

of feces. Some of the vitamins, amino acids, and other growth factors produced by these bacteria are absorbed by the gut lining. In this way, the bacteria perform a service to us.

Water is considered unsafe for swimming when the coliform bacterial count reaches a certain level. A high count is an indication of the amount of fecal material that has entered the water. The more fecal material present, the greater the possibility that pathogenic or disease-causing organisms are also present.

Human Issue

Ulcers, we are told, are more apt to occur in individuals who have a limited ability to cope with stress. Although formerly the medical profession was inclined to consider only physiological symptoms, more and more doctors have also begun to take into account the psychological state of their patients. Is it the duty of the physicians to consider the "whole" patient, or is it sufficient to treat only the obvious symptoms of disease? Some would argue that there are plenty of psychologists and psychiatrists to deal with psychological symptoms, and it is too time consuming for physicians to consider these. Others feel that the practice of medicine is incomplete if all of a patient's symptoms are not considered. What do you think?

Diarrhea and Constipation

Two common complaints associated with the large intestine are diarrhea and constipation.

The major causes of *diarrhea* are infection of the lower tract and nervous stimulation. In the case of infection, such as food poisoning caused by eating contaminated food, the intestinal wall becomes irritated, and peristalsis increases. Water is

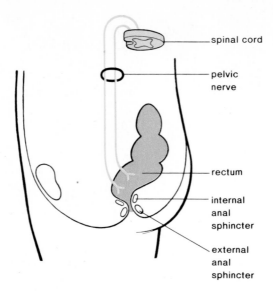

spinal cord

pelvic nerve

rectum

internal anal sphincter

external anal sphincter

Figure 4.10   Defecation reflex. The accumulation of feces in the rectum causes it to stretch, which initiates a reflex action resulting in rectal contraction and expulsion of the fecal material.

not absorbed as a protective measure, and the diarrhea that results serves to rid the body of the infectious organisms. In nervous diarrhea, the nervous system stimulates the intestinal wall, and diarrhea results. Prolonged diarrhea can lead to dehydration because of water loss and to disturbances in the heart's contraction due to an imbalance of salts in the blood.

When a person is constipated, the feces are dry and hard. One cause of this condition is that socialized persons have learned to inhibit defecation to the point that the desire to defecate is ignored. Two components of the diet can help to prevent constipation: water and fiber (roughage). Water intake prevents drying out of the feces, and fiber provides the bulk needed for elimination. It is possible that regularity helps to prevent colon cancer because feces are in contact with the membrane of the colon for less time. Even so, the frequent use of laxatives is discouraged, but if it should be necessary to take a laxative, a bulk laxative is the most natural because, like fiber, it produces a soft mass of cellulose in the colon. Lubricants, like mineral oil, make the colon slippery, and saline laxatives, like milk of magnesia, act osmotically—they prevent water from being absorbed and may even cause water to enter the colon, depending on the dosage. Some laxatives are irritants; they increase peristalsis to the degree that the contents of the colon are expelled.

Chronic constipation is associated with the development of hemorrhoids, a condition that is discussed on page 110.

The large intestine does not produce digestive enzymes; it does absorb water and salts. In diarrhea, too little water has been absorbed; in constipation, too much water has been absorbed.

Accessory Organs

The pancreas and the liver are the accessory organs of digestion. Figure 4.1 shows the ducts that conduct pancreatic juice from the pancreas and bile from the liver to the duodenum.

Pancreas

The **pancreas** lies deep in the abdominal cavity, resting on the posterior abdominal wall. It is an elongated and somewhat flattened organ that has both an endocrine function and an exocrine function. We now are interested in its exocrine function—most of its cells produce pancreatic juice that contains digestive enzymes for carbohydrate, protein, and fat. In other words, the pancreas secretes enzymes for the digestion of all types of food. The enzymes travel by way of the pancreatic duct and the common duct to the duodenum of the small intestine (fig. 4.1).

Liver

The **liver,** which is the largest gland in the body, lies mainly in the upper right portion of the abdominal cavity, under the diaphragm.

Bile Production   **Bile** is a yellowish green fluid because it contains the bile pigments bilirubin and biliverdin, which come from the breakdown of hemoglobin, the pigment found in red blood cells. Bile also contains bile salts (derived from cholesterol), which emulsify fat in the duodenum of the small intestine. When fat is emulsified, it breaks up into droplets that can be acted upon by a digestive enzyme from the pancreas (p. 78).

Up to 1,500 ml of bile are produced by the liver each day. This bile is sent by way of bile ducts to the gallbladder, where it is stored. The **gallbladder** is a pear-shaped, muscular sac attached to the undersurface of the liver. Here, water is absorbed so that bile becomes a thick, mucuslike material. Bile leaves the gallbladder by the common bile duct and proceeds to the duodenum by the common duct (fig. 4.1).

Other Functions of the Liver   In some ways, the liver acts as the gatekeeper to the blood. Once nutrient molecules have been absorbed by the small intestine, they enter the **hepatic portal vein,** pass through the blood vessels of the liver, and then enter the hepatic vein. The arrangement (fig. 4.11) is like this:

small intestine—hepatic portal vein—liver—hepatic vein

As blood passes through the liver, it removes poisonous substances and works to keep the contents of the blood constant. For example, excess glucose present in the hepatic portal vein is removed and is stored by the liver as glycogen:

$$glucose \longrightarrow glycogen + H_2O$$

Between eating periods, when the glucose level of the blood falls below 0.1 percent, glycogen is broken down to glucose, which enters the hepatic vein. In this way, the glucose content of the blood remains near 0.1 percent. It is interesting to note that glycogen sometimes is called animal starch because both starch and glycogen are made up of glucose molecules (p. 21).

If, by chance, the supply of glycogen or glucose runs short, the liver converts amino acids to glucose molecules:

$$amino\ acids \longrightarrow glucose + amino\ groups$$

Recall that amino acids contain nitrogen in the form of amino groups, whereas glucose contains only carbon, oxygen, and hydrogen. Therefore, before amino acids can be converted to glu-

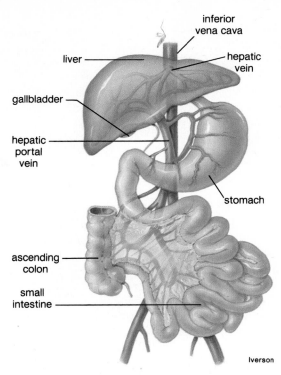

Figure 4.11 Hepatic portal system. The hepatic portal vein takes the products of digestion from the digestive system to the liver, where they are processed before they enter the circulatory system proper.

Figure 4.12 Gallstones. After removal, this gallbladder was cut open to show its contents—numerous gallstones. The dime was added later to indicate the size of the stones.

There are two accessory organs of digestion that send secretions to the duodenum via ducts. The pancreas produces pancreatic juice, which contains digestive enzymes for carbohydrate, protein, and fat. The liver produces bile, which is stored in the gallbladder.

cose molecules, **deamination,** or the removal of amino groups from the amino acids, must take place. By an involved metabolic pathway, the liver converts these amino groups to urea:

$$H_2N - \overset{\overset{\displaystyle O}{\|}}{C} - NH_2$$

Urea is the usual nitrogenous waste product of humans; after its formation in the liver, it is transported to the kidneys for excretion.

The liver also makes blood proteins from amino acids. These proteins are not used as food for cells; rather, they serve important functions within the blood itself.

Altogether, we have mentioned the following functions of the liver:

1. Destroys old red blood cells and converts hemoglobin to the breakdown products (bilirubin and biliverdin) excreted along with bile salts in bile.
2. Produces bile, which is stored in the gallbladder before entering the small intestine, where it emulsifies fats.
3. Stores glucose as glycogen after eating and breaks down glycogen to glucose to maintain the glucose concentration of the blood between eating.
4. Produces urea from the breakdown of amino acids.
5. Makes blood proteins.
6. Detoxifies the blood by removing and metabolizing poisonous substances.

Liver Disorders When a person is *jaundiced,* there is a yellowish tint to the skin due to an abnormally large amount of bilirubin in the blood. In one type of jaundice, called *hemolytic jaundice,* red blood cells are broken down in such quantity that the liver cannot excrete the bilirubin fast enough, and the excess spills over into the bloodstream. In *obstructive jaundice,* there is an obstruction of the bile duct or damage to the liver cells, which causes an increased amount of bilirubin to enter the bloodstream. Obstructive jaundice often occurs when crystals of cholesterol precipitate out of bile and form **gallstones,** which on occasion also contain calcium carbonate. The stones may be so numerous that passage of bile along the bile duct is blocked, and the gallbladder must be removed (fig. 4.12). In the meantime, the bile leaves the liver by way of the blood, and a jaundiced appearance results.

Jaundice also can result from hepatitis, an inflammation of the liver sometimes due to viral infection. *Hepatitis A* (infectious hepatitis caused by HAV) usually is contracted after consuming food, drink, or shellfish containing HAV. The virus often causes only grippelike symptoms although abdominal discomfort, weakness, and even jaundice can develop. *Hepatitis B* (serum hepatitis caused by HBV) is a viral disease transmitted via blood transfusions, unsterile needles (of drug addicts), or any body secretion, such as saliva, semen, or milk. Although symptoms are somewhat the same as those for hepatitis A, they can last much longer. To recover from a severe attack, a long recuperation period commonly is required, during which time

Table 4.2 Comparison of Enzymes

Enzyme	Source	Optimum pH	Type of food digested	Product
salivary amylase	saliva	neutral	starch	maltose
pepsin	stomach	acidic	protein	peptides
pancreatic amylase	pancreas	basic	starch	maltose
trypsin	pancreas	basic	protein	peptides
lipase	pancreas	basic	fat	glycerol; fatty acids
peptidases	intestine	basic	peptides	amino acids
maltase	intestine	basic	maltose	glucose

the patient is in a very weakened condition. *Hepatitis C* (non-A, non-B caused by HCV), like hepatitis B, is transmitted by virus-contaminated blood, but the symptoms are more likely mild. Because the HBV and HBC remain in the blood even for years, carriers (people who can pass on the disease even though they have no symptoms) often are seen. Evidence suggests that all three types of hepatitis can be spread by sexual contact.

Cirrhosis is a chronic disease of the liver in which the organ first becomes fatty. Liver tissue then is replaced by inactive fibrous scar tissue. In alcoholics, who often get cirrhosis of the liver, the condition most likely is caused by the excessive amounts of alcohol the liver is forced to break down. When alcohol, a 2-carbon compound, is metabolized, active acetate (AA) results, and these molecules can be synthesized to fatty acids. To accomplish this synthesis, smooth ER increases dramatically in the liver. This may be the first step toward cirrhosis.

The liver is a very critical organ. Any malfunction is a matter of considerable concern.

Digestive Enzymes

The digestive enzymes are **hydrolytic enzymes** that catalyze degradation by the introduction of water at specific bonds (fig. 1.8). Digestive enzymes are no different from any other enzyme of the body. For example, they are proteins having a particular shape that fits their substrate. They have a preferred pH. This pH maintains their shape, enabling them to speed up their specific reaction.

The various digestive enzymes are present in the digestive juices mentioned previously. We now consider each of the enzymes listed in table 4.2 as we discuss the digestion of starch, protein, and fat, the major components of food.

In the mouth, saliva from the salivary glands has a neutral pH and contains **salivary amylase,** an enzyme that acts on starch:

$$\text{starch} + H_2O \xrightarrow{\text{salivary amylase}} \text{maltose}$$

In this equation, salivary amylase is written above the arrow to indicate that it is neither a reactant nor a product in the reaction. It merely speeds up the reaction in which its substrate,

starch, is digested to many molecules of maltose. Maltose is not one of the small molecules that can be absorbed by the gut lining. However, additional digestive action in the small intestine converts maltose to glucose.

In the stomach, gastric juice secreted by gastric glands has a very low pH—about 2—because it contains hydrochloric acid (HCl). Pepsinogen, a precursor that is converted to the enzyme **pepsin** when exposed to hydrochloric acid, is also present in gastric juice. Pepsin acts on protein to produce peptides:

$$\text{protein} + H_2O \xrightarrow{\text{pepsin}} \text{peptides}$$

Peptides vary in length, but they always consist of a number of linked amino acids. Peptides are too large to be absorbed by the gut lining; however, they are later broken down to amino acids in the small intestine.

Pancreatic juice, which enters the duodenum, is basic because it contains sodium bicarbonate ($NaHCO_3$). It also contains digestive enzymes for all types of food. One pancreatic enzyme, called **pancreatic amylase,** digests starch:

$$\text{starch} + H_2O \xrightarrow{\text{pancreatic amylase}} \text{maltose}$$

Trypsin is a pancreatic enzyme that digests protein:

$$\text{protein} + H_2O \xrightarrow{\text{trypsin}} \text{peptides}$$

Trypsin is secreted as trypsinogen, which is converted to trypsin in the duodenum.

Lipase digests fat droplets after they have been emulsified by bile salts:

$$\text{fat} \xrightarrow{\text{bile salts}} \text{fat droplets}$$

$$\text{fat droplets} + H_2O \xrightarrow{\text{lipase}} \text{glycerol} + \text{fatty acids}$$

The end products of lipase digestion, glycerol and fatty acids, are small enough to cross the cells of the intestinal villi, where absorption takes place. As mentioned previously, glycerol and fatty acids enter the cells of the villi, and within these cells, they rejoin to give fat, which enters the lacteals (fig. 4.8).

Table 4.3 Digestive Enzymes

Reaction	Enzyme	Gland	Site of action
starch + $H_2O \longrightarrow$ maltose	salivary amylase pancreatic amylase	salivary pancreas	mouth small intestine
maltose + $H_2O \longrightarrow$ glucose*	maltase	intestinal	small intestine
protein + $H_2O \longrightarrow$ peptides	pepsin trypsin	gastric pancreas	stomach small intestine
peptides + $H_2O \longrightarrow$ amino acids*	peptidases	intestinal	small intestine
fat + $H_2O \longrightarrow$ glycerol + fatty acids*	lipase	pancreas	small intestine

*Absorbed by villi.

Note: Food is made up largely of carbohydrate (starch), protein, and fat. These very large macromolecules are broken down by digestive enzymes to small molecules that can be absorbed by intestinal villi. This table indicates the steps needed for carbohydrate digestion (starch and maltose), protein digestion (protein and peptides), and fat digestion (fat) and shows that they are all hydrolytic reactions.

Peptidases and maltase produced by the epithelial cells of the intestinal villi complete the digestion of protein and starch to small molecules that cross into the cells of the villi. Peptides, which result from the first step in protein digestion, are digested to amino acids by peptidases:

$$\text{peptides} + H_2O \xrightarrow{\text{peptidases}} \text{amino acids}$$

Maltose, which results from the first step in starch digestion, is digested to glucose by maltase:

$$\text{maltose} + H_2O \xrightarrow{\text{maltase}} \text{glucose}$$

Other disaccharides, each of which has its own enzyme, are digested in the small intestine. The absence of any one of these enzymes can cause illness. For example, many people, including as many as 75 percent of American blacks, cannot digest lactose, the sugar found in milk, because they do not produce lactase, an enzyme that converts lactose to its components, glucose and galactose. Drinking untreated milk often gives these individuals the symptoms of *lactose intolerance* (diarrhea, gas, cramps) caused by a large quantity of undigested lactose in the intestine. In most areas, it is possible to purchase milk made lactose-free by the addition of lactase.

Table 4.3 lists the enzymes needed for the digestion of each of the major components of food. Two enzymes are needed for the digestion of starch to maltose, and two enzymes are needed for the digestion of protein to amino acids, but only one enzyme is required for the digestion of fat. (Not all digestive enzymes are listed; for example, there are intestinal nucleases that digest RNA and DNA to nucleotides.)

Nutrition

The body requires many different types of organic molecules and a smaller number of various types of inorganic ions and compounds from the diet each day. Nutrition involves an interaction between food and the living organism, and a *nutrient* is a substance in food that is used by the body for the maintenance of health. In order to be sure the diet contains all the essential nutrients, it is recommended that we eat a balanced diet. A *balanced diet* is ensured by eating a variety of food from the four food groups pictured in figure 4.13.

In order to get a daily supply of essential nutrients, it is necessary to have a balanced diet.

Food consists largely of proteins, carbohydrates, and lipids (fats and cholesterol). Therefore, we begin by considering these substances.

Proteins

Foods rich in protein include red meat, fish, poultry, dairy products, legumes, nuts, and cereals. Following digestion of protein, amino acids enter the bloodstream and are transported to the tissues. Most of these amino acids are incorporated into structural proteins found in muscles, skin, hair, and nails. Others are used to synthesize such proteins as hemoglobin, plasma proteins, enzymes, and hormones.

Protein formation requires 20 different types of amino acids. Of these, nine are required from the diet because the body is unable to produce them. These are termed the **essential amino acids.** The body produces the other 11 amino acids by simply transforming one type into another type. Some protein sources, such as meat, are *complete;* they provide all types of amino acids. Vegetables and grains supply us with amino acids, but each vegetable or grain alone is an *incomplete* protein source because at least one of the essential amino acids is absent. However, it is possible to combine foods in order to acquire all of the essential amino acids. For example, the combinations of cereal with milk, or beans with rice provide all the essential amino acids.

A complete source of protein is absolutely necessary to ensure a sufficient supply of the essential amino acids.

Even though in this country we emphasize protein intake, it does not take very much protein to meet the daily requirement. In the United States, the *required dietary (daily) allow-*

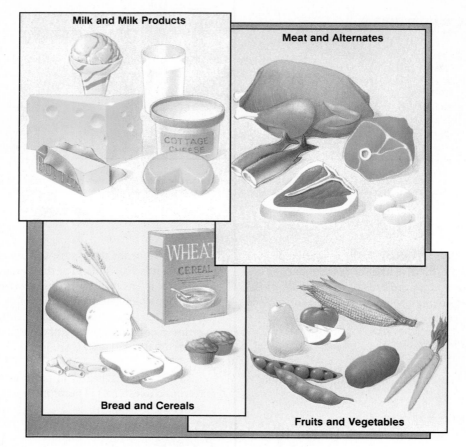

Figure 4.13   The 4 food groups. A variety of foods from each group should be eaten daily.

ances (RDAs) are determined by the National Research Council, a part of the National Academy of Sciences. The RDA for the reference woman (120 lb) is 44 g of protein a day. For the reference man (154 lb), it is 56 g of protein a day. A single serving of roast beef (3 oz) provides 25 g of protein, and a cup of milk provides 8 g.

Unfortunately, in this country, the manner in which we meet our daily requirement for protein is not always the most healthful way. The foods that are richest in protein are apt to be richest in fat.

Foods richest in protein also tend to be richest in fat.

While it is very important to meet the RDA for protein, consuming more actually can be detrimental. Calcium loss in the urine has been noted when dietary protein intake is over twice the RDA. Everything considered, it is probably a good idea to depend on protein from plant origins (e.g., whole-grain cereals, dark breads, legumes) to a greater extent than is the custom in this country.

Carbohydrates

The quickest, most readily available source of energy for the body is carbohydrates, which can be complex as in breads and cereals or simple as in candy, ice cream, and soft drinks. As mentioned previously, starches are digested to glucose, which is stored by the liver in the form of glycogen. Between eating, blood glucose is maintained at about 0.1 percent by the breakdown of glycogen or by the conversion of amino acids to glucose. If necessary, these amino acids are taken from the muscles, even from heart muscle. To avoid this situation, it is suggested that the daily diet contain at least 100 g of carbohydrate. As a point of reference, a slice of bread contains approximately 14 g of carbohydrate.

Carbohydrates are needed in the diet to maintain the blood glucose level.

Actually, the dietary guidelines produced jointly by the U.S. Department of Agriculture and the Department of Health and Human Services recommend that we increase the proportion of carbohydrates per total energy content of the diet:

	Typical Diet (%)	Recommended Diet (%)
Proteins	12	12
Carbohydrates	46	58
Fats	42	30

Further, it is assumed that these carbohydrates are complex and not simple (fig. 4.14). Simple carbohydrates (e.g., sugars) are labeled "empty calories" by some dieticians because they con-

How Does Your Diet Rate for Variety?

Check the blank that best describes your eating habits.	Seldom or never	1 or 2 times a week	3 to 4 times a week	Almost daily
How often do you eat	____	____	____	____
1. at least six servings of bread, cereals, rice, crackers, pasta, or other foods made from grains (a serving of one slice of bread or a half-cup cereal, rice, etc.) per day?				
2. foods made from whole grains?	____	____	____	____
3. three different kinds of vegetables per day?	____	____	____	____
4. cooked dry beans or peas?	____	____	____	____
5. a dark-green leafy vegetable, such as spinach or broccoli?	____	____	____	____
6. two kinds of fruit or fruit juice per day?	____	____	____	____
7. two servings (three if teenager, pregnant, or breastfeeding) of milk, cheese, or yogurt per day?	____	____	____	____
8. two servings of lean meat, poultry, fish, or alternates, such as eggs, dry beans, or nuts per day?	____	____	____	____

Source: United States Department of Agriculture, *Human Nutrition Information Service Home and Garden Bulletin No. 232–1*, April 1986.
Note: See page 92 for answers.

Figure 4.14　Complex carbohydrates. To meet our energy needs, dieticians recommend complex carbohydrates like those shown here rather than simple carbohydrates like candy and ice cream. The latter are more likely to cause weight gain than the complex ones displayed.

Table 4.4　Reducing Dietary Sugar

To reduce dietary sugar, the following suggestions are recommended.
1. Eat fewer sweets, such as candy, soft drinks, ice cream, and pastry.
2. Eat fresh fruits or fruits canned without heavy syrup.
3. Use less sugar—white, brown, or raw—and less honey and syrups.
4. Avoid sweetened breakfast cereals.
5. Eat less jelly.
6. Drink pure fruit juices, not imitations.
7. When cooking, use spices like cinnamon to flavor foods instead of using sugar.
8. Do not put sugar in tea or coffee.

tribute to energy needs and weight gain and are not a part of foods that supply other nutritional requirements. Table 4.4 gives suggestions on how to cut down on your consumption of dietary sugars (simple carbohydrates).

In contrast to simple sugars, complex carbohydrates are likely to be accompanied by a wide range of other nutrients and by fiber, which is nondigestible plant material. Insoluble fiber, such as that found in wheat bran, has a laxative effect and therefore may reduce the risk of colon cancer. Soluble fiber, such

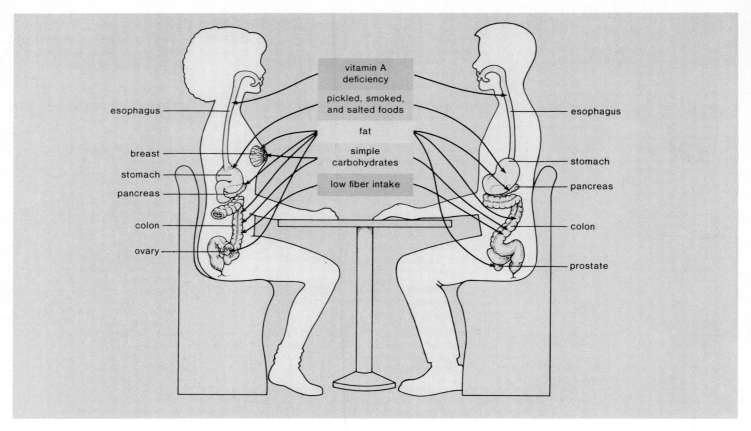

Figure 4.15   Diet and cancer. Evidence is growing to suggest that the noted dietary factors can contribute to the development of cancer.

as that found in oat bran, may possibly reduce cholesterol in the blood because it combines with cholesterol in the gut and prevents it from being absorbed.

While the diet should have an adequate amount of fiber, a high-fiber diet can be detrimental. Some evidence suggests that the absorption of iron, zinc, and calcium is impaired by a high-fiber diet.

Complex carbohydrates along with fiber are considered beneficial to health.

Carbohydrates usually provide most of the dietary Kcalories, even though they have fewer Kcalories[1] per gram than fats:

Food Component	Kcal/g
Carbohydrate	4.1
Protein	4.1
Fat	9.3

A Kcalorie is a measurement unit used to indicate the energy content of food. All types of foods can be used as energy sources in the body. Because carbohydrate makes up the bulk of the diet, it provides most of the Kcalories in the diet.

---

[1]The amount of heat required to raise 1g of water 1° C

Lipids

Our discussion of lipids is divided into two parts: fats and cholesterol.

Fats   Fats are present not only in butter, margarine, and oils, but also in foods high in protein. After being absorbed, the products of fat digestion are transported by the lymph and the blood to the tissues. The liver can alter ingested fats to suit the body's needs, except it is unable to produce the fatty acid linoleic acid. Since this is required for phospholipid production, linoleic acid is considered an essential fatty acid.

Fats have the highest Kcaloric content, but they should not be avoided entirely because they contain the essential fatty acid linoleic acid.

While we need to be sure to ingest some fat in order to satisfy our need for linoleic acid, recent dietary guidelines (p. 80) suggest that we should reduce the amount of fat per total energy content of the diet from 40 to 30 percent. Dietary fat has been implicated in cancer of the colon, the pancreas, the prostate, and the breast (fig. 4.15). Many animal studies have shown that a high-fat diet stimulates the development of mammary tumors, while a low-fat diet does not. It also has been found that women who have a high-fat diet are more likely to develop breast cancer. Surprisingly, it has been discovered that a reduction in the amount of linoleic acid in the diet helps to prevent breast cancer.

Table 4.5 Reducing Dietary Fat

To reduce dietary fat, the following suggestions are recommended.
1. Choose lean red meat, poultry, fish, or dry beans and peas as a protein source.
2. Trim fat off meat and remove skin from poultry before cooking.
3. Cook meat or poultry on a rack so that fat will drain off.
4. Broil, boil, or bake, rather than fry.
5. Limit your intake of butter, cream, hydrogenated oils, shortenings, and coconut and palm oil.*
6. Use herbs and spices to season vegetables instead of butter, margarine, or sauces. Use lemon juice instead of salad dressing.
7. Drink skim milk instead of whole milk, and use skim milk in cooking and baking.

*Although coconut and palm oils are from vegetable sources, they are saturated fats.

Linoleic acid is found in corn, safflower, sunflower, and other common plant oils but is not abundant in olive oil or in fatty fishes and marine animals.

There is very strong evidence that women who have a diet high in fat are more apt to develop breast cancer.

Fat is the component of food that has the highest energy content (9.3 Kcal/g compared to 4.1 Kcal/g for carbohydrate). Raw potatoes, which contain roughage, have about 0.9 Kcalories per gram, but when they are cooked in fat, the number of Kcalories jumps to 6 Kcalories per gram. Another problem for those trying to limit their Kcaloric intake is that fat is not always highly visible; butter melts on toast or potatoes. Table 4.5 gives suggestions for cutting down on the amount of fat in the diet.

As a nation, we have increased our consumption of fat from vegetable sources and have decreased our consumption from animal sources, such as red meat and butter. Most likely, this is due to recent studies linking diets high in saturated fats and cholesterol to hypertension and heart attack.

Cholesterol

The risk of cardiovascular disease includes many factors that are discussed on page 108. One of these factors, according to the National Heart, Lung, and Blood Institute, is a cholesterol blood level of 240 mg/100 ml or higher. If the cholesterol level is this high, additional testing can determine how much of each of two important subtypes of cholesterol is in the blood. Cholesterol is carried from the liver to the cells (including the endothelium of the arteries) by plasma proteins called *low-density lipoprotein (LDL),* and is carried away from the cells to the liver by *high-density lipoprotein (HDL).* LDL-cholesterol apparently contributes to the formation of plaque, which can clog the arteries, while HDL-cholesterol protects against the development of clogged arteries. It has been found that a diet low in saturated fats prevents the accumulation of LDL-cholesterol in the body. Therefore, the suggestions in table 4.5 are also recommended for cholesterol management. It may also be helpful to restrict the intake of cholesterol itself. To do this, white fish,

poultry, and shellfish are recommended; cheese, egg yolks, and liver are not recommended. Egg whites can be substituted for egg yolks for both cooking and eating.

Human Issue

Particularly in regard to foods, people seem to follow fads as witnessed by the recent craze about eating oat bran. Eating oat bran instead of wheat bran might lower the cholesterol blood level in someone with a high risk of heart attack, but such a diet might only increase the chances of colon cancer in someone with a low risk of heart attack. In other words, it is misleading for advertisements to encourage all persons to increase their intake of a particular food. What should be done about this? The government attempts to regulate the health claims of food processors but should it also be up to the individual to research these matters for themselves?

The intake of soluble fiber is believed to possibly reduce the cholesterol blood level by combining with cholesterol in the gut and carrying it out of the body. Foods high in soluble fiber are oat bran, oatmeal, beans, corn, and certain fruits, such as apples, citrus fruits, and cranberries.

Vitamins and Minerals

Vitamins

Vitamins are organic compounds (other than carbohydrate, fat, and protein) that the body is unable to produce but uses for metabolic purposes. Therefore, vitamins must be present in the diet; if they are lacking, various symptoms develop (fig. 4.16). There are many substances that are advertised as vitamins, but in reality there are only 13 vitamins (table 4.6). In general, carrots, squash, turnip greens, and collards are good sources of vitamin A. Citrus fruits and other fresh fruits and vegetables are natural sources of vitamin C. Sunshine and irradiated milk are primary sources of vitamin D, and whole grains are good sources of B vitamins.

It is not difficult to acquire sufficient vitamins if your diet is balanced because each vitamin is needed in small amounts only. Many vitamins are portions of coenzymes, or enzyme helpers. For example, niacin is part of the coenzyme NAD (p. 41), and riboflavin is part of another dehydrogenase, FAD. Coenzymes are needed in only small amounts because each can be used over and over again.

The National Academy of Sciences suggests that we eat more fruits and vegetables in order to acquire a good supply of vitamins C and A because these two vitamins may help to guard against the development of cancer. Nevertheless, they discourage the intake of excess vitamins by way of pills because this practice possibly can lead to illness. For example, excess vitamin C can cause kidney stones, and this excess is converted to oxalic acid, a molecule that is toxic to the body. Vitamin A taken in excess over long periods can cause hair loss, bone and joint pains, and loss of appetite. Excess vitamin D can cause an overload of calcium in the blood, which in children, leads to loss of appetite and retarded growth. Megavitamin therapy always should be supervised by a physician.

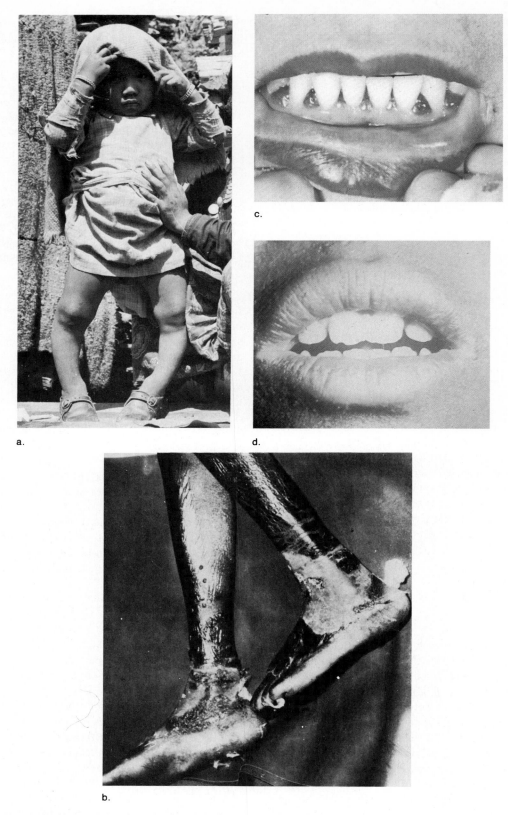

Figure 4.16   Illnesses due to vitamin deficiency. *a.* Bowing of bones (rickets) due to vitamin D deficiency. *b.* Dermatitis of areas exposed to light (pellagra) due to niacin deficiency. *c.* Bleeding of gums (scurvy) due to vitamin C deficiency. *d.* Fissures of lips (cheilosis) due to riboflavin deficiency.

Table 4.6 Vitamins: Their Role in the Body and Food Sources

Vitamins	Role in body	Good food sources
Fat soluble		
vitamin A	assists in the formation and maintenance of healthy skin, hair, and mucous membranes; aids in the ability to see in dim light (night vision); essential for proper bone growth, tooth development, and reproduction	deep yellow/orange and dark green vegetables and fruits (carrots, broccoli, spinach, cantaloupe, sweet potatoes); cheese, milk, and fortified margarines
vitamin D	aids in the formation and maintenance of bones and teeth; assists in the absorption and use of calcium and phosphorus	milks fortified with vitamin D; tuna, salmon, or cod liver oil; also made in the skin when exposed to sunlight
vitamin E	protects vitamin A and essential fatty acids from oxidation; prevents cell membrane damage	vegetable oils and margarine, nuts, wheat germ and whole grain breads and cereals, green leafy vegetables
vitamin K	aids in synthesis of substances needed for clotting of blood; helps maintain normal bone metabolism	green leafy vegetables, cabbage, and cauliflower; also made by bacteria in intestines of humans, except for newborns
Water soluble		
vitamin C	important in forming collagen, a protein that gives structure to bones, cartilage, muscle, and vascular tissue; helps maintain capillaries, bones, and teeth; aids in absorption of iron; helps protect other vitamins from oxidation	citrus fruits, berries, melons, dark green vegetables, tomatoes, green peppers, cabbage, and potatoes
thiamin	helps in release of energy from carbohydrates; promotes normal functioning of nervous system	whole-grain products, dried beans and peas, sunflower seeds, nuts
riboflavin	helps body transform carbohydrate, protein, and fat into energy	nuts, yogurt, milk, whole-grain products, cheese, poultry, leafy green vegetables
niacin	helps body transform carbohydrate, protein, and fat into energy	nuts, poultry, fish, whole-grain products, dried fruit, leafy greens, beans; can be formed in the body from tryptophan, an essential amino acid found in protein
vitamin B_6	aids in the use of fats and amino acids; aids in the formation of protein	sunflower seeds, beans, poultry, nuts, leafy green vegetables, bananas, dried fruit
folic acid	aids in the formation of hemoglobin in red blood cells; aids in the formation of genetic material	dark green leafy vegetables, nuts, beans, whole-grain products, fruit juices
pantothenic acid	aids in the formation of hormones and certain nerve-regulating substances; helps in the metabolism of carbohydrate, protein, and fat	nuts, beans, seeds, dark green leafy vegetables, poultry, dried fruit, milk
biotin	aids in the formation of fatty acids; helps in the release of energy from carbohydrate	occurs widely in foods, especially eggs. Made by bacteria in the human intestine
vitamin B_{12}	aids in the formation of red blood cells and genetic material; helps the functioning of the nervous system	milk, yogurt, cheese, fish, poultry, and eggs; not found in plant foods unless fortified (such as in some breakfast cereals)

From David C. Nieman, et al., *Nutrition.* Copyright © 1990 Wm. C. Brown Publishers, Dubuque, Iowa. All Rights Reserved. Reprinted by permission.

A properly balanced diet includes all the vitamins and the minerals needed by most individuals to maintain health.

Minerals

In addition to vitamins, various **minerals** are required by the body (table 4.7). Minerals are divided into macrominerals, which are recommended in amounts more than 100 mg per day, and microminerals (trace elements), which are recommended in amounts less than 20 mg per day. The macrominerals sodium, magnesium, phosphorus, chlorine, potassium, and calcium serve as constituents of cells and body fluids and as structural components of tissues. For example, calcium is needed for the construction of bones and teeth and for nerve conduction and muscle contraction.

The microminerals seem to have very specific functions. For example, iron is needed for the production of hemoglobin, and iodine is used in the production of thyroxin, a hormone produced by the thyroid gland. As research continues, more and more elements are added to the list of microminerals considered to be essential. During the past three decades, molybdenum, selenium, chromium, nickel, vanadium, silicon, and even arsenic have been found to be essential to good health in very small amounts.

Occasionally, it has been found that individuals do not receive enough iron, calcium, magnesium, or zinc in their diet. Adult females need more iron in the diet than males (RDA of 18 mg compared to 10 mg) because they lose hemoglobin each month during menstruation. Stress can bring on a magnesium deficiency, and due to its high-fiber content, a vegetarian diet may make zinc less available to the body. However, a varied and complete diet usually supplies the mineral RDAs.

Table 4.7   Minerals: Their Role in the Body and Food Sources

Minerals	Role in body	Good food sources
Major minerals		
calcium	used for building bones and teeth and maintaining bone strength; also involved in muscle contraction, blood clotting, and maintenance of cell membranes	all dairy products, dark green leafy vegetables, beans, nuts, sunflower seeds, dried fruit, molasses, canned fish
phosphorus	used to build bones and teeth; release energy from carbohydrate, proteins, and fats; and form genetic material, cell membranes, and many enzymes	beans, sunflower seeds, milk, cheese, nuts, poultry, fish, lean meats
magnesium	used to build bones, produce proteins, release energy from muscle carbohydrate stores (glycogen), and regulate body temperature	sunflower and pumpkin seeds, nuts, whole-grain products, beans, dark green vegetables; dried fruit, lean meats
sodium	regulates body-fluid volume and blood acidity; aids in transmission of nerve impulses	most of the sodium in the American diet is added to food as salt (sodium chloride) in cooking, at the table, or in commercial processing. Animal products contain some natural sodium
chloride	is a component of gastric juice and aids in acid-base balance	table salt, seafood, milk, eggs, meats
potassium	assists in muscle contraction, the maintenance of fluid and electrolyte balance in the cells, and the transmission of nerve impulses; also aids in the release of energy from carbohydrate, proteins, and fats	widely distributed in foods, especially fruits and vegetables, beans, nuts, seeds, and lean meats
Minor minerals		
iron	involved in the formation of hemoglobin in the red blood cells of the blood and myoglobin in muscles; also a part of several enzymes and proteins	molasses, seeds, whole-grain products, fortified breakfast cereals, nuts, dried fruits, beans, poultry, fish, lean meats
zinc	involved in the formation of protein (growth of all tissues), wound healing, and prevention of anemia; a component of many enzymes	whole-grain products, seeds, nuts, poultry, fish, beans, lean meats
iodine	integral component of thyroid hormones	table salt (fortified), dairy products, shellfish, and fish
fluoride	maintenance of bone and tooth structure	fluoridated drinking water is the best source; also found in tea, fish, wheat germ, kale, cottage cheese, soybeans, almonds, onions, milk
copper	vital to enzyme systems and in manufacturing red blood cells. Needed for utilization of iron	nuts, oysters, seeds, crab, wheat germ, dried fruit, whole grains, legumes
selenium	functions in association with vitamin E and may assist in protecting tissues and cell membranes from oxidative damage; may also aid in preventing cancer	nuts, whole grains, lean pork, cottage cheese, milk, molasses, squash
chromium	required for maintaining normal glucose metabolism; may assist insulin function	nuts, prunes, vegetable oils, green peas, corn, whole grains, orange juice, dark green vegetables, legumes
manganese	needed for normal bone structure, reproduction, and the normal functioning of the central nervous system; is a component of many enzyme systems	whole grains, nuts, seeds, pineapple, berries, legumes, dark green vegetables, tea
molybdenum	component of enzymes and may help prevent dental caries	tomatoes, wheat germ, lean pork, legumes, whole grains, strawberries, winter squash, milk, dark green vegetables, carrots

From David C. Nieman, et al., *Nutrition*. Copyright © 1990 Wm. C. Brown Publishers, Dubuque, Iowa. All Rights Reserved. Reprinted by permission.

Calcium   There has been much interest in the dietary addition of calcium supplements (fig. 4.17) to counteract osteoporosis, a degenerative bone disease that afflicts an estimated one-fourth of older men and one-half of older women in the United States. These individuals have porous bones that break easily because they lack sufficient calcium. In 1984, a National Institutes of Health conference on osteoporosis advised postmenopausal women to increase their daily intake of calcium to 1,500 mg and all others to 1,000 mg (compared with the RDA of 800 mg).

However, studies have shown that calcium supplements cannot prevent osteoporosis after menopause even when the dosage is 3,000 mg a day. In postmenopausal women, bone-eating cells called osteoclasts are known to be more active than bone-forming cells called osteoblasts (p. 216). Until now, the most effective defense against osteoporosis in older women has been exercise, which encourages the work of osteoblasts, and estrogen replacement. Recently, however, studies have shown that the drug etidronate disodium, which inhibits osteoclast activity, is effective in osteoporotic women when administered at the proper dosage.

Young women can guard against osteoporosis by forming strong, dense bones before menopause. Eighteen-year-old women are apt to get only 679 mg of calcium a day when the RDA is

Figure 4.17   Calcium in the diet. Many over-the-counter calcium supplements are now available to boost calcium content in the diet.

Table 4.8   Reducing Dietary Sodium

To reduce dietary sodium, the following suggestions are recommended.
1. Use spices instead of salt to flavor foods.
2. Add little or no salt to foods at the table, and add only small amounts of salt when you cook.
3. Eat unsalted crackers, pretzels, potato chips, nuts, and popcorn.
4. Avoid frankfurters, ham, bacon, luncheon meat, smoked salmon, sardines, and anchovies.
5. Avoid processed cheese and canned or dehydrated soups.
6. Avoid prepared catsup and horseradish.
7. Avoid brine-soaked foods, such as pickles and olives.

800 mg (or 1,000 mg according to the NIH). They should consume more calcium-rich foods, such as milk and dairy products. Taking calcium supplements may not be as effective; a cup of milk supplies 270 mg of calcium, while a 500-mg tablet of calcium carbonate provides only 200 mg. The excess supplemental calcium is not taken up by the body; it is not in a form that is *bioavailable*. However, an excess of bioavailable calcium can lead to kidney stones.

Dietary calcium and exercise, plus drug therapy if needed, are the best safeguards against osteoporosis.

Sodium   The recommended amount of sodium intake per day is 400–3,300 mg, and the average American takes in 4,000–4,700 mg. In recent years, this imbalance has caused concern because high sodium intake has been linked to hypertension in some people. About one-third of the sodium we consume occurs naturally in foods; another one-third is added during commercial processing; and we add the last one-third either during home cooking or at the table in the form of table salt.

Clearly, it is possible for us to cut down on the amount of sodium in the diet. Table 4.8 gives recommendations for doing so.

Excess sodium in the diet may lead to hypertension; therefore, excess sodium intake should be avoided.

Dieting

Figure 4.18 indicates that a loss of weight will occur if the caloric intake is reduced while still maintaining the same level of activity. According to nutritionists, the best way to lose weight is to modify our behavior and to lose weight slowly. Behavior modification requires that we examine our eating behaviors, identify the situations that cause us to snack unnecessarily, and work at changing these. For example, if you are used to having cake and ice cream for dessert, substitute fruit. If you pass an ice cream shop every day and tend to stop in, then change your route and do not go by this shop.

Physical activity is a very important addition to your daily routine. Aside from helping to firm up the body after weight loss, exercise puts you in the mood to keep the weight off, and it burns off calories.

The reading on page 89 lists and discusses some of the types of fad diets that are popular. Several are dangerous to your health and do not work because the weight loss is simply regained once the diet ceases. Our body is not adapted to rapid weight loss. First, if Kcalories are severely and suddenly restricted, the body burns protein rather than fat. This protein is removed from the muscles, even the heart muscle. Second, the body apparently has a "set point" for its usual amount of fat. If the amount of stored fat falls below this point, the fat cells are believed to signal the brain by the release of a chemical substance. In response, the metabolic rate is lowered so that fewer Kcalories are needed to stay at the same weight. This set-point hypothesis also explains why people tend to immediately regain any weight that has been lost through dieting. There is some evidence that exercise and avoidance of fatty foods lowers the set point for body fat.

The overall conclusion, then, is that in order to keep unwanted pounds off, a long-term program is needed. This program should include regular exercise and a balanced diet that avoids fatty foods.

Fad diets can be dangerous to your health. A balanced diet containing a reduced number of Kcalories along with an exercise program offer long-term weight control.

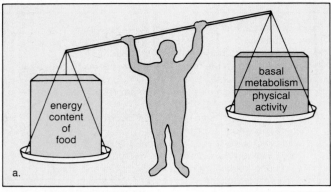

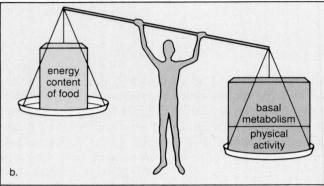

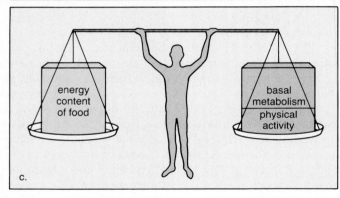

Figure 4.18   Diagram illustrating the relationship between caloric intake and weight gain or loss. In each instance, energy needs are divided between basal metabolism (basal metabolic rate) and physical activity. *a.* Energy content of food is greater than energy needs of the body—weight gain occurs. *b.* Energy content of food is less than energy needs of the body—weight loss occurs. *c.* Energy content of food equals energy needs of the body—no weight change occurs.

Eating Disorders

Authorities recognize three primary eating disorders; obesity, bulimia, and anorexia nervosa. Although they exist in a continuum as far as body weight is concerned, there is much overlap between them.

Obesity is defined as a body weight of more than 20 percent above the ideal weight. It most likely is caused by a combination of factors, including endocrinal, metabolic, and social factors. The social factors include the eating habits of other family members. Obese individuals need to consult a physician if they want to bring their body weight down to normal and to keep it there permanently.

Obesity has many complex causes that possibly can be detected by a physician.

Bulimia can coexist with either obesity or anorexia nervosa. People, usually young women, who are afflicted have the habit of eating to excess and then purging themselves by some artificial means, such as vomiting or laxatives. These individuals usually are depressed, but whether the depression causes or is caused by the bulimia cannot be determined. While individual psychological help does not seem to be effective, there is some indication that group therapy helps. The possibility of a hormonal disorder has not been ruled out, however.

Anorexia nervosa is diagnosed when an individual is extremely thin but still claims to "feel fat" and continues to diet. It is possible these individuals have an incorrect body image that makes them think they are fat. It is also possible they have various psychological problems, including a desire to suppress their sexuality. Menstruation ceases in very thin women.

Both bulimia and anorexia nervosa are serious disorders that require the assistance of competent medical personnel.

What's Your Diagnosis?

Jim is a college freshman living on campus for the first time. During the first several weeks of the fall semester, he stays confined to his dormitory room, when not attending class. Jim has developed the habit of eating mainly take-out meals that he orders from local fast-food restaurants. He supplements these meals with snacks such as candy bars and ice cream.

By the end of his first month on campus, Jim has gained six pounds. He tells his physician that he feels tired constantly, with little energy for any extracurricular activities. It has become a chore to simply carry out the daily routine activities at college. Jim starts to doubt his ability to succeed in school and also begins to miss classes by his second month on campus.

1. What is your diagnosis of Jim's problem? 2. How are the symptoms of Jim's condition consistent with your diagnosis? 3. What remedies, nutritional and behavioral, would you recommend to help Jim?

"Crash" Diets

Many people end up willing to try any gimmick that is supposed to lead to permanent weight loss. The diets that have been promoted in recent years range from the merely ineffective to the outright dangerous, with a scattering of reasonably sound ones. Most weight-loss aids, schemes, and plans fall into one of the following categories.

Pills. Some of the widely marketed over-the-counter drugs sold for dieting are supposed to suppress the appetite; others claim to "burn" fat. Appetite suppressants may work at first, but weight-control researchers say that to lead to permanent weight loss, a pill would have to be taken for life, like blood-pressure pills. Doctors say most diet pills are useless, and some—in particular, the amphetamines and similar prescription drugs—are addictive. So-called starch-blocker pills, which bind to starch and make it indigestible, are now illegal in the United States but still are sold on the black market. They cause a number of adverse side effects, including abdominal cramps.

Low-carbohydrate diets. Severely cutting back on carbohydrate upsets the body's chemical balance in such a way that fluids are depleted. While this gives the illusion of weight loss, fat is not lost, and the water weight eventually will be regained. Besides, carbohydrate is the body's prime source of energy. Despite common opinion, starches are not fattening—rather, fat is fattening.

Fasting. Some years ago, a popular diet had people foregoing food entirely and living on liquid protein drinks and vitamins. A few people on this regime died, probably because their bodies were forced to digest so much muscle that their heart muscles failed.

Liquid diets. While some liquid diets provide sufficient protein and vitamins, they may restrict the dieter to 400 Kcalories or less a day. Because the body cannot burn fat quickly enough to compensate for so few calories, however, muscle also is digested. Most doctors do not recommend cutting Kcalories to fewer than 1,200 a day on any diet.

Single-category diets. These programs call for a diet restricted entirely to one kind of food, such as fruits or vegetables or rice alone. However, no single category of food provides enough nutrients to maintain healthy body tissue. Some dieters in recent years had a dramatic revelation of the inadequacy of such diets—their hair and fingernails fell out!

High-carbohydrate, low-fat diets. Most of the sound diets fall into this category. High-carbohydrate, high-fiber foods are good sources of energy and nutrients, and most are low in fat. Combined with exercise, this kind of diet promotes gradual loss of body fat. Still, the number of Kcalories allotted per day must be sufficient to prevent the body from consuming muscle tissue. A few high-carbohydrate diet plans on the market cut Kcalories too severely.

Reprinted from the 1988 *Medical and Health Annual,* copyright 1987, with the permission of Encyclopaedia Britannica, Inc., Chicago, Illinois.

Nutrition and weight loss. Nutritionists advise that it is not necessary to purchase diet foods. Simply exercise and eat a variety of wholesome foods in small quantities. Dieting should be considered a lifelong project rather than a short-term one.

Summary

In the mouth, food is chewed and starch is acted upon by salivary amylase. After swallowing, peristaltic action moves the food along the esophagus to the stomach. Here, pepsin, in the presence of hydrochloric acid (HCl), acts on protein. In contrast, the small intestine has a basic pH environment. Here, fat is emulsified by bile salts to fat droplets before being acted upon by pancreatic lipase. Protein is digested by pancreatic trypsin, and starch is digested by pancreatic amylase. The cells that line the intestinal wall produce intestinal enzymes to finish the digestion of protein and carbohydrate. Only nondigestible material passes from the small intestine to the large intestine. The large intestine absorbs water from this material. It also contains a large population of bacteria that can use the material as food. In the process, the bacteria produce vitamins that can be absorbed and used by our body.

The walls of the small intestine have fingerlike projections called villi within which are blood capillaries and a lymphatic lacteal. Amino acids and glucose enter the blood; glycerol and fatty acids re-form to give fat before entering the lacteal. The blood from the small intestine moves into the hepatic portal vein, which goes to the liver—an organ that monitors and contributes to blood composition.

A balanced diet is required for good health. Food should provide us with all necessary vitamins, amino acids, fatty acids, and an adequate amount of energy. If the Kcaloric value of food consumed is greater than that needed for body functions and activity, weight gain occurs.

Study Questions

1. List the parts of the digestive tract, anatomically describe them, and state the contribution of each to the digestive process. (p. 69).
2. Discuss the absorption of the products of digestion into the circulatory system. (p. 74)
3. What is the common intestinal bacterium? What do these bacteria do for us? (p. 75)
4. List the accessory glands, and describe the part they play in the digestion of food. (p. 76)
5. List six functions of the liver. How does the liver maintain a constant glucose level in the blood? (p. 77)
6. What is jaundice? (p. 77) Cirrhosis of the liver? (p. 78)
7. Discuss the digestion of starch, protein, and fat, listing all the steps that occur to bring about digestion of each of these. (p. 78)
8. A balanced diet includes food from what four food groups? (p. 79)
9. Give reasons why carbohydrates, fats, proteins, vitamins, and minerals are all necessary to good nutrition. (pp. 79–87)
10. What factors determine how many Kcalories should be ingested? (p. 87)

Objective Questions

1. In the mouth, salivary _____ digests starch to _____ .
2. When swallowing, the _____ covers the opening to the larynx.
3. The _____ takes food to the stomach, where _____ is primarily digested.
4. The gallbladder stores _____ , a substance that _____ fat.
5. The pancreas sends digestive juices to the _____ , the first part of the small intestine.
6. Pancreatic juice contains _____ for digesting protein, _____ for digesting starch, and _____ for digesting fat.
7. Whereas pepsin prefers a _____ pH, the enzymes found in pancreatic juice prefer a _____ pH.
8. The products of digestion are absorbed into the cells of the _____ , fingerlike projections of the intestinal wall.
9. After eating, the liver stores glucose as _____ .
10. The diet should include a complete protein source, one that includes all the _____ .

Answers to Objective Questions

1. amylase, maltose 2. epiglottis 3. esophagus, protein 4. bile, emulsifies 5. duodenum 6. trypsin, pancreatic amylase, lipase 7. strongly acidic, slightly basic 8. villi 9. glycogen 10. essential amino acids

Label this Diagram.
See figure 4.1 (p. 70) in text.

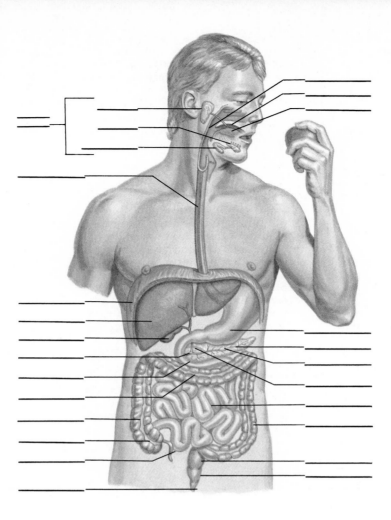

Critical Thinking Questions

1. Why would you expect hormones rather than the nervous system to control the release of digestive enzymes?

2. Reexamine the digestive tract illustration (figure 4.1) and give examples to show that the structure of individual portions along the tract suit their particular functions.

3. What is the hepatic portal system and what purpose does it serve in humans?

Selected Key Terms

amylase (am′ĭ-lās)   a starch-digesting enzyme secreted by the salivary glands (salivary amylase) and the pancreas (pancreatic amylase).   78

colon (ko′lon)   the section of the large intestine between the cecum and the rectum.   75

epiglottis (ep′′ĭ-glot′is)   a structure that covers the glottis during the process of swallowing.   72

esophagus (ē-sof′ah-gus)   a tube that transports food from the mouth to the stomach.   72

gastric gland (gas′trik gland)   gland within the stomach wall that secretes gastric juice.   73

glottis (glot′is)   slitlike opening between the vocal cords.   72

hard palate (hard pal′at)   anterior portion of the roof of the mouth that contains several bones.   71

hydrolytic enzyme (hi-dro-lit′ik en′zīm)   an enzyme that catalyzes a reaction in which the substrate is broken down with the addition of water.   78

lipase (lip′ās)   a fat-digesting enzyme secreted by the pancreas.   78

lumen (loo′men)   the cavity inside any tubular structure, such as the lumen of the gut.   72

pepsin (pep′sin)   a protein-digesting enzyme secreted by gastric glands.   78

peristalsis (per′′ĭ-stal′sis)   a rhythmic contraction that moves the contents along in tubular organs, such as the digestive tract.   72

pharynx (far'ingks)   a common passageway (throat) for both food intake and air movement.   71

salivary gland (sal'i-ver-e gland)   a gland that makes saliva and secretes it into the mouth.   71

soft palate (soft pal'at)   entirely muscular posterior portion of the roof of the mouth.   71

sphincter (sfingk'ter)   a muscle that surrounds a tube and closes or opens the tube by contracting and relaxing.   73

trypsin (trip'sin)   a protein-digesting enzyme secreted by the pancreas.   78

villus (vil'us)   fingerlike projection that lines the small intestine and functions in absorption.   74

vitamin (vi'tah-min)   essential requirement in the diet, needed in small amounts, that is often a part of coenzymes.   83

Answer Box—How Does Your Diet Rate for Variety?

Compare your answers to the best answer listed below.

1. **ALMOST DAILY.** Many people believe that eating breads and cereals will make you fat. That's not true for most of us. Extra calories often come from the fat and/or sugar you MAY eat with them. Both whole grain and enriched breads and cereals provide starch and essential vitamins.

2. **ALMOST DAILY.** Whole grain breads and cereals contain vitamins, minerals, and dietary fiber that are low in the diets of many Americans. Select whole grain cereals and bakery products—those with a whole grain listed first on the ingredient label, or make your own and use whole wheat flour.

3. **ALMOST DAILY.** Vegetables vary in the amounts of vitamins and minerals they contain, so it's important to include several kinds every day.

4. **3 to 4 TIMES A WEEK.** Dry beans and peas fit into two food groups because of the nutrients they provide. They can be used as an alternate to meat, poultry, and fish. They are also an excellent vegetable choice.

5. **3 to 4 TIMES A WEEK.** Popeye gulped down spinach to build his superior strength. Although this effect of spinach was exaggerated, spinach and other dark green leafy vegetables are excellent sources of some nutrients that are low in many diets.

6. **ALMOST DAILY.** Fruits are nature's sweets. They taste good and are good for you. Choose several different kinds each day.

7. **ALMOST DAILY.** Adults as well as children need the calcium and other nutrients found in milk, cheese, and yogurt.

8. **ALMOST DAILY.** Most Americans include some meat, poultry, or fish in their diets regularly. Dry beans and peas, peanuts (including peanut butter), nuts and seeds, and eggs can be used as alternates.

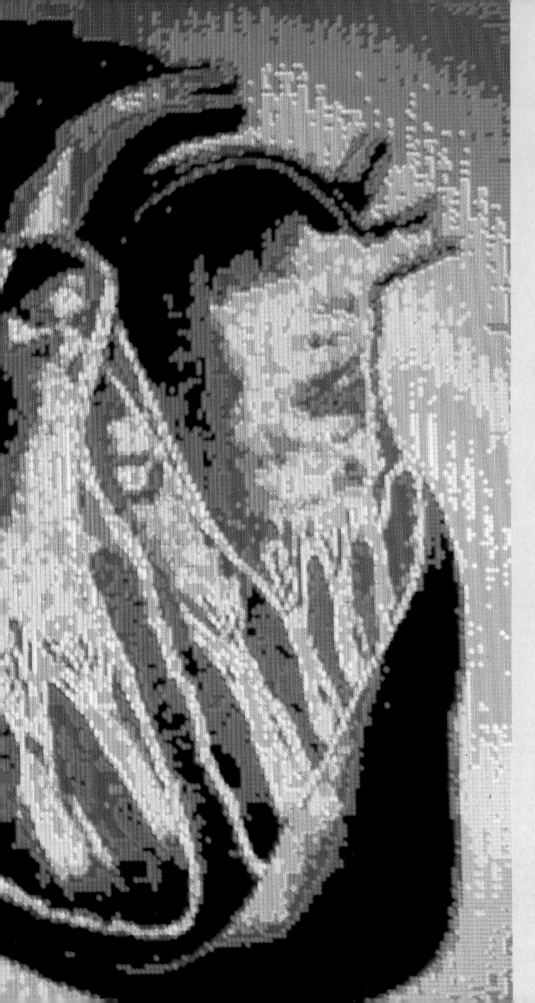

Chapter Five

Circulation

Chapter Concepts

1 In human beings, blood, kept in motion by the pumping of the heart, circulates through a series of vessels.

2 The heart is actually a double pump: the right side pumps blood to the lungs, and the left side pumps blood to the rest of the body.

3 The purpose of circulation is to deliver blood to the capillaries, where exchange of molecules takes place.

4 Although the circulatory system is very efficient, it is still subject to various degenerative disorders.

This is computer-generated art. How did the artist do? Can you identify any of the four chambers of the heart and the attached vessels?

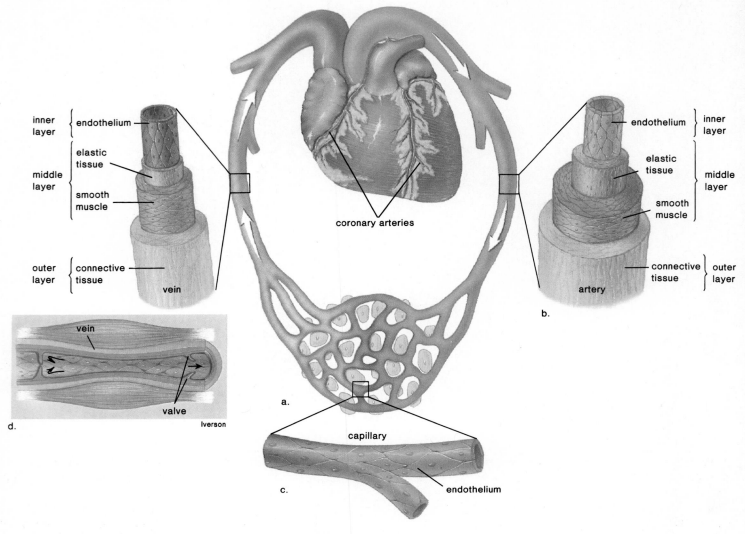

inner layer — endothelium
middle layer — elastic tissue, smooth muscle
outer layer — connective tissue
vein

coronary arteries

inner layer — endothelium
middle layer — elastic tissue, smooth muscle
outer layer — connective tissue
artery
b.

vein
valve
d.
Iverson

a.

capillary
c.
endothelium

Figure 5.1   Blood vessels. *a.* Blood leaving the heart moves from an artery to arterioles to capillaries to venules and then returns to the heart by way of a vein. *b.* Arteries have well-developed walls with a thick middle layer of elastic tissue and smooth muscle. *c.* Capillary walls are one cell thick. *d.* Veins have flabby walls, particularly because the middle layer is not as thick. Veins have valves that point toward the heart.

Life is dependent on the proper functioning of the coronary blood vessels (fig. 5.1), little tubes that serve the needs of cardiac muscle. A common circulatory problem today is blockage of the coronary arteries so that they are unable to function properly. Although coronary heart disease develops slowly over the years, a heart attack may come on quite suddenly. Evidence is growing that coronary heart disease (CHD) may be preventable in part, but there is no quick fix. A lifetime of devotion to these little vessels is required, and good health habits are a necessity. A now-famous study for over 38 years of 6,000 residents in the city of Framingham, Massachusetts, has helped investigators determine that cigarette smoking, elevated blood cholesterol, and the presence of hypertension all predispose one to CHD. Other important factors are lack of exercise, obesity, stress, and a family history of coronary artery disease.

Circulatory System

Blood Vessels

The blood vessels are arranged so that they continually carry blood from the heart to the tissues and then return it from the tissues to the heart. Blood vessels are of three types: the **arteries** (and **arterioles**) carry blood away from the heart; the **capillaries** exchange material with the tissues; and the **veins** (and **venules**) return blood to the heart.

Arteries and Arterioles

Arteries have thick walls (fig. 5.1*b*) because in addition to an inner endothelial layer and an outer fibrous connective tissue

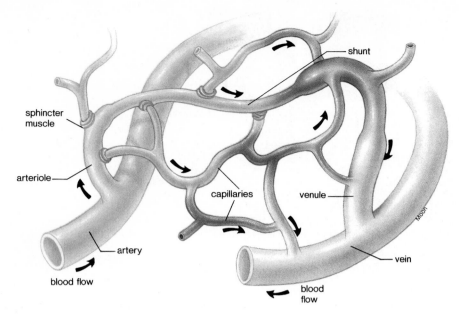

Figure 5.2 Anatomy of a capillary bed. Capillary beds form a maze of vessels that lie between an arteriole and a venule. Blood can move directly between the arteriole and the venule by way of a shunt. When sphincter muscles are closed, blood flows through the shunt. When sphincter muscles are open, the capillary bed is open, and blood flows through the capillaries. Except in the lungs, as blood passes through a capillary, it gives up its oxygen (O_2). Therefore, blood goes from being oxygenated in the arteriole (red color) to being deoxygenated (blue color) in the vein.

layer, they have a thick middle layer that contains elastic and muscular tissues. The elastic tissue is loose connective tissue that is rich in elastic fibers, and the muscular tissue is smooth muscle. The elastic fibers enable an artery to expand and to accommodate the sudden increase in blood volume that results after each heartbeat. Arterial walls are so thick that the walls themselves are supplied with blood vessels.

Arterioles are small arteries just visible to the naked eye. The middle layer of arterioles has some elastic tissue but is composed mostly of smooth muscle, the fibers of which encircle the arteriole. If these muscle fibers contract, the lumen of the arteriole gets smaller; if the fibers relax, the lumen of the arteriole enlarges. Whether arterioles are constricted or dilated affects blood pressure. The greater the number of vessels dilated, the lower the blood pressure.

Capillaries

Arterioles branch into small vessels called capillaries. Each capillary is an extremely narrow, microscopic tube with a wall composed of only one layer of endothelial cells (fig. 5.1c). *Capillary beds* (networks of many capillaries) are present in almost all regions of the body; consequently, a cut to any body tissue draws blood. The capillaries are the most important part of a closed circulatory system because an exchange of nutrient and waste molecules takes place across their thin walls. Oxygen and nutrients diffuse out of a capillary into the tissue fluid that surrounds cells, and carbon dioxide and other wastes diffuse into the capillary (see fig. 6.6). Some water also leaves a capillary;

any excess is picked up by lymphatic vessels, which return it to the blood circulatory system. The lymphatic system is discussed in chapter 7.

Since the capillaries serve the cells, the heart and the other vessels of the circulatory system can be thought of as a means by which blood is conducted to and from the capillaries. Not all capillary beds (fig. 5.2) are open or in use at the same time. After eating, the capillary beds of the digestive tract are usually open; during muscular exercise, the capillary beds of the skeletal muscles are open. Most capillary beds have a shunt through which blood moves directly from arteriole to venule when the capillary bed is closed. Sphincter muscles encircle the entrance to each capillary. When these are constricted, the bed is closed, and when they are relaxed, the bed is open. As expected, the larger the number of capillary beds open, the lower the blood pressure.

Veins and Venules

Veins and venules take blood from the capillary beds to the heart. First, the venules drain the blood from the capillaries and then join to form a vein (fig. 5.1d). The wall of a venule has the same three layers as an artery, but the wall is much thinner than that of an artery because the middle layer of elastic and smooth muscle tissues is poorly developed. Within some veins, especially in the major veins of the arms and the legs, there are **valves** that allow blood to flow only toward the heart when they are open and prevent the backward flow of blood when they are closed.

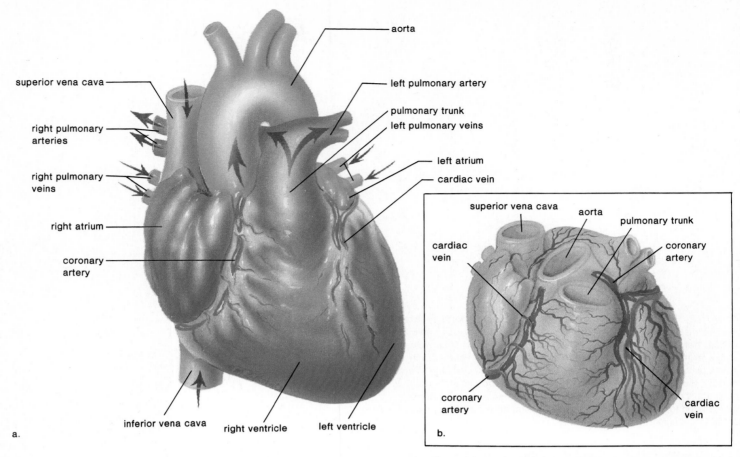

Figure 5.3 External heart anatomy. *a.* The venae cavae bring deoxygenated blood to the right side of the heart from the body, and the pulmonary arteries take this blood to the lungs. The pulmonary veins bring oxygenated blood from the lungs to the left side of the heart, and the aorta takes this blood to the body. *b.* The coronary arteries and cardiac veins pervade cardiac muscle. The coronary arteries are the first blood vessels to branch off the aorta. They bring oxygen and nutrients to cardiac cells.

At any given time, more than half of the total blood volume is in the veins and the venules. If a loss of blood occurs, for example due to hemorrhaging, nervous stimulation causes the veins to constrict, providing more blood to the rest of the body. In this way, the veins act as a blood reservoir.

Arteries and arterioles carry blood away from the heart; veins and venules carry blood to the heart; and capillaries join arterioles to venules.

Heart

The **heart** is a cone-shaped (fig. 5.3), muscular organ about the size of a fist. It is located between the lungs directly behind the sternum and is tilted so that the apex is directed to the left. The major portion of the heart is cardiac muscle tissue called the **myocardium.** The muscle fibers of the myocardium are branched and joined so tightly that prior to studies with the electron microscope, it was thought they were continuous. Now it is known they are individual fibers. The inner surface of the heart is lined with endothelial tissue called *endocardium,* which resembles squamous epithelium. The outside of the heart is surrounded by a membrane called *pericardium,* and this also forms a sac, called the pericardial sac, within which the heart is located. Normally, this sac contains a small quantity of liquid to lubricate the heart.

Internally (fig. 5.4), the heart has a right side and a left side separated by the **septum.** The heart has four chambers: two upper, thin-walled **atria** (singular, atrium), sometimes called auricles, and two lower, thick-walled **ventricles.** The atria are much smaller than the strong, muscular ventricles.

The heart also has **valves** that direct the flow of blood and prevent its backward flow. The valves that lie between the atria and the ventricles are called the **atrioventricular valves.** These valves are supported by strong fibrous strings called chordae tendineae. These cords, which are attached to muscular projections of the ventricular walls, support the valves and prevent them from inverting. The atrioventricular valve on the right side is called the tricuspid valve because it has three cusps, or flaps. The valve on the left side is called the bicuspid, or mitral, because it has two flaps. There are also **semilunar valves,** which resemble half moons, between the ventricles and their attached vessels.

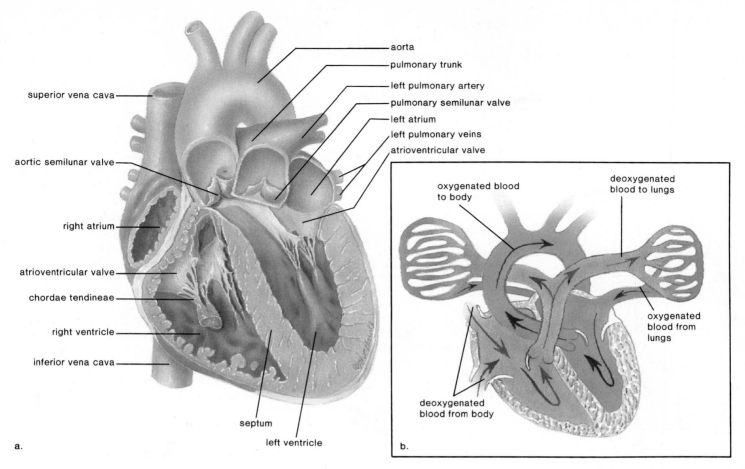

superior vena cava

aortic semilunar valve

right atrium

atrioventricular valve

chordae tendineae

right ventricle

inferior vena cava

septum

left ventricle

a.

aorta
pulmonary trunk
left pulmonary artery
pulmonary semilunar valve
left atrium
left pulmonary veins
atrioventricular valve

oxygenated blood to body

deoxygenated blood to lungs

oxygenated blood from lungs

deoxygenated blood from body

b.

Figure 5.4   Internal view of the heart. *a.* The right side of the heart contains deoxygenated blood. The venae cavae empty into the right atrium, and the pulmonary trunk and arteries leave the right ventricle. The left side of the heart contains oxygenated blood. The pulmonary veins enter the left atrium, and the aorta leaves from the left ventricle.

b. A diagrammatic representation of the heart that allows you to trace the path of the blood. On the right: venae cavae, right atrium, right ventricle, pulmonary arteries to lungs. On the left: pulmonary veins, left atrium, left ventricle, aorta to body. Restate this and give the name of the valves where appropriate.

Humans have a four-chambered heart (two atria and two ventricles), and the right side is separated from the left side by a septum.

Path of Blood in the Heart

We can trace the path of blood through the heart (fig. 5.4*b*) in the following manner:

The superior (anterior) **vena cava** and the inferior (posterior) vena cava carrying deoxygenated blood (low in oxygen and high in carbon dioxide) enter the right atrium.

The right atrium sends blood through an atrioventricular valve (the tricuspid valve) to the right ventricle.

The right ventricle sends blood through the pulmonary semilunar valve into the pulmonary trunk and the pulmonary arteries to the lungs.

The pulmonary veins carrying oxygenated blood (high in oxygen and low in carbon dioxide) from the lungs enter the left atrium.

The left atrium sends blood through an atrioventricular valve (the bicuspid, or mitral, valve) to the left ventricle.

The left ventricle sends blood through the aortic semilunar valve into the aorta to the body proper.

From this description, you can see that deoxygenated blood never mixes with oxygenated blood and that blood must pass through the lungs in order to pass from the right side to the left side of the heart. In fact, the heart is a *double pump* because the right side of the heart sends blood through the lungs, and the left side sends blood throughout the body (fig. 5.5).

There are two circular pathways (circuits) for blood in the body: one from the heart to the lungs and back to the heart, and another from the heart to the body and back to the heart. Since the left ventricle has the harder job of pumping blood to the entire body, its walls are thicker than those of the right ventricle, which pumps blood to the lungs.

The right side of the heart pumps blood to the lungs, and the left side of the heart pumps blood to the tissues.

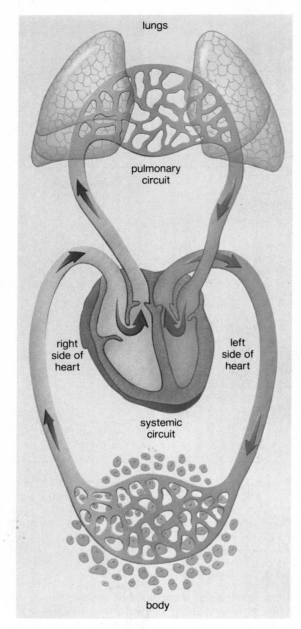

Figure 5.5 Diagram of pulmonary and systemic circuits. The blue-colored vessels carry deoxygenated blood, while the red-colored vessels carry oxygenated blood. Notice that the blood cannot move from the right side of the heart to the left side without passing through the lungs.

Heartbeat

Cardiac Cycle

From this description of the path of blood through the heart, it might seem that the right and left sides of the heart beat independently of one another, but actually, they contract together. First, the two atria contract simultaneously; then the two ventricles contract at the same time. The word **systole** refers to contraction of heart muscle, and the word **diastole** refers to relaxation of heart muscle. The heart contracts, or beats, about

70 times a minute, and each heartbeat lasts about 0.85 second. Each heartbeat, or *cardiac cycle* (fig. 5.6), consists of the following elements:

Time	Atria	Ventricles
0.15 sec	Systole	Diastole
0.30 sec	Diastole	Systole
0.40 sec	Diastole	Diastole

This shows that while the atria contract, the ventricles relax, and vice versa, and that all chambers rest at the same time for 0.40 second. The short systole of the atria is appropriate since the atria send blood only into the ventricles. It is the muscular ventricles that actually pump blood out into the circulatory system proper. When the word *systole* is used alone, it usually refers to the left ventricular systole.

The heartbeat is divided into three phases. First, the atria contract, and then the ventricles contract. (When the atria are in systole, the ventricles are in diastole, and vice versa.) Finally, all chambers are in diastole.

Heart Sounds

When the heart beats, the familiar lub-DUPP sound is heard as the valves of the heart close. The lub is caused by vibrations of the heart when the atrioventricular valves close, and the DUPP is heard when vibrations occur due to the closing of the semilunar valves. Heart murmurs, or a slight slush sound after the lub, are often due to ineffective valves that allow blood to pass back into the atria after the atrioventricular valves have closed. Rheumatic fever resulting from a bacterial infection is one cause of a faulty valve, particularly the bicuspid valve. If operative procedures are unable to open and/or restructure the valve, it can be replaced with an artificial valve.

The heart sounds are due to the closing of the heart valves.

Cardiac Conduction System

The heart will beat independently of any nervous stimulation. In fact, it is possible to remove the heart of a small animal, such as a frog, and watch it undergo contraction in a petri dish. The reason for this is the presence of nodal tissue, which has both muscular and nervous characteristics. The **SA (sinoatrial) node,** is found in the upper dorsal wall of the right atrium and the **AV (atrioventricular) node** is found in the base of the right atrium very near the septum (fig. 5.7a). The SA node initiates the heartbeat and automatically sends out an excitation impulse every 0.85 second to cause the atria to contract. When the impulse reaches the AV node, it signals the ventricles to contract by way of specialized fibers called Purkinje fibers. The SA node is called the **pacemaker** because it usually keeps the heartbeat regular. If the SA node fails to work properly, the heart still beats, but irregularly. To correct this condition, it is possible to implant in the body an artificial pacemaker that automatically gives an electric shock to the heart every 0.85 second. This causes the heart to beat regularly again.

During the first phase of
the cardiac cycle (0.15 sec),
the atria are in systole and
pumping blood into the
ventricles.

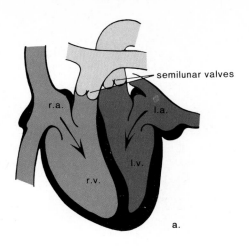

a.

During the second phase of
the cardiac cycle (0.30 sec),
the ventricles are in systole.
Deoxygenated blood goes to the
lungs by way of the pulmonary
arteries and oxygenated blood
goes to the body by way of the
aorta.

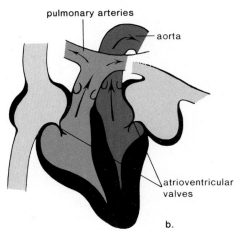

b.

During the third phase of the
cardiac cycle, all chambers
are in diastole (0.40 sec).
Deoxygenated blood enters the
right side of the heart from
the vena cavae. Oxygenated
blood enters the left side
of the heart from the
pulmonary veins.

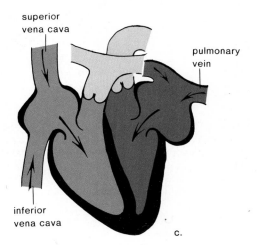

c.

Figure 5.6   Stages in the cardiac cycle. Note when the semilunar
valves and the atrioventricular valves are open or closed.

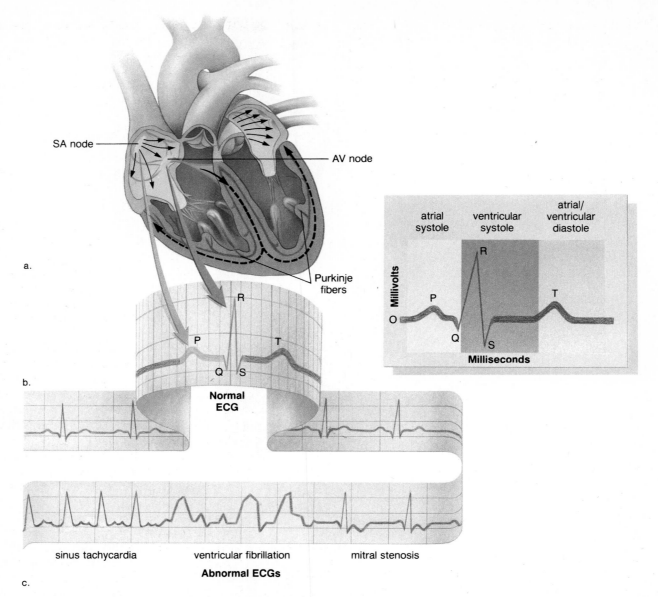

a.

b.

Normal ECG

sinus tachycardia   ventricular fibrillation   mitral stenosis

Abnormal ECGs

c.

Figure 5.7   Control of the heart cycle. *a.* The SA node sends out a stimulus that causes the atria to contract. When this stimulus reaches the AV node, it signals the ventricles to contract by way of the Purkinje fibers. *b.* A normal ECG indicates that the heart is functioning properly. The P wave occurs as the atria contract; the QRS wave occurs as the ventricles contract; and the T wave occurs when the ventricles are recovering from contraction. *c.* Abnormal ECGs: sinus tachycardia is an abnormally fast heartbeat due to a fast pacemaker; ventricular fibrillation is irregular heartbeat due to irregular stimulation of the ventricles; and mitral stenosis occurs because the bicuspid (mitral) valve is obstructed.

The SA node is the natural pacemaker that keeps the heart beating regularly.

Electrocardiogram (ECG)   With the contraction of any muscle, including the myocardium, ionic changes occur that can be detected by recording devices that measure voltage changes. (Voltage, which in this case is measured in millivolts, is the difference in polarity between two electrodes attached to the body.) The record that results, called an **electrocardiogram** (fig. 5.7*b*),

has three waves. The first wave, called the P wave, represents the excitation and contraction of the atria. The second wave, or the QRS wave, occurs during ventricular excitation and contraction. The third wave, or T wave, is caused by the recovery of the ventricles. An examination of the electrocardiogram indicates whether the heartbeat has a normal or an irregular pattern.

Ventricular fibrillation caused by uncoordinated contraction of the myocardium is of special interest because it can be caused by an injury or drug overdose. It is the most common cause of sudden cardiac death in a seemingly healthy person. Once the ventricles are fibrillating, they have to be defibrillated by applying a strong electric current for a short period of time. Then the SA node may be able to reestablish a coordinated beat.

The conduction system of the heart includes the SA node, the AV node, and the Purkinje fibers. With an ECG, it is possible to determine if the conduction system, and therefore the beat of the heart, is regular.

Nervous Control of the Heartbeat

The rate of the heartbeat is also under nervous control. A cardiac center in the medulla oblongata (p. 202) of the brain can alter the beat of the heart by way of the *autonomic nervous system* (p. 199). This system is made up of two divisions: the *parasympathetic system,* which promotes those functions we tend to associate with normal activities, and the *sympathetic system,* which brings about those responses we associate with times of stress. For example, the parasympathetic system causes the heartbeat to slow down, and the sympathetic system increases the heartbeat. Various factors, such as the relative need for oxygen or blood pressure, determine which of these systems is activated.

The heart rate is regulated largely by the autonomic nervous system.

Vascular Pathways

The cardiovascular system, which is represented in figure 5.8, includes two circuits: the **pulmonary circuit,** which circulates blood through the lungs, and the **systemic circuit,** which serves the needs of body tissues.

Pulmonary Circuit

The path of blood through the lungs can be traced as follows. Blood from all regions of the body first collects in the right atrium and then passes into the right ventricle, which pumps it into the pulmonary trunk. The pulmonary trunk divides into the *pulmonary arteries,* which divide into the arterioles of the lungs. The arterioles take blood to the pulmonary capillaries, where carbon dioxide and oxygen are exchanged. The blood then enters the pulmonary venules that lead back through the *pulmonary veins* to the left atrium. Since the blood in the pulmonary arteries is deoxygenated but the blood in the pulmonary veins is oxygenated, it is not correct to say that all arteries carry oxygenated blood and all veins carry deoxygenated blood. It is just the reverse in the pulmonary circuit.

The pulmonary arteries take deoxygenated blood to the lungs, and the pulmonary veins return oxygenated blood to the heart.

Systemic Circuit

The systemic circuit includes all of the other arteries and veins shown in figure 5.8. The largest artery in the systemic circuit is the **aorta,** and the largest veins are the superior and inferior venae cavae. The **superior vena cava** collects blood from the head, the chest, and the arms, and the **inferior vena cava** collects blood from the lower body regions. Both enter the right atrium. The aorta and the venae cavae serve as the major pathways for blood in the systemic circuit.

The path of systemic blood to any organ in the body begins in the left ventricle, which pumps blood into the aorta. Branches from the aorta go to the major body regions and organs. For example, the path of blood to the kidneys can be traced as follows:

> left ventricle—aorta—renal artery—renal arterioles, capillaries, venules—renal vein—inferior vena cava— right atrium

To trace the path of blood to any organ in the body, you need only mention the aorta, the proper branch of the aorta, the organ, and the vein returning blood to the vena cava. In most instances, the artery and the vein that serve the same organ are given the same name (fig. 5.8). In the systemic circuit, unlike the pulmonary system, arteries contain oxygenated blood and have a bright red color, but veins contain deoxygenated blood and appear a purplish color.

The **coronary arteries** (fig. 5.3b), which are a part of the systemic circuit, are extremely important because they serve the heart muscle itself. (The heart is not nourished by the blood in its chambers.) The coronary arteries arise from the aorta just above the aortic semilunar valve. They lie on the exterior surface of the heart, where they branch off in various directions into the arterioles. The coronary capillary beds join to form venules. The venules converge to form the cardiac veins, which empty into the right atrium. The coronary arteries have a very small diameter and may become blocked, as discussed on page 108.

The body has a portal system, the hepatic portal system (see fig. 4.11), that is associated with the liver. A portal system begins and ends in capillaries; in this instance, the first set of capillaries occurs at the villi of the small intestine and the second occurs in the liver. Blood passes from the capillaries of the villi into venules that join to form the *hepatic portal vein,* a vessel that connects the villi of the intestine with the liver. The *hepatic vein* leaves the liver and enters the inferior vena cava.

While figure 5.8 is helpful in tracing the path of the blood, remember that all parts of the body receive both arteries and veins, as illustrated in figure 5.9.

The systemic circuit takes blood from the left ventricle of the heart to the right atrium of the heart. It serves the body proper.

Features of the Circulatory System

When the left ventricle contracts, the blood is sent into the aorta under pressure.

Pulse

The surge of blood entering the arteries causes their elastic walls to swell, but then they almost immediately recoil. This alternating expansion and recoil of an arterial wall can be felt as a

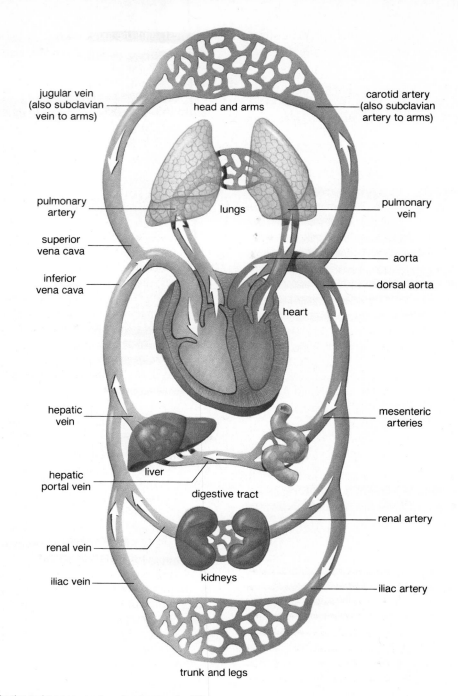

jugular vein
(also subclavian
vein to arms)

head and arms

carotid artery
(also subclavian
artery to arms)

pulmonary
artery

lungs

pulmonary
vein

superior
vena cava

aorta

inferior
vena cava

dorsal aorta

heart

hepatic
vein

mesenteric
arteries

hepatic
portal vein

liver

digestive tract

renal artery

renal vein

iliac vein

kidneys

iliac artery

trunk and legs

Figure 5.8   Blood vessels in the pulmonary and systemic circuits. The
blue-colored vessels carry deoxygenated blood, and the red-colored
vessels carry oxygenated blood; the arrows indicate the flow of blood.
Compare this diagram, useful for learning to trace the path of blood, to
figure 5.9 in order to realize that both arteries and veins go to all parts
of the body. Also, there are capillaries in all parts of the body. No cell is
located far from a capillary.

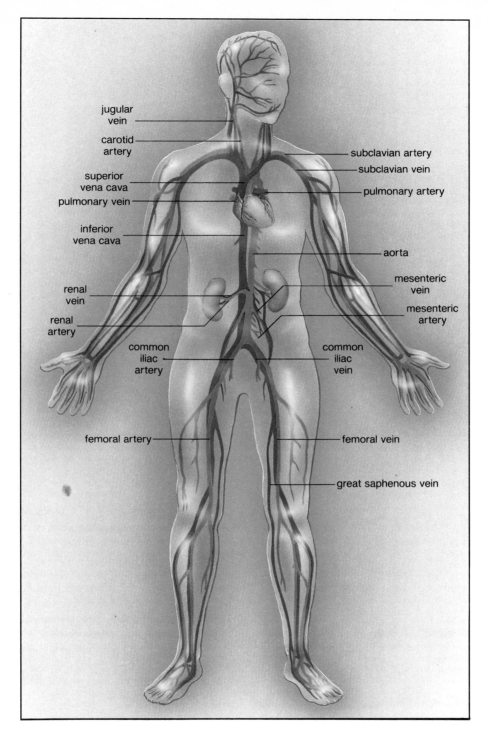

jugular
vein

carotid
artery

superior
vena cava

pulmonary vein

inferior
vena cava

renal
vein

renal
artery

common
iliac
artery

femoral artery

subclavian artery

subclavian vein

pulmonary artery

aorta

mesenteric
vein

mesenteric
artery

common
iliac
vein

femoral vein

great saphenous vein

Figure 5.9   Human circulatory system. A more realistic representation
of major blood vessels in the body shows that arteries and veins go to
all parts of the body. The superior and inferior venae cavae take their
names from their relationship to which organ?

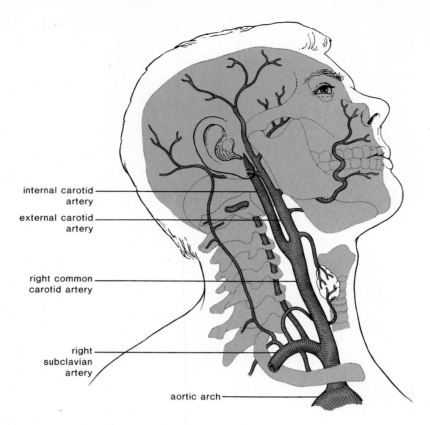

internal carotid
artery

external carotid
artery

right common
carotid artery

right
subclavian
artery

aortic arch

Figure 5.10　The common carotid artery. This artery is located in the neck region and can be used to take the pulse. The pulse indicates how rapidly the heart is beating.

pulse in any artery that runs close to the body's surface. It is customary to feel the pulse by placing several fingers on the radial artery, which lies near the outer border of the palm side of the wrist. The carotid artery is another good location to feel the pulse (fig. 5.10). The pulse rate, normally 70 times a minute, indicates the rate of the heartbeat because the arterial walls pulse whenever the left ventricle contracts.

The pulse rate indicates the heartbeat rate.

Blood Pressure

Blood pressure is the pressure of the blood against the wall of a blood vessel.

Measurement of Blood Pressure

A sphygmomanometer is used to measure blood pressure (fig. 5.11). This instrument consists of a hollow cuff connected by tubing to a compressible bulb and to a manometer, or pressure gauge. The cuff is placed about the upper arm over the brachial artery and is inflated with air. Eventually, the brachial artery is squeezed shut, and when a stethoscope is placed in the crook of the elbow, no sounds are heard. Air is slowly released, and the cuff is deflated until a sharp sound is heard through the stethoscope. The examiner glances at the pressure gauge and notes the pressure. This is the systolic blood pressure, the highest arterial pressure due to ejection of blood from the heart. The systolic pressure causes the blood to flow in the artery despite the pressure exerted by the cuff.

As the cuff is further deflated, tapping sounds become louder. Abruptly, the sounds become dull and muffled before they disappear. The examiner again notes the pressure. This is diastolic blood pressure, the lowest arterial pressure. Diastolic pressure occurs while the heart ventricles are relaxing.

Normal resting blood pressure for a young adult is said to be 120 mm of mercury (Hg) over 80 mm, or simply 120/80. The higher number is the systolic pressure, and the lower number is the diastolic pressure.

Blood Pressure throughout the Body

Actually, 120/80 is the expected blood pressure in the brachial artery of the arm; blood pressure decreases with distance from the left ventricle. Blood pressure is, therefore, higher in the arteries than in the arterioles. Further, there is a sharp drop in blood pressure when the arterioles reach the capillaries. The decrease can be correlated with the increase in the total cross-sectional area of the vessels as blood moves through arteries, arterioles, and then into capillaries. There are more arterioles than arteries and many more capillaries than arterioles (fig. 5.12).

Figure 5.11   This technician is in the act of applying a stethoscope to the left arm, which is enclosed by a sphygmomanometer. She will inflate the cuff with air, and as she gradually reduces the pressure, she will listen by means of the stethoscope for sounds that indicate the blood in an artery is moving past the cuff. A blood pressure gauge on the cuff is used to tell the systolic and diastolic blood pressure.

Blood pressure steadily decreases from the aorta to the veins.

Velocity of Blood Flow

The velocity of blood flow varies in different parts of the circulatory system (fig. 5.12). Blood pressure accounts for the velocity of the blood flow in the arterial system and therefore, as blood pressure decreases due to the increased cross-sectional area of the arterial system, so does velocity. The blood moves more slowly through the capillaries than it does through the aorta. This is important because the slow progress allows time for the exchange of molecules between the blood and the tissues.

Blood pressure cannot account for the movement of blood through the venules and the veins since they lie on the other side of the capillaries. Instead, movement of the blood through the venous system is due to skeletal muscle contraction. When the skeletal muscles contract, they press against the weak walls of the veins. This causes the blood to move past a *valve* (fig. 5.13). Once past the valve, the blood cannot return. The importance of muscle contraction in moving blood in the venous system can be demonstrated by forcing a person to stand rigidly still for a number of hours. Frequently, fainting occurs because the blood collects in the limbs, robbing the brain of oxygen. In this case, fainting is beneficial because the resulting horizontal position aids in getting blood to the head. Blood flow gradually increases in the venous system (fig. 5.12) due to a progressive reduction in the cross-sectional area as small venules join to form veins.

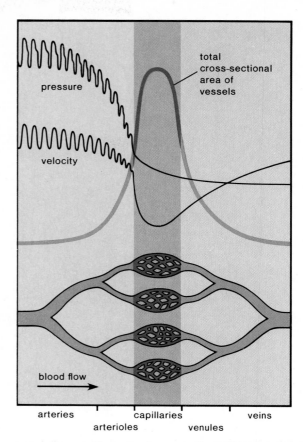

Figure 5.12   Diagram illustrating how velocity and blood pressure are related to the total cross-sectional area of blood vessels. Capillaries have the greatest cross-sectional area and the least pressure and velocity. Skeletal muscle contraction, not blood pressure, accounts for the velocity of blood in the veins.

The two venae cavae together have a cross-sectional area only about double that of the aorta. The blood pressure is lowered in the thoracic cavity whenever the thorax expands during inspiration. This also aids the flow of venous blood into the thoracic cavity because blood flows in the direction of reduced pressure.

Blood pressure accounts for the flow of blood in the arteries and the arterioles; skeletal muscle contraction accounts for the flow of blood in the venules and the veins.

Circulatory Disorders

During the past 30 years, the number of deaths due to cardiovascular disease has declined more than 30%. Even so, more than 50% of all deaths in the United States still are attributable to cardiovascular disease. The number of deaths due to hypertension, stroke, and heart attack is greater than the number due to cancer and accidents combined.

Cardiovascular disease is the number one killer in the United States.

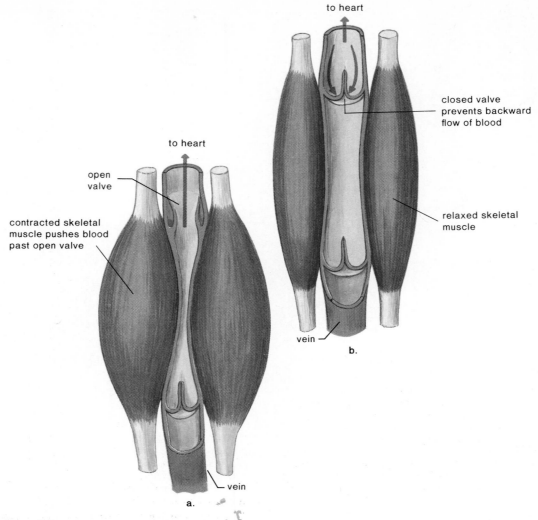

Figure 5.13 Skeletal muscle contraction moves blood in veins. *a.* Muscle contraction exerts pressure against the vein, and blood moves past the valve. *b.* Once blood has moved past the valve, it cannot return.

Hypertension

It is estimated that about 20% of all Americans suffer from *hypertension,* high blood pressure indicated by a blood pressure reading. Women of any age are considered to have hypertension if their blood pressure reading is 160/95 or above. For a man under age 45, a reading above 130/90 is hypertensive, and beyond age 45, a reading above 140/95 is considered hypertensive. While both systolic and diastolic pressures are considered important, it is the diastolic pressure that is emphasized when medical treatment is being considered.

The reasons for the development of hypertension are various. One possible scenario is described in figure 5.14. Blood pressure normally rises with excitement or alarm due to the involvement of the sympathetic nervous system (p. 201), which causes arterioles to constrict and the heart to beat faster. When arterioles are constricted, reduced blood flow to the kidneys causes these organs to release renin, a molecule that brings about vasoconstriction and sodium retention, as described in chapter

9. Vasoconstriction and sodium retention lead to high blood pressure. Excess sodium in the blood causes an increase in blood volume due to water retention.

Medical treatment can control hypertension with the following drugs. Sympathetic-blocking agents act at arrow 1 in figure 5.14 and prevent action of the sympathetic nervous system. Two types of drugs that act at arrow 2 are available. Calcium channel blockers are new drugs that prevent constriction of arteries because calcium is needed for muscle contraction (p. 231). Use of vasodilators results in relaxation of the muscles in arterial walls. Another new drug is an enzyme inhibitor that works at arrow 3 to prevent renin from producing its normal effects (p. 181). Finally, diuretics act at arrow 4 and cause the kidneys to excrete excess salts and fluids.

Since these drugs can have side effects that are at best unpleasant, it is wise to adopt a life-style that protects against the development of hypertension. The reading on page 108 discusses good health habits that lower the risk of cardiovascular disorders.

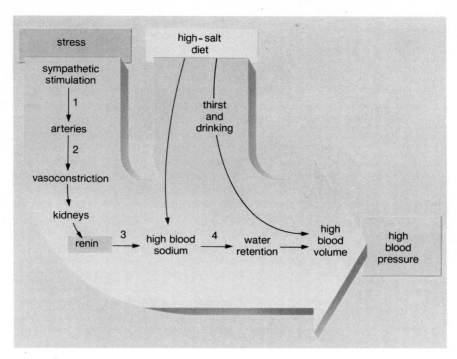

Figure 5.14 A scheme that explains the development of high blood pressure due to either stress or a high-salt diet. The numbers (1–4) indicate the site of action of antihypertensive drugs, as discussed in the text.

Diet, stress, and kidney involvement are implicated in the development of hypertension in some persons.

Human Issue

Sometimes a medical drug has unwanted side effects. For example, a study found that the popular diuretic hydrochlorothiazide, prescribed for high blood pressure, tends to decrease the ability to metabolize the sugar glucose and tends to raise the blood cholesterol level. Should doctors tell their patients about possible side effects of drugs and leave it up to the patients whether to take the drug or not? Should they do this even if the patient doesn't have enough knowledge to appreciate the risks? Or should doctors assume complete responsibility for prescribing the drug?

Atherosclerosis

Hypertension also is seen in individuals who have *atherosclerosis* (formerly called arteriosclerosis), an accumulation of soft masses of fatty materials, particularly cholesterol, beneath the inner linings of arteries. Such deposits are called *plaque*, and as it develops, it tends to protrude into the vessel and interferes with the flow of blood. Atherosclerosis begins in early adulthood and develops progressively through middle age, but symptoms may not appear until an individual is 50 or older. To prevent its onset and development, a diet low in saturated fat and choles-

terol is recommended by the American Heart Association and other organizations, as discussed in the reading on page 108.

Plaque can cause a clot to form on the irregular arterial wall. As long as the clot remains stationary, it is called a *thrombus*, but when and if it dislodges and moves along with the blood, it is called an *embolus*. If *thromboembolism* is not treated, complications can arise, as mentioned in the following section.

Development of atherosclerosis, which is associated with a high blood cholesterol level, can lead to thromboembolism.

Stroke and Heart Attack

Both strokes and heart attacks are associated with hypertension and atherosclerosis. A *stroke* (cerebrovascular accident) occurs when a portion of the brain dies due to a lack of oxygen. A stroke, characterized by paralysis or death, often results when a small arteriole bursts or is blocked by an embolus. A person sometimes is forewarned of a stroke by a feeling of numbness in the hands or the face, difficulty in speaking, or temporary blindness in one eye.

A *heart attack* (myocardial infarct) occurs when a portion of the heart muscle dies because of a lack of oxygen. Due to atherosclerosis, the coronary artery may be partially blocked. The individual may then suffer from *angina pectoris*, characterized by a radiating pain in the left arm. When a coronary artery is completely blocked, perhaps because of thromboembolism, a heart attack occurs.

Cardiovascular Disease

Because sudden cardiac death happens once every 72 seconds in the United States, it is well to identify the factors that predispose an individual to cardiovascular disease. The risk factors for cardiovascular disease include the following:

Male sex
Family history of heart attack under age 55
Smoking more than 10 cigarettes a day
Severe obesity (30% or more overweight)
Hypertension
Unfavorable HDL and LDL blood cholesterol levels
Impaired circulation to the brain or the legs
Diabetes mellitus

Hypertension is well recognized as a major factor in cardiovascular disease, and two controllable behaviors contribute to hypertension. Smoking cigarettes, including filtered cigarettes, causes hypertension, as does obesity. It is best to never take up the habit of smoking cigarettes, but most of the detrimental effects can be reversed if you stop smoking. Since it is very difficult for obese individuals to lose weight, it is recommended that weight control be a lifelong endeavor. Investigators have identified several behaviors that may help to reduce the possibility of heart attack and stroke. Exercise seems to be critical. Sedentary individuals have a risk of cardiovascular disease that is about double that of those who are very active. One physician, for example, recommends that his patients walk for one hour, three times a week. Stress reduction also is desirable. The same investigator recommends everyday meditation and yoga-like stretching and breathing exercises to reduce stress.

Another behavior that is much in the news of late is the adoption of a diet that is low in saturated fats and cholesterol (see p. 83) because such a diet is believed by many to protect against the development of cardiovascular disease. Cholesterol is ferried in the blood by two types of plasma proteins called LDL (low-density lipoprotein) and HDL (high-density lipoprotein). LDL (called "bad" lipoprotein) takes cholesterol to the tissues from the liver, and HDL (called "good" lipoprotein) transports cholesterol out of the tissues to the liver. When the blood LDL level is abnormally high or the HDL level is abnormally low, cholesterol accumulates in the cells. When cholesterol-laden cells line the arteries, plaque develops, which interferes with circulation (see figure).

Cholesterol guidelines have been established by the National Heart, Lung, and Blood Institute. According to the institute, everyone should know his or her blood cholesterol level. Individuals with a borderline-high cholesterol (200–239 mg/100 ml) should be tested further if they already have heart disease or if they have two known risk factors for cardiovascular disease (see list). Individuals with a high blood cholesterol level (240 mg/100 ml) always should be tested further. Persons with an LDL cholesterol level of over 130 mg/100 ml should be treated if they have other risk factors, and those with an LDL cholesterol level of 160 mg/100 ml should be treated even if this is the only risk factor.

Persons with a total-to-HDL cholesterol ratio higher than 4.5 also are considered to be at risk. Heart attack has occurred in individuals who have a normal total cholesterol level, but who also have an unfavorable total-to-HDL cholesterol ratio. For example, if a person's total cholesterol blood level is 200, but the HDL level is only 25 mg/100 ml, then the total-to-HDL cholesterol ratio is 8.0, and circulatory difficulties most likely will develop.

First and foremost, treatment for unfavorable cholesterol levels consists of adopting a diet that is low in saturated fat and cholesterol (see p. 83). Although the prescribed diet does not lower cholesterol blood level in all persons, it is expected to do so for most individuals. If diet alone does not bring down the blood cholesterol level, drugs can be prescribed. Some of the drugs act in the intestine to remove cholesterol, and others act in the body to prevent its production. These drugs reduce the blood cholesterol level, but

---

Stroke and heart attack are associated with both hypertension and atherosclerosis.

Medical and Surgical Treatment

Medical and surgical treatments now are available for blocked coronary arteries.

Thrombolytic Therapy Medical treatment for thromboembolism includes two drugs that can be given intravenously to dissolve a clot: streptokinase and tPA. Both drugs convert plasminogen, a molecule found in blood, into plasmin, an enzyme that can dissolve a blood clot. In fact, tPA, which stands for tissue plasminogen activator, is the body's very own way of converting plasminogen to plasmin. Streptokinase and tPA are used particularly when it is known that a clot is present.

If a person has symptoms of angina or thromboembolism, then an anticoagulant drug, such as aspirin, may be given. Aspirin works by inhibiting platelets, cell fragments that trigger clot formation. Aspirin reduces the stickiness of platelets and thus lowers the probability that a clot will form. There is evidence that aspirin protects against first heart attacks, but there is no clear support for taking aspirin every day to prevent strokes in symptom-free people. Physicians warn that long-term use of aspirin might have harmful effects, including bleeding in the brain.

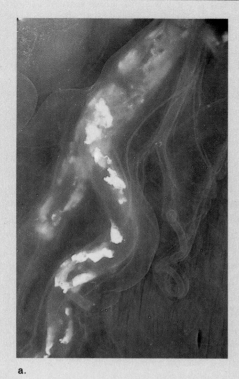

a.

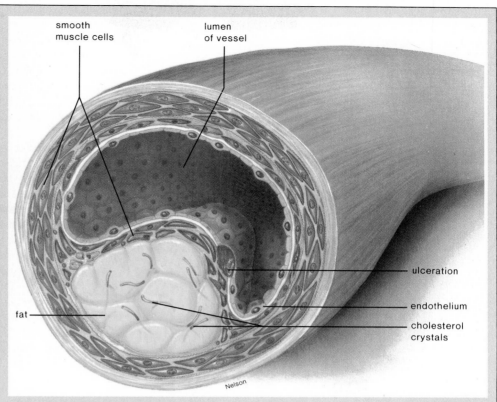

smooth
muscle cells

lumen
of vessel

ulceration

endothelium

cholesterol
crystals

fat

Nelson

b.

Plaque. *a.* Plaque (yellow) in the coronary artery of a heart patient. *b.* Cross section of plaque shows its composition and indicates how it bulges out into the lumen of an artery, obstructing blood flow.

their long-term side effects are not completely known and may be serious. Considering this, some investigators do not recommend these drugs to lower blood cholesterol levels. A diet low in saturated fat and cholesterol can lower the total blood cholesterol level and the LDL level of some individuals, but this diet most likely will not raise the HDL level. Aside from certain drugs that apparently can raise HDL level, exercise is sometimes effective.

There is nothing that can be done about some of the cardiovascular risk factors, such as male gender and family history. However, other risk factors likely can be controlled if the individual believes it is worth the effort. It is clear that the four great admonitions for a healthy life—heart-healthful diet, regular exercise, proper weight maintenance, and refraining from smoking—all contribute to acceptable blood pressure and blood cholesterol levels.

Arterial Plaque Surgical procedures are available to clear clogged arteries. In one procedure, a plastic tube is threaded into an artery of an arm or a leg and is guided through a major blood vessel toward the heart. When the tube reaches the region of plaque in a coronary or carotid artery (fig. 5.10), a balloon attached to the end of the tube inflates, forcing the vessel open. The problem with this procedure is the vessel may not remain open, and worse, it may cause clots to form. Therefore, thousands of persons each year have *coronary bypass* surgery. During this operation, surgeons take a segment of another blood vessel, often from a large vein in the leg, and stitch one end to the aorta and the other end to a coronary artery past the point of obstruction (fig. 5.15).

Once the bypass is complete, some physicians use lasers to open up clogged coronary vessels. One day, it may be possible to use lasers independently, without first opening the thoracic cavity.

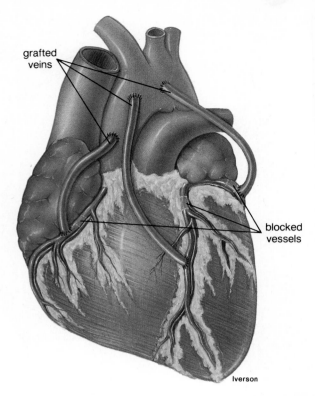

grafted veins

blocked vessels

Iverson

Figure 5.15 Coronary bypass operation. During this operation, the surgeon grafts segments of a leg vein between the aorta and the coronary vessels, bypassing areas of blockage. Patients who are ill enough to require surgery often receive 2 or 3 bypasses in a single operation.

Donor Heart Transplants and Artificial Heart Implants
Persons with weakened hearts eventually may suffer from _congestive heart failure,_ meaning the heart no longer is able to pump blood adequately. These individuals, depending on their age, are candidates for a donor heart transplant. The difficulties with a donor heart transplant are, first, one of availability and, second, the tendency of the body to reject foreign organs. Sometimes, it is possible to repair the heart instead of replacing it. For example, a back muscle can be wrapped around a weak heart. An artificial implant pacemaker causes the muscle to contract regularly and helps to pump the blood.

On December 2, 1982, Barney Clark became the first person to receive an artificial heart. The heart's two polyurethane ventricles were attached to Clark's own atria and blood vessels by way of Dacron fittings. Two long tubes were stretched between the artificial heart and an external machine that periodically sent bursts of air into the ventricles, forcing the blood out into the aorta and the pulmonary trunk. As yet, no one has survived for longer than a few months after receiving an artificial heart due to complications, such as formation of blood clots. For this reason, artificial hearts are not implanted except on a temporary basis in patients awaiting a transplant.

Varicose Veins and Phlebitis _Varicose veins_ are abnormal and irregular dilations in superficial (near the surface) veins, particularly those in the lower legs. Varicose veins in the rectum, however, are commonly called piles, or more properly, _hemorrhoids._ Varicose veins develop when the valves of the veins become weak and ineffective due to a backward pressure of the blood. The problem can be aggravated when venous blood flow is obstructed by crossing the legs or by sitting in a chair so that its edge presses against the backs of the knees.

Phlebitis, or inflammation of a vein, is a more serious condition, particularly when a deep vein is involved. Blood in the inflamed vessel may clot, in which case thromboembolism occurs. An embolus that originates in a systemic vein eventually may come to rest in a pulmonary arteriole, blocking circulation through the lungs. This condition, termed _pulmonary embolism,_ can result in death.

What's Your Diagnosis?

Mary has started a new job, serving as an editor for a major publisher. For two months she has worked 10 to 12 hours daily, five or six days per week. She feels that she is constantly under pressure to meet the deadlines of her job. Mary's work environment certainly is a stressful one as she continues to adjust to the challenges of her new occupation. Lately, Mary has been developing headaches late in the workday.

Upon visiting a physician, Mary, age 48, has her blood pressure measured. The reading is 170/100. Both figures are significantly higher than previous recordings at any stage of her life. Her resting pulse rate is also consistently around 95 beats per minute.

1. What is your diagnosis of Mary's condition? 2. How are the symptoms of Mary's condition consistent with your diagnosis? 3. What medical and behavioral treatment would you recommend to help Mary?

Summary

The movement of blood in the circulatory system is dependent on the beat of the heart. During the cardiac cycle, the SA node (pacemaker) initiates the beat and causes the atria to contract. The AV node picks up the stimulus and initiates contraction of the ventricles. The heart sounds, lub-DUPP, are due to the closing of the atrioventricular valves, followed by the closing of the semilunar valves.

The circulatory system is divided into the pulmonary and systemic circuits. In the pulmonary circuit, the pulmonary artery takes blood from the right ventricle to the lungs, and the pulmonary veins return it to the left atrium. To trace the path of blood in the systemic circuit, start with the aorta from the left ventricle. Follow its path until it branches to an artery going to a specific organ. It can be assumed that the artery divides into arterioles and capillaries, and that the capillaries lead to venules. The vein that takes blood to the vena cava most likely has the same name as the artery.

Blood pressure accounts for the flow of blood in the arteries, but because blood pressure drops off after the capillaries, it cannot cause blood flow in the veins. Skeletal muscle contraction pushes blood past a venous valve, which then shuts, preventing backward flow. The velocity of blood flow is slowest in the capillaries, where exchange of nutrient and waste molecules takes place.

Hypertension and atherosclerosis are two circulatory disorders that lead to heart attack and to stroke. Medical and surgical procedures are available to control cardiovascular disease, but the best policy is prevention by following a heart-healthy diet, getting regular exercise, maintaining a proper weight, and not smoking cigarettes.

Study Questions

1. What types of blood vessels are there? Discuss their structure and function. (p. 94)
2. Trace the path of blood in the heart, mentioning the vessels attached to and the valves within the heart. (p. 97)
3. Describe the cardiac cycle (using the terms systole and diastole), and explain the heart sounds. (p. 98)
4. Describe the cardiac conduction system and an ECG. Tell how an ECG is related to the cardiac cycle. (p. 100)

5. Trace the path of blood in the pulmonary circuit as it travels from and returns to the heart. (p. 101)
6. Trace the path of blood from the mesenteric arteries to the aorta, indicating which of the vessels are in the systemic circuit and which are in the pulmonary circuit. (p. 101)
7. What is blood pressure, and where in the body is the average normal arterial blood pressure about 120/80? (p. 104)

8. In which type of vessel is blood pressure highest? Lowest? Velocity is lowest in which type of vessel, and why is it lowest? Why is this beneficial? What factors assist venous return of the blood? (p. 104)
9. What is atherosclerosis? (p. 107) Name two illnesses associated with hypertension and thromboembolism. (p. 107) Discuss the medical and surgical treatment of cardiovascular disease. (p. 108)

Objective Questions

1. Arteries are blood vessels that take blood _____ from the heart.
2. When the left ventricle contracts, blood enters the _____ .
3. The pulmonary veins carry blood _____ in oxygen.
4. The right side of the heart pumps blood to the _____ .
5. The _____ node is known as the pacemaker.

6. The arteries that serve the heart are the _____ arteries.
7. The pressure of blood against the walls of a vessel is termed _____ .
8. Blood moves in arteries due to _____ and in veins due to _____ .
9. Reducing the amount of _____ and _____ in the diet may reduce the chances of plaque buildup in arteries.

10. Varicose veins develop when _____ become weak and ineffective.

Answers to Objective Questions

1. away 2. aorta 3. high 4. lungs 5. SA 6. coronary 7. blood pressure 8. blood pressure, skeletal muscle contraction 9. saturated fat, cholesterol 10. valves

Label this Diagram.
See figure 5.4 (p. 97) in text.

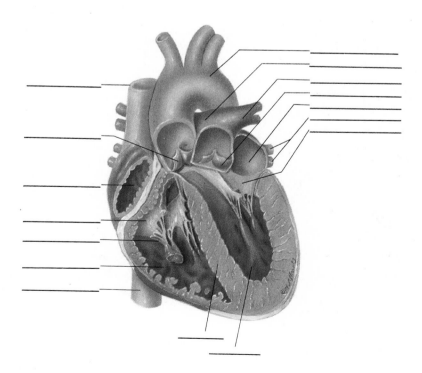

Critical Thinking Questions

1. The ancients did not know that the blood circulates. Instead, they thought that the veins and arteries were two different, unrelated systems of tubes. What evidence can you provide to prove that the blood does indeed circulate?

2. Fishes have only a two-chambered heart (one atrium and one ventricle); amphibians have a three-chambered heart (two atria and one ventricle); humans have a four-chambered heart. Why is this complexity of value in the human heart?

3. What organs mentioned in the chapter contribute to maintaining the blood pressure, and how do they contribute?

Selected Key Terms

aorta (a-or′tah)   major systemic artery that receives blood from the left ventricle.   101

arteriole (ar-te′re-ōl)   vessel that takes blood from arteries to capillaries.   94

artery (ar′ter-e)   vessel that takes blood away from the heart; characteristically possessing thick elastic and muscular walls.   94

atrium (a′tre-um)   chamber; particularly an upper chamber of the heart that lies above the ventricles.   96

AV node (a-ve nōd)   a small region of neuromuscular tissue that transmits impulses received from the SA node to the ventricular walls.   98

capillary (kap′ĭ-lar′′e)   microscopic vessel connecting arterioles to venules through the thin walls of which molecules either exit or enter the blood.   94

coronary artery (kor′ō-na-re ar′ter-e)   artery that supplies blood to the wall of the heart.   101

diastole (di-as′to-le)   relaxation of heart chamber.   98

pulmonary circuit (pul′mo-ner′′e ser′kit)   that part of the circulatory system that takes deoxygenated blood to and oxygenated blood away from the lungs.   101

SA node (es a nōd)   small region of neuromuscular tissue that initiates the heartbeat. Also called the pacemaker.   98

systemic circuit (sis-tem′ik ser′kit)   that part of the circulatory system that serves body parts other than the gas-exchanging surfaces in the lungs.   101

systole (sis′to-le)   contraction of a heart chamber.   98

valve (valv)   membranous extension of a vessel or heart wall that opens and closes, ensuring one-way flow.   95

vein (vān)   vessel that takes blood to the heart; characteristically having nonelastic walls.   94

vena cava (ve′nah ka′vah)   large systemic vein that returns blood to the right atrium of the heart; the superior vena cava draws blood from the head and shoulders, whereas the inferior vena cava draws blood from the rest of the body.   97

ventricle (ven′tri-k′l)   cavity in an organ, such as a lower chamber of the heart.   96

venule (ven′ūl)   vessel that takes blood from capillaries to veins.   94

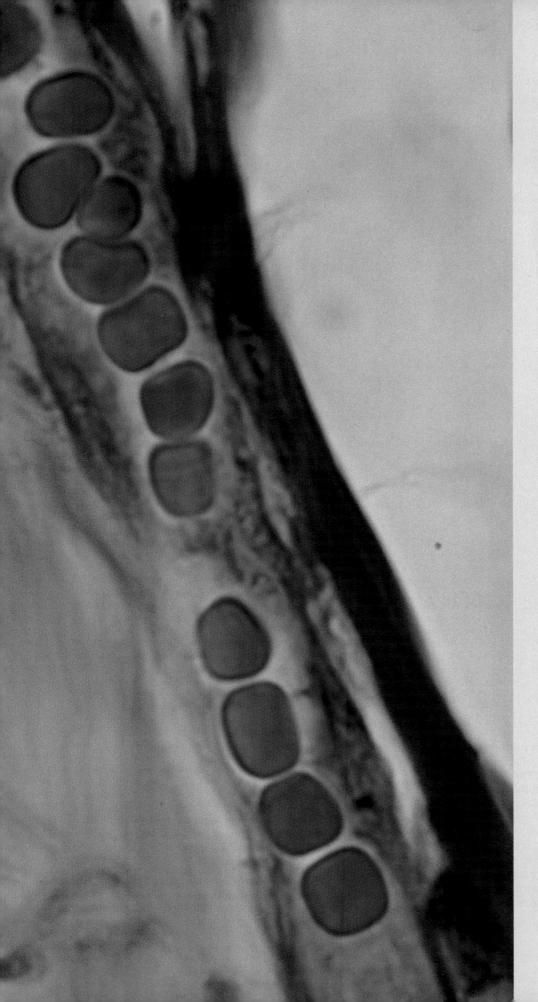

Chapter Six

Blood

Chapter Concepts

1 Blood, which is composed of cells and a fluid containing many inorganic and organic molecules, has three primary functions: transporting, clotting, and infection fighting.

2 Exchange of molecules between blood and tissue fluid takes place across capillary walls.

3 Blood is typed according to the antigens present on the red blood cells.

4 All of the functions of blood can be correlated with the ability of the body to maintain a constant environment.

Red blood cells in single file within a capillary of the scalp. These tiny cells carry oxygen throughout the body, enabling tissue cells to carry out aerobic cellular respiration and to acquire the ATP energy they need to function.

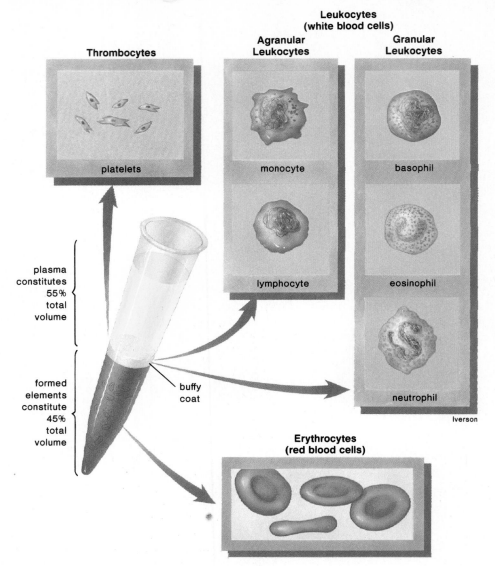

Thrombocytes

platelets

**Leukocytes
(white blood cells)**

**Agranular
Leukocytes**

monocyte

lymphocyte

**Granular
Leukocytes**

basophil

eosinophil

neutrophil

Iverson

plasma
constitutes
55%
total
volume

formed
elements
constitute
45%
total
volume

buffy
coat

**Erythrocytes
(red blood cells)**

Figure 6.1   Composition of blood. When blood is transferred to a test
tube and is centrifuged, it forms two layers. The transparent yellow top
layer is plasma, the liquid portion of blood. The formed elements are in
the bottom layer: the white blood cells (leukocytes) and the platelets
(thrombocytes) are in a narrow buff-colored band above the red blood
cells (erythrocytes). The white blood cells are either granular leukocytes
or agranular leukocytes.

**In persons with sickle cell disease, the red blood cells aren't
biconcave disks (fig. 6.1) like normal red blood cells; they are
elongated. In fact, they are sickle shaped. The defect is caused
by an abnormal hemoglobin that piles up inside the cells. Because the sickle-shaped cells can't pass along narrow capillary
passageways like disk-shaped cells, they clog the vessels and
break down. No wonder, then, that persons with sickle cell disease suffer from poor circulation, anemia, and poor resistance
to infection. Internal hemorrhaging leads to further complications like jaundice, episodic pain of the abdomen and joints,
and damage to internal organs. The importance of a normal
structure and function of blood components is dramatically illustrated by sickle cell disease.**

Transport Function of Blood

If blood is transferred from a person's vein to a test tube and is
prevented from clotting, it separates into two layers (fig. 6.1).
The lower layer consists of red blood cells (erythrocytes), white
blood cells (leukocytes), and blood platelets (thrombocytes).
Collectively, these are called the **formed elements.** Formed elements make up about 45% of the total volume of whole blood.
The upper layer, called plasma, contains a variety of inorganic
and organic substances dissolved or suspended in water. Plasma
accounts for about 55% of the total volume of whole blood.

Table 6.1 Components of Blood

Blood	Function	Source
Formed elements		
red blood cells	transport oxygen (O_2)	bone marrow
platelets	clotting	bone marrow
white blood cells	fight infection	bone marrow
*Plasma**		
water	maintain blood volume; transport molecules	absorbed from intestine
plasma proteins	maintain blood osmotic pressure and pH	
albumin	transport	liver
fibrinogen	clotting	liver
globulins		
alpha and beta	transport	liver
gamma	fight infection	lymphocytes
gases		
oxygen (O_2)	aerobic cellular respiration	lungs
carbon dioxide (CO_2)	end product of metabolism	tissues
nutrients		
fats, glucose, amino acids, etc.	food for cells	absorbed from intestinal villi
salts	maintain blood osmotic pressure and pH; aid metabolism	absorbed from intestinal villi
wastes		
urea, uric acid, ammonia	end products of metabolism	tissues and liver
hormones, vitamins, etc.	aid metabolism	varied

*Plasma is 90–92% water, 7–8% plasma proteins, and not quite 1% salts; all other components are present in even smaller amounts.

Table 6.1 lists the components of blood—some of which participate in its transport function. The transport function of blood helps maintain the constancy of tissue fluid. Blood transports oxygen (O_2) from the lungs and the nutrients from the intestine to the capillaries of the body where they enter tissue fluid. Here, blood also takes up carbon dioxide (CO_2) and other waste (molecules) given off by the cells and transports them away. Carbon dioxide exits blood at the lungs. Urea, a substance formed by the liver following deamination of amino acids, travels by way of the bloodstream to the kidneys and is excreted. Figure 6.2 diagrams the transport function of blood, indicating the manner in which this function helps to keep the internal environment relatively constant.

Homeostasis is only possible because blood brings nutrients to the cells and removes their wastes.

Plasma

Plasma (table 6.1) is the liquid portion of blood. Small organic molecules, such as glucose and urea, simply dissolve in plasma, but large organic molecules combine with proteins for transport.

Plasma Proteins

Plasma proteins make up 8–9% of plasma. These molecules assist in transporting large organic molecules in blood. For example, the molecule bilirubin, a breakdown product of hemoglobin, is transported by **albumin,** and the **alpha** and **beta globulins**[1] transport hormones and fat-soluble molecules. The lipoproteins that transport cholesterol are globulins.

The movement of blood in the arteries is dependent upon blood pressure, but blood needs a certain *viscosity,* or thickness, in order to exert pressure. This property of blood is largely dependent on plasma proteins and red blood cells. Maintenance of blood viscosity is, therefore, another way that proteins contribute to the transport of molecules.

Blood also needs a certain *volume* in order to exert a pressure. Because plasma proteins are too large to pass through a capillary wall, the fluid within capillaries is always an area of lesser concentration of water compared to tissue fluid, and water, therefore, passes into capillaries. Maintenance of osmotic pressure is associated particularly with albumin, the smallest and most plentiful plasma protein (table 6.2).

Certain plasma proteins have specific functions that are not duplicated by any other proteins. We will discuss how fibrinogen is necessary to blood clotting. We also will discuss the gamma globulins, which are antibodies that help to fight infection.

---

[1]When globulins undergo electrophoresis (are put in an electrical field), they separate into three major components called alpha globulin, beta globulin, and gamma globulin. The gamma globulin fraction contains antibodies.

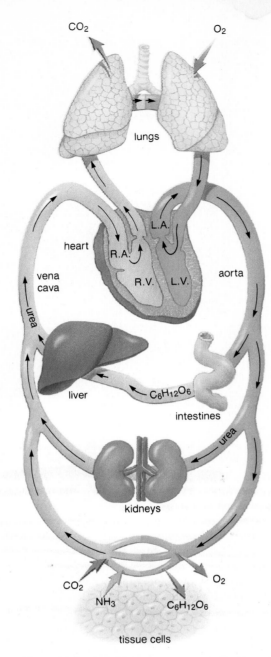

Figure 6.2 Diagram illustrating the transport function of blood. Oxygen (O_2) is transported from the lungs to the tissues, and carbon dioxide (CO_2) is transported from the tissues to the lungs. Urea, a nitrogenous end product made in the liver, is excreted by the kidneys. Glucose ($C_6H_{12}O_6$) is absorbed by the gut and may be stored temporarily in the liver as glycogen before it is transported to the tissues.

Table 6.2 Plasma Proteins

Name	Function
albumin	maintains osmotic pressure and transports bilirubin
globulins	
alpha and beta	transport hormones and fat-soluble molecules
gamma	fight infection
fibrinogen	causes blood clotting

Plasma proteins assist in the transport function of blood. They serve as carriers for some molecules, and they help to maintain the viscosity and the volume of blood.

Human Issue

The ancients believed that the blood carried vital spirits and the heart was the center of the emotions. In other words, the cardiovascular system was seen as possessing a life-force and contributing to our human nature. Today there is an artificial heart, and there may soon be artificial blood. Other structures in the body can also be successfully replaced by artificial structures. This raises an interesting question: How much of a human body could consist of artificial substances and parts before we considered the individual a robot rather than a human?

Red Blood Cells (Erythrocytes)

Red blood cells also are involved in the transport function of blood. They contain the respiratory pigment **hemoglobin,** which carries oxygen (O_2) (fig. 6.3). Since hemoglobin is a red pigment, the cells appear red, and their color also makes the blood red. There are between 4 and 6 million red blood cells per mm^3 of whole blood, and each of these cells contains about 200 million hemoglobin molecules. If this much hemoglobin were suspended within the plasma rather than enclosed within the cells, blood would be so thick the heart would have difficulty pumping it.

Each hemoglobin molecule (fig. 6.3c) contains four polypeptide chains that make up the protein **globin,** and each chain is associated with **heme,** a complex iron-containing group. Iron combines loosely with oxygen, and in this way, oxygen is carried in blood.

Humans are active, warm-blooded animals; the brain and the muscles often require much oxygen within a short period of time. Plasma carries only about 0.3 ml of oxygen per 100 ml, but whole blood carries 20 ml of oxygen per 100 ml. This shows that hemoglobin increases the carrying capacity of blood more than 60 times. Although the iron portion of hemoglobin carries oxygen, the equation for oxygenation of hemoglobin is usually written as

$$Hb + O_2 \underset{\text{tissues}}{\overset{\text{lungs}}{\rightleftarrows}} HbO_2$$

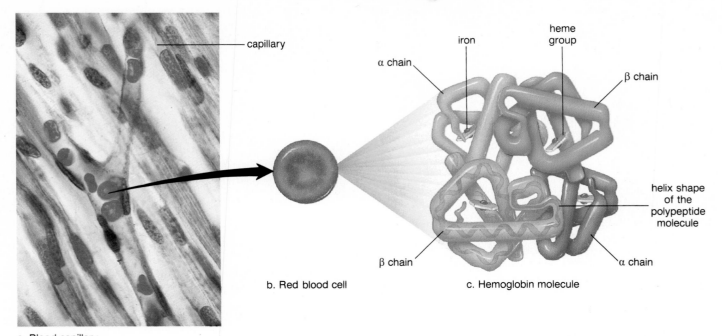

capillary

iron

heme group

α chain

β chain

β chain

α chain

helix shape of the polypeptide molecule

b. Red blood cell

c. Hemoglobin molecule

a. Blood capillary

Figure 6.3 Physiology of red blood cells. *a.* Red blood cells move single file through the capillaries. *b.* Each red blood cell is a biconcave disk containing many molecules of hemoglobin, the respiratory pigment. *c.* Hemoglobin contains four polypeptide chains, two of which are alpha (*α*) chains and two of which are beta (*β*) chains. The plane in the center of each chain represents an iron-containing heme group. Oxygen combines loosely with iron when hemoglobin is oxygenated.

The hemoglobin on the right, which is combined with oxygen, is called *oxyhemoglobin.* Oxyhemoglobin forms in the lungs and has a bright red color. The hemoglobin on the left, which has given up oxygen to tissue fluid, has a dark purple color.

Hemoglobin is an excellent oxygen carrier because it combines loosely with oxygen in the cool, neutral conditions of the lungs. It then readily gives up oxygen under the warm and more acidic conditions of the tissues.

Carbon monoxide (CO), present in automobile exhaust and cigarette smoke, combines with hemoglobin more readily than oxygen, and it stays combined for several hours, regardless of the environmental conditions. Accidental death or suicide from carbon monoxide poisoning occurs because the hemoglobin of blood is not available for oxygen transport. This transport function of blood is so important that life can be sustained temporarily by giving a patient a hemoglobin substitute transfusion when whole blood is not available or cannot be given. The reading on page 118 discusses the possible benefits of this "artificial blood."

Oxygen is transported to the tissues in combination with hemoglobin, a pigment found in red blood cells.

Red blood cells that are not engaged in oxygen transport assist in transporting carbon dioxide (CO_2). First, hemoglobin (Hb) combines with carbon dioxide to form *carbaminohemoglobin:*

$$Hb + CO_2 \underset{\text{lungs}}{\overset{\text{tissues}}{\rightleftharpoons}} HbCO_2$$

However, this combination with hemoglobin actually represents only a small portion of the carbon dioxide in the blood. Most of the carbon dioxide is transported as the *bicarbonate ion.* This ion forms after carbon dioxide combines with water. Carbon dioxide combined with water is carbonic acid (H_2CO_3), which dissociates (breaks down) to a hydrogen (H^+) ion and a bicarbonate ion (HCO_3^-):

$$CO_2 + H_2O \underset{\text{lungs}}{\overset{\text{tissues}}{\rightleftharpoons}} H_2CO_3 \underset{\text{lungs}}{\overset{\text{tissues}}{\rightleftharpoons}} H^+ + HCO_3^-$$

An enzyme within red blood cells, called *carbonic anhydrase,* speeds up this reaction. Most of the released hydrogen ions, are absorbed by the globin portions of hemoglobin, and the bicarbonate ions diffuse out of red blood cells to be carried in the plasma. Hemoglobin, which combines with a hydrogen ion called *reduced hemoglobin,* can be symbolized as HHb. Reduced hemoglobin plays a vital role in maintaining the pH of blood.

Once systemic venous blood reaches the lungs, the reaction just described takes place in the reverse: the bicarbonate ion joins with a hydrogen ion to form carbonic acid, and this splits into carbon dioxide and water. The carbon dioxide diffuses

Artificial Blood

There are risks to having a blood transfusion, such as an immune reaction and acquiring an infection. A cross-matching test between the donor's blood and the recipient's blood usually detects if the recipient has antibodies in the plasma that will react against antigens on the membrane of the donor's red blood cells and vice versa. Blood also is screened for the presence of 2 viruses that are especially troublesome. They are the hepatitis B virus and the AIDS virus. Blood donors are questioned carefully, and their blood is tested for the presence of these viruses. Despite the care that is taken to avoid immune reactions and the transference of disease, it would be advantageous to develop an artificial blood that has neither of these risks.

Investigation is proceeding in three directions. Enrico Bucci, a blood-substitute specialist at the University of Maryland School of Medicine, is working with modifying hemoglobin itself. He links several hemoglobin molecules so that they do not become "lost" from the circulatory system and uses chemicals to modify the complexes' oxygen affinity so that they are more likely to give up the oxygen when needed. Many other problems remain, however. The hemoglobin is taken from whole blood, and it alone may contain infectious material. Modified hemoglobin also seems to cause a generalized constriction of the body's blood vessels, making oxygenation of tissues more difficult.

Anthony Hunt and colleagues at the University of California at San Francisco are working with artificial red blood cells called neohemocytes (NHCs). To make the cells, purified human hemoglobin taken from outdated donor blood is encapsulated in a lipid bilayer membrane. The artificial cells are much

View inside a capillary in which blood has been replaced with a 25 percent suspension of hemoglobin-containing synthetic neohemocytes (NHCs). A normal red blood cell is shown for scale. Unlike living red blood cells, neohemocytes are nearly spherical and can have one, two, or three chambers.

smaller than normal human red blood cells, and they do not contain as much hemoglobin. However, when tested in rats, the animals survived until they were sacrificed for gross toxicity studies. The investigators believe that the tests are successful enough to warrant further study. They want to improve the stability and the vascular retention time of the cells since they are removed from the bloodstream and are broken down at a faster rate than normal cells.

George Groveman is director of new-products marketing at Alpha Therapeutic, a subsidiary of a Japanese firm that is based in Los Angeles. His firm is working on a third possibility, a substance called perfluorocarbon emulsion oil (PFC) that can be transfused and can carry oxygen much like hemoglobin does. This substance has served as a blood substitute for

humans in emergency situations in which only the oxygen-carrying function of blood was required. FDA approval has been sought by a Japanese corporation to market PFC under the trade name of Fluosol-DA, but thus far the FDA has denied permission on the grounds that the clinical trials were not successful enough. There is some hope, though, that PFC will be approved for localized use. For example, it may be helpful to administer Fluosol-DA when a person is suffering a heart attack or is undergoing thrombolytic therapy (p. 108).

Although researchers have been working for 20 years to produce a blood substitute for general use, the prospects are still in the future. Anthony Hunt says, "Physiological systems always turn out to be more complicated than we thought."

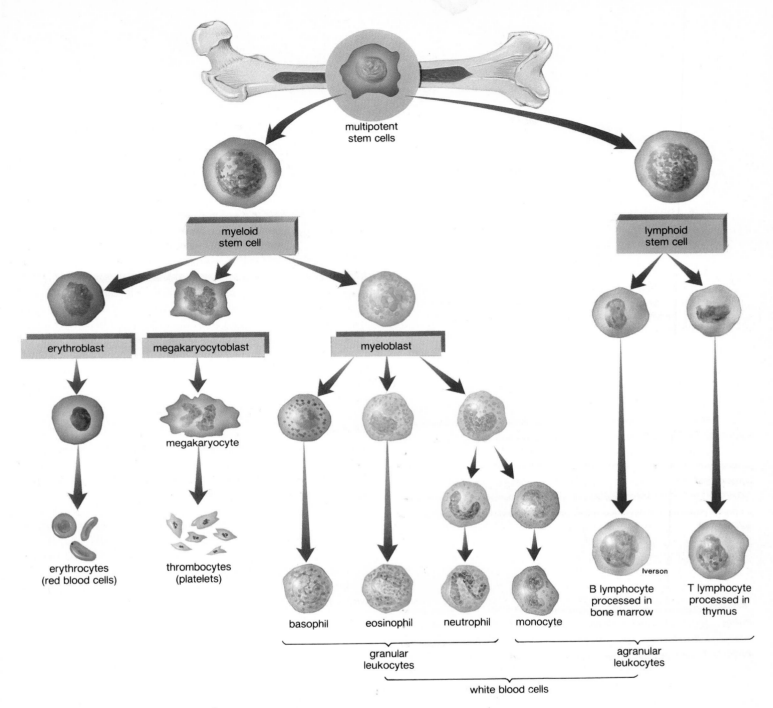

Figure 6.4   Formation of formed elements in red bone marrow. Multipotent stem cells give rise to specialized stem cells. The myeloid stem cell gives rise to still other cells that become red blood cells, platelets, and all the white blood cells except lymphocytes. The lymphoid stem cell gives rise to the lymphocytes.

out of blood into the lungs for expiration. Now, hemoglobin is ready again to transport oxygen.

Hemoglobin participates in the transport of carbon dioxide in blood and helps to buffer the blood.

Life Cycle of Red Blood Cells

Red blood cells (erythrocytes) are manufactured continuously in the bone marrow of the skull, the ribs, the vertebrae, and the ends of the long bones. The number produced increases whenever arterial blood carries a reduced amount of oxygen (O_2), as happens when an individual first takes up residence at a high altitude. Under these circumstances, the kidneys (and probably other organs as well) produce a growth factor called erythropoietin that stimulates cell division and differentiation of erythrocyte stem cells in the bone marrow (fig. 6.4). Erythropoietin, now mass-produced through biotechnology (p. 391), is sometimes abused by athletes in order to raise their red cell count.

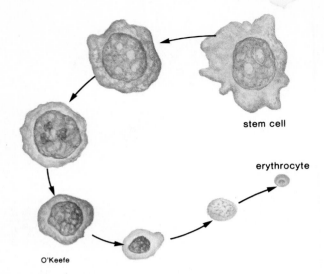

stem cell

erythrocyte

O'Keefe

Figure 6.5   Maturation of red blood cells (erythrocytes). Red blood cells are made in the bone marrow, where stem cells continuously divide. During the maturational process, a red blood cell loses its nucleus and gets much smaller.

Before they are released from the bone marrow into blood, red blood cells lose their nucleus and acquire hemoglobin (fig. 6.5). Possibly because they lack a nucleus, red blood cells only live about 120 days. They are destroyed chiefly in the *liver* and the *spleen,* where they are engulfed by large phagocytic cells. When red blood cells are broken down, the hemoglobin is released. The iron is recovered and is returned to the bone marrow for reuse. The heme portion of the molecule undergoes chemical degradation and is excreted by the liver in the bile as bile pigments. These bile pigments contribute to the color of feces.

Anemia

When there is an insufficient number of red blood cells or the cells do not have enough hemoglobin, the individual suffers from **anemia** and has a tired, run-down feeling. In iron-deficiency anemia, the hemoglobin blood level is low. It may be that the diet does not contain enough iron. Certain foods, such as raisins and liver, are rich in iron, and the inclusion of these in the diet can help to prevent this type of anemia.

In another type of anemia, called pernicious anemia, the digestive tract is unable to absorb enough vitamin B_{12}. This vitamin is essential to the proper formation of red blood cells; without it, immature red blood cells tend to accumulate in the bone marrow in large quantities. A special diet and administration of vitamin B_{12} by injection is an effective treatment for pernicious anemia.

Illness (anemia) results when the blood has too few red blood cells and/or not enough hemoglobin.

Capillary Exchange within the Tissues

Arterial End of the Capillary

When arterial blood enters the tissue capillaries (fig. 6.6), it is bright red because red blood cells are carrying oxygen. It is also rich in nutrients that are dissolved in the plasma. At the arterial end of the capillary, blood pressure (40 mm Hg) is higher than the osmotic pressure of the blood (15 mm Hg). Blood pressure, you recall, is created by the pumping of the heart; the osmotic pressure is caused by the presence of salts and, in particular, by the plasma proteins that are too large to pass through the wall of the capillary. Since the blood pressure is higher than the osmotic pressure, fluid together with oxygen and nutrients (glucose and amino acids) exit from the capillary. This is a *filtration* process because large substances, such as red blood cells and plasma proteins, remain, but small substances, such as water and nutrient molecules, leave the capillaries. Tissue fluid, created by this process, consists of all the components of plasma except the proteins.

Midsection

Along the length of the capillary, molecules follow their concentration gradient as diffusion occurs. Diffusion, you recall, is the movement of molecules from an area of greater concentration to an area of lesser concentration. The area of greater concentration for nutrients is always blood because after these molecules have passed into the tissue fluid, they are taken up and metabolized by the tissue cells. The cells use glucose ($C_6H_{12}O_6$) and oxygen (O_2) in the process of cellular respiration, and they use amino acids for protein synthesis. Following cellular respiration, the cells give off carbon dioxide (CO_2) and water (H_2O). Carbon dioxide and waste products of metabolism, leave the cell by diffusion. Since tissue fluid is always the area of greater concentration for these waste materials, they diffuse into the capillary.

Oxygen and nutrient molecules (e.g., glucose and amino acids) exit a capillary near the arterial end; waste molecules (e.g., carbon dioxide) enter a capillary near the venous end.

Venous End of the Capillary

At the venous end of the capillary, blood pressure is much reduced (10 mm Hg), as can be verified by reviewing figure 5.12

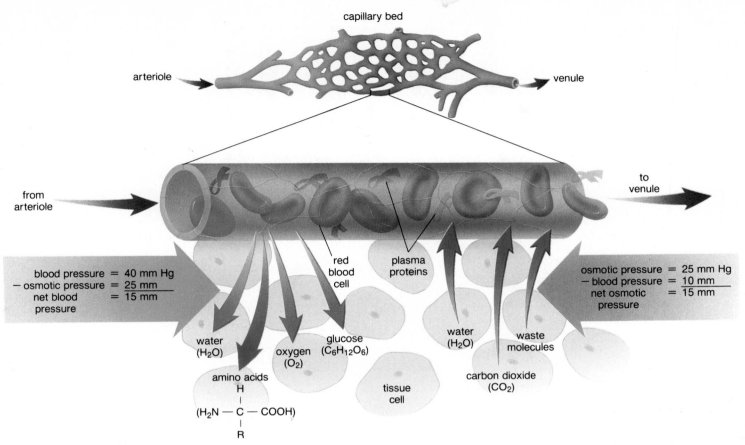

Figure 6.6 Diagram of a capillary illustrating the exchanges that take place and the forces that aid the process. At the arterial end of a capillary, the blood pressure is higher than the osmotic pressure; therefore, water (H_2O), oxygen, (O_2), amino acids, and glucose ($C_6H_{12}O_6$) tend to leave the bloodstream. At the venous end of a capillary, the osmotic pressure is higher than the blood pressure; therefore, water (H_2O), carbon dioxide (CO_2), and other wastes tend to enter the bloodstream. Notice that the red blood cells and the plasma proteins are too large to exit from a capillary.

(p. 105). However, there is no reduction in osmotic pressure (25 mm Hg), which tends to pull fluid into the capillary. As water enters a capillary, it brings with it additional waste molecules. Blood that leaves the capillaries is deep purple in color because red blood cells contain reduced hemoglobin. Carbon dioxide is being carried in the plasma as the bicarbonate ion (HCO_3^-).

Retrieving fluid by means of osmotic pressure is not completely effective. There is always some fluid that is not picked up at the venous end. This excess tissue fluid enters the lymphatic capillaries (fig. 6.7). Lymph is tissue fluid contained within lymphatic vessels. Lymph is returned to the systemic venous blood when the major lymphatic vessels enter the subclavian veins (p. 132).

Lymphatic capillaries lie in close proximity to blood capillaries, where they collect excess tissue fluid.

Blood-Clotting Function of Blood

When an injury to a blood vessel occurs, **clotting,** or coagulation, of blood takes place. This is obviously a protective mechanism to prevent excessive blood loss. As such, blood clotting is another mechanism by which blood components maintain homeostasis.

If blood is allowed to clot in a test tube, a yellowish fluid develops above the clotted material (fig. 6.8). This fluid is called

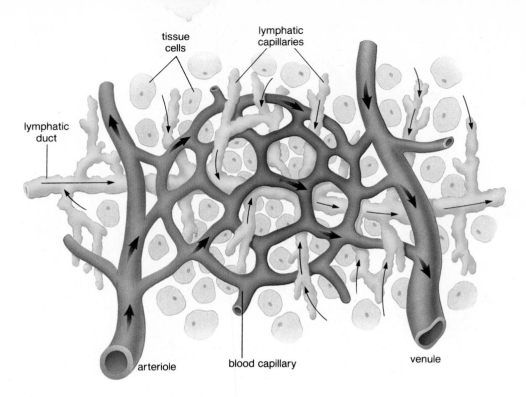

Figure 6.7   Lymphatic vessels. Arrows indicate that lymph is formed when lymphatic capillaries take up excess tissue fluid. Lymphatic capillaries lie near blood capillaries.

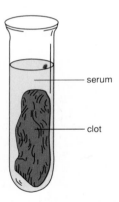

Figure 6.8   Serum formation. When blood clots, serum is squeezed out as a solid plug is formed. In a blood vessel, this plug can help to prevent further blood loss.

Table 6.3   Body Fluids

Name	Composition
blood	formed elements and plasma
plasma	liquid portion of blood
serum	plasma minus fibrinogen
tissue fluid	plasma minus proteins
lymph	tissue fluid within lymphatic vessels

serum, and it contains all the components of plasma except fibrinogen. Because we now have used a number of different terms to refer to various body fluids related to blood, table 6.3 reviews these terms for you.

There are at least 12 clotting factors in the blood that participate in the formation of a blood clot. We will discuss the roles played by platelets, prothrombin, and fibrinogen. **Platelets** result from fragmentation of certain large cells, called megakaryocytes, in the bone marrow (fig. 6.4). These cells are produced at a rate of 200 billion a day, and the bloodstream carries more than a trillion. **Fibrinogen** and **prothrombin** are proteins

manufactured and deposited in blood by the liver. Vitamin K is necessary to the production of prothrombin, and if by chance this vitamin is missing from the diet, hemorrhagic disorders develop.

When a blood vessel in the body is damaged, platelets clump at the site of the puncture and partially seal the leak. They and the injured tissues release a clotting factor called prothrombin activator that converts prothrombin to thrombin. This reaction requires calcium (Ca^{++}) ions. **Thrombin,** in turn, acts as an enzyme that severs two short amino acid chains from each fibrinogen molecule. These activated fragments then join end to end, forming long threads of **fibrin.** Fibrin threads wind around the platelet plug in the damaged area of the blood vessel and provide the framework for the clot. Red blood cells also are trapped within the fibrin threads; these cells make a clot appear red. The steps necessary for blood clotting upon injury are quite complex and are summarized in figure 6.9.

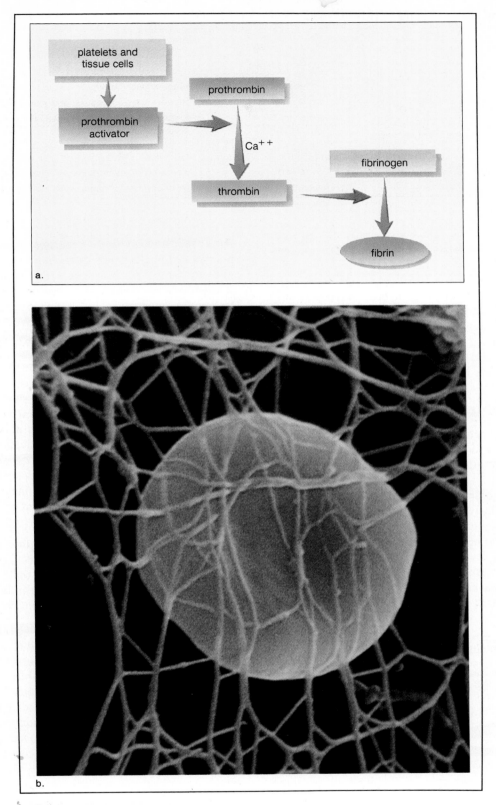

Figure 6.9   *a.* Fibrin threads form from activated fibrinogen, a normal component of blood plasma. *b.* Scanning electron micrograph showing a red blood cell (erythrocyte) caught in the fibrin threads of a clot.

Table 6.4 White Blood Cells (Leukocytes)

Granular leukocytes (polymorphonuclear)			Function
	Size	Granules stain	
neutrophils	9–12 μm	lavender	phagocytize primarily bacteria
eosinophils	9–12 μm	red	phagocytize and destroy antigen-antibody complexes
basophils	9–12 μm	deep blue	congregate in tissues; release histamine when stimulated
Agranular leukocytes			
	Size	Type of nucleus	
lymphocytes	8–10 μm	indented	B type produce antibodies in blood and lymph
monocytes	12–20 μm	irregular	become macrophages—phagocytize bacteria and viruses

A blood clot consists of platelets and red blood cells entangled within fibrin threads.

A fibrin clot is only temporarily present. As soon as blood vessel repair is initiated, an enzyme called plasmin destroys the fibrin network and restores the fluidity of plasma. This is a protective measure because a blood clot can act as a thrombus or an embolus. A blood clot interferes with circulation and can cause the death of tissues in the area.

Infection-Fighting Function of Blood

The body defends itself against microorganisms in various ways (see chapter 7). Two components of blood, white blood cells and antibodies, contribute to this defense. Their roles are explored briefly here.

Diseases and disorders that affect the quantity of white blood cells are particularly threatening. *Leukemia* is a form of cancer characterized by uncontrolled production of abnormal white blood cells. These cells accumulate in the bone marrow, the lymph nodes, the spleen, and the liver so that these organs are unable to function properly. Since the initiation of "total therapy," the combined use of chemotherapy with radiation therapy, a large number of leukemia patients are able to remain in remission for years.

White blood cells fight infection. They attack microorganisms that have invaded the body.

White Blood Cells (Leukocytes)

White blood cells differ from red blood cells in that they are usually larger, have a nucleus, lack hemoglobin, and without staining, appear white in color. White blood cells are not as numerous as red blood cells, with only 7,000–8,000 cells per mm³.

Table 6.4 lists the different types of white blood cells, and figure 6.10 shows two in detail. On the basis of structure, it is possible to divide white blood cells into the granular leukocytes and the agranular leukocytes. The **granular leukocytes** (neutrophils, eosinophils, and basophils) have granules in the cyto-

plasm that contain powerful digestive enzymes. Some have a many-lobed nucleus joined by nuclear threads; therefore, they also are called polymorphonuclear. Granular leukocytes are formed and mature in the bone marrow (fig. 6.4). The **agranular leukocytes** (monocytes and lymphocytes) do not have prominent granules in the cytoplasm and have a nucleus that is mononuclear. Agranular leukocytes are also produced in the bone marrow, but certain of the lymphocytes mature in the thymus. Agranular leukocytes are stored in the spleen, the lymph nodes, the tonsils, and other lymphoid organs.

Infection fighting by white blood cells is dependent primarily on the neutrophils, which comprise 60–70% of all leukocytes, and the lymphocytes, which make up 25–30% of the leukocytes. Neutrophils are phagocytic; they destroy many bacteria by traveling to the site of invasion and engulfing the foe. Some lymphocytes produce immunoglobulins or **antibodies,** that combine with foreign substances to inactivate them. Neutrophils and lymphocytes can be compared in the following manner:

Neutrophils	*Lymphocytes*
Granules in cytoplasm	No granules in cytoplasm
Polymorphonuclear	Mononuclear
Produced in the bone marrow	Produced in the bone marrow and accumulate in lymphoid tissue
Phagocytic	Make gamma globulins (antibodies)

Ordinarily, the total number of white blood cells increases when there is infection. Sometimes, when only one type of white blood cell increases in number, it is possible to help to diagnose the illness by doing a differential white blood cell count, involving microscopic examination of a blood sample and counting the number of each type of white blood cell up to a total of 100 cells. For example, in *infectious mononucleosis,* the characteristic finding is a large number of lymphocytes of the B type that are atypical in appearance. This condition, caused by an Epstein-Barr viral infection, takes its name from the fact that lymphocytes are mononuclear. On the other hand, the *AIDS* (acquired immunodeficiency syndrome) virus attacks the T type of lymphocyte, and the patient eventually has a reduced number of this type of lymphocyte in his or her blood (p. A–4). Infection is usually diagnosed by testing for the presence of antibodies against the virus in blood.

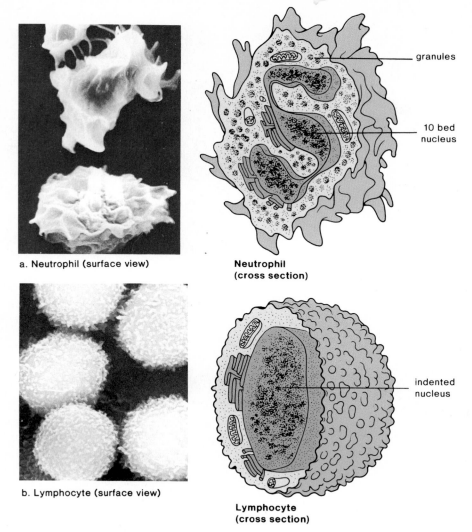

a. Neutrophil (surface view)

Neutrophil (cross section)

granules

10 bed nucleus

b. Lymphocyte (surface view)

Lymphocyte (cross section)

indented nucleus

Figure 6.10 Two leukocytes of interest. *a.* Scanning electron micrograph and an artist's representation of a neutrophil. *b.* Scanning electron micrograph and an artist's representation of a lymphocyte.

Antibodies

Bacteria and their toxins cause lymphocytes of the B type to produce antibodies because they contain antigens. **Antigens** are protein, sometimes polysaccharide, molecules that are foreign to the individual. When activated, each B lymphocyte produces only one type of antibody that is specific for one type of antigen. *Antibodies* combine with their antigen in such a way that the antigen is rendered harmless. Sometimes antibodies cause agglutination (clumping) of the antigen or simply prepare it for phagocytosis. In any case, keep in mind that the antigen is the foreigner, and the antibody is the molecule prepared by the body.

The antigen-antibody reaction is a lock-and-key reaction in which the molecules fit together like a lock and key. Specificity is possible because each type of antibody has variable regions, unique sequences of amino acids that result in receptor sites capable of combining with one type of antigen. In other words, this particular sequence of amino acids shapes a site where the antibody fits a specific antigen (fig. 7.7).

Immunity

An individual is actively immune when the body has antibodies that can react to a disease-causing antigen. Blood in these individuals contains lymphocytes that are capable of producing the necessary antibodies for a length of time. Exposure to the antigen, either naturally or by way of a vaccine, can cause active immunity to develop. Chapter 7 deals with immunity and explores the topic in detail.

Lymphocytes are responsible for immunity. B lymphocytes produce antibodies that specifically combine with disease-causing antigens.

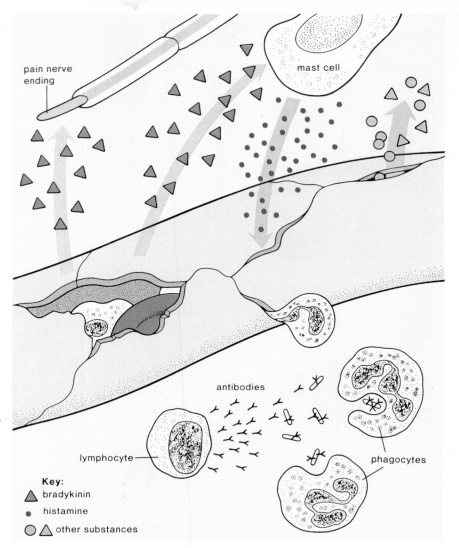

pain nerve
ending

mast cell

antibodies

lymphocyte

phagocytes

Key:
▲ bradykinin
• histamine
◯ △ other substances

Figure 6.11   Inflammatory reaction. When a blood vessel is injured,
release of bradykinin stimulates the pain nerve ending and the mast
cells that secrete histamine. Histamine dilates blood vessels.
Neutrophils congregate at the injured site. These amoeboid cells
squeeze through the capillary wall and begin to phagocytize bacteria,
especially those that have been attacked by antibodies.

Inflammatory Reaction

Whenever the skin is broken due to a minor injury, a series of
events occurs that is known as the **inflammatory reaction** because there is swelling and reddening at the site of the injury.
Figure 6.11 illustrates the participants in the inflammatory reaction. One participant, the mast cells, possibly is derived from
basophils, a type of white blood cell that takes up residence in
the tissues.

When an injury occurs, a capillary and several tissue cells
are apt to rupture and to release certain precursors that lead to
the presence of **bradykinin,** a molecule that (1) initiates nerve
impulses resulting in the sensation of pain, and (2) stimulates

mast cells to release **histamine,** another molecule that together
with bradykinin (3) causes the capillary to dilate and to become
more permeable. The enlarged capillary causes the skin to
redden, and its increased permeability allows proteins and fluids
to escape so that swelling results.

Any break in the skin allows bacteria and viruses to enter
the body. In figure 6.11, a B lymphocyte is releasing antibodies
that attack the bacteria, preparing them for **phagocytosis.** When
a neutrophil phagocytizes a bacterium, an intracellular vacuole
is formed. The engulfed bacterium is destroyed by hydrolytic
enzymes when the vacuole combines with a granule.

Also present in tissues are monocyte-derived **macrophages,** large phagocytic cells that are able to devour a hundred

invaders and still survive. Some tissues, particularly connective tissue, have resident macrophages that routinely act as scavengers, devouring old blood cells, bits of dead tissue, and other debris. Macrophages also are capable of bringing about an explosive increase in the number of leukocytes by liberating a growth factor that passes by way of the blood to the bone marrow, where it stimulates the production and the release of white blood cells, usually neutrophils.

As the infection is being overcome, some neutrophils die. These, along with dead tissue, cells, and bacteria and living white blood cells, form **pus,** a thick, yellowish fluid. Pus indicates that the body is trying to overcome the infection.

The inflammatory reaction is a "call to arms"—it marshals phagocytic white blood cells to the site of invasion by bacteria.

Human Issue

When people undergo operations, they often require a blood transfusion. Some people today are storing blood that they themselves might use later. That's appropriate if they expect to undergo an operation sometime soon, but there is not enough space for blood banks to store blood for each individual indefinitely. Therefore, it seems that we do need a blood bank for everyone's use. Do you think that all should be required to give blood to a blood bank or is it all right to expect to be given blood even when none has been contributed?

Blood Typing

ABO System

The cell membrane of red blood cells contains molecules that may differ from one individual to the next. When the blood of one individual is given to another, certain molecules can act as antigens in the recipient. These antigens are known as type A and type B, and blood is typed as described in table 6.5 according to whether or not these antigens are present on a person's red blood cells. For example, if a person has type A blood, the A antigen is on his or her red blood cells. This molecule is not an antigen to this individual, although it can be an antigen to a recipient of his or her blood.

In the simplified ABO system, there are four types of blood: A, B, AB, and O. Type O blood has neither the A antigen nor the B antigen on red blood cells; the other types of blood are designated by the antigen(s) present on red blood cells.

Within the plasma, there are antibodies to the antigens that are *not* present on the person's red blood cells. Therefore, for example, type A blood has an antibody called anti-B in the plasma. Type AB blood has neither anti-A nor anti-B antibodies because both antigens are on the red blood cells. This is reasonable because if these antigens were present, **agglutination,** or clumping of red blood cells, would occur. Agglutination of red blood cells can cause blood to stop circulating in small blood

Table 6.5 Blood Groups

Type	Antigen on red blood cells	Antibody in plasma	U.S. black (%)*	U.S. caucasian (%)*
A	A	anti-B	25	41
B	B	anti-A	20	7
AB	A,B	none	4	2
O	none	anti-A, anti-B	51	50

*Blood type frequency for other races is not available.

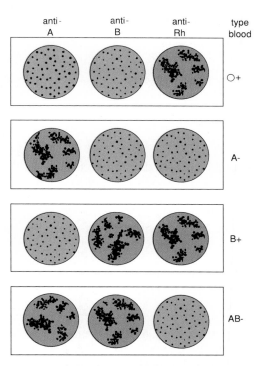

Figure 6.12   Blood typing. The standard test to determine ABO and Rh blood type consists of putting a drop of anti-A antibodies, anti-B antibodies, and anti-Rh antibodies on a slide. To each of these, a drop of the person's blood is added. If agglutination occurs, the person has this antigen on red blood cells. Shown are several possible results.

vessels, and this leads to organ damage. It also is followed by hemolysis, which brings about the death of the individual.

For a recipient to receive blood from a donor, the recipient's plasma must not have an antibody that causes the donor's cells to agglutinate. For this reason, it is important to determine each person's blood type. Figure 6.12 demonstrates a way to use the antibodies derived from plasma to determine the blood type. If clumping occurs after a sample of blood is exposed to a particular antibody, the person has that type of blood.

Rh System

Another important antigen in matching blood types is the **Rh factor.** Eighty-five percent of the U.S. population has this particular antigen on the red blood cells and are Rh+ (Rh positive).

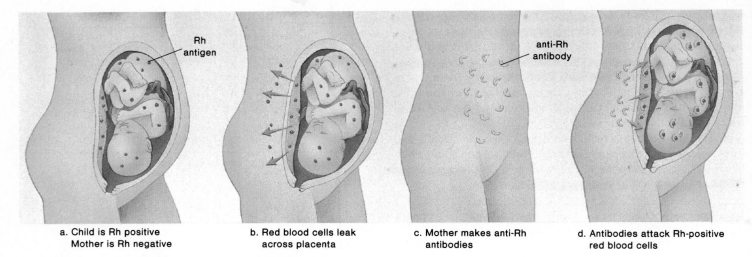

a. Child is Rh positive
   Mother is Rh negative

b. Red blood cells leak
   across placenta

c. Mother makes anti-Rh
   antibodies

d. Antibodies attack Rh-positive
   red blood cells

Figure 6.13   Diagram (*a–d*) describing the development of hemolytic disease of the newborn.

Fifteen percent do not have this antigen and are Rh⁻ (Rh negative). Rh⁻ individuals normally do not have antibodies to the Rh factor, but they may make them when exposed to the Rh factor. It is possible to use anti-Rh antibodies for blood testing. When Rh⁺ blood is mixed with anti-Rh antibodies, agglutination occurs.

The designation of blood type usually also includes whether the person has the Rh factor (Rh⁺) or does not have the Rh factor (Rh⁻) on the red blood cells.

During pregnancy (fig. 6.13), if the mother is Rh⁻ and the father is Rh⁺, the child may be Rh⁺. The Rh⁺ red blood cells may begin leaking across the placenta into the mother's circulatory system, as placental tissues normally break down before and at birth. This causes the mother to produce anti-Rh antibodies. In this or a subsequent pregnancy with another Rh⁺ baby, anti-Rh antibodies (but usually not anti-A and anti-B antibodies discussed earlier) may cross the placenta and destroy the child's red blood cells. This is called hemolytic disease of the newborn (HDN).

The Rh problem has been solved by giving Rh⁻ women an Rh immunoglobulin injection either midway through the first pregnancy or no later than 72 hours after giving birth to any Rh⁺ child. This injection contains anti-Rh antibodies that attack any of the baby's red blood cells in the mother's blood before

these cells can stimulate her immune system to produce her own antibodies. The injection is not beneficial if the woman has already begun to produce antibodies; therefore, the timing of the injection is most important.

The possibility of hemolytic disease of the newborn exists when the mother is Rh⁻ and the father is Rh⁺.

What's Your Diagnosis?

Mary is a 19-year-old college student. On a recent trip to the mountains, she had trouble breathing at high altitudes. In general, she complains of a tired and "run-down" feeling and fatigue after mild strenuous exercise.

A battery of laboratory tests were requested by her physician. The percentage of formed elements was 35%. The red cell count and concentration of hemoglobin were measured and found to be low.

1. What is your diagnosis of Mary's condition? 2. Why are the laboratory results consistent with your diagnosis and with Mary's symptoms? 3. What treatment might you recommend for Mary?

Summary

Nutrients and wastes are transported in plasma, but oxygen is combined with hemoglobin within red blood cells. The end result of transport is capillary exchange with tissue fluid, regulated by blood pressure and osmotic pressure. Blood clotting requires a series of enzymatic reactions involving platelets, prothrombin, and fibrinogen. In the final reaction, fibrinogen becomes fibrin threads that entrap red blood cells.

White blood cells and gamma globulin proteins are required to fight infections. The two most prevalent white blood cells are the neutrophils and the lymphocytes. Neutrophils are involved in the inflammatory reaction, and lymphocytes are involved in immunity development.

Blood transfusions require compatible blood types. Of consideration are the antigens (A and B) on the red blood cells and the antibodies (anti-A and anti-B) in the plasma. Another important antigen is the Rh antigen, particularly because an Rh⁻ mother may produce anti-Rh antibodies that will attack the red blood cells of an Rh-positive fetus.

Study Questions

1. State the major components of plasma. Name the plasma proteins, and tell their common function as well as their specific functions. (p. 115)
2. Give the equation for the oxygenation of reduced hemoglobin. Where does this reaction occur? Where does the reverse reaction occur? (p. 116)
3. Give an equation that indicates how carbon dioxide (CO_2) commonly is carried in blood. Indicate the direction of the reaction in the tissues and in the lungs. In what ways does hemoglobin aid the process of transporting carbon dioxide? (p. 117)
4. Discuss the life cycle of red blood cells. (p. 119)
5. What forces operate to facilitate exchange of molecules across the capillary wall? (pp. 120–121)
6. Name the steps that take place when blood clots. Which substances are present in blood at all times, and which appear during the clotting process? (p. 122)
7. Define blood, plasma, tissue fluid, lymph, and serum. (p. 122)
8. Name and discuss two ways that blood fights infection. Associate each of these with a particular type of white blood cell. (p. 124)
9. Describe the inflammatory reaction, and give a role for each type of cell and chemical that participates in the reaction. (p. 126)
10. What are the four ABO blood types in humans? For each, state the antigen(s) on the red blood cells and the antibody(ies) in the plasma. (p. 127)
11. Problems can arise during childbearing if the mother is which Rh type and the father is which Rh type? Explain why this is so. (p. 128)

Objective Questions

1. The liquid part of blood is called _____ .
2. Red blood cells carry _____ , and white blood cells _____ _____ .
3. Hemoglobin that is carrying oxygen is called _____ .
4. Human red blood cells lack a _____ and only live about _____ days.
5. When a blood clot occurs, fibrinogen has been converted to _____ threads.
6. The most common granular leukocyte is the _____ , a phagocytic white blood cell.
7. B lymphocytes produce _____ that react with antigens.
8. At a capillary, _____ , _____ , and _____ leave the arterial end, and _____ and _____ enter the venous end.
9. Type AB blood has the antigens _____ and _____ on red blood cells and _____ antibodies in plasma.
10. Hemolytic disease of the newborn can occur when the mother is _____ and the father is _____ .

Critical Thinking Questions

1. Multicellular animals have exchange areas with the external environment in order to maintain the internal environment. Explain with reference to figure 6.2.
2. Blood coagulation is a series of enzymatic reactions that lead from one to the other. Of what value is such a complicated system?
3. In what way is the inflammatory reaction a lifesaving event?

Selected Key Terms

agglutination (ah-gloo''ti-na'shun)   clumping of cells, particularly in reference to red blood cells involved in an antigen-antibody reaction.   127

agranular leukocyte (ah-gran'u-lar loo'ko-sīt)   white blood cell that does not contain distinctive granules.   124

antibody (an'ti-bod''e)   a protein produced in response to the presence of some foreign substance in blood or the tissues.   124

antigen (an-ti-jen)   a foreign substance, usually a protein, that stimulates the immune system to produce antibodies.   125

clotting (klot'ing)   process of blood coagulation, usually when injury occurs.   121

fibrinogen (fi-brin'o-jen)   plasma protein that is converted into fibrin threads during blood clotting.   122

formed element (form'd el'ē-ment)   a constituent of blood that is either cellular (red blood cells and white blood cells) or at least cellular in origin (platelets).   114

granular leukocyte (gran'u-lar loo'ko-sīt)   white blood cell that contains distinctive granules.   124

inflammatory reaction (in-flam'ah-to''re re-ak'shun)   a tissue response to injury that is characterized by dilation of blood vessels and accumulation of fluid in the affected region.   126

macrophage (mak'ro-fāj)   an enlarged monocyte that ingests foreign material and cellular debris.   126

phagocytosis (fag''o-si-to'sis)   the taking in of bacteria and/or debris by engulfing.   126

platelet (plāt'let)   a formed element that is necessary to blood clotting.   122

prothrombin (pro-throm'bin)   plasma protein that is converted to thrombin during the process of blood clotting.   122

pus (pus)   thick, yellowish fluid composed of dead phagocytes, dead tissue, and bacteria.   127

serum (se'rum)   light-yellow liquid left after clotting of blood.   122

thrombin (throm'bin)   an enzyme that converts fibrinogen to fibrin threads during blood clotting.   122

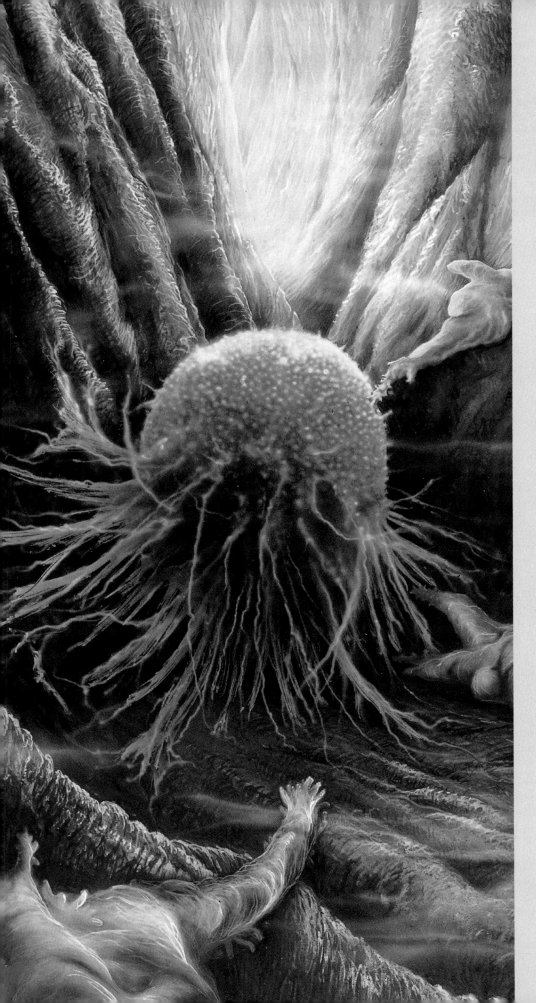

Chapter Seven

Lymphatic System and Immunity

Chapter Concepts

1 The lymphatic vessels form a one-way system that transports lymph from the tissues and fat from the lacteals to certain cardiovascular veins.

2 Lymphocytes are produced in the bone marrow and accumulate in other lymphoid organs.

3 The body has various general (nonspecific) ways to protect itself from disease.

4 Immunity is specific and requires two types of lymphocytes, B lymphocytes and T lymphocytes. Both of these are produced in the bone marrow.

5 While immunity preserves our existence, it also is responsible for certain undesirable effects, such as tissue rejection, allergies, and autoimmune diseases.

6 Immunotherapy involves the use of vaccines to achieve long-lasting immunity and the use of antibodies to provide temporary immunity.

A disruptive cancer cell coursing through a lymphatic vessel is a threat to the body's integrity. The lymphatic system aids in the body's defense but may, on occasion, help spread cancer in the manner depicted here.

Immunization (fig. 7.1) protects children from diseases. The success of immunization is witnessed by the fact that the vaccination against smallpox is no longer required because the disease has been eradicated. However, parents today often fail to get their children immunized because they don't realize the importance of doing so. Therefore, one occasionally reads in the newspapers about an outbreak of measles at a college or hospital in the United States because adults were never optimally immunized as children. The United Nations has established the goal of universal immunization of the world's children against diphtheria, tetanus, pertussis, polio, measles, and tuberculosis. Safe and effective vaccines are available against these infections and are being included in a worldwide program.

Lymphatic System

The **lymphatic system** consists of lymphatic vessels (fig. 7.2) and the lymphoid organs. This system, which is closely associated with the cardiovascular system, has three main functions: (1) lymphatic vessels take up excess tissue fluid and return it to the bloodstream; (2) lymphatic capillaries absorb fats at the intestinal villi and transport them to the bloodstream (p. 122); and (3) the lymphatic system helps to defend the body against disease.

Lymphatic Vessels

Lymphatic vessels are quite extensive; every region of the body is supplied richly with lymphatic capillaries. The construction of the larger lymphatic vessels is similar to that of cardiovascular veins, including the presence of valves. Also, the movement of lymph within these vessels is dependent upon skeletal muscle contraction. When the muscles contract, the lymph is squeezed past a valve that closes, preventing the lymph from flowing backwards.

The lymphatic system begins with lymphatic capillaries that lie near blood capillaries. These capillaries take up fluid that has diffused from and has not been reabsorbed by the blood capillaries (fig. 6.7). Once tissue fluid enters the lymphatic vessels, it is called **lymph.** The lymphatic capillaries join to form lymphatic vessels that merge before entering one of two ducts: the thoracic duct or the right lymphatic duct.

The *thoracic duct* is much larger than the right lymphatic duct. It serves the lower extremities, the abdomen, the left arm, and the left side of the head and the neck. In the thorax, the left thoracic duct enters the left subclavian vein. The *right lymphatic duct* serves only the right arm and the right side of the head and the neck. It enters the right subclavian vein.

The lymphatic system is a one-way system. Lymph flows from a capillary to ever-larger lymphatic vessels and finally to a lymphatic duct, which enters a subclavian vein.

Edema

Edema is localized swelling caused by the accumulation of tissue fluid. Tissue fluid accumulates if too much of it is being made and/or not enough of it is being drained away. Pulmonary edema

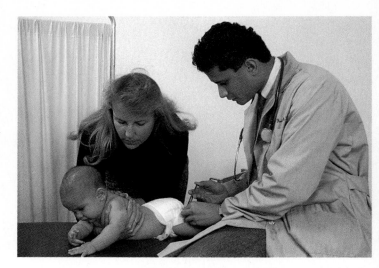

Figure 7.1   All babies should be immunized according to the schedule recommended by U.S. Centers for Disease Control (see page 142).

is a life-threatening condition that can complicate the recovery of a patient suffering from congestive heart failure. In congestive heart failure, the heart is not pumping and keeping the blood flowing adequately. Blood backs up in the pulmonary circuit, and the increase in blood pressure causes excess tissue fluid and an increase in interstitial pressure to the point that the walls of the air sacs in the lungs rupture. When fluid accumulates in the lungs, the patient can suffocate.

The malfunction of the lymphatic system causes tissue fluid to accumulate because it is not being drained away. During an operation for breast cancer, lymph nodes and lymphatic vessels sometimes are removed because they can be involved in the spread of cancerous cells. This procedure results in an inability of the body to collect tissue fluid; edema results.

Edema is characterized by localized swelling due to an abnormal accumulation of tissue fluid. Either too much tissue fluid is being made and/or not enough is being drained.

Lymphoid Organs

The lymphoid organs include the bone marrow, the lymph nodes, the spleen, and the thymus (fig. 7.3).

Bone Marrow

In the adult, bone marrow is present only in the bones of the skull, the sternum, the ribs, the clavicle, the spinal column, and the ends of the femur and the humerus (p. 216). The bone marrow consists of a network of connective tissue fibers, called reticular fibers, that are produced by cells called reticular cells. These and the cells that are developing into the blood cells are packed about thin-walled venous sinuses. Differentiated blood cells enter the bloodstream at these sinuses.

Radioactive tracer studies have shown that the bone marrow is the site of origination for all the types of blood cells (fig. 6.4), including both the granular and agranular leukocytes.

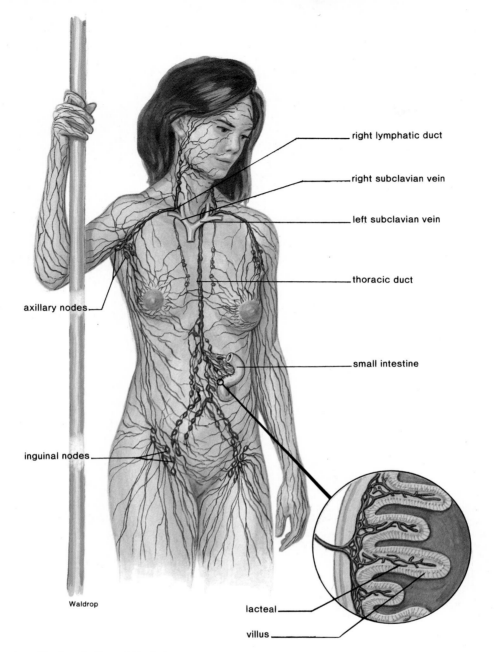

right lymphatic duct

right subclavian vein

left subclavian vein

thoracic duct

axillary nodes

small intestine

inguinal nodes

Waldrop

lacteal

villus

Figure 7.2   Lymphatic system. The lymphatic system drains excess fluid from the tissues and returns it to the cardiovascular system. The thoracic duct and the right lymphatic duct are the major lymphatic vessels. The enlargement shows the lymphatic vessels, called lacteals, that are present in the intestinal villi.

In other words, the bone marrow contains lymphoid tissue that produces lymphocytes. The B (for bone marrow) lymphocytes mature in the bone marrow, and the T (for thymus) lymphocytes mature in the thymus gland. The structure and the function of B and T lymphocytes are discussed in the following section.

The bone marrow also contains monocytes that have developed into resident macrophages. These large phagocytic cells help to cleanse the marrow and the adjacent blood sinuses.

Lymph Nodes

At certain points along lymphatic vessels, small (about 2.5 cm), ovoid or round structures called lymph nodes occur. A lymph node has a fibrous connective tissue capsule. Connective tissue also divides a node into nodules. Each nodule contains a sinus (open space) filled with many lymphocytes and macrophages. As lymph passes through the sinuses, it is purified of infectious organisms and any other debris.

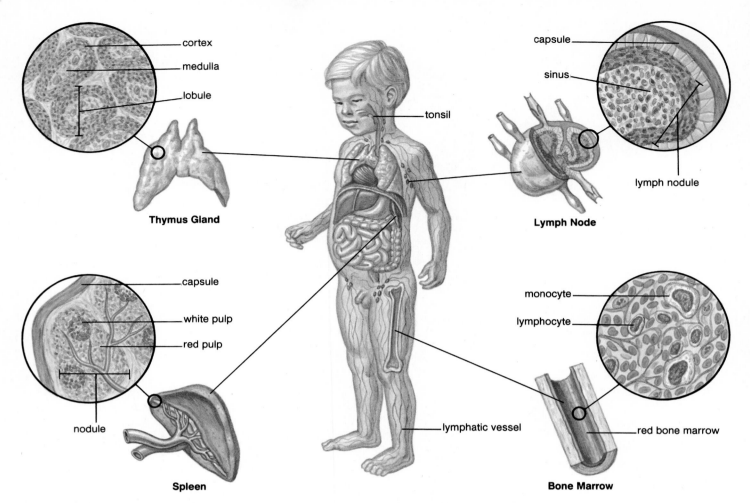

Thymus Gland

cortex
medulla
lobule

tonsil

Lymph Node

capsule
sinus
lymph nodule

Spleen

capsule
white pulp
red pulp
nodule

Bone Marrow

monocyte
lymphocyte
red bone marrow

lymphatic vessel

Figure 7.3   Lymphoid organs. The bone marrow is the site of lymphocyte and monocyte (macrophage) production and B cell maturation. The thymus is the site of T cell maturation. (A child is shown because the thymus is larger in children than in adults.) The lymph nodes are the sites for lymphocyte and macrophage accumulation. (The tonsils are modified lymph nodes.) The spleen is a site for lymphocyte, macrophage, and red blood cell accumulation.

While nodules usually occur within lymph nodes, they also can occur singly or in groups. The *tonsils* are composed of partly encapsulated lymph nodules. There are also nodules called *Peyer's patches* within the intestinal wall.

The lymph nodes occur in groups in certain regions of the body. For example, the inguinal nodes are in the groin and the axillary nodes are in the armpits.

Lymph nodes are divided into sinus-containing lobules, where the lymph is cleansed by phagocytes.

Spleen

The spleen is located in the upper left abdominal cavity just beneath the diaphragm. The spleen is constructed the same as a lymph node. Outer connective tissue divides the organ into lobules that contain sinuses. In the spleen, however, the sinuses are filled with blood instead of lymph. Especially since the blood vessels of the spleen can expand, this organ serves as a blood reservoir and makes blood available in times of low pressure or when the body needs extra oxygen in the blood.

A spleen nodule contains red pulp and white pulp. Red pulp contains red blood cells, lymphocytes, and macrophages. The white pulp contains only lymphocytes and macrophages. Both types of pulp help to purify the blood that passes through the spleen. If the spleen ruptures due to injury, it can be removed. Although the functions of the spleen are duplicated by other organs, the individual is expected to be slightly more susceptible to infections and may have to take antibiotic therapy indefinitely.

The spleen is divided into sinus-containing lobules, where the blood is cleansed by phagocytes.

Thymus

The thymus is located along the trachea behind the sternum in the upper thoracic cavity. This gland varies in size, but it is larger in children than in adults. The thymus also is divided into lobules by connective tissue. The T lymphocytes mature in these lobules. Those in the medulla are more mature than those in the cortex of the thymus.

The thymus secretes thymosin, a molecule that is believed to be an inducing factor; that is, it causes pre-T cells to become T cells. Thymosin also may have other functions in immunity.

The thymus is divided into lobules, where T lymphocytes mature.

Immunity

The body is prepared to protect itself from foreign substances and cells, including infectious microbes. The *first line of defense* is available immediately because it involves mechanisms that are nonspecific. The *second line of defense* takes a little longer to act because it is highly specific and contains mechanisms that are tailored to a particular threat.

General Defense

The environment contains many types of organisms that are able to invade and to infect the body. There are three general defense mechanisms that are useful against all types of organisms: barriers to entry, phagocytic white blood cells, and protective proteins.

Barriers to Entry

The skin and the mucous membrane lining the respiratory and digestive tracts serve as mechanical barriers to entry by bacteria and viruses. The secretions of the oil glands in the skin contain chemicals that weaken or kill bacteria. The respiratory tract is lined by cells that sweep mucus and trapped particles up into the throat, where they can be swallowed. The stomach has an acidic pH that inhibits the growth of many types of bacteria. A mix of bacteria that normally reside in the intestine and other organs, such as the vagina, prevents pathogens from taking up residence.

Phagocytic White Blood Cells

If microbes do gain entry to the body, as described in the section on the inflammatory response on page 126, other nonspecific forces come into play. For example, neutrophils and monocyte-derived macrophages (fig. 7.4) are phagocytic white blood cells that engulf some bacteria upon contact. Infections may be accompanied by fever, which is a protective response because phagocytes function better at a higher-than-normal body temperature.

Protective Proteins

The **complement system** is a series of proteins produced by the liver that are present in the plasma. When the first protein is activated, a cascade of reactions occurs. Every protein molecule in the series activates another in a predetermined sequence. In the end, certain proteins form pores in bacterial cell walls and membranes. These pores allow fluids and salt to enter the bacterial cell, until it bursts (fig. 7.5).

If complement is unable to destroy the microbes directly, some complement proteins still can coat them and other proteins attract phagocytes to the scene. Although complement is

a general defense mechanism, it also plays a role in specific defense, as we will see later.

When viruses infect a tissue cell, the infected cell usually produces and secretes interferon. **Interferon** binds to receptors on noninfected cells, and this action causes these cells to prepare for possible attack by producing substances that interfere with viral replication.

A cell that has bound interferon is protected against any type of virus; therefore, interferon should be very useful in preventing viral infection. However, interferon is specific to the species; only human interferon can be used in humans. It used to be quite a problem to collect enough interferon for clinical and research purposes, but now interferon is made by recombinant DNA technology.

The first line of defense against disease is nonspecific. It consists of barriers to entry, phagocytic white blood cells, and protective proteins.

Specific Defense

Sometimes, we are threatened by an invasion by microorganisms that cannot be counteracted successfully by the general defense mechanisms. In such cases, it is necessary to develop immunity against a specific antigen. **Antigens** are usually protein (or polysaccharide) molecules that specific lymphocytes recognize as foreign to the body. Antigens occur on bacteria and viruses, but they also can be part of a foreign cell or a cancerous cell. Ordinarily, we do not become immune to our body's own normal cells; therefore, it is said that the immune system is able to distinguish self from nonself.

Immunity usually lasts for some time. For example, once we recover from the measles, we usually cannot be infected by the measles virus a second time. Immunity is primarily the result of the action of the B lymphocytes and the T lymphocytes. As mentioned earlier, B (for bone marrow) lymphocytes mature in the bone marrow, and T (for thymus) lymphocytes mature in the thymus gland. **B lymphocytes,** also called B cells, become plasma cells that produce **antibodies,** proteins that are capable of combining with and inactivating antigens. These antibodies are secreted into the blood and the lymph. In contrast, **T lymphocytes,** also called T cells, do not produce antibodies. Instead, certain T cells directly attack cells bearing antigens they recognize. Other T cells regulate the immune response.

Lymphocytes are capable of recognizing an antigen because they have receptor molecules on their surface. The shape of the receptors on any particular lymphocyte is complementary to a portion of one specific antigen. It is often said that the receptor and the antigen fit together like *a lock and a key*. It is estimated that during our lifetime, we encounter a million different antigens so we need the same number of different lymphocytes for protection against those antigens. It is remarkable that so much diversification occurs during the maturation process that in the end there is a different type of lymphocyte for each possible antigen. Despite this great diversity, none of the lymphocytes ordinarily attacks the body's own cells. It is

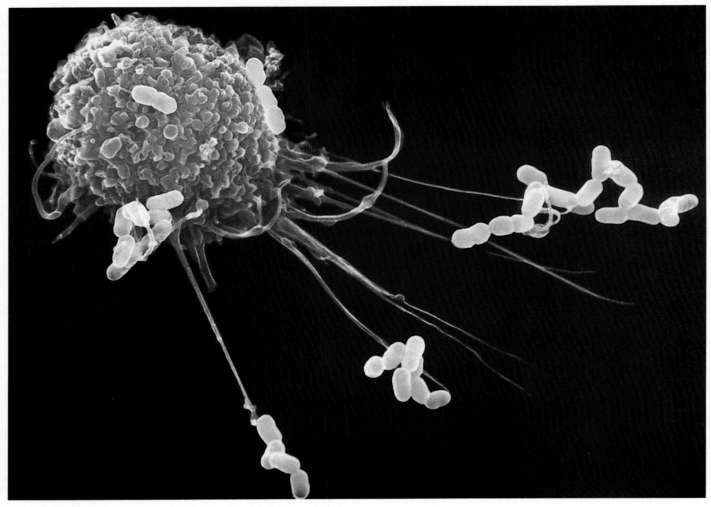

Figure 7.4   Macrophage (red) engulfing bacteria (green). Monocyte-derived macrophages are the body's scavengers. They phagocytize microbes and debris in the body's fluids and tissues.

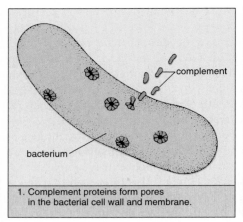

1. Complement proteins form pores in the bacterial cell wall and membrane.

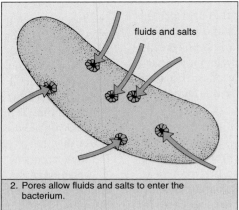

fluids and salts

2. Pores allow fluids and salts to enter the bacterium.

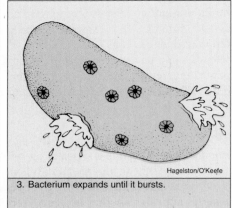

Hagelston/O'Keefe

3. Bacterium expands until it bursts.

Figure 7.5   Action of the complement system. The complement system consists of a number of proteins always present in the plasma. When activated, some of these proteins form pores in bacterial cell wall membranes that allow fluids and salts to enter until the cell eventually bursts.

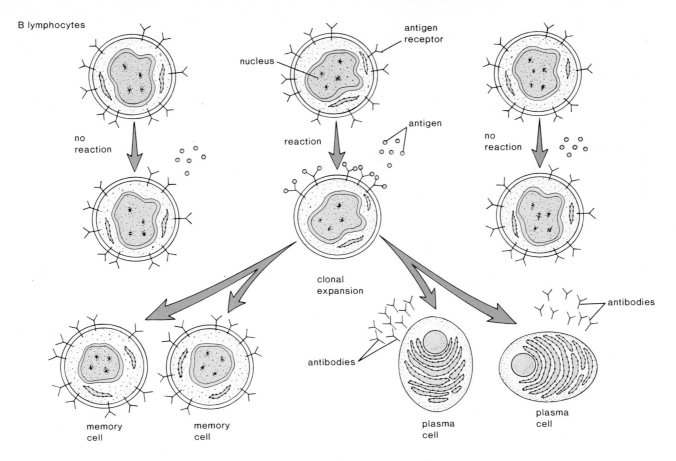

Figure 7.6 Clonal selection theory as it applies to B cells. An antigen activates the appropriate B cell, which clones if stimulated by a helper T cell (fig. 7.10). During the cloning process, many plasma cells that produce specific antibodies against this antigen are produced by the fifth day. Memory cells that retain the ability to secrete these antibodies at a future time are also produced.

believed that if by chance a lymphocyte arises that is equipped to respond to the body's own proteins, it normally is suppressed and develops no further.

There are two types of lymphocytes. B cells produce and secrete antibodies that combine with antigens. Certain T cells directly attack antigen-bearing cells, and others regulate the immune response.

Action of B Cells

The receptor on a B cell is called a membrane-bound antibody because it is structured like an antibody. When a B cell encounters a bacterial cell or a toxin bearing an appropriate antigen, it is activated; that is, it has the potential to produce many **plasma cells** that will secrete antibodies against this antigen (fig. 7.6). (For the B cell to realize this potential, it must be stimulated by a helper T cell [p. 138]). All of the plasma cells derived from one parent lymphocyte are called a clone, and a clone produces the same type of antibody. Notice that a B cell does not clone until its antigen is present. The *clonal selection theory* states that the antigen selects which B cell will produce a clone of plasma cells.

Once antibody production is sufficient, the antigen disappears from the system, and the development of plasma cells ceases. However, some members of a clone do not participate in antibody production; instead, they remain in the bloodstream as **memory B cells.** Memory B cells are capable of producing the antibody specific to a particular antigen for some time. As long as these cells are present, the individual is said to be actively immune: future antibody production is possible because the memory cells can produce more plasma cells if the same antigen invades the system again.

Defense by B cells is called *antibody-mediated immunity* because B cells produce antibodies. It also is called humoral immunity because these antibodies are present in the bloodstream.

B cells are responsible for antibody-mediated immunity. After they recognize an antigen, they (if stimulated by a helper T cell) divide to produce both antibody-secreting plasma cells and memory B cells.

Antibodies　　The most common type of antibody (IgG) is a Y-shaped protein molecule having two arms. Each arm has a long "heavy" chain and a short "light" chain of amino acids.

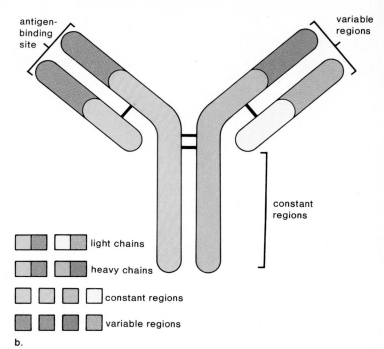

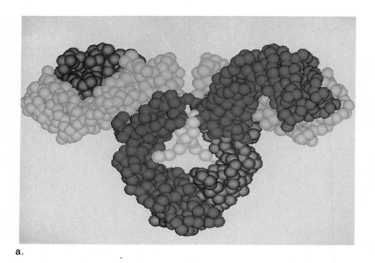

Figure 7.7 Structure of the most common antibody (IgG).
a. Computer model of an antibody. *b.* Schematic drawing. Each arm of an antibody contains a light (short) chain and a heavy (long) chain. The amino acid chains make up constant and variable regions. In the variable regions, the amino acid sequence varies so that there is a specific antibody for any particular antigen. This antigen binds to the variable regions in a lock-and-key manner.

These chains have *constant regions,* where the sequence of amino acids is set, and *variable regions,* where the sequence of amino acids varies (fig. 7.7). The antigen binds to the antibody at the variable regions of one arm in a lock-and-key manner. In other words, the variable regions form an antibody-binding site that is specific for a particular antigen.

The constant regions are not identical among all the antibodies. Instead, they are the same for different classes of antibodies. Most antibodies found in the blood belong to the class IgG (immunoglobulin G) (table 7.1). Antibodies are found in the gamma globulin portion of the blood.

The antigen-antibody reaction can take several forms, but quite often the antigen-antibody reaction produces complexes of antigens combined with antibodies (fig. 6.11). Such an antigen-antibody complex, sometimes called the immune complex, marks the antigen for destruction by other forces. For example, the complex may be engulfed by neutrophils or macrophages or it may activate a portion of blood serum called complement. As mentioned earlier, complement refers to a series of different proteins with various functions (p. 135).

An antibody combines with its antigen in a lock-and-key manner. The antibody-antigen reaction can lead to complexes that contain several molecules of antibody and antigen.

Actions of T Cells

There are four different types of T cells: cytotoxic T cells, helper T cells, memory T cells, and suppressor T cells. All four types look alike but can be distinguished by their functions.

Table 7.1 Antibodies

Classes	Description
IgG	main antibody type in circulation; attacks microorganisms and their toxins
IgA	main antibody type in secretions, such as saliva and milk; attacks microorganisms and their toxins
IgE	antibody type on mast cells; responsible for allergic reactions
IgM	antibody type found in circulation; largest antibody, with 5 subunits
IgD	antibody type found primarily as a membrane-bound Ig

Cytotoxic T (T_C) cells sometimes are called killer T cells. In immune individuals, they attack and destroy cells bearing a foreign antigen, such as virus-infected or cancerous cells. These T cells have storage vacuoles that contain a chemical called perforin because it perforates cell membranes. The perforin molecules form a pore in the membrane that allows water and salts to enter. The cell under attack then swells and eventually bursts (fig. 7.8).

It often is said that T cells are responsible for *cell-mediated immunity,* characterized by destruction of antigen-bearing cells. Of all the T cells, only T_C cells are involved in this type of immunity.

Helper T (T_H) cells regulate immunity by enhancing the response of other immune cells. In response to an antigen, they

Figure 7.8   Cell-mediated immunity. *a.* Scanning electron micrograph showing cytotoxic T cells attacking a cancer cell. *b.* The cancer cell now is destroyed. *c.* During the killing process, the vacuoles in a T cell fuse with the cell membrane and release units of the protein perforin. These units combine to form pores in the target cell membrane. Thereafter, fluid and salts enter so that the target cell eventually bursts.

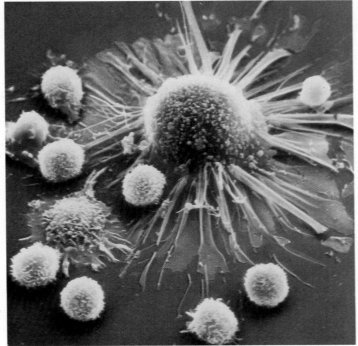

a.

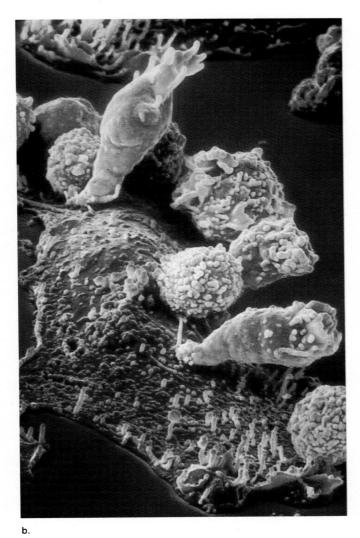

b.

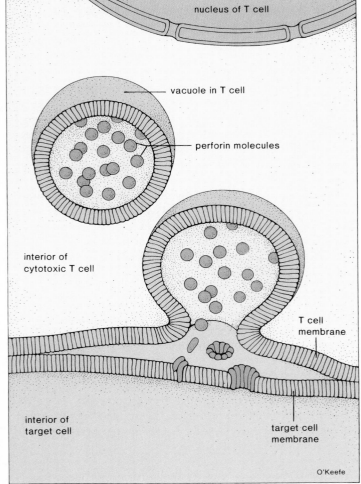

nucleus of T cell

vacuole in T cell

perforin molecules

interior of cytotoxic T cell

T cell membrane

interior of target cell

target cell membrane

O'Keefe

c.

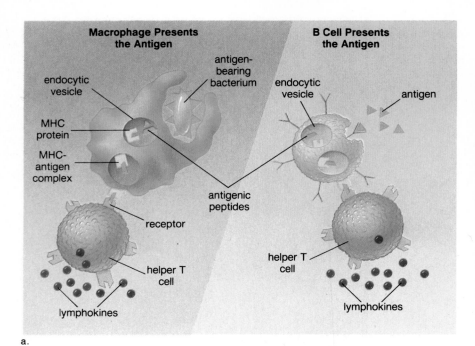

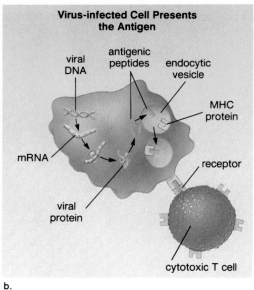

a.

b.

Figure 7.9   T cell activation. *a*. Either a macrophage or a B cell presents an antigen to a helper T cell. To accomplish this, the antigen has to be digested to peptides that are combined with an MHC protein. The complex is presented to the T cell. In return, the helper T cell produces and secretes lymphokines that stimulate T cells and other immune cells. *b*. Cells infected with a virus present one of the viral proteins along with an MHC protein to a cytotoxic T cell. This causes the cytotoxic T cell to attack and destroy any cell infected with the same virus. (see fig. 7.8).

enlarge and secrete lymphokines, including interferon and the interleukins. **Lymphokines** are stimulatory molecules that cause T_H cells to clone and other immune cells to perform their functions. For example, T_H cells stimulate macrophages to phagocytize and B cells to manufacture antibodies.[1] Because the AIDS virus attacks T_H cells, it inactivates the immune response. Following page 310, the AIDS supplement discusses this disease in detail.

When an activated T_H cell divides, the clone contains **suppressor T (T_S) cells** and **memory T (T_M) cells.** Once there is a sufficient number of T_S cells, the immune response ceases. Following suppression, a population of T_M cells persists, perhaps for life. These cells are able to secrete lymphokines and to stimulate macrophages and B cells whenever the same antigen enters the body once again.

T_C cells are responsible for cell-mediated immunity; T_H cells promote the immune response; T_S cells suppress the immune response; and T_M cells maintain immunity.

Activation of Cytotoxic and Helper T Cells   T cells have receptors just as B cells do. Unlike B cells, however, T_C cells and T_H cells are unable to recognize an antigen that simply is present in lymph or blood. Instead, the antigen must be pre-

sented to them by an *antigen-presenting cell (APC)*. When an APC, such as a macrophage, engulfs a bacterium or a virus, the APC enzymatically breaks it down to peptide fragments that are antigenic (have the properties of an antigen). The antigenic peptide fragment is linked to an **MHC** (major histocompatibility complex) **protein,** and together they are displayed to a T cell at the cell membrane. The importance of MHC proteins first was recognized when it was discovered that they contribute to the specificity of tissues and make it difficult to transplant tissue from one person to another. In other words, the donor and the recipient must be histo-(tissue) compatible (the same or nearly so) for a transplant to be successful without the administration of immunosuppressive drugs.

Figure 7.9*a* shows a macrophage and also a B cell presenting an antigen to a T_H cell. Once a T_H cell recognizes an antigen, it undergoes clonal expansion, producing T_S cells and T_M cells that also can recognize this same antigen. While well-known APCs are macrophages and B cells, actually any cell in the body can be an APC. For example, figure 7.9*b* shows a virus-infected cell presenting an antigen to a T_C cell. Once a T_C cell recognizes an antigen, it attacks and destroys any cell that is infected with the same virus.

In order for T_C and T_H cells to recognize an antigen, the antigen, along with an MHC protein, must be presented to them by an APC.

Table 7.2 and figure 7.10 summarize our discussion of B cells and T cells.

---

[1]The lymphokine secreted by helper T cells that stimulates B cells once was called B cell growth factor. Now it is called interleukin-4.

Table 7.2 Some Properties of B Cells and T Cells

Property	B cells	T cells	
		Cytotoxic	**Helper**
type of immunity	antibody mediated	cell mediated	—————
antigen recognition	direct recognition	must be presented by APC	must be presented by APC
response	become antibody-producing plasma cells	search for and destroy antigen-bearing cells	secrete lymphokines; stimulate other immune cells
final response	memory cells	—————	suppressor and memory cells

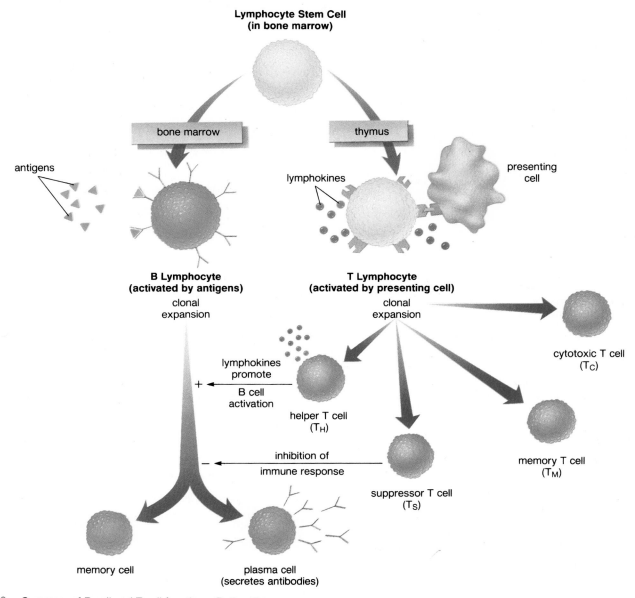

Figure 7.10   Summary of B cell and T cell functions. Both cells are produced in the bone marrow, but only B cells mature there. T cells mature in the thymus. An antigen activates the appropriate B cell, which divides and differentiates to give a clone of antibody-producing plasma cells and memory cells. An antigen activates a T cell (either cytotoxic T or helper T) if it is presented with an MHC by an APC. Cytotoxic T cells then attack and destroy antigen-bearing cells; helper T cells divide to give a clone of helper T cells that produce lymphokines and also suppressor T cells and memory T cells.

Immunotherapy

The immune system can be manipulated to help people avoid or recover from diseases. Some of these techniques have been utilized for a long time, and some are relatively new.

Induced Immunity

Active immunity, which provides long-lasting protection against a disease-causing organism, develops after an individual is infected with a virus or a bacterium. In many instances today, however, it is not necessary to suffer an illness to become immune because it is possible to be artificially immunized against a disease. One recommended immunization schedule for children is given in figure 7.11. The importance of following a recommended immunization schedule is being demonstrated at this time. There have been outbreaks of childhood communicable diseases among college-age people because they were not immunized properly when they were younger.

Immunization requires the use of **vaccines,** which are traditionally bacteria and viruses (antigens) that have been treated so that they are no longer virulent (able to cause disease). New methods of producing vaccines are being developed. For example, it is possible to use the recombinant DNA technique to mass-produce a protein that can be used as a vaccine. This method is being used to prepare a vaccine against hepatitis B.

After a vaccine is given, it is possible to determine the amount of antibody present in a sample of serum—this is called the *antibody titer.* After the first exposure to an antigen, a primary response occurs. For a period of several days, no antibodies are present; then, there is a slow rise in the titer, followed by a gradual decline (fig. 7.12). After a second exposure, a secondary response may occur. If so, the titer rises rapidly to a level much greater than before. The second exposure in that case often is called the "*booster*" because it boosts the antibody titer to a high level. The antibody titer now may be high enough to prevent disease symptoms even if the individual is exposed to the disease. If so, the individual is now immune to that particular disease. A good secondary response can be related to the number of plasma cells and memory cells in the serum. Upon the second exposure, these cells already are present, and antibodies can be produced rapidly.

Vaccines can be used to make people actively immune.

Passive immunity occurs when an individual is given antibodies (gammaglobulins) to combat a disease. Since these antibodies are not produced by the individual's B cells, passive immunity is short-lived. For example, newborn infants possess passive immunity because antibodies have crossed the placenta from the mother's blood. These antibodies soon disappear, however, so that within a few months, infants become more susceptible to infections. Breast feeding (fig. 7.13) prolongs the passive immunity an infant receives from the mother because there are antibodies in the mother's milk.

Even though passive immunity does not last, it sometimes is used to prevent illness in a patient who has been unexpectedly exposed to an infectious disease. Usually, the person receives an

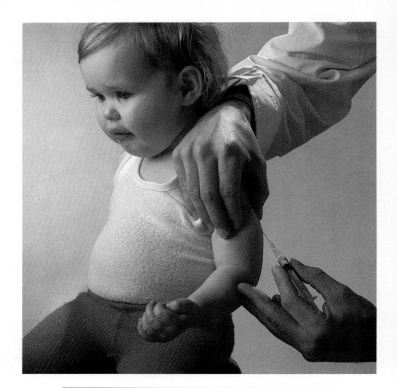

Age	Vaccines
2 months old	1st DTP[a] 1st polio[b]
4 months old	2nd DTP 2nd polio
6 months old	3rd DTP
9 months old	Measles[c]
15 months old	MMR[d] 4th DTP 3rd polio
18 months old	HbCV[e]
4–6 years old	5th DTP 4th polio
14–16 years old	Td[f]

[a]Diphtheria, tetanus, and pertussis (whooping cough).

[b]Trivalent oral containing poliovirus types 1, 2, and 3.

[c]Measles only vaccine should be administered in high-risk areas.

[d]Measles, mumps, and rubella (German measles).

[e]Haemophilus influenzae b conjugated to a protein carrier.

[f]Compared to DTP, contains same amount of tetanus but reduced dose of diphtheria. Repeat every 10 years.

Figure 7.11 Suggested immunization schedule for infants and young children. Children who are not immunized are subject to childhood diseases that can cause serious health consequences.

Source: Data from U.S. Department of Health and Human Services, *Morbidity and Mortality Weekly Report,* April 7, 1989, page 5.

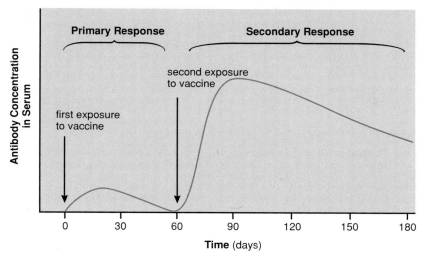

Figure 7.12   Development of active immunity due to immunization. The primary response after the first exposure of a vaccine is minimal, but the secondary response that may occur after the second exposure shows a dramatic rise in the amount of antibody present in serum.

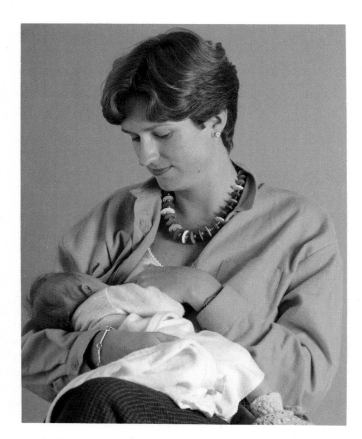

Figure 7.13   Example of passive immunity. Breast feeding is believed to provide a newborn with antibodies during the period of time when the child is not yet producing antibodies.

injection of a serum containing antibodies. This may have been taken from donors who have recovered from the illness. In the past, horses were immunized, and serum was taken from them to provide the needed antibodies. Horses were used to produce antibodies against diphtheria, botulism, and tetanus. Occasionally, a patient who received these antibodies became ill because the serum contained proteins that the individual's immune system recognized as foreign. This was called serum sickness.

Passive immunity is short-lived because the antibodies are administered to and not made by the individual.

Lymphokines

Lymphokines are being investigated as possible adjunct therapy for cancer and AIDS because they stimulate white blood cell formation and/or function. Both interferon and various interleukins have been used as immunotherapeutical drugs, particularly to potentiate the ability of the individual's own T cells (and possibly B cells) to fight cancer.

Interferon is a substance produced by leukocytes, fibroblasts, and probably most cells in response to a viral infection. When it is produced by T cells, it is called a lymphokine. Interferon still is being investigated as a possible cancer drug, but thus far it has proven to be effective only in certain patients, and the exact reasons as yet cannot be discerned. For example, interferon has been found to be effective in up to 90% of patients with a type of leukemia known as hairy-cell leukemia (because of the hairy appearance of the malignant cells).

When and if cancerous cells carry an altered protein on their cell surface, by all rights they should be attacked and destroyed by T_C cells. Whenever cancer develops, it is possible that the T_C cells have not been activated. In that case, the use of lymphokines might awaken the immune system and lead to the destruction of the cancer. In one technique being investigated, researchers first withdraw T cells from the patient and activate the cells by culturing them in the presence of an *interleukin* (IL-2). The cells then are reinjected into the patient, who is given doses of the interleukin to maintain the killer activity of the T cells.

Blood Cell Growth Factors

Growth factors are hormones that stimulate the proliferation of cells. The various interleukins (IL-1 through IL-11) are blood cell growth factors, some of which are specific for white blood cells. Other white blood cell growth factors have also been discovered. The best known is *GM-CSF* (granulocyte-macrophage colony-stimulating factor), which enhances the activity of the myeloid stem cell (fig. 6.4). When GM-CSF was given to AIDS patients, it temporarily raised their white cell count. In addition, scientists recently announced the discovery of *SCF* (stem cell growth factor), which can stimulate all types of stem cells. Since growth factors can be mass produced by genetic engineering techniques (chapter 19), there is great hope that SCF will prove helpful to patients with immunity problems. Growth factors can be used as adjuncts for vaccines, for treatment of chronic infectious diseases, and for rebuilding immunity following radiation treatment or chemotherapy.

Human Issue

John Moore contributed to the research preliminary to the mass production of GM-CSF today. Because he was suffering from a type of leukemia called hairy-cell leukemia, doctors removed his spleen, which had swollen to many times its normal size. They then used a sliver of his cancerous spleen to establish an immortal cell line of T cells which continuously produce GM-CSF. Only because of this was mass production then possible. Moore's doctor has been handsomely reimbursed for his contribution to the endeavor. When Moore signed a release form allowing his tissue to be used for research, he had no idea of its significance. Should patients be required to sign such forms? Should Moore have been compensated for the use of his spleen? If so, when should he have been compensated—when it was removed or when it was found to be valuable?

The interleukins and other blood cell growth factors show some promise of potentiating the individual's own immune system.

Monoclonal Antibodies

Every plasma cell derived from the same B cell secretes antibodies against a specific antigen, as previously discussed. These are **monoclonal antibodies** because all of them are the same type (mono) and because they are produced by plasma cells derived from the same B cell (clone).

One method of producing monoclonal antibodies in vitro (in laboratory glassware) is depicted in figure 7.14. B lymphocytes are removed from the body (today, usually a mouse) and are exposed to a particular antigen. Then they are fused with a myeloma cell (a malignant plasma cell) because these cells, unlike normal plasma cells, live and divide indefinitely. The fused cells are called hybridomas; *hybrid* because they result from the fusion of two different cells and *oma* because one of the cells is a cancer cell.

At present, monoclonal antibodies are being used for quick and certain diagnosis of various conditions. For example, a particular hormone is present in the urine of a pregnant woman. A monoclonal antibody can be used to detect the hormone and so indicate that the woman is pregnant. Monoclonal antibodies also are used for identifying infections (fig. 7.15). And because they can distinguish between cancerous and normal tissue cells, they are used to carry radioactive isotopes or toxic drugs only to tumors so these can be destroyed selectively.

Monoclonal antibodies are considered to be a biotechnology product because the production process makes use of a living system to mass-produce the product.

Monoclonal antibodies are produced in pure batches—they react specifically with just one type of molecule (antigen); therefore, they can distinguish one cell, or even one molecule, from another.

Immunological Side Effects and Illnesses

The immune system protects us from disease because it can tell self from nonself. Sometimes, however, the immune system is underprotective, as when an individual develops cancer, or is overprotective, as when an individual has allergies.

Allergies

Allergies are caused by an overactive immune system that forms antibodies to substances that usually are not recognized as foreign substances. Unfortunately, allergies usually are accompanied by coldlike symptoms, or even at times, by severe systemic reactions, such as anaphylactic shock, a sudden drop in blood pressure.

Of the five varieties (table 7.1) of antibodies—IgA, IgD, IgE, IgG, and IgM—IgE antibodies cause allergies. IgE antibodies are found in the bloodstream, but they, unlike the other types of antibodies, also reside in the membrane of *mast cells* found in the tissues. Some investigators contend that mast cells are basophils that have left the bloodstream and have taken up residence in the tissues. In any case, when the *allergen,* an antigen that provokes an allergic reaction, attaches to the IgE antibodies on mast cells, these cells release histamine and other substances that cause mucus secretion and airway constriction, resulting in the characteristic symptoms of allergy. On occasion, basophils and other white blood cells release these chemicals into the bloodstream. The increased capillary permeability that results from this can lead to fluid loss and shock.

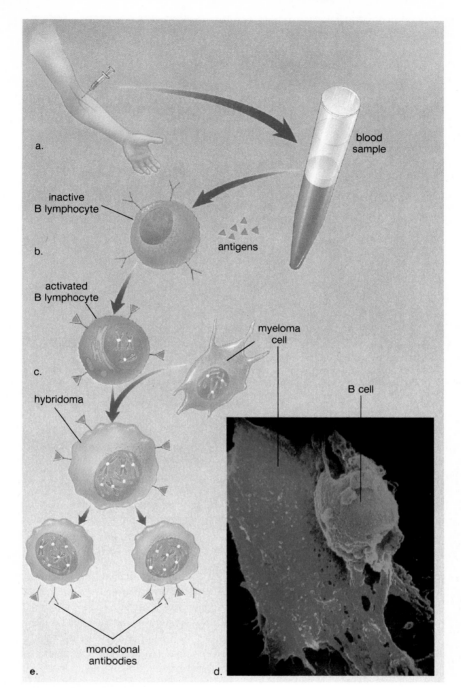

Figure 7.14 One possible method of producing human monoclonal antibodies. *a.* Blood sample is taken from a patient. *b.* Inactive B lymphocytes from the sample are exposed to an antigen. *c.* Activated lymphocytes are fused with myeloma cells. *d.* Scanning electron micrograph of cells fusing. *e.* Resulting hybridomas divide repeatedly, yielding many cells that produce monoclonal antibodies.

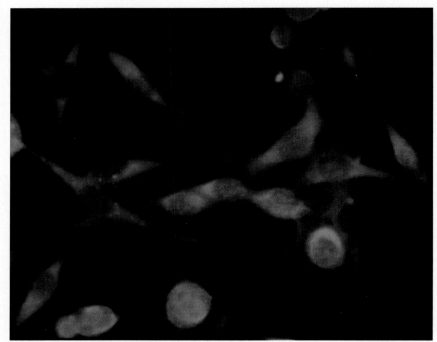

Figure 7.15   Monoclonal antibody use. These cells are infected with herpes simplex type 1 virus (HSV-1), which can be detected by a monoclonal antibody specific for the virus and tagged with a fluorescent dye. When the cells fluoresce, it shows that the virus is inside the cells.

Allergy shots sometimes prevent the onset of allergic symptoms. Injections of the allergen cause the body to build up high quantities of IgG antibodies, and these combine with allergens received from the environment before they have a chance to reach the IgE antibodies located in the membrane of mast cells.

Allergic symptoms are caused by the release of histamine and other substances from mast cells.

Tissue Rejection

Certain organs, such as skin, the heart, and the kidneys, could be transplanted easily from one person to another if the body did not attempt to *reject* them. Rejection occurs because the transplanted organ is foreign to the individual. Both T_C cells and/or antibodies can bring about disintegration of the foreign tissue.

Organ rejection can be controlled in two ways: careful selection of the organ to be transplanted and the administration of immunosuppressive drugs. It is best if the transplanted organ has the same type of MHC proteins as those of the recipient, because T_C cells learn to recognize foreign MHC proteins. The immunosuppressive drug cyclosporine has been in use for some years. A new experimental drug, FK-506, eventually may replace cyclosporine as the drug of choice for transplant patients.

In more than 100 patients taking FK-506, the rate of organ rejection was one-sixth that of cyclosporine.

When an organ is rejected, the immune system is attacking cells that bear different MHC proteins from those of the individual.

Autoimmune Diseases

Certain human illnesses are believed to be due to the production of antibodies that act against an individual's own tissues. In myasthenia gravis, autoantibodies attack the neuromuscular junctions so that the muscles do not obey nervous stimuli. Muscular weakness results. In MS (multiple sclerosis), antibodies attack the myelin of nerve fibers, causing various neuromuscular disorders. A person with SLE (systemic lupus erythematosus) forms various antibodies to different constituents of the body, including the DNA of the cell nucleus. The disease sometimes results in death, usually due to kidney damage. In rheumatoid arthritis, the joints are affected. When an autoimmune disease occurs, a viral infection of tissues often has set off an immune reaction to the body's own tissues. There is evidence to suggest that type I diabetes is the result of this sequence of events, as well as heart damage following rheumatic fever.

Autoimmune diseases seem to be preceded by a viral infection that fools the immune system into attacking the body's own tissues.

Summary

The lymphatic system consists of lymphatic vessels and lymphoid organs. The lymphatic vessels collect excess tissue fluid and fat molecules at lacteals and carry these to the cardiovascular system. Lymphocytes are produced in the bone marrow and accumulate in the other lymphoid organs (bone marrow, thymus, lymph nodes, and spleen).

The body is prepared to defend itself in both a generalized and a specific manner. Barriers to entry, phagocytic white blood cells, and protective chemicals react to any threat. The immune response is specific to each particular antigen and requires two types of lymphocytes, both of which are produced in the bone marrow. B cells mature in the bone marrow, and T cells mature in the thymus.

B cells directly recognize an antigen and give rise to antibody-secreting plasma cells if stimulated to do so by T_H cells. In order for the T cell to recognize an antigen, the antigen must be presented by an APC. There are four types of T cells. T_C cells kill cells on contact; T_H cells stimulate other immune cells; T_S cells suppress the immune response. There are also memory T and memory B cells that remain in the body and provide long-lasting immunity.

Immunity can be fostered by immunotherapy. Vaccines are available to promote active immunity, and antibodies sometimes are available to provide an individual with short-term passive immunity. Lymphokines, notably interferon and interleukins, plus other blood cell growth factors, are used to promote the body's ability to recover from cancer and AIDS.

Immunity has certain undesirable side effects. Allergies are due to an overactive immune system that forms antibodies to substances not normally recognized as foreign. T_C cells attack transplanted organs, although immunosuppressive drugs are available. Autoimmune illnesses occur when antibodies against the body's own cells form.

Study Questions

1. What is the lymphatic system, and what are its three functions? (p. 132)
2. Describe the structure and the function of the bone marrow, lymph nodes, the spleen, and the thymus. (p. 132)
3. What are the general defense mechanisms of the body? (p. 135)
4. B cells are responsible for which type of immunity? What is the clonal selection theory? (p. 137)
5. Describe the structure of an antibody, including the terms *variable regions* and *constant regions*. (p. 137)
6. Name the four types of T cells, and state their functions. (p. 138)
7. Explain the process by which a T cell is able to recognize an antigen. (p. 140)
8. How is active immunity achieved? How is passive immunity achieved? (p. 142)
9. What are lymphokines, and how are they used in immunotherapy? (p. 143)
10. How are monoclonal antibodies produced, and what are their applications? (p. 144)
11. Discuss allergies, tissue rejection, and autoimmune diseases as they relate to the immune system. (p. 144)

Objective Questions

1. Lymphatic vessels collect excess _____ and return it to the _____ veins.
2. The function of lymph nodes is to _____ the lymph.
3. T lymphocytes have passed through the _____ .
4. A stimulated B cell produces antibody-secreting _____ cells and _____ cells that are ready to produce the same type of antibody at a later time.
5. B cells are responsible for _____-mediated immunity.
6. In order for a T cell to recognize an antigen, it must be presented by an _____ along with an MHC protein.
7. T cells produce _____ , which are stimulatory chemicals for all types of immune cells.
8. Cytotoxic T cells are responsible for _____-mediated immunity.
9. Allergic reactions are associated with the release of _____ from mast cells.
10. The body recognizes foreign cells because they bear different _____ proteins than the body's cells.
11. Immunization with _____ brings about active immunity.
12. Hybridomas produce _____ antibodies.

Answers to Objective Questions

1. tissue fluid, subclavian 2. purify 3. thymus 4. plasma, memory 5. antibody 6. APC 7. lymphokines 8. cell 9. histamine 10. MHC 11. vaccines 12. monoclonal

Critical Thinking Questions

1. Argue both for and against the suggestion that the lymphatic system should be considered part of the blood circulatory system.
2. There are more cells in the immune system than any other system. What does this tell us about the environment of organisms?
3. The immune system has various ways of attacking foreign antigens. Does this seem to be a duplication of effort or a necessity?

Selected Key Terms

B lymphocyte (lim′fo-sīt) lymphocyte that matures in the bone marrow and reacts to antigens by producing antibodies. 135

complement system (kom′plĕ-ment sis′tem) a series of proteins in plasma that are involved in both general and specific immune effects. 135

cytotoxic T cell (si′′to-tok′sik te sel) T lymphocyte that attacks cells bearing foreign antigens. 138

helper T cell (hel′per te sel) T lymphocyte that stimulates other immune cells to perform their respective function. 138

lymph (limf) fluid having the same composition as tissue fluid and carried in lymphatic vessels. 132

lymphatic system (lim-fat′ik sis′tem) vascular system that takes up excess tissue fluid and transports it to the bloodstream. 132

lymphokine (lim′fo-kīn) molecule secreted by T lymphocytes that has the ability to affect the characteristics of lymphocytes and monocytes. 143

memory B cell (mem′o-re be sel) cell derived from a B lymphocyte that is ever present within the body and that produces a specific antibody and accounts for the development of active long-lasting immunity. 137

MHC protein (pro′tēn) major histocompatibility protein; a surface molecule that serves as a genetic marker. 140

monoclonal antibody (mon′′o-klōn′al an′ti-bod′′e) one of many antibodies of the same type produced by cells that are derived from a single lymphocyte that is fused with a cancerous cell. 144

plasma cell (plaz′mah sel) cell derived from B cell lymphocyte that is specialized to mass-produce antibodies. 138

suppressor T cell (sŭ-pres′or te sel) T lymphocyte that suppresses certain other T and B lymphocytes from continuing to perform their respective function. 140

T lymphocyte (lim′fo-sīt) lymphocyte that matures in the thymus and exists as 4 types: cytotoxic, helper, suppressor, and memory T cells. 135

vaccine (vak′sēn) antigens prepared in such a way that they can promote active immunity without causing disease. 142

Chapter Eight

Respiration

Chapter Concepts

1 As air passes along the respiratory tract, it is filtered, warmed, and saturated with water before gas exchange takes place across a very extensive moist surface.

2 Breathing brings in oxygen needed by the cells for cellular respiration and rids the body of carbon dioxide, a by-product of cellular respiration.

3 The respiratory pigment hemoglobin combines with oxygen in the lungs and releases oxygen in the tissues. It also aids in the transport of carbon dioxide, largely by its ability to buffer.

4 The respiratory tract is especially subject to disease because it can serve as an entry way for infectious agents. Polluted air contributes to two major lung disorders—emphysema and cancer.

Cilia that project from the epithelium of the trachea keep the lungs clean by sweeping mucus and debris into the throat. This scanning electron micrograph shows the three-dimensional shapes of the cilia and the cells.

Fish take their oxygen from the water. Why must we take our oxygen with us when we descend into the depths (fig. 8.1)? For one thing, we are warm blooded and therefore require more oxygen intake per unit time than do fish. Water contains 5–7 ppt (parts of oxygen per thousands parts of water), but air contains 210 ppt. If our oxygen demands were lower, could we breathe water? The answer seems to be yes, particularly if the water is cold. Scientists have studied the mammalian diving reflex, which allows air-breathing whales and seals to remain submerged in cold water. When frigid water splashes over the forehead and nose, nerves signal the brain to divert oxygen-rich blood from the limbs to the heart and brain. Water entering the lungs further cools the body, slowing down the metabolic rate, thereby reducing the brain's oxygen requirements. It was not viewed as astonishing, then, when physicians were able to revive a toddler who fell into an ice-cold creek near Salt Lake City and was not found for 66 minutes. Once her temperature rose from 66° F to 77° F, her heart began beating and her pupils reacted to light. Her only complication was a slight tremor which continues to improve.

Breathing

Breathing is more eminently necessary than eating. While it is possible to stop eating altogether for several days, it is not possible to remain alive for longer than several minutes without breathing. Breathing supplies the body with oxygen needed for cellular respiration, as indicated in the following equation:[1]

$$36 \text{ ADP} + 36 \text{ } \textcircled{P} \longrightarrow 36 \text{ ATP}$$

$$C_6H_{12}O_6 + 6O_2 \longrightarrow 6H_2O + 6CO_2$$

As glucose is broken down to carbon dioxide and water, 36 ATP are formed.

Breathing not only draws oxygen into the body, it also pushes carbon dioxide out. It is only the first step of respiration, which can be said to include the following steps (fig. 8.2):

1. **Breathing:** entrance and exit of air into and out of the lungs
2. **External respiration:** exchange of the gases oxygen (O_2) and carbon dioxide (CO_2) between air and blood
3. **Internal respiration:** exchange of the gases (oxygen and carbon dioxide) between blood and the tissue fluid
4. **Aerobic cellular respiration:** production of ATP in cells

The normal breathing rate is about 14 to 20 times per minute. The more energy expended, the greater the breathing rate. The average young adult male utilizes about 250 ml of oxygen per minute in a basal, or restful, state. Exercise and the digestion of food raise the need for oxygen. The average amount of oxygen needed with mild exercise is 500 ml of oxygen per minute.

[1]The body requires oxygen (O_2) for the respiration of fats and amino acids as well as for glucose.

Figure 8.1   When individuals scuba dive, they carry a tank that supplies them with air. One of the major problems with deep dives is due to the increased pressure under water. Nitrogen dissolves in the blood and then begins to bubble out and collect in the joints when the diver rises to the surface. To prevent this so-called decompression sickness, it is necessary to decompress slowly so that the nitrogen is expired by the respiratory system.

Passage of Air

During both **inspiration** (inhalation) and **expiration** (exhalation), air is conducted toward or away from the lungs by a series of cavities, tubes, and openings, which are listed in order in table 8.1 and illustrated in figure 8.3.

As air moves in along the air passages, it is filtered, warmed, and moistened. Filtering is accomplished by coarse hairs and cilia in the region of the nostrils and by cilia alone in the rest of the nose and the windpipe. In the nose, the hairs and the cilia act as a screening device. In the trachea, cilia beat upward, carrying mucus, dust, and occasional bits of food that "went down the wrong way" into the pharynx, where the accumulation can be swallowed or expectorated. The air is warmed by heat given off by the blood vessels lying close to the surface of the lining of the air passages, and it is moistened by the wet surface of these passages.

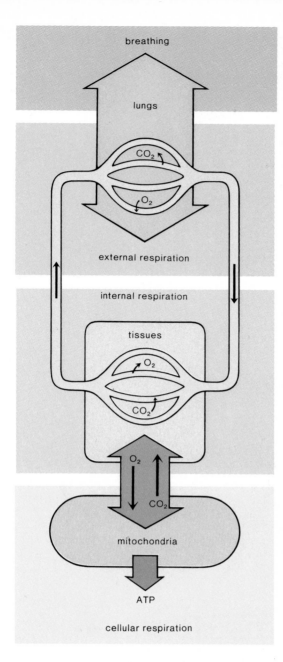

Figure 8.2 Respiration is divided into 4 components: breathing brings oxygen into the lungs; external respiration is the exchange of gases in the lungs; internal respiration is the exchange of gases in the tissues; and cellular respiration is the production of ATP in cells—an oxygen-requiring process.

On the other hand, as air moves out during expiration, it cools and loses its moisture. As the gas cools, it deposits its moisture on the lining of the windpipe and the nose, and the nose may even drip as a result of this condensation. The air still retains so much moisture, however, that upon expiration on a cold day, it condenses and forms a small cloud.

Air is warmed, filtered, and moistened as it moves from the nose toward the lungs.

Table 8.1 Path of Air

Structure	Function
nasal cavities	filter, warm, and moisten air
nasopharynx	passage of air from nose to throat
pharynx (throat)	connection to surrounding regions
glottis	passage of air into larynx
larynx (voice box)	sound production
trachea (windpipe)	passage of air to bronchi
bronchi	passage of air to each lung
bronchioles	passage of air to each alveolus
alveoli	air sacs for gas exchange

Each portion of the air passage has its own structure and function, as described in the sections that follow.

Nose

The nose contains two *nasal cavities*, narrow canals with convoluted lateral walls that are separated from one another by a median septum. In the narrow upper recesses of the nasal cavities are special ciliated cells (see fig. 12.4) that act as odor receptors. Nerves lead from these cells to the brain, where the impulses are interpreted as smell.

The tear (lacrimal) glands drain into the nasal cavities by way of tear ducts. For this reason, crying produces a runny nose.

The nasal cavities also communicate with the cranial sinuses, air-filled spaces in the skull that are lined with mucous membrane. If these membranes are inflamed due to a cold or an allergic reaction, mucus can accumulate in the sinuses and cause a sinus headache.

The nasal cavities empty into the nasopharynx, a chamber just beyond the soft palate. The *eustachian tubes* lead from the nasopharynx to the middle ears (see fig. 12.12).

The nasal cavities, which receive air, open into the nasopharynx.

Pharynx

The **pharynx** is in the back of the throat; therefore, air taken in by either the nose or the mouth enters the pharynx. In the pharynx, the air passage and the food passage temporarily join. The trachea, which lies in front of the esophagus, is normally open, allowing the passage of air, but the esophagus normally is closed and opens only when swallowing occurs. The larynx lies at the top of the trachea.

Air from either the nose or the mouth enters the pharynx, as does food. The passage of air continues in the larynx and the trachea.

Larynx

The **larynx** can be imagined as a triangular box whose apex, the Adam's apple, is located at the front of the neck. At the top of the larynx is a variable-sized opening called the **glottis.** When food is being swallowed, the glottis is covered by a flap of tissue

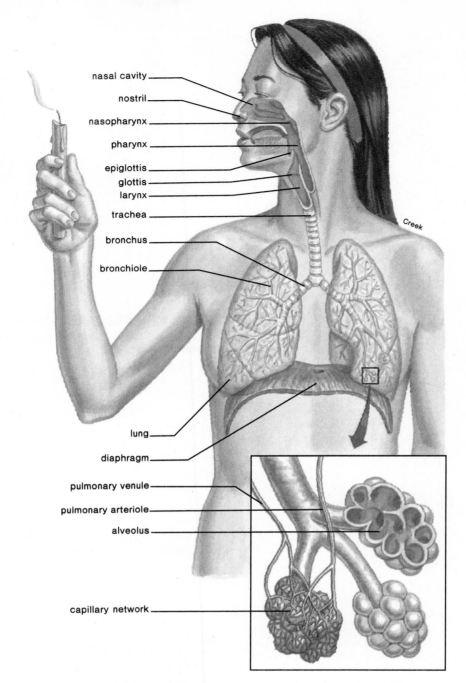

nasal cavity

nostril

nasopharynx

pharynx

epiglottis

glottis

larynx

trachea

bronchus

bronchiole

Creek

lung

diaphragm

pulmonary venule

pulmonary arteriole

alveolus

capillary network

Figure 8.3   Diagram of the human respiratory tract, with internal structure of one lung revealed in an enlargement of a section of the lung. Gas exchange occurs in the alveoli, which are surrounded by a capillary network. Notice that the pulmonary arteriole carries deoxygenated blood (colored blue) and the pulmonary venule carries oxygenated blood (colored red). (See page 101.)

called the **epiglottis** so that no food passes into the larynx. If, by chance, food or some other substance does gain entrance to the larynx, reflex coughing usually occurs to expel the substance. If this reflex is not sufficient, it may be necessary to resort to the Heimlich maneuver (fig. 8.4).

At the edges of the glottis and embedded in mucous membrane are elastic ligaments called the **vocal cords** (fig. 8.5). These cords, stretching from the back to the front of the larynx at the sides of the glottis, vibrate when air is expelled past them through the glottis. Vibration of the vocal cords produces sound. The

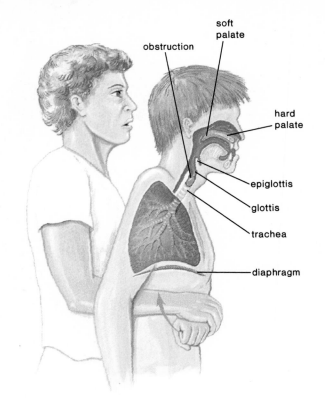

Figure 8.4   The Heimlich maneuver. More than 8 Americans choke to death each day on food lodged in the trachea. A simple process termed the abdominal thrust (Heimlich) maneuver can save the life of a person who is choking. The abdominal thrust maneuver is performed as follows. If the victim is standing or sitting: (1) Stand behind the victim or the victim's chair, and wrap your arms around his or her waist; (2) grasp your fist with your other hand, and place the fist against the victim's abdomen, slightly above the navel and below the rib cage; (3) press your fist into the victim's abdomen with a quick upward thrust; (4) repeat several times if necessary. If the victim is lying down: (1) Position the victim on his or her back; (2) face the victim, and kneel on his or her hips; (3) with one of your hands on top of the other, place the heel of your bottom hand on the abdomen, slightly above the navel and below the rib cage; (4) press into the victim's abdomen with a quick upward thrust; (5) repeat several times if necessary. If you are alone and choking, use anything that applies force just below your diaphragm. Press into a table or a sink, or use your own fist.

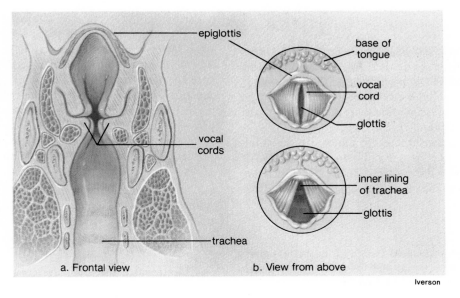

Iverson

Figure 8.5   Operation of vocal cords. The vocal cords lie at the edges of the glottis. When air is expelled from the trachea, the cords vibrate, producing the voice. *a.* Frontal view. *b.* View from above.

high or low pitch of the voice depends upon the length, the thickness, and the degree of elasticity of the vocal cords and the tension at which they are held. The loudness, or intensity, of the voice depends upon the amplitude of the vibrations, or the degree to which vocal cords vibrate.

At the time of puberty, the growth of the larynx and the vocal cords is much more rapid and accentuated in the male than in the female, causing the male to have a more prominent Adam's apple and a deeper voice. The voice "breaks" in the young male due to his inability to control the longer vocal cords.

The larynx is the voice box because it contains the vocal cords at the sides of the glottis, an opening sometimes covered by the epiglottis.

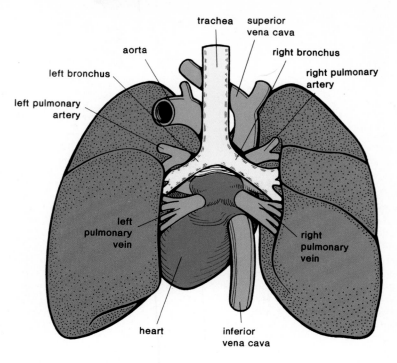

labels in figure:
trachea
superior vena cava
aorta
right bronchus
left bronchus
right pulmonary artery
left pulmonary artery
left pulmonary vein
right pulmonary vein
heart
inferior vena cava

Figure 8.6 Posterior view of the heart and the lungs showing the relationship of the pulmonary vessels to the trachea and the bronchial tubes. Trace the path of air to the left lung, and trace the path of blood from the heart to the left lung and return. Why are the arteries blue and the veins red?

Trachea

The **trachea** is a tube held open by C-shaped cartilaginous rings. Ciliated mucous membrane (see fig. 3.3) lines the trachea, and normally these cilia keep the windpipe free of debris. Smoking is known to destroy the cilia, and consequently the soot in cigarette smoke collects in the lungs. Smoking is discussed more fully at the end of this chapter.

If the trachea is blocked because of illness or accidental swallowing of a foreign object, it is possible to insert a tube by way of an incision made in the trachea. This tube acts as an artificial air intake and exhaust duct. The operation is called a *tracheostomy.*

Bronchi

The trachea divides into two **bronchi** (singular, bronchus) that enter the right and left lungs and then branch into a great number of smaller passages called **bronchioles.** The two bronchi resemble the trachea in structure, but as the bronchial tubes divide and subdivide, their walls become thinner and the small rings of cartilage are no longer present. Each bronchiole terminates in an elongated space enclosed by a multitude of air pockets, or sacs, called **alveoli** (singular, alveolus) (fig. 8.3), which make up the lungs.

Lungs

Within the lungs, each alveolar sac consists of only one layer of squamous epithelium surrounded by blood capillaries. Gas exchange occurs between air in the alveoli and blood in the capillaries.

A film of lipoprotein that lines the alveoli of mammalian lungs lowers the surface tension and prevents them from closing. The lungs collapse in some newborn babies, especially premature infants, who lack this film. This condition, called infant respiratory distress syndrome, often results in death.

There are approximately 300 million alveoli, having a total cross-sectional area of 50–70 m². This is about 40 times the surface area of the skin. Because of their many air spaces, the lungs are very light; normally, a piece of lung tissue dropped in a glass of water floats.

Air moves from the trachea and the two bronchi, which are held open by cartilaginous rings, into the lungs. The lungs are composed of air sacs called alveoli.

Externally, the lungs are cone-shaped organs that lie on both sides of the heart in the thoracic (chest) cavity. Each lung has a narrow and rounded apex that approaches the neck. The base of each lung is broad and concave in shape so as to fit upon the convex surface of the diaphragm. The other surfaces of the lungs follow the contours of the ribs and the organs in the thoracic cavity.

Figure 8.6 shows the relationship of the pulmonary vessels to the trachea and the bronchial tubes. The branches of the pulmonary artery, carrying deoxygenated blood, accompany the bronchial tubes and form a mass of capillaries around the alveoli. The four pulmonary veins, carrying oxygenated blood, collect blood from these capillaries. These veins then empty into the left atrium of the heart.

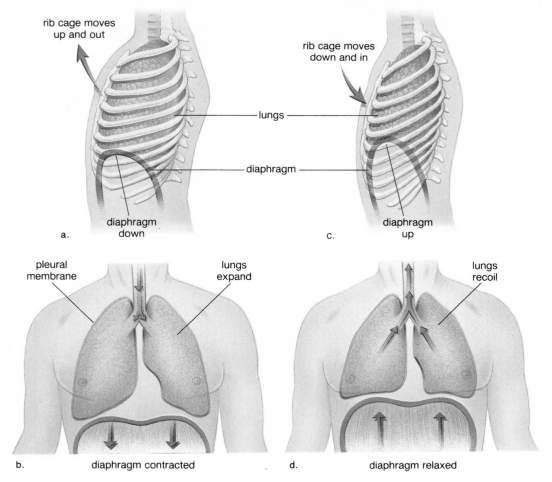

a.

c.

pleural membrane

lungs expand

lungs recoil

b. diaphragm contracted

d. diaphragm relaxed

rib cage moves up and out

rib cage moves down and in

lungs

diaphragm

diaphragm down

diaphragm up

Figure 8.7 Inspiration versus expiration. *a.* When the rib cage lifts up and outward and the diaphragm lowers, the lungs expand so that *b.* air is drawn in. This sequence of events is only possible because the pressure within the intrapleural space, containing a thin film of fluid, is less than atmospheric pressure. *c.* When the rib cage lowers and the diaphragm rises *d.* the lungs recoil so that air is forced out.

Mechanism of Breathing

In order to understand **ventilation,** the manner in which air is drawn into and expelled out of the lungs, it is necessary to remember first that when you are breathing, there is a continuous column of air from the pharynx to the alveoli of the lungs; that is, the air passages are open.

Secondly, note that the lungs lie within the sealed-off thoracic cavity. The **ribs,** hinged to the vertebral column at the back and to the *sternum* (breastbone) at the front, along with the muscles that lie between them, make up the top and the sides of the thoracic cavity. The **diaphragm,** a dome-shaped horizontal muscle, forms the floor of the thoracic cavity. The lungs themselves are enclosed by the **pleural membranes** (fig. 8.7). The outer pleural membrane adheres closely to the thoracic cavity wall and the diaphragm, and the inner membrane is fused to the lungs. The two pleural layers lie very close to one another, separated only by a thin film of fluid. Normally, the intrapleural pressure is less than atmospheric pressure. The importance of

this reduced pressure is demonstrated when, by design or accident, air enters the intrapleural space. The lungs collapse, and inspiration is impossible.

The lungs are completely enclosed and by way of the pleural membranes, adhere to the thoracic cavity walls.

Inspiration - phase 1

Carbon dioxide (CO_2) and hydrogen (H^+) ions are the primary stimuli that cause us to breathe. When the concentration of carbon dioxide and subsequently the concentration of hydrogen ions (see following sections) reach a certain level in blood, the *respiratory center* in the medulla oblongata, the stem portion of the brain, is stimulated. This center is not affected by low oxygen (O_2) levels. There also are chemoreceptors in the *carotid bodies,* located in the carotid arteries, and in the *aortic bodies,* located in the aorta, that respond primarily to hydrogen

Inspiration requires that the phrenic nerve and these structures be active.

Expiration requires that the vagus nerve and these structures be active.

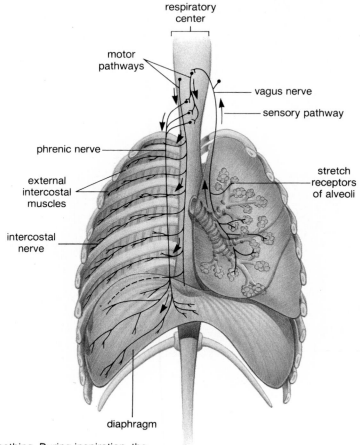

respiratory center

motor pathways

vagus nerve

sensory pathway

phrenic nerve

external intercostal muscles

stretch receptors of alveoli

intercostal nerve

diaphragm

Figure 8.8   Nervous control of breathing. During inspiration, the respiratory center stimulates the intercostal (rib) muscles and the diaphragm to contract by way of the phrenic nerve. Nerve impulses from the expanded alveoli traveling by way of the vagus nerve then inhibit the respiratory center. Lack of stimulation causes the rib muscles and the diaphragm to relax, and expiration follows.

ion concentration [H^+] but also to the level of carbon dioxide and oxygen in blood. These bodies communicate with the respiratory center.

In humans, the breathing rate is regulated by the amount of carbon dioxide in blood.

When the breathing center is stimulated, a nerve impulse goes out by way of nerves to the diaphragm and the rib cage (fig. 8.8). In its relaxed state, the *diaphragm* is dome shaped,

but upon stimulation, it contracts and lowers. When the rib muscles contract, the *rib cage* moves upward and outward. Both of these contractions serve to increase the size of the thoracic cavity. As the thoracic cavity increases in size, the lungs expand. When the lungs expand, air pressure within the enlarged alveoli lowers and is immediately rebalanced by air rushing in through the nose or the mouth.

Inspiration (fig. 8.7a and b) is the active phase of breathing. During this time, the diaphragm contracts, the rib muscles contract, the lungs are pulled open, and air comes

Table 8.2 Breathing Process

Inspiration	Expiration
Medulla sends stimulatory message to diaphragm and rib muscles.	Stretch receptors in lungs send inhibitory message to medulla.
Diaphragm contracts and flattens.	Diaphragm relaxes and resumes a dome position.
Rib cage moves up and out.	Rib cage moves down and in.
Lungs expand.	Lungs recoil.
Negative pressure builds in lungs.	Positive pressure builds in lungs.
Air is pulled in.	Air is forced out.

rushing in. Note that air comes in because the lungs already have opened up; air does not force the lungs open. This is why it sometimes is said that *humans breathe by negative pressure*. The creation of a partial vacuum sucks air into the lungs.

Stimulated by nervous impulses, the rib cage lifts up and out and the diaphragm lowers to expand the thoracic cavity and the lungs, allowing inspiration to occur.

Expiration — phase 2

When the lungs are expanded, stretching of the alveoli stimulates special receptors in the alveolar walls, and these receptors initiate nerve impulses that travel from the inflated lungs to the breathing center. When the impulses arrive at the medulla oblongata, the center is inhibited and stops sending signals to the diaphragm and the rib cage. The *diaphragm* relaxes and resumes its dome shape (fig. 8.7c and d). The abdominal organs press up against the diaphragm. The *rib cage* moves down and inward. The elastic lungs recoil, and air is pushed out.

Table 8.2 summarizes the events that occur during inspiration and expiration. It is clear that while inspiration is an active phase of breathing, normally expiration is passive because the breathing muscles automatically relax following contraction. It is possible, in deeper and more rapid breathing, for both phases to be active because there is another set of rib muscles whose contraction can forcibly cause the thoracic wall to move downward and inward. Also, when the abdominal wall muscles are contracted, there is an increase in pressure that helps to expel air.

When nervous stimulation ceases, the rib cage lowers and the diaphragm rises, allowing the lungs to recoil and expiration to occur.

Lung Capacities

When we breathe, the amount of air moved in and out with each breath is called the **tidal volume.** Normally, the tidal volume is about 500 ml, but we can increase the amount inhaled and exhaled by deep breathing. The total volume of air that can be moved in and out during a single breath is called the **vital capacity** (fig. 8.9). First, we can increase inspiration by as much as 3,100 ml of air. This is called the *inspiratory reserve volume.* Similarly, we can increase expiration by contracting the thoracic muscles. This is called the *expiratory reserve volume* and measures approximately 1,400 ml of air. Vital capacity is the sum of tidal, inspiratory reserve, and expiratory reserve volumes.

Note in figure 8.9 that even after very deep breathing, some air (about 1,000 ml) remains in the lungs; this is called the **residual volume.** This air is no longer useful for gas exchange purposes. In some lung diseases, such as emphysema (p. 162), the residual volume builds up because the individual has difficulty emptying the lungs. This means that the lungs tend to be filled with useless air, and as you can see from examining figure 8.9, the vital capacity is reduced.

Some of the inspired air never reaches the lungs; instead it fills the conducting airways (fig. 8.10). These passages are not used for gas exchange and therefore are said to contain dead space filled with *dead air.* To ventilate the lungs, then, it is better to breathe more slowly and deeply because it ensures that a greater percentage of the tidal volume reaches the lungs.

If we breathe through a tube, we increase the amount of dead air that never reaches the lungs. Any device that increases the amount of dead space beyond maximal inhaling capacity spells death for the individual because the air inhaled never reaches the alveoli.

The manner in which we breathe has physiological consequences.

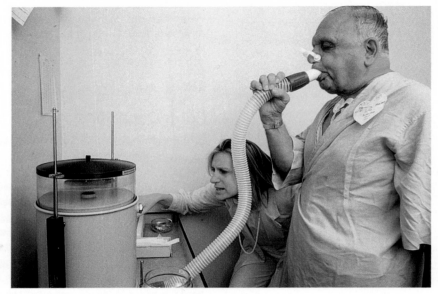

a.

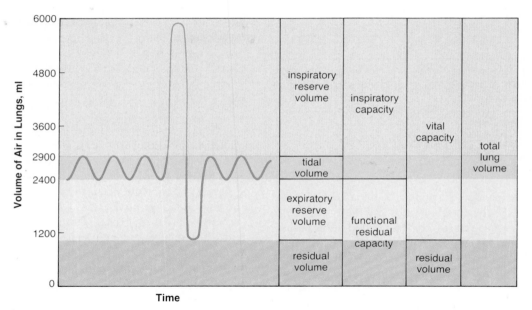

b.

Figure 8.9   Vital capacity. *a.* This individual is using a spirometer that measures the amount of air that can be maximally inhaled and exhaled. When he inspires, a pen moves up, and when he expires, a pen moves down. *b.* The resulting pattern, such as the one shown here, is called a spirogram.

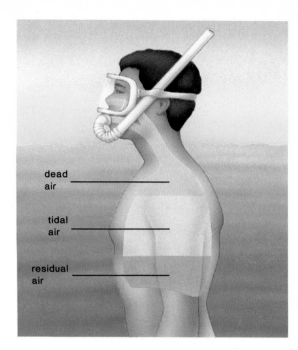

dead
air _____

tidal
air _____

residual
air _____

Figure 8.10   Distribution of air in lungs. Dead air never reaches the lungs and residual air is used air. Only tidal air brings additional oxygen for respiration.

External and Internal Respiration

External Respiration

The term *external respiration* refers to the exchange of gases between air in the alveoli of the lung and blood in the pulmonary capillaries (fig. 8.11). The wall of an alveolus consists of a thin, single layer of cells, and the wall of a blood capillary also consists of such a layer. Since neither wall offers resistance to the passage of gases, *diffusion* alone governs the exchange of oxygen (O_2) and carbon dioxide (CO_2) between alveolar air and blood. Active cellular absorption and secretion do not appear to play a role. Rather, the direction in which the gases move is determined by the pressure or tension gradients between blood and inspired air.

Atmospheric air contains little carbon dioxide, but blood flowing into the lung capillaries is almost saturated with the gas. Therefore, *carbon dioxide diffuses out of blood into the alveoli of the lung.* The pressure pattern is the reverse for oxygen. Blood coming into the pulmonary capillaries is deoxygenated, and alveolar air is oxygenated; therefore, *oxygen diffuses into the capillary.* Breathing at high altitudes is less effective than at low altitudes because the air pressure is lower, making the concentration of oxygen (and other gases) lower than normal; therefore, less oxygen diffuses into the blood. Breathing problems do not occur in airplanes because the cabin is pressurized to maintain an appropriate pressure. Emergency oxygen is available in case the pressure is, for one reason or another, reduced.

As blood enters the pulmonary capillaries (fig. 8.11), most of the carbon dioxide is being carried as the bicarbonate ion (HCO_3^-). As the little remaining free carbon dioxide begins to diffuse out, the following reaction is driven to the right:

$$H^+ + HCO_3^- \longrightarrow H_2CO_3 \longrightarrow H_2O + CO_2\uparrow$$

(bicarbonate
ion)

"Up" arrow indicates carbon dioxide is leaving the body.

The enzyme carbonic anhydrase (p. 117), present in red blood cells, speeds up the reaction. As the reaction proceeds, the respiratory pigment **hemoglobin** gives up the hydrogen (H^+) ions it has been carrying; HHb becomes Hb.

Now, hemoglobin more readily takes up oxygen and becomes oxyhemoglobin.

$$Hb + O_2 \longrightarrow HbO_2\downarrow$$

(oxyhemoglobin)

"Down" arrow indicates that oxygen is entering the body.

It is remarkable that at the partial pressure[2] of oxygen in the lungs (P_{O_2} = about 100 mm Hg), hemoglobin is about 95% saturated (fig. 8.12). Hemoglobin takes up oxygen in increasing amounts as the P_{O_2} increases and likewise gives it up as the P_{O_2} decreases. The curve begins to level off at about 90 mm Hg. This means that hemoglobin easily retains oxygen in the lungs but tends to release it in the tissues. This effect is potentiated by the fact that hemoglobin takes up oxygen more readily in the cool temperature (fig. 8.12*a*) and neutral pH (fig. 8.12*b*) of the lungs. On the other hand, it gives up oxygen more readily at the warmer temperature and more acidic pH of the tissues.[3]

External respiration, the exchange of oxygen for carbon dioxide between air within alveoli and blood in pulmonary capillaries, is dependent on the process of diffusion.

---

[2]Air exerts pressure, and the amount of pressure each gas exerts in air is called its partial pressure, symbolized by a capital P.

	pH	Temperature
[3]Lungs	7.40	37° C (98.6° F)
Body	7.38	38° C (100.4° F)

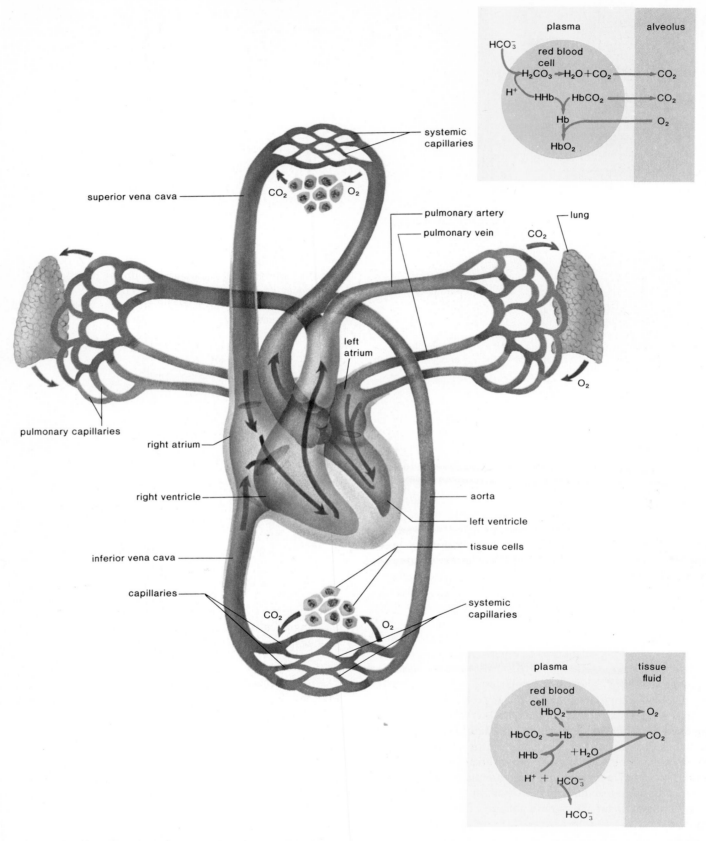

plasma | alveolus

HCO_3^-

red blood cell

$H_2CO_3 \rightarrow H_2O + CO_2 \longrightarrow CO_2$

H^+ HHb $HbCO_2 \longrightarrow CO_2$

Hb $\longrightarrow O_2$

HbO_2

systemic capillaries

superior vena cava

CO_2 O_2

pulmonary artery

pulmonary vein

lung

CO_2

left atrium

pulmonary capillaries

right atrium

O_2

right ventricle

aorta

left ventricle

tissue cells

inferior vena cava

capillaries

CO_2 O_2

systemic capillaries

plasma | tissue fluid

red blood cell

$HbO_2 \longrightarrow O_2$

$HbCO_2 \longleftarrow Hb$ CO_2

HHb $+H_2O$

$H^+ + HCO_3^-$

HCO_3^-

Figure 8.11   Diagram illustrating external and internal respiration. During external respiration in the lungs, carbon dioxide (CO_2) leaves blood and oxygen (O_2) enters blood. During internal respiration in the tissues, oxygen leaves blood and carbon dioxide enters blood. Steps necessary for gas exchange are shown for the lungs (*above*) and for the tissues (*below*).

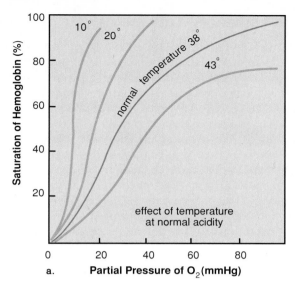

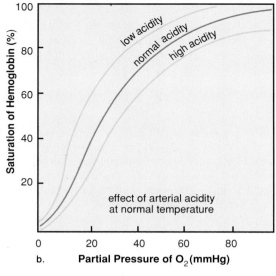

Figure 8.12 Hemoglobin saturation curves. Percent saturation of hemoglobin with oxygen (O_2) varies with the partial pressure of oxygen. Hemoglobin is more saturated in the lungs, where the temperature and acidity are lower, than in the tissues, where the temperature and acidity are higher. *a.* These data show that it takes a lower partial pressure of oxygen to saturate hemoglobin when the temperature is 10° C than when the temperature is 43° C. *b.* These data show that it takes a lower partial pressure of oxygen to saturate hemoglobin when the acidity is low than when the acidity is high.

Internal Respiration

The term *internal respiration* refers to the exchange of gases between blood in systemic capillaries and tissue fluid. When blood enters the systemic capillaries (fig. 8.11), oxyhemoglobin gives up oxygen and it diffuses out of blood into the tissues:

$$HbO_2 \longrightarrow Hb + O_2$$

(oxyhemoglobin) (oxygen)

Diffusion of oxygen out of blood into the tissues occurs because the oxygen concentration of tissue fluid is low—the cells continuously use up oxygen in cellular respiration.

Diffusion of carbon dioxide into blood from the tissues occurs because carbon dioxide concentration of tissue fluid is high. Carbon dioxide is produced continuously by cells, and it collects in tissue fluid. Carbon dioxide enters the blood and then the red blood cells, where a small amount is taken up by hemoglobin, forming carbaminohemoglobin (p. 117). Most carbon dioxide combines with water to form carbonic acid, which dissociates to hydrogen ions and the bicarbonate ion. The enzyme carbonic anhydrase, present in red blood cells, speeds up the reaction:

$$CO_2 + H_2O \rightleftharpoons H_2CO_3 \rightleftharpoons H^+ + HCO_3^-$$

carbon dioxide — water — carbonic acid — hydrogen ion — bicarbonate ion

The globin portion of hemoglobin combines with excess hydrogen ions produced by the reaction, and Hb becomes reduced hemoglobin, HHb. In this way, the pH of blood remains fairly constant. The bicarbonate ion diffuses out of the red blood cells and is carried in the plasma.

Internal respiration, the exchange of oxygen for carbon dioxide between blood in the tissue capillaries and tissue fluid, is dependent on the process of diffusion.

Respiration and Health

We have seen that the full length of the respiratory tract is lined with a warm, wet mucous membrane lining that is constantly exposed to environmental air. The quality of this air, determined by the pollutants and the germs it contains, can affect our health.

Germs frequently spread from one individual to another by way of the respiratory tract. Droplets from one single sneeze can be loaded with billions of bacteria or viruses. The mucous membranes are protected by the production of mucus and by the constant beating of the cilia, but if the number of infective agents is large and/or our resistance is reduced, respiratory infections such as colds and influenza (flu) can result. Other more serious infections and/or disorders are discussed here.

Human Issue

Many Americans are fighting high home heating costs by switching to wood-burning stoves and furnaces. Like car exhaust fumes, smoke from burning wood contains hydrocarbons and other cancer-causing chemicals. Approximately 80% of the hydrocarbons found in recent wintertime air samples in a New Mexico city were generated by burning wood. Also, nearly 60% of the cancer-causing potential of this city's winter air was attributable to wood smoke. After weighing the economic and medical pro's and con's, do you think people should be using wood-burning stoves and furnaces?

¥ characteristics of each

Bronchitis

Viral infections can spread from the nasal cavities to the sinuses (sinusitis), to the middle ears (otitis media), to the larynx (laryngitis), and to the bronchi (bronchitis). Acute bronchitis (fig. 8.13) usually is caused by a secondary bacterial infection of the bronchi, resulting in a heavy mucous discharge with much coughing. Acute bronchitis usually responds to antibiotic therapy. Chronic bronchitis is not necessarily due to infection. It often is caused by constant irritation of the lining of the bronchi, which as a result undergo degenerative changes, including the loss of cilia and their normal cleansing action. There is frequent coughing, and the individual is more susceptible to respiratory infections. Chronic bronchitis is most often seen in cigarette smokers.

Strep Throat

Strep throat is a very severe throat infection caused by the bacterium *Streptococcus pyogenes*. Swallowing may be difficult, and there is fever. Unlike a viral infection, strep throat should be treated with antibiotics. If not treated, it can lead to complications, such as rheumatic fever, in which the heart valves may be permanently affected.

Pneumonia

Most forms of pneumonia are caused by either a bacterium or a virus that has infected the lungs. However, AIDS patients are susceptible to a particularly rare form of pneumonia caused by the protozoan *Pneumocystis carinii*. Sometimes, pneumonia is localized in specific lobes of the lungs. These lobes become inoperative as they fill with mucus and pus. Obviously, the more lobes involved, the more serious the infection.

Pulmonary Tuberculosis

Tuberculosis is caused by the tubercle bacillus. An infection causes the alveoli to burst and be replaced by inelastic connective tissue. It is possible to tell if a person has ever been exposed to tuberculosis with a skin test in which a highly diluted extract of the bacilli is injected into the skin of the patient. A person who has never been in contact with the bacillus shows no reaction, but one who has developed immunity to the organism shows an area of inflammation that peaks in about 48 hours. If these bacilli invade the lung tissue, the cells build a protective capsule about the foreigners to isolate them from the rest of the body. This tiny capsule is called a *tubercle*. If the resistance of the body is high, the imprisoned organisms die, but if the resistance is low, the organisms eventually can be liberated. If a chest X ray detects tubercles, the individual is put on appropriate drug therapy to ensure the localization of the disease and the eventual destruction of any live bacterial organisms.

Emphysema

Emphysema refers to the destruction of lung tissue, with accompanying ballooning or inflation of the lungs due to trapped air. The trouble stems from the destruction and the collapse of the bronchioles. When this occurs, the alveoli are cut off from renewed oxygen supply and the air within them is trapped. The trapped air very often causes the alveolar walls to rupture and a fibrous thickening of associated blood vessel walls. The victim is breathless and may have a cough. Since the surface area for gas exchange is reduced, not enough oxygen reaches the heart and the brain. Even so, the heart works furiously to force more blood through the lungs, which can lead to a heart condition. Lack of oxygen to the brain can make the person feel depressed, sluggish, and irritable.

Chronic bronchitis and emphysema are two conditions most often caused by smoking.

Pulmonary Fibrosis

Inhaling particles, such as silica (sand), coal dust, and asbestos, can lead to pulmonary fibrosis in which fibrous connective tissue builds up in the lungs. Breathing capacity can be seriously impaired, and the development of cancer is common. Since asbestos has been used so widely as a fireproofing and insulating agent, unwarranted exposure has occurred.

Lung Cancer

Lung cancer used to be more prevalent in men than in women, but recently lung cancer has surpassed breast cancer as the leading cause of death in women. This can be linked to an increase in the number of women who smoke today. Autopsies on

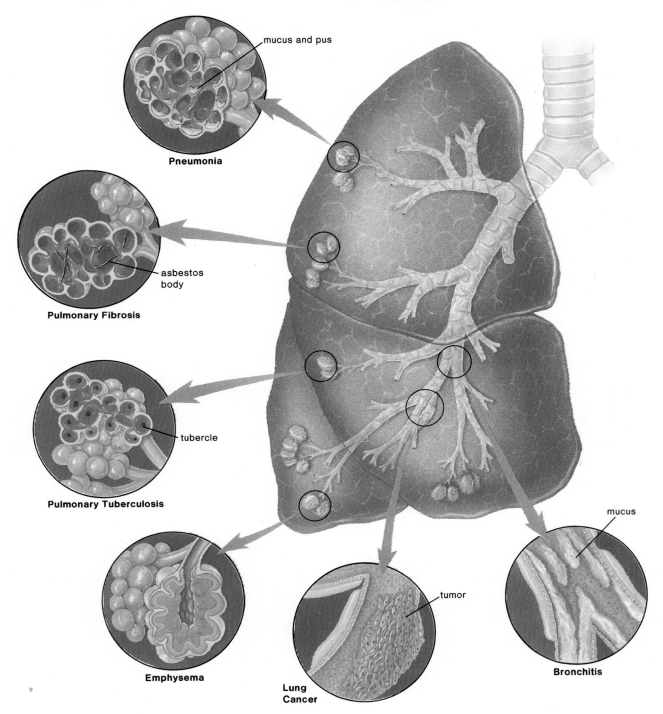

Figure 8.13   Common bronchi and lung infectious diseases and disorders. Exposure to infectious microbes and/or polluted air, including cigarette and cigar smoke, causes the diseases and disorders pictured here.

Risks of Smoking versus Benefits of Quitting

Based on available statistics, the American Cancer Society informs us that smoking carries a high risk. Among the risks of smoking are the following:

Shortened life expectancy   A twenty-five-year-old who smokes two packs of cigarettes a day has a life expectancy 8.3 years shorter than a nonsmoker. The greater the number of packs smoked, the shorter the life expectancy.

Lung cancer   Smoking cigarettes is the major cause of lung cancer in both men and women. Lung cancer now surpasses breast cancer as the leading cause of death in women because more women are smoking.

Cancer of the larynx, mouth, esophagus, bladder, and pancreas   The chances of developing these cancers are from 2 to 17 times higher in cigarette smokers than in nonsmokers.

Leukemia   Smokers of more than 25 cigarettes a day for 15 years have 2½ to 3 times the risk of developing leukemia even after they have stopped smoking.

Bronchitis and emphysema   Cigarette smokers have 4 to 25 times the risk of death from these diseases as nonsmokers. Damage is seen in lungs of even young smokers.

Coronary heart disease   Cigarette smoking is the major factor in 120,000 additional U.S. deaths from coronary heart disease each year.

Reproductive effects   Smoking mothers have more stillbirths and low-birthweight babies who are more vulnerable to disease and death. Children of smoking mothers are smaller and underdeveloped physically and socially even seven years after birth.

In the same manner, the American Cancer Society informs smokers of the benefits of quitting. These benefits include the following:

Risk of premature death is reduced   Do not smoke for 10 to 15 years, and the risk of death due to any one of the cancers mentioned approaches that of the nonsmoker.

Health of respiratory system improves   The cough and excess sputum disappear during the first few weeks after quitting. As long as cancer has not yet developed, all the ill effects mentioned can reverse themselves and the lungs can become healthy again. In patients with emphysema, the rate of alveoli destruction is reduced and lung function may improve.

Coronary heart disease risk sharply decreases   After only one year the risk factor is greatly reduced, and after 10 years an exsmoker's risk is the same as that of those who never smoked.

The increased risk of having stillborn children and underdeveloped children disappears   Even for women who do not stop smoking until the fourth month of pregnancy, such risks to infants are decreased.

People who smoke must ask themselves if the benefits of quitting outweigh the risks of smoking.

Text reproduced by permission of the American Cancer Society, Inc.

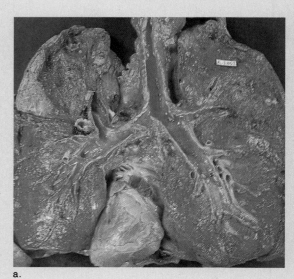
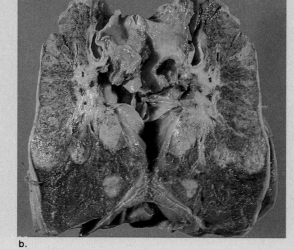

a.

b.

Normal lung versus cancerous lung
a. Normal lung with heart in place. Notice the healthy red color. *b.* Lungs of a heavy smoker. Notice how black the lungs are except where cancerous tumors have formed.

smokers have revealed the progressive steps by which the most common form of lung cancer develops. The first event appears to be thickening or callusing of the cells that line the bronchi. (Callusing occurs whenever cells are exposed to irritants.) Then there is a loss of cilia so that it is impossible to prevent dust and dirt from settling in the lungs. Following this, cells with atypical nuclei appear in the callused lining. A disordered collection of cells with atypical nuclei is considered to be cancer in situ (at one location). A final step occurs when some of these cells break loose and penetrate the other tissues, a process called metastasis. Now the cancer has spread. The tumor may grow until the bronchus is blocked, cutting off the supply of air to that lung. The lung then collapses, and the secretions trapped in the lungs spaces become infected, with pneumonia or a lung abscess occurring. The only treatment that offers a possibility of cure, before secondary growths have had time to form, is to remove the lung completely. This operation is called *pneumonectomy*.

The incidence of lung cancer is much higher in individuals who smoke than in those who do not smoke.

A recent finding is that *involuntary smoking,* simply breathing in air filled with cigarette smoke, also can cause lung cancer and other illnesses associated with smoking. The reading on page 164 lists these various illnesses associated with smoking. If a person stops both voluntary and involuntary smoking and if the body tissues are not already cancerous, they return to normal.

Human Issue

On the basis that passive smoking is detrimental to their health, nonsmokers have been slowly able to restrict those areas where smokers are permitted to smoke. A new law in New York prohibits smoking altogether in enclosed public places such as taxis, stores, and rest rooms. Smoking in federal buildings across the country has recently been restricted to specific, confined areas designated for smokers. Then, too, some employers are now refusing to hire smokers because they raise the insurance premiums too high, and statistics show that they take too many sick days. Smokers feel that their rights are being denied. Do you think that smokers are being unfairly discriminated against?

Summary

Air enters and exits the lungs by way of the respiratory tract (table 8.2). Inspiration begins when the breathing center in the medulla oblongata sends excitatory nerve impulses to the diaphragm and the rib cage. As they contract, the diaphragm lowers and the rib cage moves up and out; the lungs expand, creating a partial vacuum that causes air to rush in. Nerves within the expanded lungs then send inhibitory impulses to the breathing center. As the diaphragm relaxes, it resumes its dome shape, and as the rib cage retracts, air is pushed out of the lungs during expiration.

External respiration occurs when carbon dioxide (CO_2) leaves blood and oxygen (O_2) enters blood at the alveoli. Oxygen is transported to the tissues in combination with hemoglobin. Internal respiration occurs when oxygen leaves the blood and carbon dioxide enters the blood at the tissues. Carbon dioxide is primarily carried to the lungs in the form of the bicarbonate ion (HCO_3^-).

There are a number of illnesses associated with the respiratory tract. In addition to colds and flu, pneumonia and tuberculosis are serious lung infections. Two illnesses that have been attributed to breathing polluted air are emphysema and lung cancer.

Study Questions

1. Name and explain the four parts of respiration. (p. 150)
2. List the parts of the respiratory tract. What are the special functions of the nasal cavity, the larynx, and the alveoli? (p. 151)
3. What are the steps in inspiration and expiration? How is breathing controlled? (p. 155)
4. Why can we not breathe through a very long tube? (p. 157)
5. What physical process is believed to explain gas exchange? (p. 159)
6. What two equations are needed to explain external respiration? (p. 159)
7. How is hemoglobin remarkably suited to its job? (p. 159)
8. What two equations are needed to explain internal respiration? (p. 161)
9. Name and discuss some infections of the respiratory tract. (p. 162)
10. What are emphysema and pulmonary fibrosis, and how do they affect a person's health? (p. 162)
11. By what steps is cancer believed to develop in a person who smokes? (p. 162)

Objective Questions

1. In tracing the path of air, the _____ immediately follows the pharynx.
2. The lungs contain air sacs called _____ .
3. The breathing rate is primarily regulated by the amount of _____ and _____ in blood.
4. Air enters the lungs after they have _____ .
5. Carbon dioxide (CO_2) is carried in blood as the _____ ion.
6. The hydrogen (H^+) ions given off when carbonic acid (H_2CO_3) dissociates are carried by _____ .
7. Gas exchange is dependent on the physical process of _____ .
8. Reduced hemoglobin becomes oxyhemoglobin in the _____ .
9. The most likely cause of emphysema and chronic bronchitis is _____ .
10. Most cases of lung cancer actually begin in the _____ .

Answers to Objective Questions

1. larynx 2. alveoli 3. CO_2, H^+ 4. expanded 5. bicarbonate 6. globin portion of hemoglobin 7. diffusion 8. lungs 9. cigarette smoking 10. bronchi

Label this Diagram.
See figure 8.3 (p. 152) in text.

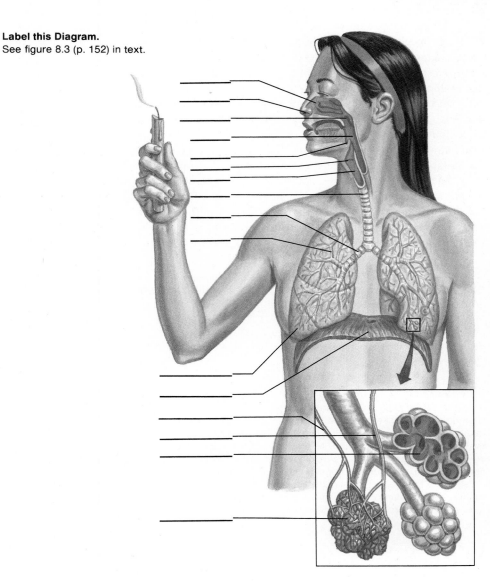

Critical Thinking Questions

1. If the human respiratory system were a flow-through system, it would have an opening both for intake and outtake. How would this affect various features of the respiratory system?

2. Fishes live in the water and expose their gills to the external environment. We live on land and have our lungs deep within the body. Explain why this is appropriate.

3. Newborn infants have poorly developed chest muscles and therefore they rely primarily on movements of the diaphragm to ventilate their lungs. Explain why an infant's body appears to lengthen during inhalation and shorten during exhalation.

Selected Key Terms

alveolus (al-ve'o-lus) (pl. alveoli)   saclike structure that is an air sac of a lung.   154

bronchiole (brong'ke-ōl)   the smaller air passages in the lungs.   154

bronchus (brong'kus) (pl. bronchi)   one of the two major divisions of the trachea leading to the lungs.   154

diaphragm (di'ah-fram)   a sheet of muscle that separates the thoracic cavity from the abdominal cavity.   155

expiration (eks''pĭ-ra'shun)   process of expelling air from the lungs; exhalation.   150

glottis (glot'is)   slitlike opening in the larynx between the vocal cords.   151

hemoglobin (he'mo-glo''-bin)   a red, iron-containing pigment in blood that combines with and transports oxygen.   159

inspiration (in''spi-ra'shun)   the act of breathing in.   150

larynx (lar'ingks)   the structure that contains the vocal cords; voice box.   151

rib (rib)   bone hinged to the vertebral column and sternum that, with muscle, defines the top and sides of the thoracic cavity.   155

trachea (tra'ke-ah)   a tube that is supported by C-shaped cartilaginous rings that lies between the larynx and the bronchi; the windpipe.   154

ventilation (ven''ti-la'shun)   breathing; the process of moving air into and out of the lungs.   155

vocal cord (vo'kal kord)   fold of tissue within the larynx that creates vocal sounds when it vibrates.   152

Chapter Nine

Excretion

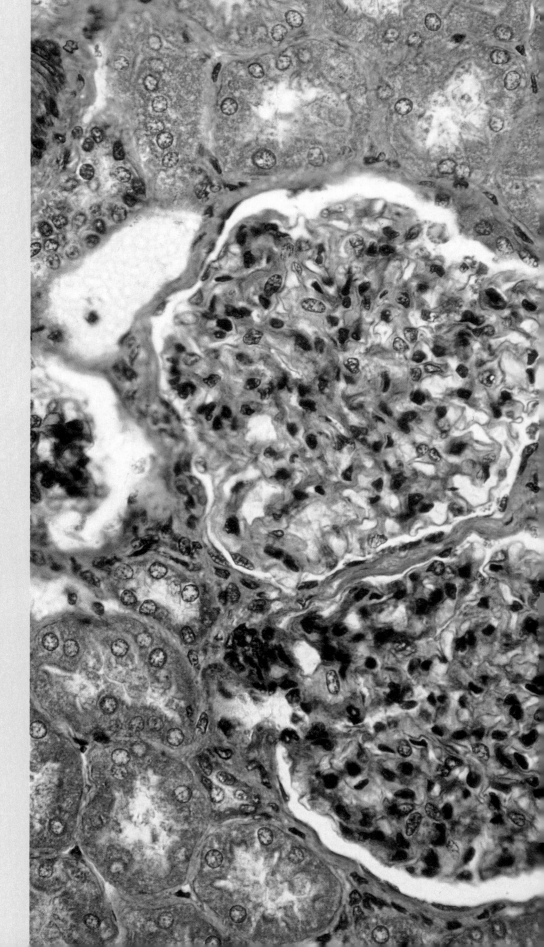

Chapter Concepts

1 Excretion rids the body of unwanted substances, particularly the end products of metabolism.

2 Several organs assist in the process of excretion, but the kidneys, which are a part of the urinary system, are the primary organs of excretion.

3 The formation of urine by the more than 1 million nephrons present in each kidney serves not only to rid the body of nitrogenous wastes but also to regulate the salt/water content and the pH of blood.

4 The kidneys, the malfunction of which causes illness and perhaps death, are important organs of homeostasis.

Microscopic anatomy of the kidneys. This cross section shows two capillary tufts surrounded by numerous tubules. The tubules carry away impurities filtered from the capillary tufts.

Figure 9.1   Unlike seabirds that have a special nasal gland to excrete excess salt, humans can't drink seawater and remain alive. Their kidneys need more water than is available in seawater in order to flush out the excess salt.

Everyone knows that if you cross a desert without a supply of fresh water, death is likely. Dehydration occurs because the body loses water and salts through sweating and urination. But what if you were lost at sea (fig. 9.1)? Can you drink salt water? I'm afraid not. Humans have no special way to rid the body of excess salt, and the kidneys would have to excrete more liquid than was consumed in order to wash out the excess salt. If there is no fresh water available, don't drink at all and don't eat salty fish either!

Excretory Substances and Organs

Excretion rids the body of metabolic wastes. The organs of excretion are shown in figure 9.2; they excrete the toxic substances listed in table 9.1. In addition, salts and water are constantly excreted.

Several of the end products excreted by humans are related to nitrogen metabolism since amino acids, nucleotides, and creatine all contain nitrogen.

Nitrogenous End Products

Ammonia (NH_3) arises upon deamination, or removal of amino groups from amino acids. It is extremely toxic, and only animals living in water, who continually flush out their body with water, largely excrete ammonia (fig. 9.3).

Urea is the primary nitrogenous end product of amino acid metabolism in human beings. Deamination of amino acids occurs in the liver, and the amino groups do not become ammonia; instead they enter a complicated series of reactions called the urea cycle. In this cycle, carrier molecules also take up carbon dioxide (CO_2) to finally release urea:

$$H_2N - \overset{\overset{\displaystyle O}{\|}}{C} - NH_2 \rightarrow urea$$

Uric acid is the usual nitrogenous end product of many terrestrial animals that need to conserve water (figure 9.4). In humans, uric acid only occurs when nucleotides are broken

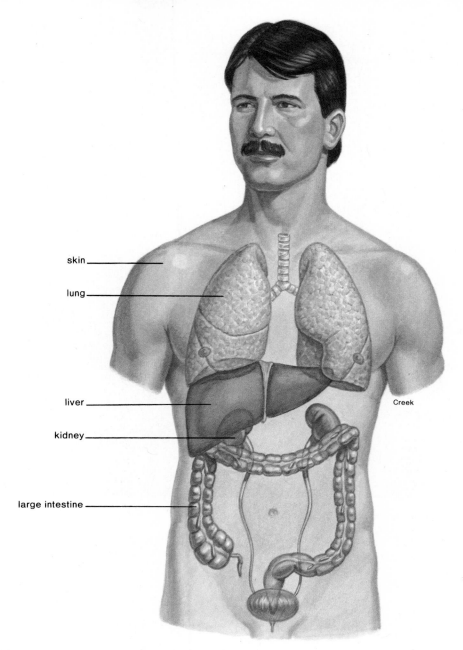

skin

lung

liver

kidney

large intestine

Creek

Figure 9.2   Organs of excretion. The lungs excrete carbon dioxide (CO_2); the liver excretes hemoglobin breakdown products in bile; the large intestine excretes certain heavy metals; the skin excretes perspiration; and the kidneys excrete urine. Excretion, ridding the body of metabolic wastes, should not be confused with defecation, ridding the body of nondigestable remains.

Table 9.1 Some Metabolic End Products

Name	End product of	Primarily excreted by
Nitrogenous wastes		
ammonia	amino acid metabolism	kidneys
urea	amino acid metabolism	kidneys and skin
uric acid	nucleotide metabolism	kidneys
creatinine	creatine phosphate metabolism	kidneys
Other wastes		
bile pigments	hemoglobin metabolism	liver
carbon dioxide	cellular respiration	lungs
Metals		
iron	hemoglobin metabolism	large intestine
calcium	muscle and nerve metabolism	large intestine

Figure 9.4 Seabirds on cliff. Birds excrete uric acid, a solid material, as their nitrogenous waste. It is mixed with fecal material in a common repository for the urinary, digestive, and reproductive systems. Seabirds congregate in such numbers that their droppings build up to give a nitrogen-rich substance called guano. At one time, guano was harvested for natural fertilizer.

Figure 9.3 Marine fishes. Most aquatic animals, including most bony fishes, excrete ammonia as their nitrogenous waste. Marine fishes, such as these, excrete excess salt by way of their gills.

down metabolically; if uric acid is present in excess, it precipitates out of the plasma. Crystals of uric acid sometimes collect in the joints, producing a painful ailment called gout.

Creatinine is an end product of muscle metabolism. It results when **creatine phosphate,** a molecule that serves as a reservoir of high-energy phosphate, breaks down.

Other Excretory Substances

Other excretory substances are bile pigments, carbon dioxide (CO_2), ions (salts), and water (H_2O).

Bile Pigments

Bile pigments are derived from the heme portion of hemoglobin and are incorporated into bile within the liver (fig. 9.5). Although the liver produces bile, it is stored in the gallbladder before passing into the small intestine by way of ducts. If for any reason a bile duct is blocked, bile spills out into the blood, producing a condition called jaundice in which the skin is discolored (p. 77).

Carbon Dioxide

The lungs are the major organs of *carbon dioxide* (CO_2) excretion, although the kidneys are also important. The kidneys excrete bicarbonate ions (HCO_3^-), the form in which carbon dioxide is carried in the blood.

Ions

Various ions (salts) that have participated in metabolism are excreted. The blood level of these ions is important to the pH, the osmotic pressure, and the electrolyte balance of blood. The balance of potassium (K^+) ions and sodium (Na^+) ions is important to nerve conduction. The level of calcium (Ca^{++}) ions in blood affects muscle contraction; iron (Fe^{++}) ions take part in hemoglobin metabolism; and magnesium (Mg^{++}) ions help many enzymes to function properly.

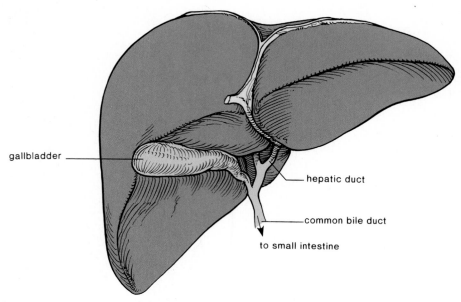

Figure 9.5   The liver, an organ of excretion. The liver breaks down hemoglobin and converts the products to bile pigments. Bile is stored in the gallbladder before being sent to the small intestine by way of ducts.

Water

Water (H_2O) is an end product of metabolism; it also is taken into the body when food and liquids are consumed. The amount of fluid in the blood helps to determine blood pressure. Treatment of hypertension sometimes includes the administration of a diuretic drug that increases the excretion of sodium and water by the kidneys.

Urea, salts, and water are the primary constituents of human urine. Carbon dioxide is excreted as a gas in the lungs and as the bicarbonate ion in the kidneys.

Organs of Excretion

The kidneys are the primary excretory organs, but there are other organs that also function in excretion (fig. 9.2), such as those described in the discussion that follows.

Skin

The sweat glands in the skin (see fig. 3.13) excrete perspiration, which is a solution of water, salt, and some urea. A sweat gland is made up of a coiled tubule portion in the dermis and a narrow, straight duct that exits from the epidermis. Although perspiration is an excretion, we perspire not so much to rid the body of waste as to cool the body. The body cools because heat is lost as perspiration evaporates. Sweating keeps the body temperature within normal range during muscular exercise or when the outside temperature rises. In times of renal failure, more urea than usual may be excreted by the sweat glands, to the extent that a so-called urea frost is observed on the skin.

Liver

The liver excretes bile pigments that are incorporated into bile, a substance stored in the gallbladder before it passes into the small intestine by way of ducts (fig. 9.5). The yellow pigment found in urine, called urochrome, also is derived from the breakdown of heme, but this pigment is deposited in blood and subsequently is excreted by the kidneys.

Human Issue

Are family members obligated to donate organs to other family members? Doctors can perform kidney, pancreas, bone-marrow, and now even liver transplants taken from a living donor. Recently, 29-year-old Teresa Smith gave the left lobe of her liver to her 21-month-old daughter Alyssa. During the operation, surgeons had to remove Teresa's gallbladder in order to gain access to the liver and when they tore her spleen, it had to be removed. Now Teresa may have to take antibiotics for the rest of her life. Do doctors have the right to ask parents to donate vital organs to their children and do parents have the right to refuse? What about sibling-to-sibling or child-to-parent transplants?

Lungs

The process of expiration (breathing out) not only removes carbon dioxide (CO_2) from the body, it also results in the loss of water (H_2O). The air we exhale contains moisture, as demonstrated by blowing onto a cool mirror.

Table 9.2　Composition of Urine

Water	95%
Solids	5%
Organic wastes	(per 1,500 ml of urine)
urea	30 g
creatinine	1–2 g
ammonia	1–2 g
uric acid	1 g
Ions (salts)	25 g

Positive ions	*Negative ions*
sodium	chlorides
potassium	sulfates
magnesium	phosphates
calcium	

Large Intestine

Certain salts, such as those of iron and calcium, are excreted directly into the cavity of the large intestine by the epithelial cells lining it. These salts leave the body in the feces.

At this point, it is helpful to remember that the term *defecation,* and not *excretion,* is used to refer to the elimination of feces from the body. Substances that are excreted are waste products of metabolism. Undigested food and bacteria, which make up feces, have never been a part of the functioning of the body, but salts that are passed into the gut are excretory substances because they were once metabolites in the body.

Kidneys

The kidneys excrete urine, which ordinarily contains organic wastes and salts (table 9.2). The kidneys are a part of the urinary system.

There are various organs that excrete metabolic wastes, but only the kidneys consistently rid the body of urea.

Urinary System

The urinary system includes the structures illustrated in figure 9.6 and listed in table 9.3. The organs are listed in order, according to the path of urine.

Path of Urine

Urine is made by the **kidneys,** two bean-shaped, reddish-brown organs, each about the size of a fist. One kidney is found on either side of the vertebral column just below the diaphragm. The kidneys lie in depressions against the deep muscles of the back beneath the serous membrane of the abdominal cavity, where they also receive some protection from the lower rib cage. Each is covered by a tough fibrous capsule of connective tissue overlaid by adipose tissue.

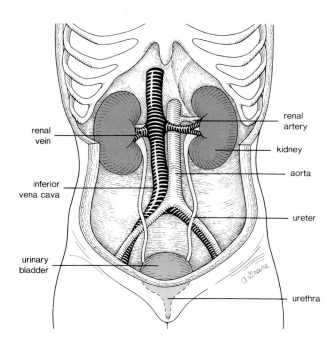

Figure 9.6　The urinary system. Urine is found only within the kidneys, the ureters, the urinary bladder, and the urethra.

Table 9.3　Urinary System

Organ	Function
kidneys	produce urine
ureters	transport urine
urinary bladder	stores urine
urethra	eliminates urine

The **ureters** are muscular tubes that convey the urine from the kidneys toward the bladder by peristaltic contractions. Urine enters the bladder by peristaltic contractions, in jets that occur at the rate of five per minute.

The **urinary bladder,** which can hold up to 600 ml of urine, is a hollow, muscular organ that gradually expands as urine enters. In the male, the urinary bladder lies ventral to the rectum, the seminal vesicles, and the vas deferens. In the female, the urinary bladder is ventral to the uterus and the upper vagina.

The **urethra,** which extends from the urinary bladder to an external opening, differs in length in the female and the male. In the female, the urethra lies ventral to the vagina and is only about 2.5 cm long. The short length of the female urethra invites bacterial invasion and explains why the female is more prone to urinary infections. In the male, the urethra averages 15 cm when the penis is relaxed. As the urethra leaves the urinary bladder, it is encircled by the prostate gland (see fig. 14.2). In older men, enlargement of the prostate gland can prevent urination, a condition that usually can be corrected surgically.

There is no connection between the genital (reproductive) and urinary systems in females, but there is a connection in males

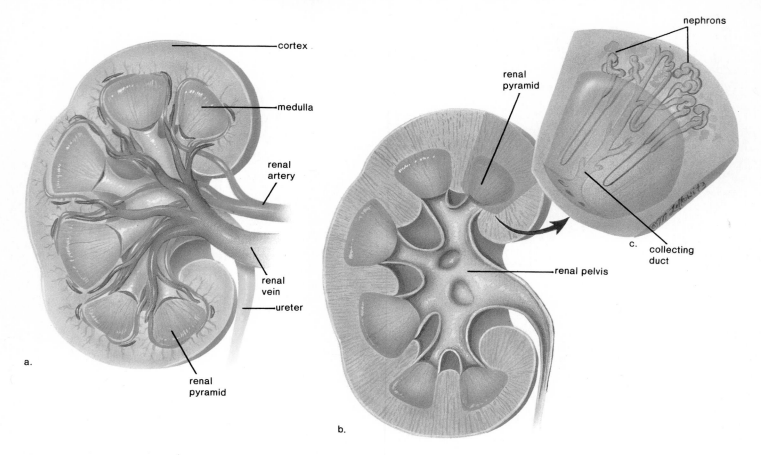

Figure 9.7 Gross anatomy of the kidney. *a.* A longitudinal section of the kidney showing the blood supply. Note that the renal artery divides into smaller arteries that frame the pyramids of the medulla. Smaller veins join to form the renal vein. *b.* Same section without the blood supply. Now it is easier to make out the cortex, the medulla, and the renal pelvis, which connects with the ureter. *c.* An enlargement of one pyramid showing the placement of nephrons.

(compare figs. 14.2 and 14.7). This double function does not alter the path of urine, and it is important to realize that urine is found only in those structures listed in table 9.3.

Urination

When the urinary bladder fills with urine, stretch receptors send nerve impulses to the spinal cord; nerve impulses leaving the cord then cause the urinary bladder to contract and the sphincters to relax so that urination is possible. In older children and adults, it is possible for the brain to control this reflex, delaying urination until a suitable time.

Only the urinary system, consisting of the kidneys, the urinary bladder, the ureters, and the urethra, ever hold urine.

Kidneys

On the concave side of each kidney, there is a depression where the renal blood vessels and the ureters enter (fig. 9.7*a*). When a kidney is sliced lengthwise, it is possible to make out three

regions: (1) an outer granulated layer called the **cortex,** which dips down in between (2) a radially striated or lined layer called the **medulla,** and (3) an inner space, or cavity, called the renal **pelvis** (fig. 9.7*b*), which is continuous with the ureter.

Upon closer examination, you can see that the medulla contains conical masses of tissue called renal pyramids. At the tip of each pyramid, there is a tube that joins with others to form the renal pelvis.

Nephrons

Microscopically, the kidney is composed of over 1 million **nephrons,** sometimes called renal or kidney tubules (figs. 9.7*c* and 9.8). Each nephron is made up of several parts. The blind end of the nephron is pushed in on itself to form a cuplike structure called **Bowman's capsule.** The outer layer of Bowman's capsule is composed of squamous epithelial cells; the inner layer is composed of specialized cells that allow easy passage of molecules. Next, there is a **proximal** (meaning near the Bowman's capsule) **convoluted tubule** in which the cells are cuboidal, with many mitochondria and an inner brush border (tightly packed microvilli). Then the cells become flat and the tube narrows and makes

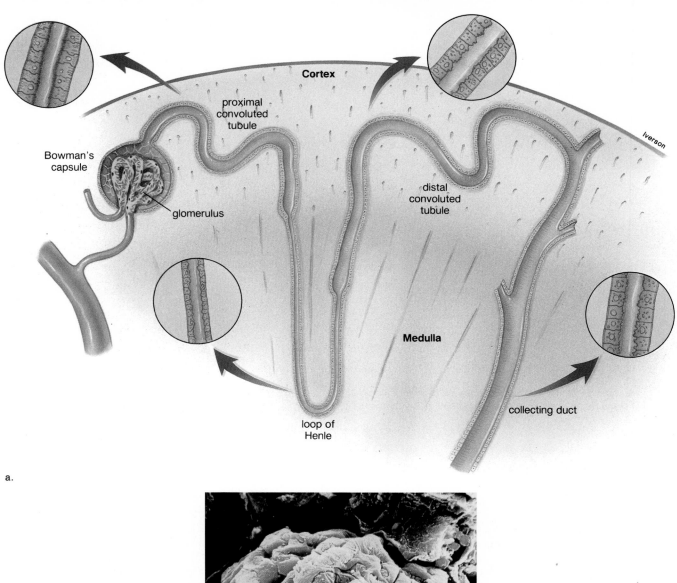

a.

b.

Figure 9.8   Nephron gross and microscopic anatomy. *a.* A nephron is made up of Bowman's capsule, the proximal convoluted tubule, the loop of Henle, the distal convoluted tubule, and the collecting duct. Note the portions that are in the cortex and the portions that are in the medulla. The blowups show the types of tissue at these different locations. *b.* Scanning electron micrograph of the glomerulus, a capillary tuft found in Bowman's capsule.

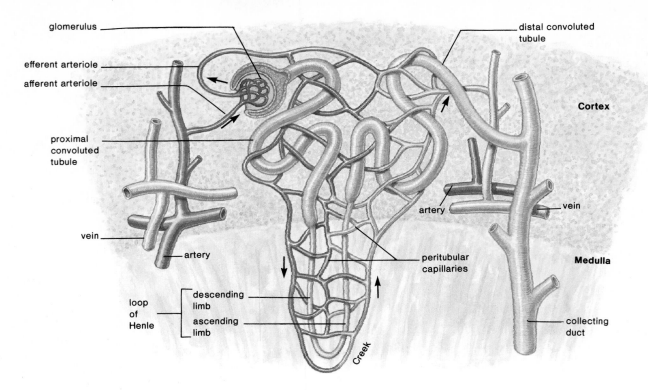

Figure 9.9   Nephron anatomy. You can trace the path of blood about the nephron by following the arrows. Use table 9.4 to help you trace the path of blood about a nephron.

Table 9.4   Circulation about a Nephron

Name of structure	Comment
afferent arteriole	brings arteriolar blood toward Bowman's capsule
glomerulus	capillary tuft enveloped by Bowman's capsule
efferent arteriole	takes arteriolar blood away from Bowman's capsule
peritubular capillary network	capillary bed that envelops the rest of the tubule
venule	takes venous blood away from the tubule

a U-turn to form the portion of the tubule called the **loop of Henle.** This leads to the **distal** (far from Bowman's capsule) **convoluted tubule,** where the cells are cuboidal, again with mitochondria but no brush border. The distal convoluted tubule enters the **collecting duct.**

Figure 9.8 shows that typically Bowman's capsules and convoluted tubules lie within the cortex and account for its granular appearance. Loops of Henle and collecting ducts lie within the triangular-shaped pyramids of the medulla. Because the loops and the ducts are longitudinal structures, they account for the striped appearance of the renal pyramids.

Urine Formation

Each nephron has its own blood supply, including two capillary regions. The **glomerulus** is a capillary tuft inside Bowman's capsule, and the **peritubular capillary** surrounds the rest of the nephron (table 9.4 and fig. 9.9). Urine formation requires the movement of molecules between these capillaries and the nephron. Three steps are involved: *pressure filtration, selective reabsorption,* and *tubular excretion.*

The pattern of blood flow about the nephron is critical to urine formation.

Pressure Filtration

Figure 9.10 gives a simple overview of urine formation. Whole blood, of course, enters an afferent arteriole and a glomerulus (fig. 9.10*b*). Under the influence of glomerular blood pressure, which is usually about 60 mm Hg, small molecules move from the glomerulus to the inside of Bowman's capsule across the thin walls of each. This is a **pressure filtration** process because large molecules and formed elements are unable to pass through. In effect, then, blood that enters the glomerulus is divided into two portions: the filterable components and the nonfilterable components.

Filterable Blood Components	*Nonfilterable Blood Components*
Water	Formed elements (blood cells
Nitrogenous wastes	and platelets)
Nutrients	Proteins
Ions (salts)	

The filterable components form a filtrate called the **glomerular filtrate** that contains small dissolved molecules in approximately the same concentration as plasma. The filtrate stays inside of Bowman's capsule, and the nonfilterable components leave the glomerulus by way of an efferent arteriole.

A consideration of the preceding filterable substances leads us to conclude that if the composition of urine were the same as that of glomerular filtrate, the body would continually lose nutrients, water, and salts. Death from dehydration, starvation, and low blood pressure would quickly follow. Therefore, we can assume that the composition of the filtrate must be altered as this fluid passes through the remainder of the tubule.

During pressure filtration, water, salts, nutrient molecules, and waste molecules move from the glomerulus to the inside of Bowman's capsule. The filtered substances are called the glomerular filtrate.

Selective Reabsorption

Both passive and active reabsorption of molecules from the nephron to the blood of the peritubular capillary occur as the filtrate moves along the **proximal convoluted tubule.**

Because of *passive reabsorption,* even some urea is reabsorbed (table 9.5). However, we are particularly interested in the passive reabsorption of water (H_2O). Two factors aid this process. The nonfilterable proteins remain in blood, and salt is returned to blood. Following active reabsorption of sodium (Na^+) ions, chloride (Cl^-) ions follow passively, as does water. Therefore, water moves from the area of greater concentration in the filtrate to the area of lesser concentration in blood because of the osmolarity difference created largely by salt reabsorption. This process occurs along the length of the nephron, until eventually nearly all water and sodium ions have been reabsorbed (table 9.5).

Table 9.5 Reabsorption from Nephron

Substance	Amount filtered per day	Amount excreted per day	Reabsorption (%)
water (L)	180	1.8	99.0
sodium (g)	630	3.2	99.5
glucose (g)	180	0.0	100.0
urea (g)	54	30.0	44.0

From A. J. Vander, et al., *Human Physiology*, 4th ed. Copyright © 1985 McGraw-Hill Publishing Company, New York, NY. Reprinted by permission of McGraw-Hill, Inc.

The cells that line the proximal convoluted tubule are anatomically adapted for *active reabsorption* (fig. 9.11). These cells have numerous microvilli, each about 1 μm in length, that increase the surface area for reabsorption. In addition, the cells contain numerous mitochondria, which produce the energy necessary for active transport. Reabsorption by active transport is **selective reabsorption** because only molecules recognized by carrier molecules move across the membrane. After passing through the tubule cells, the molecules enter the blood of the peritubular capillary.

Glucose is an example of a molecule that ordinarily is reabsorbed completely (table 9.5). Such molecules are actually reabsorbed until their threshold level is obtained; thereafter, they appear in the urine. For example, the threshold level of glucose is about 180 mg of glucose per 100 ml of blood. After this amount is reabsorbed, any excess molecules present in the filtrate appear in the urine. In diabetes mellitus (sugar diabetes, p. 266), the filtrate contains excess glucose molecules because the liver fails to store glucose as glycogen.

We have seen that the filtrate that enters the proximal convoluted tubule is divided into two portions: the components that are reabsorbed from the tubule into blood and the components that are nonreabsorbed.

Reabsorbed Filtrate Components	*Nonreabsorbed Filtrate Components*
Most water	Some water
Nutrients	Much nitrogenous waste
Required ions (salts)	Excess ions (salts)

The substances that are not reabsorbed become the tubular fluid that enters the loop of Henle.

During selective reabsorption, nutrient and salt molecules are actively reabsorbed from the proximal convoluted tubule into the peritubular capillary, and water follows passively.

Tubular Excretion

The distal convoluted tubule continues the work of the proximal convoluted tubule in that salt and water are both reabsorbed. As before, sodium (Na^+) ions are actively reabsorbed into the blood capillary, and thereafter chloride (Cl^-) ions and water follow passively. In this region of the nephron also, substances are added to the urine by a process called tubular excretion, or

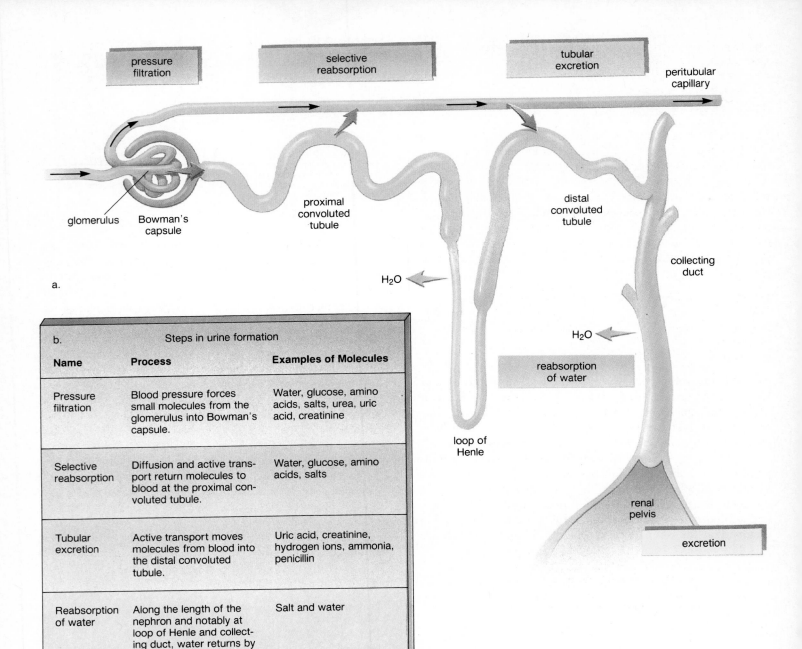

| pressure filtration | selective reabsorption | tubular excretion | peritubular capillary |

glomerulus   Bowman's capsule   proximal convoluted tubule   distal convoluted tubule   collecting duct

a.

H_2O

H_2O

reabsorption of water

loop of Henle

renal pelvis

excretion

b. Steps in urine formation

Name	Process	Examples of Molecules
Pressure filtration	Blood pressure forces small molecules from the glomerulus into Bowman's capsule.	Water, glucose, amino acids, salts, urea, uric acid, creatinine
Selective reabsorption	Diffusion and active transport return molecules to blood at the proximal convoluted tubule.	Water, glucose, amino acids, salts
Tubular excretion	Active transport moves molecules from blood into the distal convoluted tubule.	Uric acid, creatinine, hydrogen ions, ammonia, penicillin
Reabsorption of water	Along the length of the nephron and notably at loop of Henle and collecting duct, water returns by osmosis following active reabsorption of salt.	Salt and water
Excretion	Urine formation rids body of metabolic wastes.	Water, salts, urea, uric acid, ammonia, creatinine

Figure 9.10   Steps in urine formation (simplified). *a.* The steps are noted at the location of the nephron where they occur. *b.* The molecules involved in the steps and processes are listed.

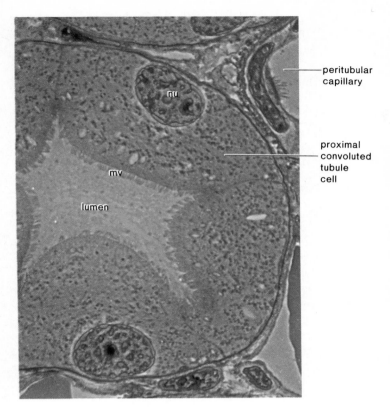

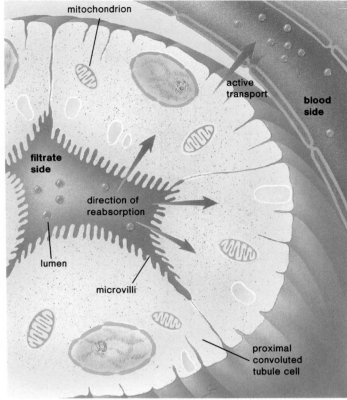

a.

b.

Figure 9.11   The cells that line the lumen (inside) of the proximal convoluted tubule, where selective reabsorption takes place. *a.* This photo shows that the cells have a brushlike border composed of microvilli (*mv*) that greatly increases the surface area exposed to the lumen. The peritubular capillary surrounds the cells (Nu = nucleus). *b.* Each cell has many mitochondria that supply the energy needed for active transport, the process that moves molecules from the lumen to the capillary.

augmentation. The cells that line this portion of the tubule have numerous mitochondria because **tubular excretion** is an active process just like selective reabsorption, although the molecules are moving in the opposite direction. Some molecules that are actively excreted are uric acid, creatinine, hydrogen (H^+) ions, ammonia, and penicillin.

During tubular excretion, certain molecules are actively secreted from the peritubular capillary into the fluid of the distal convoluted tubule. These molecules are found in urine.

Reabsorption of Water

Water is reabsorbed along the whole length of the nephron, but the excretion of a hypertonic urine (one that is more concentrated than blood) is dependent upon the action of the loop of Henle and the collecting duct.

The **loop of Henle,** which typically lies in the medulla (fig. 9.9*a*) is made up of a *descending* (going down) limb and an

ascending (going up) limb. Salt (Na^+Cl^-) passively diffuses out of the lower portion of the ascending limb, but the upper thick portion of the limb actively transports salt out into the tissue of the outer medulla (fig. 9.12). Less and less salt is available for transport from the tubule as fluid moves up the thick portion of the ascending limb. In the end, there is an osmotic gradient within the tissues of the medulla: the concentration of salt is greater in the direction of the inner medulla. (Note that water cannot leave the ascending limb because the limb is impermeable to water.)

If you examine figure 9.12 carefully, you can see that the *innermost* portion of the inner medulla has the highest concentration of solutes. This cannot be due to salt because active transport of salt does not start until the thick portion of the ascending limb. Some urea is believed to leave the lower portion of the collecting duct, and it is this molecule that contributes to the high solute concentration of the inner medulla. This urea passes back into the loop of Henle and thereby eventually reaches the collecting duct once again.

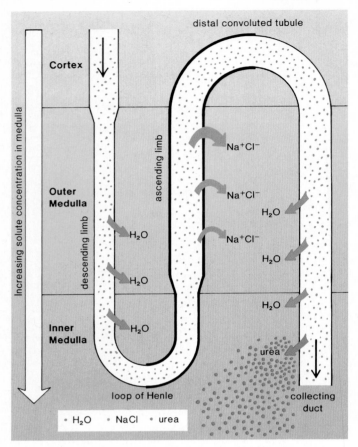

Figure 9.12   Reabsorption of water at the loop of Henle and the collecting duct. Salt (Na⁺Cl⁻) diffuses and is extruded by the ascending limb of the loop of Henle into the medulla; also, urea is believed to leave the collecting duct and to enter the tissues of the medulla. This creates a hypertonic environment that draws water out of the descending limb and the collecting duct. The urea, but not water, passes back into the loop of Henle.

Because of the solute concentration gradient of the medulla, water leaves the descending limb of the loop of Henle along its length. This is a *countercurrent mechanism*—the increasing concentration of solute encounters the decreasing number of water molecules in the descending limb, ensuring that water continues to leave the descending limb from the top to the bottom.

Fluid entering the **collecting duct** comes from the distal convoluted tubule. This fluid is isotonic to the cells of the cortex. This means that to this point, the net effect of reabsorption of water and salt is the production of a fluid that has the same tonicity as blood. Now, however, the collecting duct passes through the medulla, which is increasingly hypertonic, as previously explained (fig. 9.12). Therefore, water diffuses out of the collecting duct into the medulla, and the urine within the collecting duct becomes hypertonic to blood plasma.

Urine, the composition of which is listed in table 9.2, now passes out of the collecting duct into the pelvis of the kidney.

Urine contains all the molecules that were not reabsorbed and the ones that underwent tubular excretion at the distal convoluted tubule.

Water diffuses from the descending limb of the loop of Henle and the collecting duct due to an increasingly hypertonic kidney medulla. Urine (table 9.2) formation is complete.

Regulatory Functions of the Kidneys

Adjustment of Blood pH and Ion Balance

The kidneys help to maintain the pH level of blood within a narrow range, and the whole nephron takes part in this process. The excretion of hydrogen (H^+) ions and ammonia (NH_3), together with the reabsorption of sodium (Na^+) and bicarbonate ions (HCO_3^-), is adjusted to keep the pH within normal bounds. If blood is acidic, hydrogen ions are excreted in combination with ammonia, while sodium and bicarbonate ions are reabsorbed. This restores the pH because sodium ions promote the formation of hydroxide (OH^-) ions, while bicarbonate ions take up hydrogen ions when carbonic acid (H_2CO_3) is formed. If blood is basic, fewer hydrogen ions are excreted and fewer sodium and bicarbonate ions are reabsorbed.

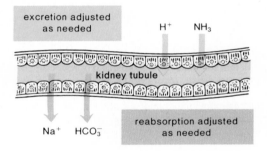

These examples also show that the kidneys regulate the ion balance in blood by controlling the excretion and the reabsorption of various ions. Sodium (Na^+) is an important ion in plasma that must be regulated, but the kidneys also excrete or reabsorb other ions, such as the bicarbonate ion, potassium ions, and magnesium ions, as needed.

Blood Volume

Maintenance of blood volume and ion balance is under the control of hormones. **ADH (antidiuretic hormone)** is a hormone secreted by the posterior pituitary that primarily maintains blood volume. ADH increases the permeability of the collecting duct so that more water can be reabsorbed. In order to understand the function of this hormone, consider its name. *Diuresis* means increased amount of urine, and *antidiuresis* means decreased amount of urine. When ADH is present, more water is reabsorbed and a decreased amount of urine results. This hormone is secreted according to whether blood volume needs to be in-

Table 9.6 Antidiuretic Hormone (ADH)

| increase in ADH | increased reabsorption of water | less urine |
| decrease in ADH | decreased reabsorption of water | more urine |

creased or decreased. When water is reabsorbed at the collecting duct, blood volume increases, and when water is not reabsorbed, blood volume decreases. In practical terms (table 9.6), if an individual does not drink much water on a certain day, the *posterior lobe of the pituitary* releases ADH, more water is reabsorbed, blood volume is maintained at a normal level, and, consequently, there is less urine. On the other hand, if an individual drinks a large amount of water and does not perspire much, the posterior lobe of the pituitary does not release ADH, more water is excreted, blood volume is maintained at a normal level, and a greater amount of urine is formed.

Drinking alcohol causes diuresis because it inhibits the secretion of ADH. The dehydration that follows is believed to contribute to the symptoms of a "hangover." Drugs called diuretics often are prescribed for high blood pressure. The drugs cause salts and water to be excreted; therefore, they reduce blood volume and blood pressure. Concomitantly, any *edema* (p. 132) present also is reduced.

Aldosterone, secreted by the adrenal cortex, is a hormone that primarily maintains sodium (Na^+) and potassium (K^+) ion balance. It causes the distal convoluted tubule to reabsorb sodium ions and to excrete potassium ions. The increase of sodium ions in the blood causes water to be reabsorbed, leading to an increase in blood volume and blood pressure.

Blood pressure is constantly monitored by the afferent arteriole cells within the juxtaglomerular apparatus. The juxtaglomerular apparatus (fig. 9.13) occurs at a region of contact between the afferent arteriole and the distal convoluted tubule. The afferent arteriole cells in the region secrete renin when blood pressure is insufficient to promote efficient filtration in the glomerulus. *Renin* is an enzyme that cleaves a large plasma protein, angiotensinogen. This process releases angiotensin I (fig. 9.13c). This molecule undergoes further cleavage by an enzyme called converting enzyme, which is present in the lining of the pulmonary capillaries of the lungs. Converting enzyme changes angiotensin I to angiotensin II, and angiotensin II stimulates the adrenal cortex to release aldosterone. Now, blood pressure rises.

A possible mechanism also has been suggested for the suppression of renin secretion. Perhaps the distal convoluted tubule cells in the juxtaglomerular apparatus are sensitive to sodium ion concentration in urine, and when there is a high sodium ion concentration in the distal convoluted tubule, they inhibit the afferent arteriole cells from secreting renin. If so, this mechanism may be faulty in certain individuals. The renin-angiotensin-aldosterone system always seems to be active in some

patients with hypertension. In response to this possibility, there is a new medicine for hypertension that inhibits converting enzyme.

The kidneys contribute to homeostasis by excreting urea. They also maintain both the pH and the ion balance of blood and regulate the volume of blood, three very important functions.

Problems with Kidney Function

Because of the great importance of the kidney to the maintenance of body fluid homeostasis, renal failure is a life-threatening event. There are many types of illnesses that cause progressive renal disease and renal failure.

Infections of the urinary tract themselves are a fairly common occurrence, particularly in the female since the urethra is considerably shorter than that of the male. If the infection is localized in the urethra, it is called *urethritis*. If it invades the urinary bladder, it is called *cystitis*. Finally, if the kidneys are affected, the infection is called *pyelonephritis*. Glomerular damage sometimes leads to blockage of the glomeruli so that no fluid moves into the tubules, or it can cause the glomeruli to become more permeable than usual. This is detected when a **urinalysis** is done. If the glomeruli are too permeable, albumin, white blood cells, or even red blood cells appear in the urine. Trace amounts of protein in the urine is not a matter of concern, however.

When glomerular damage is so extensive that more than two-thirds of the nephrons are incapacitated, waste substances accumulate in blood. This condition is called *uremia* because urea is one of these substances that accumulate. Although nitrogenous wastes can cause serious damage, the retention of water and salts is of even greater concern. The latter causes edema, fluid accumulation in the body tissues. Imbalance in the ionic composition of body fluids even can lead to loss of consciousness and to heart failure.

Kidney Replacement

Kidney Transplant

Patients with renal failure sometimes can undergo a kidney transplant operation, during which a functioning kidney from a donor is received. As with all organ transplants, there is the possibility of organ rejection. Receiving a kidney from a close relative has the highest chance of success. The current one-year survival rate is 97% if the kidney is received from a relative and 90% if it is received from a nonrelative.

Dialysis

If a satisfactory donor cannot be found for a kidney transplant, which is frequently the case, the patient can undergo dialysis treatments, utilizing either a kidney machine or continuous ambulatory peritoneal (abdominal) dialysis, or CAPD. Dialysis is defined as the diffusion of dissolved molecules through a

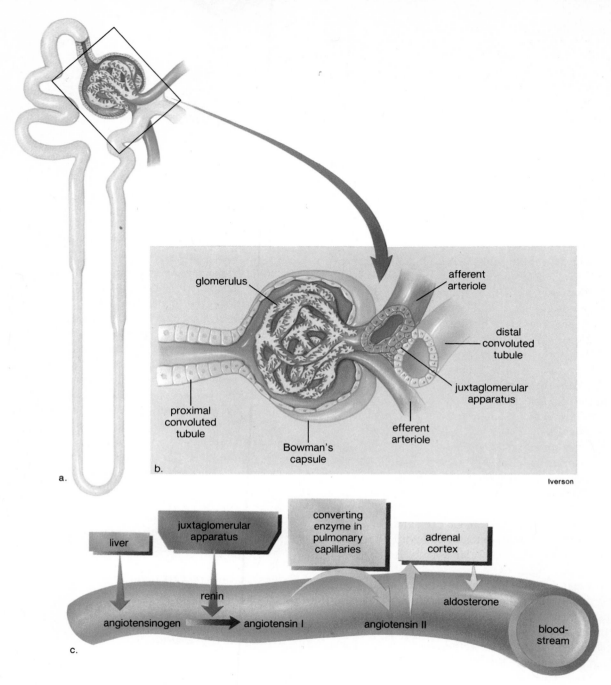

glomerulus

afferent
arteriole

distal
convoluted
tubule

juxtaglomerular
apparatus

proximal
convoluted
tubule

efferent
arteriole

Bowman's
capsule

a.

b.

Iverson

liver

juxtaglomerular
apparatus

converting
enzyme in
pulmonary
capillaries

adrenal
cortex

renin

aldosterone

angiotensinogen

angiotensin I

angiotensin II

blood-
stream

c.

Figure 9.13   Juxtaglomerular apparatus. *a*. This drawing shows how it is possible for the afferent arteriole and the distal convoluted tubule to lie next to one another. The juxtaglomerular apparatus occurs where they touch. *b*. Cross section shows exact location of the juxtaglomerular apparatus, which releases renin if the blood pressure in the afferent arteriole falls. *c*. Renin is an enzyme that cleaves angiotensinogen, a plasma protein made by the liver. This releases angiotensin I, which is changed to angiotensin II by a converting enzyme found in the lining of the pulmonary (lung) capillaries. Angiotensin II stimulates the adrenal cortex to release aldosterone into blood. Now, the blood pressure rises because sodium (Na^+) ions are reabsorbed to a greater extent.

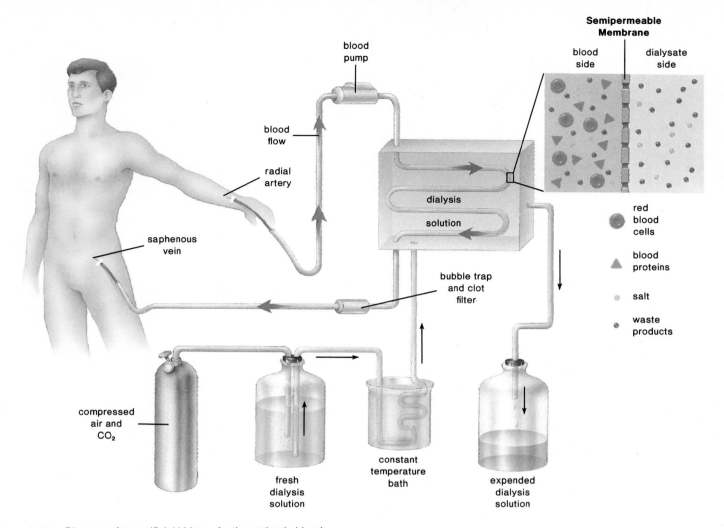

Figure 9.14 Diagram of an artificial kidney. As the patient's blood circulates through dialysis tubing, it is exposed to a solution. Wastes exit from blood into the solution because of a preestablished concentration gradient. In this way, blood is cleansed and the pH also can be adjusted.

semipermeable membrane. These molecules, of course, move across a membrane from the area of greater concentration to one of lesser concentration.

Kidney machine dialysis is more properly called hemodialysis (fig. 9.14) because the patient's blood is directly cleaned as it passes through a semipermeable membranous tube in contact with a balanced salt (dialysis) solution. Substances more concentrated in blood diffuse into the dialysis solution, also called the dialysate. Conversely, substances more concentrated in the dialysate diffuse into blood. Accordingly, the artificial kidney can be utilized either to extract substances from blood, including waste products or toxic chemicals and drugs, or to add substances to blood, for example, bicarbonate ions (HCO_3^-) if blood is acidic. In the course of a 6-hour hemodialysis, from 50 to 250 grams of urea can be removed from a patient, which greatly exceeds the urea clearance of normal kidneys. Therefore, a patient need undergo treatment only about twice a week.

In the case of CAPD, a fresh amount of dialysate is introduced directly into the abdominal cavity from a bag attached to a permanently implanted plastic tube. Waste and water molecules pass into the dialysate from the surrounding organs before the fluid is collected four or eight hours later. The individual can go about his or her normal activities during CAPD, unlike during hemodialysis.

Human Issue

Imagine a situation in which (1) a patient is not doing well on dialysis and doctors believe that only a kidney transplant will allow an improvement in health and (2) there is in the same hospital another patient who is being kept alive only by machines. Under what circumstances—if any—would it be proper to disconnect the machines so that a kidney of the patient (2) could be donated to this other patient (1)?

Kidney transplants and dialysis are available procedures for persons who have suffered renal failure.

Summary

The end products of metabolism are, for the most part, nitrogenous wastes, such as ammonia, urea, uric acid, and creatinine, all of which are excreted primarily by the urinary system. The urinary system contains the kidneys, whose macroscopic anatomy is dependent on nephrons. Urine formation requires three steps: during pressure filtration, small components of plasma pass into Bowman's capsule from the glomerulus due to blood pressure; during selective reabsorption, nutrients and sodium are actively reabsorbed from the proximal convoluted tubule back into blood; during tubular excretion, a few types of substances are actively secreted into the distal convoluted tubule from blood.

Water is reabsorbed along the length of the nephron, but it is the loop of Henle and the collecting duct that allow us to secrete a hypertonic urine. ADH, a hormone produced by the posterior pituitary, controls the reabsorption of water directly, and aldosterone from the adrenal cortex controls it indirectly by affecting sodium (Na^+) ion reabsorption. The whole nephron participates in maintaining the pH of blood by regulating the pH of urine. In practice, hydrogen (H^+) ions are excreted and sodium bicarbonate (HCO_3^-) ions are reabsorbed to maintain the pH.

Various problems can lead to kidney failure. In such cases, the person either can receive a kidney from a donor or undergo dialysis treatments by means of the kidney machine or CAPD.

Study Questions

1. Name four nitrogenous end products, and explain how each is formed in the body. (p. 169)
2. Name several organs of excretion and the substances they excrete. (p. 172)
3. What is the composition of urine? (p. 173)
4. Give the path of urine. (p. 173)
5. Name the parts of a nephron. (p. 174)
6. Trace the path of blood about the nephron. (p. 176)
7. Describe how urine is made by telling what happens at each part of the tubule. (p. 178)
8. Explain these terms: *pressure filtration, selective reabsorption,* and *tubular excretion.* (p. 177)
9. How does the nephron regulate the blood volume and the pH of blood? (p. 180)
10. Explain how the artificial kidney machine and CAPD work. (p. 181)

Objective Questions

1. The primary nitrogenous end product of humans is _____ .
2. The large intestine is an organ of excretion because it rids the body of

    _____ .
3. Urine leaves the urinary bladder in the

    _____ .
4. The capillary tuft inside Bowman's capsule is called the _____ .
5. _____ is a substance that is found in the filtrate, is reabsorbed, and is still in urine.
6. _____ is a substance that is found in the filtrate, is minimally reabsorbed, and is concentrated in the urine.
7. Tubular excretion takes place at the _____ , a portion of the nephron.
8. Reabsorption of water from the collecting duct is regulated by the hormone

    _____ .
9. In addition to excreting nitrogenous wastes, the kidneys adjust the

    _____ and _____ of blood.
10. Persons who have nonfunctioning kidneys often have their blood cleansed by a _____ machine.

Answers to Objective Questions

1. urea 2. certain salts, e.g., calcium and iron 3. urethra 4. glomerulus 5. Water 6. Urea 7. distal convoluted tubule 8. ADH 9. volume, pH 10. hemodialysis

Label this Diagram.
See figure 9.9 (p. 176) in text.

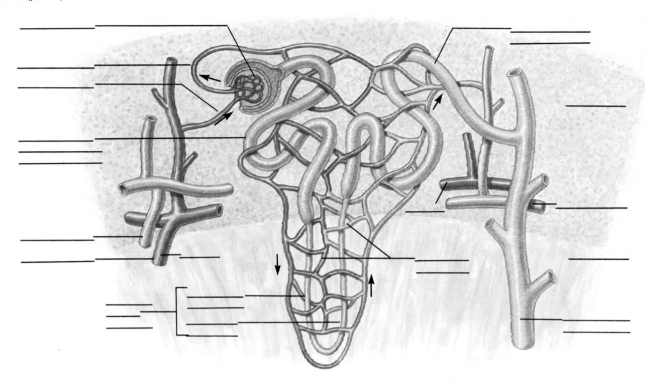

Critical Thinking Questions

1. Would animals that live in fresh water or on land have well-developed loops of Henle? Why?

2. Why is the term filtration used both in regard to a capillary in the tissues and to the glomerulus in the nephron? What differences are there between these two processes?

3. Physicians recognize that the presence of protein in the urine is the single most important indicator of kidney disease. How does kidney disease cause plasma proteins to be lost via the kidneys? What are the bodily consequences of losing plasma proteins in this way?

Selected Key Terms

antidiuretic hormone (an''tĭ-dĭ''u-ret'ik hor'mōn)   ADH; sometimes called vasopressin, a hormone secreted by the posterior pituitary that controls the rate at which water is reabsorbed by the kidneys. 180

Bowman's capsule (bo'manz kap'sūl)   a double-walled cup that surrounds the glomerulus at the beginning of the nephron. 174

collecting duct (kō-lekt'ing dukt)   a tube that receives urine from several distal convoluted tubules. 176

distal convoluted tubule (dis'tal kon'vo-lūt-ed too'būl)   highly coiled region of a nephron that is distant from Bowman's capsule because it occurs after the loop of Henle. 176

excretion (eks-kre'shun)   removal of metabolic wastes. 173

glomerular filtrate (glo-mer'u-lar fil'trāt)   the filtered portion of blood that is contained within Bowman's capsule. 177

glomerulus (glo-mer'u-lus)   the cluster of capillaries surrounded by Bowman's capsule in a nephron. 176

kidney (kid'ne)   one of two organs in the urinary system that form, concentrate, and excrete urine. 173

nephron (nef'ron)   the anatomical and functional unit of the vertebrate kidney; kidney tubule. 174

pelvis (pel'vis)   a hollow chamber in the kidney that lies inside the medulla and receives freshly prepared urine from the collecting ducts. 174

peritubular capillary (per''ĭ-tu'bu-lar kap'ĭ-lar''e)   capillary that surrounds a nephron and functions in reabsorption during urine formation. 176

pressure filtration (presh'ur fil-tra'shun)   the movement of small molecules from the glomerulus into Bowman's capsule due to the action of blood pressure. 177

proximal convoluted tubule (prok'sĭ-mal kon'vo-lūt-ed too'bul)   highly coiled region of a nephron near Bowman's capsule. 174

selective reabsorption (sĕ-lek'tiv re''ab-sorp'shun)   the movement of nutrient molecules, as opposed to waste molecules, from the contents of the nephron into blood at the proximal convoluted tubule. 177

tubular excretion (too'bu-lar eks-kre'shun)   the movement of certain molecules from blood into the distal convoluted tubule so that they are added to urine. 179

urea (u-re'ah)   primary nitrogenous waste of humans derived from amino acid breakdown. 169

ureter (u-re'ter)   one of two tubes that take urine from the kidneys to the urinary bladder. 173

urethra (u-re'thrah)   tube that takes urine from the bladder to outside. 173

uric acid (u'rik as'id)   waste product of nucleotide breakdown. 169

urinary bladder (u'rĭ-ner''e blad'der)   an organ where urine is stored before being discharged by way of the urethra. 173

Further Readings for Part II

Ada, G. L., and Sir Gustav Nossal, August 1987. The clonal-selection theory. *Scientific American.*

Buisseret, P. D. August 1982. Allergy. *Scientific American.*

Cohen, I. R. April 1988. The self, the world and autoimmunity. *Scientific American.*

Doolittle, R. F. December 1981. Fibrinogen and fibrin. *Scientific American.*

Gallo, R. C. January 1987. The AIDS virus. *Scientific American.*

Grey, Howard M., Alessandro Sette, and Soren Buus. November 1989. How T-cells see antigen. *Scientific American.*

Guyton, A. C. 1979. *Physiology of the human body.* 5th ed. Philadelphia: W. B. Saunders.

Hole, J. W. 1990. *Human anatomy and physiology.* 5th ed. Dubuque, IA: Wm. C. Brown Publishers.

Human nutrition: Readings from Scientific American. 1978. San Francisco: W. H. Freeman.

Jarvik, R. K. January 1981. The total artificial heart. *Scientific American.*

Kessel, R. G., and R. H. Kardon. 1979. *Tissues and organs: A text-atlas of scanning electron microscopy.* San Francisco: W. H. Freeman.

Mills, John, and Henry Masur. August 1990. AIDS-related infections. *Scientific American.*

Moog, F. November 1981. The lining of the small intestine. *Scientific American.*

Perutz, M. F. December 1978. Hemoglobin structure and respiratory transport. *Scientific American.*

Rennie, John. December 1990. The body against itself. *Scientific American.*

Rose, N. R. February 1981. Autoimmune diseases. *Scientific American.*

Schmidt-Nielson, K. May 1981. Countercurrent systems in animals. *Scientific American.*

Scientific American. October 1988. Entire issue devoted to AIDS.

Scrimshaw, N. S., and V. R. Young. September 1976. The requirements of human nutrition. *Scientific American.*

Smith, Kendall A. March 1990. Interleukin-2. *Scientific American.*

Tonegawa, S. October 1985. The molecules of the immune system. *Scientific American.*

Vander, A. J. 1980. *Human physiology: The mechanisms of body function.* 4th ed. New York: McGraw-Hill.

Wurtman, R. M. and J. J. Wortman. January 1989. Carbohydrates and depression. *Scientific American.*

Young, J. D. and Z. A. Cohn. January 1988. How killer cells kill. *Scientific American.*

Zucker, M. B. June 1980. The functioning of blood platelets. *Scientific American.*

Part Three

Integration and Coordination

in Humans

The nervous and hormonal systems help maintain a relatively constant internal environment, homeostasis, by coordinating the functions of the body's other systems. The nervous system acts quickly but provides short-lived regulation, and the hormonal system acts more slowly but provides a more sustained regulation of body parts.

Organisms must also be able to react with the external environment in order to survive. The sense receptors are the organs that inform the organism about the outside environment. This information is then processed by the nervous system, and the individual responds through the muscular system. These physiological events allow humans to maintain an external environment that is compatible with maintenance of an internal environment.

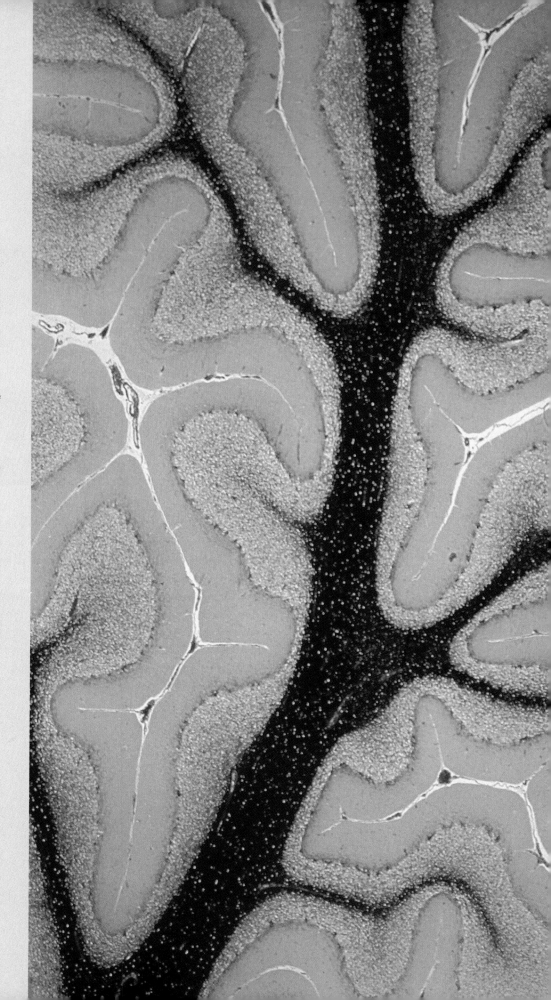

Chapter Ten

Nervous System

Chapter Concepts

1 The nervous system is made up of cells called neurons that are specialized to carry nerve impulses. A nerve impulse is an electrochemical change.

2 Transmission of impulses between neurons is accomplished by means of chemicals called neurotransmitter substances.

3 The nervous system consists of the central and peripheral nervous systems. The two systems are joined when a reflex occurs.

4 The central nervous system, made up of the spinal cord and the brain, is highly organized. Consciousness is a function only of the cerebrum, which is most highly developed in humans.

5 Drugs that affect the psychological state of the individual, such as alcohol, marijuana, cocaine, and heroin, are abused, often to the detriment of the body.

Convoluted gray matter of the cerebellum, the portion of the brain responsible for skeletal muscle coordination. Trapeze artists, skateboard riders, and even ice skaters need a cerebellum that functions well.

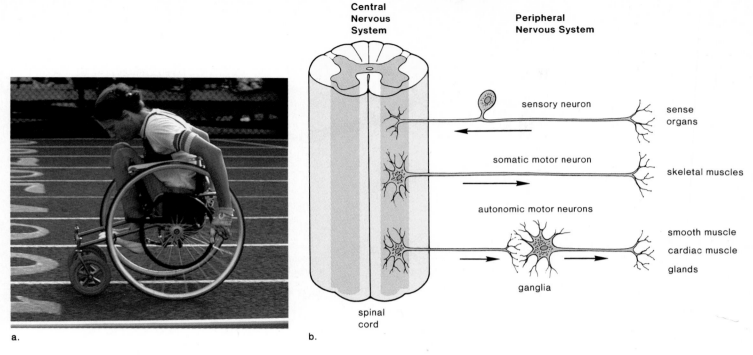

Central Nervous System

Peripheral Nervous System

sensory neuron

sense organs

somatic motor neuron

skeletal muscles

autonomic motor neurons

smooth muscle

cardiac muscle

glands

ganglia

spinal cord

a.

b.

Figure 10.1 *a.* In paraplegics, nerve impulses no longer flow between the peripheral nervous system (the legs) and the central nervous system (the brain and spinal cord). *b.* In the peripheral nervous system (PNS), the nerves contain sensory nerve fibers that take messages from sense organs to the central nervous system (CNS), and motor nerve fibers that take messages from the CNS to muscles and glands.

Paraplegia, paralysis of the legs, and tetraplegia, paralysis of all four limbs, occur when the spinal cord is partially or wholly severed. Not until this decade did scientists believe there was hope for a cure. Now they have learned to coax nerve regrowth within the central nervous system (fig. 10.1*b*). In hamsters with severed optic nerves, researchers not only got the optic nerves to regrow, they detected nerve signals in the visual center of the brain after flashing a light in front of the hamsters' eyes. They reported the work demonstrates that "one can make functional connections between neurons that are widely separated by injury."

Neurons

The nerve cell is called a neuron, and it is neurons that carry nerve impulses (messages) within the nerves of the peripheral nervous system (PNS) and within the central nervous system (CNS). All neurons (fig. 10.2) have three parts: dendrite(s), cell body, and axon. A **dendrite** conducts nerve impulses toward the **cell body,** the part of a neuron that contains the nucleus and other organelles. An **axon** conducts nerve impulses away from the cell body. There are three types of neurons: sensory neuron, motor neuron, and interneuron. A **sensory neuron** takes a message from a **receptor,** a sense organ, to the CNS and has a long dendrite and a short axon. A **motor neuron** takes a message away from the CNS to an **effector,** a muscle fiber or a gland, and has short dendrites and a long axon. Because motor neurons cause muscle fibers and glands to react, they are said to **innervate** these structures. Sometimes a sensory neuron is referred to as the *afferent neuron,* and the motor neuron is called the *efferent neuron.* These words, which are derived from Latin, mean running to and running away from, respectively. Obviously, they refer to the relationship of these neurons to the CNS.

An **interneuron** (also called association neuron or connector neuron) always is found completely within the CNS and

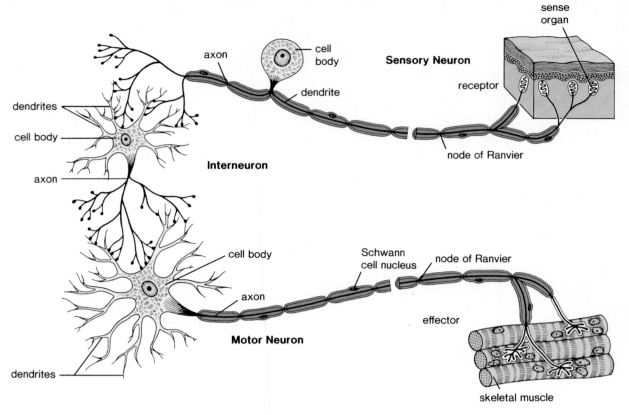

Figure 10.2   Types of neurons. A sensory neuron, an interneuron, and a motor neuron are drawn here to show their arrangement in the body. How does this arrangement correlate with the function of each neuron?

conveys messages between parts of the system. An interneuron has short dendrites and a long or a short axon. Table 10.1 summarizes the three types of neurons that also are illustrated in figure 10.2.

Although all neurons have the same three parts, each is specialized in structure and in function. Specialization is dependent on the location of the neuron in relation to the CNS.

The dendrites and the axons of neurons sometimes are called fibers, or processes. Most long fibers, whether dendrites or axons, are covered by tightly packed spirals of *Schwann cells* (fig. 10.3). Schwann cells first encircle an axon, and then as they wrap themselves around the axon many times, they lay down several layers of cellular membrane containing myelin, a lipid substance that is an excellent insulator. Myelin gives nerve fibers their white, glistening appearance. Because of the manner in which Schwann cells wrap themselves about nerve fibers, two sheaths are formed (fig. 10.3). The outermost sheath is called **neurilemma,** or the cellular sheath, and the inner one is called the **myelin sheath.** The neurilemma plays an important role in nerve regeneration. If a nerve fiber is accidently severed, the part on the far side of the cell body degenerates except for the neurilemma, which can serve as a passageway for new growth stemming from the remaining portion of the nerve fiber. The myelin sheath is involved in nerve conduction and is discussed in the following section.

Table 10.1   Neurons

Neuron	Structure	Function
sensory neuron (afferent)	long dendrite, short axon	carry nerve impulses (messages) from periphery to the CNS*
motor neuron (efferent)	short dendrites, long axon	carry nerve impulses (messages) from the CNS to periphery
interneuron	short dendrites, long or short axon	carry nerve impulses (messages) within the CNS

*CNS = central nervous system

Schwann cells are one of several types of glial cells in the nervous system. Glial cells service the neurons—they have supportive and nutritive functions.

Nerve Impulse

A neuron is specialized to conduct nerve impulses. The nature of a **nerve impulse** has been studied using giant axons from the squid and an instrument called a voltmeter. Voltage is a measure of the electrical potential difference between two points, which in this case are the inside and the outside of the axon.

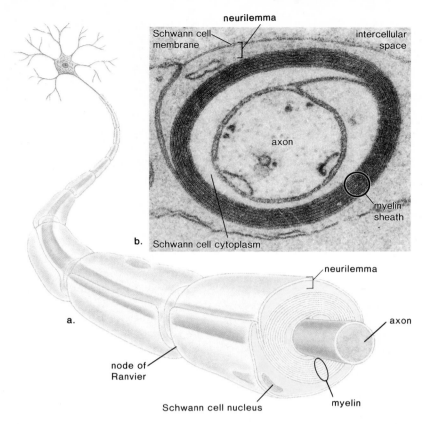

Figure 10.3   Neurilemma and myelin sheath. *a.* Axon of a motor neuron ending in a cross section of neurilemma and myelin sheath that encloses the long fibers of all neurons. The myelin sheath is composed of many layers of Schwann cell membrane and has a white, glistening appearance in the body. *b.* Electron micrograph of a cross section of an axon surrounded by neurilemma and myelin sheath.

The change in voltage is displayed on an *oscilloscope,* an instrument with a screen that shows a trace, or pattern, indicating a change in voltage with time (fig. 10.4).

Resting Potential

In the experimental setup shown in figure 10.5, an oscilloscope is wired to two electrodes, one inside and one outside a giant axon of the squid. The axon is essentially a membranous tube filled with cytoplasm, or in this case, axoplasm. When the axon is not conducting an impulse, the oscilloscope records a *membrane potential* (potential difference across a membrane) equal to about −65 mV (millivolts). This reading indicates that the inside of the neuron is negative compared to the outside. This is called the **resting potential** because the axon is not conducting an impulse.

The existence of this polarity (charge difference) can be correlated with a difference in ion distribution on either side of the axomembrane (cell membrane of the axon). As figure 10.6*a* shows, there is a higher concentration of sodium (Na^+) ions outside the axon and a higher concentration of potassium (K^+)

Figure 10.4   Scientist working at an oscilloscope, the electrical recording device that measures changes in voltage wherever an electrode is placed on or inserted in a neuron.

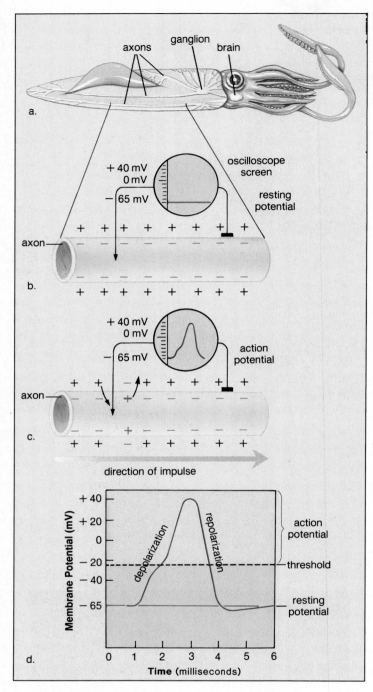

Figure 10.5 Nerve impulse. *a.* The squid axons shown produce rapid muscular contraction so that the squid can move quickly. *b.* These axons are so large (about 1 mm in diameter) that a microelectrode can be inserted inside. When the axon is not conducting a nerve impulse, the electrode registers and the oscilloscope records a resting potential of −65 mV. *c.* When the axon is conducting a nerve impulse, the *threshold* for an action potential is achieved, and there is a rapid change in potential from −65 mV to +40 mV (called depolarization) followed by a return to −65 mV (called repolarization). *d.* Enlargement of action potential of nerve impulse.

ions inside the axon. The unequal distribution of these ions is due to the action of the *sodium-potassium pump.* This is an active transport system in the cell membrane that pumps sodium ions out of and potassium ions into the axon. The work of the pump maintains the unequal distribution of sodium and potassium ions across the axomembrane.

The pump is always working because the membrane is somewhat permeable to these ions and they tend to diffuse toward their lesser concentration. Since the membrane is more permeable to potassium than to sodium, there are always more positive ions outside the axomembrane than inside; this accounts for the polarity recorded by the oscilloscope. There are also large, negatively charged proteins in the axoplasm, which are termed immobile in figure 10.6*a* because they are too large to cross the axomembrane.

Action Potential

If the axon is stimulated to conduct a nerve impulse by an electric shock, by a sudden difference in pH, or by a pinch, a trace appears on the oscilloscope screen. This pattern, caused by rapid polarity changes and called the **action potential,** has an upswing and a downswing.

Sodium Gates Open As the action potential swings up from −65 mV to +40 mV, sodium (Na$^+$) ions rapidly move across the axomembrane to the inside of the axon. The stimulation of the axon has caused the gates of the sodium channels to open temporarily, allowing sodium to flow into the axon. This sudden permeability of the axomembrane causes the oscilloscope to record a *depolarization:* the charge inside of the fiber changes from negative to positive as sodium ions enter (fig. 10.6*b*).

Potassium Gates Open As the action potential swings down from +40 mV to at least −65 mV, potassium (K$^+$) ions rapidly move from the inside to the outside of the axon. The axomembrane suddenly has become permeable to potassium because the potassium gates of the potassium channels have opened temporarily, allowing potassium ions to flow out of the axon. The oscilloscope records a *repolarization* as the charge inside the axon becomes negative again (fig. 10.6*c*).

Sodium-Potassium Pump A fiber can conduct a volley of nerve impulses because only a small number of ions are exchanged with each impulse. When the fiber rests, however, there is a refractory period during which the sodium-potassium pump continues to return sodium (Na$^+$) to the outside and potassium (K$^+$) ions to the inside of the axon (fig. 10.6*d*). During the refractory period, a neuron is unable to conduct a nerve impulse.

All neurons, whether sensory or motor, transmit the same type of nerve impulse—an electrochemical change that is propagated along the nerve fiber(s).

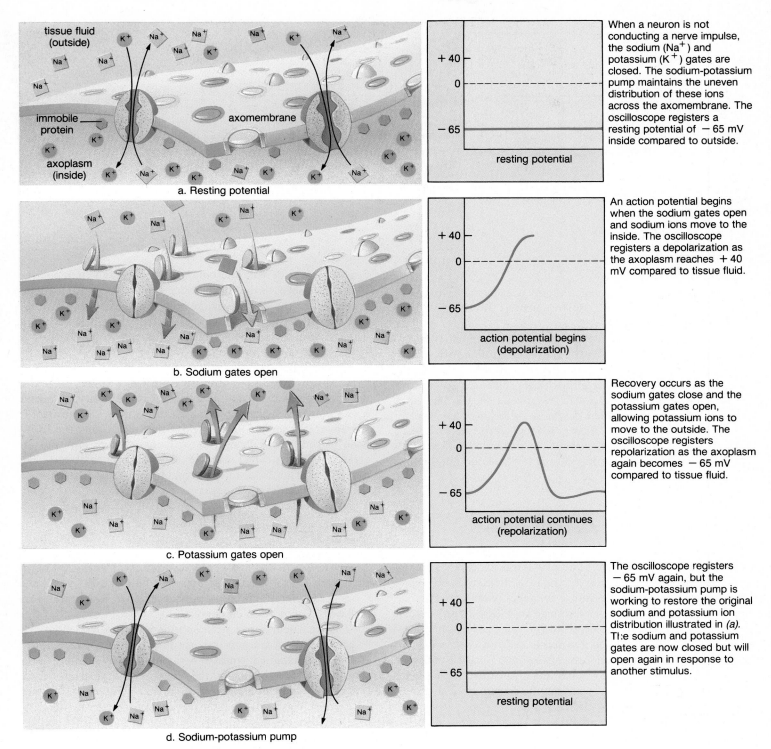

tissue fluid (outside)

immobile protein

axomembrane

axoplasm (inside)

a. Resting potential

When a neuron is not conducting a nerve impulse, the sodium (Na$^+$) and potassium (K$^+$) gates are closed. The sodium-potassium pump maintains the uneven distribution of these ions across the axomembrane. The oscilloscope registers a resting potential of −65 mV inside compared to outside.

resting potential

b. Sodium gates open

An action potential begins when the sodium gates open and sodium ions move to the inside. The oscilloscope registers a depolarization as the axoplasm reaches +40 mV compared to tissue fluid.

action potential begins (depolarization)

c. Potassium gates open

Recovery occurs as the sodium gates close and the potassium gates open, allowing potassium ions to move to the outside. The oscilloscope registers repolarization as the axoplasm again becomes −65 mV compared to tissue fluid.

action potential continues (repolarization)

d. Sodium-potassium pump

The oscilloscope registers −65 mV again, but the sodium-potassium pump is working to restore the original sodium and potassium ion distribution illustrated in (a). The sodium and potassium gates are now closed but will open again in response to another stimulus.

resting potential

Figure 10.6   Action and resting potential. The action potential is the result of an exchange of sodium (Na$^+$) and potassium (K$^+$) ions and is recorded as a change in polarity by an oscilloscope (as shown on the right). So few ions are exchanged for each action potential that it is possible for a nerve fiber to repeatedly conduct nerve impulses. Whenever the fiber rests, the sodium-potassium pump restores the original distribution of ions.

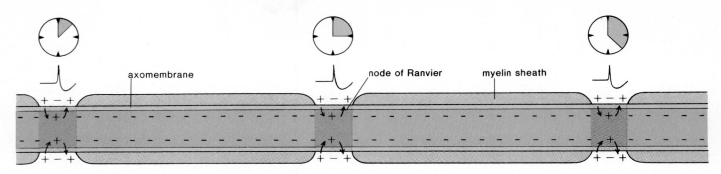

Figure 10.7   Longitudinal section of a vertebrate axon illustrating the manner by which the nerve impulse travels down a long nerve fiber. The speed of the impulse is due to the fact that it jumps from one node of Ranvier to the next.

Speed of Conduction

The oscilloscope records changes at only one location in a nerve fiber. Actually, however, the action potential travels along the length of a fiber (fig. 10.7).

In invertebrates, nerve fibers do not have a myelin sheath as they do in vertebrates. The speed of conduction in invertebrate nerve fibers can reach 20 m per second, but the speed of conduction in myelinated vertebrate fibers can reach 200 m per second. Notice in figure 10.2 that the myelin sheath has gaps between Schwann cells called the *nodes of Ranvier*. The speed of conduction in myelinated fibers is much faster because the action potential jumps from one node of Ranvier to the next (fig. 10.7). This is called saltatory (saltatory means jumping) conduction.

Transmission across a Synapse

The mechanism by which an action potential passes from one neuron to another is not the same as the mechanism by which an action potential is conducted along a neuron. Each axon branches into many fine terminal branches, each of which is tipped by a small swelling, or terminal knob (fig. 10.8*a*). Each knob lies very close to the dendrite (or cell body) of another neuron. This region is called a **synapse,** and the knob is called a **synaptic ending.** The membrane of the knob is called the **presynaptic membrane,** and the membrane of the next neuron just beyond the knob is called the **postsynaptic membrane.** The small gap between is the **synaptic cleft.**

Transmission of nerve impulses across a synaptic cleft is carried out by **neurotransmitter substances,** which are stored in synaptic vesicles (fig. 10.8*b*) before their release. When nerve impulses traveling along an axon reach a synaptic ending, the axomembrane becomes permeable to calcium (Ca^{++}) ions. These

ions then interact with microfilaments, causing them to pull the synaptic vesicles to the inner surface of the presynaptic membrane. When the vesicles merge with this membrane, a neurotransmitter substance is discharged into the synaptic cleft. The neurotransmitter molecules diffuse across the cleft to the postsynaptic membrane, where they bind with a receptor in a lock-and-key manner (fig. 10.8*c*).

Neurotransmitter substances can be excitatory or inhibitory. If the neurotransmitter substance is excitatory, the membrane potential of the postsynaptic membrane decreases, the sodium (Na^+) ion channels open at that locale, and the likelihood of the neuron firing (transmitting a nerve impulse) increases. If the neurotransmitter substance is inhibitory, the membrane potential of the postsynaptic membrane increases as the inside becomes more negative, and the likelihood of a nerve impulse decreases.

Neurotransmitter Substances

Acetylcholine (ACh) and **norepinephrine (NE)** are well-known excitatory neurotransmitters active in both the PNS and the CNS. Examples of inhibitory substances, so far discovered only in the CNS, are given on page 206.

Once a neurotransmitter substance has been released into a synaptic cleft, it has only a short time to act. In some synapses, the cleft contains enzymes that rapidly inactivate the neurotransmitter. For example, the enzyme **acetylcholinesterase (AChE),** or simply cholinesterase, breaks down ACh. In other synapses, the synaptic ending rapidly absorbs the neurotransmitter substance, possibly for repackaging in synaptic vesicles or for chemical breakdown. The enzyme monoamine oxidase breaks down NE after it is absorbed. The short existence of neurotransmitters in the synapse prevents continuous stimulation (or inhibition) of postsynaptic membranes.

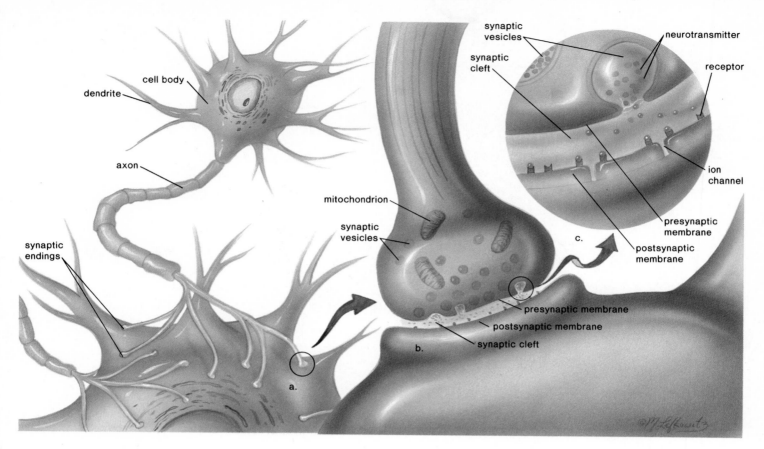

Figure 10.8 Diagrammatic representation of 3 different magnifications. *a.* Typically, a synapse is located wherever an axon is close to a dendrite or a cell body. Drawing based on a photomicrograph shows that there are several synaptic endings per axon because of terminal branching of the axon. *b.* Drawing based on low-power electron micrographs shows that a branch ends in a terminal knob having numerous synaptic vesicles, each filled with a neurotransmitter substance. This drawing makes it clear that a synapse contains a cleft (space) between the axon and the dendrite or the cell body. *c.* Drawing based on high-power electron micrographs shows that a synapse consists of a presynaptic membrane, the synaptic cleft, and the postsynaptic membrane. When a nerve impulse reaches the synaptic vesicles, the synaptic vesicles move toward and fuse with the presynaptic membrane and they discharge their contents. The neurotransmitter substance diffuses across the cleft and combines with a receptor. A nerve impulse may follow.

Transmission of nerve impulses across a synapse is dependent on a neurotransmitter substance that changes the permeability of the postsynaptic membrane.

Summation and Integration

A dendrite or a cell body is on the receiving end of many synapses. Whether or not a neuron fires depends on **summation,** the net effect of all the excitatory and the inhibitory neurotransmitters received. If enough sodium (Na^+) ion channels open, excitation is sufficient to raise the membrane potential above threshold level (fig. 10.5*d*), and the neuron fires. Otherwise, it does not fire.

The CNS *integrates* (sums up) the information it receives from all over the body. Summation in a neuron is integration at the cellular level. Integration in the brain allows us to make decisions about the body in general.

Because a neuron either fires or does not fire, it is said to obey an all-or-none law. A **nerve** does not obey the all-or-none law because it contains many fibers, any number of which can be carrying nerve impulses. Therefore, a nerve can have degrees of performance.

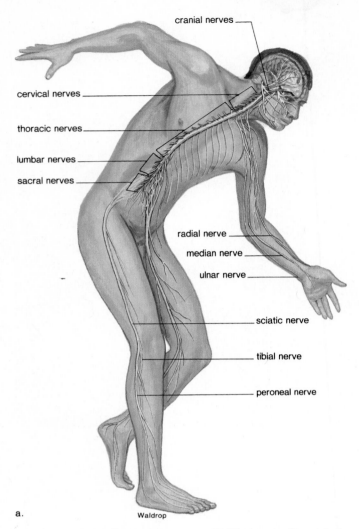

cranial nerves

cervical nerves

thoracic nerves

lumbar nerves

sacral nerves

radial nerve

median nerve

ulnar nerve

sciatic nerve

tibial nerve

peroneal nerve

a.

Waldrop

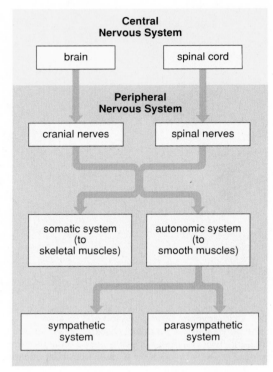

Central Nervous System

brain

spinal cord

Peripheral Nervous System

cranial nerves

spinal nerves

somatic system (to skeletal muscles)

autonomic system (to smooth muscles)

sympathetic system

parasympathetic system

b.

Figure 10.9   Peripheral nervous system (PNS) compared to central nervous system (CNS). *a.* The CNS lies in the center of the body, and the PNS lies to either side. *b.* The CNS contains the brain and the spinal cord, and the PNS contains the nerves. There are cranial and spinal nerves within the somatic and autonomic nervous systems. The nerves of the autonomic nervous system belong to the sympathetic system or the parasympathetic system.

Peripheral Nervous System

The **peripheral nervous system (PNS)** is made up of nerves that are a part of either the somatic system or the autonomic system (fig. 10.9). The autonomic system is further divided into the sympathetic and parasympathetic systems.

Nerves

Nerves are structures that contain only long dendrites and/or long axons. This is so because neuron cell bodies are found only in the brain, spinal cord, and ganglia. **Ganglia** are collections of cell bodies within the PNS.

There are three types of nerves (table 10.2). **Sensory nerves** contain only the long dendrites of sensory neurons; **motor nerves**

Table 10.2   Nerves

Type of nerve	Consists of	Function
sensory nerve	long dendrites of sensory neurons only	carries message from receptors to CNS
motor nerve	long axons of motor neurons only	carries message from CNS to effectors
mixed nerve	both long dendrites of sensory neurons and long axons of motor neurons	carries message in dendrite to CNS and away from CNS in axons

Note: Compare this table to table 10.1.

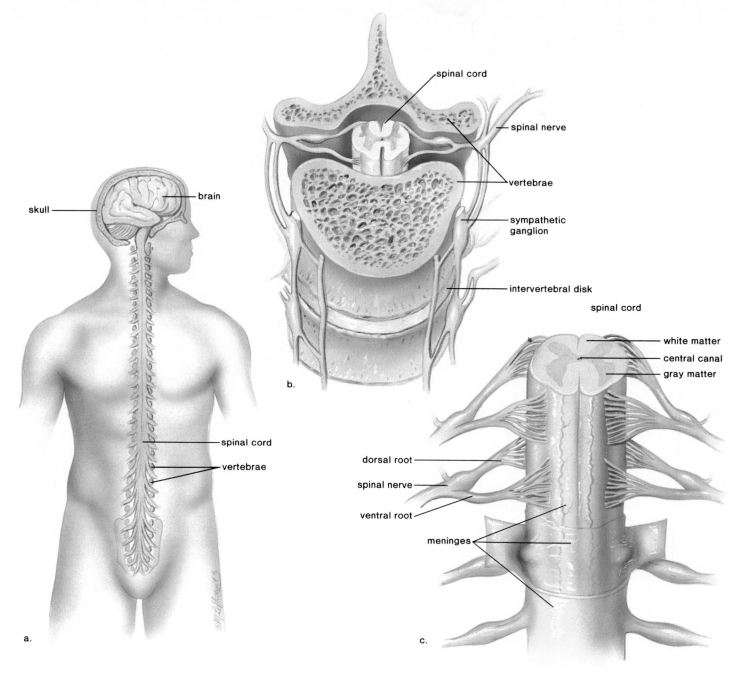

Figure 10.10 The anatomy of the spinal cord. *a*. The CNS consists of the brain and the spinal cord. The brain is protected by the skull, and the spinal cord is protected by the vertebrae. *b*. Cross section of the spine, showing spinal nerves. The human body has a total of 31 pairs of spinal nerves. *c*. This cross section of the spinal cord shows that a spinal nerve has a dorsal and a ventral root. Also, the cord is protected by 3 layers of tissue called the meninges. Spinal meningitis is an infection of these layers.

contain only the long axons of motor neurons; **mixed nerves,** however, contain both the long dendrites of sensory neurons and the long axons of motor neurons. Each nerve fiber within a nerve is surrounded by myelin (fig. 10.3), and therefore nerves have a white, shiny, glistening appearance.

Cranial Nerves

Humans have twelve pairs of **cranial nerves** attached to the brain (fig. 10.9). Some of these are sensory, some are motor, and others are mixed. Notice that although the brain is a part of the CNS, the cranial nerves are a part of the PNS. All cranial nerves, except the vagus, are concerned with the head, neck, and facial regions of the body, but the vagus nerve has many branches to serve the internal organs.

Spinal Nerves

Humans have 31 pairs of **spinal nerves.** Each spinal nerve emerges from the spinal cord (fig. 10.10) by two short branches, or roots, which lie within the vertebral column. The *dorsal root* can be identified by the presence of an enlargement called the dorsal root ganglion. This ganglion contains the cell bodies of the sensory neurons whose dendrites conduct impulses toward the cord. The ventral root of each spinal nerve contains the axons of motor neurons that conduct impulses away from the cord. These two roots join just before a spinal nerve leaves the vertebral column. Therefore, all spinal nerves are mixed nerves that contain many sensory dendrites and motor axons. Each spinal nerve serves the particular region of the body in which it is located.

In the PNS, cranial nerves take impulses to and/or from the brain, and spinal nerves take impulses to and from the spinal cord.

Human Issue

What is the best approach to the treatment of pain? Should people take pills for pain, or should they try to endure the pain? There are different thresholds for pain; what is unbearable pain to one is not unbearable to another. Even so, should people be cautioned against taking drugs for pain? Would it be better to carefully study the psychological as well as the physical reasons why a person has pain? Or would this simply make the person feel guilty for having pain?

Somatic Nervous System

The **somatic nervous system** includes all nerves that serve the musculoskeletal system and the exterior sense organs, including those in the skin. Exterior sense organs are receptors that receive environmental stimuli and then initiate nerve impulses.

Muscle fibers are effectors that bring about a reaction to the stimulus. Receptors are studied in chapter 12, and muscle effectors are studied in chapter 11.

Reflex Arc

Reflexes are automatic, involuntary responses to changes occurring inside or outside the body. In the somatic nervous system, outside stimuli often initiate a reflex action. Some reflexes, such as blinking the eye, involve the brain, while others, such as withdrawing the hand from a hot object, do not necessarily involve the brain. Figure 10.11 illustrates the path of the second type of reflex action. Whenever a person touches a very hot object, a *receptor* in the skin generates nerve impulses that move along the dendrite of a *sensory neuron* toward the cell body and the CNS. The cell body of a sensory neuron is located in the **dorsal root ganglion** just outside the cord. From the cell body, the impulses travel along the axon of the sensory neuron and enter the cord by the dorsal root of a spinal nerve. The impulses then pass to many interneurons, one of which connects with a motor neuron. The short dendrites and the cell body of the *motor neuron* lead to the axon, which leaves the cord by way of the ventral root of a spinal nerve. The nerve impulses travel along the axon to *muscle fibers* that then contract so that the hand is withdrawn from the hot object. (See table 10.3 for a listing of these events.)

Various other reactions usually accompany a reflex response; the person may look in the direction of the object, jump back, and utter appropriate exclamations. This whole series of responses is explained by the fact that the sensory neuron stimulates several interneurons, which take impulses to all parts of the CNS, including the cerebrum, which in turn, makes the person conscious of the stimulus and his or her reaction to it.

The reflex arc is the main functional unit of the nervous system. It allows us to react to internal and external stimuli.

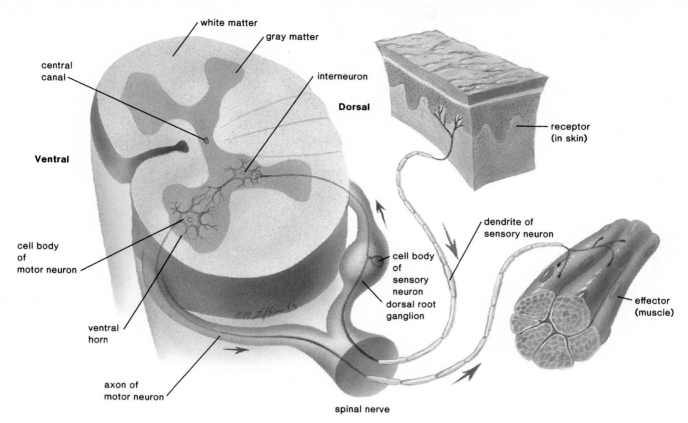

white matter

gray matter

central
canal

interneuron

Dorsal

receptor
(in skin)

Ventral

cell body
of
motor neuron

dendrite of
sensory neuron

cell body
of
sensory
neuron

dorsal root
ganglion

effector
(muscle)

ventral
horn

axon of
motor neuron

spinal nerve

Figure 10.11   Diagram of a reflex arc showing the detailed composition of a spinal nerve. When the receptors in the skin are stimulated, nerve impulses (see arrows) move along a sensory neuron to the spinal cord. (Note that the cell body of a sensory neuron is in a ganglion outside the cord.) The nerve impulses are picked up by an interneuron, which lies completely within the cord, and passed to the dendrites and cell body of a motor neuron that lies ventrally within the cord. The nerve impulses then move along the axon of the motor neuron to an effector, such as a muscle fiber that contracts. The brain receives information concerning sensory stimuli by way of other interneurons with long fibers in tracts that run up and down the cord within the white matter.

Table 10.3   Path of a Simple Reflex

1. Receptor (formulates message)*	Generates nerve impulses
2. Sensory neuron (takes message to CNS)	Impulses move along dendrite (spinal nerve)† and proceed to cell body (dorsal root ganglia) and then go from cell body to axon (spinal cord)
3. Interneuron (passes message to motor neuron)	Impulses picked up by dendrites and pass through cell body to axon (spinal cord)
4. Motor neuron (takes message away from CNS)	Impulses travel through short dendrites and cell body (spinal cord) to axon (spinal nerve)
5. Effector (receives message)	Receives nerve impulses and reacts; glands secrete and muscles contract

*Phrases within parentheses state overall function.
†Words within parentheses indicate location of structure.

Autonomic Nervous System

The **autonomic nervous system** (fig 10.12), a part of the PNS, is made up of motor neurons that control the internal organs automatically and usually without need for conscious intervention. There are two divisions of the autonomic nervous system: the sympathetic and parasympathetic systems. Both of these (1) function automatically and usually subconsciously in an involuntary manner; (2) innervate all internal organs; and (3) utilize two motor neurons and one ganglion for each impulse. The first of these two neurons has a cell body within the CNS and a **preganglionic axon.** The second neuron has a cell body within the ganglion and a **postganglionic axon.**

The autonomic nervous system controls the functioning of internal organs without need of conscious control.

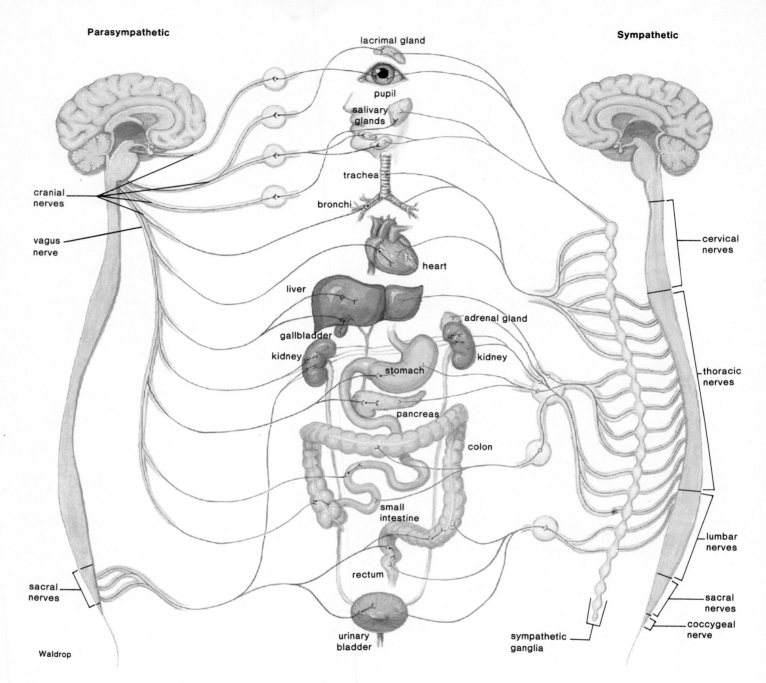

Parasympathetic

Sympathetic

lacrimal gland

pupil

salivary glands

trachea

bronchi

cranial nerves

vagus nerve

cervical nerves

heart

liver

adrenal gland

gallbladder

kidney

kidney

stomach

thoracic nerves

pancreas

colon

small intestine

lumbar nerves

rectum

sacral nerves

sacral nerves

coccygeal nerve

urinary bladder

sympathetic ganglia

Waldrop

Figure 10.12   Structure and function of the autonomic nervous system. The sympathetic fibers arise from the thoracic and lumbar portion of the spinal cord; the parasympathetic fibers arise from the brain and the sacral portion of the cord. Each system innervates the same organs but has contrary effects. For example, the sympathetic system speeds up the beat of the heart and the parasympathetic system slows down the beat of the heart.

Sympathetic System

The preganglionic fibers of the **sympathetic nervous system** (fig 10.12) arise from the middle or *thoracic-lumbar portion* of the spinal cord and almost immediately terminate in ganglia that lie near the cord. Therefore, in this system, the preganglionic fiber is short, but the postganglionic fiber that makes contact with an organ is long.

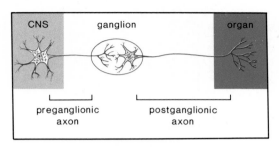

The sympathetic nervous system is especially important during emergency situations and is associated with "fight or flight." For example, it inhibits the digestive tract, but it dilates the pupil, accelerates the heartbeat, and increases the breathing rate. It is not surprising, then, that the neurotransmitter released by the postganglionic axon is norepinephrine (NE), a chemical close in structure to adrenalin, a well-known heart stimulant.

The sympathetic nervous system brings about those responses we associate with "fight or flight."

Parasympathetic System

A few cranial nerves, including the vagus nerve, together with fibers that arise from the bottom portion of the cord form the **parasympathetic nervous system** (fig. 10.12). Therefore, this system often is referred to as the *craniosacral portion* of the autonomic nervous system. In the parasympathetic nervous system, the preganglionic fiber is long and the postganglionic fiber is short because the ganglia lie near or within the organ.

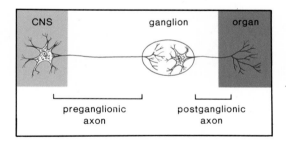

The parasympathetic system, sometimes called the "housekeeper system," promotes all the internal responses we associate with a relaxed state; for example, it causes the pupil of the eye to contract, it promotes digestion of food, and it retards the heartbeat. The neurotransmitter utilized by the parasympathetic system is acetylcholine (ACh).

Table 10.4 Sympathetic System versus Parasympathetic System

Sympathetic system	Parasympathetic system
Fight or flight	Normal activity
Norepinephrine is neurotransmitter	Acetylcholine is neurotransmitter
Postganglionic fiber is longer than preganglionic fiber	Preganglionic fiber is longer than postganglionic fiber
Preganglionic fiber arises from middle portion of cord	Preganglionic fiber arises from brain and lower portion of cord

Figure 10.12 contrasts the sympathetic and parasympathetic systems, and table 10.4 lists all the differences we have noted between these two systems.

The parasympathetic nervous system brings about the responses we associate with normally restful activities.

Central Nervous System

The **central nervous system (CNS)** consists of the spinal cord and the brain. As figures 10.10 and 10.13 illustrate, the CNS is protected by bone: the brain is enclosed within the skull and the spinal cord is surrounded by vertebrae. Also, both the brain and the spinal cord are wrapped in three protective membranes known as **meninges;** meningitis is an infection of these coverings. The spaces between the meninges are filled with **cerebrospinal fluid,** which cushions and protects the CNS. A small amount of this fluid sometimes is withdrawn for laboratory testing when a spinal tap (i.e., lumbar puncture) is done. Cerebrospinal fluid also is contained within the **central canal** of the spinal cord and the **ventricles** of the brain. The latter are interconnecting spaces that produce and serve as a reservoir for cerebrospinal fluid.

Spinal Cord

The spinal cord lies along the middorsal line of the body (fig. 10.10). It has two main functions: (1) it is the center for many reflex actions, and (2) it provides a means of communication between the brain and the spinal nerves that leave the cord.

The path of a spinal reflex passes through the gray matter of the cord (fig. 10.11). It is gray because it contains short fibers and cell bodies that are unmyelinated. In cross section, the gray matter looks like a butterfly or the letter H. The axons of sensory neurons are found in the dorsal regions (horns) of the gray matter, and the dendrites and cell bodies of motor axons are found in the ventral regions (horns) of the gray matter. Short interneurons connect sensory to motor neurons on the same sides and opposite sides of the cord.

The white matter of the cord is found in between the regions of the gray matter (fig. 10.10c). The white matter of the

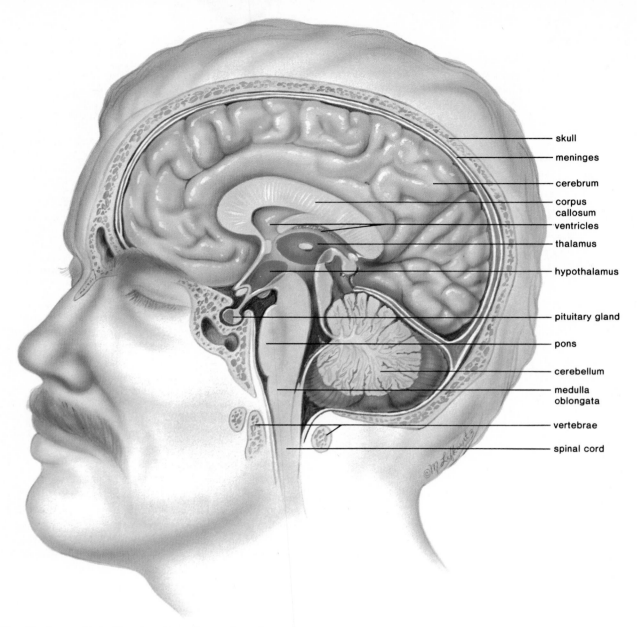

skull
meninges
cerebrum
corpus callosum
ventricles
thalamus
hypothalamus
pituitary gland
pons
cerebellum
medulla oblongata
vertebrae
spinal cord

Figure 10.13 The human brain. Note how large the cerebrum is compared to the rest of the brain.

cord is white because it contains myelinated long fibers of interneurons that run together in bundles called tracts. These tracts connect the cord to the brain. Dorsally, there are primarily ascending tracts that take information to the brain, and ventrally, there are primarily descending tracts that carry information down from the brain. Because the tracts at one point cross over, the left side of the brain controls the right side of the body and the right side of the brain controls the left side of the body.

The CNS lies in the midline of the body and consists of the brain and the spinal cord, where sensory information is received and motor control is initiated.

Brain

The largest and most prominent portion of the human brain (fig. 10.13) is the cerebrum. Consciousness resides only in the cerebrum; the rest of the brain functions below the level of consciousness. In addition to the portions mentioned, remember that the unconscious brain contains many tracts that relay messages to and from the spinal cord.

Unconscious Brain

The **medulla oblongata** lies closest to the spinal cord and contains centers for heartbeat, respiration, and vasoconstriction (blood pressure). It also contains reflex centers for vomiting, coughing, sneezing, hiccuping, and swallowing.

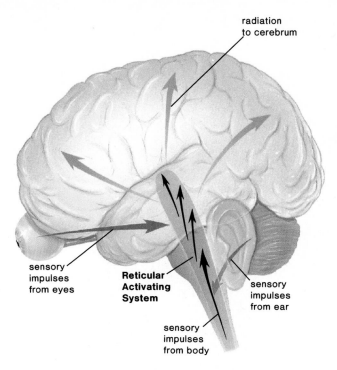

radiation to cerebrum

sensory impulses from eyes

Reticular Activating System

sensory impulses from body

sensory impulses from ear

Figure 10.14   The reticular activating system. Sensory information received by the reticular activating system is sorted out by the thalamus before nerve impulses are sent to the cerebrum.

Figure 10.15   Skateboarding. This sport requires muscular coordination and balance, movements that are controlled by the cerebellum.

The **hypothalamus** is concerned with homeostasis, or the constancy of the internal environment, and contains centers for hunger, sleep, thirst, body temperature, water balance, and blood pressure. The hypothalamus controls the pituitary gland and thereby serves as a link between the nervous and endocrine systems.

The medulla oblongata and the hypothalamus both are concerned with control of the internal organs.

The *midbrain* and the *pons* contain tracts that connect the cerebrum with other parts of the brain. In addition, the pons functions with the medulla to regulate breathing rate, and the midbrain has reflex centers concerned with head movements in response to visual and auditory stimuli.

The **thalamus** is a central relay station for sensory impulses traveling upward from other parts of the cord and the brain to the cerebrum. It receives all sensory impulses (except those associated with the sense of smell) and channels them to appropriate regions of the cerebrum. In other words, it is the last portion of the brain for sensory input before the cerebrum.

The thalamus has connections to various parts of the brain by way of nerve fibers that radiate from the upper part of the *reticular activating system (RAS)* (fig. 10.14). The RAS, which extends from the medulla to the thalamus, sorts out stimuli received from sense organs including the eyes and ears, passing on only those that require immediate attention. The thalamus sometimes is called the gatekeeper to the cerebrum because it alerts the cerebrum to only certain sensory input. We are unaware of many of the sensory impulses received by the CNS.

The thalamus receives sensory impulses from other parts of the CNS and channels only certain of these to the cerebrum.

The **cerebellum,** a bilobed butterfly-shaped structure is the second largest portion of the brain. It functions in muscle coordination (fig. 10.15), integrating impulses received from higher centers to ensure that all the skeletal muscles work together to produce smooth and graceful motions. The cerebellum also is responsible for maintaining normal muscle tone and transmitting impulses that maintain posture. It receives information from the inner ear indicating the position of the body and sends impulses to those muscles whose contraction maintains or restores balance.

The cerebellum controls balance and complex muscular movements.

Conscious Brain

The **cerebrum**, the only area of the brain responsible for consciousness, is the largest portion of the brain in humans. The outer layer of the cerebrum, called the *cerebral cortex,* is gray in color and contains cell bodies and short fibers. The cerebrum is divided into halves known as the right and left **cerebral hemispheres.** Each hemisphere contains four types of superficial lobes: **frontal, parietal, temporal,** and **occipital** (fig. 10.16).

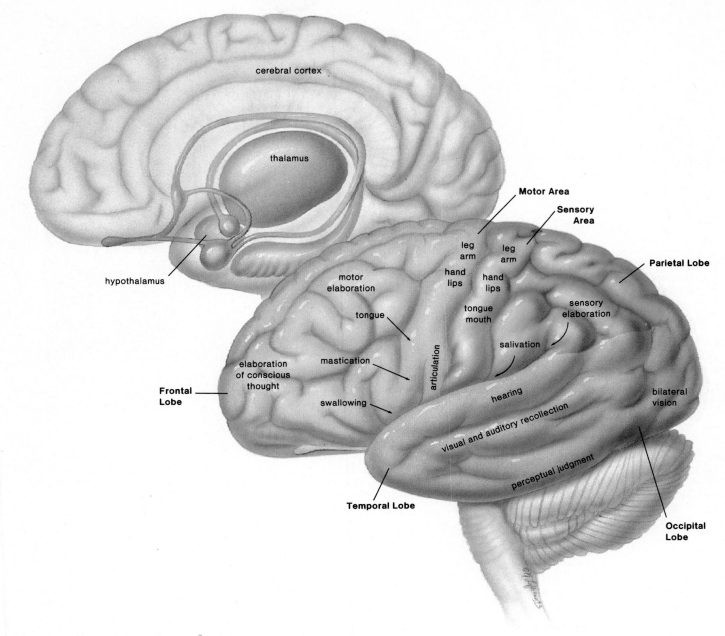

cerebral cortex

thalamus

Motor Area

Sensory Area

leg
arm
hand
lips
tongue
mouth

leg
arm
hand
lips

hypothalamus

motor
elaboration

Parietal Lobe

sensory
elaboration

tongue

salivation

elaboration
of conscious
thought

mastication

articulation

hearing

bilateral
vision

Frontal
Lobe

swallowing

visual and auditory recollection

Temporal Lobe

perceptual judgment

Occipital
Lobe

Figure 10.16   Two views of the cerebrum. (*Rear*) The limbic system, which includes portions of the cerebrum, the thalamus, and the hypothalamus, is concerned mainly with emotion and memory. (*Front*) The convoluted cortex of the cerebrum is divided into 4 lobes; frontal, temporal, parietal, and occipital. It is possible to map the cerebrum since each area has a particular function.

Certain areas of the cerebral cortex have been "mapped" in great detail (table 10.5). For example, we know which portions of the frontal lobe control various parts of the body and which portions of the parietal lobe receive sensory information from these same parts. Each of the four lobes of the cerebral cortex contains association areas that receive information from the other lobes and integrate it into higher, more complex levels of consciousness. These areas are concerned with intellect, artistic and creative abilities, learning, and memory.

Consciousness is the province of the cerebrum, the most highly developed portion of the human brain. The cerebrum is responsible for higher mental processes, including the interpretation of sensory input and the initiation of voluntary muscular movements.

There has been a great deal of testing to determine whether the right and left halves of the cerebrum serve different functions. These studies tend to suggest that the left half of the brain

Table 10.5 Functions of the Cerebral Lobes

Lobe	Functions
frontal lobes	Motor areas control movements of voluntary skeletal muscles. Association areas carry on higher intellectual processes, such as those required for concentration, planning, complex problem solving, and judging the consequences of behavior.
parietal lobes	Sensory areas are responsible for the sensations of temperature, touch, pressure, and pain from the skin. Association areas function in the understanding of speech and in using words to express thoughts and feelings.
temporal lobes	Sensory areas are responsible for hearing and smelling. Association areas are used in the interpretation of sensory experiences and in the memory of visual scenes, music, and other complex sensory patterns.
occipital lobes	Sensory areas are responsible for vision. Association areas function in combining visual images with other sensory experiences.

From John W. Hole, Jr., *Human Anatomy and Physiology*, 5th ed. Copyright © 1990 Wm. C. Brown Publishers, Dubuque, Iowa. All Rights Reserved. Reprinted by permission.

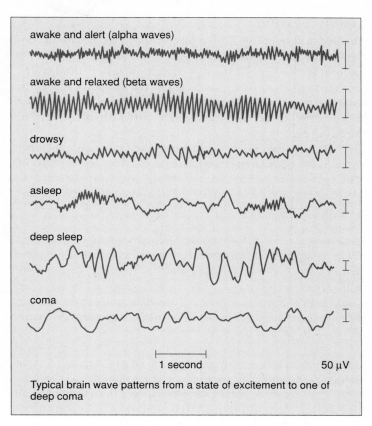

awake and alert (alpha waves)

awake and relaxed (beta waves)

drowsy

asleep

deep sleep

coma

1 second   50 µV

Typical brain wave patterns from a state of excitement to one of deep coma

Figure 10.17 Electroencephalograms (EEGs), brain wave patterns from a state of excitement to coma. The alpha waves, which appear when the subject is awake and alert, are the most common. Second most common are the beta waves, which are recorded when the subject is awake but relaxed. Note the progressive change in the pattern from a state of drowsiness to comatose state.

is the verbal (word) half and the right half of the brain is the visual (spatial relation) and artistic half. However, other results indicate that such a strict dichotomy does not always exist between the two halves. In any case, the two cerebral hemispheres normally share information because they are connected by a horizontal tract called the **corpus callosum** (fig. 10.13).

Severing the corpus callosum can control severe epileptic seizures but results in a person with two brains, each with its own memories and thoughts. Today, use of the laser permits more precise treatment without this side effect. *Epilepsy* is caused by a disturbance of the normal communication between the RAS and the cortex. In a grand mal or epileptic seizure, the cerebrum is extremely excited. Due to a reverberation of signals within the RAS and the cerebrum, the individual loses consciousness, even while convulsions are occurring. Finally, the neurons fatigue and the signals cease. Following an attack, the brain is so fatigued the person must sleep for a while.

EEG The electrical activity of the brain can be recorded in the form of an **electroencephalogram (EEG).** Electrodes are taped to different parts of the scalp, and an instrument called the electroencephalograph records the so-called brain waves (fig. 10.17).

When the subject is awake, two types of waves are usual: *alpha waves,* with a frequency of about 6–13 per second and a potential of about 45 mV, predominate when the person is alert, and *beta waves,* with higher frequencies but lower voltage, appear when the person is relaxed.

During an 8-hour sleep, there are usually five times when the brain waves become slower and larger than alpha waves. During each of these times, there are irregular flurries as the eyes move back and forth rapidly. When subjects are awakened

during the latter, called **REM** (rapid eye movement) **sleep,** they always report that they were dreaming. The significance of REM sleep still is being debated, but some studies indicate that REM sleep is needed for memory to occur.

The EEG is a diagnostic tool; for example, an irregular pattern can signify epilepsy or a brain tumor. A flat EEG signifies lack of electrical activity of the brain, or brain death, and thus it can be used to determine the precise time of death.

Human Issue

Modern medical technology is able to prolong the life of terminally ill patients, and this raises thorny ethical issues. For example, doctors in California removed the feeding tubes from a hopelessly brain-damaged patient who subsequently died. Later the doctors were charged with murder, even though they had acted at the family's request. Presently, a number of states and the District of Columbia recognize "living wills" that request hospitals not take extraordinary measures to prolong the life of the person who made up the will. Do you think this is appropriate, or are there better solutions to the "right-to-die" problem? If you do support the use of such wills, should each of us be required to decide whether or not we want our lives to be prolonged?

Limbic System

The **limbic system** (fig. 10.16) involves portions of both the unconscious brain and the conscious brain. It lies just beneath the cerebral cortex (fig. 10.16) and contains neural pathways that connect portions of the frontal lobes, the temporal lobes, the thalamus, and the hypothalamus. Several masses of gray matter that lie deep within each hemisphere of the cerebrum, termed the *basal nuclei,* are also a part of the limbic system.

Stimulation of different areas of the limbic system causes the subject to experience rage, pain, pleasure, or sorrow. By causing pleasant or unpleasant feelings about experiences, the limbic system apparently guides the individual into behavior that is likely to increase the chance of survival.

Learning and Memory   The limbic system also is involved in the processes of learning and memory. Learning requires memory, but just what permits memory to occur is not known definitely. Experimentation with invertebrates, such as slugs and snails, indicates that learning is accompanied by an increase in the number of synapses in the brain while forgetting involves a decrease in the number of synapses.

Some investigators study short-term and long-term memory in monkeys. An example of short-term memory in humans is the ability to recall a telephone number long enough to dial it; an example of long-term memory is the ability to recall the events of the day. It is believed that for both types of memory, impulses move within the limbic circuit, but eventually long-term memories are stored in sensory areas. The involvement of the limbic system certainly explains why emotionally charged events result in our most vivid memories and why sensory stimulation can awaken a complex memory.

The limbic system is involved particularly in emotions and in memory and learning.

Neurotransmitters in the Brain

The neurotransmitters ACh and NE are found in the brain as well as serotonin and dopamine, which are also excitatory transmitters. Serotonin and dopamine are associated with behavioral states such as mood, sleep, attention, learning, and memory. Gamma-aminobutyrate (GABA) and glycine are two inhibitory transmitters in the brain.

Endorphins are called the body's own natural opioids because morphine and heroin utilize their receptors in the CNS. When endorphins are present, neurons do not release substance P, a neurotransmitter that brings about the sensation of pain. Exercise has been associated with the presence of endorphins, and this may account for the so-called "runner's high."

Neurotransmitter Disorders   *Parkinson disease,* characterized by a wide-eyed, unblinking expression, an involuntary tremor of the fingers and the thumbs, muscular rigidity, and a shuffling gait, is due to dopamine deficiencies. *Huntington disease* (p. 366) is believed to be due to a malfunction of GABA and now it seems that *Alzheimer disease,* a severe form of sen-

ility, may be due to a deterioration of cells in the basal nuclei that use ACh as a transmitter. It's hypothesized that a gene located on chromosome number 21, which normally is active only during development and which directs the production of a protein associated with neuron death, inexplicably has been turned on in these patients.

Some neurological illnesses are associated with the deficiency of a particular neurotransmitter in the brain.

Drug Abuse

A wide variety of chemicals can alter the functioning of the body, but our discussion will center on four drugs most commonly abused because they alter the mood and/or emotional state: alcohol, marijuana, cocaine, and heroin. *Drug abuse* is evident when a person takes a drug at a dose level and under circumstances that increase the potential for a harmful effect.

Drug abusers are apt to display a *physical dependence* on the drug (formerly called an addiction to the drug). Dependence has developed when the person (1) spends much time thinking about the drug or arranging to get it; (2) often takes more of the drug than was intended; (3) is *tolerant* to the drug—that is, must increase the amount of the drug to get the same effect; (4) has *withdrawal symptoms* when he or she stops taking the drug; and (5) has an ongoing desire to cut down on use.

Drug Action

Drugs that affect the nervous system have two general effects: (1) they affect the RAS (p. 203) and the limbic system, and (2) they either promote or decrease the action of a particular neurotransmitter. There are a number of different ways drugs can influence the transmission of neurotransmitters, some of which are described in figure 10.18. It is clear, as outlined in table 10.6, that stimulants can either enhance the action of an excitatory neurotransmitter or block the action of an inhibitory neurotransmitter. Depressants can either enhance the action of an inhibitory neurotransmitter or block the action of an excitatory neurotransmitter.

Drug abuse often results in physical dependence on the drug because the drug interferes with normal neurotransmitter function in the brain.

Table 10.6   Drug Action

Drug action	Neurotransmitter	Result
blocks	excitatory	depression
enhances	excitatory	stimulation
blocks	inhibitory	stimulation
enhances	inhibitory	depression

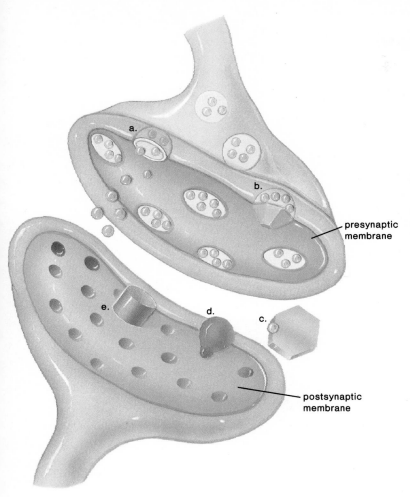

Figure 10.18   Drug action at synapses. *a.* Drug stimulates release of neurotransmitter. *b.* Drug blocks release of neurotransmitter. *c.* Drug combines with neurotransmitter, preventing its breakdown or reuptake. *d.* Drug mimics neurotransmitter. *e.* Drug blocks receptor so that neurotransmitter cannot be received.

presynaptic membrane

postsynaptic membrane

Alcohol

The type of alcohol consumed is ethanol, the production of which is discussed on page 44. While it is possible to drink alcohol in moderation, the drug often is abused. Alcohol use becomes "abuse," or an illness, when alcohol ingestion impairs an individual's social relationships, health, job efficiency, or ability to avoid legal difficulties (fig. 10.19). Table 10.7 lists some of the questions that are used to identify the alcohol-dependent person.

Alcohol effects on the brain are biphasic: after consuming several drinks, blood alcohol concentration rises rapidly and the drinker reports feeling "high" and happy (euphoric). After 90 minutes and lasting until some 330–400 minutes after consumption, the drinker feels depressed and unhappy (dysphoric) (fig. 10.20). On the other hand, if the drinker continues to drink in order to maintain a high blood level of the alcohol, he or she will experience everincreasing loss of control. Coma and death are even possible if a substantial amount of alcohol (1¼ pints of whiskey) is consumed within an hour.

Figure 10.19   Drug abuse involvement in accidents. Those who abuse drugs, including alcohol, are more likely to be involved in automobile accidents. Unfortunately, others often suffer the consequences.

Mode of Action and Associated Illnesses

Recent research indicates that alcohol acts on the GABA (an inhibitory neurotransmitter) receptor and potentiates GABA's ability to increase chloride (Cl^-) ion uptake. Most likely, it does this by disordering membrane lipids.

Cirrhosis of the Liver   The stomach and the liver contain the enzyme alcohol dehydrogenase, which begins the breakdown of alcohol to acetic acid. A new study reports that women have less of this enzyme in their stomach, and this may explain why women show a greater sensitivity to alcohol, including a greater chance of liver damage. Acetic acid becomes active acetate and, therefore, it can be used in the liver to produce energy, but the calories provided are termed "empty" because they contribute to energy needs and weight gain without supplying any other nutritional requirements. Worse still, the molecules (glucose and fatty acids) that the liver ordinarily uses as an energy source are converted to fats, and the liver cells become engorged with fat droplets. After a few years of being overtaxed, the liver cells begin to die, causing an inflammatory condition known as alcoholic hepatitis. Finally, scar tissue appears in the liver, and it no longer is able to perform its vital functions. This condition is called cirrhosis of the liver, a frequent cause of death among drinkers. Brain impairment and generalized deterioration of other vital organs also are seen in heavy drinkers.

Table 10.7 Some Questions to Identify the Alcohol-Dependent Person

1. Do you occasionally drink heavily after a disappointment, a quarrel, or when the boss gives you a bad time?
2. When you are having trouble or feel under pressure, do you always drink more heavily than usual?
3. Have you noticed that you are able to handle more liquor than you did when you were first drinking?
4. Did you ever wake up the "morning after" and discover that you could not remember part of the evening before, even though your friends tell you that you did not "pass out"?
5. When drinking with other people, do you try to have a few extra drinks when others will not know it?
6. Are there certain occasions when you feel uncomfortable if alcohol is not available?
7. Have you recently noticed that when you begin drinking you are in more of a hurry to get the first drink than you used to be?
8. Do you sometimes feel a little guilty about your drinking?
9. Are your secretly irritated when your family or friends discuss your drinking?
10. Have your recently noticed an increase in the frequency of your memory "blackouts"?
11. When you are sober, do you often regret things you have done or said while drinking?
12. Have you often failed to keep the promises you have made to yourself about controlling or cutting down on your drinking?
13. Do more people seem to be treating you unfairly without good reason?
14. Do you eat very little or irregularly when you are drinking?
15. Do you get terribly frightened after you have been drinking heavily?

Source: National Council on Alcoholism.

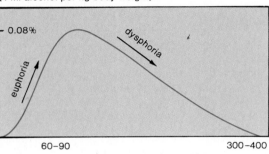

Typical blood alcohol curve for a normal drinker (1 ml alcohol per kg body weight)

Blood Alcohol Concentration (ml per kg)

0.08%

euphoria

dysphoria

60–90   300–400

Elapsed Time (minutes)

Figure 10.20   Typical blood alcohol curve for a normal drinker after intake of 1 ml of alcohol per kg of body weight. As blood alcohol concentration increases, the user often feels euphoric (happy), but as blood alcohol concentration declines, the user feels dysphoric (unhappy). In most states, a person is considered to be legally drunk when the blood alcohol content is 0.1%. This usually requires imbibing 3 mixed drinks within one and one-half hours.

It should be stressed that the early signs of deterioration can be reversed if the habit of drinking to excess is given up.

Alcohol is the most abused drug in the United States. Its abuse often results in well-recognized illnesses and early death.

Marijuana

The dried flowering tops, leaves (fig. 10.21), and stems of the Indian hemp plant *Cannabis sativa* contain and are covered by a resin that is rich in THC (tetrahydrocannabinol). The names *cannabis* and *marijuana* can apply to the plant or to THC.

The effects of marijuana differ depending upon the strength and the amount consumed, the expertise of the user, and the setting in which it is taken. Usually, the user reports

Figure 10.21   *Cannabis sativa.* This plant, which contains marijuana, is often smoked in the same manner as tobacco.

experiencing a mild euphoria along with alterations in vision and judgment that result in distortions of space and time. The inability to concentrate and to speak coherently and motor incoordination also can be involved.

Intermittent use of low-potency marijuana generally is not associated with obvious symptoms of toxicity, but heavy use can produce chronic intoxication. Intoxication is recognized by the presence of hallucinations, anxiety, depression, rapid flow of ideas, body image distortions, paranoid reactions, and similar psychotic symptoms. The terms *cannabis psychosis* and *cannabis delirium* refer to such reactions.

Mode of Action

Marijuana is classified as an hallucinogen. It is possible that, like LSD (lysergic acid diethylamide), it has an effect on the action of serotonin, an excitatory neurotransmitter.

The use of marijuana does not seem to produce physical dependence, but a psychological dependence on the euphoric and sedative effects can develop. Craving or difficulty in stopping use also can occur as a part of regular heavy use.

Marijuana has been called a *gateway drug* because adolescents who have used marijuana also tend to try other drugs. For example, in a study of 100 cocaine abusers, 60% had smoked marijuana for more than 10 years.

Associated Illnesses

Usually marijuana is smoked in a cigarette form called a joint. Since this allows toxic substances, including carcinogens, to enter the lungs, chronic respiratory disease and lung cancer are considered dangers of long-term, heavy use. Some researchers claim that marijuana use leads to long-term brain impairment. Others report that males and females suffer reproductive dysfunctions. *Fetal cannabis syndrome,* which resembles fetal alcohol syndrome (p. 322), has been reported.

Some psychologists are very concerned about the use of marijuana among adolescents because it can be used as a means to avoid coming to grips with the many personal problems that often develop during this maturational phase.

Although marijuana does not produce physical dependence, it does produce psychological dependence.

Cocaine

Cocaine is an alkaloid derived from the shrub *Erythroxylum cocoa. Cocaine* is sold in powder form and as *crack,* a more potent extract (fig. 10.22). Users often use the word *rush* to describe the feeling of euphoria that follows intake of the drug. Snorting (inhaling) produces this effect in a few minutes; injection, within 30 seconds; and smoking, in less than 10 seconds. Persons dependent upon the drug are, therefore, most likely to smoke cocaine. The rush only lasts a few seconds and then is

a.

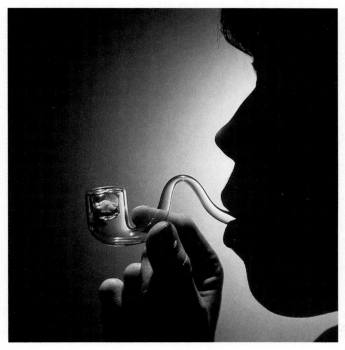

b.

Figure 10.22 Cocaine use. *a.* Crack, the ready-to-smoke form of cocaine that is more potent and more deadly than the powder. *b.* Users often smoke crack in a glass water pipe. The high produced consists of a ''rush'' lasting a few seconds, followed by a few minutes of euphoria. Continuous use makes the user extremely dependent on the drug.

replaced by a state of arousal that lasts from 5 to 30 minutes. Then the user begins to feel restless, irritable, and depressed. To overcome these symptoms, the user is apt to take more of the drug, repeating the cycle over and over again until there is no more drug left. A binge of this sort can go on for days, after which the individual suffers a *crash*. During the binge period, the user is hyperactive and has little desire for food or sleep, but has an increased sex drive. During the crash period, the user is fatigued, depressed, irritable, has memory and concentration problems, and displays no interest in sex. Indeed, men are often impotent. Other drugs, such as marijuana, alcohol, or heroin, often are taken to ease the symptoms of the crash.

Mode of Action

Cocaine affects the concentration of dopamine, an excitatory neurotransmitter, in brain synapses. After release, dopamine ordinarily is withdrawn into the presynaptic cell for recycling and reuse. Cocaine prevents the reuptake of dopamine, and this causes an excess of dopamine in the synaptic cleft so that the user experiences the sensation of a rush. The adrenalin-like effects of dopamine account for the state of arousal that lasts for some minutes after the rush experience.

With continued cocaine use, the body begins to make less dopamine as a compensation for a seemingly excess supply. The user, therefore, now experiences *tolerance* (always needing more of the drug for the same effect), *withdrawal* (symptoms described previously when a drug is not taken), and an intense *craving* for cocaine. These are indications that the person is highly dependent upon the drug or, in other words, that cocaine is extremely addictive.

Human Issue

What is the solution to the drug abuse problem? Would education, legalization, or substitution alone or in any combination be especially helpful? The drug buprenorphine is an alternative to methadone, which is given to reduce the craving for heroin and to block heroin's high if the addict tries to shoot up. Unlike methadone, buprenorphine is relatively nonaddictive and carries almost no risk. Also, it seems to suppress the urge to take cocaine which is abused by many heroin addicts. Do you think it would help to make a drug like buprenorphine freely available to abusers of heroin?

Associated Illnesses

Overdosing on cocaine is a real possibility. The number of deaths from cocaine and the number of emergency room admissions for drug reactions involving cocaine have increased greatly. High doses can cause seizures and cardiac and respiratory arrest.

Individuals who snort the drug can suffer damage to the nasal tissues and even perforation of the septum between the nostrils. Whether long-term cocaine abuse causes brain damage is not yet known, but this possibility is under investigation. It is known that babies born to addicts suffer withdrawal symptoms and may suffer neurological and developmental problems.

Heroin

Heroin is derived from morphine, an alkaloid of *opium*. Heroin usually is injected. After intravenous injection, the onset of action is noticeable within one minute and reaches its peak in 3 to 6 minutes. There is a feeling of euphoria along with relief of pain. Side effects can include nausea, vomiting, dysphoria, and respiratory and circulatory depression leading to death.

Mode of Action

Heroin binds to receptors meant for the body's own opioids, the endorphins. As mentioned previously, the opioids are believed to alleviate pain by preventing the release of a neurotransmitter termed substance P from certain sensory neurons in the region of the spinal cord. When substance P is released, pain is felt, and when substance P is not released, pain is not felt. Evidence also indicates that there are opioid receptors in neurons that travel from the spinal cord to the limbic system and that stimulation of these can cause a feeling of pleasure. This explains why opium and heroin not only kill pain but also produce a feeling of tranquility.

Individuals who inject heroin become physically dependent on the drug. With time, the body's production of endorphins decreases. Now *tolerance* develops so that the user needs to take more of the drug just to prevent *withdrawal* symptoms. The euphoria originally experienced upon injection no longer is felt.

Heroin withdrawal symptoms include perspiration, dilation of pupils, tremors, restlessness, abdominal cramps, gooseflesh, defecation, vomiting, and increases in systolic pressure and respiratory rate. Those who are excessively dependent may experience convulsions, respiratory failure, and death. Infants born to women who are physically dependent also experience these withdrawal symptoms.

Cocaine and heroin produce a very strong physical dependence. An overdose of these drugs can cause death.

Designer Drugs

Designer drugs are analogues; that is, they are chemical compounds of controlled substances slightly altered in molecular structure. One such drug is MPPP (1-methyl-4-phenylprionoxypiperidine), an analogue of the narcotic fentanyl. Even small doses of the drug are very toxic; MPPP already has caused many deaths on the West Coast.

Summary

The cell bodies of nerve cells are found in the CNS and the ganglia. Axons and dendrites make up nerves. The nerve impulse is a change in permeability of the axomembrane so that sodium (Na^+) ions move to the inside of a neuron and potassium (K^+) ions move to the outside. The nerve impulse is transmitted across the synapse by neurotransmitter substances.

During a spinal reflex, a sensory neuron transmits nerve impulses from a receptor to an interneuron, which in turn, transmits impulses to a motor neuron, which conducts them to an effector. Reflexes are automatic, and some do not require involvement of the brain.

Long fibers of sensory and/or motor neurons make up cranial and spinal nerves of the somatic and autonomic divisions of the PNS. While the somatic division controls skeletal muscle, the autonomic division controls smooth muscle and the internal organs.

The CNS consists of the spinal cord and the brain. Only the cerebrum is responsible for consciousness; the other portions of the brain have their own function. The cerebrum can be mapped, and each lobe also seems to have particular functions. Neurological drugs, although quite varied, have been found to affect the RAS and the limbic system by either promoting or preventing the action of neurotransmitters.

Study Questions

1. What are the two main divisions of the nervous system? Explain why these names are appropriate. (p. 189)
2. What are the three types of neurons? How are they similar, and how are they different? (p. 189)
3. What does the term *resting potential* mean, and how is it brought about? (p. 191) Describe the two parts of an action potential and the change that can be associated with each part. (p. 192)
4. What is the sodium-potassium pump, and when is it active? (p. 192)
5. What is a neurotransmitter substance, where is it stored, how does it function, and how is it destroyed? (p. 194) Name two well-known neurotransmitters. (p. 194)
6. What are the three types of nerves, and how are they anatomically different? functionally different? Distinguish between cranial and spinal nerves. (p. 196)
7. Trace the path of a reflex action after discussing the structure and the function of the spinal cord and the spinal nerve. (p. 198)
8. What is the autonomic nervous system, and what are its two major divisions? (p. 199) Give several similarities and differences between these divisions. (p. 201)
9. Name the major parts of the brain, and give a function for each. (pp. 202–204)
10. Describe the EEG, and discuss its importance. (p. 205)
11. Describe the physiological effects and mode of action of alcohol, marijuana, cocaine, and heroin. (pp. 206–210)

Objective Questions

1. An _____ carries nerve impulses away from the cell body.
2. During the upswing of the action potential, _____ ions are moving to the _____ of the nerve fiber.
3. The space between the axon of one neuron and the dendrite of another is called the _____ .
4. ACh is broken down by the enzyme _____ after it has altered the permeability of the postsynaptic membrane.
5. Motor nerves innervate _____ .
6. The vagus nerve is a _____ nerve that controls the _____ .
7. In a reflex arc only the neuron called the _____ is completely within the CNS.
8. The brain and the spinal cord are covered by protective layers called _____ .
9. The _____ is that part of the brain that allows us to be conscious.
10. The _____ is the part of the brain responsible for coordination of body movements.

Label this Diagram.
See figure 10.2 (p. 190) in text.

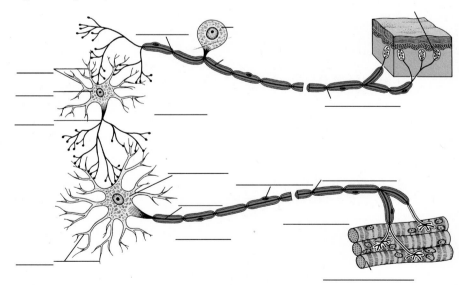

Answers to Objective Questions

1. axon 2. sodium, inside 3. synaptic cleft
4. AChE 5. muscles 6. cranial, motor, or parasympathetic; internal organs 7. interneuron
8. meninges 9. cerebrum 10. cerebellum

Critical Thinking Questions

1. Skeletal muscle fibers are innervated only by somatic motor neurons. Speculate why two different motor neurons (one from the sympathetic system and the other from the parasympathetic system) are used for internal organs and glands.

2. The limbic system is sometimes called the animal brain. Considering the function of the limbic system, why might this term be appropriate? Can you give an explanation for why we have a "primitive" brain beneath the cerebral cortex?

3. The nervous system is divided into the central nervous system (CNS) and the peripheral nervous system (PNS). In what way does this division seem awkward and artificial?

Selected Key Terms

axon (ak'son)   process of a neuron that conducts nerve impulses away from the cell body.   189

cell body (sel bod'e)   portion of a nerve cell that includes a cytoplasmic mass and a nucleus and from which nerve fibers extend.   189

central nervous system (sen'tral ner'vus sis'tem)   CNS; that portion of the nervous system that includes the brain and the spinal cord.   201

cerebral hemisphere (ser'ē-bral hem'ĭ-sfēr)   one of the large paired structures that together constitute the cerebrum of the brain.   203

dendrite (den'drīt)   process of a neuron, typically branched, that conducts nerve impulses toward the cell body.   189

effector (ef-fek'tor)   a structure, such as a muscle and a gland, that allows an organism to respond to environmental stimuli.   189

ganglion (gang'gle-on)   a collection of neuron cell bodies outside the CNS.   196

innervate (in'er-vāt)   to activate an organ, muscle, or gland by motor neuron stimulation.   189

interneuron (in''ter-nu'ron)   a neuron that is found within the CNS and that takes nerve impulses from one portion of the system to another.   189

motor neuron (mo'tor nu'ron)   a neuron that takes nerve impulses from the CNS to the effectors.   189

myelin sheath (mi'ĕ-lin shēth)   the Schwann cell membranes that cover long neuron fibers and give them a white glistening appearance.   190

nerve impulse (nerv im'puls)   an electrochemical change due to increased membrane permeability that is propagated along a neuron from the dendrite to the axon following excitation.   190

neurotransmitter substance (nu''ro-trans'mit'er sub'stans)   a chemical found at the ends of axons that is responsible for transmission across a synapse.   194

parasympathetic nervous system (par''ah-sim''pah-thet'ik ner'vus sis'tem)   that part of the autonomic nervous system that usually promotes those activities associated with a normal state.   201

peripheral nervous system (pĕrif'er-al ner'vus sis'tem)   PNS; nerves and ganglia that lie outside the CNS.   196

receptor (re-sep'tor)   a sense organ specialized to receive information from the environment.   189

sensory neuron (sen'so-re nu'ron)   a neuron that takes nerve impulses to the CNS afferent neuron.   189

somatic nervous system (so-mat'ik ner'vus sis'tem)   that portion of the PNS containing motor neurons that control skeletal muscles.   198

sympathetic nervous system (sim''pah-thet'ik ner'vus sis'tem)   that part of the autonomic nervous system that usually causes effects associated with emergency situations.   201

synapse (sin'aps)   the region between two nerve cells, where the nerve impulse is transmitted from one to the other, usually from axon to dendrite.   194

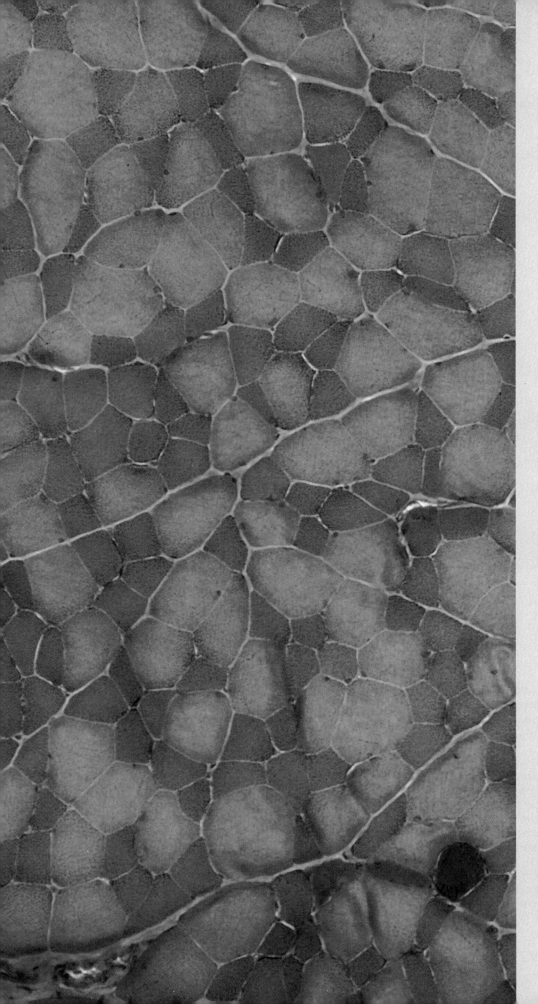

Chapter Eleven

Musculoskeletal System

Chapter Concepts

1 The skeleton, which contributes greatly to our general appearance, has various functions and is divided into the axial and appendicular skeletons.

2 Macroscopically, skeletal muscles work in antagonistic pairs and exhibit certain physiological characteristics.

3 Microscopically, muscle fiber contraction is dependent on actin filaments and myosin filaments and a ready supply of calcium (Ca^{++}) ions and ATP.

Cross section of fibers within a human skeletal muscle. The ability of a fiber to contract is dependent upon the presence of contractile filaments that appear as tiny black dots in this light micrograph. Notice that some fibers have more contractile filaments and appear darker than other fibers. Darker fibers are concentrated in "dark meat" while lighter fibers are concentrated in "light meat." This shows that a macroscopic characteristic has a microscopic explanation.

Figure 11.1 Vasily Alexeyev, a Russian weight lifter, strains to lift over 500 pounds.

Exercise causes a muscle to grow larger—there is an increase in the number and size of the protein filaments within each muscle cell and in the number of mitochondria. Strong muscles also contain more energy sources in the form of glycogen and fat, and they resist fatigue.

Apparently, however, you are born with a certain proportion of slow-twitching versus fast-twitching muscle cells (fibers), and exercise cannot drastically alter the proportion. No wonder, then, that athletes have their favorite sport at which they perform best. The slow-twitching fibers found in dark meat are generously supplied with mitochondria and capillaries. They produce a steady pulling power so necessary to swimming, running, and skiing. The fast-twitching fibers found in light meat aren't as well supplied with blood vessels nor do they have as many mitochondria. They supply bursts of energy for sprinting, weight lifting (fig. 11.1), or swinging a golf club.

Human Issue

Mental capacity is undoubtedly our highest human attribute, yet some people—particularly professional athletes—make astounding sums of money for their physical prowess. In other countries, much attention is given to other activities, such as chess and the fine arts, that require nonphysical prowess. Is it wrong for Americans to devote so much money and admiration to athletic performances rather than to activities that express mainly mental capabilities?

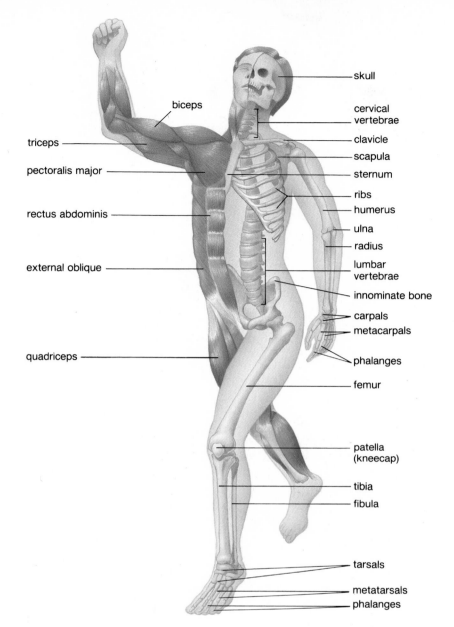

Figure 11.2 Major bones (*right*) and muscles (*left*) of the human body. The axial skeleton, composed of the skull, the vertebral column, the sternum, and the ribs, lies in the midline; the rest of the bones belong to the appendicular skeleton.

Labels (muscles, left): biceps, triceps, pectoralis major, rectus abdominis, external oblique, quadriceps

Labels (bones, right): skull, cervical vertebrae, clavicle, scapula, sternum, ribs, humerus, ulna, radius, lumbar vertebrae, innominate bone, carpals, metacarpals, phalanges, femur, patella (kneecap), tibia, fibula, tarsals, metatarsals, phalanges

Skeleton

Functions of the Skeleton

The skeleton (fig. 11.2), notably the large, heavy bones of the legs, supports the body against the pull of gravity. The skeleton also protects soft body parts. For example, the skull forms a protective encasement for the brain, as does the rib cage for the heart and the lungs. Flat bones, such as those of the skull, the ribs, and the breastbone, produce red blood cells in both adults and children. All bones are storage areas for inorganic calcium and phosphorus salts. Bones also provide sites for muscle attachment. The long bones, particularly those of the legs and the arms, permit flexible body movement.

The skeleton not only permits flexible movement, it also supports and protects the body, produces red blood cells, and serves as a storehouse for certain inorganic salts.

Bone Structure and Growth

Bone Structure

A long bone, such as the femur, illustrates principles of bone anatomy. When the bone is split open, as in figure 11.3, the longitudinal section shows that it is not solid, but has a cavity called the medullary cavity bounded at the sides by compact bone and at the ends by spongy bone. Beyond the spongy bone there is a thin shell of compact bone and finally a layer of cartilage.

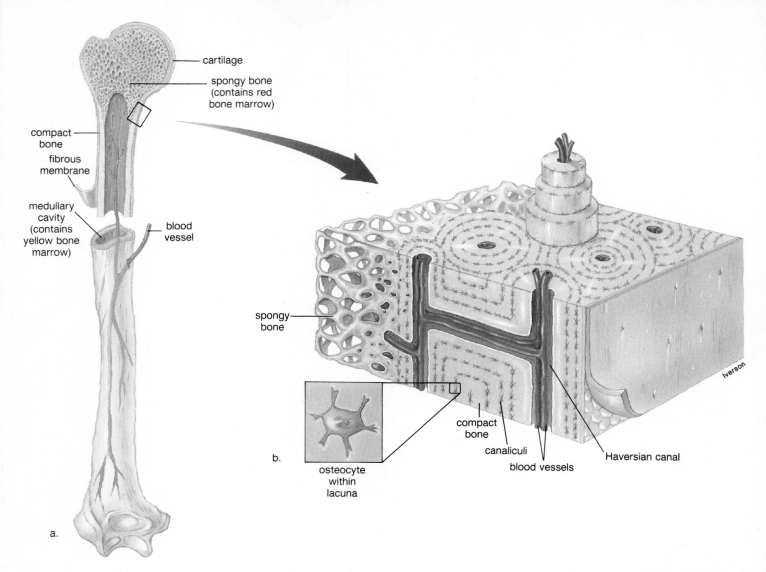

Figure 11.3 Anatomy of the long bone. *a.* A long bone is encased by fibrous membrane except at the ends, where it is covered by cartilage. The central shaft is composed of compact bone, but the ends are spongy bone, which can contain red marrow. A central medullary cavity contains yellow marrow. *b.* Close-up view of compact and spongy bone.

Compact bone, as discussed previously (p. 55), contains bone cells in tiny chambers called lacunae, arranged in concentric circles around Haversian canals. The Haversian canals contain blood vessels and nerves. The lacunae are separated by a matrix that contains protein fibers of collagen and mineral deposits, primarily of calcium and phosphorus salts.

Spongy bone contains numerous bony bars and plates separated by irregular spaces. Although lighter than compact bone, spongy bone is still designed for strength. Just as braces are used for support in buildings, the solid portions of spongy bone follow lines of stress. The spaces in spongy bone are often filled with **red marrow,** a specialized tissue that produces red blood cells. The cavity of a long bone usually contains **yellow marrow,** which is a fat-storage tissue.

Bones are not inert. They contain cells and perform various functions, including the production of blood cells.

Bone Growth

Most of the bones of the skeleton are cartilaginous during prenatal development. Later, the cartilage is replaced by bone due to the action of bone-forming cells known as **osteoblasts.** (Once they are isolated in lacunae, the cells are called osteocytes.) At first, there is only a primary ossification center at the middle of a long bone, but later, secondary centers form at the ends of the bones. There remains a *cartilaginous disk* between the primary ossification center and each secondary center, which can increase in length. The rate of growth is controlled by hormones such as growth hormones and the sex hormones. Eventually, though, the disks disappear, and the bone stops growing as the individual attains adult height.

In the adult, bone is continually being broken down and built up again. Bone-absorbing cells, called **osteoclasts,** are derived from cells carried in the bloodstream. As they break down bone, they remove worn cells and deposit calcium in the blood.

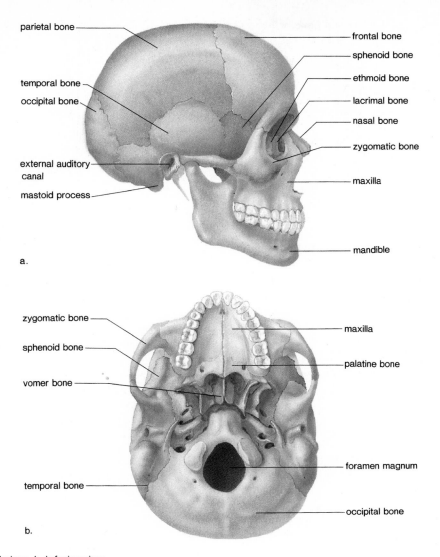

parietal bone

temporal bone

occipital bone

external auditory canal

mastoid process

frontal bone

sphenoid bone

ethmoid bone

lacrimal bone

nasal bone

zygomatic bone

maxilla

mandible

a.

zygomatic bone

sphenoid bone

vomer bone

temporal bone

maxilla

palatine bone

foramen magnum

occipital bone

b.

Figure 11.4   Skull. *a*. Lateral view. *b*. Inferior view.

Apparently after a period of about three weeks, they disappear. The destruction caused by the work of osteclasts is repaired by osteoblasts. As they form new bone, they take calcium from the blood. Eventually some of these cells get caught in the matrix they secrete and are converted to **osteocytes,** the cells found within Haversian systems (p. 55).

Bone is a living tissue, and it is always being rejuvenated.

Axial Skeleton

The skeleton may be divided into two parts: the axial skeleton and the appendicular skeleton (fig. 11.2). The **axial skeleton** lies in the midline of the body and consists of the skull, vertebral column, sternum, and ribs.

Skull

The skull is formed by the cranium and the facial bones.

Cranium   The cranium protects the brain and is composed of eight bones fitted tightly together in adults. In newborns, certain bones are not completely formed and instead are joined by membranous regions called *fontanels,* all of which usually close by the age of 16 months. The bones of the cranium contain the **sinuses,** air spaces lined by mucous membrane, which reduce the weight of the skull and give a resonant sound to the voice. Two sinuses called the mastoid sinuses drain into the middle ear. *Mastoiditis,* a condition that can lead to deafness, is an inflammation of these sinuses.

The major bones of the cranium have the same names as the lobes of the brain: frontal, parietal, temporal, and occipital. On the top of the cranium (fig. 11.4*a*), the **frontal bone** forms the forehead, the **parietal bones** extend to the sides, and the **occipital bone** curves to form the base of the skull. Here there is a large opening, the **foramen magnum** (fig. 114*b*), through which the nerve cord passes and becomes the brain stem. Below the much larger parietal bones, each **temporal bone** has an opening that leads to the middle ear. The **sphenoid bone** not only completes the sides of the skull, it also contributes to the floors and walls of the eye sockets. Likewise, the **ethmoid bone,** which lies in front of the sphenoid, is a part of the orbital wall and, in addition, is a component of the nasal septum.

The cranium contains eight bones; the frontal, two parietal, the occipital, two temporal, the sphenoid, and the ethmoid.

Facial Bones The **mandible,** or lower jaw, is the only movable portion of the skull (fig. 11.4*a*), and its action permits us to chew our food. Tooth sockets are located on this bone and on the **maxillae** (maxillary bones), the upper jaw that also forms the anterior portion of the hard palate. The **palatine bones** make up the posterior portion of the hard palate and the floor of the nasal cavity. The **zygomatic bones** give us our cheekbone prominences, and the nasal bones form the bridge of the nose. Each thin, scalelike **lacrimal bone** lies between an ethmoid bone and a maxillary bone, and the thin, flat **vomer** joins with the perpendicular plate of the ethmoid to form the nasal septum.

The facial bones include the mandible, two maxillae, two palatine, two zygomatic, two lacrimal, two nasal, and the vomer.

Vertebral Column

The **vertebral column** extends from the skull to the pelvis. Normally, the vertebral column has four curvatures that provide more resiliency and strength than a straight column could. The various vertebrae are named according to their location in the vertebral column (fig. 11.5). When the vertebrae join, they form a canal through which the spinal cord passes. The *spinous processes* of the vertebrate can be felt as bony projections along the midline of the back.

There are intervertebral disks between the vertebrae that act as a kind of padding. They prevent the vertebrae from grinding against one another and absorb shock caused by movements such as running, jumping, and even walking. Unfortunately, these disks become weakened with age and can slip or even rupture. This causes pain when the damaged disk presses up against the spinal cord and/or spinal nerves. The body may heal itself, or else the disk can be removed surgically. If the latter occurs, the vertebrae can be fused together, but this will limit the flexibility of the body. The presence of the disks allows motion between the vertebrae so that we can bend forward, backward, and from side to side.

The vertebral column, directly or indirectly, serves as an anchor for all the other bones of the skeleton (fig. 11.2). All of the twelve pairs of **ribs** connect directly to the thoracic vertebrae in the back, and all but two pairs connect either directly or indirectly via shafts of cartilage to the **sternum** (breastbone) in the front. The lower two pairs of ribs are called "floating ribs" because they do not attach to the sternum.

The vertebral column contains the vertebrae and serves as the backbone for the body.

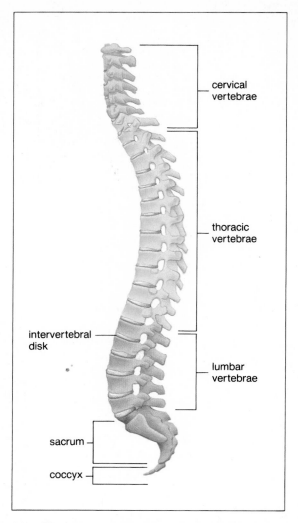

Figure 11.5 The vertebral column. The vertebrae are named according to their location in the column, which is flexible due to the intervertebral disks. Note the presence of the coccyx, the vestigial "tailbone."

Appendicular Skeleton

The **appendicular skeleton** consists of the bones within the pectoral and pelvic girdle and the attached limbs (fig. 11.2). The pectoral (shoulder) girdle and upper limbs (arms) are specialized for flexibility, but the pelvic girdle (hipbones) and lower limbs (legs) are specialized for strength.

Pectoral Girdle and Arm

The components of the **pectoral girdle** (fig. 11.6) are loosely linked together by ligaments. Each **clavicle** (collarbone) connects with the sternum in front and the **scapula** (shoulder blade) behind, but the scapula is freely movable and held in place only by muscles. This allows it to follow freely the movements of the arm. The single long bone in the upper arm (fig. 11.6), the **hu-**

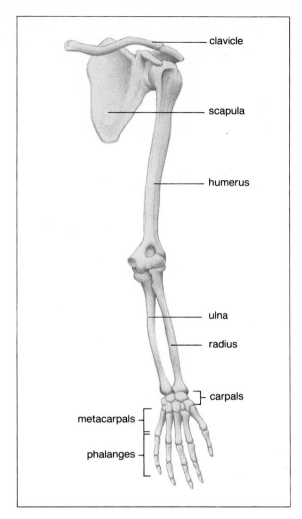

Figure 11.6   The bones of the pectoral girdle, the arm, and the hand. The humerus becomes the "funny bone" of the elbow.

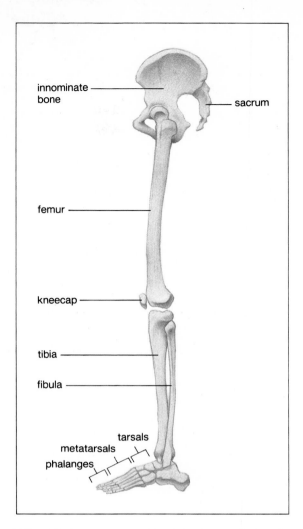

Figure 11.7   The bones of the pelvic girdle, the leg, and the foot. The femur, our strongest bone, withstands a pressure of 540 kg per 2.5 cm³ when we walk.

merus, has a smoothly rounded head that fits into a socket of the scapula. The socket, however, is very shallow and much smaller than the head. Although this means that the arm can move in almost any direction, there is little stability. Therefore, this is the joint that is most apt to dislocate. The opposite end of the humerus meets the two bones of the lower arm, the **ulna** and the **radius,** at the elbow. (The prominent bone in the elbow is the topmost part of the ulna.) When the arm is held so that the palm is turned frontward, the radius and ulna are about parallel to one another. When the arm is turned so that the palm is next to the body, the radius crosses in front of the ulna, a feature that contributes to the easy twisting motion of the forearm.

The many bones of the hand increase its flexibility. The wrist has eight **carpal** bones, which look like small pebbles. From these, five **metacarpal** bones fan out to form a framework for the palm. The metacarpal bone that leads to the thumb is placed in such a way that the thumb can reach out and touch the other digits. (**Digits** is a term that refers to either fingers or toes.) Beyond the metacarpals are the **phalanges,** the bones of the fingers and the thumb. The phalanges of the hand are long, slender, and lightweight.

Pelvic Girdle and Leg

The **pelvic girdle,** or pelvis (fig. 11.7), consists of two heavy, large **innominate** bones (hipbones). The innominate bones are anchored to the sacrum, and together these bones form a hollow cavity that is wider in females than males. The weight of the body is transmitted through the pelvis to the legs and then onto the ground. The largest bone in the body is the **femur,** or thighbone. Although the femur is a strong bone, it is doubtful that the femurs of a fairy-tale giant could support the increase in

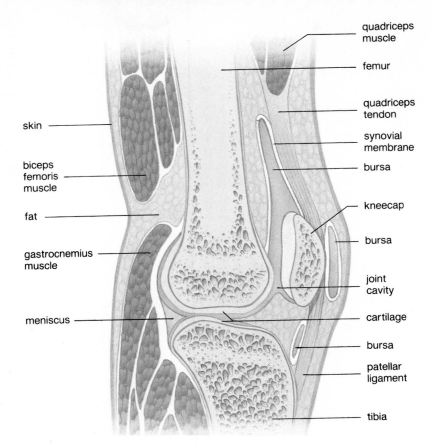

Figure 11.8   The knee joint, a freely movable synovial joint. Notice that there is a cavity between the bones that is encased by ligaments and lined by synovial membrane. The kneecap protects the joint.

weight. If a giant were ten times taller than an ordinary human being, he would also be about ten times wider and thicker, making him weigh about one thousand times as much. This amount of weight would break even giant-size femurs.

In the lower leg, the larger of the two bones, the **tibia** (fig. 11.7), has a ridge we call the shin. Both of the bones of the lower leg have a prominence that contributes to the ankle—the tibia on the inside of the ankle and the **fibula** on the outside of the ankle. Although there are seven **tarsal** bones in the ankle, only one receives the weight and passes it on to the heel and the ball of the foot. If you wear high-heeled shoes, the weight is thrown even further forward toward the front of your foot. The **metatarsal** bones form the arches of the foot. There is a longitudinal arch from the heel to the toes and a transverse arch across the foot. These provide a stable, springy base for the body. If the tissues that bind the metatarsals together become weakened, flatfeet are apt to result. The bones of the toes are called *phalanges,* just like those of the fingers, but in the foot the phalanges are stout and extremely sturdy.

The appendicular skeleton contains the bones of the girdles and limbs.

Joints

Bones are joined together at the joints, which are often classified according to the amount of movement they allow. Some bones, such as those that make up the cranium, are sutured together and are *immovable.* Other joints are *slightly movable,* such as the joints between the vertebrae. The vertebrae are separated by disks, described earlier, that increase their flexibility. Similarly, the two hipbones are slightly movable where they are ventrally joined by cartilage. Owing to hormonal changes, this joint becomes more flexible during late pregnancy, which allows the pelvis to expand during childbirth.

Most joints are *freely movable* **synovial joints,** in which the two bones are separated by a cavity. **Ligaments** composed of fibrous connective tissue bind the two bones to one another, holding them in place as they form a capsule. In a "double-jointed" individual, the ligaments are unusually loose. The joint capsule is lined by synovial membrane, which produces *synovial fluid,* a lubricant for the joint.

The knee is an example of a synovial joint (fig. 11.8). In the knee, as in other freely movable joints, the bones are capped by cartilage, although there are also crescent-shaped pieces of cartilage between the bones called **menisci.** These give added

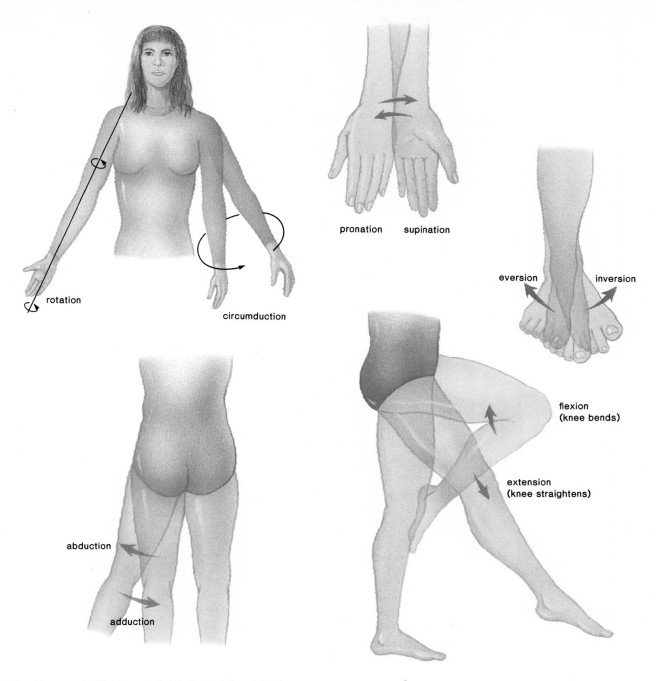

rotation

circumduction

pronation   supination

eversion   inversion

abduction

adduction

flexion
(knee bends)

extension
(knee straightens)

Figure 11.9   Movement of body parts in relation to joints. Muscles are attached to bones across a joint. One of these bones remains steady while the other bone moves.

stability, helping to support the weight placed on the knee joint. Unfortunately, athletes often suffer injury of the menisci, known as torn cartilage. The knee joint also contains thirteen fluid-filled sacs called bursae, which ease friction between tendons and ligaments and between tendons and bones. Inflammation of the bursae is called bursitis. Tennis elbow is a form of bursitis.

There are different types of movable joints. The knee and elbow joints are *hinge joints* because, like a hinged door, they largely permit movement in one direction only. More movable are the ball-and-socket joints; for example, the ball of the femur fits into a socket on the hipbone. *Ball-and-socket joints* allow movement in all planes and even a rotational movement. The various movements of body parts at joints are depicted in figure 11.9.

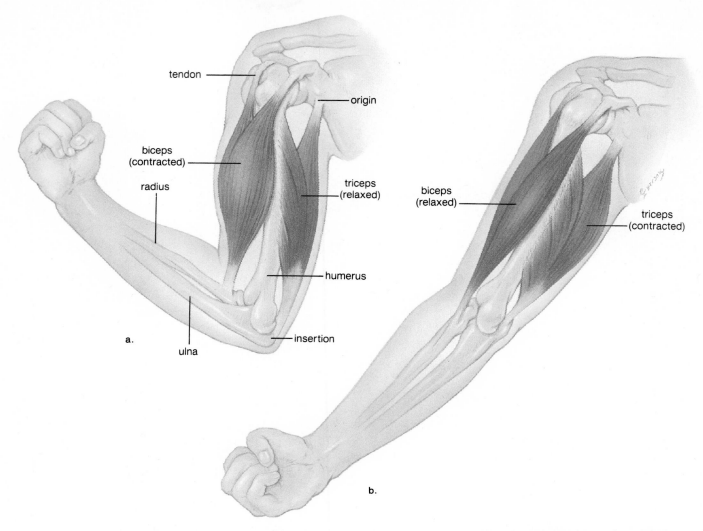

Figure 11.10 Attachment of skeletal muscles as exemplified by the biceps and the triceps. The origin of a muscle is fairly stationary; the insertion moves. These muscles are antagonistic. *a.* When the biceps contract, the lower arm is raised, and *b.* when the triceps contract, the lower arm is lowered.

Synovial joints are subject to *arthritis*. In rheumatoid arthritis, the synovial membrane becomes inflamed and thickens. Degenerative changes take place that make the joint almost immovable and painful to use. There is evidence that these effects are brought on by an autoimmune reaction. In old-age arthritis, or osteoarthritis, the cartilage at the ends of the bones disintegrates so that the two bones become rough and irregular. This type of arthritis is apt to affect the joints that have received the greatest use over the years.

Joints are classified according to the degree of movement. Some joints are immovable, some are slightly movable, and some are freely movable.

Skeletal Muscles: Macroscopic View

Muscles are effectors, which enable the organism to respond to a stimulus (p. 199). Skeletal muscles are attached to the skeleton, and their contraction accounts for voluntary movements. Involuntary muscles, both smooth and cardiac, were discussed on page 56.

Anatomy of Whole Muscle

Muscles typically are attached to bone by **tendons** made of fibrous connective tissue. Tendons most often attach muscles to the far side of a joint so that the muscle extends across the joint (fig. 11.8). When the central portion of the muscle, called the belly, contracts, one bone remains fairly stationary and the other one moves. The **origin** of the muscle is on the stationary bone, and the **insertion** of the muscle is on the bone that moves.

When a muscle contracts, it shortens. Therefore, muscles can only pull; they cannot push. Because we need both to extend and to flex at a joint, muscles generally work in *antagonistic pairs*. For example, the biceps and the triceps are a pair of muscles that move the forearm up and down (fig. 11.10). When the biceps contracts, the lower arm flexes, and when the triceps contracts, the lower arm extends.

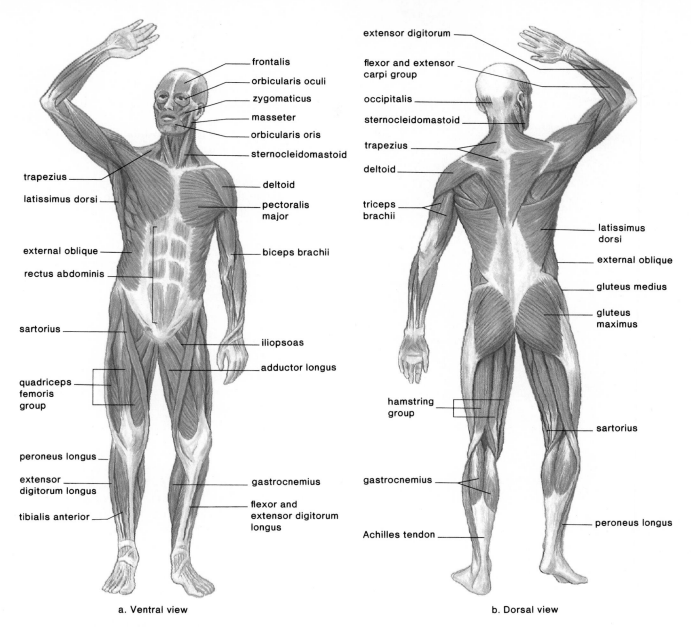

a. Ventral view

b. Dorsal view

Figure 11.11   Superficial skeletal muscles. *a*. Ventral view, *b*. dorsal view.

Figure 11.11 depicts various superficial muscles of the body.

Typically muscles are attached across a joint. When a muscle contracts, the movable end (insertion) is pulled toward the fixed end (origin), and movement occurs at the joint. This system requires that muscles work in antagonistic pairs.

Physiology of Whole Muscle

In the laboratory, it is possible to study the contraction of individual whole muscles. Customarily, a calf muscle is removed from a frog and mounted so one end is fixed and the other is movable. The mechanical force of contraction is transduced into an electrical current recorded by an apparatus called a **physiograph** (fig. 11.12*a*). The resulting pattern is called a **myogram.**

a.

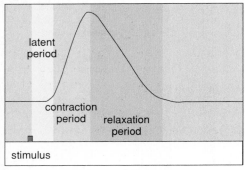

b.

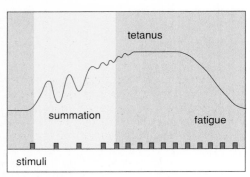

c.

Figure 11.12 Physiology of muscle contraction. *a.* A physiograph is an apparatus that can be used to record a myogram, a visual representation of the contraction of a muscle that has been dissected from an animal. *b.* Simple muscle twitch is composed of 3 periods: a latent period, a contraction period, and a relaxation period. *c.* Summation and tetanus. When a muscle is not allowed to completely relax between stimuli, the contractions increase in size. The muscle remains contracted until it fatigues.

All-or-None Response

A *single* muscle fiber (muscle cell, p. 56) either responds to a stimulus and contracts or it does not. At first, the stimulus may be so weak that no contraction occurs, but as soon as the strength of the stimulus reaches the *threshold stimulus,* the muscle fiber contracts completely. Therefore, a muscle fiber obeys the **all-or-none law.**

Contrary to that of an individual fiber, the strength of contraction of a whole muscle can increase according to the degree of stimulus beyond the threshold stimulus. A whole muscle contains many fibers, and the degree of contraction is dependent on the total number of fibers contracting. The *maximum stimulus* is the one beyond which the degree of contraction does not increase.

Muscle fibers obey the all-or-none law, but whole muscles do not obey this law.

Muscle Twitch

If a muscle is placed in a physiograph and is given a maximum stimulus, it contracts and then relaxes. This action—a single contraction that lasts only a fraction of a second—is called a muscle **twitch.** Figure 11.12*b* is a myogram of a twitch, which is customarily divided into the *latent period,* or the period of time between stimulation and initiation of contraction, the *contraction period,* and the *relaxation period.*

If a muscle is exposed to two maximum stimuli in quick succession, it responds to the first but not to the second stimulus. This is because it takes an instant following a contraction for the muscle fibers to recover in order to respond to the next stimulus. The very brief moment following stimulation, during which a muscle is unresponsive, is called the refractory period.

Summation and Tetanus

If a muscle is given a rapid series of threshold stimuli, it can respond to the next stimulus without relaxing completely. In this way, muscle tension summates until maximal sustained tetanic contraction called **summation** is achieved (fig. 11.12*c*). The myogram no longer shows individual twitches; rather, they are fused and blended completely into a straight line. **Tetanus** continues

Table 11.1 Characteristics of Skeletal Muscle

weight: 40% of total body weight

total number of muscles in body: over 600

number of fibers (cells) in a muscle: a few hundred to over a million

size of muscle cell: 10–100 μm diameter; up to 20 cm long

size of filaments: Thick = 15 nm diameter; thin = 5 nm diameter

time of contraction: fast fiber = 10 msec; slow fiber = 100 msec

time for contraction-relaxation cycle: 0.15 second

Table 13.1, "Characteristics of Skeletal Muscle," from *Being Human* by David W. Deamer, copyright © 1981 by Saunders College Publishing, a division of Holt, Rinehart and Winston, Inc., reprinted by permission of the publisher.

until the muscle fatigues due to depletion of energy reserve. **Fatigue** is apparent when a muscle relaxes even though stimulation continues.

Tetanic contractions occur whenever skeletal muscles are used actively. Ordinarily, however, only a portion of any particular muscle is involved—while some fibers are contracting, others are relaxing. Because of this, intact muscles rarely fatigue completely.

Muscle twitch, summation, and tetanus are related to the frequency with which a muscle is stimulated.

Muscle Tone

Skeletal muscles exhibit tone, a condition in which some fibers always are contracted. Muscle **tone** is particularly important in maintaining posture. If all the fibers within the neck, trunk, and leg muscles suddenly relaxed, the body would collapse.

Muscle spindles are receptors that are widespread throughout skeletal muscles (fig. 11.13). A muscle spindle consists of several modified muscle fibers that have sensory nerve fibers wrapped around a short, specialized region somewhere near the middle of their length. A muscle spindle generates more nerve impulses when the muscle stretches and fewer when it contracts. These nerve impulses inform the central nervous system (CNS) of the state of this particular muscle so that contraction of this muscle can be coordinated with the contraction of other muscles.

Effect of Exercise

A regular exercise program, such as the one described in table 11.A in the reading on page 227, has many benefits. Increased endurance and strength of muscles are two possible benefits. Endurance is measured by the length of time the muscle can work before fatiguing, and strength is the force a muscle (or a group of muscles) can exert against a resistance.

A regular exercise program brings about physiological changes that build endurance, such as increased muscular stores of ATP and increased tolerance to lactate buildup. Muscle strength increases as muscle enlargement occurs due to exercise. When a muscle enlarges, the number of muscle fibers does not increase; however, the protein content of the muscle increases. This happens because the contractile elements in muscles—called myofibrils—that contain the protein filaments actin and myosin increase in number.

Aside from improved endurance and strength, an exercise program also helps many other organs of the body. Cardiac muscle enlarges, and the heart can work harder than before. The resting heart rate decreases. Lung and diffusion capacity increase. Body fat decreases, but bone density increases so that breakage is less apt to occur. Fat and cholesterol blood levels decrease along with blood pressure. The reading on page 226 particularly discusses studies showing that an exercise program lowers the risk of heart attack.

Human Issue

Athletes want to win athletic competitions, and most people today aspire to appear physically fit. Pushed on by the desire for public and private admiration, some people take body-building anabolic steroids. Unfortunately, people who take steroids can suffer from changes in secondary sexual characteristics, liver cancer, kidney tumors, and circulatory disorders. One scientist took a survey of world-class athletes and asked them, "Would it be worth it to take a drug that could win you a gold medal if it meant you would be dead in five years?" Most answered yes. What would be your answer? Do you think taking steroids is worth the risk?

Exercise: Changing Perceptions

. . . If a single researcher could be considered responsible for the fitness craze, it would be Dr. Ralph Paffenbarger, a specialist in heart disease and exercise with posts at both Stanford and Harvard universities. His landmark research, involving 17,000 Harvard alumni and 6,000 San Francisco longshoremen, showed that the men who exercised vigorously and burned at least 2,000 calories a week doing so cut their risk of dying from heart disease by half. The reduction was dramatic even if the men smoked or if their parents both had died of heart disease.

The problem was the public—and many health professionals—interpreted these findings as meaning that anything less didn't do any good. If you weren't up for running 20 miles a week or an hour of tennis five times a week, or cross-country skiing for a half hour a day—well, you'd be just as well off sitting in front of the tube with a beer. The 2,000-calorie threshold was transformed into a "magic number"— above it, you were fit. Below it, you were a basket case.

Paffenbarger says he never believed that the 2,000-calorie figure was carved in stone. Neither did Dr. Arthur Leon, an epidemiologist at University of Minnesota's School of Public Health. Both were surprised, however, when Leon's research showed that a far more modest level of activity could exert a very powerful protective effect on the heart.

"People who don't want to do formal, sweaty exercise can be told that less can be beneficial," says Leon. "Just moving around more is a big help." He backs up his assertion with the results of his study, published in the *Journal of the American Medical Association* (in the fall of 1987). That study analyzed the relationship between heart attack and physical activity off the job in 12,138 men who participated in the Multiple Risk Factor Intervention Trial, a nationwide study conducted at 22 medical centers.

Fitness walkers. The enclosed shopping mall, intended to be a boon for buyers, has also become a haven for fitness walkers. A controlled climate and unimpeded passages provide a perfect environment for those determined to put in their daily mileage, rain or shine, while eliminating many of the outdoor hazards that may deter older pedestrians. Some mall walkers, like these Galleria Mall GoGetters in Glendale, California, have formed clubs. A few malls issue special walking maps, while others open on holidays, even when the stores are closed, to accommodate the local ramblers.

The trial found that men who were only moderately active—spending an average of 48 minutes a day on leisure-time physical activity—had one-third fewer heart attacks than their peers who moved around during leisure time an average of 16 minutes each day. And the moderately active group didn't spend all, or even most, of their exercise time huffing and puffing. Mostly, the report found, their activities were in the light-to-moderate range: lawn and garden work, bowling, ballroom dancing—activities we don't often even think of as exercise.

Indeed, some of the most dramatic gains are made by the sedentary folks whose initial efforts fall short of the magic 2,000 calories and whose activities never include much bouncing around. For example, after only four months of attending a twice-weekly, low-impact aerobics class at Northwestern University, men and women with rheumatoid arthritis—a chronic, severe form of joint disease—reported much less pain, swelling, fatigue, and depression than before they began exercising.

"The lower end of the spectrum gets the most benefit from its effort," says cardiologist Miller. "If you've just had a heart transplant, it takes very little—a walk around the gym—to get improvement. If you take a sedentary office worker, it doesn't take much more than adding a mile of walking a day to get improvement."

Leon and Paffenbarger point out that it's relatively easy to program greater amounts of activity into your normal day. Walk down to the accounting department at work instead of calling the accountant. Take stairs instead of elevators. Stroll during your lunchtime. A brisk walk—especially if you swing your arms—can get your heart rate up without the jouncing of jogging that many people dislike.

Leon's study showed that the benefits of exercise start to level off at a certain point—at least as far as fatal heart attack is concerned. The rate of fatal heart attack among the moderately active men was the same as the most active men in the study, who devoted an average of 134 minutes a day to leisure-time physical activity.

Does this mean the most active group was no more fit than the moderately active group? Of course not. Fitness, stresses Bud Getchell, executive director of the nonprofit National Institute for Fitness and Sport in Indianapolis, includes flexibility, muscular strength, endurance, and body composition (lean mass vs. fat), as well as cardiovascular health. To achieve the type of fitness that would allow you to take on most vigorous activities with ease—from lifting a heavy box at work, to shoveling snow without hurting your back, to enjoying a game of pickup basketball—most fitness experts say you should spend about three hours (see table 11.A), spaced out during the week, doing activities that strengthen muscles and enhance flexibility, as well as challenge the heart.

Your body will reward your efforts. Although most people think the main benefit of fitness is reducing the risk of heart disease, there's mounting evidence that other tissues and organs benefit as well. Several studies at the Institute for Aerobics Research in Dallas, Harvard, and other institutions have found that more active people have lower rates of colon, brain, kidney, and reproductive cancers, as well as leukemia, than their more sedentary counterparts. And studies that looked at exercise along with other variables found this still held true when factors like age, diet, and socioeconomic background were taken into account. Researchers speculate this may be because activity promotes the delivery of more nutrients and oxygen to these organs and tissues.

Indeed, according to Dr. Everett L. Smith, director of the Biogerontology Laboratory at the University of Wisconsin-Madison, half of the functional decline between ages 30 and 70 could be prevented if we simply used our body more. Bone density, nerve function, and kidney efficiency, as well as overall strength and flexibility, largely can be preserved into our later years simply by keeping up an active life.

First appeared in *The Boston Globe* May 1, 1988. Copyright © 1988 Sy Montgomery. Reprinted by permission.

Table 11.A A Checklist for Staying Fit

Children, 7–12	Teenagers, 13–18	Adults, 19–55	Seniors, 55 and up
vigorous activity 1–2 hours daily free play	vigorous activity 3–5 times a week	vigorous activity for one-half hour, 3 times a week	moderate exercise 3 times a week
build motor skills through team sports, dance, swimming	build muscle with calisthenics	exercise to prevent lower back pain: aerobics, stretching, yoga	plan a daily walk
encourage more exercise outside of physical education classes	plan aerobic exercise to control buildup of fat cells	take active vacations: hike, bicycle, cross-country ski	daily stretching exercises
initiate family outings: bowling, boating, camping, hiking	pursue tennis, swimming, riding—sports that can be enjoyed for a lifetime	find exercise partners: join a running club, bicycle club, outing group	learn a new sport: golf, fishing, ballroom dancing
	continue team sports, dancing, hiking, swimming		try low-impact aerobics
			before undertaking new exercises, consult your doctor

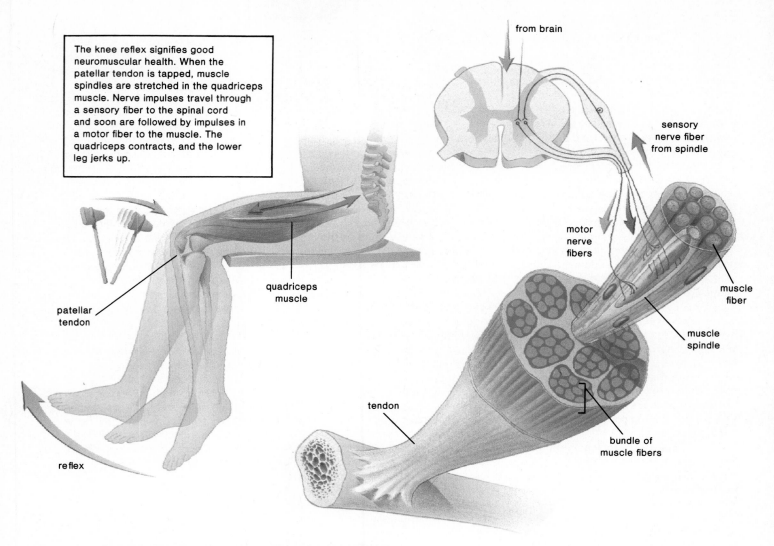

The knee reflex signifies good neuromuscular health. When the patellar tendon is tapped, muscle spindles are stretched in the quadriceps muscle. Nerve impulses travel through a sensory fiber to the spinal cord and soon are followed by impulses in a motor fiber to the muscle. The quadriceps contracts, and the lower leg jerks up.

from brain

sensory nerve fiber from spindle

motor nerve fibers

muscle fiber

muscle spindle

bundle of muscle fibers

tendon

patellar tendon

quadriceps muscle

reflex

Figure 11.13   Muscle spindle structure and function. After associated muscle fibers stretch, spindle fibers aid in the coordination of muscular contraction by sending more nerve impulses to the CNS, which then adjusts contraction of this muscle and other muscles.

Skeletal Muscles: Microscopic View

A whole skeletal muscle (fig. 11.13) is composed of a number of bundles of *muscle fibers.*

Anatomy of a Muscle Fiber

Each muscle fiber is a cell containing the usual cellular components, but special names have been assigned to some of these components, as indicated in table 11.2. A muscle fiber has some unique anatomical characteristics. For one thing, it has a T (for transverse) system (fig. 11.14). The **sarcolemma,** or cell membrane, forms *tubules* that penetrate, or dip down, into the cell so that they come into contact, but do not fuse, with expanded portions of modified ER, termed the **sarcoplasmic reticulum.** The expanded portions of the sarcoplasmic reticulum, called calcium-storage sacs, contain calcium (Ca^{++}) ions, which are essential for muscle contraction. The sarcoplasmic reticulum

Table 11.2   Muscle Cells

Component	Term
cell membrane	sarcolemma
cytoplasm	sarcoplasm
endoplasmic reticulum	sarcoplasmic reticulum

encases hundreds and sometimes even thousands of *myofibrils,* which are the contractile portions of the fibers.

Myofibrils and Sarcomeres

Myofibrils, the contractile elements in muscles, are cylindrical in shape and run the length of a muscle fiber. The light microscope shows that a myofibril has light and dark bands called *striations.* It is these striations that cause skeletal muscle to appear striated (fig. 3.11*a*). The electron microscope shows that the striations of myofibrils are formed by the placement of filaments within contractile units called **sarcomeres** (fig. 11.14*b*).

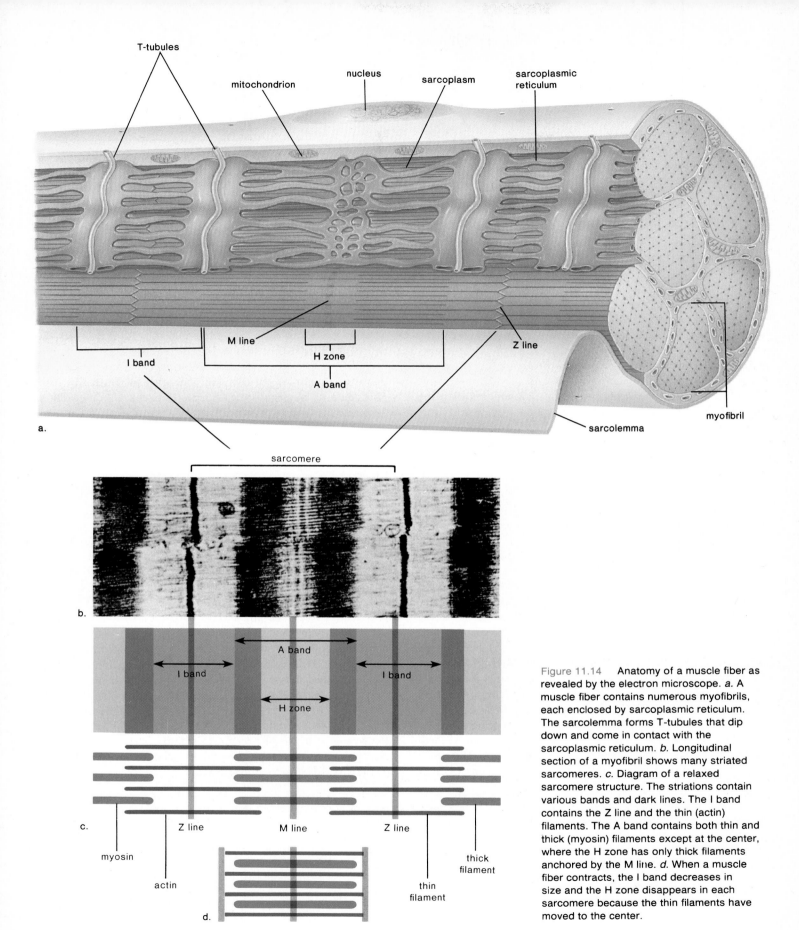

T-tubules

mitochondrion

nucleus

sarcoplasm

sarcoplasmic reticulum

M line

H zone

I band

A band

Z line

myofibril

sarcolemma

a.

sarcomere

b.

A band

I band

I band

H zone

c.

myosin

Z line

M line

Z line

thick filament

actin

thin filament

d.

Figure 11.14   Anatomy of a muscle fiber as revealed by the electron microscope. *a.* A muscle fiber contains numerous myofibrils, each enclosed by sarcoplasmic reticulum. The sarcolemma forms T-tubules that dip down and come in contact with the sarcoplasmic reticulum. *b.* Longitudinal section of a myofibril shows many striated sarcomeres. *c.* Diagram of a relaxed sarcomere structure. The striations contain various bands and dark lines. The I band contains the Z line and the thin (actin) filaments. The A band contains both thin and thick (myosin) filaments except at the center, where the H zone has only thick filaments anchored by the M line. *d.* When a muscle fiber contracts, the I band decreases in size and the H zone disappears in each sarcomere because the thin filaments have moved to the center.

A sarcomere extends between two dark lines called the Z lines (fig. 11.14c). A Z line is in a light-colored region called an I band. The center dark region of a sarcomere is called an A band. The A band is interrupted by a light center portion, the H zone. A fine, dark stripe called the M line cuts through the H zone.

The placement of *thick* and *thin* filaments in a sarcomere creates these bands and zones (fig. 11.14c). The I band is light because it contains only thin filaments; the A band is dark because it contains both thin and thick filaments except at the center in the lighter H zone, where only thick filaments are found.

The thick filaments of sarcomeres are made up of a protein called **myosin** and the thin filaments are made up of a protein called **actin** (table 11.3).

Physiology of a Muscle Fiber

A muscle fiber contracts when the sarcomeres within the myofibrils shorten. When a sarcomere shortens (fig. 11.14d), the actin (thin) filaments slide past the myosin (thick) filaments and approach one another. This causes the I band to shorten and the H zone to almost or completely disappear. The movement of actin filaments in relation to myosin filaments is called the **sliding filament theory** of muscle contraction. Notice that after the sarcomere contracts, the filaments are still the same length.

The overall formula for muscle contraction can be represented as follows:

$$\text{actin} + \text{myosin} \xrightarrow[\text{Ca}^{++}]{\text{ATP} \to \text{ADP} + \text{P}} \text{actomyosin}$$

The participants in this reaction have the functions listed in table 11.4. Even though it is the actin filaments that slide past the myosin filaments, it is the myosin filaments that do the work. In the presence of calcium (Ca^{++}) ions and ATP, portions of a myosin filament called *cross bridges* (fig. 11.15) bend backward and attach to an actin filament. (Each cross bridge binds to an actin filament at a cross-bridge binding site; actomyosin represents this in the formula above.) After attaching, the cross bridges bend forward and the actin filament is pulled along. Now, ATP is broken down by myosin, and detachment occurs. Notice that myosin is not only a structural protein, it is also an ATPase enzyme. The cross bridges attach and detach some 50–100 times as the thin filaments are pulled to the center of a sarcomere.

The sliding filament theory states that actin filaments slide past myosin filaments because myosin has cross bridges that pull the actin filaments inward.

Table 11.3   Contractile Elements

Component	Definition
myofibril	muscle cell contractile element
sarcomere	functional unit of myofibril
myosin	thick filament
actin	thin filament

Table 11.4   Muscle Contraction

Name	Function
actin filaments	slide past myosin, causing contraction
Ca^{++}	needed for myosin to bind to actin
myosin filaments	a. pull actin filaments by means of cross bridges b. are enzymatic and split ATP
ATP	supplies energy for bonding between myosin and actin

It is obvious from our discussion that ATP provides the energy for muscle contraction. In order to ensure a ready supply of ATP, muscle fibers contain **creatine phosphate** (phosphocreatine), a storage form of high-energy phosphate. Creatine phosphate does not participate directly in muscle contraction. Instead, it is used to regenerate ATP by the following reaction:

$$\text{creatine} \sim \text{P} + \text{ADP} \longrightarrow \text{ATP} + \text{creatine}$$

Oxygen Debt

When all of the creatine phosphate is depleted and no oxygen (O_2) is available for aerobic respiration, a muscle fiber can generate ATP by using fermentation, an anaerobic process (p. 43). Fermentation, which is apt to occur during strenuous exercise, can supply ATP for only a short time because lactate buildup produces muscular aching and fatigue that lasts a minute or so.

We all have had the experience of having to continue deep breathing following strenuous exercise. This continued intake of oxygen is required to complete the metabolism of lactate that has accumulated during exercise and represents an **oxygen debt** that the body must pay to rid itself of lactate. The lactate is transported to the liver, where one-fifth of it is completely broken down to carbon dioxide (CO_2) and water (H_2O) by means of the Krebs cycle and the respiratory chain (see chapter 2). The ATP gained by this respiration then is used to convert four-fifths of the lactate back to glucose.

Muscle contraction requires a ready supply of ATP. Creatine phosphate is used to generate ATP rapidly. If oxygen is in limited supply, fermentation produces ATP but results in oxygen debt.

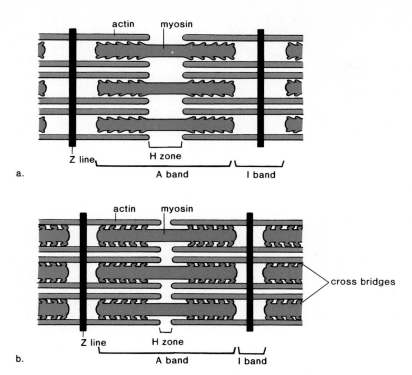

Figure 11.15 Sliding filament theory. *a*. Relaxed sarcomere. *b*. Contracted sarcomere. During contraction, the I band and the H zone decrease in size. This indicates that the thin filaments slide past the thick filaments. Even so, the thick filaments do the work by pulling the thin filaments by means of cross bridges.

Innervation

Muscles are innervated; that is, nerve impulses cause muscles to contract. Each motor axon of a nerve branches to several muscle fibers, and collectively, these muscle fibers are called a motor unit. Each branch has several terminal knobs, where there are synaptic vesicles filled with the neuromuscular transmitter acetylcholine (ACh). The region where a terminal knob lies in close proximity to the sarcolemma of a muscle fiber is called a neuromuscular junction. A **neuromuscular junction** (fig. 11.16) has the same components as a synapse: a presynaptic membrane, a synaptic cleft, and a postsynaptic membrane. Only in this case, the postsynaptic membrane is a portion of the sarcolemma of a muscle fiber.

Nerve impulses cause synaptic vesicles to merge with the presynaptic membrane and to release ACh into the synaptic cleft. When ACh reaches the sarcolemma, it is depolarized. The result is a **muscle action potential** that spreads over the sarcolemma and down the T system (fig. 11.14) to where calcium (Ca^{++}) ions are stored in calcium-storage sacs of the sarcoplasmic reticulum. When the action potential reaches a sac, calcium ions are released, and they diffuse into the sarcoplasm, where they participate in muscle contraction.

It is now necessary to consider the structure of a thin filament in more detail. Figure 11.17 shows the placement of two other proteins associated with a thin filament (the double row of twisted globular actin molecules). Threads of tropomyosin wind about a thin filament, and troponin occurs at intervals along the threads. After calcium ions are released from their storage sac, they combine with troponin. After binding occurs, the tropomyosin threads shift their position, and the cross-bridge binding sites are exposed.

The thick filament is a bundle of myosin molecules, each having a globular head. Each head is a cross bridge that has an ATP-binding site. After ATP attaches to ATP-binding sites, the cross bridges bend backward and attach to cross-bridge binding sites on the actin filaments. After attachment occurs, the cross bridges bend forward, pulling the actin filaments a short distance. Then the myosin heads break down ATP, and detachment of the cross bridges occurs. The actin filaments move nearer the center of the sarcomere each time the cycle is repeated.

The movement of the actin filaments causes muscle contraction. Contraction ceases when nerve impulses no longer stimulate the muscle fiber. With the cessation of a muscle action potential, calcium ions are pumped back into their storage sacs by active transport. Relaxation now occurs.

A neuromuscular junction functions like a synapse except that a muscle action potential causes calcium ions to be released from calcium-storage sacs, and thereafter muscle contraction occurs.

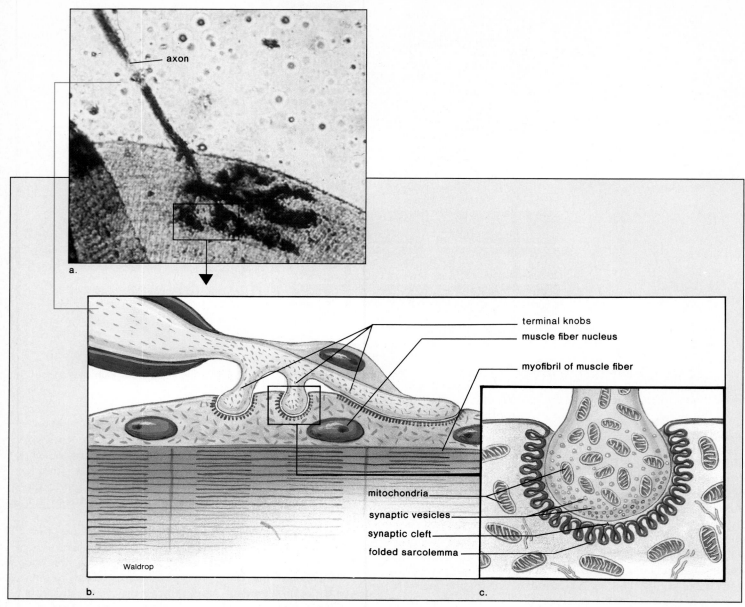

axon

terminal knobs
muscle fiber nucleus
myofibril of muscle fiber

mitochondria
synaptic vesicles
synaptic cleft
folded sarcolemma

a.

Waldrop

b.

c.

Figure 11.16   Muscle fiber innervation. *a.* Micrograph of a motor unit showing branching of an axon. *b.* A neuromuscular junction occurs where a terminal knob comes in close proximity to a muscle fiber. The knob contains synaptic vesicles (enlargement *c*) filled with ACh. When these vesicles fuse with the presynaptic membrane, ACh diffuses across the synaptic cleft to the sarcolemma. If the stimulus is sufficient, a muscle action potential begins.

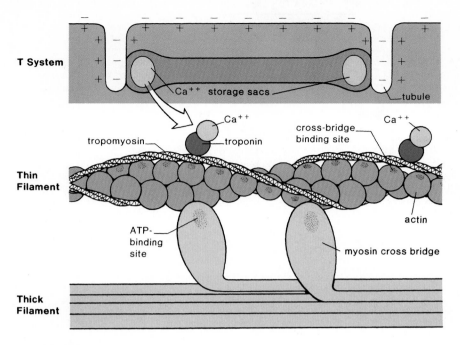

T System

Ca++ storage sacs

tubule

Ca++

tropomyosin

troponin

cross-bridge binding site

Ca++

Thin Filament

ATP-binding site

actin

myosin cross bridge

Thick Filament

Figure 11.17   Detailed structure and function of sarcomere contraction. After calcium (Ca++) ions are released from their storage sac, they combine with troponin, a protein that periodically occurs along tropomyosin threads. This causes the tropomyosin threads to shift their position so that cross-bridge binding sites are revealed along the actin filaments. The myosin filament extends its globular heads and forms cross bridges that bind to these sites. The breakdown of ATP by myosin causes the cross bridges to detach and to reattach farther along the actin. In this way, the actin filaments are pulled past the myosin filaments.

Summary

The skeleton aids movement of the body while it also supports and protects the body. Bones serve as deposits for inorganic salts, and some bones are sites for blood-cell production. The skeleton is divided into two parts: (1) the axial skeleton, which is made up of the skull, the ribs, and the vertebrae; and (2) the appendicular skeleton, which is composed of the appendages and their girdles. Joints are regions where bones are linked.

Whole skeletal muscles work in antagonistic pairs and have degrees of contraction. Muscle fibers obey the all-or-none law; it is possible to study a single contraction (muscle twitch) and sustained contraction (summation and tetanus) by using a physiograph.

Muscle fibers are cells that contain myofibrils in addition to the usual components of cells. Longitudinally, myofibrils are divided into sarcomeres, where it is possible to note the arrangement of actin filaments and myosin filaments. When a sarcomere contracts, the actin filaments slide past the myosin filaments. Myosin has cross bridges that attach to and pull the actin filaments along. ATP breakdown by myosin is necessary for detachment to occur.

Innervation of a muscle fiber begins at a neuromuscular junction. Here, synaptic vesicles release ACh into the synaptic cleft. When the sarcolemma receives ACh, a muscle action potential moves down the T system to calcium-storage sacs. When calcium ions are released, contraction occurs. When calcium ions are actively transported back into the storage sacs, muscle relaxation occurs.

Study Questions

1. Describe the anatomy of a long bone (p. 215) and how bones grow. (p. 215)
2. Distinguish between the axial and appendicular skeletons. (pp. 217, 218)
3. List the bones that form the pectoral and pelvic girdles. (pp. 218, 219)
4. How are joints classified? Describe the anatomy of a freely movable joint. (p. 220)
5. Describe how muscles are attached to bones. Why do muscles act in antagonistic pairs? (p. 222)
6. Describe the significance of threshold and maximum stimuli, muscle twitch, summation, and tetanic contraction. (p. 224)
7. How is the tone of a muscle maintained, and how do muscle spindles contribute to the maintenance of tone? (p. 225)
8. Discuss the microscopic anatomy of a muscle fiber and the structure of a sarcomere. What is the sliding filament theory? (p. 228)
9. Give the function of each participant in the following reaction:

$$ATP \longrightarrow ADP + P$$
$$actin + myosin \xrightarrow{\quad Ca^{++} \quad} actomyosin$$

   (p. 230)
10. Discuss the specific role of ATP during muscle contraction. What is creatine phosphate? (p. 230)
11. What is oxygen debt, and how is it repaid? (p. 230)
12. What causes a muscle action potential? How does the muscle action potential bring about a sarcomere and muscle fiber contraction? (p. 231)

Objective Questions

1. The skull, the ribs, and the sternum are all in the _____ skeleton.
2. The vertebral column protects the _____ cord.
3. The two bones of the lower arm are the _____ and the _____ .
4. Most joints are freely movable _____ joints in which the two bones are separated by a cavity.
5. Muscles work in _____ pairs; the biceps flexes and the triceps extends the lower arm.
6. Maximal sustained contraction of a muscle is called _____ .
7. Actin and myosin filaments are found within cell inclusions called _____ , which are divided into units called _____ .
8. The molecule _____ serves as an immediate source of high-energy phosphate for ATP production in muscle cells.
9. The juncture between axon ending and muscle cell sarcolemma is called an _____ junction.
10. A muscle action potential causes _____ ions to be released from storage sacs, and this signals the muscle fiber to contract.

Answers to Objective Questions

1. axial 2. spinal 3. radius, ulna 4. synovial 5. antagonistic 6. tetanus 7. myofibrils, sarcomeres 8. creatine phosphate 9. neuromuscular 10. calcium

Critical Thinking Questions

1. Is there any similarity between the way that nerves use ATP and the way muscles use ATP? If not, what can you say, in general, about the use of ATP?
2. What is the value of having a jointed skeleton? Demonstrate this by discussing the different types of joints?
3. Muscles and bones vary in bulk. Why do the bones and muscles of the appendages have the most bulk?

Selected Key Terms

actin (ak'tin) one of two major proteins of muscle; makes up thin filaments in myofibrils of muscle fibers. *See* myosin. 230

appendicular skeleton (ap''en-dik'u-lar skel'ĕ-ton) portion of the skeleton forming the upper extremities, the pectoral girdle, the lower extremities, and the pelvic girdle. 218

axial skeleton (ak'se-al skel'ĕ-ton) portion of the skeleton that supports and protects the organs of the head, the neck, and the trunk. 217

compact bone (kom-pakt' bōn) bone in which cells, separated by a matrix of collagen and mineral deposits, are located within Haversian systems. 216

creatine phosphate (kre'ah-tin fos'fāt) a compound unique to muscles that contains a high-energy phosphate bond. 230

insertion (in-ser'shun) the end of a muscle that is attached to a movable bone. 222

muscle action potential (mus'el ak'shun po-ten'shal) an electrochemical change due to increased sarcolemma permeability that is propagated down the T system and results in muscle contraction. 231

myofibril (mi''o-fi'bril) the contractile portion of muscle fibers. 228

myosin (mi'o-sin) one of two major proteins of muscle; makes up thick filaments in myofibrils and is capable of breaking down ATP. *See* actin. 230

neuromuscular junction (nu''ro-mus'ku-lar junk'shun) the point of contact between a nerve cell and a muscle fiber. 231

origin (or'ĭ-jin) end of a muscle that is attached to a relatively immovable bone. 222

oxygen debt (ok'sĭ-jen det) oxygen that is needed to metabolize lactate, a compound that accumulates during vigorous exercise. 230

red marrow (red mar'o) tissue located in the cavity of bones that forms blood cells. 216

synovial joint (sĭ-no've-al joint) a freely movable joint. 220

tone (tōn) The continuous partial contraction of muscle; due to contraction of a small number of muscle fibers at all times. 225

Chapter Twelve

Senses

Chapter Concepts

1 Sense organs are sensitive to environmental stimuli and therefore are termed receptors. Each receptor responds to one type of stimulus, but they all initiate nerve impulses.

2 The sensation realized is the prerogative of the region of the cerebrum receiving nerve impulses.

3 There are sense receptors that respond to mechanical stimuli, chemical stimuli, and light energy. Our knowledge of the outside world is dependent on these stimuli.

The receptors for sight are the rods and cones of the retina. In the portion of the retina featured here, the rods and cones are lined up alternately, with the rods being longer than the cones. Both initiate nerve impulses that go to the brain and make sight possible.

236

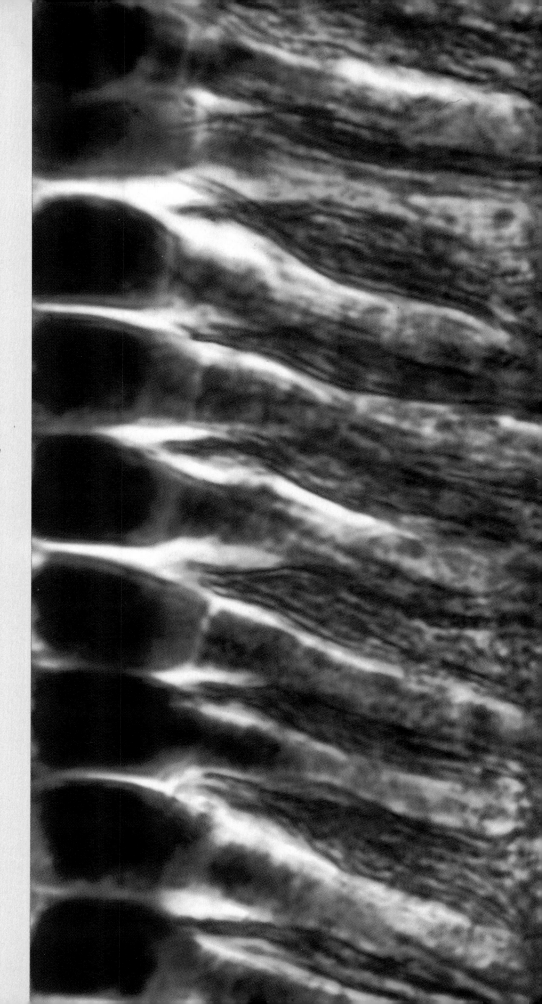

Figure 12.1   Does the man on the left appear to be smaller than the boy on the right? You are interpreting information sent to your brain by your eyes and have been fooled. The ceiling in the man's corner is actually higher than the ceiling in the boy's corner.

We view the world through our sense organs, and we can be fooled. The sense organs send information to be processed by the brain, and its interpretation is not always accurate. The boy and the man in figure 12.1 seem to be equally distant; therefore, our brain tells us that the boy is larger than the man. Actually the man's corner is farther away and has a higher ceiling than the boy's corner in this room[1] that is specially prepared to fool the mind.

General Receptors

Since sense organs receive external and internal stimuli, they are called **receptors.** Each type of receptor is sensitive to only one type of stimulus. Table 12.1 lists the receptors discussed in

[1]The room is a trapezoid with this shape:

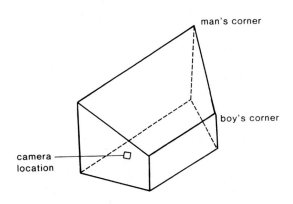

Table 12.1   Receptors

Receptors	Sense	Stimulus
General		
temperature*	hot-cold	heat flow
touch†	touch	mechanical displacement of tissue
pressure†	pressure	mechanical displacement of tissue
pain‡	pain	tissue damage
proprioceptors†	limb placement	mechanical displacement
special		
eye*	sight	light
ear†	hearing	sound waves
	balance	mechanical displacement
taste buds‡	taste	chemicals
olfactory cells‡	smell	chemicals

*Radioreceptors
†Mechanoreceptors
‡Chemoreceptors

this chapter and the stimulus to which each reacts. Receptors are the first component of a reflex arc described in chapter 10. When a receptor is stimulated, it generates nerve impulses that are transmitted to the spinal cord and/or the brain, but we are

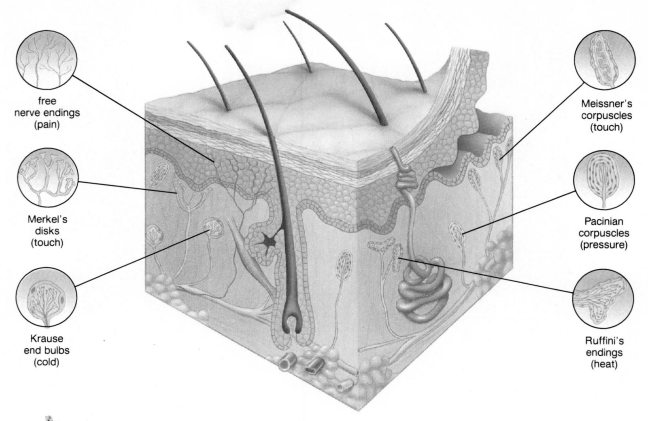

free
nerve endings
(pain)

Merkel's
disks
(touch)

Krause
end bulbs
(cold)

Meissner's
corpuscles
(touch)

Pacinian
corpuscles
(pressure)

Ruffini's
endings
(heat)

Figure 12.2 Receptors in human skin. The classical view is that each receptor has the function indicated. However, investigations in this century indicate that matters are not so clear-cut. For example, microscopic examination of the skin of the ear shows only free nerve endings (pain receptors), and yet the skin of the ear is sensitive to all sensations. Therefore, it appears that the receptors of the skin are only somewhat specialized.

conscious of a sensation only if the impulses reach the cerebrum. The sensory portion of the cerebrum can be mapped according to the parts of the body and types of sensation realized (fig. 10.16).

Receptors only initiate nerve impulses which are conducted to the cord and/or brain. Sensation is dependent upon the cerebrum.

Skin

The skin (fig. 12.2) contains receptors for touch, pressure, pain, and temperature. The skin is a mosaic of these tiny receptors, as you can determine by pressing a metal probe slowly over the skin. At certain points, there is a feeling of pressure, at others, a feeling of hot or cold (depending on the temperature of the probe). Certain parts of the skin contain more receptors for a particular sensation; for example, the fingertips have an abundance of touch receptors.

Specialized receptors in the human skin respond to temperature, touch, pressure, and pain.

Adaptation occurs when a receptor becomes so accustomed to stimuli that it stops generating impulses even though the stimulus is still present. The touch receptors are this type. They can quickly adapt to the clothing we put on so that we are not constantly aware of the feel of the clothes against our skin.

Muscles and Joints

The sense of position and movement of limbs (i.e., proprioception) is dependent upon receptors termed **proprioceptors.** Muscle spindles, discussed in chapter 11, sometimes are considered to be proprioceptors. Stretching of associated muscle fibers causes muscle spindles to increase the rate at which they fire, and for this reason, they are sometimes called **stretch receptors.** The *knee jerk* is a common example of the manner in which muscle spindles act as stretch receptors (fig. 11.13). When the legs are crossed at the knee and the tendon at the knee is tapped, both the tendon and the muscles in the thigh are stretched. Stimulated by the stretching, muscle spindles transmit impulses to the spinal cord, and thereafter the thigh muscles contract. This causes the lower leg to jerk upward in a kicking motion.

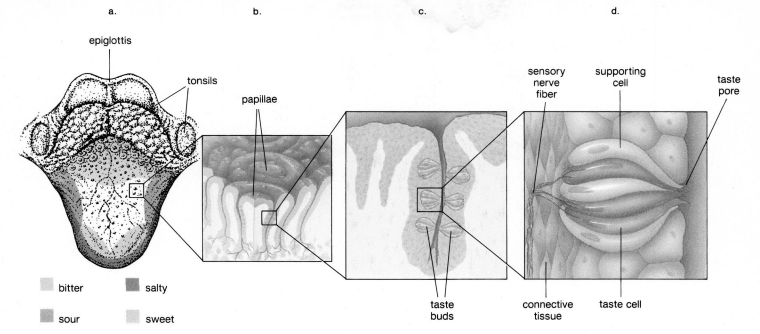

a.

epiglottis

tonsils

papillae

b.

c.

d.

sensory
nerve
fiber

supporting
cell

taste
pore

bitter   salty

sour   sweet

taste
buds

connective
tissue

taste cell

Figure 12.3   Taste buds. *a.* Elevations on the tongue indicate the presence of taste buds. The location of those receptors containing taste buds responsive to sweet, sour, salt, and bitter is indicated. *b.* Enlargement of elevations, called papillae. *c.* The taste buds occur along the walls of the papillae. *d.* Drawing shows the various cells that make up a taste bud. Taste cells in a bud end in microvilli that have receptors for the chemicals that exhibit the tastes noted in (*a*). When the chemicals combine with the receptors, nerve impulses are generated.

Proprioceptors are located in the joints and associated ligaments and tendons that respond to stretching, pressure, and pain. Nerve endings from these receptors are integrated with those received from other types of receptors so that we know the position of body parts.

Special Senses

The special senses include the chemoreceptors for taste and smell, the light receptors for sight, and the mechanoreceptors for hearing and balance.

Chemoreceptors

Taste and smell are called the *chemical senses* because these receptors are sensitive to certain chemical substances in the food we eat and in the air we breathe.

Taste buds are located primarily on the tongue (fig. 12.3). Many lie along the walls of the papillae, the small elevations visible to the naked eye. Isolated ones also are present on the palate, the pharynx, and the epiglottis.

Taste buds are pockets of cells that extend through the tongue epithelium and open at a taste pore. Taste buds have supporting cells and a number of elongated cells that end in microvilli. These cells, which have associated nerve fibers, are sensitive to chemicals. Nerve impulses most probably are generated when chemicals bind to receptor sites found on the microvilli.

It is believed that there are four types of tastes (bitter, sour, salty, sweet) and that taste buds for each are concentrated on the tongue in particular regions (fig. 12.3*a*). Sweet receptors are most plentiful near the tip of the tongue. Sour receptors occur primarily along the margins of the tongue. Salt receptors are most common on the tip and the upper front portion of the tongue. Bitter receptors are located toward the back of the tongue.

The olfactory cells (fig. 12.4) are located high in the roof of the nasal cavity. Each cell ends in a tuft of about five cilia that bear **olfactory receptor** sites for various chemicals. Research resulting in the stereochemical theory of smell suggests that different types of smell are related to the various shapes of molecules. When these combine with the receptor sites, nerve impulses are generated in olfactory nerve fibers. Within an olfactory bulb, an olfactory tract takes this sensory information to an olfactory area of the cerebral cortex. The olfactory receptors, like the touch and temperature receptors, adapt to outside stimuli. In other words, after a while, the presence of a particular chemical no longer causes the olfactory cells to generate nerve impulses, and we no longer are aware of a particular smell.

The sense of taste and the sense of smell supplement each other, creating a combined effect when interpreted by the cerebral cortex. For example, when we have a cold, we think that food has lost its taste, but actually we have lost the ability to sense its smell. This may work in reverse also. When we smell

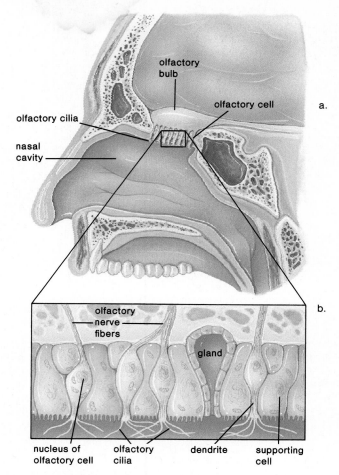

olfactory bulb

olfactory cell

a.

olfactory cilia

nasal cavity

b.

olfactory nerve fibers

gland

nucleus of olfactory cell

olfactory cilia

dendrite

supporting cell

Figure 12.4   Olfactory cell location and anatomy. *a.* The olfactory area in humans is located high in the nasal cavity. *b.* Enlargement of the olfactory cells shows they are modified neurons located between supporting cells. When olfactory cells are stimulated by chemicals in the air, nerve impulses are conducted by olfactory nerve fibers to the olfactory bulb. An olfactory tract within the bulb takes the nerve impulses to the brain.

something, some of the molecules move from the nose down into the mouth region and stimulate the taste buds there. Therefore, part of what we refer to as smell actually may be taste.

The receptors for taste (taste buds) and the receptors for smell (olfactory cilia) work together to give us our sense of taste and our sense of smell.

Photoreceptor—the Eye

The eyeball (fig. 12.5 and table 12.2), an elongated sphere about 2.5 cm in diameter, has three layers, or coats. The outer **sclera** is a white, fibrous layer except for the transparent cornea, the window of the eye. The middle, thin, dark brown layer, the **choroid,** contains many blood vessels and absorbs stray light rays. Toward the front, the choroid thickens and forms a ring-shaped structure, the ciliary body, containing the **ciliary muscle,** which

controls the shape of the lens for near and far vision. Finally, the choroid becomes a thin, circular, muscular diaphragm, the **iris,** which regulates the size of a center hole, the **pupil,** through which light enters the eyeball. The **lens,** attached to the ciliary body by ligaments, divides the cavity of the eye into two chambers. A viscous, gelatinous material, the **vitreous humor,** fills the posterior cavity behind the lens. The anterior cavity between the cornea and the lens is filled with an alkaline, watery solution secreted by the ciliary body and called the **aqueous humor.**

A small amount of aqueous humor is produced continually each day. Normally, it leaves the anterior cavity by way of tiny ducts that are located where the iris meets the cornea. When a person has glaucoma, these drainage ducts are blocked, and aqueous humor builds up. The resulting pressure compresses the arteries that serve the nerve fibers of the retina. The nerve fibers begin to die due to lack of nutrients, and the person becomes partially blind. Over time, total blindness can result. The tendency toward glaucoma is inherited in some individuals.

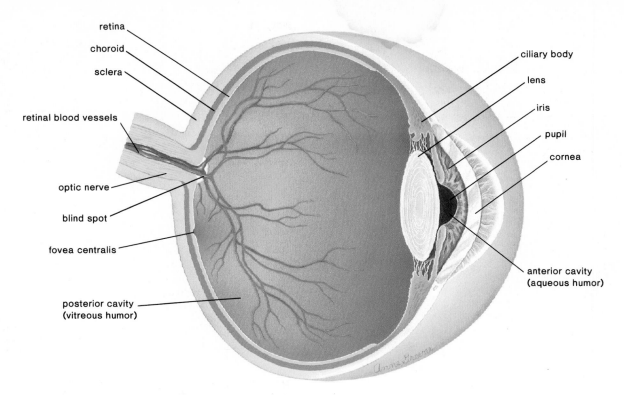

Figure 12.5 Anatomy of the human eye. Notice that the sclera becomes the cornea and that the choroid becomes the ciliary body and the iris. The retina contains the receptors for light. Vision is most acute in the fovea centralis, where there are only cones. A blind spot occurs where the optic nerve leaves the retina and where there are no receptors for light.

Table 12.2 Function of Parts of the Eye

Part	Function
sclera	protects eyeball
cornea	refracts light rays
humors	refracts light rays
iris	regulates light entrance
pupil	admits light
lens	refracts and focuses light rays
choroid	absorbs stray light
ciliary body	holds lens in place, accommodation
retina	contains receptors for sight
rods	makes black-and-white vision possible
cones	makes color vision possible
fovea centralis	contains the cones of the retina
optic nerve	transmits impulses

Retina

The inner layer of the eye, the **retina,** has three layers of cells (fig. 12.6). The layer closest to the choroid contains the sense receptors for sight, the **rods** and the **cones;** the middle layer contains bipolar cells; and the innermost layer contains ganglionic cells whose fibers become the **optic nerve.** Only the rods and the cones contain light-sensitive pigments, and therefore light must penetrate to the back of the retina before nerve impulses are generated. Nerve impulses initiated by the rods and the cones are passed to the bipolar cells, which in turn pass them to the ganglionic cells. The fibers of the ganglionic cells pass in front of the retina, forming the optic nerve, which turns to pierce the layers of the eye. Notice in figure 12.6 that there are many more rods and cones than ganglionic cells. In fact, the retina has as many as 150 million rods but only 1 million ganglionic cells and optic nerve fibers. This means that there is considerable mixing of messages and a certain amount of integration before nerve impulses are sent to the occipital lobe of the brain. There are no rods or cones where the optic nerve passes through the retina; therefore, this is a **blind spot,** where vision is impossible.

The retina contains a very special region called the **fovea centralis** (fig. 12.5), an oval, yellowish area with a depression where there are only cone cells. Vision is most acute in the fovea centralis.

The eye has three layers: the outer sclera, the middle choroid, and the inner retina. Only the retina contains the receptors for sight.

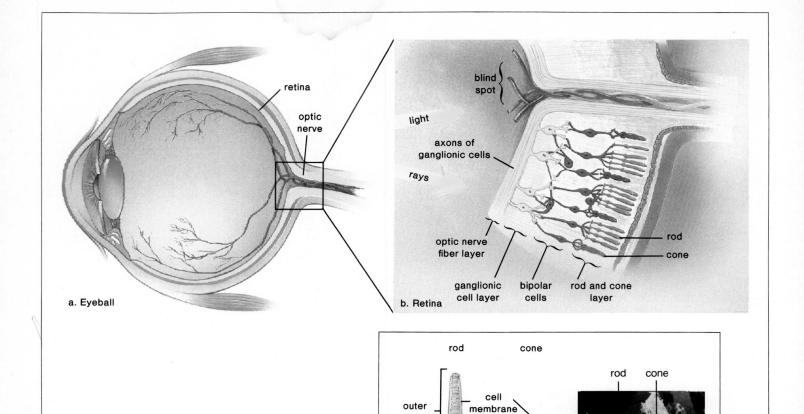

a. Eyeball

b. Retina

blind spot

light

rays

axons of ganglionic cells

optic nerve fiber layer

ganglionic cell layer

bipolar cells

rod and cone layer

rod

cone

rod

cone

outer segment

inner segment

cell body and nucleus

cell membrane

disks

connecting cilium

nucleus

fibers

synaptic endings

rod

cone

c. Drawing

d. Micrograph

Figure 12.6   Anatomy of the retina. *a.* The retina is the inner layer of the eye. *b.* Rods and cones are located at the back of the retina, followed by the bipolar and ganglionic cells, whose fibers become the optic nerve. Notice that rods share bipolar cells but cones do not. Cones, therefore, distinguish more detail. *c.* The photosensitive pigment is located in the membranous disks of the outer segment of rods and cones. *d.* Scanning electron micrograph of rods and cones. The cones are responsible for color vision; rods are responsible for night vision.

Figure 12.7   Focusing. Light rays from each point on an object are bent by the cornea and the lens in such a way that they are directed to a single point after emerging from the lens. By this process, an inverted image of the object forms on the retina.

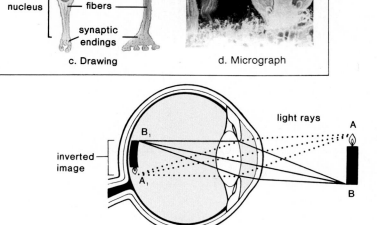

light rays

inverted image

B_1

A_1

A

B

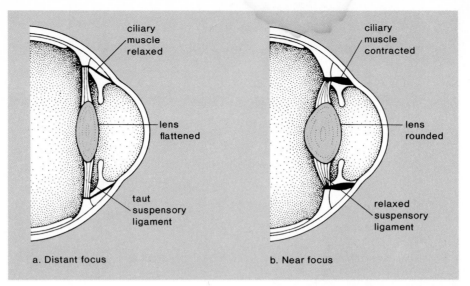

Figure 12.8   Accommodation. *a.* When the eye focuses on a distant object, the lens is flat because the ciliary muscle is relaxed and the suspensory ligament is taut. *b.* When the eye focuses on a near object, the lens is rounded because the ciliary muscle contracts, and causes the suspensory ligament to relax.

Physiology

Focusing   When we look at an object, light rays are **focused** on the retina (fig. 12.7). In this way, an *image* of the object appears on the retina. The image on the retina occurs when the rods and the cones in a particular region are excited. Obviously, the image is much smaller than the object. In order to produce this small image, light rays must be bent (refracted) and brought into focus. They are bent as they pass through the cornea. Further bending occurs as the rays pass through the lens and the humors.

Accommodation   Light rays are reflected from an object in all directions. For distant objects, only nearly parallel rays enter the eye, and the cornea alone is needed for focusing. For close objects, many of the rays are at sharp angles to one another, and additional focusing is required. The lens provides this additional focusing power as **accommodation** occurs. The lens remains flat when we view distant objects and rounds up when we view close objects. The shape of the lens is controlled by the ciliary muscle within the ciliary body. When we view a distant object, the ciliary muscle is relaxed, causing the suspensory ligaments attached to the ciliary body to be under tension; therefore, the lens remains relatively flat (fig. 12.8). When we view a close object, the ciliary muscle contracts, releasing the tension on the suspensory ligaments, and the lens rounds up due to its natural elasticity. Because close work requires contraction of the ciliary muscle, it very often causes eyestrain.

With aging, the lens loses some of its elasticity and is unable to accommodate. This usually necessitates the wearing of corrective lenses, as is discussed on page 245. The lens is also subject to *cataracts;* it can become opaque and unable to transmit rays of light. Special cells within the interior of the lens contain proteins called crystallin. Recent research suggests that cataracts develop when these proteins become oxidized, causing their

three-dimensional shape to change. If so, researchers believe that they eventually may be able to find ways to restore the normal configuration of crystallin so that cataracts can be treated medically instead of surgically. For the present, however, surgery is the only viable treatment. First, a surgeon opens the eye near the rim of the cornea. Zonulysin, an enzyme, may be used to digest away the ligaments holding the lens in place. Most surgeons then use a cryoprobe that freezes the lens for easy removal. An intraocular lens attached to the iris can be implanted in the eye so that the patient need not wear thick glasses or contact lenses.

The lens, assisted by the cornea and the humors, focuses images on the retina.

Inverted Image   The image on the retina is upside down (fig. 12.7), and it is thought that perhaps this image is righted in the brain by experience. In one experiment, scientists wore glasses that inverted the field of vision. At first, they had difficulty adjusting to the placement of the objects, but they soon became accustomed to their inverted world. Experiments such as this suggest that if we see the world upside down, the brain learns to see it right side up.

Stereoscopic Vision   We can see well with either eye alone, but the two eyes functioning together provide us with **stereoscopic vision.** Normally, the two eyes are directed by the eye muscles toward the same object, and therefore the object is focused on corresponding points of the two retinas. Each eye, however, sends its own information to the brain about the placement of the object because each forms an image from a slightly different angle. These data are pooled to produce depth perception by a two-step process. First, because the optic nerves

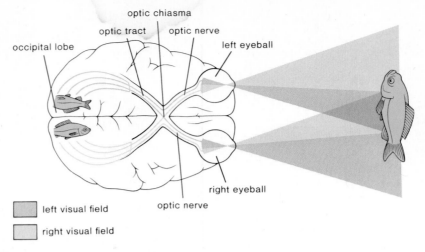

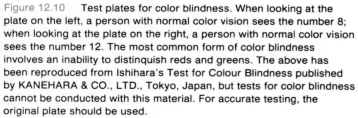

Figure 12.9   Optic chiasma. Both eyes "see" the entire object, but information from the right half of each retina goes to the right occipital lobe and information from the left half of the retina goes to the left occipital lobe. When the information is pooled, the brain "sees" the entire object in depth.

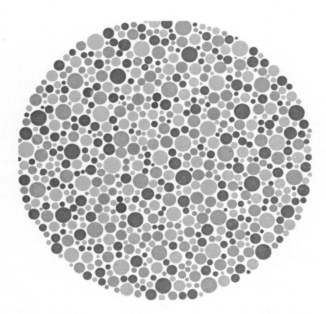

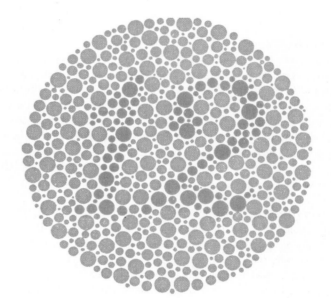

Figure 12.10   Test plates for color blindness. When looking at the plate on the left, a person with normal color vision sees the number 8; when looking at the plate on the right, a person with normal color vision sees the number 12. The most common form of color blindness involves an inability to distinguish reds and greens. The above has been reproduced from Ishihara's Test for Colour Blindness published by KANEHARA & CO., LTD., Tokyo, Japan, but tests for color blindness cannot be conducted with this material. For accurate testing, the original plate should be used.

cross at the optic chiasma (fig. 12.9), one-half of the brain receives information from both eyes about the same part of an object. Later, the two halves of the brain communicate to arrive at a complete three-dimensional interpretation of the whole object.

The anatomy and the physiology of the brain allow us to see the world right side up and in three dimensions.

Biochemistry In *dim light,* the pupils enlarge so that more rays of light can enter the eyes. As the rays of light enter, they strike the rods and the cones, but only the 150 million rods located in the periphery, or sides, of the eyes are sensitive enough to be stimulated by this faint light. The rods do not detect fine detail or color, so at night, for example, all objects appear to be blurred and to be a shade of gray. Rods do detect even the slightest motion, however, because of their abundance and position in the eyes.

Although it has been known for some time that vision generated by the rods is dependent on the presence of rhodopsin, the actual events only recently have been worked out in detail. Many molecules of rhodopsin, also called visual purple, are located within the membrane of the disks found in the outer segment (fig. 12.6c) of the rods. **Rhodopsin** is a complex molecule that contains a protein (*opsin*) and a pigment molecule called *retinal.* When light strikes rhodopsin, it breaks down to its components and this generates nerve impulses. The more rhodopsin present in the rods, the more sensitive are our eyes to dim light. Therefore, during the time required for adjustment to dim light, when we find it difficult to see, rhodopsin is being formed in the rods. Retinal is a derivative of vitamin A. Vitamin A is abundant in carrots, so the suggestion that we should eat carrots for good vision is not without foundation.

In *bright light,* the pupils get smaller so that less light enters the eyes. The cones, located primarily in the fovea centralis, are active and detect the fine detail and color of an object. In order to perceive depth, as well as to see color, we turn our eyes so that reflected light from the object strikes the fovea.

Color vision (fig. 12.10) has been shown to depend on three kinds of cones that contain pigments sensitive to either blue, green, or red light. The nerve impulses generated from one type of cone not only stimulate certain cells in the visual cortex of the brain, they also inhibit the reception of impulses from other types of cones. Complete color blindness is extremely rare. In most instances, a particular type of cone is lacking or deficient in number. The lack of red or green cones is the most common, affecting about 5% of the American population. If the eye lacks red cones, the green colors are accentuated, and vice versa.

The sense receptors for sight are the rods and the cones. The rods are responsible for vision in dim light, and the cones are responsible for vision in bright light and for color vision. When either is stimulated, nerve impulses begin and are transmitted in the optic nerve to the brain.

Corrective Lenses

The majority of people can see what is designated as a size 20 letter 20 feet away and so are said to have 20/20 vision. Persons who can see close objects but cannot see the letters from this distance are said to be nearsighted. *Nearsighted* people can see near better than they can see far. These individuals often have a steeply curved cornea or an elongated eyeball, and when they attempt to look at a far object, the image is brought to focus in front of the retina (fig. 12.11). They can see near because they can adjust the lens to allow the image to focus on the retina, but to see far, these people must wear concave lenses that diverge the light rays so that the image can be focused on the retina. The condition can also be corrected surgically. Laser corneal sculpting is an experimental alternative to the radial K procedure discussed in the human issue box. The laser blasts away one layer of cells at a time to reduce the curvature of the cornea.

Human Issue

A Russian ophthalmologist, Svyatoslav Fyodorov, has developed the eye operation known as radial keratotomy, or radial K, to correct nearsightedness. Eight to sixteen cuts are made in the cornea so that they radiate out from the center like spokes in a wheel. When the cuts heal, the cornea is flattened. Although some patients are satisfied with the result, others complain of glare and varying visual acuity. If you were an ophthalmologist, would you recommend this operation to patients who simply want to improve their appearance by not wearing glasses? Under what circumstances would you make the recommendation?

Persons who can easily see the optometrist's chart but cannot see close objects well are *farsighted;* these individuals can see far away better than they can see near. They often have a shortened eyeball, and when they try to see near, the image is focused behind the retina. When the object is far away, the lens can compensate for the short eyeball, but when the object is close, these persons must wear a convex lens to increase the bending of light rays so that the image can be focused on the retina.

When the cornea or lens is uneven, the image is fuzzy because the light rays cannot be focused evenly on the retina. This condition, called astigmatism, can be corrected by an unevenly ground lens to compensate for the uneven cornea.

Bifocals As mentioned earlier, with normal aging, the lens loses some of its ability to change shape in order to focus on close objects. Because nearsighted individuals still have difficulty seeing distant objects clearly, they must wear bifocals. In bifocals, the upper part of the lens is for distant vision and the remainder is for near vision.

The shape of the eyeball determines the need for corrective lenses; the inability of the lens to accommodate as we age also requires corrective lenses for close vision.

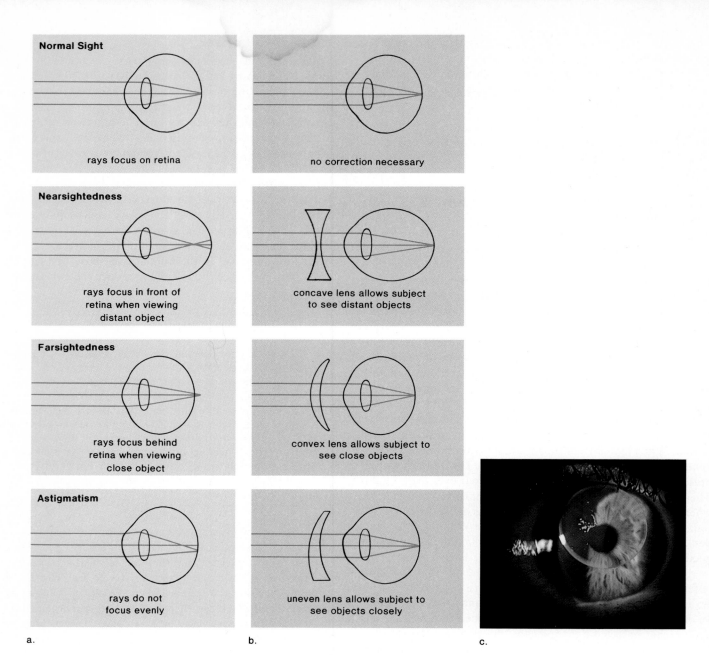

Normal Sight

rays focus on retina

no correction necessary

Nearsightedness

rays focus in front of
retina when viewing
distant object

concave lens allows subject
to see distant objects

Farsightedness

rays focus behind
retina when viewing
close object

convex lens allows subject to
see close objects

Astigmatism

rays do not
focus evenly

uneven lens allows subject to
see objects closely

a.

b.

c.

Figure 12.11   Common abnormalities of the eye and possible
corrective lenses. *a.* The cornea and the lens function in bringing light
rays (lines) to focus, but sometimes they are unable to compensate for
the shape of the eyeball or for an uneven cornea. *b.* In these instances,
corrective lenses can allow the individual to see normally. *c.*
Ophthalmologists can examine the fit of a contact lens by using a
narrow beam of light from a ''slit lamp'' while looking through a
biomicroscope.

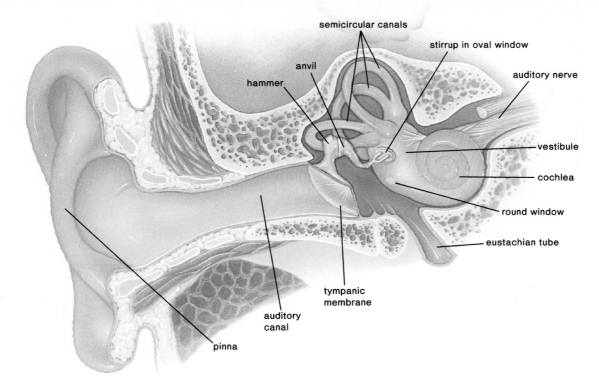

Figure 12.12   Anatomy of the human ear. In the middle ear, the hammer, the anvil, and the stirrup amplify sound waves. Otosclerosis is a condition in which the stirrup becomes attached to the inner ear and is unable to carry out its normal function. The stirrup can be replaced by a plastic piston, and thereafter the individual hears normally because sound waves are transmitted as usual to the cochlea, which contains the receptors for hearing.

Table 12.3   The Ear

	Outer ear	Middle ear	Inner ear	
			Cochlea	Sacs plus semicircular canals
Function	directs sound waves to tympanic membrane	picks up and amplifies sound waves	hearing	maintains equilibrium
Anatomy	pinna auditory canal	tympanic membrane ossicles	contains organ of Corti; auditory nerve starts here	saccule and utricle semicircular canals
Media	air	air (eustachian tube)	fluid	fluid

Path of vibration: Sound waves—vibration of tympanic membrane—vibration of hammer, anvil, and stirrup—vibration of oval window—fluid pressure waves in canals of inner ear lead to stimulation of hair cells—bulging of round window.

Mechanoreceptor—the Ear

The ear accomplishes two sensory functions: balance and hearing. The sense cells for both of these are located in the inner ear and consist of hair cells with cilia that respond to mechanical stimulation. When the cilia of any particular hair cell are displaced, the cell generates nerve impulses that are sent along a cranial nerve to the brain.

Anatomy

Figure 12.12 is a drawing of the ear and table 12.3 lists the parts of the ear. The ear has three divisions: outer, middle, and inner. The **outer ear** consists of the **pinna** (external flap) and the **auditory canal.** The opening of the auditory canal is lined with fine hairs and sweat glands. Modified sweat glands that secrete earwax, a substance that helps to guard the ear against the entrance of foreign materials, such as air pollutants, are in the upper wall of the canal.

The **middle ear** begins at the **tympanic membrane** (eardrum) and ends at a bony wall containing two small openings covered by membranes. These openings are called the **oval and round windows.** Three small bones are found between the tympanic membrane and the oval window. Collectively called the **ossicles,** individually they are the **hammer** (malleus), the **anvil** (incus), and the **stirrup** (stapes) because their shapes resemble these objects (fig. 12.12). The hammer adheres to the tympanic membrane, and the stirrup touches the oval window. The posterior wall has an opening that leads to sinuses within the mastoid portion of the temporal bone.

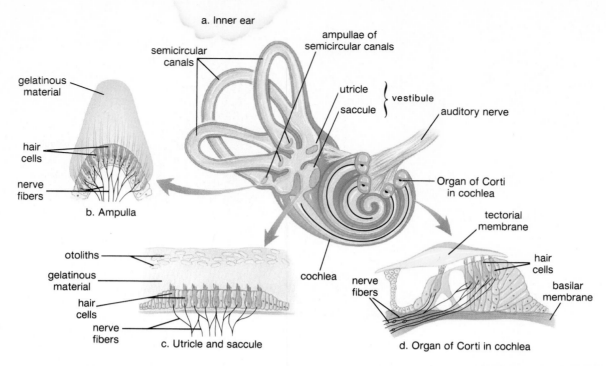

a. Inner ear

ampullae of semicircular canals

semicircular canals

gelatinous material

utricle

saccule } vestibule

auditory nerve

hair cells

nerve fibers

b. Ampulla

Organ of Corti in cochlea

tectorial membrane

hair cells

basilar membrane

otoliths

gelatinous material

hair cells

nerve fibers

c. Utricle and saccule

cochlea

nerve fibers

d. Organ of Corti in cochlea

Figure 12.13   Inner ear. *a.* The inner ear contains the semicircular canals, a vestibule, and the cochlea. The cochlea has been cut to show the location of the organ of Corti. *b.* There is an ampulla at the base of each semicircular canal that contains the receptors (hair cells) for dynamic equilibrium. *c.* The utricle and the saccule, small sacs that contain the receptors (hair cells) for static equilibrium, are in a vestibule. *d.* The sense organ for hearing, the organ of Corti, is in the cochlea. The organ of Corti also consists of hair cells.

An **eustachian tube,** which extends from each middle ear to the nasopharynx, permits equalization of air pressure to occur. Chewing gum, yawning, and swallowing in elevators and airplanes help to move air through the eustachian tubes upon ascent and descent to equalize air pressure across the tympanic membrane between the outer and middle ear.

The **semicircular canals** are arranged so that there is one in each dimension of space. The base of each canal, called the **ampulla** (fig. 12.13*b*), is slightly enlarged. Within the ampullae are little hair cells whose cilia are inserted into a gelatinous material.

A vestibule, or a chamber, lies between the semicircular canals and the cochlea. It contains two small membranous sacs called the **utricle** and the **saccule** (fig. 12.13*c*). Within both of these are little hair cells whose cilia protrude into a gelatinous material. Resting on this substance are calcium carbonate granules, or **otoliths.**

The **cochlea** resembles the shell of a snail because it spirals. Within the tubular cochlea are three canals: the vestibular canal, the **cochlear canal,** and the tympanic canal. Along the length of the basilar membrane, which forms the lower wall of the cochlear canal, are little hair cells whose cilia come into contact with another membrane called the tectorial membrane. The hair cells of the cochlear canal plus the **tectorial membrane** are called the **organ of Corti** (fig. 12.13d). When this organ sends nerve impulses to the cerebral cortex, they are interpreted as sound.

The outer ear, the middle ear, and the cochlea are necessary for hearing. The semicircular canals and the vestibule are concerned with the sense of balance.

Physiology

Balance (Equilibrium)   The sense of balance has been divided into two senses: *dynamic equilibrium,* requiring a knowledge of angular and/or rotational movement, and *static equilibrium,* requiring a knowledge of movement in one plane, either vertical or horizontal.

Dynamic equilibrium is required when the body is moving (fig. 12.14*a*). At that time, the fluid within the semicircular canals flows over and displaces the gelatinous material within the ampullae. This causes the cilia of the hair cells to bend and to initiate nerve impulses that travel to the brain. Continuous movement of the fluid in the semicircular canals causes one form of motion sickness.

When the body is still, the otoliths in the utricle and the saccule rest on the gelatinous material above the hair cells (fig. 12.14*b*). Static equilibrium is required when the body moves horizontally or vertically. At that time, the otoliths are displaced and the gelatinous material sags, bending the cilia of the hair cells beneath. Now the hair cells generate nerve impulses that travel to the brain.

Movement of the otoliths within the utricle and the saccule is important for static equilibrium. Movement of fluid within the semicircular canals contributes to our sense of dynamic equilibrium.

Hearing   The process of hearing begins when sound waves enter the auditory canal (fig. 12.15*a*). Just as ripples travel across the surface of a pond, sound travels by the successive vibrations of molecules. Ordinarily, sound waves do not carry much energy, but when a large number of waves strikes the eardrum, it moves

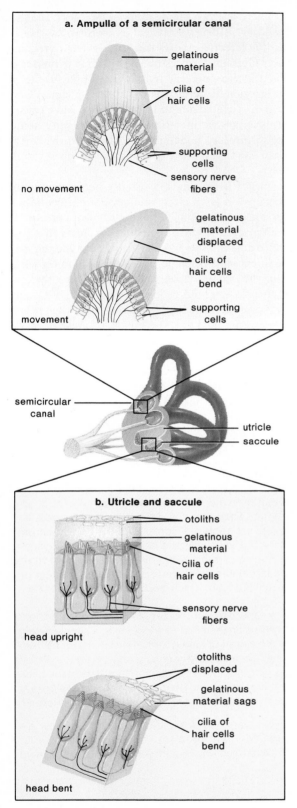

a. Ampulla of a semicircular canal

gelatinous material

cilia of hair cells

supporting cells

sensory nerve fibers

no movement

gelatinous material displaced

cilia of hair cells bend

supporting cells

movement

semicircular canal

utricle

saccule

b. Utricle and saccule

otoliths

gelatinous material

cilia of hair cells

sensory nerve fibers

head upright

otoliths displaced

gelatinous material sags

cilia of hair cells bend

head bent

Figure 12.14 Sense of balance. *a.* The ampullae of the semicircular canals contain hair cells with cilia embedded in a gelatinous material. When the head rotates, the material is displaced and the bending of the cilia initiates nerve impulses in sensory nerve fibers. This permits dynamic equilibrium. *b.* A vestibule contains the utricle and the saccule, sacs that contain hair cells with cilia embedded in a gelatinous material. When the head bends, otoliths are displaced, causing the gelatinous material to sag and the cilia to bend. This initiates nerve impulses in sensory nerve fibers. This permits static equilibrium.

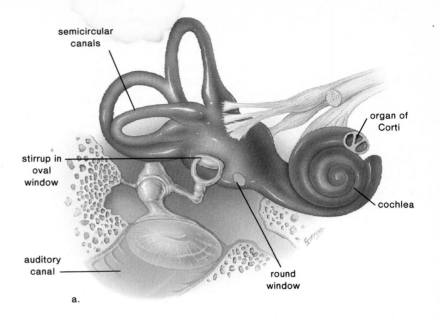

semicircular canals

organ of Corti

stirrup in oval window

cochlea

auditory canal

round window

a.

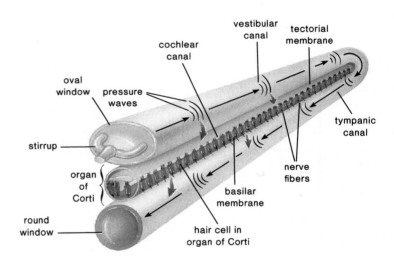

vestibular canal

tectorial membrane

cochlear canal

oval window

pressure waves

stirrup

organ of Corti

round window

tympanic canal

nerve fibers

basilar membrane

hair cell in organ of Corti

b.

Figure 12.15 Sense of hearing. *a.* The organ of Corti is located within the cochlea. *b.* In the unwound cochlea, the organ of Corti consists of hair cells resting on the basilar membrane with the tectorial membrane above. The arrows represent pressure waves that move from the oval window to the round window due to the motion of the stirrup. These pressure waves cause the basilar membrane to vibrate and the cilia of at least a portion of the 15,000 hair cells to bend against the tectorial membrane. The generated nerve impulses result in hearing.

back and forth (vibrates) ever so slightly. The hammer then takes the pressure from the inner surface of the eardrum and passes it by way of the anvil to the stirrup in such a way that the pressure is multiplied about 20 times as it moves from the eardrum to the stirrup. The stirrup strikes the oval window, causing it to vibrate, and in this way, the pressure is passed to the fluid within the inner ear.

If the cochlea is unwound, as shown in figure 12.15*b,* you can see that the vestibular canal connects with the tympanic canal and that pressure waves move from one canal to the other toward the round window, a membrane that can bulge to absorb the pressure. As a result of the movement of the fluid within the cochlea, the basilar membrane moves up and down, and the cilia of the hair cells rub against the tectorial membrane. This bending of the cilia initiates nerve impulses that pass by way of the **auditory nerve** to the temporal lobe of the brain, where the impulses are interpreted as a sound.

The organ of Corti is narrow at its base but widens as it approaches the tip of the cochlear canal. Each part of the organ is sensitive to different wave frequencies, or pitch. Near the tip, the organ of Corti responds to low pitches, such as a tuba, and near the base, it responds to higher pitches, such as a bell or a whistle. The neurons from each region along the length of the cochlea lead to slightly different areas in the brain. The pitch sensation we experience depends upon which of these areas of the brain is stimulated.

Volume is a function of the amplitude of sound waves. Loud noises cause the fluid of the cochlea to oscillate to a greater degree, and this, in turn, causes the basilar membrane to move up and down to a great extent. The resulting increased stimulation is interpreted by the brain as loudness. It is believed that tone is an interpretation of the brain based on the distribution of hair cells stimulated.

The sense receptors for sound are hair cells on the basilar membrane (the organ of Corti). When the basilar membrane vibrates, the delicate hairs touch the tectorial membrane, initiating nerve impulses that are transmitted in the auditory nerve to the brain.

Human Issue

Loss of sight is probably more critical than a hearing loss. Could this be why many young people carelessly listen to rock music that is many times louder than is safe for retention of their hearing? If not, what are the pressures that cause people to endanger their sense of hearing? The government has regulations requiring that employers protect the ears of their employees from loud noises. Should the government also make it illegal to play music so loudly at concerts and in other public places that it endangers the hearing of listeners?

Deafness   There are two major types of deafness: *conduction deafness* and *nerve deafness.* Conduction deafness can be due to a congenital defect, as those that occur when a pregnant woman contracts German measles during the first trimester of pregnancy. (For this reason every female should be sure to be immunized against rubella before the childbearing years.) Conduction deafness also can be due to infections that have caused the ossicles to fuse, restricting the ability to magnify sound waves. Because respiratory infections can spread to the ear by way of the eustachian tubes, every cold and ear infection should be taken seriously.

The wearing of a hearing aid usually helps conduction deafness. A hearing aid is an electronic device that contains a microphone for converting sound into an electrical current, an amplifier that amplifies the current, and an earphone that converts the amplified current into a louder sound than the original.

Nerve deafness most often occurs when cilia on the sense receptors within the cochlea have worn away. Since this can happen with normal aging, old people are more likely to have trouble hearing; however, nerve deafness also occurs when people listen to loud music amplified to 130 decibels. Because the usual types of hearing aids are not helpful for nerve deafness, it is wise to avoid subjecting the ears to any type of continuous loud noise. Costly cochlear implants that directly stimulate the auditory nerve are available, but those who have these electronic devices report that the speech they hear is like that of a robot.

What's Your Diagnosis?

Nancy is a woman in her mid-twenties. Over the past six months, she has noticed that it is difficult to hear sounds, including the voices of people at a normal speaking volume from more than fifteen feet away. During her teenage years, Nancy did a lot of swimming in public swimming pools. Several times she developed throat infections that spread to her right ear by way of the eustachian tube, which leads to the middle ear from the throat.

Nancy is concerned about this problem and seeks professional help. A physician tests her hearing, using a simple tuning fork. He holds the vibrating tuning fork a few inches away from her right ear. Nancy can bearly hear its humming sound. When the stem of the vibrating fork is placed on the bone at the back of her ear (temporal bone), the hearing sensation does not improve. However, when the left ear is tested, hearing is normal.

1. What is your diagnosis of Nancy's problem? 2. How are the symptoms of her problem consistent with your diagnosis? 3. What remedy would you recommend to help Nancy?

"We have an idea of what noise does to the ear," David Lipscomb (of the University of Tennessee Noise Laboratory) says. "There's a pretty clear cause-effect relationship." And these scanning electron micrographs of the cochlea's tiny structures graphically document noise trauma to the inner ear.

Hair cells transmit the mechanical energy of sound waves into those neural impulses that the brain interprets as sound. Loud noise can damage or destroy hair cells, as these scanning electron micrographs illustrate.

Hair cells come in two varieties: a single row of inner cells and a triple row of outer ones. "Outer cells degenerate before inner cells," notes Clifton Springs, New York, otolaryngologist Stephen Falk. The most subtle change wrought by noise is a development of vesicles, or blisterlike protrusions along the walls of the hair cells' cilia. Continued assault by noise leads to the rupture of the vesicles and to damage. In addition, the "cuticular plate"—base tissue supporting the cilia—may soften, followed by swelling and ultimate degeneration of hair cells.

But sensory hair cells are not the only structures at risk. Adjacent inner ear cells . . . may undergo vacuolation—development of degenerative empty spaces in cells. Even nerve fibers synapsing at the hair cells' roots may die. In the final phase of noise-induced cochlear damage, the organ of Corti—of which hair cells and supporting cells are a part—is completely denuded of its natural components and is covered by a layer of scar tissue.

Picturing the Effects of Noise

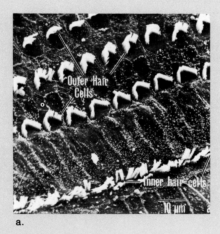

Damage to organ of Corti due to loud noise. *a.* Normal organ of Corti. *b.* Organ of Corti after 24-hour exposure to noise level typical of rock music. Note scars where cilia have worn away.

a.

b.

Summary

All receptors are the first part of a reflex arc, and they initiate nerve impulses that eventually reach the cerebrum, where sensation occurs. Among the general receptors are those located in the skin and proprioceptors located in the joints. The chemoreceptors for taste and smell are special receptors, as are the eyes and the ears.

Vision is dependent on the eye, the optic nerve, and the occipital lobe of the brain. The rods, receptors for vision in dim light, and the cones, receptors that depend on bright light and provide color and detailed vision, are located in the retina, the inner layer of the eyeball. The cornea, the humors, and especially the lens bring the light rays to focus on the retina. To see a close object, accommodation occurs as the lens rounds up. Due to the optic chiasma, both sides of the brain must function together to give us three-dimensional vision.

Hearing is a specialized sense dependent on the ear, the auditory nerve, and the temporal lobe of the brain. The outer and middle portions of the ear simply convey and magnify the sound waves that strike the oval window. Its vibrations set up pressure waves within the cochlea, which contains the organ of Corti, consisting of hair cells with the tectorial membrane above. When the hair cells strike this membrane, nerve impulses are initiated that finally result in hearing.

The ear also contains receptors for our sense of balance. Dynamic equilibrium is dependent on the stimulation of hair cells within the ampullae of the semicircular canals. Static equilibrium relies on the stimulation of hair cells by otoliths within the utricle and the saccule.

Study Questions

1. Name two factors that all receptors have in common. (p. 237)
2. What are the general receptors (p. 238) and what type are the special receptors? (p. 239)
3. Discuss the receptors of the skin and the joints. (p. 238)
4. Discuss the chemoreceptors. (p. 239)
5. Describe the anatomy of the eye (pp. 240–241), and explain focusing and accommodation. (pp. 242–243)
6. Describe sight in dim light. What chemical is responsible for vision in dim light? (p. 245) Discuss color vision. (p. 243)
7. Relate the need for corrective lenses to three possible shapes of the eye. (pp. 245–246) Discuss bifocals. (p. 245)
8. Describe the anatomy of the ear and how we hear. (pp. 247–249)
9. Describe the role of the utricle, the saccule, and the semicircular canals in balance. (p. 248)
10. Discuss the two causes of deafness, including why young people frequently suffer loss of hearing. (p. 250)

Objective Questions

1. The sense organs for position and movement are called _____ .
2. Taste buds and olfactory receptors are termed _____ because they are sensitive to chemicals in the air and food.
3. The receptors for sight, the _____ and the _____ , are located in the _____ , the inner layer of the eye.
4. The cones give us _____ vision and work best in _____ light.
5. The lens _____ for viewing close objects.
6. People who are nearsighted cannot see objects that are _____ . A _____ lens restores this ability.
7. The ossicles are _____ , the _____ and the _____ .
8. The semicircular canals are involved in our sense of dynamic _____ .
9. The organ of Corti is located in the _____ canal of the _____ .
10. Vision, hearing, taste, and smell do not occur unless nerve impulses reach the proper portion of the _____ .

Answers to Objective Questions

10. brain
anvil, stirrup 8. equilibrium 9. cochlear, cochlea
(accommodation) 6. distant, concave 7. hammer,
cones, retina 4. color, bright (day) 5. rounds up
1. proprioceptors 2. chemoreceptors 3. rods,

Label this Diagram.
See figure 12.5 (p. 241) in text.

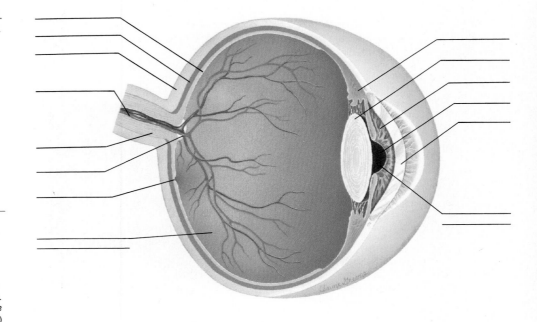

Critical Thinking Questions

1. In general, list the steps, from sense organ to brain, by which sensation occurs. Discuss the weaknesses and strengths of depending on such a method for knowledge of the outside world.
2. Question 1 assumes that the brain is responsible for sensation. For sight, why might there be some integration of data in the eye itself? Does this fact lend strength or weakness to the way by which we see?
3. How could you redesign human vision so it would be improved and more efficient?

Selected Key Terms

accommodation (ah-kom''o-da'shun) lens adjustment in order to see close objects. 243

choroid (ko'roid) the vascular, pigmented middle layer of the eyeball. 240

ciliary muscle (sil'-e-er''e mus'el) a muscle that controls the curvature of the lens of the eye. 240

cochlea (kok'le-ah) that portion of the inner ear that resembles a snail's shell and contains the organ of Corti, the sense organ for hearing. 248

cone (kōn) bright-light receptor in the retina of the eye that detects color and provides visual acuity. 241

fovea centralis (fo've-ah sen-tra'lis) region of the retina consisting of densely packed cones that is responsible for the greatest visual acuity. 241

lens (lenz) a clear membranelike structure found in the eye behind the iris. The lens brings objects into focus. 240

organ of Corti (or'gan uv kor'ti) a portion of the inner ear that contains the receptors for hearing. 248

otolith (o'to-lith) calcium carbonate granule associated with ciliated cells in the utricle and the saccule. 248

proprioceptor (pro''pre-o-sep'tor) receptor that assists the brain in knowing the position of the limbs. 238

retina (ret'i-nah) the innermost layer of the eyeball that contains the rods and the cones. 241

rhodopsin (ro-dop'sin) visual purple, a pigment found in the rods. 245

rod (rod) dim-light receptor in the retina of the eye that detects motion but no color. 241

saccule (sak'ūl) a saclike cavity that makes up part of the membranous labyrinth of the inner ear; contains receptors for static equilibrium. 248

sclera (skle'rah) white fibrous outer layer of the eyeball. 240

semicircular canal (sem''e-ser'ku-lar kah-nal') tubular structure within the inner ear that contains the receptors responsible for the sense of dynamic equilibrium. 248

tympanic membrane (tim-pan'ik mem'brān) membrane located between the outer and middle ear that receives sound waves; the eardrum. 247

utricle (u'tre-k'l) saclike cavity that makes up part of the membranous labyrinth of the inner ear; contains receptors for static equilibrium. 248

Chapter Thirteen

Hormones

Chapter Concepts

1 The endocrine system utilizes chemical messengers called hormones to bring about coordination of body parts.

2 Endocrine glands are usually ductless glands that secrete hormones directly into the bloodstream.

3 In general, secretion of hormones is controlled by a negative feedback mechanism.

4 Malfunctioning of endocrine glands can bring about a dramatic change in appearance and cause early death.

5 Environmental signals are not unique to the endocrine system and most likely are present whenever one biological element influences the behavior or the chemistry of another element.

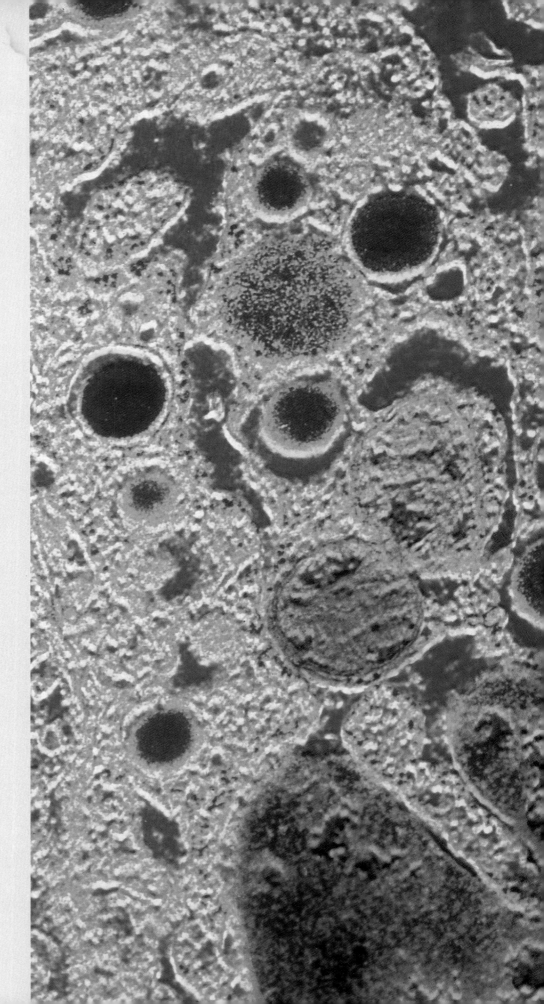

The islets of Langerhans found in the pancreas contain specialized cells. This transmission electron micrograph has been colored: the beta cells, which secrete the hormone insulin, appear green and orange, while the alpha cells, which secrete the hormone glucagon, appear dark brown.

254

Figure 13.1   Today's pregnancy tests, available in local drugstores, are based on an antigen-antibody reaction that can be detected visibly. The antigen is the hormone chorionic gonadotropin present in a pregnant woman's body and the antibody is monoclonal (p. 144).

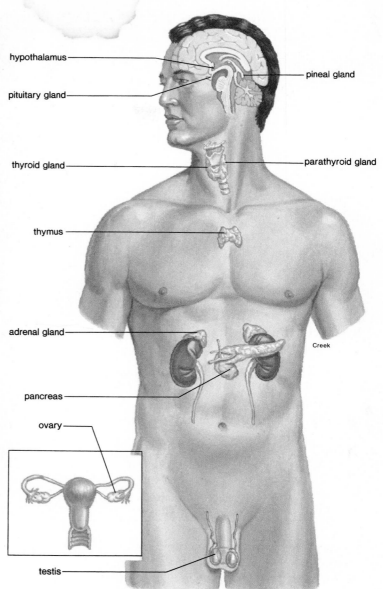

Figure 13.2   Anatomical location of major endocrine glands in the body. The hypothalamus produces the hormones secreted at the posterior pituitary and controls the anterior pituitary, which in turn controls the hormonal secretions of the thyroid, the adrenal cortex, and the sex organs (testes and ovaries). Female organs are shown in the insert.

The first pregnancy test was developed in 1928 after it was discovered that the urine of a pregnant woman contains a gonadotropic hormone that is secreted in such excess it spills over into the urine. When the urine of a pregnant woman was injected into a rabbit, characteristic changes in the animal's reproductive organs signaled pregnancy. When the urine was injected into a frog, it caused the frog to ovulate and produce eggs. The basis of a pregnancy test is the same today except that antibodies to the hormone are added to a blood or urine sample (fig. 13.1). An observable antibody-antigen reaction (called agglutination) indicates that the woman is pregnant.

Mechanism of Action

Along with the nervous system, hormones coordinate the functioning of body parts. Their presence or absence affects our metabolism, our appearance, and our behavior. **Hormones** are produced by glands (fig. 13.2) called **endocrine glands** that secrete their products internally, placing them directly in the blood. Since these glands do not have ducts for the transport of their

secretions, they are sometimes called ductless glands. All hormones are carried throughout the body by the blood, but each one affects only a specific body part or parts, appropriately termed *target organs*.

Endocrine glands secrete hormones into the bloodstream for transport to target organs.

Hormones are substances that fall into two basic categories: (1) peptides (used here to include amino acid, polypeptide, and protein hormones) and (2) steroid hormones. Steroids

Table 13.1 The Principal Endocrine Glands and Their Hormones

Endocrine gland	Hormone released	Target tissues/organ
hypothalamus	releasing and release-inhibiting hormones	anterior pituitary
anterior pituitary	thyroid-stimulating hormone (TSH, thyrotropic)	thyroid *in throat, influences metabolism*
	adrenocorticotropic hormone (ACTH)	adrenal cortex

	gonadotropic hormones
	    Follicle-stimulating (FSH)
	    Luteinizing (LH)
		gonads

	prolactin (lactogenic hormone, LTH)	mammary glands
	growth hormone (GH, somatotropin)	soft tissues, bones
	not on test → melanocyte-stimulating hormone (MSH)	melanocytes in skin
posterior pituitary (storage of hypothalamic hormones)	antidiuretic hormone (ADH, vasopressin)	kidneys
	oxytocin	uterus, mammary glands
thyroid	thyroxin	all tissues
	calcitonin	bones, kidneys, gut
parathyroids	*not on test* parathyroid hormone (PTH)	bones, kidneys, gut
adrenal cortex	glucocorticoids (cortisol)	all tissues
	mineralocorticoids (aldosterone)	kidneys
	sex hormones	sex organs, skin, muscles, bones
adrenal medulla *not on test*	norepinephrine and epinephrine	cardiac and other muscles
pancreas	insulin *peptide hormone*	liver, muscles, adipose tissue
	not on test glucagon	liver, muscles, adipose tissue
gonads testes	androgens (testosterone)	sex organs, skin, muscles, bones
ovaries	estrogen and progesterone	sex organs, skin, muscles, bones
thymus	thymosins	T lymphocytes *invaded by AIDS*
pineal gland *not on test*	melatonin	circadian rhythms

are complex rings of carbon and hydrogen atoms (fig. 1.17). The difference between steroids is due to the atoms attached to these rings. Steroid hormones are produced by the adrenal cortex, the ovaries, and the testes. All of the other glands (table 13.1) produce peptide hormones.

Peptide Hormones

Peptide hormones bind to cell membrane receptors (fig. 13.3a). The hormone-receptor complex then activates an enzyme that produces cyclic adenosine monophosphate (cAMP). cAMP is a compound made from ATP, but it contains only one phosphate group that is attached to adenosine at two locations. The cAMP now activates the enzymes of the cell to carry out their normal functions. Notice that the peptide hormone never enters the cell. Therefore, these hormones sometimes are called the *first messenger,* while cAMP, which sets the metabolic machinery in motion, is called the *second messenger.*

Steroid Hormones

Steroid hormones do not bind to cell surface receptors; they can enter the cell freely because they are relatively small, lipid sol-

uble molecules. Once inside, steroid hormones bind to receptors in the cytoplasm (fig. 13.3b). The hormone-receptor complex then enters the nucleus, where it binds with chromatin at a location that promotes activation of particular genes. Protein synthesis follows. In this manner, steroid hormones lead to protein synthesis. Steroid hormones act more slowly than peptides because it takes more time to synthesize new proteins than to activate enzymes already present in the cell.

Hormones are chemical messengers that influence the metabolism of the cell either directly or indirectly, depending on the hormone type.

Figure 13.3 ▶▶▶ Cellular activity of hormones. *a.* Peptide hormones combine with receptors located on the cell membrane. This promotes the production of cyclic AMP which in turn leads to activation of a particular enzyme. *b.* Steroid hormones pass through the cell membrane to combine with receptors; the hormone receptor complex moves into the nucleus and activates certain genes, leading to protein synthesis.

Table 13.1 Continued

Chief function of hormone	Disorders too much/too little
regulate anterior pituitary hormones	*See* anterior pituitary
stimulates thyroid	*See* thyroid ~~tested~~ through radioactive iodine
stimulates adrenal cortex	*See* adrenal cortex
controls egg and sperm production	
controls sex hormone production	*See* testis and ovary
stimulates milk production and secretion	
stimulates cell division, protein synthesis, and bone growth	giantism, acromegaly/dwarfism
regulates skin color in lower vertebrates; unknown function in humans	
stimulates water reabsorption by kidneys	diverse*/diabetes insipidus
stimulates uterine muscle contraction and release of milk by mammary glands	
increases metabolic rate; helps to regulate growth and development	exophthalmic goiter/simple goiter myxedema, cretinism
lowers blood calcium level	tetany/weak bones
raises blood calcium level	weak bones/tetany
raise blood glucose level; stimulate breakdown of protein	
stimulate kidneys to reabsorb sodium and to excrete potassium	Cushing's syndrome/Addison's disease
stimulate development of secondary sex characteristics (particularly in male)	
stimulate fight or flight reactions; raise blood glucose level	
lowers blood glucose level; promotes formation of glycogen, proteins, and fats	shock/diabetes mellitus
raises blood glucose level; promotes breakdown of glycogen, proteins, and fats	
stimulate spermatogenesis; develop and maintain secondary male sex characteristics	diverse/eunuch
stimulate growth of uterine lining; develop and maintain secondary female sex characteristics	diverse/masculinization
stimulates maturation of T lymphocytes	
involved in circadian and circannual rhythms; possibly involved in maturation of sex organs	

*The word *diverse* in this table means that the symptoms have not been described as a syndrome in the medical literature.

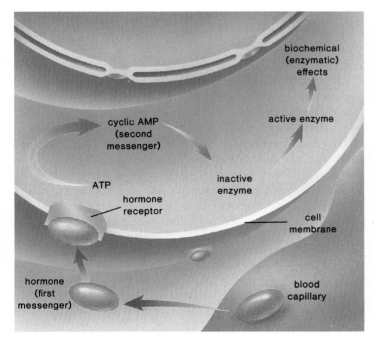

a.

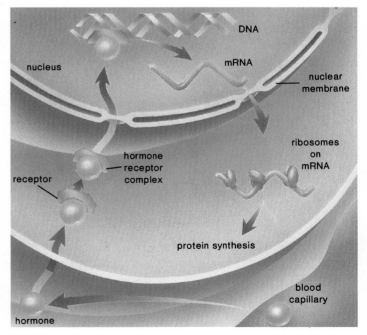

b.

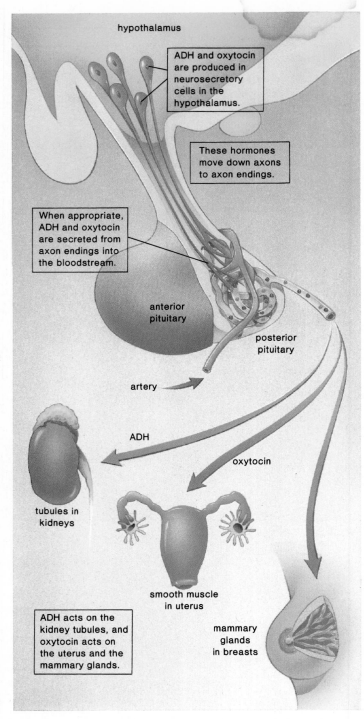

Figure 13.4   The hypothalamus produces 2 hormones, ADH and oxytocin, that are stored in and secreted by the posterior pituitary.

Hypothalamus and Pituitary Gland

The hypothalamus is a portion of the brain that regulates the internal environment. For example, the hypothalamus helps control heart rate, body temperature, and water balance, as well as the activity of the **pituitary gland.** The pituitary, a small gland about 1 cm in diameter, lies just below the hypothalamus (fig. 13.2) and is divided into two portions called the **posterior pituitary** and the **anterior pituitary.**

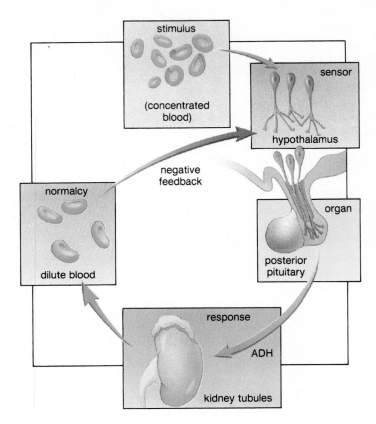

Figure 13.5   Regulation of ADH secretion. Neurons in the hypothalamus are sensitive to the osmolarity of blood. When blood is concentrated, these neurons send signals to the hypothalamus neurosecretory cells, which release ADH from their axon endings. ADH increases the permeability of the collecting ducts in the kidneys so that more water is reabsorbed. Once blood is diluted, ADH is no longer secreted. This is an example of control by negative feedback.

Posterior Pituitary

The posterior pituitary is connected to the hypothalamus by means of a stalklike structure. There are neurons in the hypothalamus called *neurosecretory cells* that both respond to neurotransmitter substances and produce the hormones that are stored in and released from the posterior pituitary (fig. 13.4).

These hormones are vasopressin, an **antidiuretic hormone (ADH)**, and oxytocin. ADH, as discussed in chapter 9, promotes the reabsorption of water from the collecting duct, a portion of the kidney tubule (nephron). When nerve cells in the hypothalamus determine that the blood is too concentrated, ADH is released from the axon endings into the bloodstream. As the blood becomes dilute, the hormone no longer is released. This is an example of control by negative feedback (fig. 13.5) because the effect of the hormone (diluted blood) acts to shut down the release of the hormone. Negative feedback mechanisms regulate the activities of most endocrine glands.

Inability to produce ADH causes **diabetes insipidus** (watery urine), in which a person produces copious amounts of urine with a resultant loss of salts from the blood. This condition can be corrected by the administration of ADH.

The Love Hormone

Your irrationally romantic après-sex sentiments may be caused by oxytocin, a primary sexual arousal hormone that triggers orgasm and may make you feel emotionally attached to your partner. Professor Niles Newton at Northwestern University Medical School was the first to hypothesize that oxytocin is "The Hormone of Love."*

Oxytocin, a hormone that stimulates the smooth muscles and sensitizes the nerves, is produced by sexual arousal—the more you're in the mood, the more oxytocin you'll produce. As you begin to make love, oxytocin is released throughout your body, making your nerves more sensitive to pleasure and giving you that turned-on feeling. As the hormone level builds, oxytocin causes the nerves in your genitals to fire spontaneously, bringing on orgasm and giving you the feeling of losing control.

However, this substance does much more than simply activate your physical response to sex. Unlike other hormones, oxytocin arousal is generated by emotional cues, in addition to physical stimuli. The look he/she gives you, the sound of his/her voice or touch of his/her hand can all start the ball rolling, making oxytocin rush through your body (did you ever notice that you sometimes feel a wave of love come over you?), and may ultimately turn what started out as pure, unadulterated lust into a full-fledged feeling of love. . . .

While oxytocin works the same way in both sexes, women seem to be more susceptible to its emotional effects. Women, it seems, are more likely to feel a need for a continued connection after sex and are more vulnerable, as the cliché goes, to "loving too much." The possible reason for this: According to a Stanford University study led by Marie Carmichael, Ph.D., women have higher levels of oxytocin than men do during sex. Apparently, it takes more oxytocin for women to achieve orgasm (one reason it could take women longer to reach orgasm than men do). . . . But keep in mind that there is one advantage: As a result of the oxytocin overload, women may be more capable than men are of having multiple orgasms and whole-body orgasms. . . .

From Charlotte Modahl, Ph.D., "The Love Hormone," November 1990. This article was originally published in *Mademoiselle*. Reprinted by permission of Dr. Charlotte Modahl.

*Newton, N. (1971) Trebly Sensuous Woman. *Psychology Today 98,* 68–71.

Newton, N. (1978) The role of the oxytocin reflexes in three interpersonal reproductive acts: coitus, birth and breastfeeding. *Clinical Psychoneuroendocrinology in Reproduction: Proceedings of the Serono Symposia 22* (L. Carenza, P. Pancheri, and L. Zichella, eds.) Academic Press.

Oxytocin is the other hormone that is made in the hypothalamus and is stored in the posterior pituitary. Oxytocin causes the uterus to contract and is used to artificially induce labor. It also stimulates the release of milk from the mother's mammary glands when a baby is nursing. Researchers now believe that oxytocin may be involved in causing orgasm, as explained in the reading "The Love Hormone."

It is appropriate to note that the neurosecretory cells in the hypothalamus provide an example of a way the nervous system and the endocrine system are joined. This topic is discussed again later.

The posterior pituitary releases two hormones, ADH and oxytocin, both of which are produced by neurosecretory cells in the hypothalamus.

Anterior Pituitary

The hypothalamus controls the anterior pituitary by producing releasing and release-inhibiting hormones. For example, there is a gonadotropic-releasing hormone (GnRH) and a gonadotropic-release-inhibiting hormone (GnRIH). The first hormone stimulates the anterior pituitary to release its gonadotropic hormones, and the second inhibits the anterior pituitary from releasing the same hormones.

Releasing and release-inhibiting hormones are transported from the hypothalamus to the anterior pituitary by way of a portal system that connects the anterior pituitary to the hypothalamus (fig. 13.6).

The anterior pituitary is controlled by releasing and release-inhibiting hormones secreted by the hypothalamus. These hormones pass to the anterior pituitary by way of a portal system.

Hormones produced by the anterior pituitary are listed in table 13.1. Three of these hormones have a direct effect on the body. **Growth hormone (GH),** or somatotropin, dramatically affects physical appearance since it determines the height of the individual (fig. 13.7). If little or no GH is secreted by the anterior pituitary during childhood, the person could become a pituitary dwarf, although of perfect proportions, quite small in stature. If too much GH is secreted, the person could become a giant. Giants usually have poor health, primarily because GH has a secondary effect on blood sugar level, promoting an illness called diabetes (sugar) mellitus, discussed following.

GH promotes cell division, protein synthesis, and bone growth. It stimulates the transport of amino acids into cells and increases the activity of ribosomes, both of which are essential to protein synthesis. In bones, it promotes growth of the cartilaginous plates and causes osteoblasts to form bone (p. 216). Evidence suggests that the effects on cartilage and bone actually may be due to hormones called somatomedins, released by the liver. GH causes the liver to release somatomedins.

If the production of GH increases in an adult after full height has been attained, only certain bones respond. These are

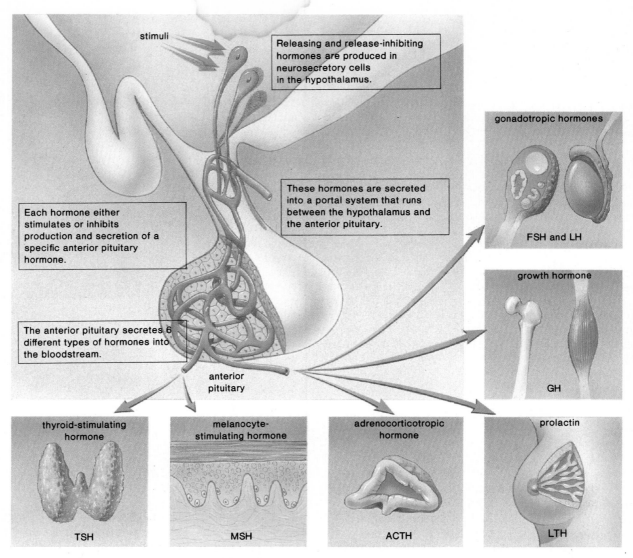

stimuli

Releasing and release-inhibiting hormones are produced in neurosecretory cells in the hypothalamus.

Each hormone either stimulates or inhibits production and secretion of a specific anterior pituitary hormone.

These hormones are secreted into a portal system that runs between the hypothalamus and the anterior pituitary.

The anterior pituitary secretes 6 different types of hormones into the bloodstream.

anterior pituitary

gonadotropic hormones

FSH and LH

growth hormone

GH

thyroid-stimulating hormone

TSH

melanocyte-stimulating hormone

MSH

adrenocorticotropic hormone

ACTH

prolactin

LTH

Figure 13.6   Hypothalamus and anterior pituitary. The hypothalamus controls the secretions of the anterior pituitary.

the bones of the jaw, the eyebrow ridges, the nose, the fingers, and the toes. When these begin to grow, the person takes on a slightly grotesque look with huge fingers and toes, a condition called **acromegaly** (fig. 13.8).

Prolactin (also called lactogenic hormone, LTH) is produced in quantity only after childbirth. It causes the mammary glands in the breasts to develop and to produce milk.

Melanocyte-stimulating hormone (MSH) causes skin color changes in lower vertebrates—but no one knows what it does in humans. However, it is derived from a molecule that is also the precursor for both adrenocorticotropic (ACTH) and the anterior pituitary endorphins. These endorphins are structurally and functionally similar to the endorphins produced in brain nerve cells.

GH and prolactin are two hormones produced by the anterior pituitary. GH influences the height of children and overproduction brings about a condition called acromegaly in adults. Prolactin promotes milk production after childbirth.

Other Hormones Produced by the Anterior Pituitary

The anterior pituitary sometimes is called the *master gland* because it controls the secretion of certain other endocrine glands (fig. 13.6). As indicated in table 13.1, the anterior pituitary secretes the following hormones, which have an effect on other glands.

Figure 13.7   Giantism. Sandy Allen, one of the world's tallest women due to a higher than usual amount of GH produced by the anterior pituitary.

1. **Thyroid-stimulating hormone (TSH)**
2. **Adrenocorticotropic hormone (ACTH),** a hormone that stimulates the adrenal cortex
3. **Gonadotropic hormones (GnH)** that stimulate the gonads, the testes in males and the ovaries in females

TSH causes the thyroid to produce thyroxin; ACTH causes the adrenal cortex to produce cortisol; and gonadotropic hormones cause the gonads to secrete sex hormones. Notice that it is now possible to indicate a three-tiered relationship between the hypothalamus, the pituitary, and the other endocrine glands. The hypothalamus produces hormones that control the anterior pituitary, and the anterior pituitary produces hormones that control the thyroid, the adrenal cortex, and the gonads. Figure 13.9 illustrates the feedback mechanism that controls the activity of all these glands.

The hypothalamus, the anterior pituitary, and the other endocrine glands controlled by the anterior pituitary all are involved in a self-regulating feedback loop.

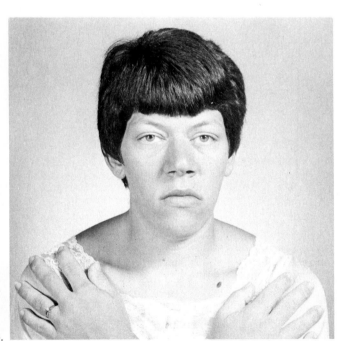

a.   b.

Figure 13.8   Acromegaly. The condition is caused by the overproduction of GH in an adult. It is characterized by an enlargement of the bones in the face and the fingers of an adult. a. At age 20, this individual was normal. b. At age 24, there is some enlargement of the nose, the jaw, and the fingers.

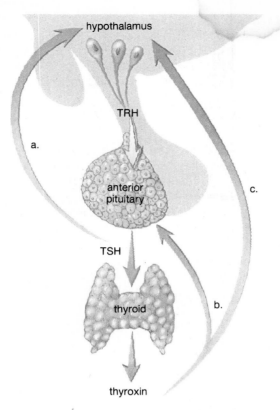

hypothalamus

TRH

a.

anterior
pituitary

c.

TSH

thyroid

b.

thyroxin

Figure 13.9   The hypothalamus-pituitary-thyroid control relationship. TRH (thyroid-releasing hormone) stimulates the anterior pituitary, and TSH (thyroid-stimulating hormone) stimulates the thyroid to secrete thyroxin. The level of thyroxin in the body is negatively controlled in 3 ways: *a.* The level of TSH exerts feedback control over the hypothalamus; *b.* the level of thyroxin exerts feedback control over the anterior pituitary; and *c.* the level of thyroxin exerts feedback control over the hypothalamus. In this way, thyroxin controls its own secretion. Cortisol and sex hormone levels are controlled in similar ways.

Thyroid and Parathyroid Glands

Thyroid Gland

The thyroid gland (fig. 13.2) is located in the neck and is attached to the trachea just below the larynx. Internally, the gland is composed of a large number of follicles filled with thyroglobulin, the storage form of thyroxin. The production of both of these requires iodine. Iodine is actively transported into the thyroid gland, where the concentration can become as much as 25 times that of blood. If iodine is lacking in the diet, the thyroid gland enlarges, producing a goiter (fig. 13.10). (Salt in the United States is iodized for this reason.) The cause of goiter becomes clear if we refer to figure 13.9. When there is a low level of thyroxin in the blood, a condition called hypothyroidism, the anterior pituitary is stimulated to produce *TSH*. TSH causes the thyroid to increase in size so that enough **thyroxin** usually is produced. In this case, enlargement continues because enough thyroxin never is produced. An enlarged thyroid that produces some thyroxin is called a **simple goiter.**

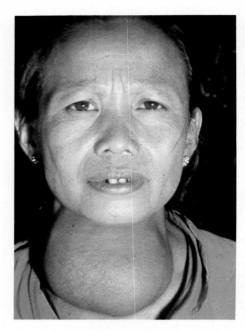

Figure 13.10   Simple goiter. An enlarged thyroid gland is often caused by a lack of iodine in the diet. Without iodine, the thyroid is unable to produce thyroxin, and continued anterior pituitary stimulation causes the gland to enlarge.

Human Issue

Simple goiter is quite rare in the United States and other developed countries, but it is not uncommon in poorer countries with inadequate health care. Even so, goiters are relatively easy to eliminate by adding iodized salt to the diet. Is the United States morally obligated to improve the health of people around the globe with respect to such medical conditions as goiter, or is this the responsibility of the individual governments? How much or how little should we do to improve the health of people in other countries?

Thyroxin

Thyroxin increases the metabolic rate. It does not have a target organ; instead, it stimulates most of the cells of the body to metabolize at a faster rate. The number of respiratory enzymes in the cell increases, as does oxygen (O_2) uptake.

If the thyroid fails to develop properly, a condition called **cretinism** results. Cretins (fig. 13.11) are short, stocky persons who have had extreme hypothyroidism since infancy and/or childhood. Thyroxin therapy can initiate growth, but unless treatment is begun within the first two months of life, mental retardation results. The occurrence of hypothyroidism in adults produces the condition known as **myxedema** (fig. 13.12), which is characterized by lethargy, weight gain, loss of hair, slowed pulse rate, decreased body temperature, and thickness and puffiness of the skin. The administration of adequate doses of thyroxin restores normal function and appearance.

In the case of hyperthyroidism (too much thyroxin), the thyroid gland is enlarged and overactive, causing a goiter to form

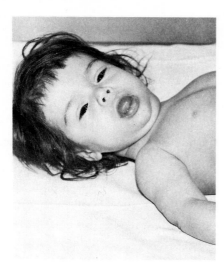

Figure 13.11   Cretinism. Cretins are individuals who have suffered from thyroxin insufficiency since birth or early childhood. Skeletal growth is usually inhibited to a greater extent than soft tissue growth; therefore, the child appears short and stocky. Sometimes, the tongue becomes so large that it obstructs swallowing and breathing.

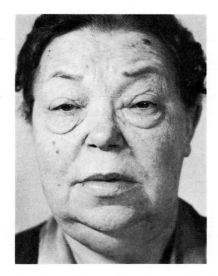

Figure 13.12   Myxedema. This condition is caused by thyroid insufficiency in the older adult and leads to swelling of the face and bagginess under the eyes.

and the eyes to protrude because of edema in the tissues of the eye sockets and swelling of muscles that move the eyes. This type of goiter is called **exophthalmic goiter** (fig. 13.13). The patient usually becomes hyperactive, nervous, irritable, and suffers from insomnia. Removal or destruction of a portion of the thyroid by means of radioactive iodine sometimes is effective in curing the condition.

Calcitonin

In addition to thyroxin, the thyroid gland also produces the hormone **calcitonin.** This hormone helps to regulate the calcium

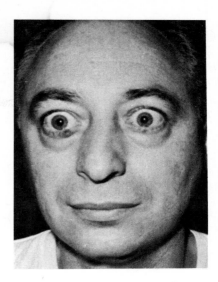

Figure 13.13   Exophthalmic goiter. Protruding eyes occur when an active thyroid gland enlarges.

level in the blood and opposes the action of parathyroid hormone. The interaction of these two hormones is discussed following.

The anterior pituitary produces TSH, a hormone that promotes the production of thyroxin by the thyroid, a gland subject to goiters. Thyroxin, which speeds up metabolism, can affect the body as a whole, as exemplified by cretinism and myxedema.

Parathyroid Glands

The parathyroid glands are embedded in the posterior surface of the thyroid gland, as shown in figure 13.14*b*. Many years ago, these four small glands sometimes were removed by mistake during thyroid surgery. Under the influence of **parathyroid hormone (PTH),** also called parathormone, the calcium level in blood increases and the phosphate level decreases. The hormone stimulates the absorption of calcium from the gut, the retention of calcium by the kidneys, and the demineralization of bone. In other words, PTH promotes the activity of osteoclasts, the bone-resorbing cells. Although this also raises the level of phosphate in the blood, PTH acts on the kidneys to excrete phosphate in the urine. When a woman stops producing the female sex hormone estrogen following menopause, she is more likely to suffer from osteoporosis, characterized by a thinning of the bones (p. 86). Whether estrogen works counter to the action of PTH has not been determined yet.

If insufficient PTH is produced, the level of calcium in blood drops, resulting in **tetany.** In tetany, the body shakes from continuous muscle contraction. The effect really is brought about by increased excitability of the nerves, which fire spontaneously and without rest. Calcium plays an important role in both nervous conduction and muscle contraction.

The level of PTH secretion is controlled by a feedback mechanism involving calcium (fig. 13.14*c*). When the calcium

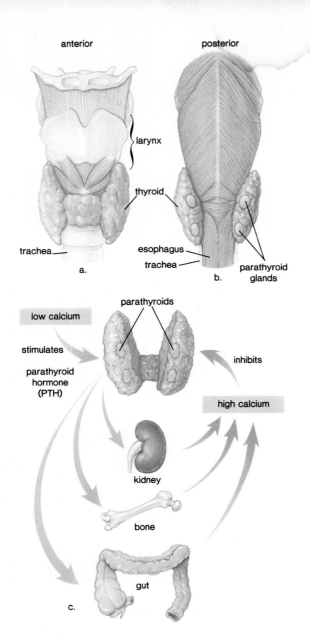

Figure 13.14   Thyroid and parathyroid glands. *a.* The thyroid is located in the neck in front of the trachea. *b.* The 4 parathyroid glands are embedded in the posterior surface of the thyroid gland. Yet, the parathyroid and thyroid glands have no anatomical or physiological connection with one another. *c.* Regulation of parathyroid hormone (PTH) secretion. A low blood level of calcium causes the parathyroids to secrete PTH, which causes the kidneys and the gut to retain calcium and osteoclasts to break down bone. The end result is an increased level of calcium in the blood. A high blood level of calcium inhibits secretion of PTH.

level rises, PTH secretion is inhibited, and when the calcium level lowers, PTH secretion is stimulated.

As mentioned previously, the thyroid secretes calcitonin, which also influences blood calcium level. Although calcitonin has the opposite effect of PTH, particularly on the bones, its action is not believed to be as significant. Still, the two hormones function together to regulate the level of calcium in the blood.

PTH maintains a high blood level of calcium by promoting its absorption in the gut, its reabsorption by the kidneys, and demineralization of bone. These actions are opposed by calcitonin produced by the thyroid.

Adrenal Glands

The adrenal glands, as their name implies (*ad* = near; *renal* = kidneys), lie atop the kidneys (fig. 13.2). Each consists of an outer portion, called the *cortex,* and an inner portion, called the *medulla.* These portions, like the anterior pituitary and the posterior pituitary, have no functional connection with one another.

Adrenal Medulla

The adrenal medulla secretes **norepinephrine** and epinephrine under conditions of stress. They bring about all those responses we associate with the "fight or flight" reaction: the blood glucose level and the metabolic rate increase, as do breathing and the heart rate. The blood vessels in the intestine constrict, and those in the muscles dilate. This increased circulation to the muscles causes them to have more stamina than usual. In times of emergency, the sympathetic nervous system *initiates* these responses, but they are maintained by secretions from the adrenal medulla.

The adrenal medulla is innervated by only one set of sympathetic nerve fibers. Recall that there are usually pre- and postganglionic nerve fibers for each stimulated organ. In this instance, what happened to the postganglionic neurons? It appears that the adrenal medulla may have evolved from a modification of the postganglionic neurons. Like the neurosecretory neurons in the hypothalamus, these neurons also secrete hormones into the bloodstream.

The adrenal medulla releases norepinephrine and epinephrine into the bloodstream. These hormones help us and other animals to cope with situations that threaten survival.

Adrenal Cortex

Although the adrenal medulla can be removed with no ill effects, the adrenal cortex is absolutely necessary to life. The two major classes of hormones made by the adrenal cortex are the *glucocorticoids* and the *mineralocorticoids.* The cortex also secretes a small amount of male sex hormone and an even smaller amount of female sex hormone. All of these hormones are steroids.

Glucocorticoids

Of the various glucocorticoids, the hormone responsible for the greatest amount of activity is **cortisol.** Cortisol promotes the hydrolysis of muscle protein to amino acids that enter the blood. This leads to an increased level of glucose when the liver converts these amino acids to glucose. Cortisol also favors metabolism of fatty acids rather than carbohydrate. In opposition to insulin, therefore, cortisol raises the blood glucose level. Cor-

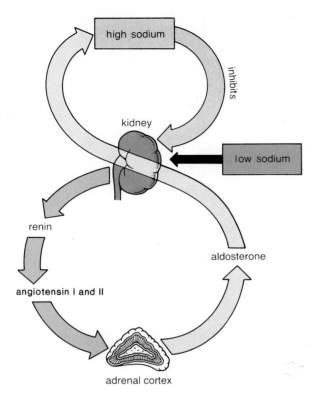

Figure 13.15 Renin-angiotensin-aldosterone system. If the blood level of sodium is low, the kidneys secrete renin. The increased renin acts via the increased production of angiotensin I and II to stimulate aldosterone secretion. Aldosterone promotes reabsorption of sodium by the kidneys; when the sodium level in the blood rises, the kidneys stop secreting renin.

high sodium inhibits **low sodium** kidney renin angiotensin I and II aldosterone adrenal cortex

tisol also counteracts the inflammatory response, which leads to the pain and the swelling of joints in arthritis and bursitis. The administration of cortisol aids these conditions because it reduces inflammation.

The secretion of cortisol by the adrenal cortex is under the control of the anterior pituitary hormone ACTH. Using the same kind of system shown in figure 13.9, the hypothalamus produces a releasing hormone (CRH) that stimulates the anterior pituitary to release ACTH. ACTH in turn stimulates the adrenal cortex to secrete cortisol, which regulates its own synthesis by negative feedback of both CRH and ACTH synthesis.

Mineralocorticoids

The secretion of mineralocorticoids, the most significant of which is **aldosterone,** is not under the control of the anterior pituitary. Aldosterone regulates the level of sodium and potassium in blood, its primary target organ being the kidney, where it promotes renal absorption of sodium and renal excretion of potassium (fig. 13.15). The level of sodium is particularly important to the maintenance of blood pressure because its concentration indirectly regulates the secretion of aldosterone. When the blood level of sodium is low, the kidneys secrete renin. Renin is an enzyme that converts the plasma protein angiotensinogen to angiotensin I, which becomes angiotensin II in the lungs. Angiotensin II stimulates the adrenal cortex to release aldosterone (see fig. 9.13). This is called the renin-angiotensin-aldosterone

system. The effect of this system is to raise the blood pressure in two ways. First, angiotensin II constricts the arteries directly, and secondly, aldosterone causes the kidneys to reabsorb sodium. When the blood level of sodium is high, water is reabsorbed, and blood volume and pressure are maintained.

Recently, it has been found that there is a hormone that acts contrary to aldosterone. This hormone is called the atrial natriuretic hormone because (1) it is produced by the atria of the heart and (2) it causes natriuresis, the excretion of sodium. Once sodium is excreted so is water; therefore, blood volume and blood pressure decrease.

Cortisol, which raises the blood glucose level, and aldosterone, which raises the blood sodium level, are two hormones secreted by the adrenal cortex.

Sex Hormones

The adrenal cortex produces a small amount of both male and female sex hormones. In males, the cortex is a source of female sex hormones, and in females, it is a source of male hormones. A tumor in the adrenal cortex can cause the production of a large amount of sex hormones, which can lead to feminization in males and masculinization in females.

Disorders

When the level of adrenal cortex hormones is low, a person begins to suffer from Addison disease. When the level of adrenal cortex hormones in the body is high, a person suffers from Cushing syndrome:

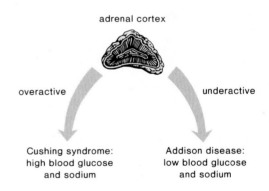

adrenal cortex

overactive — Cushing syndrome: high blood glucose and sodium

underactive — Addison disease: low blood glucose and sodium

Addison Disease   Because of the lack of cortisol, the Addison disease patient is unable to maintain the glucose level of the blood, tissue repair is suppressed, and there is a high susceptibility to any kind of stress. Even a mild infection can cause death. Due to the lack of aldosterone, the blood sodium level is low, and the person experiences low blood pressure along with acidosis and low pH. In addition, the patient has a peculiar bronzing of the skin (fig. 13.16).

Cushing Syndrome   In Cushing syndrome, a high level of cortisol causes a tendency toward diabetes mellitus, a decrease in muscular protein, and an increase in subcutaneous fat. Because of these effects, the person usually develops thin arms and

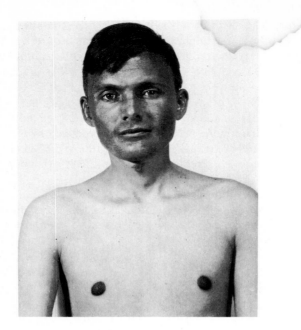

Figure 13.16 Addison disease. This condition is characterized by a peculiar bronzing of the skin, as seen in the face and the thin skin of this patient's nipples.

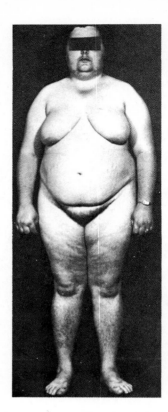

Figure 13.17 Cushing syndrome. Persons with this condition tend to have an enlarged trunk and a moonlike face. Masculinization may occur in women due to excessive male sex hormones in the body.

legs and an enlarged trunk. Due to the high level of sodium in the blood, the blood is basic and the patient has hypertension and edema of the face, which gives the face a moon shape (fig. 13.17).

Addison disease is due to adrenal cortex hyposecretion, and Cushing syndrome is due to adrenal cortex hypersecretion.

Pancreas

The **pancreas** is a long organ that lies transversely in the abdomen (fig. 13.2) between the kidneys and near the duodenum of the small intestine. It is composed of two types of tissue—exocrine, which produces and secretes *digestive* juices that go by way of the pancreatic duct and the common duct to the small intestine, and endocrine, called the **islets of Langerhans**, which produces and secretes the hormones **insulin** and **glucagon** directly into the blood.

Insulin is secreted when there is a high level of glucose in blood, which usually occurs just after eating. Insulin has three different actions: (1) it stimulates liver, fat, and muscle cells to take up and metabolize glucose; (2) it stimulates the liver and the muscles to store glucose as glycogen; and (3) it promotes the buildup of fats and proteins and inhibits their use as an energy source. Therefore, insulin is a hormone that promotes storage of nutrients so that they are on hand during leaner times. It also helps to lower the blood glucose level.

Glucagon is secreted from the pancreas between eating, and its effects are opposite to those of insulin. Glucagon stimulates the breakdown of stored nutrients and causes the blood sugar level to rise (fig. 13.18).

Diabetes Mellitus

The symptoms of **diabetes mellitus** (sugar diabetes) include the following:

Sugar in the urine
Frequent, copious urination
Abnormal thirst
Rapid loss of weight
General weakness
Drowsiness and fatigue
Itching of the genitals and the skin
Visual disturbances, blurring
Skin disorders, such as boils, carbuncles, and infection

Many of these symptoms develop because sugar is not being metabolized by the cells. The liver fails to store glucose as glycogen, and all the cells fail to utilize glucose as an energy source. This means that the blood glucose level rises very high after eating, causing glucose to be excreted in the urine. More water than usual therefore is excreted so that the diabetic is extremely thirsty.

Since carbohydrate is not being metabolized, the body turns to the breakdown of protein and fat for energy. Unfortunately, the breakdown of these molecules leads to the buildup of acids in the blood (acidosis) and to respiratory distress. It is the latter that eventually can cause coma and death of the diabetic. The symptoms that lead to coma (table 13.2) develop slowly.

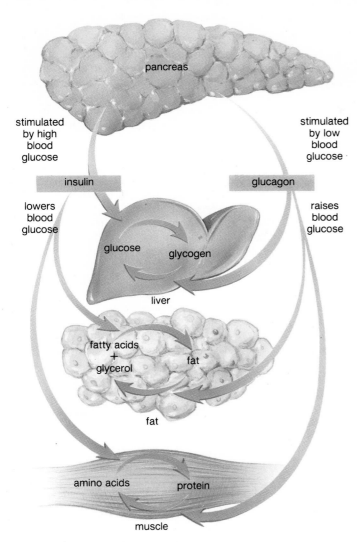

Figure 13.18 Contrary effects of insulin and glucagon. When the blood glucose level is high, the pancreas secretes insulin. Insulin promotes the storage of glucose as glycogen and the synthesis of proteins and fats. Therefore, insulin lowers the blood glucose level. When the blood glucose level is low, the pancreas secretes glucagon. Glucagon acts in opposition to insulin in all respects; therefore, glucagon raises the blood glucose level.

There are two types of diabetes. In *type I diabetes,* formerly called juvenile-onset diabetes, the pancreas is not producing insulin. Therefore, the patient must have daily insulin injections. These injections control the diabetic symptoms but still can cause inconveniences since either an overdose of insulin or the absence of regular eating can bring on the symptoms of insulin shock (table 13.2). These symptoms appear because the blood sugar level has fallen below normal levels. Since the brain requires a constant supply of sugar, unconsciousness can result. The cure is quite simple: an immediate source of sugar, such as a sugar cube or fruit juice, can counteract insulin shock immediately.

Obviously, insulin injections are not the same as a fully functioning pancreas that responds on demand to high glucose

Table 13.2 Symptoms of Insulin Shock and Diabetic Coma

Insulin shock	Diabetic coma
sudden onset	slow, gradual onset
perspiration, pale skin	dry, hot skin
dizziness	no dizziness
heart palpitation	no palpitation
hunger	no hunger
normal urination	excessive urination
normal thirst	excessive thirst
shallow breathing	deep, labored breathing
normal breath odor	fruity breath odor
confusion, disorientation, strange behavior	drowsiness and great lethargy leading to stupor
urinary sugar absent or slight	large amounts of urinary sugar
no acetone in urine	acetone present in urine

From Henry Dolger, M.D., and Bernard Seeman. *How to Live with Diabetes.* Copyright © 1972, 1965, 1958 W. W. Norton & Company, Inc., New York. Reprinted by permission.

level by supplying insulin. For this reason, some doctors advocate an islet cell transplant.

Of the 12 million people who now have diabetes in the United States, at least 10 million have *type II diabetes,* formerly called maturity-onset diabetes. In this type of diabetes, now known to occur in obese people of any age, the pancreas is producing insulin, but the cells do not respond to it. At first, the cells lack the receptors necessary to detect the presence of insulin, and later, the cells are even incapable of taking up glucose. If type II diabetes is untreated, the results can be as serious as type I diabetes. Diabetics are prone to blindness, kidney disease, and circulatory disorders, including strokes. Pregnancy carries an increased risk of diabetic coma, and the child of a diabetic is somewhat more likely to be stillborn or to die shortly after birth. It is important, therefore, to prevent or to at least control type II diabetes. The best defense is a nonfattening diet and regular exercise. If that fails, there are oral drugs that make the cells more sensitive to the effects of insulin or that stimulate the pancreas to make more of it.

Diabetes mellitus is caused by a lack of insulin or insensitivity of cells to insulin. Insulin lowers blood glucose levels by causing the cells to take up glucose and the liver to convert it to glycogen.

Other Endocrine Glands

Gonads

The gonads are the endocrine glands that produce the hormones that determine sexual characteristics. As is discussed in detail in the following chapter, the *testes* produce the androgens—the most important of which is called testosterone—which are the male sex hormones, and the *ovaries* produce estrogen and progesterone, the female sex hormones. The secretion of these hormones is under the control of the gonadotropic hormones produced by the anterior pituitary.

Side Effects of Steroids

Being a steroid user may cost an athlete far more than his or her Olympic medal: a growing body of medical evidence indicates that athletes who take steroids have experienced problems ranging from sterility to loss of libido, and the drug has been implicated in the deaths of young athletes from liver cancer and a type of kidney tumor. Steroid use has also been linked to heart disease. "Athletes who take steroids are playing with dynamite," says Robert Goldman, 29, a former wrestler and weight lifter who is now a research fellow in sports medicine at Chicago Osteopathic

Medical Center and who published a book on steroid abuse, *Death in the Locker Room* (Icarus). "Any jock who uses these drugs is taking chances not just with his health but with his life."

Anabolic steroids are essentially the male hormone testosterone and its synthetic derivatives. . . .

The great majority of physicians say the drugs upset the body's natural hormonal balance, particularly that involving testosterone, which is present, though in different amounts, in both men and women. Normally, the hypothalamus, the part of the brain that regulates many of the body's functions, "tastes" the testosterone levels; if it finds them too low, it signals the pituitary gland to trigger increased production. When the hypothalamus finds the testosterone levels too high, as it does in the case of steroid abusers, it signals the pituitary to stop production. Problems can also arise in some cases after athletes stop taking the drugs and the hypothalamus fails to get the system started again.

The results can be traumatic. Many men experience atrophy, or shrinking of the testicles, falling sperm counts,

temporary infertility and a lessening of sexual desire; some men grow breasts, while others may develop enlargement of the prostate gland, a painful condition not usually found in men under 50. Women who take too many steroids can develop male sexual characteristics. Some grow hair on their chests and faces and lose hair from their heads; many experience abnormal enlargement of the clitoris. Some cease to ovulate and menstruate, sometimes permanently.

There are several other health risks. Steroids can cause the body to retain fluid, which results in rising blood pressure. This often tempts users to fight "steroid bloat" by taking large doses of diuretics. A postmortem on a young California weight lifter who had a fatal heart attack after using steroids . . . showed that by taking diuretics he had purged himself of electrolytes, chemicals that help regulate the heart.

Steroids' effects on the body.
a. Suspected harmful effects.
b. How steroids build muscle.

The sex hormones bring about the secondary sex characteristics of males and females. Among other traits, males have greater muscular strength than females. As discussed in the reading on this page, athletes and others sometimes take so-called *anabolic steroids,* which are synthetic steroids that mimic the action of testosterone, in order to improve their strength and physique. Unfortunately, this practice is accompanied by harmful side effects.

Thymus

The **thymus** is a lobular gland that lies in the upper thorax (fig. 13.2). This organ reaches its largest size and is most active during childhood. With aging, the organ gets smaller and becomes fatty. Certain lymphocytes that originate in the bone marrow and then pass through the thymus are transformed into T cells (p. 135). The thymus produces various hormones called *thymosins,* which aid the differentiation of T cells and may

stimulate immune cells in general. There is hope that these hormones can be used in conjunction with lymphokine therapy to restore or to stimulate T cell function in patients suffering from AIDS or cancer.

Pineal Gland

The **pineal gland** produces the hormone called melatonin primarily at night. In fishes and amphibians, the pineal gland is located near the surface of the body and is a "third eye" that can receive light rays directly. In humans, the pineal gland is located in the third ventricle of the brain and cannot receive direct light signals. However, it does receive nerve impulses from the eyes by way of the optic tract.

Humans, like other animals, go through daily cycles called *circadian rhythms,* and it is believed that the secretion of melatonin may be involved in regulating these cycles, particularly the sleep cycle. Physicians at the University of Texas are con-

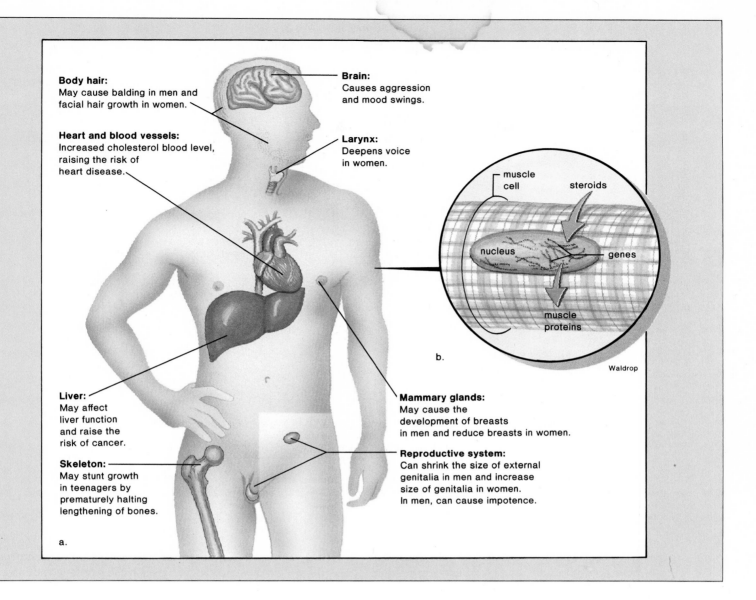

Body hair:
May cause balding in men and facial hair growth in women.

Heart and blood vessels:
Increased cholesterol blood level, raising the risk of heart disease.

Brain:
Causes aggression and mood swings.

Larynx:
Deepens voice in women.

muscle cell
steroids
nucleus
genes
muscle proteins

b.

Waldrop

Liver:
May affect liver function and raise the risk of cancer.

Skeleton:
May stunt growth in teenagers by prematurely halting lengthening of bones.

Mammary glands:
May cause the development of breasts in men and reduce breasts in women.

Reproductive system:
Can shrink the size of external genitalia in men and increase size of genitalia in women. In men, can cause impotence.

a.

ducting an experiment with a melatonin pill. They want to see if taking the pill at 2 A.M. will set a person's body clock to local time and ease the symptoms of jet lag.

Many animals also go through yearly cycles. For example, the reproductive organs in hamsters shrink in size and weight as the amount of light decreases in fall and winter. This inhibits reproduction at a time of year when food is not plentiful. In humans, it has been noted that children with brain tumors that destroy the pineal gland experience early puberty. Therefore, it is thought that the pineal gland may be involved in regulating human sexual development.

Still Other Glands

Even organs that traditionally are not considered to be endocrine glands have been found to secrete hormones. For example, as mentioned on page 265, the heart produces *atrial natriuretic hormone,* which helps to regulate the water and salt balance in the kidneys.

Human Issue

It is illegal to sell and use drugs of abuse such as heroin and cocaine. Should it also be illegal to sell and use steroids that can damage the body? Is it up to the government to regulate our use of dangerous substances or at least to educate us about their harmful effects? Or is it up to each person to learn about the dangers of using steroids in order to protect his or her own health?

A number of different types of organs and cells have been discovered to produce peptide *growth factors* that act on specific cell types. For example, we have mentioned that erythropoietin secreted by the kidneys (p. 119) stimulates red blood cell formation. Lymphokines released by helper T lymphocytes (p. 143) stimulate cytotoxic T cells. Similarly, platelets release platelet-derived growth factor and epidermal growth factor has been extracted from mouse submaxillary glands. Pituitary and

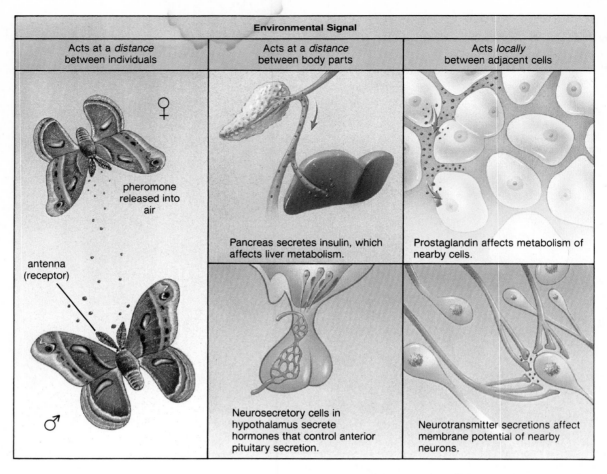

Environmental Signal		
Acts at a *distance* between individuals	Acts at a *distance* between body parts	Acts *locally* between adjacent cells

pheromone released into air

antenna (receptor)

♀

♂

Pancreas secretes insulin, which affects liver metabolism.

Prostaglandin affects metabolism of nearby cells.

Neurosecretory cells in hypothalamus secrete hormones that control anterior pituitary secretion.

Neurotransmitter secretions affect membrane potential of nearby neurons.

Figure 13.19   The 3 categories of environmental signals. Pheromones are chemical messengers that act at a distance between individuals. Endocrine hormones and neurosecretions are typically carried in the bloodstream and act at a distance within the body of a single organism. Some chemical messengers have local effects only; they pass between cells that are adjacent to one another. This, of course, includes neurotransmitter substances.

brain extracts have been found to contain a fibroblast growth factor that stimulates proliferation of cells of endodermal and mesodermal origin. Indeed, it is believed that most cells release growth factors, and the role of these in the formation of cancer will be discussed on page 404.

Prostaglandins, derived from cell membrane phospholipids, are unique in that they are produced and used locally. There are many different types of prostaglandins produced by many different tissues. In the uterus, certain prostaglandins cause muscles to contract; therefore, they are implicated in the pain and discomfort of menstruation in some women. (Antiprostaglandin therapy is useful in these cases.) On the other hand, certain prostaglandins are being used to treat ulcers because they reduce gastric secretion, to treat hypertension because they lower blood pressure, and to prevent thrombosis because they inhibit platelet aggregation. Because the different prostaglandins can have contrary effects, however, it has been very difficult to standardize their use, and in most instances, prostaglandin therapy still is considered experimental.

Environmental Signals

In this chapter, we concentrated on describing the functions of the human endocrine glands and their hormonal secretions. We already know that hormones are only one type of environmental signal or chemical messenger between cells. In fact, the concept of the environmental signal now has been broadened to include at least the following three different categories of messengers (fig. 13.19).

Environmental signals that act at a distance between individuals. Many organisms release chemical messengers, called *pheromones,* into the air or in externally deposited body fluids. These are intended to be messages for other members of the species. For example, ants lay down a pheromone trail to direct other ants to food, and the female silkworm moth releases bombykol, a sex attractant that is received by male moth antennae even several miles away. This chemical is so potent that it has been estimated that only 40 out of 40,000 receptors on the male antennae need to be activated in order for the male to respond. Mammals, too, release pheromones; the urine of dogs serves as a territorial marker, for example. Studies are being conducted to determine if humans also have pheromones.

Environmental signals that act at a distance between body parts. This category includes the endocrine secretions, which traditionally have been called hormones. It also includes the secretions of the neurosecretory cells in the hypothalamus—the production and action of ADH and oxytocin illustrate the close relationship between the nervous system and the endocrine system. Neurosecretory cells produce these hormones, which are released when these cells receive nerve impulses. As another example of the overlap between the nervous and endocrine systems, consider that endorphins on occasion travel in the bloodstream, but they act on nerve cells to alter their membrane potential. Also, norepinephrine is both a neurotransmitter and hormone secreted by the adrenal medulla.

Environmental signals that act locally between adjacent cells. Neurotransmitter substances belong in this category, as do substances like prostaglandins that are sometimes called local hormones. Also, when the skin is cut, histamine is released by mast cells and promotes the inflammatory response (p. 144).

Redefinition of a Hormone

Traditionally, a hormone was considered to be a secretion of an endocrine gland that was carried in the bloodstream to a target organ. In recent years, some scientists have broadened the definition of a hormone to include *all* types of chemical messengers. This change seemed necessary because those chemicals traditionally considered to be hormones now have been found in all sorts of tissues in the human body. For example, it is impossible for insulin produced by the pancreas to enter the brain because of the blood–brain barrier—a tight fusion of endothelial cells of the capillary walls that prevents passage of larger molecules like peptides. Yet, insulin has been found in the brain. It now appears that the brain cells themselves can produce insulin, which is used locally to influence the metabolism of adjacent cells. Also, some chemicals identical to the hormones of the endocrine system have been found in lower organisms, even in bacteria! A moment's thought about the evolutionary process helps to explain this; these regulatory chemicals may have been present in the earliest cells and only became specialized as hormones as evolution proceeded.

Summary

Hormones are chemical messengers having a metabolic effect on cells. The hypothalamus produces the hormones ADH and oxytocin, released by the posterior pituitary. In addition, the hypothalamus produces hormones that control the production of hormones by the anterior pituitary. In addition to GH and prolactin, which affect the body directly, the anterior pituitary secretes hormones that control other endocrine glands: TSH stimulates the thyroid to release thyroxin; ACTH stimulates the adrenal cortex to release glucocorticoids; gonadotropins stimulate the gonads to release the sex hormones. The secretion of hormones is controlled by negative feedback. In the case of the hormones just mentioned, this mechanism involves the hypothalamus and the anterior pituitary, in addition to the hormonal gland in question. Other hormones are PTH, the mineralocorticoids, noradrenalin, and insulin (table 13.1).

The most common illness due to hormonal imbalance is diabetes mellitus. This condition occurs when the islets of Langerhans within the pancreas fail to produce insulin. Insulin promotes the uptake of glucose by the cells and the conversion of glucose to glycogen, thereby lowering the blood glucose levels. Without the production of insulin, the blood sugar level rises, and some of it spills over into the urine. The real problem in diabetes mellitus, however, is acidosis, which may cause the death of the diabetic if therapy is not begun.

There are three categories of environmental signals: those that act at a distance between individuals (pheromones); those that act at a distance within the individual (traditional endocrine hormones and secretions of neurosecretory cells); and local messengers (such as prostaglandins and neurotransmitter substances). Since there is great overlap between these categories, perhaps the definition of a hormone now should be expanded to include all of them.

Study Questions

1. Give a definition of endocrine hormones that includes their most likely source, how they are transported in the body, and how they are received. What does "target" organ mean? (p. 255)
2. Give the location in the human body of all the major endocrine glands. Name the hormones secreted by each gland, and describe their chief function. (p. 255)
3. What two types of chemicals are hormones? Contrast how each type of hormone is received by and affects a cell. (p. 255)
4. Explain the relationship of the hypothalamus to the posterior pituitary and to the anterior pituitary. (p. 256)
5. Explain the concept of negative feedback, and give an example involving ADH. (p. 258)
6. Give an example of the three-tiered relationship between the hypothalamus, the anterior pituitary, and other endocrine glands. (p. 262)
7. Explain why the anterior pituitary can be called the master gland. (p. 260)
8. Draw a diagram to explain the contrary actions of insulin and glucagon. Use your diagram to explain the symptoms of type I diabetes mellitus. (p. 267)
9. Categorize environmental signals into three groups, and give examples of each group. (p. 270)
10. Give examples to show that there is an overlap between the mode of operation of the nervous system and that of the endocrine system. Explain why the traditional definition of a hormone may need to be expanded. (p. 271)

Objective Questions

1. The hypothalamus _____ the hormones _____ and _____, released by the posterior pituitary.
2. The _____ secreted by the hypothalamus stimulate the anterior pituitary.
3. Generally, hormone production is self-regulated by a _____ mechanism.
4. Growth hormone is produced by the _____ pituitary.
5. Simple goiter occurs when the thyroid is producing _____ (too much or too little) _____ .
6. ACTH, produced by the anterior pituitary, stimulates the _____ of the adrenal glands.

7. An overproductive adrenal cortex results in the condition called _____ .
8. Parathyroid hormone increases the level of _____ in blood.

9. Type I diabetes mellitus is due to a malfunctioning _____ , while type II diabetes is due to limited uptake of insulin by _____ .
10. Prostaglandins are not carried in _____ as are hormones secreted by the endocrine glands.

Critical Thinking Questions

1. Use figure 13.9 to explain why control by negative feedback results in a fluctuation of hormonal blood levels about a mean.

2. Name several hormones that control the concentration of blood glucose, and show that this is not an unnecessary duplication of effects.

3. Argue that it is incorrect to speak of the nervous system versus the endocrine system and that instead the two systems should be considered one system.

Selected Key Terms

acromegaly (ak″ro-meg′ah-le)   a condition resulting from an increase in GH production after adult height has been achieved.   260

cretinism (kre′tin-izm)   a condition resulting from a lack of thyroid hormone in an infant.   262

diabetes insipidus (di″ah-be′tēz in-sip′ĭ-dus)   condition characterized by an abnormally large production of urine due to a deficiency of ADH.   258

diabetes mellitus (di″ah-be′tēz mĕ-li′tus)   condition characterized by a high blood glucose level and the appearance of glucose in the urine due to a deficiency of insulin production or uptake by cells.   266

endocrine gland (en′do-krin gland)   a gland that secretes hormones directly into blood or body fluids.   255

exophthalmic goiter (ek″sof-thal′mik goi′ter)   an enlargement of the thyroid gland accompanied by an abnormal protrusion of the eyes.   263

islets of Langerhans (i′lets uv lahng′er-hanz)   distinctive groups of cells within the pancreas that secrete insulin and glucagon.   266

myxedema (mik″sĕ-de′mah)   a condition resulting from a deficiency of thyroid hormone in an adult.   262

pituitary gland (pĭ-too′ĭ-tār″e gland)   anterior portion produces six types of hormones and is controlled by hypothalamic-releasing and release-inhibiting hormones; posterior portion is connected by a stalk to the hypothalamus. 256

simple goiter (sim′p′l goi′ter)   condition in which an enlarged thyroid produces low levels of thyroxin.   262

Further Readings for Part III

Alkon, Daniel L. July 1989. Memory storage and neural systems. *Scientific American.*

Aoki, C. and P. Siekevitz. December 1988. Plasticity in brain development. *Scientific American.*

Atkinson, Mark A., and Noel K. Maclaren. July 1990. What causes diabetes? *Scientific American.*

Barlow, Robert B., Jr. April 1990. What the brain tells the eye. *Scientific American.*

Barr, M. L. 1979. *The human nervous system, an anatomical viewpoint,* 3rd. ed. New York: Harper and Row Publishers.

Bloom, F. E. October 1981. Neuropeptides. *Scientific American.*

Borg, Erik, and S. Allen Counter. August 1989. The middle-ear muscles. *Scientific American.*

Fincher, J. 1981. *The brain: mystery of matter and mind.* Washington, D.C.: U.S. News Books.

Hudspeth, A. J. January 1983. The hair cells of the inner ear. *Scientific American.*

Kalil, Ronald E. December 1989. Synapse formation in the developing brain. *Scientific American.*

Keynes, R. D. March 1979. Ion channels in the nerve cell membrane. *Scientific American.*

Koretz, J. F., and G. H. Handelman. July 1988. How the human eye focuses. *Scientific American.*

Llinas, R. R. October 1982. Calcium in synaptic transmission. *Scientific American.*

Loeb, G. E. February 1985. The functional replacement of the ear. *Scientific American.*

Mishkin, M., and Appenzeller, T. June 1987. The anatomy of memory. *Scientific American.*

Nauta, W. J. H., and M. Feirtag. September 1979. The organization of the brain. *Scientific American.*

Norman, D. A. 1982. *Learning and memory.* San Francisco: W. H. Freeman.

Norton, W. T., and P. Morell. May 1980. Myelin. *Scientific American.*

Orci, L. et al. September 1988. The insulin factory. *Scientific American.*

Rasmussen, Howard. October 1989. The cycling of calcium as an intracellular messenger. *Scientific American.*

Rubenstein, E. March 1980. Diseases caused by impaired communication among cells. *Scientific American.*

Schnapt, J. L., and D. A. Baylor. April 1987. How photoreceptor cells respond to light. *Scientific American.*

Schwartz, J. H. April 1980. The transport of substances in nerve cells. *Scientific American.*

Selim, R. D. 1982. *Muscles: the magic of motion.* Washington, D.C.: U.S. News Books.

Shashoua, V. E. July—August 1985. The role of extracellular proteins in learning and memory. *American Scientist.*

Snyder, S. H. October 1985. The molecular basis of communication between cells. *Scientific American.*

Stevens, C. F. September 1979. The neuron. *Scientific American.*

Stryer, L. July 1987. The molecules of visual excitation. *Scientific American.*

Wertenbaker, L. 1981. *The eye: Window to the world.* Washington, D.C.: U.S. News Books.

Wurtman, R. J. April 1982. Nutrients that modify brain function. *Scientific American.*

Zwislocki, J. J. 1981. Sound analysis in the ear: A history of discoveries. *American Scientist* 69:184.

Part Four

Human Reproduction

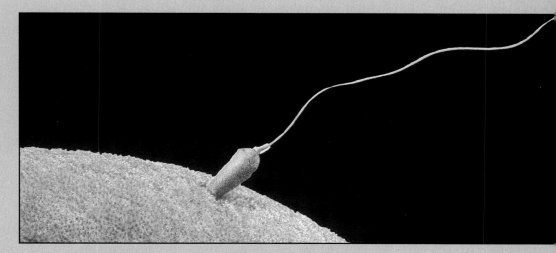

Human beings have two sexes, male and female. The anatomy of each sex functions to produce sex cells that join prior to the development of a new individual. The embryo develops into a fetus within the body of the female, and birth occurs when there is a reasonable chance for independent existence. The steps of human development can be outlined from the fertilized egg to the birth of a child.

We are in the midst of a sexual revolution. We have the freedom to engage in varied sexual practices and to reproduce by alternative methods of conception such as in vitro fertilization. With freedom comes a responsibility to be familiar with the biology of reproduction and its health consequences, not only for ourselves but for our potential offspring.

Chapter Fourteen

Reproductive

System

Chapter Concepts

1 The male reproductive system is designed for the continuous production of a large number of sperm within a fluid medium.

2 The female reproductive system is designed for the monthly production of an egg and the preparation of the uterus for possible implantation of the fertilized egg.

3 Hormones control the reproductive process and the sex characteristics of the individual.

4 Birth control measures vary in effectiveness from those that are very effective to those that are minimally effective.

5 There are alternative methods of reproduction today, including in vitro fertilization followed by artificial implantation.

Female reproductive organs superimposed on an enhanced silhouette. In the pregnant female, the uterus expands greatly to accommodate the growing fetus.

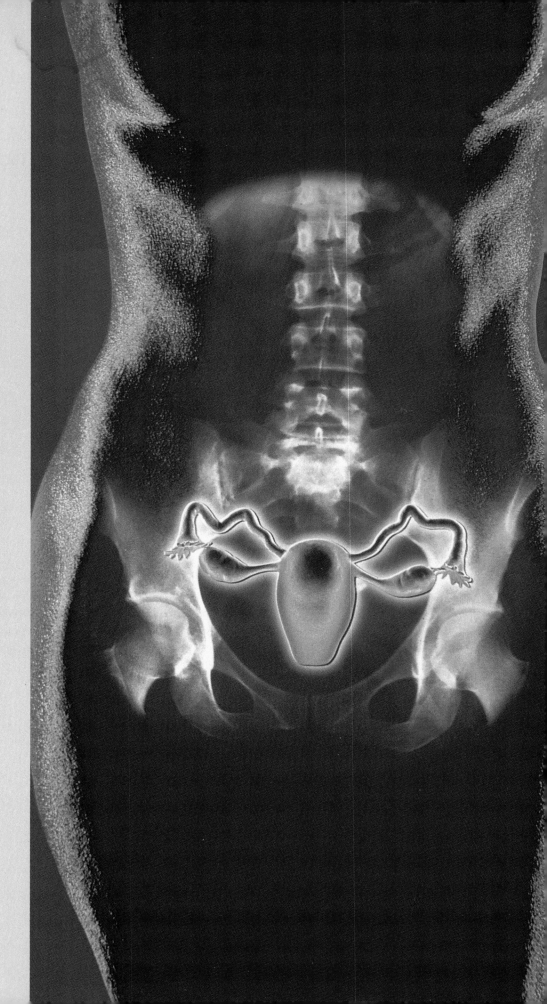

Figure 14.1 When frogs mate, they shed their eggs and sperm right in the water where fertilization takes place. The watery environment protects the gametes and zygote from drying out. When humans mate, the male deposits his sperm inside the female. Her body protects the sperm, egg, and zygote from drying out.

Organisms that reproduce in the water deposit their eggs and sperm in the water (fig. 14.1) because the aquatic environment protects them from drying out. But organisms that reproduce on the land need a mechanism to protect the gametes and developing zygote from the drying effects of the air. In humans the egg stays within the body of the female, where it is fertilized and the zygote undergoes development. The sperm pass from the male within seminal fluid, which if exposed would indeed dry out and be useless. Sexual intercourse prevents this eventuality. During sexual intercourse sperm are deposited into the female's vagina, which is lubricated by secretions. The human sex act is an adaptation to the land environment.

Human Issue

Most public schools now have some sort of sex education program, and these have wide acceptance because of our concerns over child molestation, teenage pregnancy, AIDS, and other sexually transmitted diseases. Sex education, however, raises a number of controversial issues, such as in which grade sex education should begin and how explicit the course should be. For example, should the course include a description of birth control methods and devices, and should all types of sex acts be discussed? If so, should the discussion confine itself to the medical consequences of certain acts or should it also include opinions on the morality of these behaviors? Where do you stand on these issues?

Male Reproductive System

Figure 14.2 shows the reproductive system of the male, and table 14.1 lists the anatomical parts of the system.

Testes

The **testes** lie outside the abdominal cavity of the male within the **scrotum.** The testes begin their development inside the abdominal cavity but descend into the scrotal sacs during the last two months of fetal development. If, by chance, the testes do not descend and the male is not treated or operated on to place the testes in the scrotum, sterility—the inability to produce offspring—usually follows. This is because the internal temperature of the body is too high to produce viable sperm.

Seminiferous Tubules

Fibrous connective tissue forms the wall of each testis and divides it into lobules (fig. 14.3). Each lobule contains one to three tightly coiled **seminiferous tubules** that have a combined length of approximately 250 m. A microscopic cross section through a tubule shows a lobule is packed with cells undergoing spermatogenesis (fig. 14.3c). These cells are derived from undifferentiated germ cells called spermatogonia (singular, spermatogonium) that lie just inside the outer wall and divide mitotically, always producing new spermatogonia. Some newly

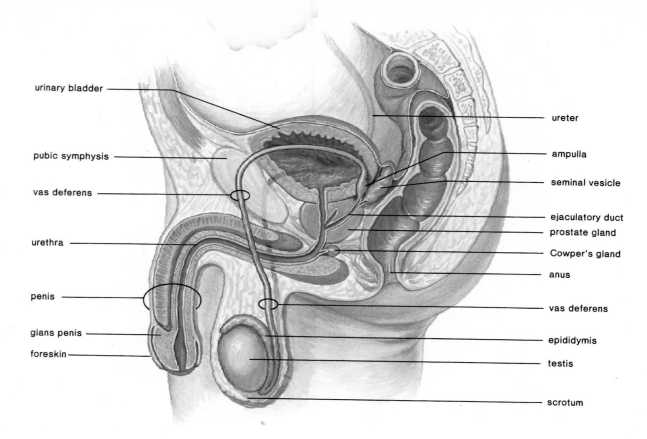

urinary bladder	ureter
pubic symphysis	ampulla
vas deferens	seminal vesicle
urethra	ejaculatory duct
	prostate gland
	Cowper's gland
penis	anus
glans penis	vas deferens
foreskin	epididymis
	testis
	scrotum

Figure 14.2   Side view of the male reproductive system. Trace the path of the genital tract from a testis to the exterior. The seminal vesicles, the Cowper's gland, and the prostate gland add secretions to seminal fluid. Notice that the penis in this drawing is not circumcised since the foreskin is present.

Table 14.1   Male Reproductive System

Organ	Function
testis	produces sperm and sex hormones
epididymis	stores sperm as they mature
vas deferens	conducts and stores sperm
seminal vesicle	contributes to seminal fluid
prostate gland	contributes to seminal fluid
urethra	conducts sperm
Cowper's gland	contributes to seminal fluid
penis	organ of copulation

formed spermatogonia move away from the outer wall to increase in size and become primary spermatocytes that undergo meiosis, a type of cell division described in chapter 4. Although these cells have 46 chromosomes, they divide to give secondary spermatocytes, each with 23 duplicated chromosomes. Secondary spermatocytes divide to give spermatids, also with 23 chromosomes, but single stranded. Spermatids then differentiate into spermatozoa, or mature sperm. Also present in the tubules are the *Sertoli* or nurse *cells,* that support, nourish, and regulate the spermatogenic cells.

Sperm   The mature sperm, or spermatozoan (fig. 14.3*d*), has three distinct parts: a head, a middle piece, and a tail. The *tail* contains the 9 + 2 pattern of microtubules typical of cilia and flagella (fig. 2.12), and the *middle piece* contains energy-producing mitochondria. The *head* contains the 23 chromosomes within a nucleus. The tip of the nucleus is covered by a cap called the **acrosome,** which is believed to contain enzymes needed for fertilization. The human egg is surrounded by several layers of cells and a mucoprotein substance. The acrosome enzymes are believed to aid the sperm in reaching the surface of the egg and allowing a single sperm to penetrate the egg.

Each acrosome may contain such a minute amount of enzyme that it requires the action of many sperm to allow just one to actually penetrate the egg. This may explain why so many sperm are required for the process of fertilization. A normal human male usually produces several hundred million sperm per day, an adequate number for fertilization. Sperm are produced continually throughout a male's reproductive life.

In males, spermatogenesis occurs within the seminiferous tubules of the testes. Sperm have a head capped by an acrosome, where 23 chromosomes reside in the nucleus, a mitochondria-containing middle piece, and a tail with a 9 + 2 pattern of microtubules.

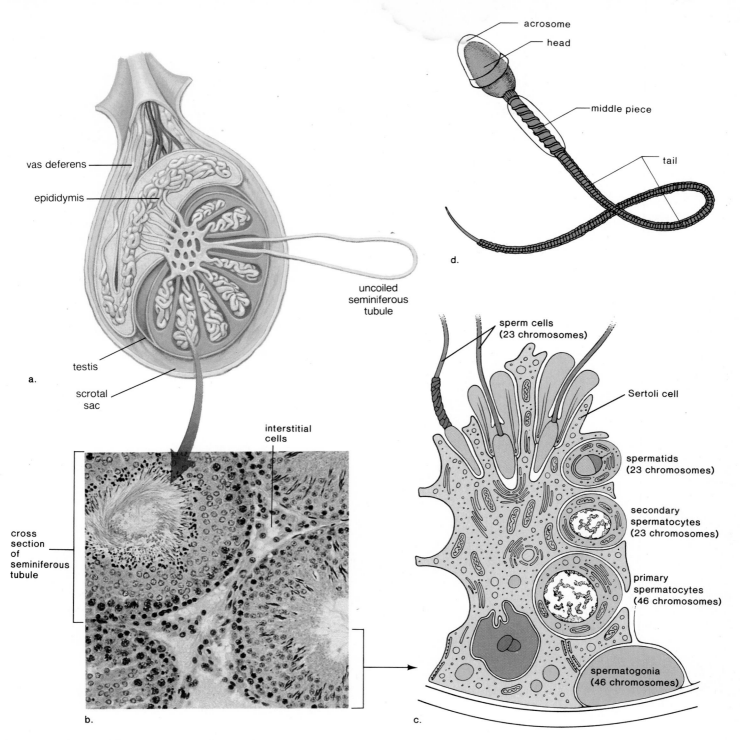

vas deferens

epididymis

testis

scrotal sac

a.

uncoiled seminiferous tubule

interstitial cells

cross section of seminiferous tubule

b.

acrosome

head

middle piece

tail

d.

sperm cells (23 chromosomes)

Sertoli cell

spermatids (23 chromosomes)

secondary spermatocytes (23 chromosomes)

primary spermatocytes (46 chromosomes)

spermatogonia (46 chromosomes)

c.

Figure 14.3   Testis and sperm. *a.* Longitudinal section of testis showing lobules containing seminiferous tubules. *b.* Light micrograph of a cross section of seminiferous tubule. *c.* Diagrammatic representation of spermatogenesis, which occurs in the wall of the tubules. *d.* Mature sperm consist of a head, a middle piece, and a tail. The head, which contains the nucleus, is capped by the enzyme-containing acrosome.

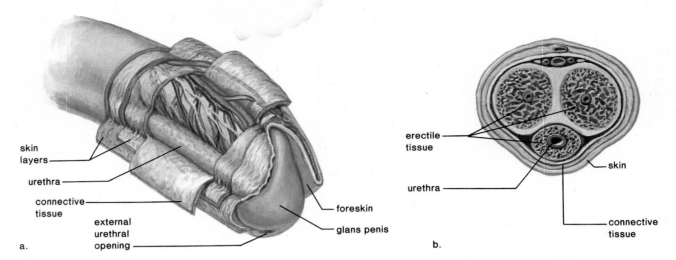

skin
layers

urethra

connective
tissue

external
urethral
opening

a.

foreskin

glans penis

erectile
tissue

urethra

skin

connective
tissue

b.

Figure 14.4   Penis anatomy. *a.* The urethra lies beneath the skin and the connective tissue, and is surrounded by erectile tissue. This tissue expands to form the glans penis, which in uncircumcised males is partially covered by the foreskin. *b.* Two other columns of erectile tissue in the penis are located dorsally.

Interstitial Cells

The male sex hormones, the androgens, are secreted by cells that lie between the seminiferous tubules. Therefore, they are called **interstitial cells** (fig. 14.3*b*). The most important of the androgens is testosterone, whose functions are discussed on page 281.

Genital Tract

Sperm are produced in the testes, but they mature in the **epididymis** (fig. 14.2), a tightly coiled tubule about 5–6 m (17 ft) in length that lies just outside each testis. During the two to four-day maturational period, the sperm develop their characteristic swimming ability. Each epididymis joins with a **vas (ductus) deferens,** which ascends through a canal called the *inguinal canal* and enters the abdominal cavity where it curves around the bladder and empties into the urethra. Sperm are stored in the first part of a vas deferens. They pass from each vas deferens into the urethra only when ejaculation (p. 280) is imminent.

The *spermatic cords* consist of fibrous connective tissue and muscle fibers that enclose the vas deferens, the blood vessels, and the nerves. The region of the inguinal canal, where the spermatic cord passes into the abdomen, remains a weak point in the abdominal wall. As such, it is frequently the site of hernias. A *hernia* is an opening or separation of some part of the abdominal wall through which a portion of an internal organ, usually the intestine, protrudes.

Seminal Fluid

At the time of ejaculation, sperm leave the penis in a fluid called **seminal fluid** (semen). Three types of glands—the seminal vesicles, the prostate gland, and the Cowper's glands—add secretion to seminal fluid. The **seminal vesicles** lie at the base of the

bladder, and each has a duct that joins with a vas deferens. The **prostate gland** is a single doughnut-shaped gland that surrounds the upper portion of the urethra just below the bladder. In older men, the prostate can enlarge and squeeze off the urethra, making urination painful and difficult. This condition can be treated medically or surgically. **Cowper's glands** are pea-sized organs that lie posterior to the prostate on either side of the urethra.

Each component of seminal fluid seems to have a particular function. Sperm are more viable in a basic solution, and seminal fluid, which is milky in appearance, has a slightly basic pH (about 7.5). Swimming sperm require energy, and seminal fluid contains the sugar fructose, which presumably serves as an energy source. Seminal fluid also contains prostaglandins, chemicals that cause the uterus to contract. Some investigators believe that uterine contractions help propel the sperm toward the egg.

Orgasm in Males

The **penis** (fig. 14.4) is the copulatory organ of males. The penis has a long shaft and an enlarged tip called the glans penis. At birth, the glans penis is covered by a layer of skin called the **foreskin,** or prepuce. Gradually, over a period of 5–10 years, the foreskin separates from the penis and may be retracted. During this time there is a natural shedding of cells between the foreskin and penis. These cells, along with an oil secretion that begins at puberty, is called smegma. In the child, no special cleansing method is needed to wash away smegma, but in the adult, the foreskin can be retracted to do so. **Circumcision** is the surgical removal of the foreskin, usually soon after birth.

When the male is sexually aroused, the penis becomes erect and ready for intercourse. **Erection** is achieved because blood sinuses within the erectile tissue of the penis fill with blood.

The discussion regarding the advisability or inadvisability of having males circumcised (see figure) has centered on the following considerations.

Religion In Islam and in Judaism, circumcision is a religious practice that represents a covenant with God made by Abraham.

Cleanliness In preteen males, a number of small glands located in the foreskin and under the corona of the glans penis begin to produce an oily secretion. This secretion along with dead skin cells forms a cheesy substance known as smegma. In the circumcised male, smegma does not build up; in the mature uncircumcised mature male, it collects and its removal requires routine hygienic care. It is only necessary to retract the foreskin and wash away the smegma. If smegma is allowed to collect, bacteria can multiply and it will develop a strong odor. Also, the bacteria can cause a variety of infections in the female vaginal tract.

Functions Whether or not the foreskin protects the glans penis or whether it increases or decreases sexual sensitivity has not been definitely determined. It retracts during sexual intercourse and no study has shown that its presence or absence has any effect on male performance or feelings.

Penile and Cervical Cancer There is evidence that uncircumcised males may be slightly more at risk for developing penile cancer. For example, penile cancer almost never occurs among Jewish males.

There is no evidence that the presence of a foreskin contributes to the development of cervical cancer in females. A study of Lebanese Muslims and Christians showed no greater incidence among the Christians compared to the Muslims who practice circumcision.

Urinary Infections It is not necessary to clean beneath the foreskin in newborn males; indeed the foreskin does not retract in newborn males and it should not be forced back. But several studies have reported a higher incidence of urinary infections among male infants. In one study, it was found that 1,661 out of 40,000 infants developed a urinary infection. Among these infants, uncircumcised males had twice the incidence of urinary tract infections compared to girls and 10 times the rate of

Circumcision

circumcised boys. It's believed that in infants who wear diapers, bacteria find a haven beneath the foreskin and, thereafter, can enter the urethra.

Surgical trauma Duration of crying and increase in heart rate suggests that circumcision is painful to newborns. Indications of pain are reduced when a local anesthetic is given. In rare instances, hemorrhage, infection, mutilation, and even death has occurred. Whether or not there is long-lasting psychological trauma has not been definitely established. Sometimes circumcision is necessary later on in life due to phimosis, a condition in which the foreskin tightens and cannot be pulled back.

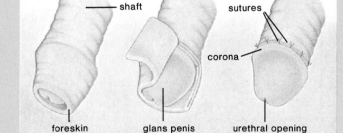

Circumcision exposes the glans penis by removing the foreskin (prepuce).

shaft

sutures

corona

foreskin

glans penis

urethral opening

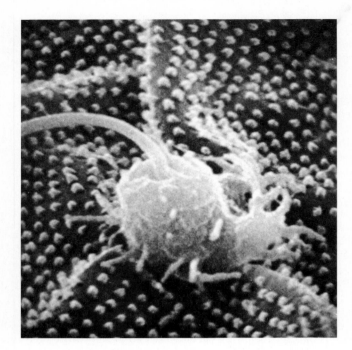

Figure 14.5   Fertilization. A single sperm enters the egg and then a new life begins. The reproductive systems of males and females are designed to bring about this union of the gametes.

Parasympathetic impulses dilate the arteries of the penis, while the veins are compressed passively so that blood flows into the erectile tissue under pressure. If the penis fails to become erect, the condition is called **impotency.** There are medical and surgical remedies for impotency.

Ejaculation

As sexual stimulation intensifies, sperm enter the urethra from each vas deferens and the glands contribute secretions to seminal fluid (semen). Once seminal fluid is in the urethra, rhythmical muscle contractions cause it to be expelled from the penis in spurts. During ejaculation, a sphincter closes off the bladder so that no urine enters the urethra. (Notice that the urethra carries either urine or semen at different times.)

The contractions that expel seminal fluid from the penis are a part of male **orgasm,** the physiological and psychological sensations that occur at the climax of sexual stimulation. The psychological sensation of pleasure is centered in the brain, but the physiological reactions involve the genital (reproductive) organs and associated muscles, as well as the entire body. Marked muscular tension is followed by contraction and relaxation.

Following ejaculation and/or loss of sexual arousal, the penis returns to its normal flaccid state. After ejaculation, a male typically experiences a period of time, called the refractory period, during which stimulation does not bring about an erection. The length of the refractory period increases with age.

There may be in excess of 400 million sperm in the 3.5 ml of semen expelled during ejaculation. The sperm count can be much lower than this, however, and fertilization (fig. 14.5) still can take place.

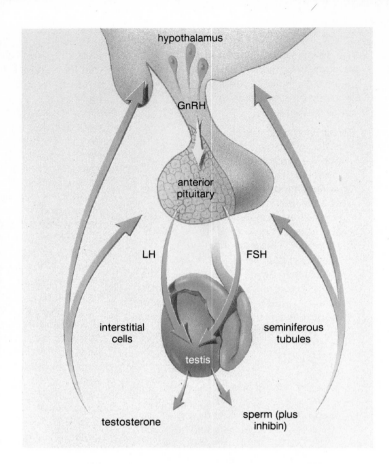

Figure 14.6   The hypothalamus–pituitary–testes control relationship. GnRH (gonadotropic-releasing hormone) stimulates the anterior pituitary to secrete the gonadotropic hormones FSH and LH. FSH stimulates the testes to produce sperm, and LH stimulates the testes to produce testosterone. Testosterone and inhibin exert negative feedback control over the hypothalamus and the anterior pituitary, and this ultimately regulates the level of testosterone in the blood.

Sperm mature in the epididymis and are stored in the vas deferens before entering the urethra just prior to ejaculation. The accessory glands (seminal vesicles, prostate gland, and Cowper's gland) add secretions to seminal fluid. Seminal fluid (semen) leaves the penis during ejaculation.

Regulation of Male Hormone Levels

The hypothalamus has ultimate control of the testes' sexual functions because it secretes gonadotropic-releasing hormone (GnRH) that stimulates the anterior pituitary to produce the gonadotropic hormones. Two gonadotropic hormones, **FSH (follicle-stimulating hormone)** and **LH (luteinizing hormone),** are named for their function in females but exist in both sexes, stimulating the appropriate gonads in each. FSH promotes spermatogenesis in the seminiferous tubules, and LH promotes the production of testosterone in the interstitial cells. Sometimes, LH in males is sometimes given the name interstitial cell-stimulating hormone (ICSH).

The hormones mentioned are involved in a feedback process (fig. 14.6) that maintains the production of testosterone at

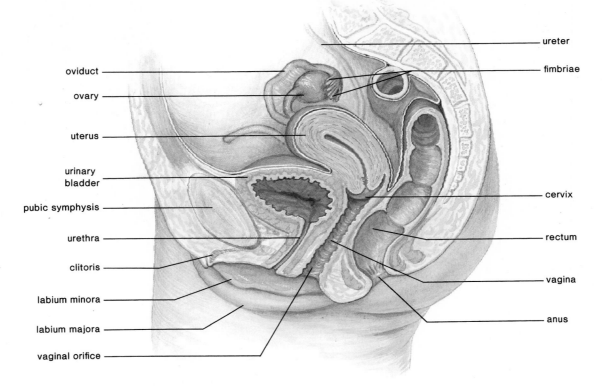

Figure 14.7   Side view of the female reproductive system. The ovaries produce one egg a month; fertilization occurs in the oviduct, and most development occurs in the uterus. The vagina is the birth canal and the organ of copulation.

a fairly constant level. For example, when the amount of testosterone in blood rises to a certain level, it causes the anterior pituitary to decrease its secretion of LH. As the level of testosterone begins to fall, the anterior pituitary increases its secretion of LH, and stimulation of the interstitial cells reoccurs. It should be emphasized that only minor fluctuations of testosterone level occur in the male and that the feedback mechanism in this case acts to maintain testosterone at a normal level. It long has been suspected that the seminiferous tubules produce a hormone that blocks FSH secretion. This substance, termed *inhibin,* recently has been isolated.

Testosterone

The male sex hormone, **testosterone,** has many functions. It is essential for the normal development and functioning of the primary sex organs, those structures we just have discussed. It is also necessary for the maturation of sperm.

Greatly increased testosterone secretion at the time of puberty stimulates the growth of the penis and the testes. Testosterone also brings about and maintains the secondary sex characteristics in males that develop at the time of puberty. Testosterone causes growth of a beard, axillary (underarm) hair, and pubic hair. It prompts the larynx and the vocal cords to enlarge, causing the voice to change. It is responsible for the great muscular strength of males, and this is the reason some

athletes take supplemental amounts of *anabolic steroids,* which are either testosterone or related chemicals. The contraindications of taking anabolic steroids are discussed in the reading on page 268. Testosterone also causes oil and sweat glands in the skin to secrete; therefore, it is largely responsible for acne and body odor. Another side effect of testosterone activity is baldness. Genes for baldness probably are inherited by both sexes, but baldness is seen more often in males because of the presence of testosterone.

Testosterone is believed to be largely responsible for the sex drive. It may even contribute to the supposed aggressiveness of males.

In males, FSH promotes spermatogenesis and LH promotes testosterone production within the testes. Testosterone stimulates growth of the male genitals during puberty and is necessary for maturation of sperm and development of secondary sex characteristics.

Female Reproductive System

Figure 14.7 illustrates the anatomical parts of the female reproductive system.

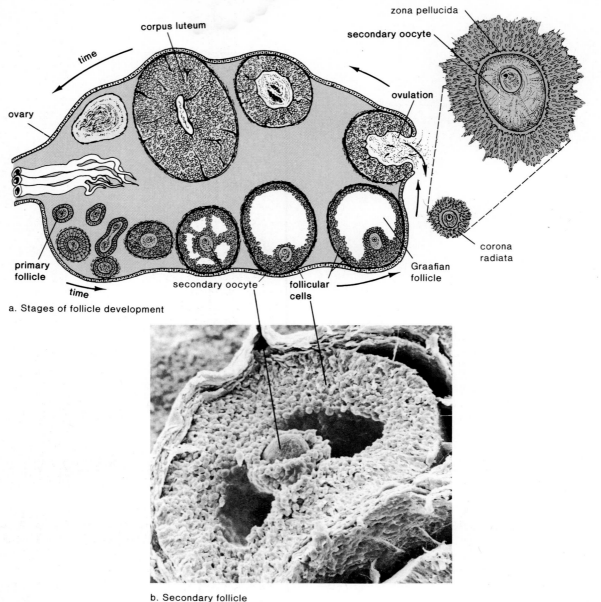

a. Stages of follicle development

zona pellucida

secondary oocyte

ovulation

corpus luteum

time

ovary

corona radiata

primary follicle

time

secondary oocyte

follicular cells

Graafian follicle

b. Secondary follicle

Figure 14.8 Anatomy of ovary and follicle. *a.* As a follicle matures, the oocyte enlarges and is surrounded by a mantle of follicular cells, called the corona radiata, and fluid. Eventually, ovulation occurs, the mature follicle ruptures, and the secondary oocyte is released. A single follicle actually goes through all stages in one place within the ovary. *b.* Scanning electron micrograph of a secondary follicle.

Ovaries

The **ovaries** lie in shallow depressions, one on each side of the upper pelvic cavity. A longitudinal section through an ovary shows that it is made up of an outer cortex and an inner medulla. There are many **follicles** in the cortex and each one contains an oocyte. A female is born with as many as 2 million follicles, but the number is reduced to 300,000—400,000 by the time of puberty. Only a small number of follicles (about 400) ever mature because a female usually produces only one egg per month during her reproductive years. Since oocytes are present at birth, they age as the woman ages. This is one possible reason why older women are more likely to produce children with genetic defects.

As the follicle undergoes maturation, it develops from a primary follicle to a secondary follicle to a **Graafian follicle** (fig. 14.8). In a primary follicle, the primary oocyte divides meiotically into two cells, each having 23 chromosomes (see fig. 17.13). One of these cells, termed the secondary oocyte, receives almost all the cytoplasm. The other is a polar body that disintegrates. A secondary follicle contains the secondary oocyte pushed to one side of a fluid-filled cavity. In a Graafian follicle,

Table 14.2 Female Reproductive System

Organ	Function
ovary	produces egg and sex hormones
oviduct (fallopian tube)	conducts egg toward uterus
uterus (womb)	houses developing fetus
cervix	contains opening to uterus
vagina	receives penis during copulation and serves as birth canal

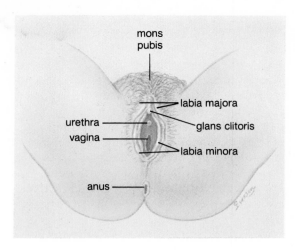

Figure 14.9 External genitalia of female. At birth, the opening of the vagina is partially occluded by a membrane called the hymen. Physical activities and sexual intercourse disrupt the hymen.

the fluid-filled cavity increases to the point that the follicle wall balloons out on the surface of the ovary and bursts, releasing the secondary oocyte (often called an egg for convenience) surrounded by a clear membrane, the zona pellucida, and follicular cells, the corona radiata. This is referred to as **ovulation.** Once a follicle has lost its egg, it develops into a **corpus luteum,** a glandlike structure. If pregnancy does not occur, the corpus luteum begins to degenerate after about 10 days. If pregnancy does occur, the corpus luteum persists for three to six months. The follicle and the corpus luteum secrete the female sex hormones estrogen and progesterone, as discussed on page 284.

In females, oogenesis occurs within the ovaries, where one follicle reaches maturity each month. This follicle balloons out of the ovary and bursts to release the egg. The ruptured follicle develops into a corpus luteum. The follicle and the corpus luteum produce the female sex hormones estrogen and progesterone.

Genital Tract

The female genital tract includes the oviducts, the uterus, and the vagina (table 14.2).

Oviducts

The oviducts, also called uterine or fallopian tubes, extend from the uterus to the ovaries. The oviducts are not attached to the ovaries; instead, they have fingerlike projections called **fimbriae** that sweep over the ovary at the time of ovulation. When the egg bursts (fig. 14.8) from the ovary during ovulation, it usually is swept up into an oviduct by the combined action of the fimbriae and the beating of cilia that line the oviducts.

Once in the oviduct, the egg is propelled slowly by cilia movement and tubular muscle contraction toward the uterus. Fertilization, the completion of oogenesis and zygote formation occurs in an oviduct because the egg only lives approximately 6 to 24 hours. The developing embryo normally arrives at the uterus after several days and then embeds, or implants, itself in the uterine lining, which has been prepared to receive it.

Uterus

The **uterus** is a thick-walled, muscular organ about the size and the shape of an inverted pear. Normally, it lies above and is tipped over the urinary bladder. The oviducts join the uterus ante-

riorly, while posteriorly, the cervix enters into the vagina nearly at a right angle. A small opening in the cervix leads to the vaginal canal. Development of the embryo normally takes place in the uterus. This organ, sometimes called the womb, is approximately 5 cm wide in its usual state but is capable of stretching to over 30 cm to accommodate the growing baby. The lining of the uterus, called the **endometrium,** participates in the formation of the placenta (p. 319), which supplies nutrients needed for embryonic and fetal development. The endometrium has two layers: a basal layer and an inner functional layer. In the nonpregnant female, the functional layer of the endometrium varies in thickness according to a monthly reproductive cycle, called the uterine cycle (p. 284).

Cancer of the cervix is a common form of cancer in women. Early detection is possible by means of a **Pap test,** which requires the removal of a few cells from the region of the cervix for microscopic examination. If the cells are cancerous, a hysterectomy may be recommended. A hysterectomy is the removal of the uterus. Removal of the ovaries in addition to the uterus is termed an ovariohysterectomy. Because the vagina remains, the woman still can engage in sexual intercourse.

Vagina

The **vagina** is a tube that makes a 45-degree angle with the small of the back. The mucosal lining of the vagina lies in folds that extend as the fibromuscular wall stretches. This capacity to extend is especially important when the vagina serves as the birth canal, and it also can facilitate intercourse, when the vagina receives the penis during copulation.

External Genitalia

The external genital organs of the female (fig. 14.9) are known collectively as the **vulva.** The vulva includes two large, hair-covered folds of skin called the **labia majora.** They extend backward from the *mons pubis,* a fatty prominence underlying the pubic hair. The **labia minora** are two small folds lying just inside the labia majora. They extend forward from the vaginal opening to encircle and form a foreskin for the *clitoris,* an organ that is homologous to the penis. Although quite small, the clitoris has a shaft of erectile tissue and is capped by a pea-shaped glans. The glans clitoris also has sense receptors that allow it to function as a sexually sensitive organ.

The *vestibule,* a cleft between the labia minora, contains the openings of the urethra and the vagina. The vagina may be partially closed by a ring of tissue called the hymen. The hymen ordinarily is ruptured by initial sexual intercourse; however, it also can be disrupted by other types of physical activities. If the hymen persists after sexual intercourse, it can be surgically ruptured.

Notice that the urinary and reproductive systems in the female are entirely separate. For example, the urethra carries only urine, and the vagina serves only as the birth canal and the organ for sexual intercourse.

The egg enters the oviducts, which lead to the uterus and then the vagina. The vagina opens into the vestibule, the location of female external genitalia.

Orgasm in Females

Sexual response in the female may be more subtle than in the male, but there are certain corollaries. The clitoris is believed to be an especially sensitive organ for initiating sexual sensations. It is possible for the clitoris to become ever so slightly erect as its erectile tissues become engorged with blood, but vasocongestion is more obvious in the labia minora, which expand and deepen in color. Erectile tissue within the vaginal wall also expands with blood, and the added pressure in these blood vessels causes small droplets of fluid to squeeze through the vessel walls and to lubricate the vagina.

Release from muscular tension occurs in females, especially in the region of the vulva and vagina but also throughout the entire body. Increased uterine motility may assist the transport of sperm toward the oviducts. Since female orgasm is not signaled by ejaculation, there is a wide range in normalcy of sexual response.

Regulation of Female Hormone Levels

Hormonal regulation in the female is quite complex, so we begin with a simplified presentation and follow with a more in-depth presentation for those who wish to study the matter in greater detail. The following glands and hormones are involved in hormonal regulation.

Hypothalamus: secretes *GnRH* (gonadotropic-releasing hormone) and *GnRIH* (gonadotropic release-inhibiting hormone)
Anterior pituitary: secretes *FSH* (follicle-stimulating hormone) and *LH* (luteinizing hormone), the gonadotropic hormones
Ovaries: secrete estrogen and progesterone, the female sex hormones

Table 14.3 Ovarian and Uterine Cycles (Simplified)

Ovarian cycle	Events	Uterine cycle	Events
follicular phase days 1–13	FSH follicle maturation estrogen	menstruation days 1–5 proliferative phase days 6–13	endometrium breaks down endometrium rebuilds
*Ovulation day 14**			
luteal phase days 15–28	LH corpus luteum progesterone	secretory phase days 15–28	endometrium thickens and glands are secretory

*Assuming a 28-day cycle

The female sex hormones have many effects on the body. In particular, estrogen secreted at the time of puberty stimulates the growth of the uterus and the vagina. Estrogen is necessary for egg maturation and is largely responsible for the secondary sex characteristics in females. For example, it is responsible for female body hair and fat distribution. In general, females have a more rounded appearance than males because of a greater accumulation of fat beneath the skin. Also, the pelvic girdle enlarges in females so that the pelvic cavity has a larger relative size compared to males; this means that females have wider hips. Both estrogen and progesterone also are required for breast development.

Hormonal Regulation (Simplified)

Ovarian Cycle The gonadotropic and sex hormones are not present in constant amounts in the female and instead are secreted at different rates during a monthly **ovarian cycle,** which lasts an average of 28 days but may vary widely in individuals. For simplicity's sake, it is convenient to emphasize that during the first half of a 28-day cycle (days 1–13, table 14.3), FSH from the anterior pituitary is promoting the development of a follicle in the ovary and that this follicle is secreting estrogen. As the estrogen blood level rises, it exerts feedback control over the anterior pituitary secretion of FSH so that this follicular phase comes to an end (fig. 14.10). The end of the follicular phase is marked by ovulation on the fourteenth day of the 28-day cycle. Similarly, it can be emphasized that during the last half of the ovarian cycle (days 15–28, table 14.3), anterior pituitary production of LH is promoting the development of a corpus luteum, which is secreting progesterone. As the blood progesterone blood level rises, it exerts feedback control over anterior pituitary secretion of LH so that the corpus luteum begins to degenerate. As the luteal phase comes to an end, menstruation occurs.

Uterine Cycle The female sex hormones cause the uterus to undergo a cyclical series of events known as the **uterine cycle** (table 14.3). A cycle that lasts 28 days is divided as follows.

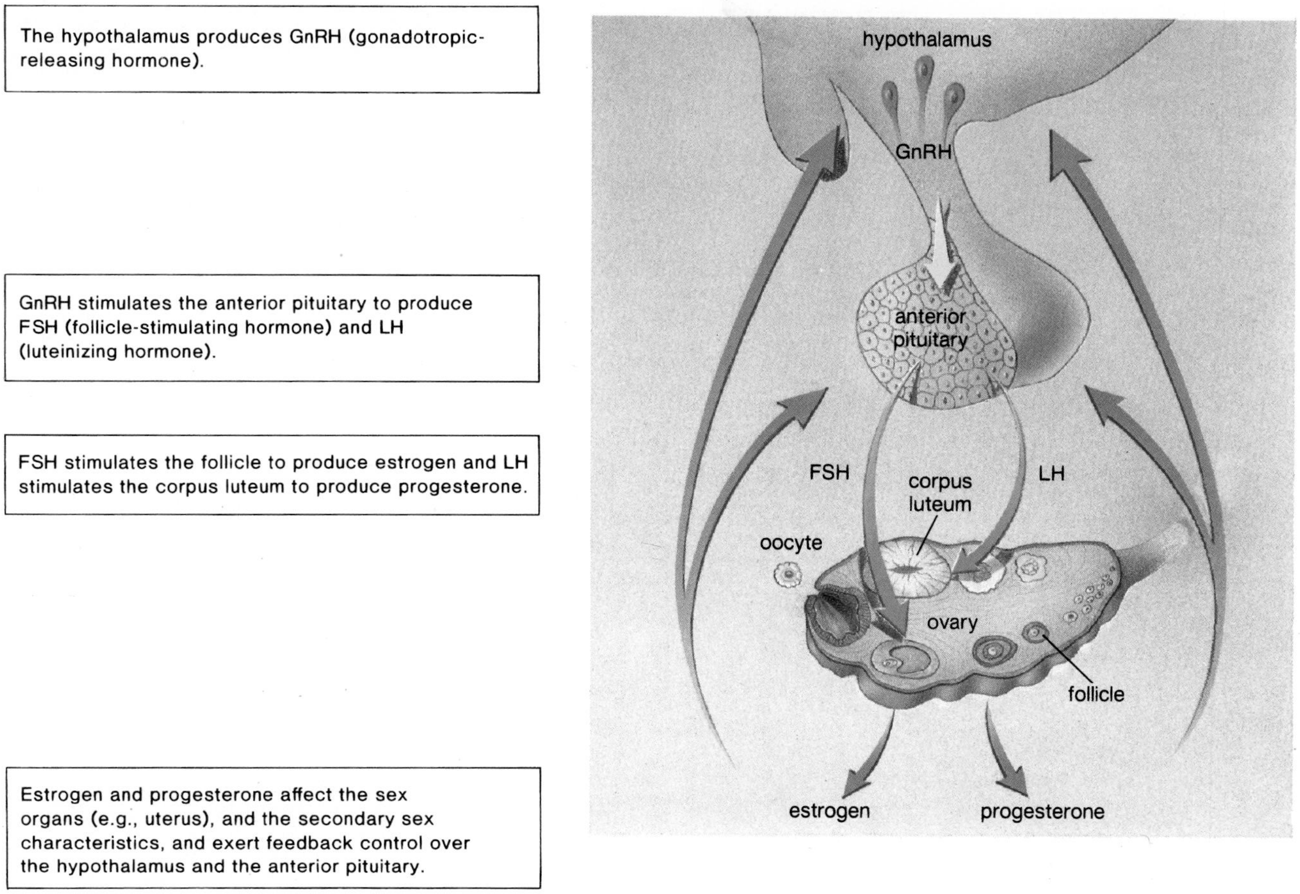

The hypothalamus produces GnRH (gonadotropic-releasing hormone).

GnRH stimulates the anterior pituitary to produce FSH (follicle-stimulating hormone) and LH (luteinizing hormone).

FSH stimulates the follicle to produce estrogen and LH stimulates the corpus luteum to produce progesterone.

Estrogen and progesterone affect the sex organs (e.g., uterus), and the secondary sex characteristics, and exert feedback control over the hypothalamus and the anterior pituitary.

Figure 14.10   The hypothalamus–pituitary–ovary control relationship.

During *days 1–5,* there is a low level of female sex hormones in the body, causing the uterine lining to disintegrate and its blood vessels to rupture. A flow of blood, known as the *menses,* passes out of the vagina during a period of **menstruation,** also known as the menstrual period.

During *days 6 –13,* increased production of estrogen by an ovarian follicle causes the endometrium to thicken and to become vascular and glandular. This is called the proliferative phase of the uterine cycle.

Ovulation usually occurs on the fourteenth day of the 28-day cycle.

During *days 15–28,* increased production of progesterone by the corpus luteum causes the endometrium to double in thickness and the uterine glands to mature, producing a thick mucoid secretion. This is called the secretory phase of the uterine cycle. The endometrium now is prepared to receive the developing embryo, but if pregnancy does not occur, the corpus luteum degenerates and the low level of sex hormones in the female body causes the uterine lining to break down. This is evident, due to the menstrual discharge that begins at this time. Even while menstruation is occurring, the anterior pituitary begins to increase its production of FSH and a new follicle begins to mature. Table 14.3 indicates how the ovarian cycle controls the uterine cycle.

Hormonal Regulation (Detailed)

Figure 14.11 shows the changes in blood concentration of all four hormones participating in the ovarian and uterine cycles. Notice that all four of these hormones (FSH, LH, estrogen, and progesterone) are present during the entire 28 days of the cycle. Therefore, in actuality, both FSH and LH are present during the follicular phase and both are needed for follicle development and egg maturation. The follicle secretes primarily estrogen and a very minimal amount of progesterone. Similarly, both LH and FSH are present in decreased amounts during the luteal phase. LH may be primarily responsible for corpus luteum formation, but the corpus luteum secretes both progesterone and estrogen. The effect that these hormones have on the endometrium has already been stated: Estrogen stimulates growth of the endometrium and readies it for reception of progesterone, which causes it to thicken and to become secretory.

Feedback Control   As the estrogen level increases during the first part of the follicular phase, FSH secretion begins to decrease due to negative feedback. However, the high level of estrogen is believed to exert *positive feedback on the hypothalamus,* causing it to secrete GnRH, after which the pituitary momentarily produces an unusually large amount of FSH and LH. It is the surge of LH that is believed to promote ovulation.

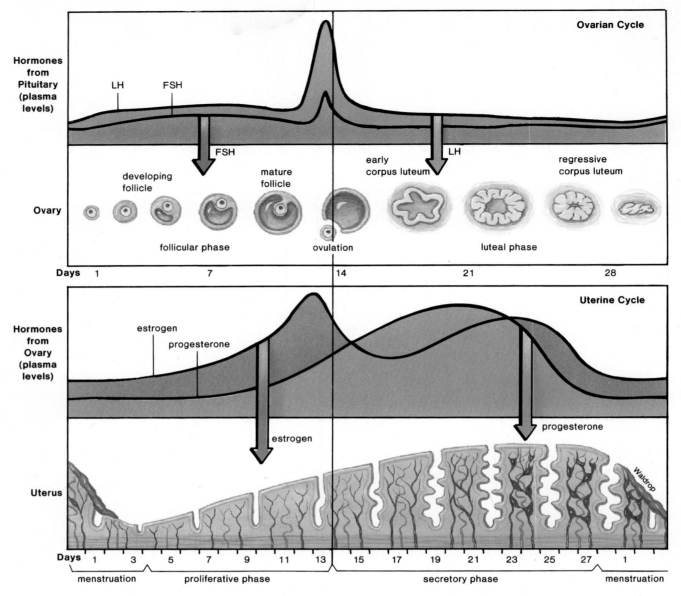

Figure 14.11   Plasma hormonal levels associated with the ovarian and uterine cycles. During the follicular phase, FSH produced by the anterior pituitary promotes the maturation of a follicle in the ovary. In the proliferative phase, the follicle produces increasing levels of estrogen, which causes the endometrial lining of the uterus to thicken.

After ovulation and during the luteal phase, LH promotes the development of the corpus luteum. In the secretory phase, the corpus luteum produces increasing levels of progesterone, which causes the endometrial lining to become secretory. Menstruation begins when progesterone production declines to a low level.

During the luteal phase, estrogen and progesterone bring about feedback inhibition as expected, and the levels of both LH and FSH decline steadily. In this way, all four hormones eventually reach their lowest levels, causing menstruation to occur. Therefore, the corpus luteum degenerates unless pregnancy occurs.

During the first half of the ovarian cycle, FSH from the anterior pituitary causes maturation of a follicle, which secretes estrogen. After ovulation and during the second half of the cycle, LH from the anterior pituitary converts the follicle into the corpus luteum, which produces progesterone. Estrogen and progesterone regulate the uterine cycle in which the endometrium builds up and then is shed during menstruation.

Pregnancy

If pregnancy occurs, menstruation does not occur. Instead, the developing embryo embeds itself in the endometrial lining several days following fertilization. Once this process, called **implantation,** is complete, a female is *pregnant*. During implantation, an embryonic membrane surrounding the embryo produces a gonadotropic hormone called **h**uman **c**horionic **go**nadotropic hormone **(HCG),** which prevents degeneration of the corpus luteum and instead causes it to secrete even larger quantities of progesterone. The corpus luteum may be maintained for as long as six months, even after the placenta is fully developed.

The **placenta** (see fig. 16.9) originates from both maternal and fetal tissue and is the region of exchange of molecules between fetal and maternal blood, although there is no mixing of the two types of blood. After its formation, the placenta continues production of HCG and begins production of progesterone and estrogen. The latter hormones have two effects: they shut down the anterior pituitary so that no new follicles mature, and they maintain the lining of the uterus so that the corpus luteum is not needed. There is no menstruation during the nine months of pregnancy.

Pregnancy Tests

Pregnancy tests, which are readily available in hospitals, clinics, and now even drug and grocery stores, are based on the fact that HCG is present in the blood and the urine of a pregnant woman.

Before the advent of monoclonal antibodies, only a hospital blood test using radioactive material was available to detect pregnancy before the first missed menstrual period. Now there is a monoclonal antibody (p. 144) test for the detection of pregnancy 10 days after conception. This test can be done on a urine sample, and the results are available within the hour.

The physical signs that oftentimes prompt a woman to have a pregnancy test are cessation of menstruation, increased frequency of urination, morning sickness, and increase in the size and the fullness of the breasts, as well as darkening of the areolae (fig. 14.12).

Female Breasts

A female breast contains 15 to 25 lobules (fig. 14.12), each with its own mammary duct that begins at the nipple and divides into numerous other ducts that end in blind sacs called *alveoli*. In a nonlactating (nonmilk-producing) breast, the ducts far outnumber the alveoli because alveoli are made up of cells that can produce milk.

Milk is not produced during pregnancy. *Prolactin* (lactogenic hormone) is needed for lactation (milk production) to begin, and the production of this hormone is suppressed because of the feedback inhibition estrogen and progesterone have on the pituitary during pregnancy. It takes a couple of days after delivery for milk production to begin, and in the meantime, the breasts produce a yellow-white fluid called *colostrum,* which differs from milk in that it contains more protein and less fat. Colostrum is a source of passive immunity for the baby.

The continued production of milk requires continued production of lactogenic hormone, which occurs as long as the woman is breast-feeding. The hormone oxytocin is necessary to milk letdown. When a breast is suckled, the nerve endings in the areola are stimulated, and nerve impulses travel to the hypothalamus, which causes oxytocin to be released by the posterior pituitary. When this hormone arrives at the breasts, it causes contraction of the lobules so that milk flows into the ducts.

Menopause

Menopause, the period in a woman's life during which the ovarian and uterine cycles cease, is likely to occur between ages 45 and 55. The ovaries are no longer responsive to the gonad-

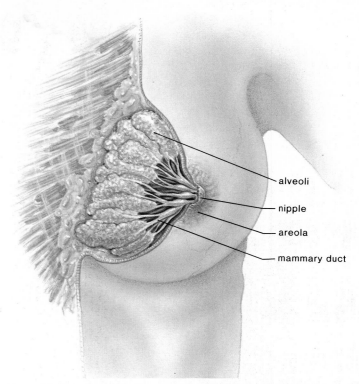

Figure 14.12   Female breast anatomy. The female breast contains lobules consisting of ducts and alveoli. The alveoli are lined by milk-producing cells in the lactating (milk-producing) breast.

otropic hormones produced by the anterior pituitary, and the ovaries no longer secrete estrogen or progesterone. At the onset of menopause, the uterine cycle becomes irregular, but as long as menstruation occurs, it is still possible for a woman to conceive. Therefore, a woman usually is not considered to have completed menopause until there has been no menstruation for a year. The hormonal changes during menopause often produce physical symptoms, such as "hot flashes" that are caused by circulatory irregularities, dizziness, headaches, insomnia, sleepiness, and depression. Again, there is a great variation among women, and any of these symptoms may be absent altogether.

Women sometimes report an increased sex drive following menopause. It has been suggested that this may be due to androgen production by the adrenal cortex.

Estrogen and to some extent progesterone affect the female genitals, promote development of the egg, and maintain the secondary sex characteristics. Prolactin causes the breasts to begin milk secretion after delivery, while another hormone, oxytocin, is responsible for milk letdown. When menopause occurs, FSH and LH are still produced by the anterior pituitary, but the ovaries no longer are able to respond.

Control of Reproduction

Birth Control

The use of birth control (contraceptive) methods decreases the probability of pregnancy but does not, except where noted, offer any protection against contracting a sexually transmitted disease such as AIDS. A common way to discuss pregnancy rate is to indicate the number of pregnancies expected per 100 women per year. For example, it is expected that 80 out of 100 young women, or 80%, who are regularly engaging in unprotected intercourse will be pregnant within a year. Another way to discuss birth control methods is to indicate their effectiveness, in which case the emphasis is placed on the number of women who will not get pregnant. For example, with the least effective method given in figure 14.13, we expect that 70 out of 100, or 70%, sexually active women will not get pregnant, and 30 women will get pregnant within a year. The very best and surest method of birth control is total abstinence.

Human Issue

Teenage pregnancy is a matter of utmost current concern. Girls who are not of an age to care for themselves are getting pregnant and either seeking abortions or having their babies. Since both of these present hardships to all concerned, solutions to the problem have been sought. Do you feel that prevention of teenage pregnancy is a private matter between parents and children, or do you feel that society should be actively involved, perhaps by making contraceptives available to young people? For example, some high schools now have birth control clinics. Do you approve of this action, or is there a better way to prevent teenage pregnancy?

Group I

Sterilization is a surgical procedure that renders the individual incapable of reproduction. Sterilization operations do not affect the secondary sex characteristics nor sexual performance.

In the male, a **vasectomy** consists of cutting and sealing the vas deferens on each side so that the sperm are unable to reach the seminal fluid that is ejected at the time of orgasm. The sperm are then largely reabsorbed. Following this operation, which can be done in a doctor's office, the amount of ejaculate remains normal because sperm account for only about 1% of the volume of semen. Also, there is no effect on the secondary sex characteristics since testosterone continues to be produced by the testes.

In the female, **tubal ligation** consists of cutting and sealing the oviducts. Pregnancy rarely occurs because the passage of the egg through the oviducts has been blocked. Whereas major abdominal surgery was formerly required for a tubal ligation, today there are simpler procedures. Using a method called *laparoscopy,* which requires only two small incisions, the surgeon inserts a small, lighted telescope to view the oviducts and a small surgical blade to sever them. An even newer method called hysteroscopy uses a telescope within the uterus to seal the tubes by means of an electric current.

Although recently developed microsurgical methods allow either a vas deferens or oviduct to be rejoined, it is still wise to view a vasectomy or tubal ligation as permanent. Even following successful reconnection, fertility is usually reduced by about 50%.

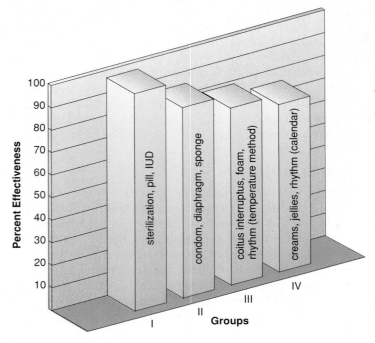

Figure 14.13 Effectiveness (the percentage of women who are not expected to be pregnant within one year) of various birth control measures. Sterilization and the pill offer the best protection, while creams, jellies, and the rhythm method offer the least protection from the occurrence of pregnancy. This graph assumes that users are properly and faithfully using the various means of birth control.

Source: Data based on Guttmacher, Alan F., *Pregnancy, Birth, and Family Planning.* New York: New American Library 1973.

The *birth control pill* (fig. 14.14*d*) is usually a combination of estrogen and progesterone that is taken for twenty-one days of a twenty-eight-day cycle (beginning at the end of menstruation). The estrogen and progesterone in the Pill effectively shut down the pituitary production of both FSH and LH so that no follicle begins to develop in the ovary; and since ovulation does not occur, pregnancy cannot take place. Both beneficial and adverse side effects have been linked to the Pill. Women report relief of menstrual discomforts and acne. They also report several minor adverse side effects such as nausea and vomiting. Less common complaints are weight gain, headaches, and chloasma (areas of darkened skin on the face). One serious side effect of the Pill is increased incidence of thromboembolism—almost exclusively in women who are over 35 and who smoke. Since there are possible side effects, those taking the Pill should always be seen regularly by a physician.

The *norplant system* is a newly available implantable device that provides birth control for up to five years. The device consists of six match-sized capsules surgically inserted under the skin of a woman's upper arm. Norplant releases low doses of the hormone levonorgestrel, a synthetic progesterone, at a constant rate. Like the birth control pill, the device prevents ovulation by disrupting the ovarian cycle.

An *IUD* (intrauterine device) (fig. 14.14*a*) is a small piece of molded plastic that is inserted into the uterus by a physician.

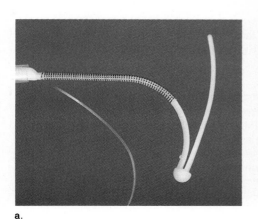

a.

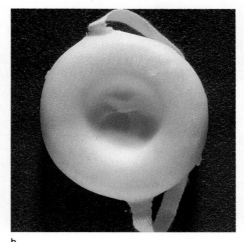

b.

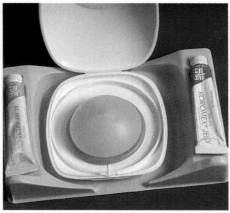

c. ✷ Sperm Trampalene

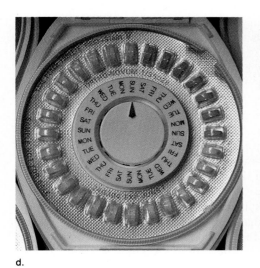

d.

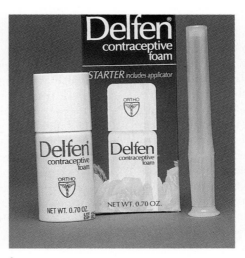

e.

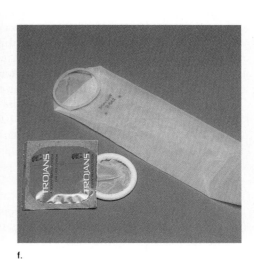

f.

Figure 14.14   Various types of birth control methods: *a.* IUD, *b.* vaginal sponge, *c.* diaphragm, *d.* birth control pills, *e.* vaginal spermicide, *f.* condom.

Two types of IUDs are now available: the copper type have a copper wire wrapped around the stem and the progesterone-releasing type have progesterone embedded in the plastic. IUDs are believed to alter the environment of the uterus and oviducts so that fertilization probably does not occur—but if it should occur, implantation cannot take place. With proper patient selection, side effects and complications of these two types of IUDs are rare. The best candidates for this use of birth control are women who have had at least one child, are of middle to older reproductive age, and who have a stable relationship with a partner who does not have a sexually transmitted disease.

Group II

The *diaphragm* (fig. 14.14c) is a soft rubber or plastic cup with a flexible rim that lodges behind the pubic bone and fits over the cervix. Each woman must be properly fitted by a physician, and the diaphragm can be inserted into the vagina two hours at most before sexual relations. It must also be used with a spermicidal jelly or cream and should be left in place for at least six hours after sexual relations. If intercourse is repeated during this time, more jelly or cream should be inserted by means of a plastic insertion tube.

The *cervical cap* is a minidiaphragm that also must be fitted by a physician and should be used with a spermicide. It is made of natural rubber or plastic and fits over the cervix like a thimble. The cap must be inserted before intercourse and remain in place for eight hours afterward. Unlike the diaphragm, the cervical cap is effective even if left in place for several days.

A *condom* (fig. 14.14f) is a thin skin (lambskin) or plastic sheath (latex) that fits over the erect penis. The ejaculate is trapped inside the sheath and, thus, does not enter the vagina.

When used in conjunction with a spermicidal foam, cream, or jelly, the protection is better than with the condom alone. Today it is possible to purchase condoms that are already lubricated with a spermicide. The condom is generally recognized as giving protection against sexually transmitted diseases such as those discussed in the following chapter.

A *vaginal sponge* (fig. 14.14*b*) permeated with spermicide and shaped to fit the cervix is a new contraceptive recently made available to the general public after seven years of testing. Unlike the diaphragm and cervical cap, the sponge need not be fitted by a physician since one size fits everyone. It is effective immediately after placement in the vagina and remains effective for twenty-four hours. Like the other means of birth control in this category, the sponge is about 85% effective in preventing pregnancy.

Group III

It is possible for the male to withdraw the penis just before ejaculation so that the semen is deposited away from the vaginal area. This method of birth control, called *coitus interruptus*, has a relatively high failure rate because a few drops of seminal fluid may escape from the penis before ejaculation takes place. Even a small amount of semen can contain numerous sperm.

Spermicidal jellies, creams, and foams (fig. 14.14*e*) contain sperm-killing ingredients and may be inserted into the vagina with an applicator up to thirty minutes before each occurrence of intercourse. Foams are considered the most effective of this group of contraceptives. When used alone, these are not highly effective means of birth control for those who have frequent intercourse. They do offer some protection against sexually transmitted disease; nonoxynol 9, a common ingredient, is a viral inhibitor giving some protection against AIDS.

Group IV

Natural family planning, formerly called the *rhythm method* of birth control, is based on the realization that a woman ovulates only once a month and that the egg and sperm are viable for a limited number of days. If the woman has a consistent twenty-eight-day cycle, then the period of "safe" days can be determined, as in figure 14.15. This method of birth control is not very effective because the days of ovulation can vary from month to month, and the viability of the egg and sperm varies perhaps monthly but certainly from person to person.

A more reliable way to practice natural family planning is to await the day of ovulation each month and then wait three more days before engaging in intercourse. They day of ovulation can be more accurately determined by noting the body temperature early each morning (body temperature rises at ovulation) or by taking the pH of the vagina each day (near the day of ovulation the vagina becomes more alkaline) or by noting the consistency of the mucus at the cervix (at ovulation the mucus is thinner and more watery). Physicians can instruct women how to do these procedures.

Only a medically recognized method of birth control such as those discussed here should be used. Douching is of little value and position of intercourse will not prevent pregnancy at all. In fact, the proximate location of the penis (at the time of ejaculation) near but not in the vagina has been known to result in pregnancy.

Numerous birth control methods and devices are available for those who wish to prevent pregnancy. The more effective methods are sterilization, the Pill, the IUD, the sponge, and the diaphragm. A condom used with a spermicidal jelly or foam is also effective. The less effective methods are spermicidal foam or jelly alone, coitus interruptus, and natural family planning.

Future Means of Birth Control

There are four areas in which birth control investigations have been directed: morning-after medication, a long-lasting method, a medication that is specifically for males, and new barrier methods.

There is a new birth control pill (Ru 486) on the market in France consisting of a synthetic steroid that prevents progesterone from acting on the uterine lining because it has a high affinity for progesterone receptors. In clinical tests, the uterine lining sloughed off within four days in 85% of women who were less than a month pregnant. To improve the success rate, the drug is administered with a small dose of prostaglandin, which causes contraction of the uterus and ejection of an embryo. The promoters of this treatment are using the term *contragestation* to describe its effects; however, it should be recognized that this medication, rather than preventing implantation, brings on an *abortion*, the loss of an implanted fetus. One day the medication might be used by many women who are experiencing delayed menstruation without knowing whether they are actually pregnant.

In this country, DES, a synthetic estrogen that affects the uterine lining making implantation difficult, sometimes is given following intercourse. Since large doses causing nausea and vomiting in the short run and possibly cancer in the long run are required, DES usually is given only for incest or rape.

Various possibilities exist for a *"male pill."* Scientists have made analogues of gonadotropic-releasing hormones that interfere with the action of this hormone and prevent it from stimulating the pituitary. The seminiferous tubules produce a hormone termed *inhibin* that inhibits FSH production by the pituitary (fig. 14.6). Testosterone and/or related chemicals can be used to inhibit spermatogenesis in males, but there are usually feminizing side effects because the excess is changed to estrogen by the body.

There has been a revival of interest in barrier methods of birth control, and a "female condom" now is being studied to determine its effectiveness against pregnancy and sexually transmitted diseases. The closed end of a large plastic tube is

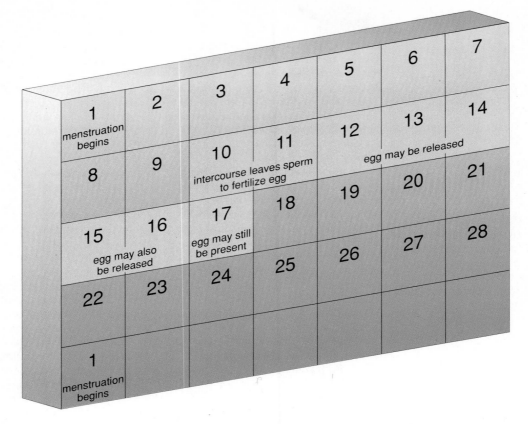

Figure 14.15   Natural family planning. The calendar shows the "safe" and "unsafe" days for intercourse. This calendar is *only* appropriate for women with regular 28-day cycles. Few women have regular cycles month after month.

anchored by a plastic ring in the upper vagina, and the open end of the tube is held in place by a thinner ring that rests just outside the vagina.

There are numerous well-known birth control methods and devices available to those who wish to prevent pregnancy. Their effectiveness varies. In addition, new methods are expected to be developed.

Infertility

Sometimes couples do not need to prevent pregnancy; conception does not occur despite frequent intercourse. The American Medical Association estimates that 15% of all couples in this country are unable to have any children and, therefore, are properly termed sterile; another 10% have fewer children than they wish and, therefore, are termed infertile. The latter assumes that the couple has been trying to become pregnant and has been unsuccessful for at least one year.

Causes of Infertility

The two major causes of infertility in females are blocked oviducts, possibly due to pelvic inflammatory disease discussed in the next chapter, and failure to ovulate due to low body weight. Endometriosis, the spread of uterine tissue beyond the uterus is also a cause of infertility as discussed in the reading on page 292. Sometimes these physical defects can be corrected surgically and/or medically. If no obstruction is apparent and body weight is normal, it is possible to give females HCG extracted from the urine of postmenopausal women. This treatment causes multiple ovulations and, sometimes, multiple pregnancies.

The most frequent causes of infertility in males is low sperm count and/or a large proportion of abnormal sperm. Disease, radiation, chemical mutagens, too much heat to the testes, and the use of psychoactive drugs can contribute to this condition.

When reproduction does not occur in the usual manner, many couples adopt a child. Others sometimes first try one of the alternative reproductive methods discussed in the following paragraphs.

Endometriosis

Each month K.C. Esperance, 31, a San Francisco nurse practitioner, suffered menstrual cramps so agonizing that she would take to her bed, curl up and pray that she would live through the next couple of days. Doctor after doctor gave her the same ineffectual advice: rest, take some codeine and bear with it.

During her teens, Maria Menna Perper, 42, a New Jersey biochemist, suffered intestinal problems around the time of her period. By her late 30s, she felt "excruciating, burning pain" in her colon every month "like clockwork." Eventually the pain became continuous, and it was impossible for her to work or even sit down.

For Anne Hicks, 29, a Portland, Ore., real estate property manager, there were no obvious signs other than her inability to become pregnant.

Despite their differing complaints, each of the women eventually discovered that she suffered from the same insidious condition: endometriosis, an often unrecognized disease that afflicts anywhere from 4 million to 10 million American women and is a major cause of infertility. The condition is caused by the spread and growth of tissue from the lining of the uterus (or endometrium) beyond the uterine walls. These endometrial cells form bandlike patches and scars throughout the pelvis and around the ovaries and the Fallopian tubes, resulting in a variety of symptoms and degrees of discomfort.

Because endometriosis has been associated with delayed childbearing, it is sometimes called the "career woman's disease." But recent studies have shown that the disorder strikes women of all socioeconomic groups and even teenagers, though those with heavier, longer, or more frequent periods may be especially susceptible. Says Dr. Donald Chatman of Chicago's Michael Reese Hospital, "Endometriosis is an equal-opportunity disease."

How the disease begins is something of a mystery. One theory ascribes it to "retrograde menstruation." Instead of flowing down through the cervix and vagina, some menstrual blood and tissue back up through the Fallopian tubes and spill out into the pelvic cavity (*see figure on page 293*). Normally this errant flow is harmlessly absorbed, but in some cases the stray tissue may implant itself outside the uterus and continue to grow. A second theory suggests that the disease arises from misplaced embryonic cells that have lain scattered around the abdominal cavity since birth. When the monthly hormonal cycles begin at puberty, says Dr. Howard Judd, director of gynecological endocrinology at UCLA Medical Center, "some of these cells get stirred up and could be a major cause of endometriosis."

If anything about endometriosis is clear, it is that once the disease has begun, it will probably get worse. Stimulated by the release of estrogen, the implanted tissue grows and spreads. Cells from the growths break away and are ferried by lymphatic fluid throughout the body, sometimes, although rarely, forming islands in the lungs, kidneys, bowel or even the nasal passages. There they respond to the menstrual cycle, causing monthly bleeding from the rectum or wherever else they have settled.

The most common symptom of endometriosis is pain, which can occur during menstruation, urination, and sexual intercourse. Unfortunately, these warnings are often overlooked by women and their doctors. Cheri Bates, 31, of Seattle, describes the cramps she suffered as "outrageous," but she assumed they were "normal." By the time her condition was discovered, scar tissue covered her reproductive organs and parts of her bladder and intestines.

To confirm that a patient has endometriosis, doctors look for the telltale tissue by peering into the pelvic cavity with a fiberoptic instrument called a laparoscope. After diagnosis, a number of treatments can be prescribed. One is pregnancy—if it is still feasible; the nine-month interruption of menstruation can help shrink misplaced endometrial tissue. Taking

Alternative Methods of Reproduction

Artificial Insemination by Donor (AID)   During **artificial insemination,** sperm are placed in the vagina by a physician. Sometimes a woman is artificially inseminated by her husband's sperm. This is especially helpful if the husband has a low sperm count—the sperm can be collected over a period of time and concentrated so that the sperm count is sufficient to result in fertilization. Often, however, a woman is inseminated by sperm acquired from a donor who is a complete stranger to her. At times, a mixture of husband and donor sperm are used.

A variation of AID is intrauterine insemination (IUI). IUI involves hormonal stimulation of the ovaries followed by placement of the donor's sperm in the uterus rather than in the vagina.

Human Issue

There are those who do not approve of alternative methods of reproduction on the grounds that (a) fertilization and pregnancy do not occur in the normal way or (b) an embryo and, therefore, a human life, might be lost during the procedure. For example, it is possible to freeze IVF embryos for use later. What should be done, however, if the mother conceives and does not want the extra embryos, or the couple divorces and the man no longer wishes to father a child? Should unwanted embryos be stored indefinitely and made available to anyone who wishes to have a child? What are the legal rights of frozen embryos that will eventually die if kept too long?

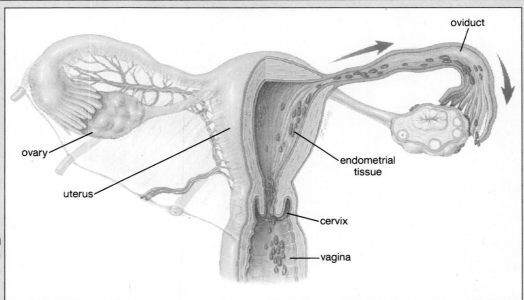

Endometriosis. It is speculated that endometriosis is caused by a backward menstrual flow as represented by the arrows in this drawing. This allows endometrial cells to enter the abdominal cavity, where they take up residence and respond to the monthly cyclic changes in hormonal levels, including those that result in menstruation.

birth-control pills may also help, but more effective is a drug called danazol, a synthetic male hormone that stops ovulation and causes endometrial tissue to shrivel. But it also can produce acne, facial-hair growth, weight gain, and other side effects.

A new experimental treatment with perhaps fewer ill effects involves a synthetic substance called nafarelin, similar to gonadotropin-releasing hormone. Normally GnRH is released in bursts by the hypothalamus gland, eventually triggering the process of ovulation. But "if the GnRH stimulation is given continuously instead of in pulses," explains Dr. Robert Jaffe of the University of California, San Francisco,

"the whole [ovulatory] system shuts off," and the endometrial implants "virtually melt away."

For severe cases of endometriosis, surgical removal of the ovaries and the uterus may be the only solution. But less extreme surgery can often help. At Atlanta's Northside Hospital, Dr. Camran Nezhat has had success with a high-tech procedure called videolaseroscopy, which employs a laparoscope rigged with a tiny video camera and a laser. The camera images, enlarged on a video screen, enable Nezhat to zero in on endometrial tissue and to vaporize it with the laser. In a study of 102 previously infertile

patients, Nezhat found that 60.7% were able to conceive within two years of videolaseroscopy treatment.

Like many other doctors who see the unfortunate consequences of endometriosis, Nezhat is concerned that a "lot of women do not seek help for this problem." Any serious pain, he notes, needs investigating. Agrees Cheri Bates (a victim), "If a doctor tells you that suffering is a woman's lot in life, get another doctor."

In Vitro Fertilization (IVF) In the case of **in vitro fertilization IVF,** hormonal stimulation of the ovaries is followed by laparoscopy. In this instance, an aspiratory tube is used for retrieving preovulatory eggs (fig. 14.8). Alternately, it is possible to place a needle through the vaginal wall and to guide it by the use of ultrasound to the ovaries where the needle is used to retrieve the eggs. This method is called transvaginal retrieval.

Concentrated sperm from the male is placed in a solution that approximates the conditions of the female genital tract. When the eggs are introduced, fertilization occurs. The resultant zygotes begin development, and after about 2 to 4 days, the embryos are inserted into the uterus of the woman, who is now in the secretory phase of her menstrual cycle. If implantation is successful, development is normal and continues to term.

Gamete Intrafallopian Transfer (GIFT) GIFT was devised as a means to overcome the low success rate (15% to 20%) of in vitro fertilization. The method is exactly the same as in vitro fertilization except the eggs and the sperm are immediately placed in the oviducts after they have been brought together. This procedure is helpful to couples whose eggs and sperm never make it to the oviducts; sometimes the egg gets lost between the ovary and the oviducts, and sometimes the sperm never reach the oviducts. GIFT has an advantage in that it is a one-step procedure for the woman—the eggs are removed and are reintroduced all in the same time period. For this reason, it is less expensive—approximately $1,500 compared with $3,000 and up for in vitro fertilization.

Figure 14.16   Mother and child. Sometimes couples utilize alternative methods of reproduction in order to experience the joys of parenthood.

Surrogate Mothers   In some instances, women have been paid to have babies by other individuals who contributed sperm (or eggs) to the fertilization process.

If all the alternative methods discussed are considered, it is possible to imagine that a baby could have five parents (1) sperm donor, (2) egg donor, (3) surrogate mother, and (4) and (5) adoptive mother and father.

Some couples are infertile due to various physical abnormalities. When corrective medical procedures fail, it is possible today to consider an alternative method of reproduction in order to be a parent (fig. 14.16).

What's Your Diagnosis?

Joan and her husband have been trying to conceive a child for the last several years, unsuccessfully. As this period of infertility continues, she begins to experience pain in her pelvic area during menstruation, urination, and sexual intercourse. Recently, Joan has also experienced discomfort and bleeding from the rectum. This has been occurring closely in time with the menstruation at the onset of her monthly uterine cycle.

With the onset of this alarming, additional problem, Joan visits a physician for an examination. He looks into the pelvic cavity with a laparoscope. He discovers the growth and spreading of tissue from the lining of the uterus, the endometrium. From these spreading cells, there are bandlike patches and scars throughout the pelvis and around the ovaries and fallopian tubes as well.

1. What is your diagnosis of Joan's condition? 2. How are the results of the physician's diagnosis consistent with your conclusion? 3. What treatment do you prescribe to help Joan?

Summary

In males, spermatogenesis occurs within the seminiferous tubules of the testes, which also produce testosterone within the interstitial cells. Sperm mature in the epididymis and are stored here and in the first part of the vas deferens before entering the urethra, along with secretions from the prostrate, seminal vesicles, and Cowper's glands. Hormonal regulation involving secretions from the hypothalamus, anterior pituitary, and the testes in the male maintains testosterone at a fairly constant level.

In females, egg production occurs within the ovaries where one follicle produces an egg each month. Fertilization, if it occurs, takes place in the oviducts, and the resulting embryo travels to the uterus where it implants itself in the uterine lining. In the nonpregnant female, hormonal regulation in the female involves the ovarian and uterine cycle, dependent upon the hypothalamus, anterior pituitary, and the female sex hormones, estrogen and progesterone.

Numerous birth control methods and devices are available for those who wish to prevent pregnancy. Infertile couples are increasingly resorting to alternative methods of reproduction.

Study Questions

1. Discuss the anatomy and physiology of the testes. (p. 275) Describe the structure of sperm. (p. 276)
2. Give the path of sperm. (p. 278)
3. What glands add secretions to seminal fluid? (p. 278)
4. Discuss the anatomy and physiology of the penis. (p. 278) Describe ejaculation. (p. 280)
5. Discuss hormonal regulation in the male. Name three functions of testosterone. (p. 280)

6. Discuss the anatomy and physiology of the ovaries. (p. 282) Describe ovulation. (p. 282)
7. Give the path of the egg. Where do fertilization and implantation occur? (p. 283) Name two functions of the vagina. (p. 283)
8. Describe the external genitalia in females. (p. 283)
9. Compare male and female orgasm. (pp. 280, 284)

10. Discuss hormonal regulation in the female, either simplified and/or detailed. (p. 284) Give the events of the uterine cycle, and relate them to the ovarian cycle. (p. 284) In what way is menstruation prevented if pregnancy occurs? (p. 286)
11. Name four functions of the female sex hormones. (p. 284) Describe the anatomy and physiology of the breast. (p. 287)
12. Aside from abstinence, discuss the various means of birth control and their relative effectiveness. (p. 287)

Objective Questions

1. If you are tracing the path of sperm, the structure that follows the epididymis is the _____ .
2. The prostate gland, Cowper's glands, and the _____ all contribute secretions to seminal fluid.
3. The primary male sex hormone is _____ .
4. An erection is caused by the entrance of _____ into the penis.
5. In the female reproductive system, the uterus lies between the oviducts and the _____ .
6. In the ovarian cycle, once each month a _____ produces an egg. In the uterine cycle, the _____ lining of the uterus is prepared to receive the zygote.
7. The female sex hormones are _____ and _____ .
8. Pregnancy in the female is detected by the presence of _____ in the blood or urine.
9. Aside from abstinence, the most effective means of birth control are _____ in males and _____ in females.
10. In vitro fertilization occurs in _____ .

Label this Diagram.
See figure 14.2 (p. 276) in text.

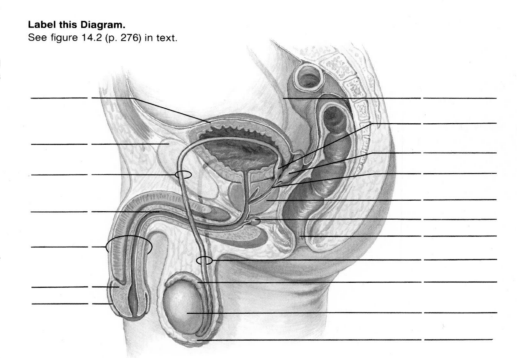

Answers

1. vas deferens 2. seminal vesicles 3. testosterone 4. blood 5. vagina 6. follicle, endometrial 7. estrogen, progesterone 8. HCG 9. vasectomy, tubal ligation 10. laboratory glassware

Critical Thinking Questions

1. As a continuation of the introduction to this chapter, state specifically the anatomical and physiological means by which humans are adapted to reproduce on land.
2. Men retain their reproductive potential much longer than women do. How is this suited to a difference in gamete production and to their different contributions to reproduction of offspring?
3. All organisms have a life that includes a reproductive strategy. For example, some insects spend much of their lives as larvae, undergo metamorphosis into winged forms, reproduce, and die all within a single season. They do not spend any time at all caring for their offspring and instead produce a large number of which a few may survive. Outline and discuss the human life strategy. Include a reference to culture, mentioned on page 4 of the text.

Selected Key Terms

endometrium (en-do-me′tre-um) the lining of the uterus that becomes thickened and vascular during the uterine cycle. 283

erection (ē-rek′shun) referring to a structure, such as the penis, that is turgid and erect as opposed to being flaccid or lacking turgidity. 278

Graafian follicle (graf′e-an fol′li-k′l) mature follicle within the ovaries that houses a developing egg. 282

implantation (im″plan-ta′shun) the attachment of the embryo to the lining (endometrium) of the uterus. 286

interstitial cells (in″ter-stish′al selz) hormone-secreting cells located between the seminiferous tubules of the testes. 278

menopause (men′o-pawz) termination of the ovarian and uterine cycles in older women. 287

menstruation (men″stroo-a′shun) a loss of blood and tissue from the uterus at the end of a uterine cycle. 285

ovarian cycle (o-va′re-an si′kl) monthly occurring changes in the ovary that affect the level of sex hormones in the blood. 284

ovaries (o′var-ez) the female gonads, the organs that produce eggs and estrogen and progesterone. 282

seminal fluid (sem′i-nal floo′id) the sperm-containing secretion of males; semen. 278

seminiferous tubules (se″mi-nif′er-us too′bulz) highly coiled ducts within the male testes that produce and transport sperm. 275

testes (tes′tēz) the male gonads, the organs that produce sperm and testosterone. 275

uterine cycle (u′ter-in si′kl) monthly occurring changes characteristic of the uterine lining. 284

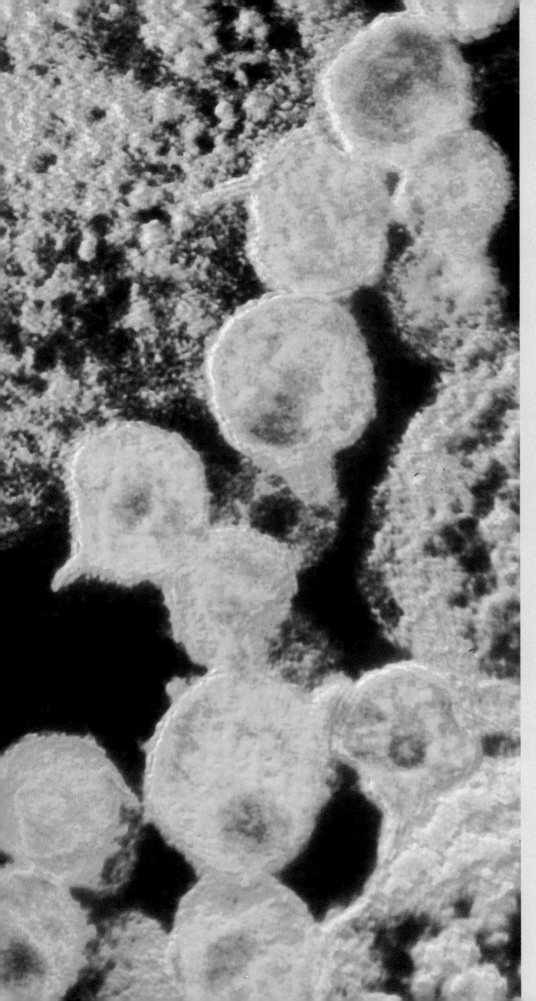

Chapter Fifteen

Sexually

Transmitted

Diseases

Chapter Concepts

1 Viruses are noncellular and have to reproduce inside a living host cell.

2 AIDS and herpes are both caused by viruses; it is difficult to find a cure.

3 Bacteria are independent cells but they lack the organelles found in human cells.

4 Gonorrhea, chlamydia, and syphilis are caused by bacteria; they are curable by antibiotic therapy.

5 There are also other fairly common sexually transmitted diseases caused by a protozoan, fungus, and even a louse.

One way to know if a microorganism causes a condition is to find it in all persons with the disease. This is a colored transmission electron micrograph of HIV (human immunodeficiency virus) positioned between two T lymphocytes, the type of cell the virus attacks. HIV is believed to be the cause of AIDS (acquired immunodeficiency syndrome).

Figure 15.1   Henry VIII had syphilis, which apparently caused all his male heirs to die from congenital defects. Luckily his daughter Elizabeth escaped the disease and went on to rule England during a period of time known as the Elizabethan era.

Can you imagine that a sexually transmitted disease could have altered the course of history? In the early part of the sixteenth century, Henry VIII (fig. 15.1) contracted syphilis just before he married Catherine of Aragon. She bore him four sons but all had congenital syphilis and were stillborn or fatally malformed. He blamed her for this tragedy and sought an annulment from the Catholic Church. When it was denied, he broke with the church so that he could divorce Catherine and take another wife. England has been a Protestant country since that time.

Human Issue

Should state and local governments make condoms available to the public in order to prevent the occurrence of sexually transmitted diseases? Do you feel that since this might promote sexual activity, it would be wrong, or do you feel that since it might in the end save tax dollars it would be worth it? Should there be a restriction as to age? If so, at what age should free condoms become available?

Viral in Origin

Sexually transmitted diseases (STDs) are contagious diseases caused by microorganisms that are passed from one human to another by sexual contact. Viruses cause numerous diseases in humans (table 15.1) including AIDS and herpes, two sexually transmitted diseases of great concern today. Since viruses are not cellular, they are incapable of independent reproduction and reproduce only inside a living cell. For this reason, they are called *obligate parasites*. A **parasite** requires a **host** organism in order to function properly, complete its life cycle, and reproduce. In the laboratory, viruses are maintained by injecting them into live chick embryos (fig. 15.2). Antibiotics which interfere with bacterial metabolism are not effective against viruses. Only recently have we been able to discover chemicals that will interfere with viral replication inside cells. Outside living cells, viruses are nonliving and can be stored just as chemicals are stored.

Viruses are tiny particles that always have at least two parts: an outer coat of protein called a capsid and an inner core

Table 15.1 Infectious Diseases Caused By Viruses

Respiratory tract	Nervous system
common colds	encephalitis
flu[a]	polio[a]
viral pneumonia	rabies[a]
Skin reactions	**Liver**
measles[a]	yellow fever[a]
German measles[a]	hepatitis A & B
chicken pox[a]	**Other**
smallpox[a]	mumps[a]
warts	cancer
Sexually transmitted	
AIDS	
herpes	

[a]Vaccines available. Yellow fever, rabies, and flu vaccines are only given if the situation requires them. Smallpox vaccinations are no longer required.

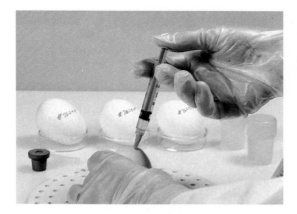

Figure 15.2   Inoculation of live chick eggs with virus particles. A virus only reproduces inside a living cell, not because it uses the cell for nutrition but because it takes over the metabolic machinery of the cell.

of nucleic acid either DNA or RNA (fig. 15.3). Many viruses also have an outer envelope that contains lipid as well as protein molecules. The lipid molecules are derived from the host's membrane, but the protein is unique to the virus.

Most viruses are extremely specific. Not only do they prefer a particular type of organism, such as humans, they also prefer a particular tissue type. This specificity is due to the ability of the virus to combine with a particular molecular configuration, such as a receptor on the cell surface. Within a half hour, the virus, or simply the nucleic acid core depending on the virus, has entered the cell. Most often the viral genes, whether they are DNA or RNA viruses, immediately take over the machinery of the cell, and the virus undergoes reproduction. These are the steps required for a DNA virus to reproduce. 1. Viral DNA replicates repeatedly, utilizing the nucleotides within the host. Multiple copies of viral DNA result. 2. Viral DNA is transcribed into mRNA, which undergoes translation. Multiple copies of coat protein result. 3. Assemblage occurs. Viral DNA

is packaged inside a coat protein. If the virus has an envelope, it is formed at the cell membrane just before the virus leaves the cell (fig. 15.3a).

There are a few types of viruses that do not immediately undergo reproduction; instead, viral DNA becomes integrated into the host DNA. RNA viruses that do this are called **retroviruses** because they have a special enzyme called *reverse transcriptase* through which their RNA is transcribed into cDNA (a DNA copy of the RNA gene) that becomes incorporated into host DNA. Sometimes the virus remains *latent,* and during this time the viral DNA is replicated along with host DNA. Eventually, viral reproduction may occur (fig. 15.3b). Certain environmental factors, such as ultraviolet radiation, can cause a latent virus to undergo reproduction.

Viruses reproduce only inside host cells. Some viruses have the ability to incorporate their DNA within host DNA. Retroviruses are RNA viruses that are able to perform reverse transcription.

Figure 15.4 gives the expected yearly increase for the most prevalent STDs we will be discussing.

AIDS

The brief discussion here may be supplemented by the insert that occurs following page 310.

The organism that causes **acquired immunodeficiency syndrome (AIDS)** is a virus called **human immunodeficiency virus (HIV).** HIV attacks the type of lymphocyte known as helper T cells (p. 138). Helper T cells, you will recall, stimulate the activities of B lymphocytes that produce antibodies. After an HIV infection sets in, helper T cells begin to decline in number and the person becomes more susceptible to other types of infections.

Symptoms

AIDS has three stages of infection. During the first stage, which may last about a year, the individual is an asymptomatic carrier. The AIDS blood test (an antibody test) is positive; the individual can pass on the infection, yet there are no symptoms. During the second stage called AIDS related complex (ARC), which may last about six to eight years, the lymph glands are swollen and there may also be weight loss, night sweats, fatigue, fever, and diarrhea. Infections like thrush (white sores on the tongue and in the mouth) and herpes (discussed on page 301) reoccur. Finally, the person may develop full-blown AIDS, especially characterized by the development of an opportunistic disease such as an unusual type of pneumonia, skin cancer, and also nervous disorders. Opportunistic diseases are ones that occur only in individuals who have little or no capability of fighting an infection. The AIDS patient usually dies about seven to nine years after infection.

Transmission

AIDS is transmitted by infected blood, semen, and vaginal secretions. In the United States, the two main affected groups are homosexual men and intravenous drug abusers (and their sexual

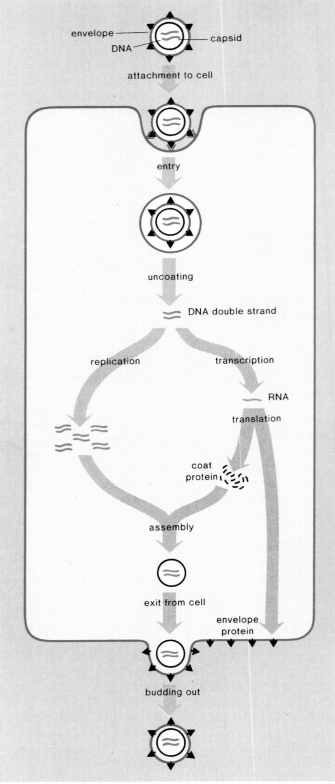

a. DNA virus

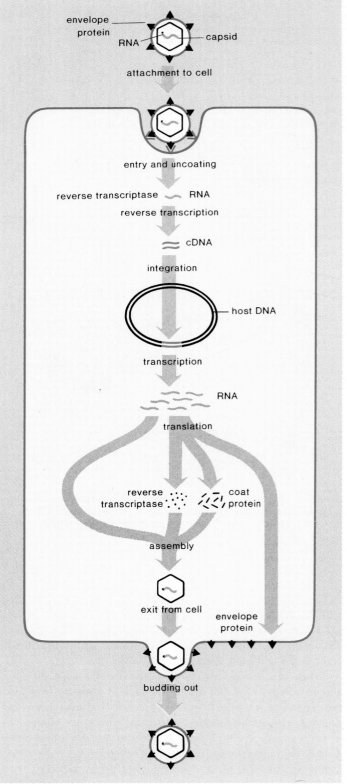

b. RNA retrovirus

Figure 15.3 Life cycles of animal viruses. *a.* DNA virus. After entering by endocytosis, the virus becomes uncoated. The DNA then codes for proteins, some of which are capsid (coat) proteins and some of which are envelope proteins. Assembly follows replication of the DNA. When the virus exits by budding, it is enclosed by an envelope made up of host cell membrane lipids and viral envelope proteins. *b.* RNA retrovirus. The life cycle includes steps not seen in *a.* The RNA genes are transcribed to cDNA (DNA copied off of RNA) that is integrated into the host DNA. Transcription produces many copies of the RNA genes, which direct the synthesis of 3 types of proteins: the enzyme reverse transcriptase, the capsid (coat) protein, and the envelope protein. Again, the virus buds from the host cell.

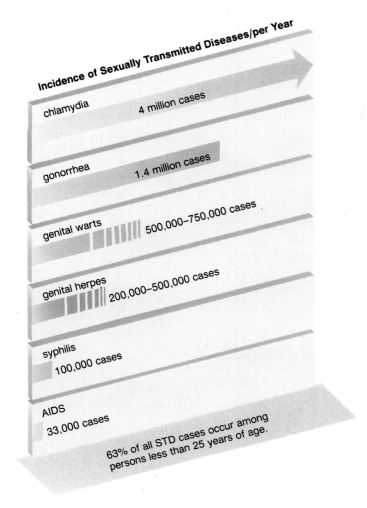

Incidence of Sexually Transmitted Diseases/per Year

chlamydia — 4 million cases

gonorrhea — 1.4 million cases

genital warts — 500,000–750,000 cases

genital herpes — 200,000–500,000 cases

syphilis — 100,000 cases

AIDS — 33,000 cases

63% of all STD cases occur among persons less than 25 years of age.

Figure 15.4 Statistics for the most common sexually transmitted diseases show that chlamydia, gonorrhea, and genital warts are all much more common than herpes, syphilis, and AIDS. Chlamydia, gonorrhea, and syphilis can be cured with antibiotic therapy.

partners). In Africa and some parts of South America, though, AIDS is apparently transmitted chiefly through heterosexual intercourse and an equal number of men and women are infected.

Certain portions of the United States have been harder hit than other portions. Even in New York City, which reports the highest level of infectivity, there are regions that have more cases than other regions. The AIDS virus can cross the placenta, and the Bronx currently has one in 43 babies born with HIV antibodies in the blood. Some of these newborns may have the antibodies without having the virus but 30% to 50% are most likely infected.

Although intravenous (IV) drug users are liable to spread the disease to the general heterosexual population, to date infection among the general population is about 4%, increasing less than 1% over the last five years. Health officials emphasize that unprotected intercourse with multiple partners or a single infected partner increases the chance of transmission. The use of a latex condom reduces the risk, but the very best preventive

measure at this time is a long-term mutually monogamous relationship with a sexual partner who is free of the disease. Casual contact with someone who is infected such as shaking hands, eating at the same table, or swimming in the same pool is not a mode of transmission.

Treatment

The drug AZT (azidothymidine) has been found to be helpful in prolonging the lives of AIDS patients. A new drug DDI (dideoxymosine), which like AZT works by preventing viral replication in cells, is undergoing clinical trials.

Investigators are trying to develop a vaccine for AIDS. The AIDS virus mutates frequently, but researchers have identified a portion of the coat envelope that they believe is relatively stable. When this is injected into the bloodstream, antibodies do develop, but it is not yet known whether such antibodies will offer protection against infection. After all, persons with AIDS do produce antibodies for several years, but for some reason the number of helper T cells still declines, and then antibody production falls off.

Human Issue

How conscientious should people be about telling potential sex partners that they have a sexually transmitted disease, especially a noncurable one like AIDS or herpes? Is it like a "buyers beware market"—it's up to the individual to inquire about the health and/or sexual practices of a potential partner? At the other extreme, should people with AIDS or herpes become celibates and refrain from having sex? How would you feel if you fell in love with someone and then were told that this person had AIDS or herpes?

Genital Herpes

The different herpes viruses are large DNA viruses that cause various illnesses. Chicken pox and mononucleosis, among other ailments, are due to herpes viruses. **Genital herpes** is caused by **herpes simplex virus** (fig. 15.5) of which there are two types: type 1 usually causes cold sores and fever blisters, while type 2 more often causes genital herpes. Cross-over infections do occur, however.

Transmission and Symptoms

Genital herpes is one of the more prevalent sexually transmitted diseases today; an estimated 40 million persons (17% of the United States population) have it, with an estimated 500,000 new cases appearing each year. Sometimes there are no symptoms. Or the individual may experience a tingling or itching sensation before blisters appear on the genitals within 2 to 20 days. Once the blisters rupture, they leave painful ulcers that may take as long as three weeks or as little as five days to heal. These symptoms may be accompanied by fever, pain on urination, and swollen lymph nodes.

After ulcers heal, the disease is only dormant, and blisters can reoccur repeatedly at variable intervals. Sunlight, sex, menstruation, and stress seem to cause the symptoms of genital

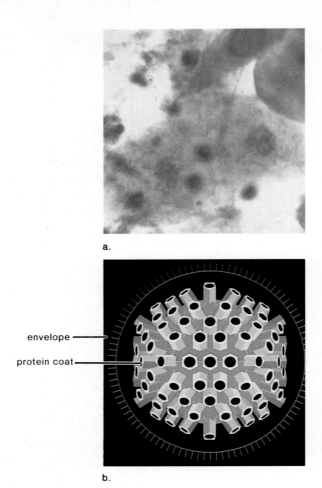

envelope

protein coat

a.

b.

Figure 15.5   *a.* Cell infected with herpes virus. *b.* Enlarged model of herpes virus.

herpes to reoccur. While the virus is latent, it resides in nerve cells near the brain and spinal cord. Herpes occasionally infects the eye, causing an eye infection that can lead to blindness, and can cause CNS infections. Type 2 formerly was thought to cause a form of cervical cancer, but this is no longer believed to be the case.

Infection of the newborn can occur if the child comes in contact with a lesion in the birth canal. In one to three weeks, the infant is gravely ill and can become blind, have neurological disorders including brain damage, or die. Birth by cesarean section prevents these occurrences.

Treatment

Presently there is no cure for herpes. The drugs vidarabine and acyclovir disrupt viral reproduction. The ointment form of acyclovir relieves initial symptoms, but the oral form prevents the occurrence of outbreaks. Work is also being done to develop a vaccine.

Genital Warts

Genital warts are caused by a *human papillomavirus (HPV),* which is a cuboidal DNA virus that reproduces in the nuclei of skin cells. Plantar warts and common warts are also caused by HPVs.

Transmission and Symptoms

Some HPVs are sexually transmitted. Quite often, carriers do not have any sign of warts, although flat lesions may be present. When present, the warts commmonly are seen on the penis and foreskin of males and near the vaginal opening in females. If the warts are removed, they may reoccur.

HPVs, rather than genital herpes, now are associated with cancer of the cervix, as well as tumors of the vulva, the vagina, the anus, and the penis. Some researchers believe that the viruses are involved in 90% to 95% of all cases of cancer of the cervix. Teenagers with multiple sex partners seem to be partic-

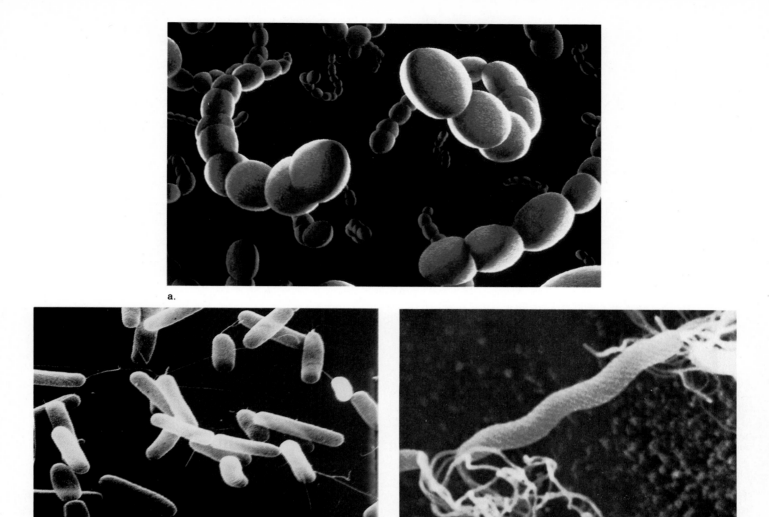

a.

b.

c.

Figure 15.6   Scanning electron micrographs of bacteria. *a.* Spherical-shaped bacteria. *b.* Rod-shaped bacteria. *c.* Spiral-shaped bacteria that use flagella for locomotion. See figure 15.7 for a generalized drawing of a bacteria.

ularly susceptible to HPV infections. More cases of cancer of the cervix are being seen among this age group.

Treatment

Presently, there is no cure for an HPV infection, but it can be treated effectively by surgery, freezing, application of an acid, or laser burning, depending on severity. A suitable medication to treat genital warts before cancer occurs is being sought, and efforts also are underway to develop a vaccine.

Bacterial in Origin

Although **bacteria** are generally larger than viruses, they are still quite small. Since they are microscopic, it is not always obvious that they are abundant in the air, water, soil, and on most objects. It has even been suggested that the combined weight of all bacteria would exceed that of any other type of organism on earth. Bacteria occur in three basic shapes (fig. 15.6): rod (bacillus); round, or spherical (coccus); and spiral (a curved shape called a spirillum). Some bacteria can locomote by means of flagella.

The bacterial cell is termed a *prokaryotic* (meaning, before the nucleus) cell to distinguish it from a *eukaryotic* (meaning, true nucleus) cell. Human cells are eukaryotic and contain numerous organelles in addition to a nucleus. Prokaryotic cells lack these organelles, except for ribosomes, but still they are functioning cells. They do have DNA; it is just not contained within a nuclear envelope, and although they have no mitochondria,

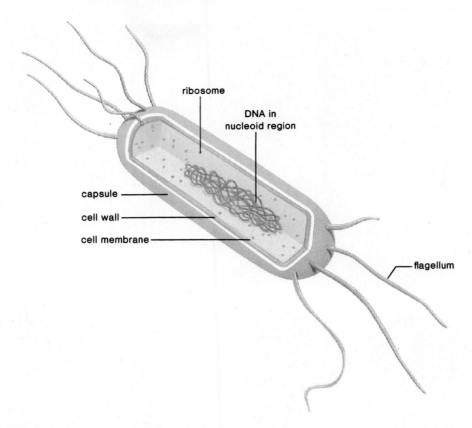

ribosome

DNA in
nucleoid region

capsule

cell wall

cell membrane

flagellum

Figure 15.7 Bacteria are prokaryotic cells and lack the organelles found in eukaryotic (e.g., human) cells.

they do have respiratory enzymes located in the cytoplasm. Notice in figure 15.7 that the bacterial cell is surrounded by a cell wall in addition to a cell membrane. Some bacteria are also surrounded by a polysaccharide or polypeptide capsule that enhances their **virulence** (that is, ability to cause disease). If they are motile, they have flagella (flagellum, singular).

Bacteria reproduce asexually by **binary fission.** First, the single chromosome duplicates, and then the two chromosomes move apart into separate areas. Next the cell membrane and cell wall grow inward and partition the cell into two daughter cells, each of which now has its own chromosome (fig. 15.8). Under favorable conditions, growth may be very rapid with cell division occurring as often as every 12 to 15 minutes. When faced with unfavorable environmental conditions, some bacteria can form **endospores.** During spore formation, the cell shrinks, rounds up within the former cell membrane, and secretes a new, thicker wall inside the old one. Endospores are amazingly resistant to extreme temperatures, drying out, and harsh chemicals, including acids and bases. When conditions are again suitable for growth, the spore absorbs water, breaks out of the inner shell, and becomes a typical bacterial cell.

Most bacteria are free-living **saprophytes** that perform many useful services in the environment. Saprophytes send out digestive enzymes into the environment to break down large molecules into small molecules that can be absorbed across the cell membrane. Most bacteria are aerobic and require a constant supply of oxygen as we do, but a few are anaerobic, even

being killed by the presence of oxygen. Table 15.2 lists the human diseases caused by bacteria; only a few serious illnesses are caused by anaerobic bacteria, such as botulism, gas gangrene, and tetanus. These bacteria and others produce toxins, chemicals that seriously interfere with the normal functioning of the body. Sometimes just a bacterial toxin is used to make a vaccine.

Bacteria have long been used by humans to produce various products commercially. Chemicals, such as ethyl alcohol, acetic acid, butyl alcohol, and acetone are produced by bacteria. Bacterial action is involved in the production of butter, cheese, sauerkraut, rubber, cotton, silk, coffee, and cocoa. By means of gene splicing, bacteria are now used to produce human insulin and interferon, as well as other types of proteins (p. 390). Even certain antibiotics are produced by bacteria.

General cleanliness is the first step toward preventing the spread of infectious bacteria. Disinfectants and antiseptics also help reduce the number of infectious bacteria. **Sterilization,** a process that kills all living things, even endospores, is used whenever all bacteria must be killed. Sterilization can be achieved by use of an autoclave (fig. 15.9), a container that admits steam under pressure. If bacteria do invade our bodies and cause an infection, antibiotic therapy is often helpful.

Bacteria are prokaryotic cells capable of independent existence. Most are free-living, but a few cause human diseases that can be cured by antibiotic therapy.

Figure 15.8   Reproduction in bacteria. Above, the single chromosome is attached to the cell membrane where it is replicating. As the cell membrane and cell wall lengthen, the two chromosomes separate. Once fission has taken place, each bacterium has its own chromosome.

cell wall
cell membrane
chromosome
cytoplasm

Table 15.2   Infectious Diseases Caused by Bacteria

Respiratory tract	**Nervous system**
strep throat	tetanus[a]
pneumonia	botulism
whooping cough[a]	meningitis
diphtheria[a]	
tuberculosis[a]	**Digestive tract**
	food poisoning
Skin reactions	typhoid fever[a]
staph (pimples and boils)	cholera[a]
gas gangrene[a] (wound infections)	
	Sexually transmitted
	gonorrhea
	syphilis
	chlamydia

[a]Vaccines are available. Tuberculosis vaccine is not used in this country. Typhoid fever, cholera, and gas gangrene vaccines are given if the situation requires it. Others are routinely given.

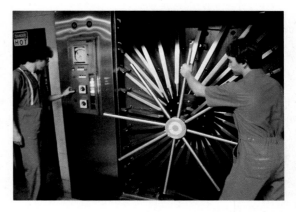

Figure 15.9   Sterilization by autoclaving. Hospital employees are closing the door and operating a large sterilizer. Sterilization permits surgical procedures to be done with reduced fear of subsequent infection. During autoclaving, steam under pressure kills bacterial cells and endospores.

An **antibiotic** is a chemical that selectively kills bacteria when it is taken into the body as a medicine. Most antibiotics are produced naturally by soil microorganisms. Penicillin is made by the fungus *Penicillium;* streptomycin, tetracycline, and erythromycin are all produced by the bacterium, *Streptomyces.* Sulfa, an analog of a bacterial growth factor, can be produced in the laboratory.

Antibiotics are metabolic inhibitors specific for bacterial enzymes. This means that they poison bacterial enzymes without harming host enzymes. Penicillin blocks the synthesis of the bacterial cell wall; streptomycin, tetracycline, and erythromycin block protein synthesis; and sulfa prevents the production of a coenzyme.

There are problems associated with antibiotic therapy. Some patients are allergic to antibiotics and the reaction may even be fatal. Antibiotics not only kill off disease-causing bacteria, they also reduce the number of beneficial bacteria in the intestinal tract. The use of antibiotics sometimes prevents natural immunity from occurring, leading to the necessity for reoccurring antibiotic therapy. Most important, perhaps, is the growing resistance of certain strains of bacteria. Tetracycline and penicillin, long used to cure gonorrhea, now have a failure rate of more than 20% against certain strains of *Gonococcus*.

Most physicians believe that antibiotics should only be administered when absolutely necessary, and some believe that if this is not done then resistant strains of bacteria will completely replace present strains and antibiotic therapy will no longer be effective at all.

Gonorrhea

Gonorrhea is caused by the bacterium *Neisseria gonorrhaeae* which is a diplococcus, meaning that generally the two spherical cells stay together (fig. 15.10).

Symptoms

The diagnosis of gonorrhea in the male is not difficult as long as he displays typical symptoms (as many as 20% of males may be asymptomatic). The patient complains of pain on urination and has a thick, greenish yellow urethral discharge three to five days after contact. In the female, the bacteria may first settle within the urethra or near the cervix, from which they may

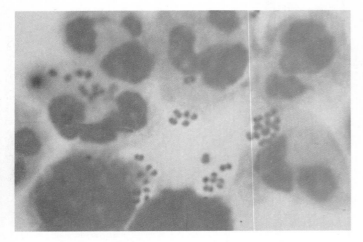

Figure 15.10   Gonorrheal bacteria (*Neisseria gonorrheae*) in male urethral discharge. If you look carefully, you will notice that the round bacteria occur in pairs; for this reason, they are called diplococci.

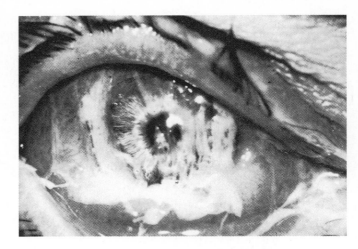

Figure 15.11   Gonorrhea infection of the eye is possible whenever the bacteria comes in contact with the eyes. This can happen when the newborn passes through the birth canal. Manual transfer from the genitals to the eyes is also possible.

spread to the oviducts, causing **pelvic inflammatory disease (PID).** PID from gonorrhea is especially apt to occur in the female using an intrauterine device as a birth control measure (p. 289). As the inflamed tubes heal, they may become partially or completely blocked by scar tissue. As a result, the female is sterile or, at best, subject to ectopic pregnancy. Similarly, there may be inflammation in untreated males followed by scarring of the vas deferens. Unfortunately, 60% to 80% of females are asymptomatic until they develop severe pains in the abdominal region due to PID. PID affects about one million women a year in the United States.

Homosexual males develop gonorrhea proctitis, or infection of the anus, with symptoms including anal pain and blood or pus in the feces. Oral sex can cause infection of the throat and the tonsils. Gonorrhea also can spread to other parts of the body, causing heart damage or arthritis. If, by chance, the person touches infected genitals and then his or her eyes, a severe eye infection can result (fig. 15.11).

Eye infection leading to blindness can occur as a baby passes through the birth canal. Because of this, all newborn infants receive eye drops containing antibacterial agents, such as silver nitrate, tetracycline, or penicillin, as a protective measure.

Transmission and Treatment

Gonococci live for a very short time outside the body; therefore, most infections are spread by intimate contact, usually sexual intercourse. A female has a 50% to 60% risk of contracting the disease after a single exposure to an infected male, whereas a male has a 20% risk after exposure to an infected female.

Blood tests for gonorrhea are being developed, but in the meantime, it is necessary to diagnose the condition by microscopically examining the discharge of males or by growing a culture of the bacterium from both males and females to positively identify the organism (fig. 15.12). Because there is no blood test, it is very difficult to find asymptomatic carriers who

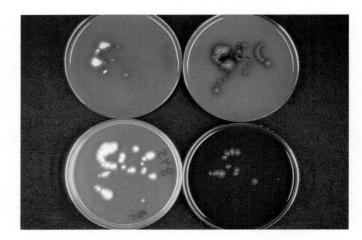

Figure 15.12   Culture plates with bacterial colonies. Visual examination and biochemical tests allow medical personnel to determine the type of bacteria growing on culture plates.

are capable of passing on the condition without realizing it. If the infection is diagnosed, however, gonorrhea can be treated using antibiotics. There is no vaccine for gonorrhea, and immunity does not seem possible; therefore, it is possible to contract the disease many times over.

Gonorrhea is one of the oldest known and most common of the sexually transmitted diseases. Untreated, an infection can cause sterility in either sex. Appropriate antibiotic therapy will cure the condition.

Chlamydia

Chlamydia is named for the tiny bacterium that causes it (*Chlamydia trachomatis*). For years, chlamydiae were considered to be more closely related to viruses than to bacteria, but today it is known that they are prokaryotic cells. Even so, they are obligate parasites due to their inability to produce ATP molecules. After a cell phagocytizes them, they develop inside the phagocytic vacuole which eventually bursts and liberates many new infective chlamydiae.

New chlamydial infections occur at an even faster rate than gonorrheal infections. They are the most common cause of NGU, **nongonococcal urethritis.** About 8 to 21 days after infection, men experience a mild burning sensation on urination and a mucoid discharge. Women may have a vaginal discharge along with the symptoms of a urinary tract infection. Unfortunately, a physician mistakenly may diagnose a gonorrheal or urinary infection and prescribe the wrong type of antibiotic, or the person may never seek medical help. In either case, the infection can cause PID and sterility or ectopic pregnancy.

If a newborn comes in contact with chlamydia during delivery, inflammation of the eyes or pneumonia can result. Some believe that chlamydial infections increase the possibility of premature and stillborn births.

Detection and Treatment

New and faster laboratory tests are now available for chlamydia detection. Their expense sometimes prevents public clinics from using them, however. It's been suggested that these criteria could help physicians decide which women should be tested: no more than 24 years old; having had a new sex partner within the preceding two months; having a cervical discharge; bleeding during parts of the vaginal exam; and using a nonbarrier method of contraception. Some doctors, however, are routinely prescribing additional antibiotic appropriate to treating chlamydia for anyone who has gonorrhea because 40% of females and 20% of males with gonorrhea also have chlamydia.

As with AIDS, condoms serve as a protection against both gonorrheal and chlamydia infections. The concomitant use of a spermicide containing nonoxynol 9 gives added protection.

PID and sterility are common effects of a chlamydia infection in the female. This condition often accompanies gonorrheal infection.

Syphilis

Syphilis is caused by a bacterium called *Treponema pallidum,* an actively motile, corkscrewlike organism that is classified as a spirochete. Because this bacterium is difficult to stain, it shows up only when viewed with a dark-field microscopic field (fig. 15.13). Syphilis is less common than gonorrhea (fig. 15.14), but it is the more serious of the two infections.

Syphilis has three stages, which can be separated by latent stages in which the bacteria are resting before multiplying again. During the primary stage, a hard chancre (ulcerated sore with hard edges) indicates the site of infection (fig. 15.14*a*). The chancre can go unnoticed, especially since it usually heals spontaneously, leaving little scarring. During the secondary stage,

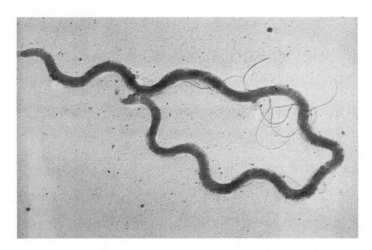

Figure 15.13　*Treponema pallidum,* the cause of syphilis. Notice the characteristic spiral shape and the presence of flagella.

proof that bacteria have invaded and spread throughout the body is evident when the victim breaks out in a rash (fig. 15.14b). Curiously, the rash does not itch and is seen even on the palms of the hands and the soles of the feet. There can be hair loss and infectious gray patches on the mucous membranes, including the mouth. These symptoms disappear of their own accord.

During a tertiary stage, which lasts until the patient dies, syphilis may affect the cardiovascular system, and weakened arterial walls (aneurysms) are seen, particularly in the aorta. In other instances, the disease may affect the nervous system. An infected person may become mentally retarded, become blind, walk with a shuffle, or show signs of insanity, for example. **Gummas,** large destructive ulcers (fig. 15.14*c*) may develop on the skin or within the internal organs.

Congenital syphilis is caused by syphilic bacteria crossing the placenta. The child is stillborn, or blind, with many other possible anatomical malformations.

Transmission and Treatment

The syphilis bacterium is not present in the environment; it is only present within human beings. Only close intimate contact such as sexual intercourse transmits the condition from one person to another.

Diagnosis of syphilis can be made by dark-field microscopic examination of fluids from lesions for the bacterium which is actively motile and has a corkscrew shape. There are also blood tests which are not positive until at least six weeks after initial infection.

Syphilis is a very devastating disease. Control of syphilis depends on prompt and adequate treatment of all new cases; therefore it is very important for all sexual contacts to be traced so that they can be treated. Use of condoms can prevent exposure. As with other STDs, the incidence of syphilis is rising in most parts of the world. One hundred thousand cases are reported annually and the highest incidence of these cases is among persons 29 to 39 years of age.

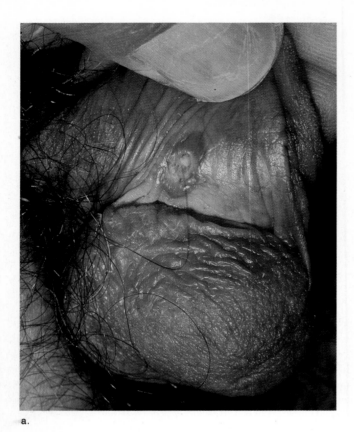

a.

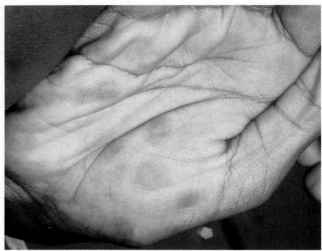

b.

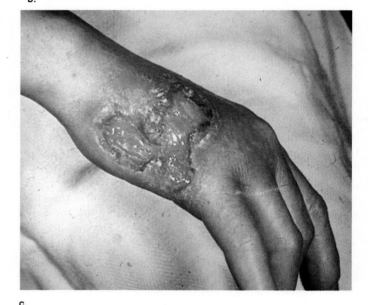

c.

Figure 15.14   The three stages of syphilis. *a.* The first stage is a hard chancre where the bacterium enters the body. *b.* The second stage is a body rash that occurs even on the palms of the hands and soles of the feet. *c.* In the tertiary stage, gummas may appear on skin or internal organs.

Human Issue

If a woman has AIDS or syphilis, the fetus can become infected while yet in the womb. Should women who have been treated for these conditions be discouraged from having children? If so, how should this be effectively handled?

Herpes, chlamydia, and gonorrhea can infect the newborn as the child passes through the birth canal. Should women who have been treated for these conditions insist on a cesarean birth so that their babies are sure to be protected? What would your reaction be if your spouse did not tell you that he or she had herpes and then a child was infected during birth?

Other Diseases

Vaginitis

Two other types of organisms are of interest: protozoans and fungi. **Protozoans** are eukaryotes that usually exist as single cells. Each type of protozoan has its own mode of location. There are some that move by extensions of the cytoplasm, called pseudopodia, some that move by cilia, and some that have flagella. Protozoans are most often found in an aquatic environment such as freshwater ponds and the ocean simply teems with them. All protozoans require an outside source of nutrients, but only the parasitic ones take this nourishment from their host.

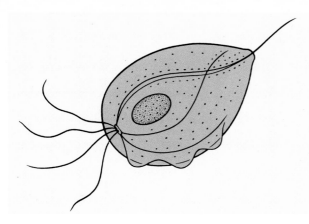

Figure 15.15 *Trichomonas vaginalis.* This protozoan is pear-shaped and uses flagella to move about.

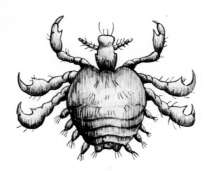

Figure 15.16 *Phthirus pubis,* the parasitic louse that infects the pubic hair of humans.

Most **fungi** are nonpathogenic saprophytes, as are bacteria, but a few are parasitic. Usually the body of a fungus is made up of filaments called hyphae, but yeasts are an exception since they are single cells. Almost everyone is familiar with yeast. It is used to make bread rise and to produce wine, beer, and whiskey because of its ability to ferment.

Females very often have vaginitis or infection of the vagina, that is caused by the flagellated protozoan *Trichomonas vaginalis* (fig. 15.15) or the yeast *Candida albicans.* The protozoan infection causes a frothy white or yellow foul-smelling discharge accompanied by itching, and the yeast infection causes a thick, white, curdy discharge, also accompanied by itching. *Trichomonas* is most often acquired through sexual intercourse, and the asymptomatic male is usually the reservoir of infection. *Candida albicans,* however, is a normal organism found in the vagina; its growth simply increases beyond normal under certain circumstances. Women taking the birth control pill are sometimes prone to yeast infections, for example. Also, the indiscriminate use of antibiotics can alter the normal balance of organisms in the vagina so that a yeast infection will flare up.

Pubic Lice (Crabs)

Small lice, animals that look like crabs under low magnification infect the pubic hair, underarm hair, and even occasionally the eyebrows of humans. The females lay their eggs around the base of the hair, and these eggs hatch within a few days to produce a larger number of animals that suck blood from their host and cause severe itching, particularly at night.

The crab louse (fig. 15.16) can be contracted by direct contact with an infected person or by contact with his or her clothing or bedding. In contrast to most types of sexually transmitted diseases, self-treatment that does not require shaving is possible. The lice can be killed by gamma benzene hexachloride, marketed in shampoo, lotion, or cream under the trade name Kwell.

What's Your Diagnosis?

Joe, who has a drug dependency, injects the drug intravenously into his body on a daily basis. He shares the syringes for these drug injections with several friends. Joe has had this dependency for several years. About a year ago, he noticed the development of constant swelling of his lymph nodes. The tenderness and swelling of the nodes in his neck were particularly prominent. Recently he has developed new symptoms including weight loss, fatigue, constant fevers, diarrhea, and periods of prolonged sweating at night. Joe cannot recount any group of abnormalities rivaling this during his entire life. Only when this recent group of symptoms surfaces does he seek medical diagnosis and treatment. Tests conducted on Joe's blood reveal the presence of the HIV viruses and a reduced number of T cells.

1. What is your diagnosis of Joe's condition? 2. How are Joe's symptoms consistent with your diagnosis? 3. What treatment do you recommend for Joe?

Summary

Human beings are subject to sexually transmitted diseases (STDs). Some of these are viral in origin (AIDS, genital herpes, genital warts). Viruses are tiny particles that always have an outer coat of protein and an inner core of nucleic acid that may be DNA or RNA. DNA viruses are apt to undergo reproduction inside a cell immediately, while RNA viruses may be retroviruses that incorporate cDNA into host DNA for some time before reproduction. Some STDs are bacterial in origin (gonorrhea, chlamydia, syphilis). Bacteria are prokaryotic cells that reproduce by binary fission and may form endospores. Most bacteria are aerobic saprophytes that perform useful services for humans, but a few cause disease. Other STDs are caused by a protozoan (vaginitis), a fungus (yeast infection), or animals (crabs).

Study Questions

1. Describe the structure and life cycle of viruses, including those that reproduce immediately and those that undergo a latent period. (p. 299)
2. What are the symptoms for the three stages leading up to and including full-blown AIDS? (p. 299)
3. What are two main groups of individuals having AIDS in the United States, and how might transmission be prevented? (p. 299)
4. Give the cause and the expected yearly increase in the number of cases of herpes, gonorrhea, chlamydia, and syphilis. (p. 301)
5. Describe the progressive symptoms of a herpes infection. (p. 301)
6. List the three shapes of bacteria, and describe the structure of a prokaryotic cell. (p. 303)
7. Describe the symptoms of a gonorrheal infection in the male and in the female. What is PID, and how does it affect reproduction? (p. 306)
8. Describe the three stages of syphilis. (p. 307)
9. How does the newborn acquire an infection of AIDS, herpes, gonorrhea, chlamydia, or syphilis? What effects do these infections have on infants? (pp. 301, 302, 306, 307)
10. List other common sexually transmitted diseases, and describe the associated symptoms. (p. 308)

Objective Questions

1. All viruses have an inner core of _____ and a coat of _____ .
2. AIDS is caused by a type of RNA virus known as a _____ .
3. Although it is a sexually transmitted disease, intravenous drug abusers get AIDS because the virus lives in _____ cells.
4. Herpes simplex virus type 1 causes _____ , and type 2 causes _____ .
5. Bacterial cells have one of three shapes: _____ , _____ , and _____ .
6. Females are often asymptomatic for gonorrhea until they develop _____ .
7. The use of a _____ can reduce the risk of acquiring a STD.
8. In the tertiary stage of syphilis, there may be large sores called _____ .
9. These three sexually transmitted diseases are curable by antibiotic therapy: _____ , _____ , and _____ .
10. Women who take the birth control pill are subject to a _____ infection.

Answers

1. nucleic acid, protein 2. retrovirus 3. blood
4. cold sores, genital herpes 5. rod, spherical, spiral 6. PID 7. condom 8. gummas
9. gonorrhea, chlamydia, syphilis 10. yeast

Critical Thinking Questions

1. Parasites reproduce inside living hosts, and they all must have a means of transmission from host to host. Explain the necessity for this stage in the organism's life cycle.
2. Reexamine figure 15.3, and list all the steps in the life cycle of a retrovirus where a drug (medication) might be designed to interfere with the cycle. Discuss in specific terms.
3. From your knowledge of the human body, discuss the mode of entry by parasites in general. Give specific examples that may have been mentioned in previous chapters of the text.

Selected Key Terms

AIDS (ādz) acquired immunodeficiency syndrome, a disease caused by HIV and transmitted via body fluids; characterized by failure of the immune system. 299

bacteria (bak-te're ah) prokaryotes that lack the organelles of eukaryotic cells. 303

binary fission (bi'na-re fish'un) reproduction by division into two equal parts by a process that does not involve a mitotic spindle. 304

chlamydia (klah-mid'e-ah) an organism that causes a sexually transmitted disease particularly characterized by urethritis. 306

endospore (en'do-spōr) a resistant body formed by bacteria when environmental conditions worsen. 304

fungus (fung'gus) an organism, usually composed of strands called hyphae, that lives chiefly on decaying matter; e.g., mold and yeasts. 309

genital (jen'i-tal) pertaining to the external sexual organs as in genital herpes and genital warts. 301, 302

gonorrhea (gon''o-re'ah) contagious sexually transmitted disease caused by bacteria and leading to inflammation of the urogenital tract. 305

gummas (gum'ahz) large unpleasant sores that may occur during the tertiary stage of syphilis. 307

herpes simplex virus (her'pēz sim'plex vi'rus) a virus of which type I causes cold sores and type II causes genital herpes. 301

HIV viruses responsible for AIDS; human immunodeficiency virus. 299

host (hōst) an organism on or in which another organism lives. 298

parasite (par'ah-sīt) an organism that resides externally on or internally within another organism and does harm to this organism. 298

pelvic inflammatory disease (pel'vik in-flam'ah-to''re di-zēz) PID; a disease state of the reproductive organs caused by an organism that is sexually transmitted. 306

retrovirus (ret''ro-vi'rus) virus capable of integrating its genes into those of the host by utilizing RNA to DNA transcription. 299

saprophyte (sap'ro-fīt) a heterotrophic organism such as bacteria and fungi that externally breaks down dead organic matter before absorbing the products. 304

sterilization (ster''i-li-za'shun) the inability to reproduce; a surgical procedure eliminating reproductive capability; the absence of living organisms due to exposure to environmental conditions that are unfavorable to sustain life. 304

syphilis (sif'i-lis) chronic, contagious sexually transmitted disease caused by a bacterium that is a spirochete. 307

AIDS Supplement

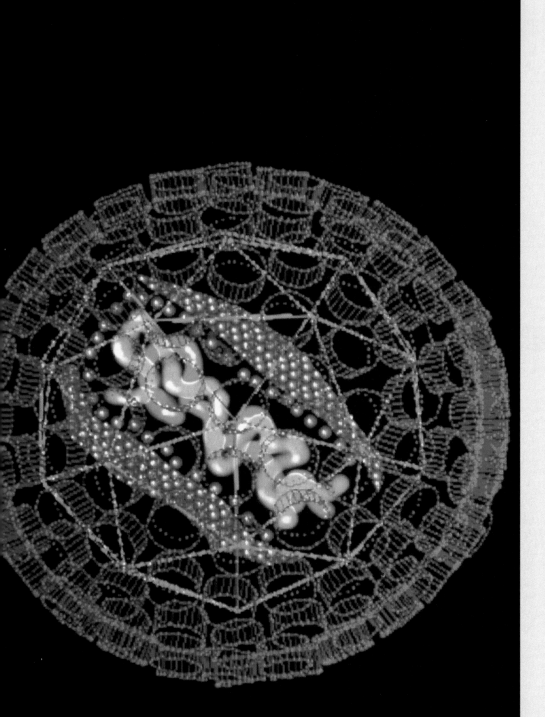

Computer graphics illustration of the structure of an AIDS virus. The spherical viral envelope is studded with glycoproteins (red and blue cylinders). Inside, a protein coat (green) surrounds the central core (blue spheres). Inside the core, the orange and light blue snakelike structure represents genetic information in the form of RNA and the enzyme reverse transcriptase.

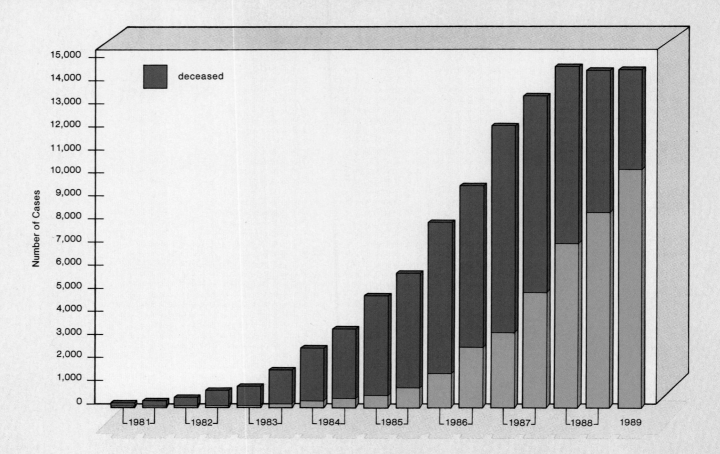

Figure A.1   Number of reported AIDS cases in the U.S. each year since 1981. As of mid-1989, over 100,000 cases had been reported; about 60% of those have already died.

Sources: Data from W. L. Heyward and J. W. Curran, "The Epidemiology of AIDS in the U.S." in *Scientific American*, October 1988; and *HIV/AIDS Surveillance*, Year-End Edition, January 1990, U.S. Department of Health and Human Services.

AIDS

AIDS (acquired immunodeficiency syndrome) is caused by a group of related retroviruses known as HIV (human immunodeficiency viruses). The full name of AIDS can be explained in this way: acquired means that the condition is caught rather than inherited; immune deficiency means that the virus attacks the immune system so there is greater susceptibility to certain opportunistic infections and cancer; and syndrome means that some fairly typical infections and cancers occur in the infected person.

Origin of AIDS

The origin of the AIDS virus has not yet been determined. It has been suggested that HIV originated in Africa and then spread to Europe and the United States. Even today, there are monkeys in Africa infected with immunosuppressive viruses which could have mutated to HIV after humans ate monkey meat.

Most likely, HIV entered the United States on numerous occasions as early as the 1950s. Presently, the first documented case is a 15-year-old male who died in Missouri in 1969 with skin lesions now known to be characteristic of an AIDS-related cancer. Doctors froze some of his tissues because they could not identify the cause of death. Recently, these tissues were examined and found to be infected with HIV. Researchers also want to test the preserved tissue samples of a 49-year-old Haitian who died in New York during 1959 of a type of pneumonia now known to be AIDS related.

British scientists were recently able to show by examining preserved tissues that a Manchester seaman most probably died in 1959 of AIDS. This may be one of the first cases of AIDS because scientists believe the immunosuppressive monkey virus may have evolved into HIV during the late 1950s.

During the 1960s, it was the custom to list leukemia as the cause of death in immunodeficient patients. Most likely some of these people actually died of AIDS. Since HIV is not extremely infectious, it took several decades for the number of AIDS cases to increase to the point that AIDS became recognizable as a specific and separate disease.

Prevalence of AIDS Today

Because there is no cure for AIDS, the number of diagnosed cases ever increases (fig. A.1). AIDS is not distributed equally throughout the country: at the present time, New York City,

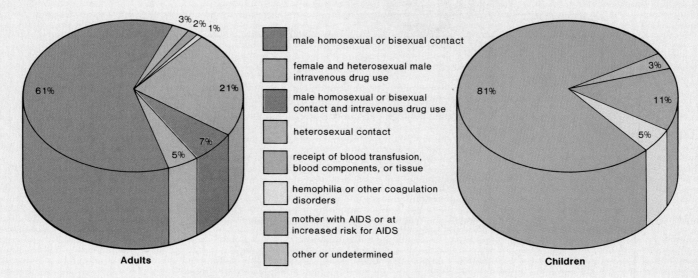

Figure A.2 Pie charts showing the distribution of AIDS among adults (left) and children (right). Homosexual or bisexual men and intravenous drug abusers account for 89% of all adult cases. Most children acquired the disease from a mother who was a carrier or who had AIDS.

Sources: Data from W. L. Heyward and J. W. Curran, "The Epidemiology of AIDS in the U.S." in *Scientific American*, October 1988; and *HIV/AIDS Surveillance*, Year-End Edition, January 1990, U.S. Department of Health and Human Services.

San Francisco, and Los Angeles alone have over one-third of the cases. However, AIDS is now spreading to other cities and rural areas as well.

Although over half of AIDS patients are white, the blacks and hispanics have a higher proportionate number of cases compared to whites. Presently, 90% of all AIDS patients are males and only 10% are female. These statistics are understandable when one realizes that homosexuals are a high-risk group for AIDS; the other high-risk group is intravenous (IV) drug users. The behavior of these two groups places them at risk because of the mode of transmission of HIV.

Transmission of AIDS

A person has AIDS after HIV enter the blood and infect helper T lymphocytes (p. 138) of a type known as T4 cells. As the disease progresses, the number of T4 cells decline in number.

Since HIV are present in the blood of an infected person, they can be transmitted when intravenous drug abusers share contaminated needles. About 28% of total AIDS cases are intravenous drug abusers (fig. A.2).

A small number of persons (about 3% of total cases) have contracted AIDS by receiving infected blood during a blood transfusion. Particularly, hemophiliacs who require more frequent blood transfusions than most or routine injections of a blood product called factor-VIII have been known to get AIDS in this manner. Since 1985, blood donations in the United States have been tested for contamination with HIV. For various reasons, the testing process is not absolutely perfect and, therefore, the slight possibility of getting AIDS from a blood transfusion

remains. Persons who are about to undergo an operation can predonate their own blood and/or have noninfected friends do so.

AIDS is a sexually transmitted disease. Semen can contain the virus or more likely an infected lymphocyte. Anal intercourse is common among homosexual and bisexual men who account for about 70% of the total cases in the United States. The inner lining of the rectum is a thin, single-cell layer that is easily torn during intercourse. Viruses or infected lymphocytes within semen deposited in the rectum can therefore pass into the blood. The vagina has a thicker lining of cells than the rectum and it does not as easily allow the passage of the virus or infected lymphocyte into the blood. The uterus is a different matter, however. Its thinner, more vulnerable walls do allow the passage of a virus or infected lymphocyte into the blood. Once a female is infected, her vaginal secretions might very well contain the virus and/or infected lymphocytes.

The use of latex condoms can help prevent the spread of AIDS. If infected semen is trapped within the condom, the virus and/or infected lymphocyte obviously cannot pass into the blood and infect a partner. Unfortunately, however, mishaps with condoms are more likely to occur with anal intercourse than with vaginal intercourse. The use of a spermicidal jelly that contains nonoxynol-9 along with a condom is an added protection because this spermicide also kills the virus and infected lymphocytes.

From this discussion, it is apparent that male to male, and male to female transmission, is considered more likely than female to male. Nevertheless, female to male transmission does occur and may become more prevalent as more females become

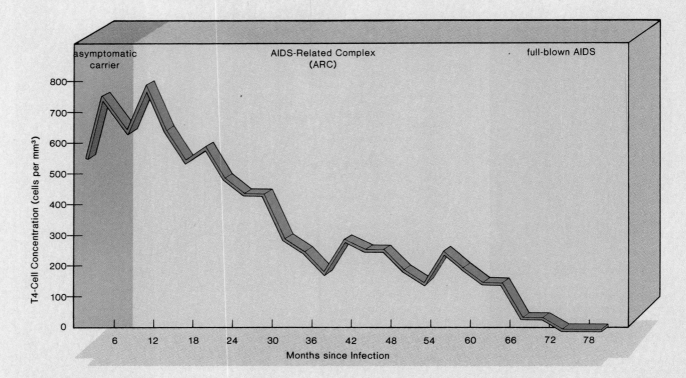

Figure A.3   T4-cell concentration in a young man whose HIV infection followed a typical course. After about 3 months, the concentration of T4 cells increases as the body attempts to fight the infection. During this time, the individual is an asymptomatic carrier. Once the T4-cell concentration begins to decline, the individual develops the symptoms of ARC. Finally, full-blown AIDS is diagnosed when the patient has one or more of the opportunistic infections. By this time, the T4-cell concentration is below 100 per mm³ and the individual soon dies.

Source: Data from R. R. Redfield and D. S. Burke, "HIV Infection: The Clinical Picture" in *Scientific American*, October 1988.

infected. One unhappy side effect to female infection is the fact that viruses and infected lymphocytes can pass to a fetus via the placenta or to an infant via mother's milk. Presently, infected infants account for about 1% of all AIDS cases.

Symptoms of AIDS

Most often, when discussing the symptoms of AIDS, the discussion centers around the symptoms of full-blown AIDS. This is a mistake because it does not take into account the complex course of an HIV infection.

First of all, it is important to realize that exposure to HIV, and not membership in a risk group, is the first step towards AIDS. Thereafter, it is possible to relate the progression of the disease to the number of T4 cells in the body as is done in figure A.3 based on the data from a particular individual. For the sake of our discussion, figure A.3 is divided into three stages:

Name of Stage	Months
Asymptomatic carrier	First 9 months or longer
AIDS Related Complex (ARC)	10–63 months
Full-blown AIDS	64–83 months

Asymptomatic Carrier

Usually people do not have any symptoms at all after initial infection. A few (1% to 2%) do have mononucleosis-like symp-

toms that may include fever, chills, aches, swollen lymph glands, and an itchy rash. These symptoms disappear, however, and there are no other symptoms for quite some time.

During the asymptomatic-carrier stage, the individual exhibits no symptoms and yet is highly infectious. According to figure A.4, first the blood contains a high amount of HIV, and then the body begins to produce antibodies against the virus. The standard HIV blood test tests for the presence of antibody, and not for the presence of HIV itself. This means that an individual is infectious before the HIV test becomes positive.

A small percentage of false-negative and false-positive test results do occur with the HIV-antibody test. A positive result can be verified with a more expensive test that has a higher accuracy. Some people are against routine testing for HIV antibodies because it could subject those who test positive to unfair social discrimination; however, it is possible that routine testing could also help prevent the spread of AIDS.

It is important to realize that AIDS can spread among persons who do not realize that they are infected. Most infected young people are in this category; they have not yet developed any symptoms of infection and do not know they are infected.

Most infected persons will go on to develop the symptoms of ARC and then full-blown AIDS.

Aids Related Complex (ARC)

Several months to several years after infection, the individual may begin to show some symptoms. The most common symptom

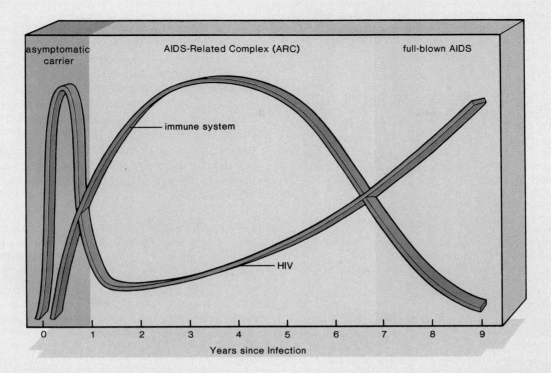

Figure A.4   Balance of power between HIV and the immune system during the course of an HIV infection. During the time that a person is an asymptomatic carrier, the concentration of HIV in the body is high. Then the immune system becomes active, and antibodies are produced. While an infected person has ARC, the immune system begins to lose the battle and eventually the battle is lost when a person develops full-blown AIDS.

Source: Data from R. R. Redfield and D. S. Burke, "HIV Infection: The Clinical Picture" in *Scientific American*, October 1988.

of ARC is swollen lymph glands in the neck, armpits, or groin that persist for three months or more. Swollen lymph glands are believed to occur because B cells in the lymph glands are hyperactivated and producing a flood of antibodies in an effort to control the infection (fig. A.4).

Still, during this time, the number of T4 cells continues to decline. Symptoms that indicate that HIV infection is present are severe fatigue not related to exercise or drug use; unexplained persistent or recurrent fevers, often with night sweats; persistent cough not associated with smoking, a cold, or the flu; and persistent diarrhea. Also possible are signs of nervous system impairment, including loss of memory, inability to think clearly, loss of judgment, and/or depression.

When the individual develops non-life-threatening and recurrent infections, it is a signal that full-blown AIDS will occur shortly. One possible infection is thrush, a fungal infection that is identified by the presence of white spots and ulcers on the tongue and inside the mouth. The fungus may also spread to the vagina, resulting in a chronic infection there. Another frequent infection is herpes simplex with painful and persistent sores in the skin surrounding the anus, the genital area, and/or the mouth.

Full-Blown AIDS

Whereas it was believed previously that a percentage of persons with ARC would never progress to full-blown AIDS, it is now thought that almost every person with ARC will eventually develop full-blown AIDS. The number of persons with full-blown AIDS represents only the "tip of the iceberg" when one considers the total number of persons who are infected with HIV (fig. A.5).

In this final stage of an HIV infection, the AIDS patient who is now suffering from "slim disease" (as AIDS is called in Africa)—severe weight loss and weakness due to persistent diarrhea and coughing—will most likely succumb to one of the so-called opportunistic infections. These infections are called opportunistic because they are caused by microbes that cannot ordinarily start an infection, but they have the opportunity in AIDS patients because of the severely impaired immune system. (fig. A.3). Some of the opportunistic infections are:

Pneumocystis carinii pneumonia. The lungs become useless as they fill with fluid and debris due to an infection with this organism. There is not a single documented case of *P. carinii* pneumonia in a person with normal immunity; therefore its presence serves to establish the diagnosis of AIDS in about 60% of patients.

Toxoplasmic encephalitis is caused by a one-cell parasite that lives in cats and other animals as well as humans. Many persons harbor a latent infection in the brain or muscle, but in AIDS patients the infection leads to loss of brain cells, seizures, weakness, or decreased sensation on one side of the body.

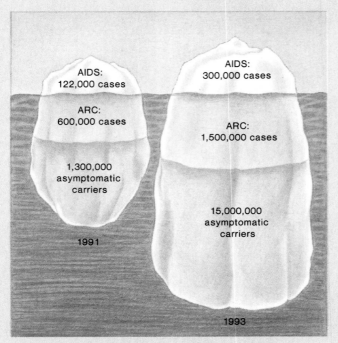

Figure A.5 Figure shows that most persons who have an HIV infection are asymptomatic carriers. They can pass the infection to others without knowing that they are infected. A much smaller proportion of persons with an HIV infection have ARC and know that they are infected but are still able to carry on a fairly normal life. A still smaller proportion of those infected have been diagnosed as having AIDS. This diagnosis is based on the presence of an opportunistic disease.

Sources: From Frank D. Cox, *The AIDS Booklet*. Copyright © 1989 Wm. C. Brown Publishers, Dubuque, Iowa; and *HIV/AIDS Surveillance*, Year-End Edition, January 1990, U.S. Department of Health and Human Services.

Mycobacterium avium infection affects many organs.

Infection of the bone marrow can contribute to a decrease in red cells, white cells, and platelets in AIDS patients. Kaposi's sarcoma is an unusual cancer of blood vessels, which gives rise to reddish-purple, coin-size spots and lesions on the skin.

Although drugs are being developed to deal with opportunistic diseases in AIDS patients, death usually follows in two to four years. Some may continue to lead a fairly normal life for some months, but eventually they are hospitalized due to weight loss, constant fatigue, and multiple infections.

Treatment

Presently there is no cure for AIDS. Investigators have developed drug treatments that will prolong the lives of those infected and are attempting to develop vaccines that will prevent infection in the first place.

The drug zidovudine formerly known as azidothymidine (AZT) has been shown to be effective in those with full-blown AIDS and also seems to prevent the progression of the disease in infected persons who have fewer than 500 T4 cells per mm^3, but who exhibit no symptoms. A new drug called dideoxyinosine (DDI), which like AZT works by preventing viral replication in cells, is available and thus far shows fewer side effects. Researchers are hopeful that a protease inhibitor drug will soon be available. Proteases are the enzymes HIV needs to reproduce. Also, new drugs are being developed to fight the opportunistic infections that AIDS patients get. Controlling these infections could prolong the lives of AIDS patients. The current hope is to manage AIDS as a chronic disease with increasingly effective drugs.

Most AIDS cases in the United States are caused by HIV-1 and, therefore, the vaccines being developed are directed against this virus. Notice that the virus has an outer envelope molecule called gp120 (fig. A.6). When gp120 combines with a CD4 molecule that projects from helper T lymphocyte, the virus enters helper T cells. Two experimental vaccines for HIV-1 designed to stimulate antibodies to gp120 are being studied clinically. Both of these vaccines utilize only gp120, but one uses it directly and the other uses a treated vaccinia (cowpox) virus that has gp120 gene inserted into it. The hope is that this gene will go on producing gp120 molecules after entering host-cell blood. An entirely different approach is being taken by Jonas Salk, who developed the polio vaccine. His vaccine utilizes whole HIV-1 killed by treatment with chemicals and radiation. So far, this vaccine has been found to be effective against experimental HIV-1 infection in a few chimpanzees.

Future of AIDS

Currently at least 1 million and probably more persons in the United States are infected with human immunodeficiency virus (HIV). Although there was a decline in the rate of increase of reported cases in 1987, the total number of cases reported is expected to continue to increase for at least the next several years. Figure A.5 gives the projected number of cases for the year 1993. The incidence among women and heterosexuals is expected to increase.

AIDS Prevention

We have indicated that AIDS is transmitted by sexual contact, or by sharing contaminated needles or blood. Shaking hands, hugging, social kissing, coughing or sneezing, and swimming in the same pool will not transmit the AIDS virus. You cannot get AIDS from inanimate objects such as toilets, doorknobs, telephones, office machines, or household furniture.

The following behaviors are ones that will help prevent the spread of AIDS and decrease the projected number of new cases:

1. Do not use alcohol or drugs in a way that may prevent you from being able to control your behavior. Especially do not take up the habit of injecting drugs into veins.
2. If you are already a drug abuser and cannot stop your behavior, then always use a new sterile needle for

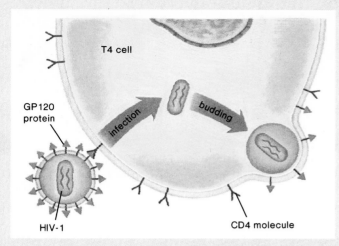

Figure A.6   An HIV-1 has an envelope molecule called GP120 that allows it to attach to CD4 molecules that project from a T4 cell. Infection of the T4 cell follows and viruses eventually bud from the infected T4 cell. If the immune system can be trained by the use of a vaccine to attack and destroy all cells that bear GP120, a person would not be able to be infected with HIV-1.

injection or one that has been cleaned in bleach.

3. Either abstain from sexual intercourse or develop a long-term monogamous (always the same partner) sexual relationship with a partner who is free of HIV and is not an intravenous drug abuser.

4. If you do not know for certain that your partner has been free of HIV for the past five years, always use a latex condom during sexual intercourse. Be sure to follow the directions supplied by the manufacturer. Use of a spermicide containing nonoxynol-9 in addition to the condom can offer further protection because nonoxynol-9 also kills the virus and virus-infected lymphocytes.

5. Refrain from multiple sex partners especially with homosexual or bisexual men, or intravenous drug abusers of either sex. The risk of contracting AIDS is greater in persons who already have another STD.

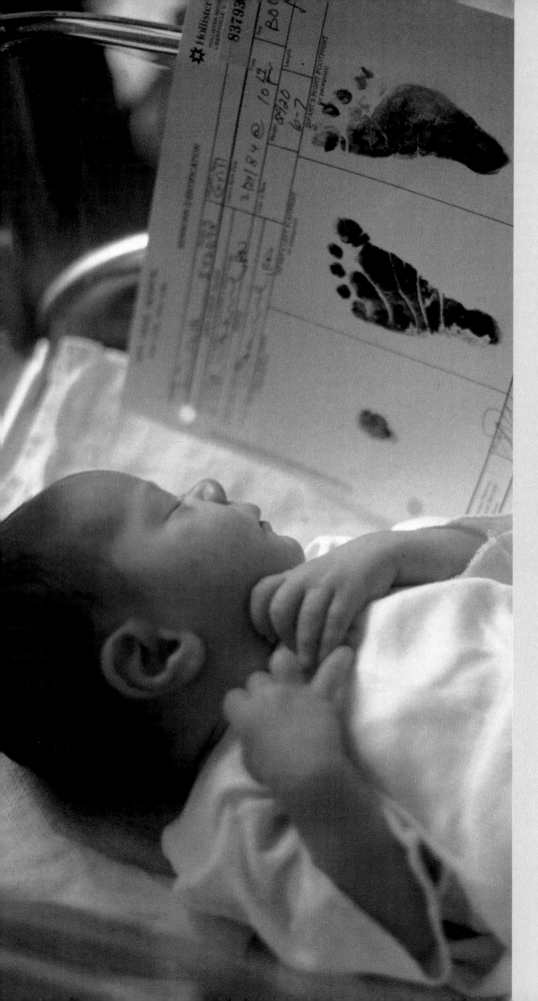

Chapter Sixteen

Development

Chapter Concepts

1 The first stages of human development lead to the establishment of the embryonic germ layers.

2 Induction, or the ability of one tissue to influence the development of another, can explain differentiation, or specialization of parts, and can account for the orderliness of development.

3 Human embryos are dependent on the presence of the extraembryonic membranes, the chorion and amnion, to nourish and protect them during development.

4 Human development is divided into the embryonic period when structures first arise and fetal development when there is a refinement of these structures.

Development has a cellular basis. Union of the sperm and egg begins cell division, growth, and differentiation of cells that result in a baby identified by footprints and thumbprints.

How do you define environment? Illustrations in this chapter show the human embryo surrounded by a watery sac as it develops within the uterus of its mother (p. 321). No doubt this is the embryo's environment, but is the mother's environment also the embryo's environment? In many ways it is. If the mother is x-rayed, has AIDS, or takes a harmful drug, the embryo can be harmed irrevocably. On the other hand, if the mother takes care of herself and receives adequate nutrition, the child is more likely to be both physically and mentally healthy.

Well, what about psychological assaults on the mother? Can exposure to stressful situations, for example, be harmful to the embryo and later to the fetus? Perhaps so. One study shows that during and after World War II, the incidence of birth defects in Germany was markedly increased. The possibility exists that this was due to mental anguish as well as to inadequate prenatal care. It has even been suggested that the personality of the child and adult is also affected by prenatal environmental factors including those within the womb itself. Perhaps we are psychologically affected by whether the womb is large or small, or whether or not the womb contains another developing embryo, for example.

Human Issue

It is well known that a woman's adverse health habits—ingestion of alcohol, caffeine, and other drugs, infection by sexually transmitted diseases—can harm the embryo developing in her uterus. Even before a woman knows she is pregnant, the development of embryonic organs is well advanced and subject to damage. Should sexually active young women, whether or not they are practicing birth control, constantly watch what they eat, where they go, and how much sleep they get in order to make sure any offspring are physically and mentally healthy? What should young men do to promote the birth of healthy babies? What sort of role—if any—should the government play in promoting good health habits during pregnancy?

Developmental Processes

Fertilization occurs in the upper third of an oviduct (fig. 16.1), and development begins even as the embryo passes down this tube to the uterus. For two months the developing human being is called an **embryo** because the organ systems are forming. Only when we can recognize that this will be a human being is the term **fetus** used. The fetal period is largely a time of maturation and growth of structures that have already formed.

While development is occurring, these processes take place.

1. **Cleavage:** Within 30 hours after fertilization, the zygote begins to divide so that at first there are 2, then 4, 8, 16, and 32 cells, and so forth. Since increase in size does not accompany these divisions, the embryo is at first no larger than the zygote (the fertilized egg) was. Cell division during cleavage is mitotic, and each cell receives a full complement of chromosomes and genes.

2. **Morphogenesis:** Morphogenesis refers to the shaping of the embryo and is first evident when certain cells are seen to move, or migrate, in relation to other cells. By these movements, the embryo begins to assume various shapes.

3. **Differentiation:** When a cell takes on a specific structure and function, differentiation occurs. The first system to become visibly differentiated is the nervous system.

4. **Growth:** Later cell division is accompanied by an increase in size of the daughter cells, and growth in the true sense of the term takes place.

Embryonic Development

First Week

By the time the embryo has reached the uterus on the third day it is a **morula,** a ball of many cells. The morula is not much larger than the zygote because there has been no growth. By about the fifth day, the morula has been transformed into the **blastocyst,** a hollow ball of about one hundred cells. The blastocyst has a single layer of outer cells called the trophoblast and an inner cell mass. The **trophoblast,** which will give rise to the **chorion,** is not to be part of the embryo, and for that reason it is called an **extraembryonic membrane.** The **inner cell mass** will eventually become the fetus. Each cell within the inner cell mass has the genetic capability of becoming a complete individual. Sometimes during human development, the inner cell mass splits, and two embryos start developing rather than one. These two embryos will be *identical twins* (fig. 16.2) because they have inherited exactly the same chromosomes. *Fraternal twins,* who arise when two different eggs are fertilized by two different sperm, do not have identical chromosomes. It has even been known to happen that these "twins" have different fathers.

During the first week, the embryo undergoes cleavage, and then the morula becomes the blastocyst having two main parts, the outer trophoblast (to become the chorion) and the inner cell mass (to become the fetus).

Second Week

At the end of the first week, the embryo begins the process of *implanting* in the wall of the uterus. The trophoblast secretes enzymes to digest away some of the tissue and blood vessels of

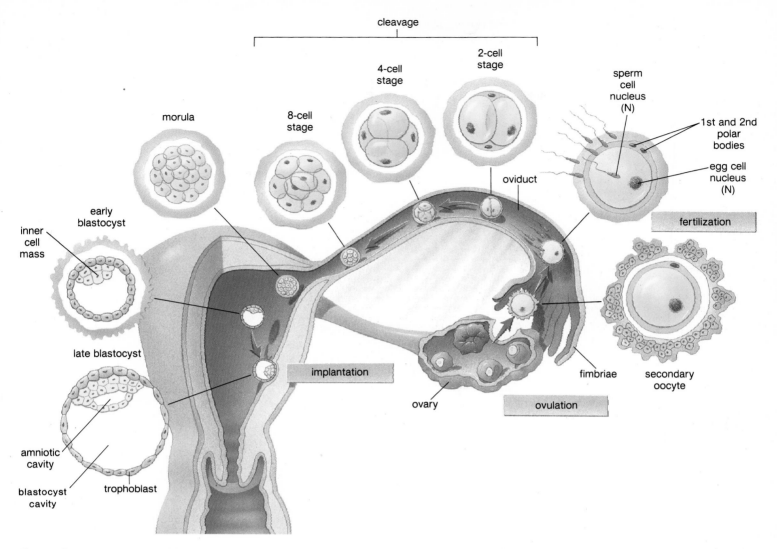

Figure 16.1 Human development before implantation. Structures and events proceed counterclockwise. At ovulation, the secondary oocyte leaves the ovary. Fertilization occurs in the oviduct. As the zygote moves along the oviduct, it undergoes cleavage to produce a morula. The blastocyst forms and implants itself in the uterine lining.

the uterine wall (fig. 16.1). The embryo is now about the size of the period at the end of this sentence. The trophoblast begins to secrete *HCG* (human chorionic gonadotropin), the hormone that is the basis for the pregnancy test and that serves to maintain the corpus luteum past the time it normally disintegrates. Because of this, the endometrium is maintained and menstruation does not occur.

As the week progresses, the inner cell mass detaches itself from the trophoblast, and two more extraembryonic mem-

branes form. The **yolk sac,** which forms below the embryo, has no nutritive function, but it is the first site of blood cell formation. However, the **amnion** and its cavity are where the embryo (and then the fetus) develops. The amniotic fluid acts as an insulator against cold and heat and also absorbs any shock, such as a blow to the mother's abdomen.

Gastrulation occurs during the second week. The inner cell mass now has flattened into the *embryonic disk* composed of two layers of cells (table 16.1): *ectoderm* above and *endoderm*

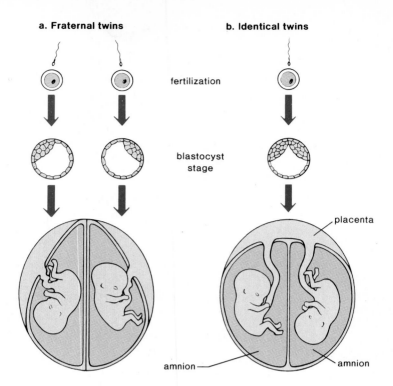

a. Fraternal twins

b. Identical twins

fertilization

blastocyst
stage

placenta

amnion

amnion

Figure 16.2   Conception of fraternal versus identical twins.
a. Fraternal twins are formed when 2 eggs are released and fertilized.
Fraternal twins receive a different genetic inheritance from both the
mother and father. They can even have different fathers. *b.* Identical
twins occur when the embryo breaks in two during an early stage of
development. Identical twins have the exact same genetic inheritance
from both the mother and father.

Table 16.1   Germ Layers and Organ Development

Ectoderm	Mesoderm	Endoderm
skin epidermis including hair, nails, and sweat glands	all muscles	lining of digestive tract, trachea, bronchi, lungs, gallbladder, and urethra
nervous system including brain, spinal cord, ganglia, and nerves	dermis of skin	liver
retina, lens, and cornea of eye	all connective tissue including bone, cartilage, and blood	pancreas
inner ear	blood vessels	thyroid, parathyroid, and thymus glands
lining of nose, mouth, and anus	kidneys	urinary bladder
tooth enamel	reproductive organs	

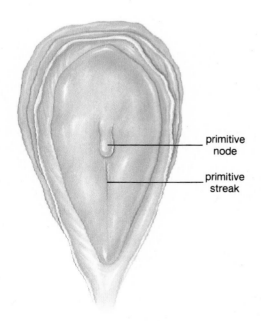

primitive
node

primitive
streak

Figure 16.3   Human embryo at 16 days. The primitive node marks the
extent of the primitive streak, where invagination occurs to establish
the germ layers—ectoderm, endoderm, and mesoderm.

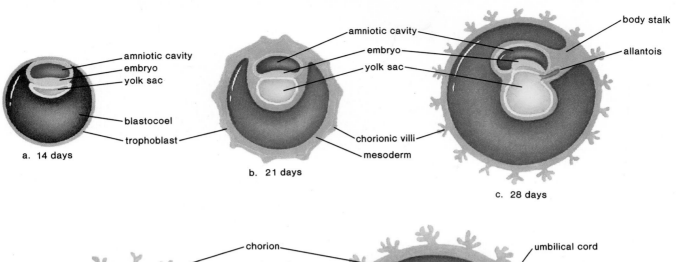

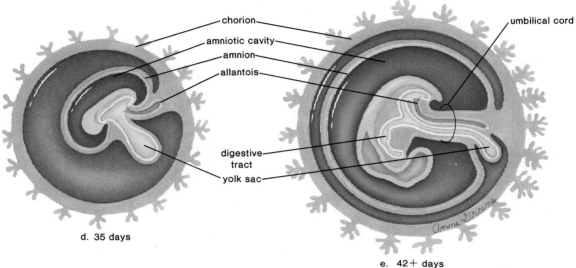

Figure 16.4   Stages showing the early appearance of the extraembryonic membranes and the formation of the umbilical cord in the human embryo. *a.* At 14 days, the amniotic cavity appears. *b.* At 21 days, the chorion and the yolk sac are apparent. *c.* At 28 days, the body stalk and the allantois form. *d.* At 35 days, the embryo begins to take shape as the umbilical cord forms. *e.* Eventually, the umbilical cord is fully formed.

below. Once the embryonic disk elongates to become the *primitive streak* (fig. 16.3), a third layer of cells, the *mesoderm,* forms by invagination of cells along the streak. The trophoblast is reinforced by mesoderm and becomes the chorion (fig. 16.4).

It is possible to relate the development of future organs to these so-called **germ layers.** In general, ectoderm becomes the nervous system, the skin, the hair, and the nails; endoderm produces the inner linings of the digestive, respiratory, and urinary tracts; and mesoderm produces the muscles, the skeleton, and the circulatory systems (fig. 16.5).

The germ layers (ectoderm, endoderm, and mesoderm) are laid down during the third week of development. Development of the organs can be related to these germ layers.

Ectopic Pregnancy

An **ectopic** (out of place) **pregnancy** occurs whenever the embryo implants itself in a location other than the uterus. In most of these instances, the embryo implants itself in the oviduct, but implantation in an ovary or even the abdominal cavity has been known to occur. The oviduct is unable to expand sufficiently to permit the pregnancy to continue, and it will rupture if the embryo is not removed.

PID (p. 306) due to a sexually transmitted disease is the most common cause of ectopic pregnancy because the resultant scarring entraps the embryo and prevents it from traveling to the uterus. Other conditions that predispose a woman to ectopic pregnancies are an infection following childbirth or abortion, a history of ectopic pregnancy, endometriosis, tubal ligation, and the use of fertility drugs. Fertility drugs cause multiple ovulations, and not all of the fertilized eggs may make it to the uterus.

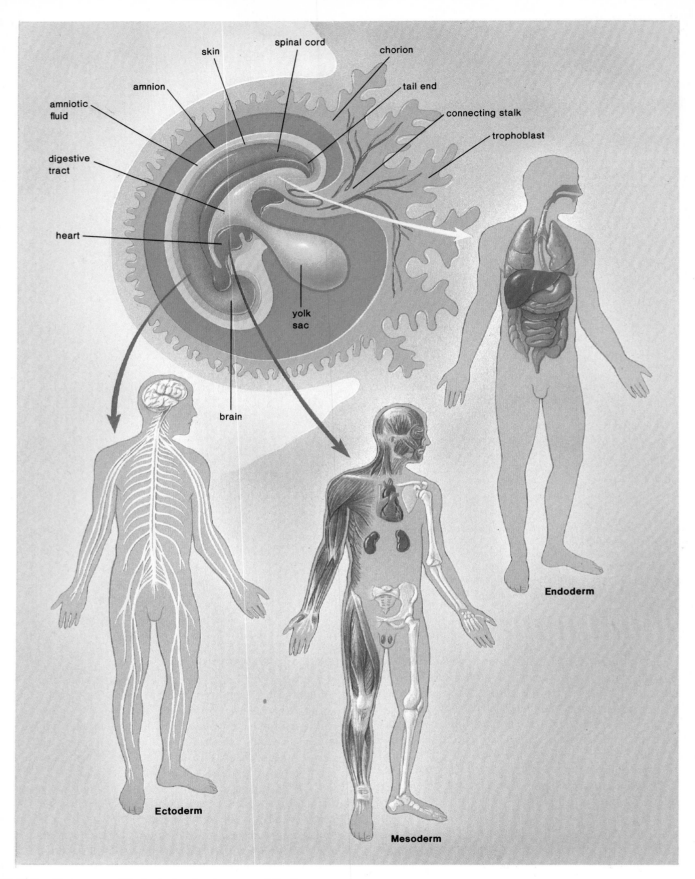

Figure 16.5   The organs of the body develop from one of the three
germ layers.

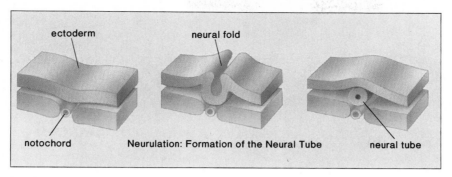

ectoderm   neural fold

notochord   Neurulation: Formation of the Neural Tube   neural tube

a.

The classical signs of ectopic pregnancy are lower abdominal pain, often more noticeable on one side of the abdomen, a missed menstrual period accompanied by irregular vaginal bleeding, and the presence of a tender mass on one side of the pelvis. A positive pregnancy test along with an ultrasound scan that does not indicate uterine pregnancy is also highly suggestive of an ectopic pregnancy.

It is very important for women to be aware of and consider the possibility of an ectopic pregnancy because undiagnosed cases can lead to oviduct rupture, internal bleeding, shock, and possible death.

Third Week

Two important organ systems make their appearance during the third week. The nervous system is the first organ system to be visually evident. At first, a thickening appears along the entire dorsal length of the embryo, and then invagination occurs as neural folds appear. When the neural folds meet at the midline, the neural tube, which later develops into the brain and the nerve cord, is formed (fig. 16.6a).

The notochord is replaced later by the vertebral column; the neural tube then is called the spinal cord. Figure 16.6b shows a human embryo at the end of the third week, after neurulation has begun. A cross section of this embryo (fig. 16.6c) indicates that the **notochord** derived from mesoderm lies beneath the developing neural tube. The notochord is a supporting rod that will later be replaced by the vertebral column.

Experiments, particularly in the frog, have shown that if the presumptive (potential) nervous system lying just above the notochord is cut out and transplanted to another region of the embryo, it will not form a neural tube. On the other hand, if the presumptive notochord is cut out and transplanted beneath what would be belly ectoderm, this ectoderm now differentiates into neural tissue. These experiments indicate that notochord mesoderm causes the overlying ectoderm to form the nervous system, and it is said that the dorsal mesoderm induces the formation of the neural tube. The process of **induction** can help explain the orderly development of the embryo. One tissue is induced by another, and this in turn induces another tissue, and so on, until development is complete. Most likely the inducing tissue produces chemical messengers that are received by the tissue being induced.

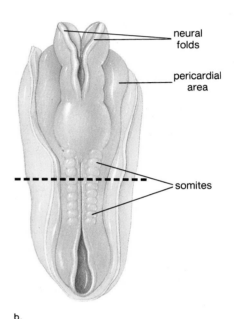

neural folds

pericardial area

somites

b.

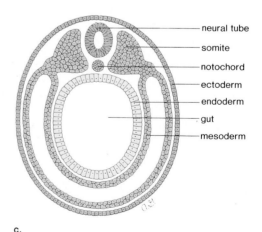

neural tube
somite
notochord
ectoderm
endoderm
gut
mesoderm

c.

Figure 16.6   *a.* Closure of the neural folds is taking place in the manner shown. *b.* Human embryo at 21 days. The pericardial area contains the primitive heart, and the somites are the precursors of the muscles. *c.* Cross section of *b.* where the dotted line appears. Notice that the notochord lies just beneath the neural tube. In this colored drawing, ectoderm-derived structures are in blue, mesoderm-derived structures are in red, and endoderm-derived structures are in yellow.

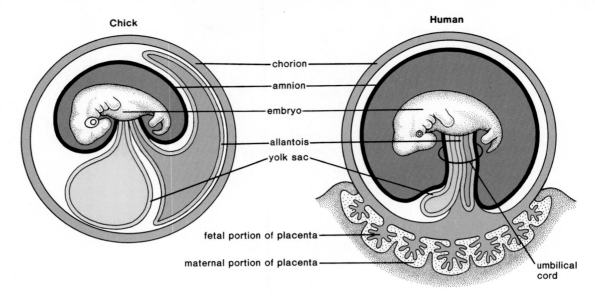

Chick

Human

chorion
amnion
embryo
allantois
yolk sac
fetal portion of placenta
maternal portion of placenta
umbilical cord

Figure 16.7 Extraembryonic membranes are not part of the embryo. These membranes are found during the development of chicks and humans, where each has a specific function. In the chick, the chorion lies just beneath the shell and performs gas exchange. The allantois collects nitrogenous wastes. The yolk sac provides nourishment, and the amnion provides a watery environment. In humans, only the chorion and amnion have comparable functions to those of the chick. The chorion forms the fetal half of the placenta, where exchange occurs with mother's blood, and the amnion provides a watery environment. The yolk sac has no nutritive function but is the first site of blood cell production; the allantois gives rise to the umbilical vessels.

Development of the Heart

Development of the heart also begins in the third week and continues into the fourth week. At first there are right and left heart tubes; when these fuse, the heart begins pumping blood even though the chambers of the heart are not yet fully formed. The veins enter posteriorly and the arteries exit anteriorly from this largely tubular heart, but later the heart will twist so that all major blood vessels enter or exit anteriorly.

During the third week, major organs, such as the nerve cord and the heart, first make their appearance.

Fourth and Fifth Weeks

At four weeks, the embryo is barely larger than the height of this print. There is a bridge of mesoderm called the body stalk that connects the caudal end of the embryo with the chorion (fig. 16.4c). The fourth extraembryonic membrane, the **allantois** (fig. 16.7), is contained within this stalk, and its blood vessels become the umbilical blood vessels. Then the head and the tail lift up, and the body stalk moves anteriorly by constriction (fig. 16.4d). Once this process is complete, the **umbilical cord** that connects the developing embryo to the placenta is fully formed (fig. 16.4e).

Little flippers called limb buds appear later; the arms and the legs develop from the limb buds, and even the hands and the feet become apparent. At the same time, during the fifth week, the head becomes larger than previously, and the sense organs become more prominent. It is possible to make out the developing eyes, ears, and even nose.

During the fourth and fifth weeks, human features, like the head, the arms, and the legs, begin to make their appearance.

Sixth through Eighth Weeks

There is a remarkable change in external appearance during the sixth through eighth weeks of development (fig. 16.8), from a form that is difficult to recognize as human to one that easily is recognizable as human. Concurrent with brain development, the head achieves its normal relationship with the body as a neck region develops. The nervous system is developed well enough to permit reflex actions, such as a startle response to being touched. At the end of this period, the embryo is about 38 mm (1½ inches) long and weighs no more than an aspirin tablet, even though all organ systems are established.

Placenta

The *placenta* begins formation once the embryo is implanted fully. Treelike extensions of the chorion called **chorionic villi** project into the maternal tissues. Later, these disappear in all areas except where the placenta develops. By the tenth week, the placenta (fig. 16.9) is formed fully and begins to produce

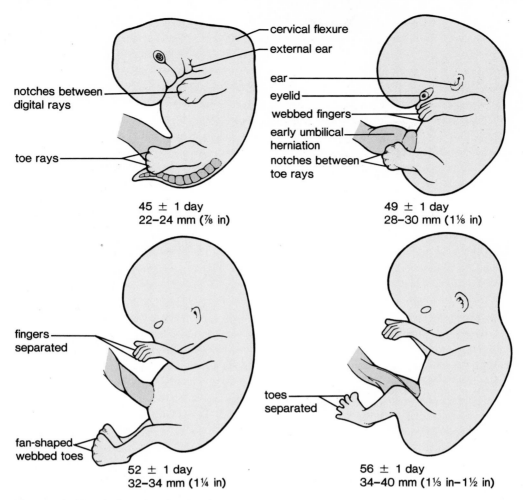

notches between
digital rays

toe rays

cervical flexure
external ear

45 ± 1 day
22–24 mm (⅞ in)

ear
eyelid
webbed fingers
early umbilical
herniation
notches between
toe rays

49 ± 1 day
28–30 mm (1⅛ in)

fingers
separated

fan-shaped
webbed toes

52 ± 1 day
32–34 mm (1¼ in)

toes
separated

56 ± 1 day
34–40 mm (1⅓ in–1½ in)

Figure 16.8   Human embryo at the days indicated and at the size
noted. Note that the drawings are for a 10-day span.

progesterone and estrogen (fig. 16.10). These hormones have two effects: due to their negative feedback effect on the hypothalamus and the anterior pituitary, they prevent any new follicles from maturing, and they maintain the lining of the uterus—now the corpus luteum is not needed. There is no menstruation during pregnancy.

The placenta has a fetal side contributed by the chorion and a maternal side consisting of uterine tissues. Notice in figure 16.9 how the chorionic villi are surrounded by maternal blood sinuses; yet, the blood of the mother and the fetus never mix since exchange always takes place across cell membranes. Carbon dioxide and other wastes move from the fetal side to the maternal side, and nutrients and oxygen move from the maternal side to the fetal side of the placenta. The umbilical cord stretches between the placenta and the fetus. Although it may seem that the **umbilical cord** travels from the placenta to the intestine, actually the umbilical cord simply is taking fetal blood

to and from the placenta. The umbilical cord is the lifeline of the fetus because it contains the umbilical arteries and vein, which transport waste molecules (carbon dioxide and urea) to the placenta for disposal and take oxygen and nutrient molecules from the placenta to the rest of the fetal circulatory system.

Harmful chemicals also can cross the placenta, and this is of particular concern during the embryonic period, when various structures first are forming. Each organ or part seems to have a sensitive period during which a substance can alter its normal function. The reading on page 322 concerns the origination of birth defects and explains ways to detect or predict genetic defects before birth.

At the end of the embryonic period, all organ systems are established and there is a mature and fully functioning placenta. The embryo is only about 38 mm (1½ inches) long.

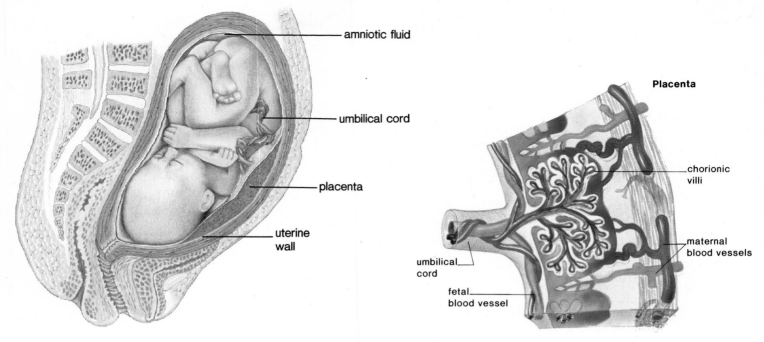

Figure 16.9 Anatomy of the placenta. The placenta is composed of both fetal and maternal tissues. Chorionic villi penetrate the uterine lining and are surrounded by maternal blood. Exchange of molecules between fetal and maternal blood takes place across the walls of the villi. Oxygen and nutrient molecules enter fetal blood; carbon dioxide and urea exit from fetal blood.

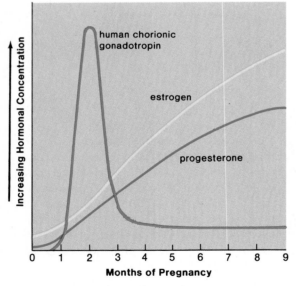

Figure 16.10 Hormones during pregnancy. Human chorionic gonadotropin is secreted by the trophoblast during the first 3 months of pregnancy. This maintains the corpus luteum, which continues to secrete estrogen and progesterone. At about 5 weeks of pregnancy, the placenta begins to secrete estrogen and progesterone in increasing amounts as the corpus luteum degenerates.

Human Issue

All sorts of operations may be available soon to correct fetal abnormalities. When Blake Schultz was a fetus, he had a hole in his diaphragm that allowed his stomach, spleen, and intestines to move into his thoracic cavity and put pressure on his lungs. Surgeons cut into his mother's abdominal wall and uterus and pulled him out enough so that they could open his left side, push the organs back into their natural place, before repairing his diaphragm and side. Do you think Blake's mother was obligated to undergo this operation? One out of 2,200 fetuses have a similar defect. Should all pregnant women be required to undergo sonograms so that fetal malformations can be detected and operated on before birth? Where do you draw the line between fetal rights and the rights of the mother?

Fetal Development

Third and Fourth Months

Figure 16.11 shows that an embryo is not recognizably human, while a fetus is recognizably human. At the beginning of the third month, the fetal head is still very large, the nose is flat, the eyes are far apart, and the ears are distinctively present. Head growth now begins to slow down as the rest of the body increases in length. Epidermal refinements such as eyelashes, eyebrows, hair on head, fingernails, and nipples appear.

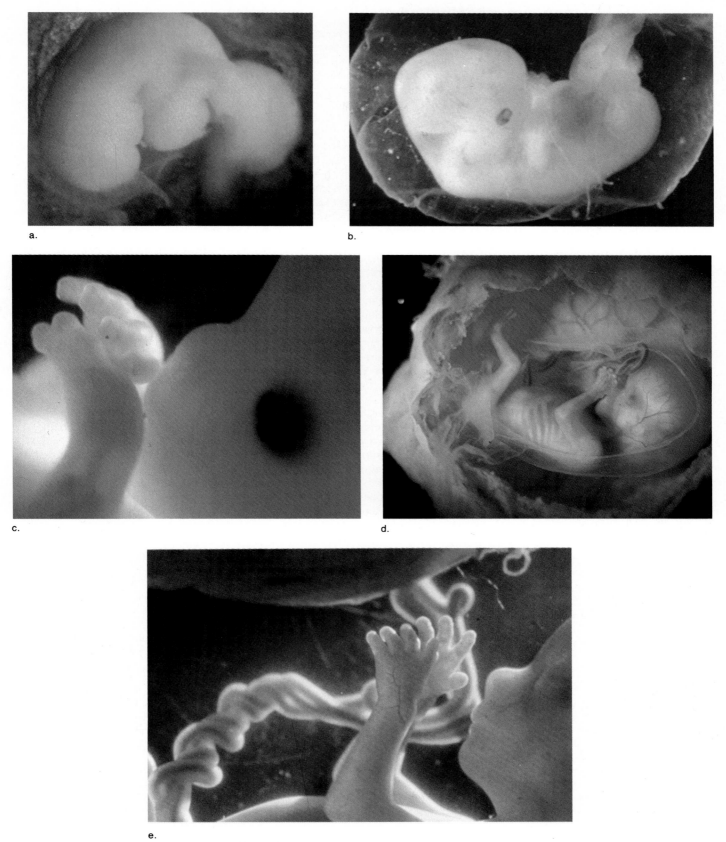

a.

b.

c.

d.

e.

Figure 16.11   Human development from the fourth to the sixteenth week. *a.* In the 4-week-old embryo, the body is flexed and C shaped. *b.* At the end of 6 weeks, the head becomes disproportionately large. *c.* In the 8-week-old embryo, the nose is flat, the eyes are far apart, and the eyelids are fused. *d.* Surrounded by the extraembryonic membranes, this 12-week-old fetus appears to be sucking its thumb. *e.* At 16 weeks, the blood vessels are easily visible through the transparent skin.

Birth Defects

It is believed that at least 1 in 16 newborns has a birth defect, either minor or serious, and the actual percentage may be even higher. Most likely, only 20% of all birth defects are due to heredity. Those that are due to heredity sometimes can be detected before birth. Amniocentesis allows the fetus to be tested for abnormalities of development. Chorionic villi sampling allows the embryo to be tested, and just recently, a method has been developed for screening eggs to be used for in vitro fertilization (see figure on page 323).

Treatment of the fetus in the womb is a rapidly developing area of medical expertise. Biochemical defects sometimes can be treated by giving the mother appropriate medicines. For example, if a baby is unable to synthesize vitamin B and/or is unable to use biotin efficiently, the mother can take these substances in doses large enough to prevent any untoward effects. Structural defects sometimes can be corrected by surgery. For example, if the fetus has water on the brain or is unable to pass urine, tubes that temporarily allow the fluid to pass out into the amniotic fluid can be inserted even while the fetus is still in the womb. Physicians are hopeful that eventually all sorts of structural defects can be corrected by lifting the fetus from the womb long enough for corrective surgery to be done.

It is recommended that all females take everyday precautions to protect any future and/or presently developing embryos and fetuses from defects that are not due to heredity. X-ray diagnostic therapy should be avoided during pregnancy because X rays are mutagenic to a developing embryo or fetus. Children born to women who received X-ray treatment are apt to have birth defects and/or to develop leukemia later on. Fetotoxic chemicals, such as pesticides and many organic industrial chemicals, are also mutagenic. Cigarette smoke not only contains carbon monoxide but also some of these very same fetotoxic chemicals. Babies born to smokers are often underweight and are subject to convulsions.

Pregnant Rh-negative women should receive a RhoGam injection to prevent the production of Rh antibodies. These antibodies can cause nervous system and heart defects.

Sometimes, birth defects are caused by microorganisms. Females can be immunized before the childbearing years for rubella (German measles), which causes birth defects such as deafness. Immunization for sexually transmitted diseases is not possible. The AIDS virus can cross the placenta, and over 1,500 babies who contracted AIDS while in their mother's womb are now mentally retarded. When a mother has herpes, gonorrhea, or chlamydia, newborns can become infected as they pass through the birth canal and might become blind or develop other mental and physical defects. Birth by cesarean section could prevent these occurrences.

Drugs of all types should be avoided. Certainly illegal drugs, like marijuana, cocaine, and heroin, should be completely avoided. "Cocaine babies" now make up 60% of drug-affected babies. Severe fluctuations in blood pressure accompany the use of cocaine and temporarily deprive the developing brain of oxygen. Cocaine babies have visual problems, lack coordination, and are mentally retarded. The drugs aspirin, caffeine (present in coffee, tea, and cola), and alcohol should be severely limited. It is not unusual for babies of drug addicts and alcoholics to display withdrawal symptoms and to have various abnormalities. Babies born to women who have about 45 drinks a month and as many as 5 drinks on one occasion are apt to have FAS (fetal alcohol syndrome). These babies have decreased weight, height, and head size, with malformation of the head and face. Mental retardation is common in FAS infants.

Medications can also sometimes cause problems. When the synthetic hormone DES was given to pregnant women to prevent miscarriage, their daughters showed various abnormalities of the reproductive organs and an increased tendency toward cervical cancer. Other sex hormones, including birth control pills, possibly can cause abnormal fetal development, including abnormalities of the sex organs. The tranquilizer thalidomide is well known for having caused deformities of the arms and legs in children born to women who took the drug. Therefore, a woman has to be very careful about taking medications while pregnant.

Now that physicians and laypeople are aware of the various ways in which birth defects can be prevented, it is hoped that the incidence of birth defects will decrease in the future.

Three methods for genetic defect testing before birth. *a.* Amniocentesis cannot be done until the sixteenth week of pregnancy. A long needle is passed through the abdominal wall to withdraw a small amount of amniotic fluid along with fetal cells. Since there are only a few cells in the amniotic fluid, testing must be delayed for 4 weeks until a cell culture produces enough cells for testing purposes.

b. Chorionic villi sampling can be done as early as the fifth week of pregnancy. The doctor inserts a long, thin tube through the vagina into the uterus. With the help of ultrasound, which gives a picture of the uterine contents, the tube is placed between the lining of the uterus and the chorion. Suction is then used to remove a sampling of the chorionic villi cells. Chromosome analysis and biochemical tests for several different genetic defects can be done immediately on these cells.

c. Screening eggs for genetic defects is a new technique. Preovulation eggs are removed by aspiration after a laparoscope (a telescope with a cold light source), is inserted into the abdominal cavity through a small incision in the region of the navel. The prior administration of FSH ensures that several eggs are available for screening. Only the chromosomes within the first polar body are tested because if the woman is heterozygous for a genetic defect and it is found in the polar body, then the egg must be normal. Normal eggs undergo *in vitro* fertilization and are placed in the prepared uterus. At present, only 1 in 10 attempts results in a birth, but it is known ahead of time that the child will be normal.

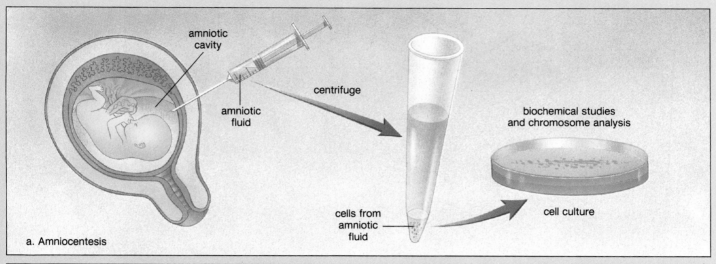

amniotic
cavity

amniotic
fluid

centrifuge

biochemical studies
and chromosome analysis

cells from
amniotic
fluid

cell culture

a. Amniocentesis

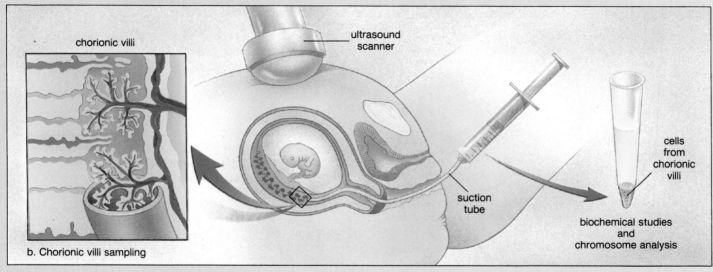

chorionic villi

ultrasound
scanner

cells
from
chorionic
villi

suction
tube

biochemical studies
and
chromosome analysis

b. Chorionic villi sampling

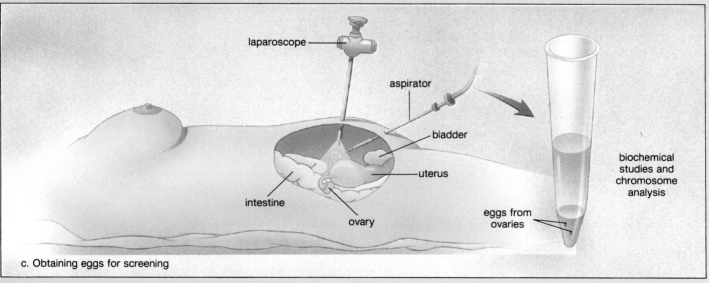

laparoscope

aspirator

bladder

uterus

intestine

ovary

biochemical
studies and
chromosome
analysis

eggs from
ovaries

c. Obtaining eggs for screening

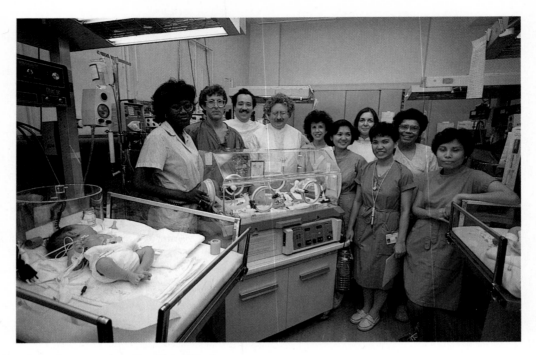

Figure 16.12   Newborn intensive care unit staff at Mount Sinai Hospital in New York. Six nurses, a newborn specialist (neonatologist), a resident, and 2 psychologists care for each baby and the parents. Only babies who have low birth weight (below 5.5 lbs.) are placed in intensive care.

Cartilage is replaced by *bone* as ossification centers appear in most of the bones. Cartilage remains at the ends of the long bones, and ossification is not complete until age 18 or 20. The skull has six large membranous areas called *fontanels* that permit a certain amount of flexibility as the head passes through the birth canal and allow rapid growth of the brain during infancy. The fontanels disappear by two years of age.

Sometime during the third month, it is possible to distinguish males from females. Apparently, the Y chromosome has a gene called the testis determining factor gene (TDF) that triggers the differentiation of gonads into testes. Once the testes differentiate, they produce androgens, the male sex hormones. The androgens, especially testosterone, stimulate the growth of the male external genitalia. In the absence of androgens, female genitalia form. Fetal ovaries do not produce estrogen because there is plenty of it circulating in the mother's bloodstream.

At this time, both testes and ovaries are located within the abdominal cavity, but later, in the last trimester of fetal development, the testes descend into the scrotal sacs (scrotum). Sometimes the testes fail to descend, and in that case, an operation may be done later to place them in their proper location.

During the fourth month, the fetal heartbeat is loud enough to be heard when a physician applies a stethoscope to the mother's abdomen. By the end of this month, the fetus is less than 140 mm (6 in) in length and weighs a little more than 200 g (½ lb).

During the third and fourth months, it is obvious that the skeleton is becoming ossified. The sex of the individual is now distinguishable.

Fifth through Seventh Months

During the fifth through seventh months, the mother begins to feel movement. At first, there is only a fluttering sensation, but as the fetal legs grow and develop, kicks and jabs are felt. The fetus, though, is in the fetal position with the head bent down and in contact with the flexed knees.

The wrinkled, translucent, pink-colored skin is covered by a fine down called **lanugo.** This in turn is coated with a white, greasy, cheeselike substance called **vernix caseosa,** which probably protects the delicate skin from the amniotic fluid. The eyelids now are open fully, however.

At the end of this period, weight has increased to almost 1,350 gm (3 lb) and the length to almost 300 mm (12 in). It is possible that if born now, the baby will survive.

Premature Babies

About 7% of all newborns in the United States weigh less than 5½ pounds. Most of these are premature babies (fig. 16.12) who face the following difficulties.

Respiratory Distress Syndrome (Hyaline Membrane Disease)
The lungs don't produce enough of a chemical surfactant that helps the alveoli stay open. Therefore, the lungs tend to collapse instead of expanding to be filled with air.

Retinopathy of Prematurity The high level of oxygen needed to insure adequate gas exchange by the immature lungs can lead to proliferation of blood vessels within the eyes with ensuing blindness.

Intracranial Hemorrhage The delicate blood vessels in the brain are apt to break, causing swelling and inflammation of the brain. If not fatal, this can lead to brain damage.

Jaundice The immature liver fails to excrete the waste product bilirubin; instead, it builds up in the blood, possibly causing brain damage.

Infections There is a low level of antibodies in the body, and the various medical procedures performed could possibly introduce germs. Also, an infection of the bowels is common, along with perforation, bleeding, and shock.

Circulatory Disorders Fetal circulation, discussed in the following section, has two features—the oval opening between the atria, and the arterial duct that allows blood to bypass the lungs. If these features persist in the newborn, a mixing of oxygenated blood with deoxygenated blood will result and circulation of the blood will be impaired, perhaps leading to blue baby syndrome or heart failure.

The reasons for premature birth have been investigated, and it's been concluded that prenatal care, including good nutrition and the willingness to refrain from excessive drinking of alcohol and smoking cigarettes, could reduce the incidence of premature birth and/or low birth weight.

Fetal Circulation

As figure 16.13 shows, the fetus has four circulatory features that are not present in adult circulation.

1. **Oval opening,** or *foramen ovale,* an opening between the two atria. This opening is covered by a flap of tissue that acts as a valve.
2. **Arterial duct,** or *ductus arteriosus,* a connection between the pulmonary artery and the aorta.
3. **Umbilical arteries** and **vein,** vessels that travel to and from the placenta, leaving waste and receiving nutrients.
4. **Venous duct,** or *ductus venosus,* a connection between the umbilical vein and the inferior vena cava.

All of these features can be related to the fact that the fetus does not use its lungs for gas exchange since it receives oxygen and nutrients from the mother's blood by way of the placenta.

To trace the path of blood in the fetus, begin with the right atrium (fig. 16.13). From the right atrium, the blood may pass directly into the left atrium by way of the oval opening or it may pass through the atrioventricular valve into the right ventricle. From the right ventricle, the blood goes into the pulmonary artery, but because of the arterial duct, most of the blood

then passes into the aorta. Therefore, by whatever route the blood takes, most of the blood reaches the aorta instead of the lungs.

Blood within the aorta travels to the various branches, including the iliac arteries that connect to the umbilical arteries leading to the placenta. Exchange between maternal blood and fetal blood takes place at the placenta. It is interesting to note that the blood in the umbilical arteries, which travels to the placenta, is low in oxygen, but the blood in the umbilical vein, which travels from the placenta, is high in oxygen. The umbilical vein enters the venous duct, which passes directly through the liver. The venous duct then joins with the inferior vena cava, a vessel that contains deoxygenated blood. The vena cava returns this "mixed blood" to the heart.

The most common of all cardiac defects in the newborn is the persistence of the oval opening. With the tying of the cord and the expansion of the lungs, blood enters the lungs in quantity. Return of this blood to the left side of the heart usually causes a flap to cover the opening. Incomplete closure occurs in nearly one out of four individuals, but even so, passage of the blood from the right atrium to the left atrium rarely occurs because either the opening is small or it closes when the atria contract. In a small number of cases, the passage of impure blood from the right side to the left side of the heart is sufficient to cause a "blue baby." Such a condition now can be corrected by open-heart surgery.

The arterial duct closes because endothelial cells divide and block off the duct. Remains of the arterial duct and parts of the umbilical arteries and vein later are transformed into connective tissue.

Eighth and Ninth Months

As the time of birth approaches, the fetus rotates so that the head is pointed toward the cervix (fig. 16.14). If the fetus does not turn, then the likelihood of a breech birth (rump first) may call for a cesarean section. It is very difficult for the cervix to expand enough to accommodate this form of birth, and asphyxiation of the baby is more likely.

At the end of this time period, the fetus is about 530 mm (21 in) long and weighs about 3,400 gm (7½ lb). Weight gain is due largely to an accumulation of fat beneath the skin.

From the fifth to the ninth month, the fetus continues to grow and to gain weight. Babies born after six or seven months may survive but are subject to various illnesses that can have lasting effects or cause an early death.

Birth

The uterus characteristically contracts throughout pregnancy. At first, light, often indiscernible contractions lasting about 20–30 seconds occur every 15–20 minutes, but near the end of pregnancy, they become stronger and more frequent so that the woman may think falsely that she is in labor. The onset of true

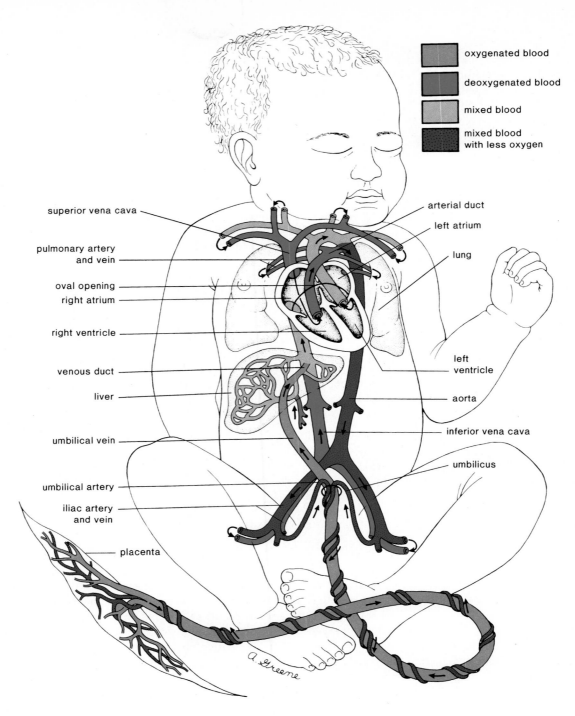

oxygenated blood

deoxygenated blood

mixed blood

mixed blood
with less oxygen

superior vena cava

pulmonary artery
and vein

oval opening

right atrium

right ventricle

venous duct

liver

umbilical vein

umbilical artery

iliac artery
and vein

placenta

arterial duct

left atrium

lung

left
ventricle

aorta

inferior vena cava

umbilicus

a. Greene

Figure 16.13   Fetal circulation. Oxygenated blood becomes mixed
with deoxygenated blood when the umbilical vein joins with the inferior
vena cava via the venous duct. This mixed blood is routed to the left
ventricle by way of the oval opening and then passes to the aorta and
brain. Deoxygenated blood from the superior vena cava is routed to the
aorta via the arterial duct; therefore, blood in the dorsal aorta is mixed
blood with less oxygen.

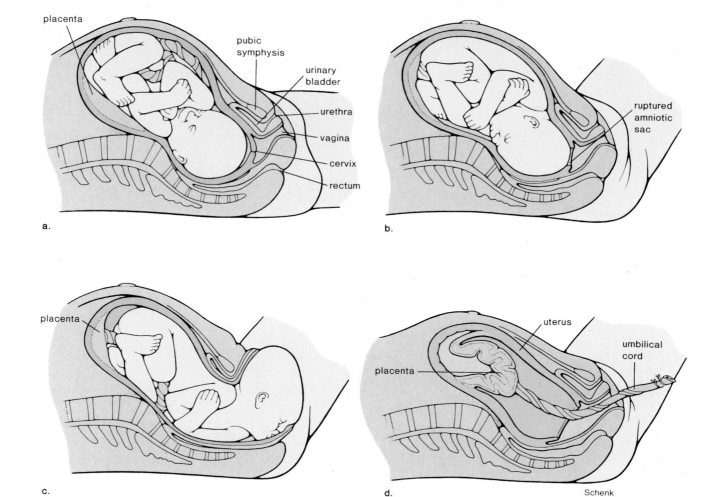

Figure 16.14 Three stages of parturition. *a.* Position of fetus just before birth begins. *b.* Dilation of cervix. *c.* Birth of baby. *d.* Expulsion of afterbirth.

labor is marked by uterine contractions that occur regularly every 15–20 minutes and last for 40 seconds or more. Birth, also called **parturition** or labor, has three stages.

The events that cause parturition still are not known entirely, but there is now evidence suggesting the involvement of prostaglandins. It may be, too, that the prostaglandins cause the release of oxytocin from the maternal posterior pituitary. Both prostaglandins and oxytocin cause the uterus to contract, and either hormone can be given to induce parturition.

Stages

During the *first stage* of parturition, the cervix dilates; during the *second,* the baby is born; and during the *third,* the afterbirth is expelled.

Stage 1

Prior to or concomitant with the first stage of parturition, there can be a "bloody show" caused by the expulsion of a mucous plug from the cervical canal. This plug prevents bacteria and sperm from entering the uterus during pregnancy.

Uterine contractions during the first stage of labor occur in such a way that the cervical canal slowly disappears (fig.

16.14*b*) as the lower part of the uterus is pulled upward toward the baby's head. This process is called *effacement,* or "taking up the cervix." With further contractions, the baby's head acts as a wedge to assist cervical dilation. The baby's head usually has a diameter of about 10 cm; therefore, the cervix has to dilate to this diameter in order to allow the head to pass through. If it has not occurred already, the amniotic membrane is apt to rupture now, releasing the amniotic fluid, which escapes out the vagina. The first stage of labor ends once the cervix is dilated completely.

Stage 2

During the second stage of parturition, the uterine contractions occur every one to two minutes and last about one minute each. They are accompanied by a desire to push, or bear down. As the baby's head gradually descends into the vagina, the desire to push becomes greater. When the baby's head reaches the exterior, it turns so that the back of the head is uppermost (fig. 16.14*c*). Since the vagina may not expand enough to allow passage of the head without tearing, an *episiotomy* often is performed. This incision, which enlarges the opening, is stitched later and heals more perfectly than a tear. As soon as the head is delivered, the baby's shoulders rotate so that the baby faces

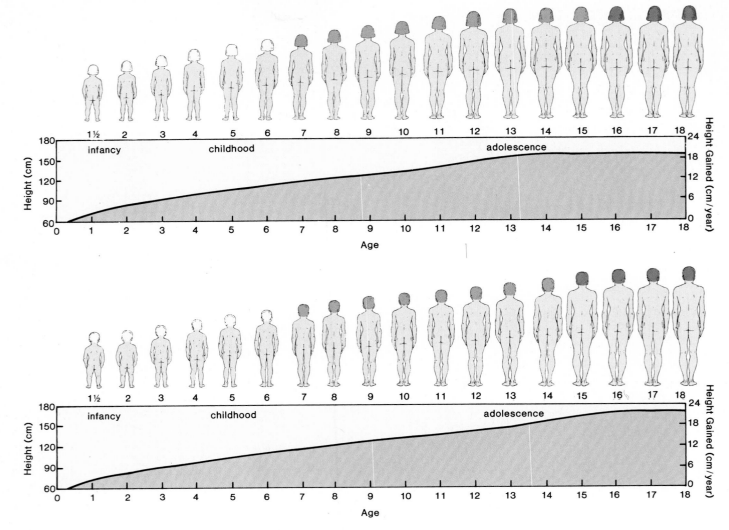

Figure 16.15 The average height of females (*above*) and males (*below*) from infancy through adolescence. Each sex has an adolescent growth spurt, but the male's growth spurt occurs about 2 years after the female's.

either to the right or the left. The physician may hold the head at this time and guide it downward, while one shoulder and then the other emerges. The rest of the baby follows easily.

Once the baby is breathing normally, the umbilical cord is cut and tied, severing the child from the placenta. The stump of the cord shrivels and leaves a scar, which is the navel.

Stage 3

The placenta, or *afterbirth,* is delivered during the third stage of labor (fig. 16.14*d*). About 15 minutes after delivery of the baby, uterine muscular contractions shrink the uterus and dislodge the placenta. The placenta then is expelled into the vagina. As soon as the placenta and its membranes are delivered, the third stage of labor is complete.

During the first stage of birth, the cervix dilates; during the second, the child is born; and during the third, the afterbirth is expelled.

Human Development after Birth

Development does not cease once birth has occurred but continues throughout the stages of life: infancy, childhood, adolescence, and adulthood.

Infancy lasts until about two years of age. It is characterized by tremendous growth and sensorimotor development. During *childhood,* the individual grows, and the body proportions change (fig. 16.15). *Adolescence* begins with *puberty,* when the secondary sex characteristics appear and the sexual organs

Figure 16.16 Aging is a slow process during which the body undergoes changes that eventually will bring about death even if no marked disease or disorder is present. Although the human life span probably cannot be expanded, it is possible to expand the health span, the length of time the body functions normally.

become functional. At this time, there is an acceleration of growth leading to changes in height, weight, fat distribution, and body proportions. Males commonly experience a growth spurt later than females; therefore, they grow for a longer period of time. Males are generally taller than females and have broader shoulders and longer legs relative to their trunk length.

Developmental changes keep occurring throughout infancy, childhood, adolescence, and adulthood.

Adulthood and Aging

Young adults are at their physical peak in muscle strength, reaction time, and sensory perception. The organ systems at this time are best able to respond to altered circumstances in a homeostatic manner. From now on, there is an almost imperceptible, gradual loss in certain of the body's abilities. **Aging** encompasses these progressive changes that contribute to an increased risk of infirmity, disease, and death (fig. 16.16).

Today, there is great interest in **gerontology,** the study of aging, because there are now more older individuals in our so-

ciety than ever before and the number is expected to rise dramatically. In the next half-century, those over age 75 will rise from the present 13 million to 35 to 45 million, and those over age 80 will rise from 3 million to 6 million individuals. The human life span is judged to be a maximum of 110 to 115 years. The present goal of gerontology is not necessarily to increase the life span but to increase the health span, the number of years that an individual enjoys the full functions of all body parts and processes.

Theories of Aging

There are many theories about what causes aging. Three of these are considered here.

Genetic in Origin   Several lines of evidence indicate that aging has a genetic basis. (1) The children of long-lived parents tend to live longer than those of short-lived parents. Perhaps the genes are programmed to control aging and the time of death. The maximum life span of animals is species-specific; for humans it is about 110 years. (2) The number of times a cell divides is also species-specific. The maximum number of times human cells divide is around 50. Perhaps as we grow older, more and more cells are unable to divide any longer, and instead they undergo degenerative changes and die. Senescence (old-age) genes have been identified in cultures of human skin cells. (3) Some cell lines may become nonfunctional long before the maximum number of divisions has occurred. Whenever DNA replicates, mutations can occur, and this can lead to the production of nonfunctional proteins. Eventually, the number of inadequately functioning cells can build up, and this contributes to the aging process.

There is most likely a genetic basis for most of the illnesses associated with old age, such as the thinning of the bones and an impaired immune system. Researchers expect to discover the genes that cause people to become infirm.

Human Issue

Almost 30% of today's federal health-care budget is spent on medical care for the elderly. With the oncoming demographic shift toward elderly people, can we afford to take care of ourselves when we become old? A former Colorado governor bluntly suggested that "we all have the duty to die" when we grow old. This may be an extreme view, but it points to several age-related issues that are becoming increasingly common. Should the government pay most of the health-care costs of the elderly? Should physicians not prescribe expensive treatments if a person is past a certain age limit, say 75, and cannot afford to pay? Should the elderly not be considered for organ transplants if the organs are needed by younger individuals?

Whole Body Processes   A decline in the hormonal system can affect many different organs of the body. For example, type II diabetes is common in older individuals. The pancreas makes insulin, but the cells lack the receptors that enable them to respond. Menopause in women occurs for a similar reason. There is plenty of FSH in the bloodstream, but the ovaries do not respond. Perhaps aging results from the loss of hormonal activities and a decline in the functions they control.

The immune system, too, no longer performs as it once did, and this can affect the body as a whole. The thymus gland gradually decreases in size, and eventually most of it is replaced by fat and connective tissue. The incidence of cancer increases among the elderly, which may signify that the immune system is no longer functioning as it should. This idea is substantiated, too, by the increased incidence of autoimmune diseases in older individuals.

It is possible, though, that aging is not due to the failure of a particular system that can affect the body as a whole, but to a specific type of tissue change that affects all organs and even the genes. It has been noticed for some time that proteins—such as collagen, which makes up the white fibers (p. 54) and is present in many support tissues—become increasingly cross-linked as people age. Undoubtedly, this cross-linking contributes to the stiffening and the loss of elasticity characteristic of aging tendons and ligaments. It also may account for the inability of organs, such as the blood vessels, the heart, and the lungs, to function as they once did. Some researchers now have found that glucose has the tendency to attach to any type of protein, which is the first step in a cross-linking process that ends with the formation of advanced glycosylation end products (AGEs).

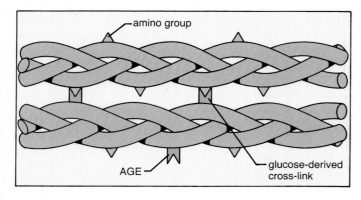

amino group

AGE

glucose-derived cross-link

AGE-derived cross-links not only explain why cataracts develop, they also may contribute to the development of atherosclerosis and to the inefficiency of the kidneys in diabetics and older individuals. Even DNA-associated proteins seem capable of forming AGE-derived cross-links, and perhaps this increases the rate of mutations as we age. These researchers presently are experimenting with the drug aminoguanidine, which can prevent the development of AGEs.

Extrinsic Factors   The current data about the effects of aging often are based on comparisons of the characteristics of the elderly to younger age groups, but perhaps today's elderly were not as aware when they were younger of the importance of, for example, diet and exercise to general health. It is possible, then, that much of what we attribute to aging is instead due to years of poor health habits.

For example, osteoporosis is associated with a progressive decline in bone density in both males and females so that fractures are more likely to occur after only minimal trauma. Osteoporosis is common in the elderly—by age 65, one-third of women will have vertebral fractures, and by age 81, one-third of women and one-sixth of men will have suffered a hip fracture. While there is no denying that there is a decline in bone mass as a result of aging, certain extrinsic factors are also important. The occurrence of osteoporosis itself is associated with cigarette smoking, heavy alcohol intake, and perhaps inadequate calcium intake. Not only is it possible to eliminate these negative factors by personal choice, it also is possible to add a positive factor. A moderate exercise program has been found to slow down the progressive loss of bone mass.

Rather than collecting data on the average changes observed between different age groups, it might be more useful to note the differences within any particular age group. If this type of comparison is done, extrinsic factors that contribute to a decline and extrinsic factors that promote the health of an organ can be identified.

Effect of Aging on Body Systems

Keeping in mind that we want to accept such data with reservations, we will still discuss in general the effects of aging on the various systems of the body. Figure 16.17 compares the percentage of function of various organs in a 75 to 80-year-old person to that of a 20-year-old person whose organs are assumed to function at 100% capacity. When making this comparison, we should keep in mind that the body has a vast functional reserve; it still can perform well even when not at 100% capacity.

The Skin   As aging occurs, the skin becomes thinner and less elastic because the number of elastic fibers decreases and the collagen fibers undergo cross-linking as discussed previously. Also, there is less adipose tissue in the subcutaneous layer; therefore, older people are more likely to feel cold. The loss of thickness accounts for sagging and wrinkling of the skin.

There are fewer hair follicles so that the hair on the scalp and the extremities thins out. The number of sebaceous glands is reduced, and the skin tends to crack. There is a decrease in the number of melanocytes, making the hair turn gray and the skin to become pale. In contrast, some of the remaining pigment cells are larger, and pigmented blotches appear in the skin.

Processing and Transporting   Cardiovascular disorders are the leading cause of death among the elderly. The heart shrinks because there is a reduction in cardiac muscle cell size. This leads to loss of cardiac muscle strength and reduced cardiac output. Still, it is observed that the heart, in the absence of disease, is able to meet the demands of increased activity. It can increase its rate to double or triple the amount of blood pumped each minute even though the maximum possible output declines.

Because the middle coat of arteries contains elastic fibers that most likely are subject to cross-linking, the arteries become more rigid with time, and their size is further reduced by plaque (p. 108). Therefore, blood pressure readings gradually rise. Such changes are common in individuals living in Western industrial-

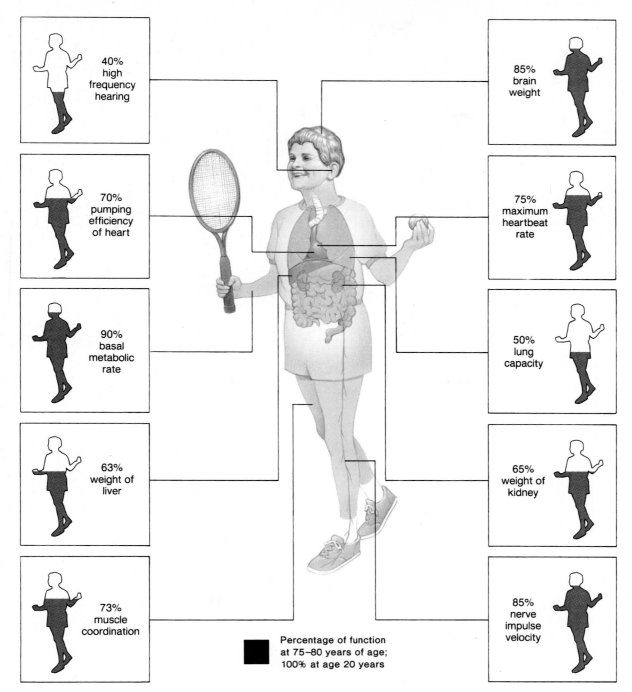

40%
high
frequency
hearing

70%
pumping
efficiency
of heart

90%
basal
metabolic
rate

63%
weight of
liver

73%
muscle
coordination

85%
brain
weight

75%
maximum
heartbeat
rate

50%
lung
capacity

65%
weight of
kidney

85%
nerve
impulse
velocity

Percentage of function
at 75–80 years of age;
100% at age 20 years

Figure 16.17   Percentage of organ function remaining in a 75 to
80-year-old person as measured against that of a 20-year-old person.

ized countries, but not in agricultural societies. As mentioned earlier, diet has been suggested as a way to control degenerative changes in the cardiovascular system (p. 83).

There is reduced blood flow to the liver, and this organ does not metabolize drugs as efficiently as before. This means that as a person gets older, less medication is needed to maintain the same level in the bloodstream.

Circulatory problems often are accompanied by respiratory disorders and vice versa. Growing inelasticity of lung tissue means that ventilation is reduced. Because we rarely use the entire vital capacity, these effects are not noticed unless there is increased demand for oxygen.

There is also reduced blood supply to the kidneys. The kidneys become smaller and less efficient at filtering wastes. Salt and water balance are difficult to maintain, and the elderly dehydrate faster than younger people. Difficulties involving urination include incontinence and the inability to urinate. In men, the prostate gland may enlarge and reduce the diameter of the urethra, making urination so difficult that surgery often is needed.

The loss of teeth, which often is seen in elderly people, is more apt to be the result of long-term neglect than a result of aging. The digestive tract loses tone and secretion of saliva and gastric juice is reduced, but there is no indication of reduced absorption. Therefore, an adequate diet, rather than vitamin and mineral supplements, is recommended. There are common complaints of constipation, increased amount of gas, and heartburn, but gastritis, ulcers, and cancer can also occur.

Integration and Coordination   It often is mentioned that while most tissues of the body regularly replace their cells, some at a faster rate than others, the brain and the muscles do not. No new nerve or skeletal muscle cells are formed in the adult. However, contrary to previous opinion, recent studies show that few neural cells of the cortex are lost during the normal aging process. This means that cognitive skills remain unchanged even though there is characteristically a loss in short-term memory. Although the elderly learn more slowly than the young, they can acquire new material and remember it as well as the young. It is noted that when more time is given for the subject to respond, age differences in learning decrease.

Neurons are extremely sensitive to oxygen deficiency, and if neuron death does occur, it may not be due to aging itself but to reduced blood flow in narrowed blood vessels. Specific disorders, such as depression, Parkinson disease, and Alzheimer disease (p. 206), sometimes are seen, but they are not common. Reaction time, however, does slow, and more stimulation is needed for hearing, taste, and smell receptors to function as before. After age 50, there is a gradual reduction in the ability to hear tones at higher frequencies, and this can make it difficult to identify individual voices and to understand conversation in a group. The lens of the eye does not accommodate as well and also may develop a cataract. Glaucoma is more likely to develop because of a reduction in the size of the anterior chamber of the eye.

Loss of skeletal muscle mass is not uncommon, but it can be controlled by a regular exercise program. There is a reduced capacity to do heavy labor, but routine physical work should be no problem. A decrease in the strength of the respiratory muscles and inflexibility of the rib cage contribute to the inability of the lungs to expand as before, and reduced muscularity of the urinary bladder contributes to difficulties in urination.

As noted before, aging is accompanied by a decline in bone density. Osteoporosis, characterized by a loss of calcium and minerals from bone, is not uncommon, but there is evidence that proper health habits can prevent its occurrence. Arthritis, which restricts the motility of joints, also is seen. In arthritis, as the articular cartilage deteriorates, ossified spurs develop. These cause pain upon movement of the joint.

Weight gain occurs because the metabolic rate decreases and inactivity increases. Muscle mass is replaced by stored fat and retained water.

Reproductive System   Females undergo menopause, and thereafter the level of female sex hormones in blood falls markedly. The uterus and the cervix are reduced in size, and there is a thinning of the walls of the oviducts and the vagina. The

Figure 16.18   The aim of gerontology is to allow the elderly to enjoy living. This requires studying the possible debilities that can occur as one ages and then making recommendations as to how to forestall or prevent their occurrences.

external genitals become less pronounced. In males, the level of androgens falls gradually over the age span 50 to 90, but sperm production continues until death.

It is of interest that as a group, females live longer than males. Although their health habits may be poorer, it is also possible that the female sex hormone estrogen offers to women some protection against circulatory disorders when they are younger. Males suffer a marked increase in heart disease in their forties, but an increase is not noted in females until after menopause. Then women lead men in the incidence of stroke. Men are still more likely than women to have a heart attack, however.

Conclusion

We have listed many adverse effects due to aging, but it is important to emphasize that while such effects are seen, they are not a necessary occurrence (fig. 16.18). We must discover any extrinsic factors that precipitate these adverse effects and guard against them. Just as it is wise to make the proper preparations to remain financially independent when older, it is also wise to realize that biologically successful old age begins with the health habits developed when you are younger.

Exercise and dieting are two behaviors that reverse the gradual decay of the body. It is possible that much of the decline in older people is due to physical inactivity. And those who are overweight are less inclined to exercise than those who are slim. In any case, it seems that those who are thin live longer. Studies are underway to confirm these impressions.

Summary

During development, cleavage, growth, morphogenesis, and differentiation occur. Embryonic development occurs during the first eight weeks, and fetal development occurs during the third to the ninth month. The first three weeks are marked by the development of the extraembryonic membranes and the germ layers from which the various organs are derived. The central nervous system and the heart are the first organs to appear, but by the end of the eighth week all organ systems are present, and externally the fetus has a human appearance.

The fetus is dependent upon the placenta for gas exchange and as a source of nutrient molecules. During this fetal period, there is a refinement of all organs, and a bony skeleton replaces the cartilaginous skeleton. A suitable weight increase is important to survival after birth. Parturition or birth has three stages: during the first stage, the cervix dilates; during the second, the child is born; and during the third, the afterbirth is expelled.

Development continues after birth. Infancy lasts until two years of age, and by this time the child has developed all sorts of motor skills and is usually able to form sentences. Puberty marks the division between childhood and adolescence. Females have a growth spurt at about age 12 while boys have theirs at age 15. Aging encompasses progressive changes, from about age 20 on, that contribute to an increased risk of infirmity, disease, and death.

Study Questions

1. List and discuss the processes of development. (p. 312)
2. List the events of embryonic development in a sequential manner. (pp. 312–319)
3. What are the three germ layers? What structures are associated with each germ layer? (p. 315)
4. What is induction? Discuss an experiment that was done to show that induction takes place. (p. 317)
5. Draw a generalized cross section of a four-week embryo, and label the parts. (p. 317)
6. What are the extraembryonic membranes of the chick and human embryo? What are their respective functions? (p. 318)
7. Give several ways in which birth defects can be prevented. (p. 322)
8. Trace the path of blood in the fetus from the umbilical vein to the aorta using two different routes. (p. 326)
9. Describe the three stages of parturition. (p. 327)
10. List and discuss the stages of development after birth. (p. 328)
11. Explain the current theories in regard to the causes of aging. (p. 329)
12. What are the major changes in body systems that have been observed as people age? (p. 330)

Objective Questions

1. When cells take on a specific structure and function, _____ occurs.
2. The morula becomes the _____, a structure that contains the inner cell mass.
3. The _____ membranes include the chorion, _____, yolk sac, and allantois.
4. The blastocyst _____ itself in the uterine lining.
5. The notochord _____ the formation of the nervous system.
6. During embryonic and fetal development, gas exchange occurs at the _____.
7. During development, there is a connection between the pulmonary artery and the aorta called the _____.
8. Fetal development begins with the _____ month.
9. The fontanels are commonly called _____.
10. In most births, the _____ appears before the rest of the body.
11. As we age, the proteins in the body undergo _____, a process that causes body parts to become stiff and rigid.
12. Most deaths are due to failure of the _____ system.

Label this Diagram.
See figure 16.7 (p. 318) in text.

Human

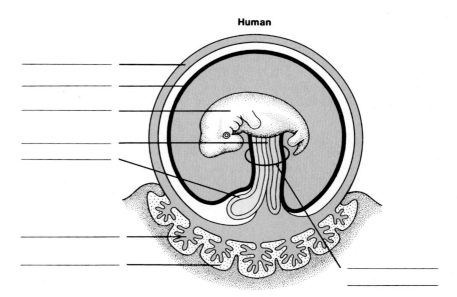

Answers

1. differentiation 2. blastocyst 3. extraembryonic, amnion 4. implants 5. induces 6. placenta 7. arterial duct 8. third 9. soft spots 10. head 11. cross-linking 12. cardiovascular

Critical Thinking Questions

1. Very often evolution makes use of previously evolved structures for a different purpose. Why does this seem reasonable, and what example of this methodology has been given in this chapter?

2. The theories on aging have been divided into genetic and whole body. Develop an overall theory that would draw from both of these types of theories.

3. Poor health habits contribute to loss of body functions as a person ages. What do you think people can do to improve their health and reduce loss of body functioning as they age?

Selected Key Terms

allantois (ah-lan'to-is)   one of the extraembryonic membranes; in reptiles and birds a pouch serving as a repository for nitrogenous waste; in mammals a source of blood vessels to and from the placenta. 318

amnion (am'ne-on)   an extraembryonic membrane; a sac around the embryo containing fluid. 313

chorion (ko're-on)   an extraembryonic membrane that forms an outer covering around the embryo; in reptiles and birds it functions in gas exchange; in mammals it contributes to the formation of the placenta. 312

differentiation (dif''er-en''she-a'shun)   the process and developmental stages by which a cell becomes specialized for a particular function. 312

extraembryonic membranes (eks''trah-em''bre-on'ik mem'brānz)   membranes that are not a part of the embryo but are necessary to the continued existence and health of the embryo. 312

gerontology (jer''on-tol'o-je)   the study of aging; those progressive changes that contribute to an increased risk of infirmity, disease, and death. 329

induction (in-duk'shun)   a process by which one tissue controls the development of another, as when the embryonic notochord induces the formation of the neural tube. 317

lanugo (lah-nu'go)   downy hair with which a fetus is born; fetal hair. 324

morphogenesis (mor''fo-jen'ĕ sis)   the movement of cells and tissues to establish the shape and structure of an organism. 312

parturition (par''tu-rish'un)   birth of a human and the expulsion of the extraembryonic membranes through the terminal portion of the female reproductive tract. 327

trophoblast (trof'o-blast)   the outer membrane that surrounds the human embryo and, when thickened by a layer of mesoderm, becomes the chorion, an extraembryonic membrane. 312

umbilical cord (um-bil'i-kal kord)   cord connecting the fetus to the placenta through which blood vessels pass. 319

vernix caseosa (ver'niks ka''se-o'sah)   cheeselike substance covering the skin of the fetus. 324

yolk sac (yōk sak)   one of the extraembryonic membranes within which yolk is found; in mammals, the first site of blood-cell formation in the embryo. 313

Further Readings for Part IV

Baconsfield, P., G. Birdwood, and R. Baconsfield. August 1980. The placenta. *Scientific American.*

Cerami, A. et al. May 1987. Glucose and aging. *Scientific American.*

DeRobertis, E. M., and J. B. Gurdon. December 1979. Gene transplantation and the analysis of development. *Scientific American.*

Frisch, R. E. March 1988. Fatness and fertility. *Scientific American.*

Goldberg, S., and B. DeVitto. 1983. *Born too soon: Preterm birth and early development.* San Francisco: W. H. Freeman and Company.

Guttmacher, A. F. 1973. *Pregnancy, birth, and family planning.* New York: New American Library.

Leach, P. 1980. *Your baby and child from birth to age five.* New York: Alfred A. Knopf.

Lein, A. 1979. *The cycling female: Her menstrual rhythm.* San Francisco: W. H. Freeman.

Mader, S. S. 1991. *Human reproductive biology,* 2d ed. Dubuque, IA: Wm. C. Brown Publishers.

Nilsson, L. 1977. *A child is born,* rev. ed. New York: Delacorte Press.

Oldstone, Michael B. A. August 1989. Viral alteration of cell function. *Scientific American.*

Rugh, R., et al. 1971. *From conception to birth: The drama of life's beginnings.* New York: Harper & Row, Publishers.

Scientific American. October 1988. Entire issue devoted to AIDS.

Ulmann, Andre, Georges Teutsch, and Daniel Philibert. June 1990. RU 486. *Scientific American.*

Vander, A. J., et al. 1980. *Human physiology,* 3d ed. New York: McGraw-Hill.

Volpe, E. P. 1983. *Biology and human concerns,* 3d ed. Dubuque, IA: Wm. C. Brown Publishers.

Wantz, M. S., and J. E. Gay. 1981. *The aging process: A health perspective.* Cambridge, MA: Winthrop.

Wassarman, P. M. December 1988. Fertilization in mammals. *Scientific American.*

Part Five

Human Genetics

Human beings practice sexual reproduction, which requires gamete production, fertilization, and zygote development. Gametes carry half the total number and different combinations of chromosomes. Therefore except for identical twins, no child receives exactly the same combination of chromosomes and genes as another. It is sometimes possible to determine the chances of an offspring receiving a particular gene and, therefore, inheriting a genetic disorder.

Genes, now known to be constructed of DNA, control not only the metabolism of the cell, but also, ultimately, the characteristics of the individual. DNA contains a code for the sequence of amino acids in proteins, which are synthesized at the ribosomes. The step-by-step procedure by which the DNA code is transcribed and translated to ensure the formation of a particular protein has been discovered.

It is possible to make recombinant DNA, a combination of DNA from two different sources. Because of this, it is possible to bioengineer bacteria to produce a human protein or bioengineer plants to have new characteristics, for example. This is the basis of the new and burgeoning field of biotechnology.

Cancer is a cellular disease brought on by DNA mutations that transform a normal cell into a cancer cell. Oncogenes or cancer-causing genes are derived from normal genes that keep cell division under control. Because environmental factors play a major role in causing genes to become oncogenes, the possibility exists of reducing the incidence of cancer. Knowledge about the detection and treatment of cancer is improving daily.

Chapter Seventeen

Chromosomes and Chromosomal Inheritance

Chapter Concepts

1 The nucleus contains the gene-bearing chromosomes, which occur as homologous pairs.

2 Ordinary cell division ensures that each new cell will receive a full complement of chromosomes and genes.

3 Reduction division is required to produce the sex cells, which contain half the full number of chromosomes. When the sperm (male sex cell) fertilizes the egg (female sex cell), the full number of chromosomes is restored.

4 Chromosomal inheritance includes the autosomal and sex chromosomes. Females are XX and males are XY.

5 The genes are on the chromosomes, and all possible combinations of chromosomes and genes occur among the gametes.

Each and every cell in your body has the same number and same kinds of chromosomes, a copy of the very ones inherited from your parents. How is this possible? When a cell divides, a copy of every chromosome goes to each daughter cell. Therefore, in this photo, there are two bunches of chromosomes, one for each daughter cell.

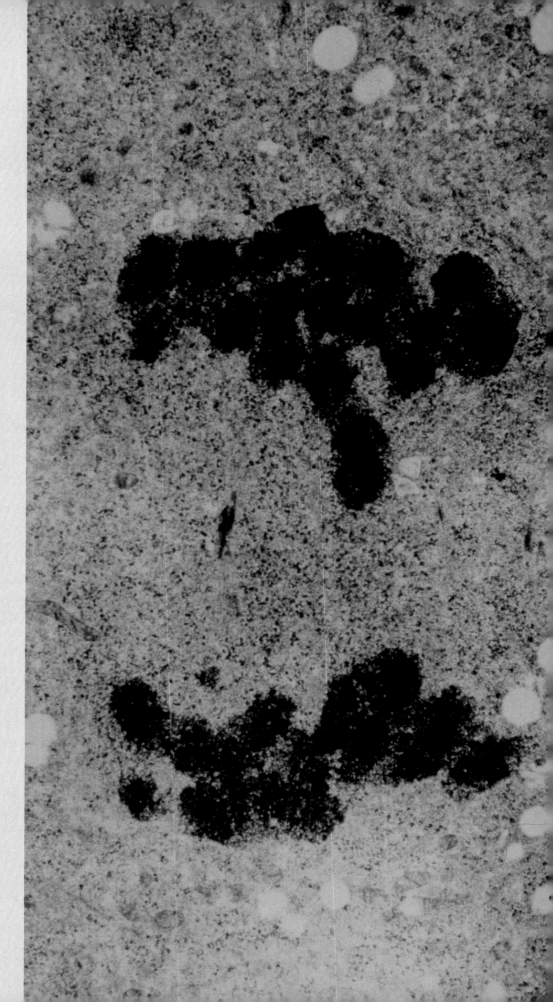

Figure 17.1   Most mothers are devoted to their children. Is this because children inherit half their chromosomes from their mothers?

Mother's love (fig. 17.1) has been extolled by poets and religious leaders for centuries as selfless devotion. However, behavioral geneticists have a possible alternative explanation. First of all, a woman's chances of having children are not as great as a man's—she produces only one egg a month for a limited number of years, but a man produces millions of sperm each day even into old age. Children have some of the same chromosomes as their mother, and therefore a child represents a way for her to be immortal in a biological sense. Then, too, you have to consider that a woman is certain in a way that no man can be that a child is hers. Although this explanation of motherly love is based on a certain amount of selfishness, the result is the same. Most mothers will do all in their power to protect and teach their children so that they can become productive members of society.

Chromosomes

Genes carried on chromosomes determine what the cell is like and what the individual is like. An examination of the body cells of a multicellular organism shows that all the nuclei have the same number of chromosomes. This number is characteristic of

the organism—corn plants have twenty chromosomes, houseflies have twelve, and humans have forty-six.

Karyotype

In a nondividing cell, there is indistinct and diffuse *chromatin* (fig. 2.1), but in a dividing cell chromatin becomes the short and thick *chromosomes*. A cell may be photographed just prior to division so that a picture of the chromosomes is obtained. The chromosomes may be cut out of the picture and arranged by pairs (fig. 17.2) of homologous chromosomes. **Homologous chromosomes** are recognized by the fact that each member of a pair is of the same size and has the same general appearance. The resulting display of pairs of chromosomes is called a **karyotype.** Although both males and females have twenty-three pairs of chromosomes, one of these pairs is of unequal length in males. The larger chromosome of this pair is called the X and the smaller is called the Y. Females have two X chromosomes in their karyotype. The X and Y chromosomes are called the **sex chromosomes** because they contain the genes that determine sex. The other chromosomes, known as **autosomes,** include all of the

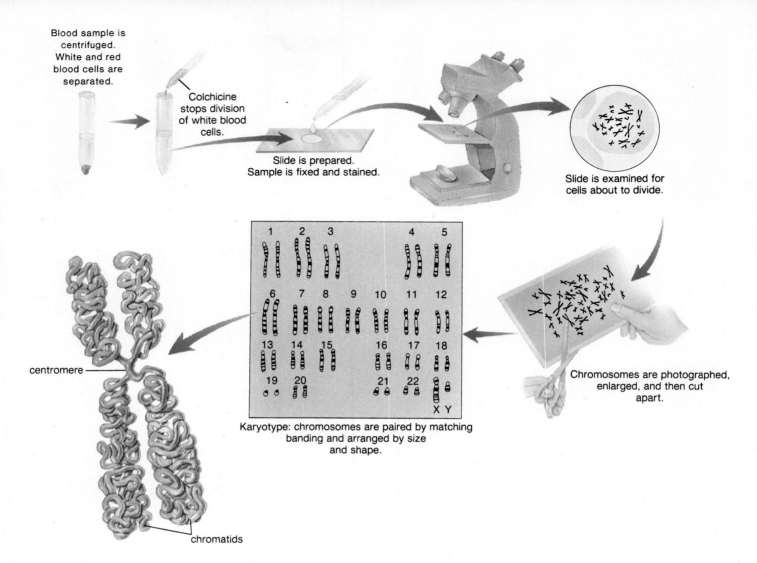

Blood sample is centrifuged. White and red blood cells are separated.

Colchicine stops division of white blood cells.

Slide is prepared. Sample is fixed and stained.

Slide is examined for cells about to divide.

Chromosomes are photographed, enlarged, and then cut apart.

Karyotype: chromosomes are paired by matching banding and arranged by size and shape.

centromere

chromatids

Figure 17.2 Human karyotype preparation. As illustrated here, the stain used can result in chromosomes with a banded appearance. The bands help researchers to identify and to analyze the chromosomes. The enlargement of a chromosome on the far left shows that the chromosomes in a karyotype consist of 2 chromatids held together by a centromere. This is the appearance of chromosomes just before they divide.

pairs of chromosomes except the X and Y chromosomes. Each pair of autosomes in the human karyotype is numbered.

Notice, as further illustrated in figure 17.2, that prior to division each chromosome is composed of two identical parts, called **chromatids.** These two sister (twin) chromatids are genetically identical and contain the same *genes,* the units of heredity that control the cell. The chromatids are held together at a region called the **centromere.**

A karyotype shows the individual's total number of chromosomes arranged by homologous pairs.

Life Cycle

Two types of cell division occur during the human life cycle: *mitosis* and *meiosis* (fig. 17.3). Meiosis (reduction division) occurs as a part of gametogenesis, the production of **gametes,** which is a term used collectively to mean the sperm and egg. Because of meiosis, the sperm contains twenty-three chromosomes and the egg contains twenty-three chromosomes. A new individual comes into existence when the sperm of the male fertilizes the egg of the female. The resulting **zygote** contains twenty-three pairs of chromosomes; one of each pair was contributed by the father, and one of each pair was contributed by the mother. As the zygote grows to become the adult, mitosis occurs so that each and every cell has forty-six chromosomes. In this way, each body cell contains the full complement of chromosomes and genes. The full complement of chromosomes is called the **diploid,** or *2N,* number of chromosomes.

Meiosis occurs only in the sex organs, or **gonads**—the testes in males and the ovaries in females. Here, diploid germ cells develop into the gametes that have half the total number, called the **haploid,** or *N,* number of chromosomes. The haploid number always has one of each kind of chromosome. When a haploid

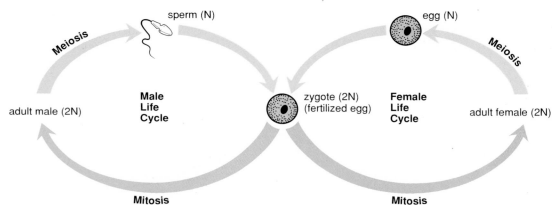

Figure 17.3   Life cycle of humans. Notice that there are 2 types of cell division—mitosis and meiosis. For the male life cycle, read as follows: After the sperm fertilizes the egg, the diploid zygote divides mitotically as the individual becomes the adult male. Then, meiosis occurs during the production of haploid sperm. For the female life cycle, read as follows: After the sperm fertilizes the egg, the zygote divides mitotically as the individual becomes the adult female. Then, meiosis occurs during the production of haploid eggs.

Table 17.1   Mitosis versus Meiosis

Cell type	Cell division	Description	Result
somatic or body cells	mitosis	2N (diploid) → 2N (diploid)	more body cells = growth
germ cells in gonads of animals	meiosis	2N (diploid) → N (haploid)	gamete or sex cell production

sperm fertilizes a haploid egg, the new individual has the diploid number of chromosomes, half of which came from the father and half of which came from the mother. Table 17.1 summarizes the major differences between mitosis and meiosis.

The life cycle of humans requires two types of cell divisions: mitosis and meiosis. Mitosis is responsible for growth and repair, while meiosis is required for gamete production.

Mitosis

Overview

Mitosis[1] *is cell division in which the daughter cells retain the same number and kinds of chromosomes as the mother cell.* Therefore, the newly formed cells are genetically identical. The

[1]The term *mitosis* technically refers only to nuclear division but for convenience is used here to refer to division of the entire cell.

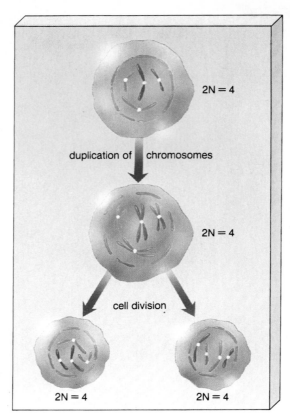

Figure 17.4 Mitosis overview. Following duplication, each chromosome in the mother cell contains 2 sister chromatids. During mitotic division, the sister chromatids separate so that daughter cells have the same number and kinds of chromosomes as the mother cell.

mother cell is the cell that divides, and the *daughter cells* are the resulting cells. Although *humans have forty-six chromosomes,* each cell in figure 17.4 contains only four chromosomes for simplicity's sake. (In determining the number of chromosomes, it is necessary to count only the number of independent centromeres.) A cell prepares for mitosis by replication of the genetic material contained within each chromosome. **Replication** is the process by which DNA makes a copy of itself, as is described in detail in chapter 19. Because of replication, each chromosome contains two sister chromatids. During mitosis, the sister chromatids separate, ensuring that each new cell will receive a copy of each chromosome rather than two copies of one chromosome and none of another. Different genes are on different chromosomes, and it is necessary for every cell to have a copy of each chromosome in order to have a full complement of genes. As an aid in describing the events of mitosis, the process has been divided into four phases: prophase, metaphase, anaphase, and telophase (fig. 17.5).

Mitosis ensures that each cell in the body is genetically identical. At the time of division, a chromosome consists of sister chromatids. When these separate, each newly forming cell receives the same number and kinds of chromosomes as the original cell.

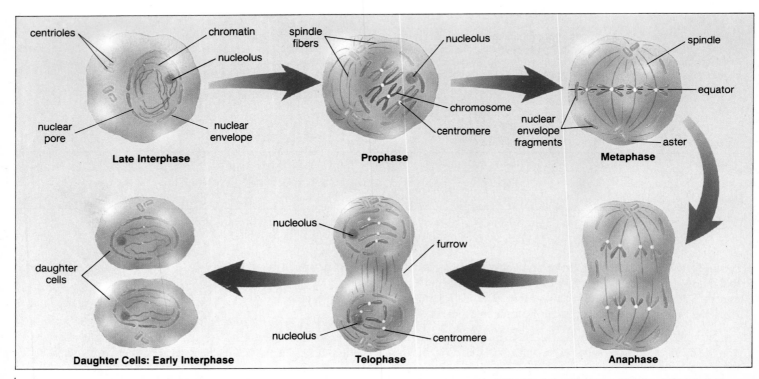

Figure 17.5 Mitotic cell division. Mitosis has 4 stages: prophase, metaphase, anaphase, and telophase. Interphase, which occurs between divisions, is not part of mitosis. Notice that the centrioles duplicate during interphase so that there are 2 pairs at the start of mitosis (compare early interphase to late interphase).

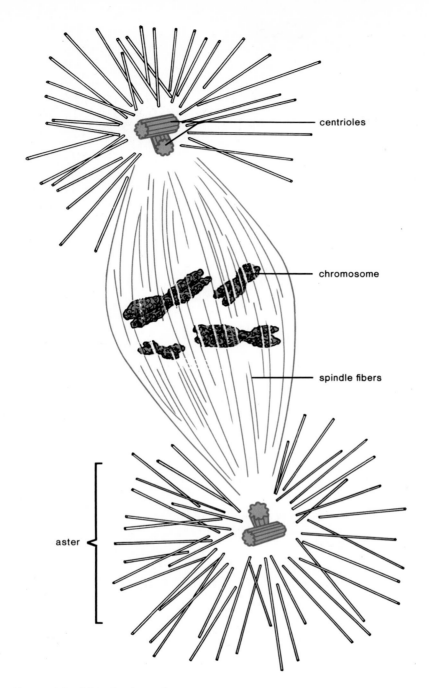

centrioles

chromosome

spindle fibers

aster

Figure 17.6   The spindle apparatus consists of the structures shown. Each spindle fiber is a bundle of microtubules. Some believe centrioles are involved in the production of microtubules and, therefore, of the spindle.

Stages of Mitosis

Prophase

It is apparent during **prophase** that cell division is about to occur. The chromosomes are now visible, and there are two pairs of centrioles outside the nucleus. As the pairs of centrioles begin separating and moving toward opposite ends of the nucleus, **spindle fibers** appear between them. The nuclear envelope begins to fragment, and the nucleolus begins to disappear.

Toward the end of prophase, the chromosomes are randomly placed even though the spindle appears to be fully formed.

The Function of Centrioles   The entire spindle apparatus is shown in figure 17.6. It consists of asters, spindle fibers, and centrioles. Both the short **asters** radiating from the centrioles and the long spindle fibers are composed of microtubules. It is known that microtubules are capable of assembling and disassembling, which accounts for the appearance and the disappearance of asters and spindle fibers. It is possible that the centrioles are a part of organizing centers for spindle formation, but it also could be that their location at the poles simply ensures that each daughter cell receives a pair of centrioles.

Metaphase

During **metaphase,** the nuclear envelope is fragmented and the spindle occupies the region formerly occupied by the nucleus. Each chromosome is attached to the spindle and moves to the equator (center) of the spindle. Metaphase is characterized by a fully formed spindle, with the chromosomes, each composed of two chromatids, aligned at the equator (fig. 17.7*a*). At the close of metaphase, the centromeres uniting the chromatids split.

Anaphase

During **anaphase,** the chromatids separate (fig. 17.7*b*). *Once separated, the chromatids are called chromosomes.* The chromosomes now move up (*ana* means up) to the opposite poles of the spindle. Separation of the sister chromatids ensures that each cell receives a copy of each type of chromosome and thereby has a full complement of genes. As the newly formed chromosomes move to opposite poles, the entire cell elongates (fig. 17.5).

Telophase

Telophase begins when the chromosomes arrive at the poles. During **telophase** the spindle disappears, possibly due to disassembly of the microtubules making up the spindle fibers. As the nuclear envelopes re-form and the nucleoli reappear in each daughter cell, the chromosomes become indistinct chromatin again. Following nuclear division, cytoplasmic division, sometimes called **cytokinesis,** usually occurs. In animal cells, _furrowing,_ or an indentation of the membrane between the two daughter cells, divides the cytoplasm. Furrowing is complete when each daughter cell has a complete membrane enclosing it. Microfilaments are believed to take part in the furrowing process since they are always in the vicinity.

Table 17.2 is a summary of the stages of mitosis.

Cell Cycle

The **cell cycle** (fig. 17.8) includes interphase and the four stages of division. During interphase, a cell that remains nonspecialized will resemble figure 2.1; however, some cells become differentiated and then they resemble specialized cells. Each type of specialized cell has a characteristic life span. For example, red blood cells live about 120 days, but many nerve cells live as long as the individual does.

During interphase, nonspecialized cells prepare to divide again; that is, they prepare to complete the cell cycle. DNA replicates as the chromosomes duplicate and the centrioles also duplicate at this time. There is a limit to the number of times any human cell will divide before death follows degenerative changes. Most will divide about 50 times, and only cancer cells retain the ability to divide repeatedly. Cancer cells have abnormal chromosomes and other abnormalities of cell structure. Aging of cells in the individual appears to be a normal process most likely controlled by the nucleus. In the laboratory, aging of cells can be delayed by environmental circumstances such as reduced temperature, but it cannot be postponed indefinitely.

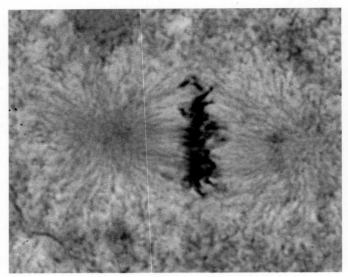

a.

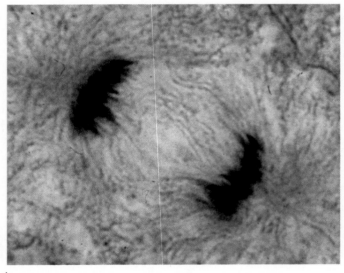

b.

Figure 17.7 Photomicrographs of cells undergoing mitosis. *a.* During metaphase, the chromosomes are lined up along the equator of the spindle. *b.* During anaphase, the separation of sister chromatids results in chromosomes that are pulled by spindle fibers to opposite poles of the spindle.

Table 17.2   Stages of Mitosis

Stage	Events
prophase	Replication has occurred, and each chromosome is composed of a pair of sister chromatids.
metaphase	Chromatid pairs are at the equator of the cell.
anaphase	Sister chromatids separate, and each one is now termed a chromosome.
telophase	Each pole has the same number and kinds of chromosomes as the mother cell.

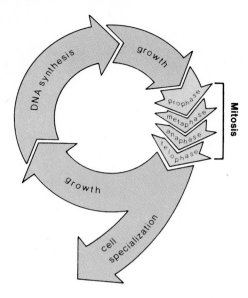

Figure 17.8   The cell cycle consists of mitosis and interphase. During interphase, there is growth before and after DNA synthesis. DNA synthesis is required for replication, the process by which DNA is duplicated. Some daughter cells "break out" of the cell cycle and become specialized cells performing a specific function.

The cell cycle includes interphase and the four stages of division: prophase, metaphase, anaphase, and telophase. Some cells break out of the cell cycle, and some prepare to enter the cycle again.

Importance of Mitosis

Mitosis assures that each body cell has the same number and kinds of chromosomes. It is important to the growth and repair of multicellular organisms. When a baby develops in its mother's womb, mitosis occurs as a component of growth. As a wound heals, mitosis occurs to repair the damage.

Mitosis also occurs during the process of asexual reproduction. In lower animals, a group of cells called a bud can give rise to an entire individual. This new individual has the same genes and is identical to the parent individual. The term *cloning* is sometimes used to refer to the asexual production of individuals from mature cells of donor animals such as mice or even humans. Thus far, only embryonic mammalian cells seem to retain the capability of beginning development again. An adult cell has an entire set of genes but is too specialized to begin development. Research is still going on to overcome this difficulty so that an animal can be cloned from an adult cell.

Human Issue

If one day it is possible for humans to make clones of themselves, should each and every person be allowed to do so? Or should only those persons who are judged to have the best of personal attributes be allowed to clone themselves? Who would make judgments about the worthiness of individuals to be cloned?

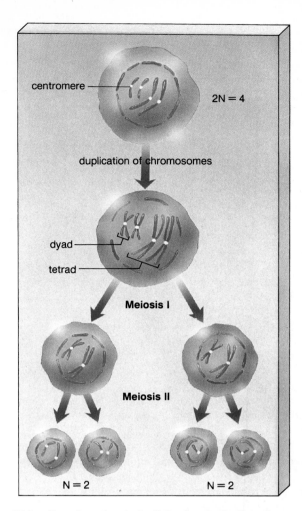

Figure 17.9   Overview of meiosis. Following duplication of chromosomes, the mother cell undergoes 2 divisions, meiosis I and meiosis II. During meiosis I, homologous chromosomes separate; during meiosis II, sister chromatids separate. The final daughter cells are haploid.

Meiosis

Overview of Meiosis

Meiosis, which requires two cell divisions, results in *four daughter cells, each having one of each kind of chromosome and therefore half the number of chromosomes as the mother cell.*[2] Figure 17.9 presents an overview of meiosis, indicating the two cell divisions, **meiosis I** and **meiosis II.** Prior to meiosis I, replication has occurred and each chromosome consists of sister chromatids held together at a centromere. Therefore, each chromosome can be called a **dyad.** During meiosis I, the homologous chromosomes come together and line up side by side due to a means of attraction still unknown. This is called **synapsis** and results in **tetrads,** an association of four chromatids that stay in close proximity during the first two phases of meiosis

---

[2]The term *meiosis* technically refers only to nuclear division but for convenience is used here to refer to division of the entire cell.

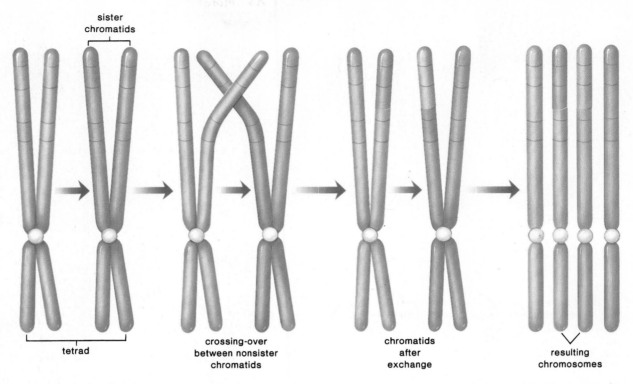

sister
chromatids

tetrad

crossing-over
between nonsister
chromatids

chromatids
after
exchange

resulting
chromosomes

Figure 17.10   Crossing over. When homologous chromosomes are in synapsis, the nonsister chromatids exchange genetic material. Following crossing over, there is a different combination of genes on the chromatids and resulting chromosomes.

I. Now an important process called **crossing over** may occur (fig. 17.10). During crossing over, nonsister chromatids of the homologous chromosome pairs exchange segments. This produces new combinations of genes on the chromatids.

Following synapsis during meiosis I, the homologous chromosomes separate. This separation means that one chromosome of every homologous pair reaches each gamete. There are no restrictions on the separation process; either chromosome of a homologous pair can occur in a gamete with either chromosome of any other pair. Therefore, any possible combination of chromosomes and genes can occur within the gametes.

Notice that at the completion of meiosis I (fig. 17.9), the chromosomes are still dyads. No replication of DNA nor duplication of chromosomes occurs between meiosis I and meiosis II. During meiosis II, the sister chromatids separate, resulting in four daughter cells. Each daughter cell has the haploid number of chromosomes. These chromosomes are singled chromosomes. You can count the number of centromeres to verify that the mother cell has the diploid number of chromosomes and each of the daughter cells has the haploid number.

Meiosis is cell division that halves the chromosome number. The process requires two divisions and results in four daughter cells, each with the haploid number of chromosomes.

Stages of Meiosis

The stages of meiosis I are diagrammed in figure 17.11a. During **prophase I,** the homologous chromosomes of each pair undergo synapsis, forming tetrads. The nuclear envelope and nucleolus do not disappear until the end of prophase I. At **metaphase I,** tetrads line up at the equator of the spindle. During **anaphase I,** the homologous chromosomes of each pair separate, and the chromosomes (each still composed of sister chromatids) move to the poles of the spindle. Each pole receives half of the total number of chromosomes. In **telophase I,** the nuclear envelope reforms and the nucleoli reappear. There is no replication of DNA between meiosis I and meiosis II.

During meiosis I, the nonsister chromatids within a tetrad exchange chromosome pieces. When the homologous chromosomes separate, each daughter cell receives one from each pair of chromosomes.

In the second division of meiosis (fig. 17.11b), the phases are referred to as prophase II, metaphase II, anaphase II, and telophase II. During **anaphase II,** centromere separation occurs, and the sister chromatids part to become independent chromosomes. At the end of **telophase II,** there are four cells. Each of these four cells is haploid; that is, the nucleus has half the number of chromosomes and half the DNA content of the mother cell nucleus.

Separation of sister chromatids during meiosis II produces a total of four daughter cells, each with the haploid number of chromosomes. Figure 17.12 contrasts mitosis to meiosis.

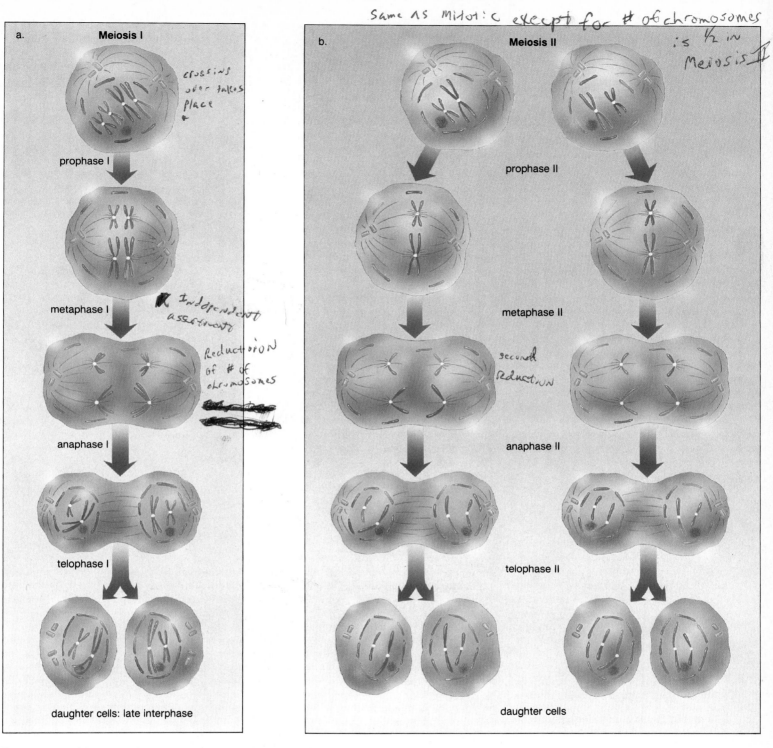

a. **Meiosis I**

crossing over takes place

prophase I

metaphase I

Independent assortment

Reduction of # of chromosomes

anaphase I

telophase I

daughter cells: late interphase

b. **Meiosis II**

Same as Mitotic except for # of chromosomes is ½ in Meiosis II

prophase II

metaphase II

second Reduction

anaphase II

telophase II

daughter cells

Figure 17.11   Meiosis. *a.* Stages of meiosis I. During meiosis I, homologous chromosomes separate so that each daughter cell has only one chromosome from each original homologous pair. For simplicity, the results of crossing over have not been depicted. *b.* Stages of meiosis II. During meiosis II, sister chromatids separate, and each daughter cell has the haploid number of chromosomes in single copy.

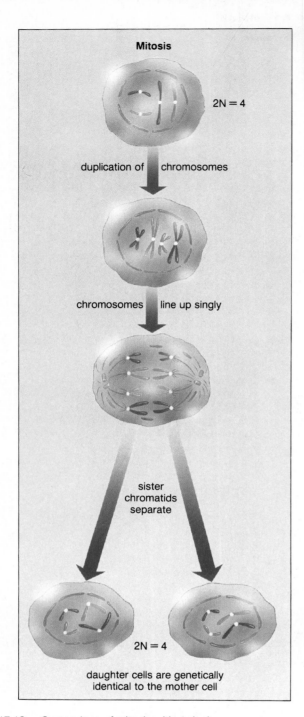

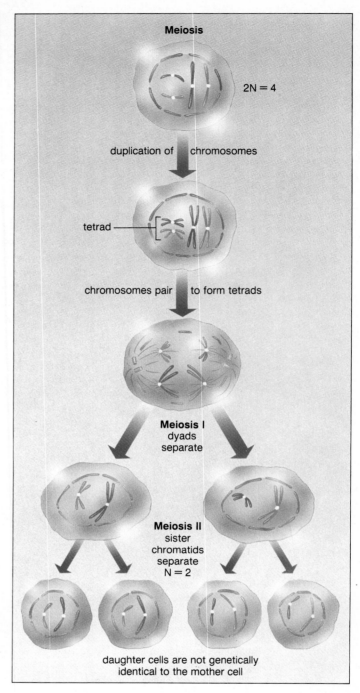

Figure 17.12   Comparison of mitosis with meiosis.

Spermatogenesis and Oogenesis

Spermatogenesis, sperm production, takes place in the testes of males. **Oogenesis,** egg production, takes place in the ovaries of females. Gamete production is different in the two sexes. Figure 17.13 shows that for each meiosis, there result four viable sperm in males. Also, spermatogenesis occurs continuously, and at the time of ejaculation, males emit as many as 400 million or more sperm.

In contrast to spermatogenesis in males, egg production occurs only once a month in females. The first meiotic division produces two cells, but one is much larger than the other. The smaller nonfunctional cell is called a **polar body** and remains attached to the large cell. At this point, ovulation occurs and the immature egg enters the oviduct (fig. 16.1) in females. The second meiotic division does not occur, and unless fertilization takes place, oogenesis is halted. If oogenesis continues, the second

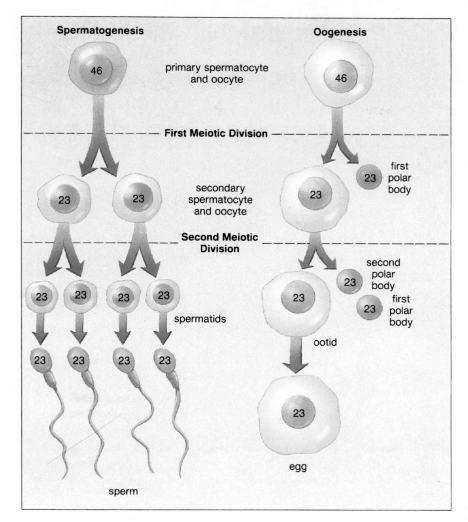

Spermatogenesis produces 4 viable sperm, and oogenesis produces one egg and at least 2 polar bodies. In humans, both sperm and egg have 23 chromosomes each; therefore, following fertilization the zygote has 46 chromosomes.

meiotic division is also unequal, so that in the end there is only one mature egg and at least two nonfunctional polar bodies that disintegrate.

Figure 17.13 shows how the sperm and egg are adapted to their function. The sperm is a tiny, flagellated cell that is adapted for swimming to the maturing egg, a large cell that contributes most of the cytoplasm and nutrients to the new individual.

Importance of Meiosis

Meiosis is nature's way of keeping the chromosome number constant from generation to generation. It assures that the next generation will have a different genetic makeup than that of the previous generation. As a result of independent assortment of chromosomes and crossing over, the gametes carry a new combination of genes. The egg carries half of the genes from the female parent and the sperm carries half of the genes from the

male parent. When the sperm fertilizes the egg, the zygote has a different combination of genes than either parent. In this way, meiosis assures genetic variation generation after generation.

Spermatogenesis in males produces four viable sperm, but oogenesis in females produces one egg and at least two polar bodies. Each gamete is specialized for the job it does; the sperm is a tiny, flagellated cell that is propelled to the cytoplasm-laden egg.

Chromosomal Inheritance

Normal Inheritance

The individual normally receives twenty-two autosomal chromosomes and one sex chromosome from each parent. The sex of the newborn child is determined by the father. If a Y-bearing

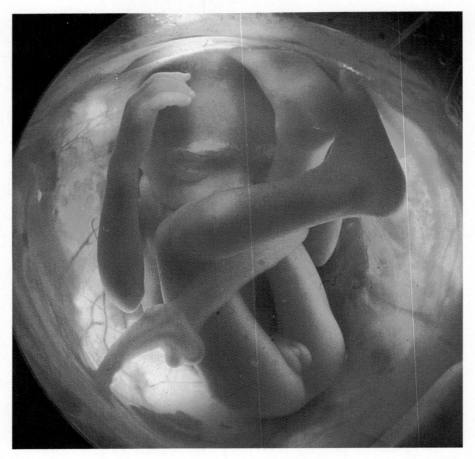

Figure 17.14   It's a girl. According to Dr. Shettles' somewhat controversial method of choosing the sex of your child, the X-bearing sperm is favored over the Y-bearing sperm if these requirements are met: (1) the vagina is acidic (a douche consisting of two tablespoons of white vinegar to a quart of water promotes this); (2) intercourse should be frequent, and penetration is shallow but ceases two or three days before ovulation. On the other hand, the Y-bearing sperm is favored over the X-bearing sperm if the following requirements are met: (1) the vagina is alkaline (a douche consisting of two tablespoons of baking soda to a quart of water promotes this; let it stand 15 minutes before using); (2) abstinence is practiced until the day of ovulation, when penetration is deep.

sperm fertilizes the egg, then the XY combination will result in the development of a male. On the other hand, if an X-bearing sperm fertilizes the egg, the XX combination results in the development of a female. All factors being equal, there is a 50% chance of having a girl or boy (fig. 17.14). It is possible to illustrate this probability by doing a **Punnett square** (fig. 17.15). In the square, all possible sperm are lined up on one side; all possible eggs are lined up on the other side (or vice versa), and every possible combination is determined. When this is done with regard to sex chromosomes, the results show one female to each male. However, for reasons that are not clear, more males than females are conceived. But from then on the death rate among males is higher; more males than females are spontaneously aborted, and this trend continues after birth until there is a dramatic reversal of the ratio of males to females (table 17.3).

The sex of a child is dependent on whether a Y-bearing or an X-bearing sperm fertilizes the X-bearing egg.

Human Issue

In a recent poll of Americans, about two-thirds of the people interviewed thought it is alright to know the sex of their child before birth, and about one-fourth thought it is justifiable to treat the father's sperm in a medical laboratory to increase the odds of conceiving a boy or a girl. Also, one person out of twenty thought abortions are justified if the child is not of the desired gender.

Should the sex of a child be of major importance? For example, is a man more manly if he has a son? Should couples simply take their chances as to the sex of their child, or is it better to predetermine the sex? What methods attempting to control the sex of children do you think should be banned?

Male Gene

For some years, it has been proposed that there is a gene located on the Y chromosome that brings about maleness. Embryos begin life with no evidence of a sex, but then along about the

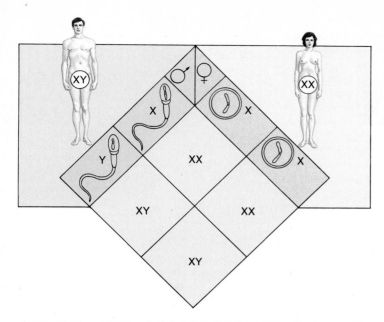

Figure 17.15 Inheritance of sex. An offspring is either male or female depending on whether an X or a Y chromosome is received from the male parent. In this Punnett square, the sperm and eggs are shown as carrying only a sex chromosome. They also carry 22 autosomes.

Table 17.3 Ratio of Males to Females in the United States

Age	Sex ratio
birth	106:100
18 years	100:100
50 years	85:100
85 years	50:100
100 years	20:100

third month of development males can be distinguished from females. Investigators have recently reported the finding of a gene they call the testis determining factor gene (TDF) on the Y chromosome. When this gene is lacking from the Y chromosome, the individual is a female even though the chromosomal inheritance is XY. On the other hand, if the gene is present in an XX individual, this person is a male. The reading on page 350 discusses the difficulty in identifying males and females by cell examination.

Abnormal Autosomal Chromosome Inheritance

Sometimes individuals are born with either too many or too few autosomal chromosomes due most likely to nondisjunction of chromosomes or sister chromatids during meiosis (fig. 17.16). It is possible also that even though there is the correct number of chromosomes, one chromosome may be defective in some way because of a chromosomal mutation (fig. 17.17). A **mutation** is a permanent change in the genetic material. *Chromosomal mutations* are known to occur after chromosomes are broken due to exposure to radiation, addictive drugs, or pesticides, for example. When the chromosomes re-form, the pieces may be rearranged. An **inversion** results when a piece of a chromosome has turned in a direction opposite to the usual. A **deletion** occurs when a piece of a chromosome is lost and the chromosome is shorter than it was formerly. In contrast, a **duplication** is the presence of a chromosome piece more than once in the same chromosome. Deletion and duplication are apt to occur between homologous chromosomes. When nonhomologous chromosomes exchange pieces, a **translocation** has taken place. The presence of a mutated chromosome can cause the individual to have reproductive problems because the abnormal chromosome can result in nonviable zygotes or a child with birth defects.

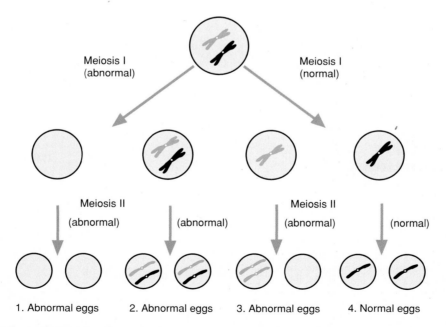

Figure 17.16 Nondisjunction during oogenesis. Nondisjunction can occur during meiosis I if the homologous chromosomes pairs fail to separate and during meiosis II if the chromatids fail to separate completely. In either case, the abnormal eggs carry an extra chromosome. Nondisjunction of the number-21 chromosome leads to Down syndrome.

Male/Female Cell Differences

On occasion, such as at athletic competitions, it is important to be able to certify that an individual is a male or a female. Since physical examinations sometimes fail—as when a male has had a sex-change operation—officials often resort to examining the cells themselves.

You could, of course, do a karyotype, but there are shorter methods. It so happens that XX females have small, darkly staining masses of condensed chromatin, called Barr bodies (named after the person who first identified them), present in their nuclei (see figure). XY males have no comparable spots of chromatin in their nuclei. It turns out that a Barr body[1] is a condensed and at least to some degree inactive X chromosome as was proposed by Mary Lyon. The validity of the *Lyon hypothesis* means that female cells function with a single X chromosome just as males do. Still, in some cells one X is condensed and in some cells the other X is condensed so that the female body is a mosaic of genetically different cells.

No doubt, you believe that observation of Barr bodies is not a guarantee of femaleness because we have already indicated that there are XX males, albeit rarely. We need another test, and there is one. There is an antigen, called an H-Y antigen, present in the cell membrane of males but not in females. It is called an antigen because females produce antibodies against it. To test for maleness, it is possible to

[1]How many Barr bodies would a person with Klinefelter syndrome have? A metafemale have? (p. 353)

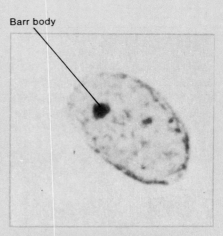

Barr body

Each female cell contains a Barr body. A Barr body is a condensed X chromosome.

suspend a sample of white blood cells in a solution that contains some of these antibodies. If the cells carry the H-Y antigen, indicating that the person is a male, the antibodies bind with them. Now, we can be certain who is a male and who is a female.

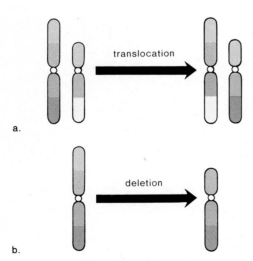

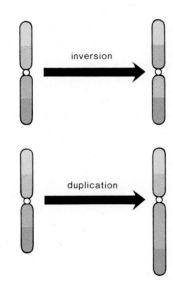

Figure 17.17   Types of chromosome mutation. *a.* Translocation is the exchange of chromosome pieces between nonhomologous parts. *b.* Deletion is the loss of a chromosome piece. *c.* Inversion occurs when a piece of a chromosome breaks loose and then rejoins in a reversed direction. *d.* Duplication occurs when the same piece is repeated within the chromosome.

a.

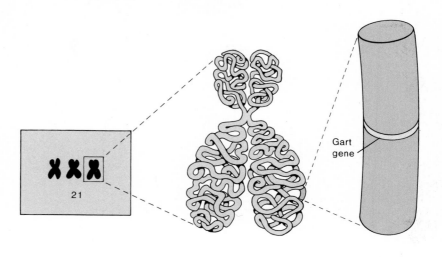

b.

Figure 17.18   Down syndrome. *a.* Common characteristics include a wide, rounded face and a fold of the upper eyelids. Mental retardation, along with an enlarged tongue, makes it difficult for persons with Down syndrome to learn to speak coherently. *b.* Karyotype of an individual with Down syndrome has an extra number-21 chromosome. More sophisticated technologies allow investigators to pinpoint the location of specific genes associated with the syndrome. The *Gart* gene, which leads to a high level of blood purines, may account for the mental retardation seen in persons with Down syndrome.

Sometimes an individual inherits too many or too few chromosomes or a defective chromosome.

Syndromes

A **syndrome** is a group or pattern of symptoms that occur together in the same individual due to the presence of an abnormal condition. We will be discussing various syndromes that result from the inheritance of chromosomal abnormalities.

Down Syndrome   The most common autosomal abnormality is seen in individuals with **Down syndrome** (fig. 17.18). This syndrome is easily recognized. Its characteristics include a short stature; an oriental-like fold of the eyelids; stubby fingers; a wide gap between the first and second toes; a large, fissured tongue; a round head; a palm crease, the so-called simian line; and, unfortunately, mental retardation that sometimes can be severe.

Persons with Down syndrome usually have three number-21 chromosomes because the egg had two number-21 chromosomes instead of one (fig. 17.18). (In 23% of the cases studied, however, the sperm had the extra number-21 chromosome.) It would appear that **nondisjunction,** a failure of the chromosome

Table 17.4   Incidence of Selected Chromosomal Abnormalities

Name	Frequency/100,000 live births
Down syndrome (general)	140
Down syndrome (mothers over age 40)	1,000
Turner syndrome	8
Metafemale	50
Klinefelter syndrome	80
XYY	100

Source: *Antenatal Diagnosis,* HEW, 1979, pp. 1–48.

pairs or chromatids to separate completely, is most apt to occur in the older female since children with Down syndrome are usually born to women over age 40 (table 17.4). If a woman wishes to know whether or not her unborn child is affected by Down syndrome, she may elect to undergo chorionic villi testing or amniocentesis, two procedures discussed in the reading on page 322. Following this procedure, a karyotype can reveal whether the child has Down syndrome. If the child has Down syndrome, she may elect to continue or to abort the pregnancy.

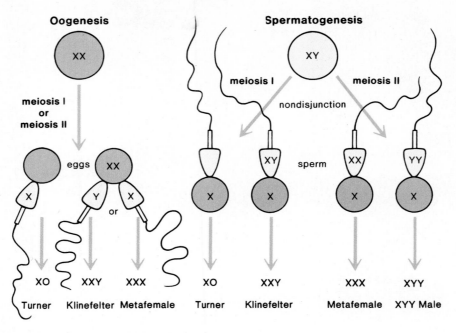

Figure 17.19 Nondisjunction of sex chromosomes during oogenesis followed by fertilization with normal sperm results in the conditions noted. Nondisjunction of sex chromosomes during spermatogenesis followed by fertilization of normal eggs results in the conditions noted.

Human Issue

Technological advancements associated with prenatal screening pose a number of ethical dilemmas for parents and their physicians. For example, recently a woman sued her obstetrician for not telling her that her child would be born with Down syndrome. She won her case, and the obstetrician has to pay for the support of the child. This illustrates one dilemma confronting physicians—are they obligated to inform a pregnant woman of any birth defects, or is it up to the woman to express her desire to know of any? Parents face potential dilemmas as well. For example, is it better for couples to remain uninformed about the results of the prenatal screening so that they don't put themselves through the agony of having to decide about an abortion, or is it better to know so that they can be mentally prepared for a child with an abnormality?

It is known that the genes that cause Down syndrome are located on the bottom third of the number-21 chromosome (fig. 17.18b), and there has been a lot of investigative work to discover the specific genes responsible for the characteristics of the syndrome. Thus far, investigators have discovered several genes that may account for various conditions seen in persons with Down syndrome. For example, they have located genes most likely responsible for the increased tendency toward leukemia, cataracts, accelerated rate of aging, and mental retardation. The latter gene, dubbed the *Gart* gene, causes an increased level of purines in the blood, a finding that is associated with mental retardation. It is hoped that it will someday be possible to find a way to control the expression of the *Gart* gene even before birth so that at least this symptom of Down syndrome will not appear.

Cri du Chat Syndrome A chromosomal deletion is responsible for **cri du chat** (cat's cry) **syndrome.** Affected individuals meow like a kitten when they cry, but more important perhaps is the fact that they tend to have a small head with malformations of the face and body and that mental defectiveness usually causes retarded development. Chromosomal analysis shows that a portion of one number-5 chromosome is missing (deleted) while the other number-5 chromosome is normal.

Down syndrome is most often due to the inheritance of an extra number-21 chromosome, and cri du chat syndrome is due to the inheritance of a defective number-5 chromosome.

Abnormal Sexual Chromosome Inheritance

Abnormal sexual chromosome constituencies (table 17.4) are also due to the occurrence of nondisjunction. Nondisjunction of the sex chromosomes during oogenesis can lead to an egg with either two X chromosomes or no X chromosomes. Nondisjunction of the sex chromosomes during spermatogenesis can result in a sperm that has no sex chromosome, both an X and a Y chromosome, two X chromosomes, or two Y chromosomes. Assuming that the other gamete is normal, the zygote could develop into an individual with one of the conditions noted in figure 17.19.

Sometimes a person inherits an abnormal combination of sex chromosomes due to nondisjunction of these chromosomes during meiosis.

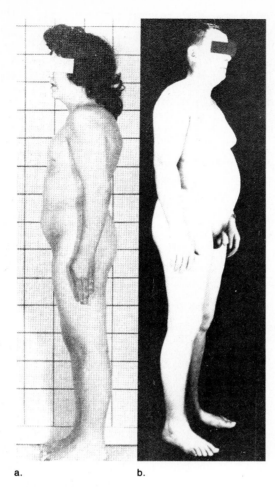

a.   b.

Figure 17.20   Abnormal sex chromosome inheritance. *a.* Female with
Turner (XO) syndrome, which is marked by a bull neck, short stature,
and immature sexual features. *b.* A male with Klinefelter (XXY)
syndrome, which is marked by immature sex organs and development
of the breasts.

Abnormalities

An XO individual with **Turner syndrome** has only one sex chromosome, an X; the O signifies the absence of the second sex chromosome. Because the ovaries never become functional, these females do not undergo puberty or menstruate, and there is a lack of breast development (fig. 17.20a). Generally, these individuals have a stocky build and a webbed neck. They also have difficulty recognizing various spatial patterns.

When an egg having two X chromosomes is fertilized by an X-bearing sperm, a **metafemale** having three X chromosomes results. It might be supposed that the XXX female with forty-seven chromosomes would be especially feminine, but this is not the case. Although there is a tendency toward learning disabilities, most metafemales have no apparent physical abnormalities and many are fertile and have children with a normal chromosome count.

When an egg having two X chromosomes is fertilized by a Y-bearing sperm, a male with **Klinefelter syndrome** results. This individual is male in general appearance, but the testes are underdeveloped and the breasts may be enlarged (fig. 17.20b). The limbs of these XXY males tend to be longer than average, body hair is sparse, and many have learning disabilities.

XYY males also occur, possibly due to nondisjunction during spermatogenesis. These males are usually taller than average, suffer from persistent acne, and tend to have barely normal intelligence. At one time, it was suggested that these men were likely to be criminally aggressive, but it has been shown that the incidence of such behavior is no greater than that among normal XY males.

Individuals are sometimes born with the sex chromosomes XO (Turner syndrome), XXX (metafemale), XXY (Klinefelter syndrome), and XYY. Individuals with a Y chromosome are always male no matter how many X chromosomes there may be; however, at least one X chromosome is needed for survival.

Summary

The life cycle of higher organisms requires two types of cell divisions, mitosis and meiosis. Mitosis assures that all cells in the body have the diploid number and same kinds of chromosomes. It is made up of four stages: prophase, metaphase, anaphase, and telophase. The cell cycle includes an additional stage termed interphase. During interphase, DNA replication causes each chromosome to have sister chromatids. When mitosis occurs, the chromatids separate, and each newly forming cell receives the same number and kinds of chromosomes as the original cell. The cytoplasm is partitioned by furrowing in human cells.

Meiosis involves two cell divisions. During meiosis I, the homologous chromosomes (following crossing over between nonsister chromatids) separate, and during meiosis II the sister chromatids separate. The result is four cells with the haploid number of chromosomes in single copy. Meiosis is a part of gamete formation in humans. Mitosis is contrasted to meiosis in figure 17.12.

An individual ordinarily receives twenty-two autosomal and one sex chromosome from each parent, but abnormalities do occur. Also, a mutated chromosome can be inherited; inversions, deletions, translocations, and duplications of chromosomal parts are known. The major inherited autosomal abnormality is Down syndrome in which the individual inherits three number-21 chromosomes due to nondisjunction during gamete formation. Examples of abnormal sex chromosome inheritance due to nondisjunction are Turner syndrome (XO), Klinefelter syndrome (XXY), XYY males, and metafemales (XXX).

Study Questions

1. Describe the normal karyotype of a human being. What is the difference between a male and a female karyotype? (p. 337)
2. Describe the makeup of a chromosome prior to the start of cell division. (p. 338)
3. How do the terms diploid (2N) and haploid (N) pertain to meiosis? (p. 338)
4. Describe the stages of mitosis, including in your description the terms *centrioles*, *nucleolus*, *spindle*, and *furrowing*. (pp. 339–342)
5. Describe the stages of meiosis I, including the terms *tetrad* and *dyad* in your description. (pp. 343–344)
6. Draw and explain a diagram illustrating crossing over. (p. 344)
7. Compare the second series of stages of meiosis to a mitotic division. (p. 344)
8. How does spermatogenesis in males compare to oogenesis in females? (p. 346)
9. What is the importance of mitosis and meiosis in the life cycle of humans? (p. 347)
10. Which parent determines the sex of the baby? What are the chances of having a boy or girl? (p. 348)
11. What is nondisjunction, and how does it occur? What is the most common autosomal chromosome abnormality? (p. 349)
12. What are four sex chromosome abnormalities? Describe each one. (p. 350)

Objective Questions

1. The arrangement of an individual's chromosomes according to homologous pairs is called a _____ .
2. The karyotype of males includes the sex chromosomes _____ , and the karyotype of females includes the sex chromosomes _____ .
3. The diploid number of chromosomes is designated the _____ number, and the haploid number is designated the _____ number.
4. If the mother cell has twenty-four chromosomes, the daughter cells following mitosis will have _____ chromosomes.
5. As the organelles called _____ separate and move to the poles, the spindle fibers appear.
6. During meiosis I, the _____ separate, and during meiosis II the _____ separate.
7. Meiosis in males is a part of _____ , and meiosis in females is a part of _____ .

8. There is a _____ chance of a newborn being a male or a female.

   For answering questions 9–15, use this key:
   a. Down syndrome
   b. Turner syndrome
   c. Klinefelter syndrome
   d. cri du chat syndrome
   e. metafemale
9. XXY _____
10. extra number-21 chromosome _____

11. XXX _____
12. deletion in number-5 chromosome _____
13. XO _____
14. due to an autosomal nondisjunction _____
15. due to a chromosomal mutation _____

Answers

1. karyotype 2. XY, XX 3. 2N, N 4. twenty-four 5. centrioles 6. homologous chromosomes, sister chromatids 7. spermatogenesis, oogenesis 8. 50% 9. c 10. a 11. e 12. d 13. b 14. a 15. d

Label this Diagram.

See figure 17.5 (p. 340) in text.

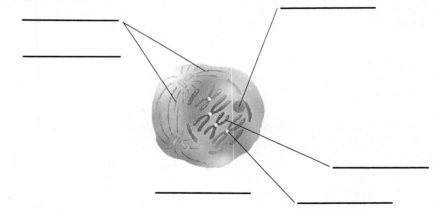

Critical Thinking Questions

1. The spindle apparatus assures the inheritance of the diploid number of chromosomes in each daughter cell. Why is there no need to assure equal distribution of the organelles to the daughter cells?
2. Both egg and sperm contribute one member of each pair of chromosomes to the new individual, but the egg contributes most of the cytoplasm. Why does this seem appropriate, considering the manner in which humans procreate, and what implications does it have for inheritance?
3. Extremely few newborns have just one autosomal chromosome of a particular pair. It's speculated that each chromosome ordinarily carries "lethal genes," genes that are so defective that if they are not counteracted by a competent gene in the other member of the homologous pair, the individual dies. Why, then, do you suppose that a person with Down syndrome can survive?

Selected Key Terms

autosomes (aw′to-sōmz) chromosomes other than sex chromosomes. 337

centromere (sen′tro-mēr) a region of a chromosome where duplicated chromosomes remain attached. 338

chromatids (kro′mah-tidz) the two identical parts of a chromosome following replication of DNA. 338

crossing over (kros′ing o′ver) the exchange of corresponding segments of genetic material between nonsister chromatids of homologous chromosomes during synapsis of meiosis I. 344

cytokinesis (si′′to-ki-ne′sis) division of the cytoplasm of a cell. 342

diploid (dip′loid) the 2N number of chromosomes; twice the number of chromosomes found in gametes. 338

Down syndrome (down sin′drōm) human congenital disorder associated with an extra number-21 chromosome. 351

gametes (gam′ets) reproductive cells that join in fertilization to form a zygote; most often an egg or sperm. 338

haploid (hap′loid) the N number of chromosomes; half the diploid number; the number characteristic of gametes that contain only one set of chromosomes. 338

homologous chromosomes (ho-mol′o-gus kro′mo-sōmz) similarly constructed; homologous chromosomes have the same shape and contain genes for the same traits. 337

karyotype (kar′e-o-tīp) the arrangement of all the chromosomes within a cell by homologous pairs in a fixed order. 337

Klinefelter syndrome (klīn′fel-ter sin′drōm) a condition caused by the inheritance of a chromosomal abnormality in number; an XXY individual. 353

meiosis (mi-o′sis) type of cell division that occurs during the production of gametes by means of which the daughter cells receive half the number of chromosomes as the mother cell. 343

metafemale (met′′ah-fe′māl) a female who has three X chromosomes. 353

mitosis (mi-to′sis) type of cell division in which daughter cells receive the exact chromosomal and genetic makeup of the mother cell; occurs during growth and repair. 339

nondisjunction (non′′dis-junk′shun) the failure of homologous chromosomes or sister chromatids to separate during the formation of gametes. 351

oogenesis (o′′o-jen′ē-sis) production of an egg in females by the process of meiosis and maturation. 346

sex chromosomes (seks kro′mo-sōmz) chromosomes responsible for the development of characteristics associated with maleness or femaleness; an X or Y chromosome. 337

spermatogenesis (sper′′mah-to-jen′ē-sis) production of sperm in males by the process of meiosis and maturation. 346

spindle fibers (spin′d′l fi′berz) microtubule bundles in cells that are involved in the movement of chromosomes during mitosis and meiosis. 341

synapsis (si-nap′sis) the attracting and pairing of homologous chromosomes during prophase I of meiosis. 343

tetrads (tet′radz) a set of four chromatids resulting from the pairing of homologous chromosomes during prophase I of meiosis. 343

Turner syndrome (tur′ner sin′drōm) a condition caused by the inheritance of an abnormality in chromosome number; an X chromosome lacks a homologous counterpart—XO. 353

Chapter Eighteen

Genes and Medical Genetics

Chapter Concepts

1 Genes, located on chromosomes, are passed from one generation to the next.

2 The Mendelian laws of genetics relate the genotype (inherited genes) to the phenotype (physical characteristics).

3 Exceptions to Mendel's laws apply to traits controlled by more than one gene and to codominant genes.

4 There are genes on the X chromosome that control traits having nothing to do with the sexual characteristics of the individual.

5 Humans are subject to many disorders due to the inheritance of faulty genes.

This boy, a participant in Special Olympics, has Down Syndrome. Investigators are only now discovering which genes on chromosome 21 bring about the characteristics of this inherited condition. The hope is that one day it will be possible to control the effects of the genes so that the characteristics will not appear.

The anatomical features of human beings are controlled by the genes. And, indeed, so are physiological and behavioral features controlled by the genes. But what are the genes? This chapter shows that it is sometimes helpful to think of genes as being units of chromosomes particularly when one wants to calculate the chances of passing on a genetic disorder. But the next chapter shows that genes are actually DNA molecules that control protein synthesis. Only now are we beginning to decipher how this function can directly affect the structure of the cell and the organism.

Simple Mendelian Inheritance

The first person to conduct a successful study of inheritance was Gregor Mendel, a Catholic monk who grew peas in a small garden plot in 1860. Mendel's studies led him to conclude that inheritance is governed by factors that exist within the individual and are passed on to offspring. Mendel said that in an individual *every trait* (for example, color of seed in peas) *is controlled by a pair of factors.* He also observed that one of the factors, controlling the same trait, can be dominant over the other, which is recessive. The individual may show the dominant characteristic (for example, yellow seed), while the recessive factor (for green seed), although present, is not expressed.

Mendel's experimental crosses made him realize that it was possible for an individual showing the dominant characteristic to pass on a factor for the recessive factor. Therefore, he concluded that while the individual has two factors for each trait, the gametes contained only one factor for each trait. This is often called Mendel's law of segregation.

Law of segregation: The factors separate when the gametes are formed, and only one factor of each pair is present in each gamete.

Today we realize that Mendel's laws are applicable not only to peas, but to all diploid individuals.

Inheritance of a Single Trait

We will use the word *trait* to mean some aspect of the individual, such as type of hair, nose, or ears. We will imagine that the genes, like the letters in the rectangles in figure 18.1 are on a chromosome. Notice that this would mean that the genes are in a particular sequence and are at particular spots, or loci, on the chromosomes. Alternate forms of a gene having the same position on a pair of chromosomes and affecting the same trait are called **alleles.** In figure 18.1, *R* is an allele of *r*, and vice versa, for example. A capital letter stands for a **dominant allele** and a lowercase letter stands for a **recessive allele.** For example, in a problem concerning hairline the *key* would be:

W = widow's peak (dominant allele)
w = continuous hairline (recessive allele)

Genotype and Phenotype

When we indicate the genes of a particular individual, two letters must be used for each trait mentioned. This is called the

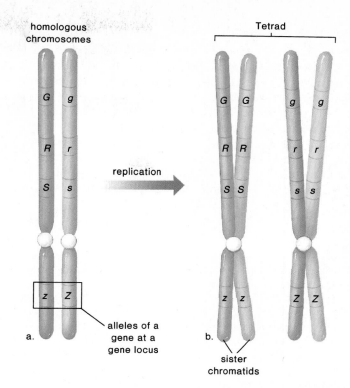

homologous chromosomes

Tetrad

replication

alleles of a gene at a gene locus

a.

b.

sister chromatids

Figure 18.1 Diagrammatic representation of a homologous pair of chromosomes before and after replication. *a.* The letters represent alleles (alternate forms of a gene). Each allelic pair, such as *Gg* or *Zz*, is located on homologous chromosomes at a particular gene locus. *b.* Following replication, each sister chromatid carries the same alleles in the same order.

Table 18.1 Genotype versus Phenotype

Genotype	Genotype versus	Phenotype
WW	homozygous dominant	widow's peak
Ww	heterozygous	widow's peak
ww	homozygous recessive	continuous hairline

genotype of the individual. The genotype may be expressed not only by using letters but also by a short descriptive phrase, as table 18.1 shows. The word **homozygous** means that the two members of the allelic pair in the zygote (*zygo*) are the same (*homo*); genotype *WW* is called *homozygous dominant* and *ww* is called *homozygous recessive*. The word **heterozygous** means that the members of the allelic pair are different (*hetero*); only *Ww* is heterozygous.

As table 18.1 also indicates, the word **phenotype** refers to the physical characteristics of the individual. What the individual actually looks like is the phenotype. (Also included in the phenotype are the microscopic and metabolic characteristics of the individual.) Notice that both homozygous dominant and heterozygous show the dominant phenotype.

Gamete Formation

Whereas the genotype has two alleles for each trait, the gametes have only one allele for each trait. This, of course, is related to the process of meiosis. The alleles are present on a homologous pair of chromosomes, and these chromosomes separate during meiosis (fig. 18.5). Therefore, the members of each allelic pair separate during meiosis, and there is only one allele for each trait in the gametes. When doing genetic problems, it should be kept in mind that no two letters in a gamete may be the same. For example, Ww would represent a possible genotype, and the possible gametes for this individual would be W, w—the comma indicating two possible gametes.

When doing genetics problems, the same alphabetic letter is used for the alleles; a capital letter indicates the dominant and a lowercase letter indicates the recessive allele. A homozygous dominant individual is indicated by two capital letters, and a homozygous recessive individual is indicated by two lowercase letters. The genotype of a heterozygous individual is indicated by a capital and a lowercase letter. Contrary to the individual, gametes have one letter of each type, either capital or lowercase as appropriate. All possible combinations of letters indicate all possible gametes.

Do Practice Problems 1 located on page 373.

Crosses

It is now possible for us to consider a particular cross. If a homozygous man with a widow's peak marries a woman with a continuous hairline, what kind of hairline will their children have?

In solving the problem, we must indicate the genotype of each parent by using letters, determine what the gametes are, and what the genotypes of the children are after reproduction. In the format that follows, **P** stands for the parental generation, and the letters in this row are the genotypes of the parents. The second row shows that each parent has only one type of gamete in regard to hairline, and therefore all the children (F_1 = first filial generation) will have a similar genotype, that is, heterozygous. Heterozygotes show the dominant characteristic, and so all the children will have a widow's peak.

P:	Widow's peak	$\times$	Continuous hairline
	WW		ww
Gametes:	W		w
F_1:		Widow's peak	
		Ww	

These individuals are **monohybrids** because they are heterozygous for only one pair of alleles. If they marry someone else with the same genotype, what type of hairline will their children have?

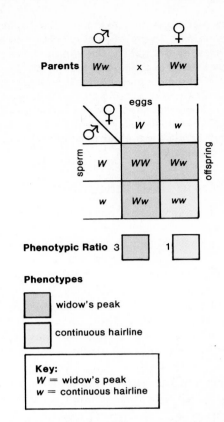

Figure 18.2   Monohybrid cross. In this cross, the parents are heterozygous for widow's peak. The chances of any child having a continuous hairline are one out of four, or 25%.

P:	Widow's peak	$\times$	Widow's peak
	Ww		Ww
Gametes:	W, w		W, w

In this problem, each parent has two possible types of gametes. In calculating F_1, it is assumed that either type of sperm has an equal chance to fertilize either type of egg. One way to assure that we have accounted for this is to use a Punnett square (p. 348).

When this is done (fig. 18.2), the results show a 3:1 phenotypic ratio; that is, three with widow's peak to one without. Such a ratio will actually be observed only if a large number of crosses of the same type take place and a large number of offspring result. Only then will all possible sperm have an equal chance to fertilize all possible eggs. It is obvious that we do not routinely observe hundreds of offspring from a single type cross in humans, and so it is customary to merely state that each child has three chances out of four to have a widow's peak, or one chance out of four to have a continuous hairline. It is important to realize that *chance has no memory;* for example, if two heterozygous parents have already had three children with a widow's peak and are expecting a fourth child, this child still has three chances out of four to have a widow's peak and only one chance out of four of having a continuous hairline (fig. 18.3). Each individual child has the same chances.

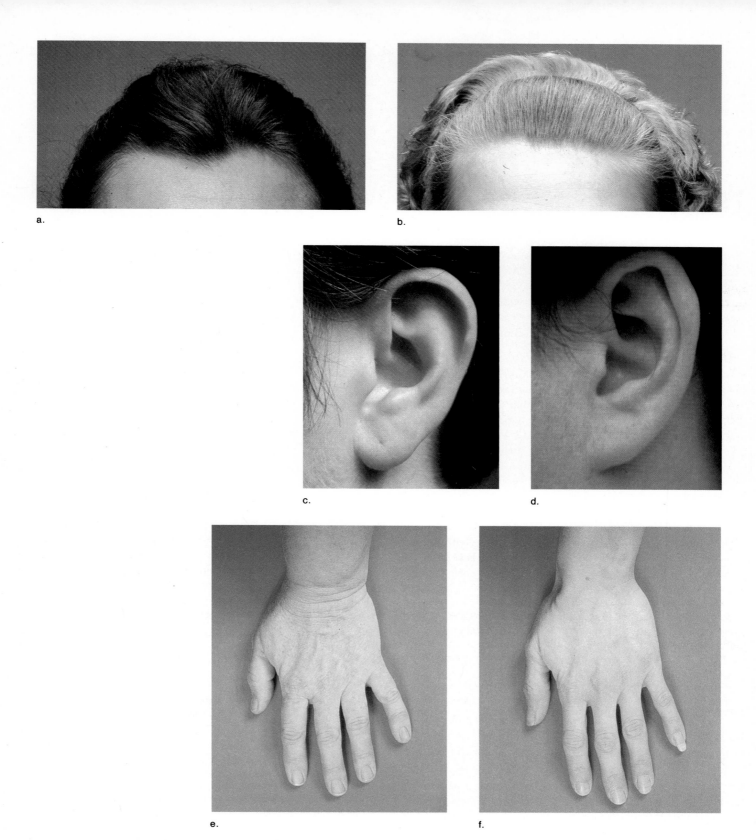

Figure 18.3   Common inherited characteristics in human beings.
Widow's peak *a.* is dominant over *b.* continuous hairline. Unattached
earlobe *c.* is dominant over *d.* attached earlobe. Short fingers *e.* are
dominant over long fingers *f.*

Probability

Another way to calculate the possible results of a cross is to realize that the chance, *or probability of receiving a particular combination of alleles, is simply the product of the individual probabilities.* In the cross just considered,

$Ww \times Ww$

the offspring have an equal chance of receiving W or w from each parent. Therefore:

Probability of W $=$ ½
Probability of w $=$ ½

and

Probability of WW $=$ ½ $\times$ ½ $=$ ¼
Probability of Ww $=$ ½ $\times$ ½ $=$ ¼
Probability of wW $=$ ½ $\times$ ½ $=$ ¼
                                     ¾ $=$ widow's peak
Probability of ww $=$ ½ $\times$ ½ $=$ ¼
                                     ¼ $=$ continuous hairline

Testcross

If an individual has the dominant phenotype, it is not possible to tell by inspection if the genotype is homozygous dominant or heterozygous. However, if the individual is crossed with a homozygous recessive, the results may indicate what the original genotype was. For example, figure 18.4 shows the different results if a man with a widow's peak is homozygous dominant, or if he is heterozygous, and married to a woman with a continuous hairline. (She must be homozygous recessive or she would not have a continuous hairline.) In the first case, the man can only sire children with widow's peaks, and in the second the chances are 2:2 or 1:1 that a child will or will not have a widow's peak. Thus, the cross of a possible heterozygote with an individual having the recessive phenotype gives the best chance of producing the recessive phenotype among the offspring. Therefore, this type of cross is called the **testcross.**

In doing an actual cross, it is assumed that all possible types of sperm fertilize all possible types of eggs. The results may be expressed as a probable phenotypic ratio; it is also possible to state the chances of an offspring showing a particular phenotype.

Do Practice Problems 2 located on page 373.

Inheritance of Multitraits

Two Traits

Although it is possible to consider the inheritance of just one trait, actually each individual passes on to his or her offspring many genes for many traits. In order to arrive at a general understanding of multitrait inheritance, we will consider the inheritance of two traits. The same principles will apply to as many traits as we might wish to consider.

When Mendel performed two-trait crosses, he formulated his second law, the law of independent assortment.

Law of independent assortment: Pairs of factors separate independently of one another to form gametes, and therefore all possible combinations of factors may occur in the gametes.

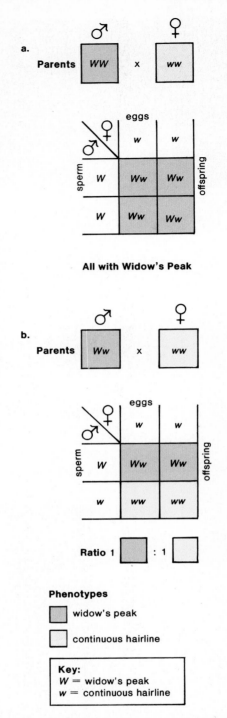

Figure 18.4 Testcross. In this example, it is impossible to tell if the male parent could be homozygous or heterozygous. The results of reproduction with a homozygous recessive, however, may determine his genotype. If he is homozygous, *a.*, none of the offspring will show the recessive characteristic, but if he is heterozygous, *b.*, there is a 50–50 chance that any offspring will show the recessive characteristic.

Figure 18.5 illustrates that the law of segregation and the law of independent assortment hold for genes on different kinds of chromosomes because of the manner in which meiosis occurs. The law of segregation is dependent on the separation of members of homologous pairs of chromosomes; and the law of independent assortment is dependent on the random arrangement

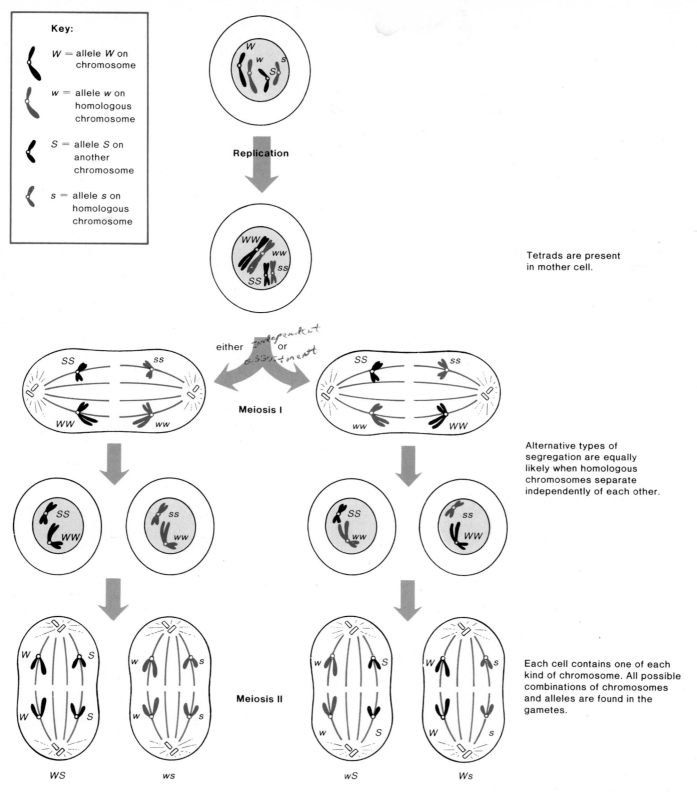

Key:

W = allele W on chromosome

w = allele w on homologous chromosome

S = allele S on another chromosome

s = allele s on homologous chromosome

Replication

Tetrads are present in mother cell.

either *independent* or *assortment*

Meiosis I

Alternative types of segregation are equally likely when homologous chromosomes separate independently of each other.

Meiosis II

Each cell contains one of each kind of chromosome. All possible combinations of chromosomes and alleles are found in the gametes.

WS   ws   wS   Ws

Figure 18.5   Chromosome basis of Mendel's laws of segregation and independent assortment. The law of segregation states that homologous chromosomes segregate during meiosis I; therefore, there is only one chromosome (and one allele) of each kind in the gametes. The law of independent assortment states that pairs of homologous chromosomes segregate independently of other homologous pairs during meiosis I, and, therefore, all possible combinations of chromosomes (and alleles) are in the gametes. During meiosis II, the chromatids separate and meiosis produces 4 daughter cells.

In this cross, the male grandparent, who is homozygous dominant, can only produce the gamete *WS*, and the female, who is homozygous recessive, can only produce the gamete *ws*.

P

Gametes

All offspring from the cross are dihybrids (*WwSs*) with widow's peak and short fingers. If dihybrids reproduce, each can produce 4 kinds of gametes: *WS, Ws, wS, ws.*

F₁

F₁ Gametes

The results of a dihybrid cross are always 9:3:3:1. Nine offspring show both dominant characteristics; 3 are mixed (i.e., widow's peak and long fingers); 3 are mixed otherwise (i.e., continuous hairline and short fingers); and 1 shows both recessive characteristics.

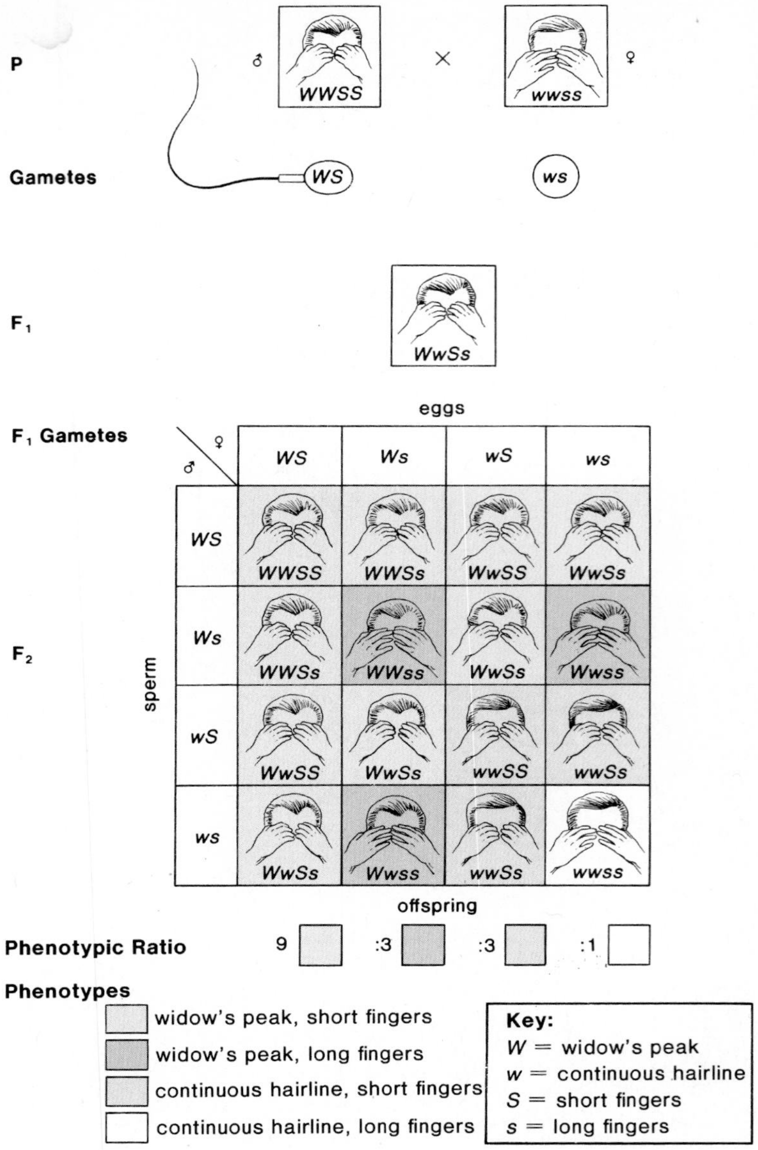

F₂

Phenotypic Ratio   9 [ ] :3 [ ] :3 [ ] :1 [ ]

Phenotypes

[ ] widow's peak, short fingers
[ ] widow's peak, long fingers
[ ] continuous hairline, short fingers
[ ] continuous hairline, long fingers

Key:
W = widow's peak
w = continuous hairline
S = short fingers
s = long fingers

Figure 18.6 A 2-trait cross that involves 2 generations. The grandparents (P) are homozygous, and the first generation (F₁) offspring are dihybrids. The second generation (F₂) offspring have a 9:3:3:1 phenotypic ratio.

of homologous pairs with respect to one another during metaphase I prior to the separation process.

Crosses

When doing a two-trait cross, we realize that the genotypes of the parents require four letters because there is an allelic pair for each trait. Second, the gametes of the parents contain one letter of each kind in every possible combination, as predicted by Mendel's law of independent assortment. Finally, in order to produce the probable ratio of phenotypes among the offspring, all possible matings are presumed to occur.

To give an example, let us cross a person homozygous for widow's peak and short fingers with a person who has a continuous hairline and long fingers. The key for such a cross is

W = Widow's peak S = Short fingers
w = Continuous hairline s = Long fingers

P: Widow's peak,   × Continuous hairline,
          Short fingers   Long fingers
          WWSS   *wwss*

Gametes: *WS*   *ws*

F₁: Widow's peak,
                Short fingers
                WwSs

In this particular cross, only one type of gamete is possible for each parent; therefore, all of the F₁ generation will have the same genotype (*WwSs*) and the same phenotype (widow's peak with short fingers). This genotype is called a **dihybrid** because the individual is heterozygous in two regards: hairline and fingers.

When a dihybrid reproduces with a dihybrid, each parent has four possible types of gametes:

F_1:	*WwSs*	×	*WwSs*
Gametes:	*WS*		*WS*
	Ws		*Ws*
	wS		*wS*
	ws		*ws*

The Punnett square (fig. 18.6) for such a cross shows the expected genotypes among sixteen offspring if all possible sperm fertilize all possible eggs. An inspection of the various genotypes in the square shows that among the offspring, *nine* will have a widow's peak and short fingers, *three* will have a widow's peak and long fingers, *three* will have a continuous hairline and short fingers, and *one* will have a continuous hairline and long fingers. This is called a 9:3:3:1 phenotype ratio, and this ratio always results when a dihybrid is mated with a dihybrid and simple dominance is present.

Probability

We can use the previous ratio to predict the chances of each child receiving a certain phenotype. For example, the possibility of getting the two dominant phenotypes together is nine out of sixteen ($9 + 3 + 3 + 1 = 16$) and that of getting the two recessive phenotypes together is one out of sixteen.

We can also calculate the chance, or probability, of these various phenotypes occurring by knowing that the *probability of combinations of independent events is the product of the probabilities of each of the events:*

Probability of widow's peak = $3/4$
Probability of short fingers = $3/4$
Probability of continuous hairline = $1/4$
Probability of long fingers = $1/4$

Therefore:

Probability of widow's peak and short fingers = $3/4 \times 3/4 = 9/16$
Probability of widow's peak and long fingers = $3/4 \times 1/4 = 3/16$
Probability of continuous hairline and short fingers = $1/4 \times 3/4 = 3/16$
Probability of continuous hairline and long fingers = $1/4 \times 1/4 = 1/16$

Testcross

An individual who shows the dominant traits can be tested for the dihybrid genotype by a mating with the recessive in both traits.

P_1:	*WwSs*	×	*wwss*
Gametes:	*WS*		*ws*
	Ws		
	wS		
	ws		

The Punnett square (fig. 18.7) shows that the resulting ratio is 1 widow's peak with short fingers : 1 widow's peak with long fingers : 1 continuous hairline with short fingers : 1 continuous hairline with long fingers, or 1:1:1:1.

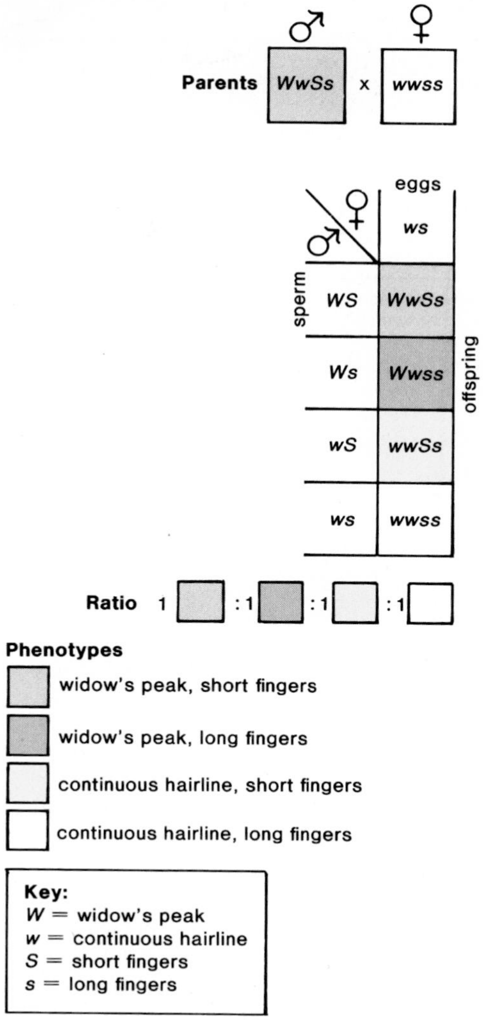

Figure 18.7 Testcross. In this example, it is impossible to tell if the male parent is homozygous dominant or if he is heterozygous for both traits. However, reproduction with a female who is recessive for both traits is likely to show which he is. If he is heterozygous, there is a 25% chance that the offspring will show both recessive characteristics and a 50% chance that they will show one or the other of the recessive characteristics.

Table 18.2 lists all of the crosses we have studied thus far, which show a frequently observed ratio. When these types of crosses are done, these ratios are observed.

Do Practice Problems 3 located on page 373.

Genetic Disorders

When studying human genetic disorders, biologists often construct **pedigree charts** that show the pattern of inheritance of a characteristic within a group of people. Let us contrast two possible patterns of inheritance in order to show how it is possible

Table 18.2 Phenotypic Ratios of Common Crosses

monohybrid × monohybrid	3:1 (dominant to recessive)
monohybrid × recessive[a]	1:1 (dominant to recessive)
dihybrid × dihybrid	9:3:3:1 (9 both dominant, 3 one dominant, 3 other dominant, 1 both recessive)
dihybrid × recessive[a]	1:1:1:1 (all possible combinations in equal number)

[a] Called a testcross because it can be used to test if the individual showing the dominant gene is homozygous or heterozygous. For a definition of all terms, see the glossary.

to determine whether the characteristic is an autosomal dominant or an autosomal recessive disorder. Autosomal disorders are caused by alleles on the autosomal chromosomes.

In both patterns, males are designated by squares and females are designated by circles. Shaded circles and squares indicate affected individuals. A line between a square and a circle represents a couple who have mated. A vertical line going downward leads, in these patterns, to a single child. (If there is more than one child, they are placed off a horizontal line.) Which pattern of inheritance do you suppose represents a dominant characteristic, and which represents a recessive characteristic?

In pattern I, the child is affected, but neither parent is; this can happen if the characteristic is recessively inherited. What are the chances that any offspring from this union will be affected? Because the parents are monohybrids, the chances are 1 in 4, or 25% (table 18.2). Notice that the parents also could be called **carriers** because they have a normal phenotype but are capable of having a child with a genetic disorder. See figure 18.8 for other ways to recognize an autosomal recessive pattern of inheritance.

In pattern II, the child is affected, as is one of the parents. When a characteristic is dominant, an affected child has at least one affected parent. Of the two patterns, this one shows a dominant pattern of inheritance. What are the chances that any offspring from this union will be affected? Because this is a monohybrid by a recessive cross, the chances are 50% (table 18.2). See figure 18.9 for other ways to recognize an autosomal dominant pattern of inheritance.

Autosomal Recessive Disorders

There are many autosomal recessive disorders that are inherited in a simple Mendelian manner, but only three of the better known are discussed. Others that are well known are albinism (lack of pigment); galactosemia (accumulation of galactose in

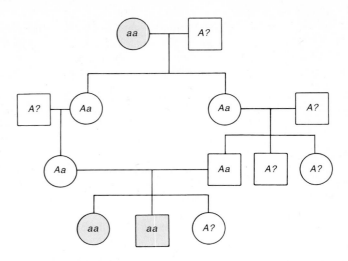

Autosomal Recessive Genetic Disorders
- Most affected children have normal parents.
- Heterozygotes have a normal phenotype.
- Two affected parents always will have affected children.
- Affected individuals who have noncarrier spouses will have normal children.
- Close relatives who marry are more likely to have affected children.
- Both males and females are affected with equal frequency.

Key:
aa = affected
Aa = carrier (appears normal)
AA = normal

Figure 18.8 Sample pedigree chart for an autosomal recessive genetic disorder. Only those affected are shaded. Go through the chart and hatch mark all those who are carriers.

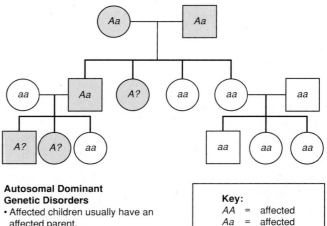

Autosomal Dominant Genetic Disorders
- Affected children usually have an affected parent.
- Heterozygotes are affected.
- Two affected parents can produce an unaffected child.
- Unaffected parents do not have affected children.
- Both males and females are affected with equal frequency.

Key:
AA = affected
Aa = affected
aa = normal

Figure 18.9 Sample pedigree chart for an autosomal dominant genetic disorder. Are there any carriers?

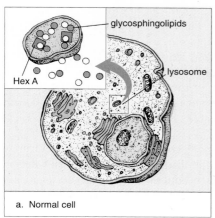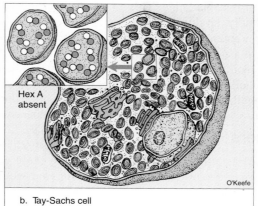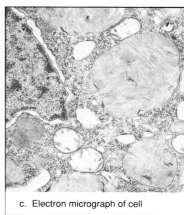

Figure 18.10   Tay-Sachs disease. *a*. When Hex A is present, glycosphingolipids are broken down. *b*. When Hex A is absent, these lipids accumulate in lysosomes and lysosomes accumulate in the cell. *c*. Electron micrograph of cell crowded with lysosomes.

the liver and mental retardation); thalassemia (production of abnormal type of hemoglobin); and xeroderma pigmentosum (inability to repair ultraviolet-induced damage). The homozygous recessive phenotype is more likely to occur among a group of people who tend to marry each other, which may explain why autosomal recessive disorders are sometimes more prevalent among members of a particular ethnic group.

Cystic Fibrosis

Cystic fibrosis is the most common lethal genetic disease among Caucasians in the United States. About 1 in 20 Caucasians is a carrier, and about 1 in 2,000 children born to this group has the disorder. In these children, the mucus in the lungs and the digestive tract is particularly thick and viscous. In the lungs, the mucus interferes with gas exchange. In the digestive tract, the thick mucus impedes the secretion of pancreatic juices, and food cannot be properly digested; large, frequent, and foul-smelling stools occur. A few individuals have been known to survive childhood, but most die from recurrent lung infections.

In the past few years, much progress has been made in our understanding of cystic fibrosis. First of all, it was discovered that chloride (Cl^-) ions fail to pass through cell membrane channels in these patients. Ordinarily, after chloride ions have passed through the membrane, water follows. It is believed that lack of water in the lungs causes the mucus to be so thick. Secondly, the cystic fibrosis gene, which is located on the number-7 chromosome, has been isolated, and soon there will be a test to identify carriers of the disease. The defective protein product has been identified, but its exact function is not yet known.

Tay-Sachs Disease

Tay-Sachs disease is the best-known genetic disease among United States Jewish people, most of whom are of central and eastern European descent. At first, it is not apparent that a baby has Tay-Sachs disease. However, development begins to slow down between four and eight months of age as neurological im-

pairment and psychomotor difficulties become apparent. Ophthalmologic examination reveals a characteristic red spot and yellowish accumulation in the region of the retina called the fovea centralis. The child gradually becomes blind and helpless, develops uncontrollable seizures, and eventually becomes paralyzed. There is no treatment or cure for Tay-Sachs disease, and most affected individuals die by the age of three or four.

So-called late-onset Tay-Sachs disease occurs in adults. The symptoms are progressive mental and motor deterioration, depression, schizophrenia, and premature death. The gene for late-onset Tay-Sachs disease has been sequenced, and this form of the disorder apparently is due to one changed pair of bases in the DNA of the number-1 chromosome.

Tay-Sachs disease results from a lack of the enzyme hexosaminidase A (Hex A) and the subsequent storage of its substrate, a glycosphingolipid, in lysosomes. Although more and more lysosomes build up in many body cells (fig. 18.10), the primary sites of storage are the cells of the nervous system, which accounts for the onset and the progressive deterioration of psychomotor functions.

There is a test to detect carriers of Tay-Sachs. The test uses a sample of serum, white blood cells, or tears to determine whether Hex A activity is present. Affected individuals have no detectable Hex A activity. Carriers have about half the level of Hex A activity found in normal individuals. Prenatal diagnosis of the disease also is possible following either amniocentesis or chorionic villi sampling.

Phenylketonuria (PKU)

Phenylketonuria (PKU) occurs in 1 in 20,000 births and so is not as frequent as the disorders previously discussed. When it does occur, the parents are very often close relatives. Affected individuals lack an enzyme that is needed for the normal metabolism of the amino acid phenylalanine, and an abnormal breakdown product, a phenylketone, accumulates in the urine. Newborns are routinely tested, and if they lack the necessary enzyme, they are placed on a diet low in phenylalanine. This diet must be continued until the brain is fully developed or else severe mental retardation develops. If a woman who is homozygous recessive for PKU wishes to have a normal child, she should resume her limited diet several months before getting pregnant; otherwise, she runs a high risk of having a microcephalic child—one with an abnormally small head.

Autosomal Dominant Disorders

There are many autosomal dominant disorders that are inherited in a simple Mendelian manner, but only two of the better known—neurofibromatosis and Huntington disease—are discussed here. Others that are well known are Marfan syndrome (connective tissue disorder); achondroplasia (dwarfism); brachydactyly (abnormally short fingers); porphyria (inability to metabolize porphyrins from hemoglobin breakdown); and hypercholesterolemia (elevated levels of cholesterol in blood).

Neurofibromatosis (NF)

Neurofibromatosis (NF), sometimes called Elephant Man disease[1], is one of the most common genetic disorders. It affects roughly 1 in 3,000 people, including an estimated 100,000 in the United States. It is seen equally in every racial and ethnic group throughout the world.

At birth or later, the affected individual may have six or more large tan spots on the skin. Such spots may increase in size and number and get darker. Small benign tumors (lumps) called neurofibromas may occur under the skin or in the muscles. Neurofibromas are made up of nerve cells and other cell types.

This genetic disorder shows *variable expressivity;* in most cases, symptoms are mild and patients live a normal life. In some cases, however, the effects are severe. Skeletal deformities, including a large head, are seen, and eye and ear tumors can lead to blindness and hearing loss. Many children with NF have learning disabilities and are overactive.

Previously researchers determined that the gene responsible for NF is located on number-17 chromosome and developed a prenatal test for diagnosing the disorder. Recently the NF gene has been isolated and found to code for a protein now called NF1 that normally suppresses abnormal cell division. Therefore the NF gene is a tumor-suppressor gene (p. 404) whose mutation leads to an ineffective NF1 protein and excessive growth of cells containing the mutated gene.

Huntington Disease

As many as 1 in 10,000 persons in the United States have **Huntington disease (HD),** a neurological disorder that affects specific regions of the brain. Most individuals who inherit the allele appear normal until middle age. Then, minor disturbances in

balance and coordination lead to progressively worse neurological disturbances. The victim becomes insane before death occurs.

Much has been learned about Huntington disease. The gene for the disease is located on the number-4 chromosome, and there is a test, of the type described in figure 19.19, to determine if the dominant gene has been inherited. Because treatment is not available, however, few may want to have this information.

Research is being conducted, though, to determine the underlying cause of the disorder. It is known that the brain of a Huntington victim produces more than the usual amount of quinolinic acid, an excitotoxin that can overstimulate certain nerve cells. It is believed to lead to the death of these cells and to the subsequent symptoms of Huntington disease. Researchers are looking for chemicals that block quinolinic acid's action or inhibit quinolinic acid synthesis.

Human Issue

One English immigrant and her 13 children gave rise to at least 432 cases of Huntington Disease in Australia. This report highlights the fact that passing on a genetic disease can have tragic consequences. Should everyone (a) be tested for and then restricted from reproducing if they have the potential for passing on a serious genetic disorder, (b) be educated about genetic disorders and the possibility of being tested, or (c) should we simply rely on medical remedies if a serious genetic disease does happen to be passed on? Where do you stand?

Some of the best-known human genetic disorders are either autosomal recessive or autosomal dominant disorders inherited in a simple Mendelian manner. Pedigree charts show the pattern of inheritance in a particular family.

Beyond Mendel's Laws

Certain traits, such as those just studied, follow the rules of simple Mendelian inheritance. There are, however, others that do not follow these rules.

Polygenic Inheritance

Two or more sets of alleles can affect the same trait, sometimes in an additive fashion. Polygenic inheritance can cause the distribution of human characteristics according to a bell-shaped curve, with most individuals exhibiting the average phenotype (fig. 18.11). The more genes that control the trait, the more continuous the distribution.

Skin Color

Just how many pairs of alleles control skin color is not known, but a range in colors can be explained on the basis of two pairs. When a black person has children by a white person, the children have medium-brown skin, but two medium-brown individ-

---

[1]Although neurofibromatosis is commonly associated with Joseph Merrick, the severely deformed nineteenth-century Londoner depicted in "The Elephant Man," researchers today believe Merrick actually suffered from a much rarer disorder called Proteus syndrome.

a.

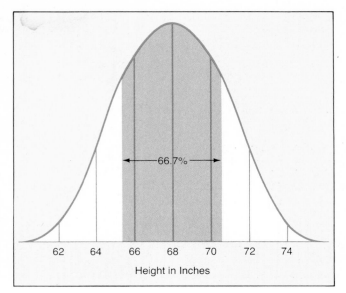

b.

Figure 18.11 Polygenic inheritance. *a.* When you record the heights of a large group of young men, the values follow a bell-shaped curve, *b.* Such distributions are seen when a trait is controlled by several sets of alleles.

Figure 18.12 Inheritance of skin color. This white husband (aabb) and his mulatto wife (AaBb) had fraternal twins, one was white and one was mulatto.

uals can produce children who range in skin color from black to white. If we *assume* that two pairs of alleles control skin color, then

Black	= *AABB*
Dark-brown	= *AABb* or *AaBB*
Medium-brown	= *AaBb* or *AAbb* or *aaBB*
Light-brown	= *Aabb* or *aaBb*
White	= *aabb*

If a medium-brown person reproduces with a white person, the very darkest individual possible is medium-brown, but a white child is also possible (fig. 18.12).

Behavioral Inheritance

Is behavior primarily inherited or is behavior shaped by environmental influences? This nature (inherited) versus nurture (environment) question has been asked for a long time, and twin studies have been employed to attempt to find the answer. Twins can be identical (derived from the same fertilized egg) or fraternal (derived from two separate eggs). Identical twins have inherited exactly the same chromosomes and genes, while fraternal twins have no more genes in common than do any other brother and sister.

Twin studies have been conducted to see to what extent behavior is inherited. It has been found that fraternal twins

Figure 18.13 Oska Stohr (right) and Jack Yufe (left) are identical twins who were reared separately but who exhibit remarkably similar behavior.

raised in the same environment are not remarkably similar in behavior whereas identical twins raised separately are sometimes remarkably similar. For example, Oskar Stohr was raised as a Catholic by his grandmother in Nazi Germany; Jack Yufe was raised by his Jewish father in the Caribbean (fig. 18.13). Yet these two men "like sweet liquers, . . . store rubber bands on their wrists, read magazines from back to front, dip buttered toast in their coffee and have highly similar personalities" . . .[2]

Responses to a questionnaire designed to provide information about behavioral traits showed that identical twins reared separately tend to have a more similar personality than fraternal twins reared together. Altogether, the data seem to show that about 50% of the *differences* in personality traits was due to polygenic inheritance and 50% was due to environmental influence.

Polygenic Disorders

A number of serious genetic disorders, such as cleft lip or palate, clubfoot, congenital dislocation of the hip, and certain neural tube defects, are traditionally believed to be controlled by a combination of genes. This belief is being challenged by researchers who studied the inheritance of cleft palate in a large family in Iceland. These researchers reported the finding of a cleft palate gene on the X chromosome.

Multiple Alleles

ABO Blood Type

Three alleles for the same gene control the inheritance of ABO blood types. These alleles determine the presence or absence of antigens on the red blood cells.

[2]Holden, C. 1980. Identical twins reared apart. *Science* 207:1323–1328.

Table 18.3 Blood Groups

Phenotype	Genotype
A	AA, AO
B	BB, BO
AB	AB
O	OO

A = type A antigen on red blood cells
B = type B antigen on red blood cells
O = no antigens on the red blood cells

Each person has only two of the three possible alleles, and both *A* and *B* are dominant over *O*. Therefore, as table 18.3 shows, there are two possible genotypes for type A blood and two possible genotypes for type B blood. On the other hand, alleles *A* and *B* are fully expressed in the presence of the other. Therefore, if a person inherits one of each of these alleles, that person will have type *AB* blood. Type O blood can only result from the inheritance of two *O* alleles.

An examination of possible matings between different blood types sometimes produces surprising results; for example,

Parents: *AO × BO*
Children: *AB, OO, AO, BO*

Therefore, from this particular mating, every possible phenotype (types AB, O, A, B blood) is possible.

Blood typing sometimes can aid in paternity suits. However, a blood test of a supposed father only can suggest that he *might* be the father, not that he definitely *is* the father. For example, it is possible, but not definite, that a man with type A blood (having genotype *AO*) is the father of a child with type O blood. On the other hand, a blood test sometimes can prove definitely that a man is not the father. For example, a man with type AB blood cannot possibly be the father of a child with type O blood. Therefore, blood tests can be used legally only to exclude a man from possible paternity.

Human Issue

A blood test done on a child born to a couple who had utilized artificial insemination (p. 292) proved that the husband was not the father of the child. Would you have this couple (1) say no more about it and raise the child as if both were the parents, (2) get a divorce on the grounds that the wife must have been having an affair, (3) sue the clinic that performed the artificial insemination on the grounds that they used the wrong semen, or (4) give up the child for adoption and try again?

The Rh Blood Factor

The Rh blood factor is inherited separately from types A, B, AB, or O type blood. In each instance, it is possible to be Rh positive (Rh+) or Rh negative (Rh-). When you are Rh positive, there is a particular antigen on the red blood cells, and when you are Rh negative, it is absent. It can be assumed that the inheritance of this antigen is controlled by a single allelic

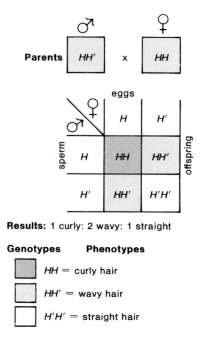

Results: 1 curly: 2 wavy: 1 straight

Genotypes	Phenotypes
HH = curly hair	
HH' = wavy hair	
$H'H'$ = straight hair	

Figure 18.14 Incomplete dominance. Among Caucasians, neither straight nor curly hair is dominant. When two wavy-haired individuals reproduce, the offspring has a 25% chance of having either straight or curly hair and a 50% chance of having wavy hair, the intermediate phenotype.

pair in which simple dominance prevails: the Rh-positive allele is dominant over the Rh-negative allele. Complications arise when an Rh-negative woman reproduces with an Rh-positive man and the child in the womb is Rh positive. Under certain circumstances (p. 128), the woman may begin to produce antibodies that will attack the red blood cells of this baby or of a future Rh-positive baby.

Degrees of Dominance

The field of human genetics also has examples of incomplete dominance and codominance. For example, when a curly-haired Caucasian reproduces with a straight-haired Caucasian, their children will have wavy hair (fig. 18.14). We already mentioned that the multiple alleles controlling blood type are codominant. An individual with the genotype AB has type AB blood. Skin color, recall, is controlled by codominant polygenes.

Sickle-Cell Disease

Sickle-cell disease is an example of a human disorder that is controlled by incompletely dominant alleles. Individuals with the genotype Hb^AHb^A are normal, those with the Hb^SHb^S genotype have sickle-cell disease, and those with the Hb^AHb^S genotype have **sickle-cell trait,** a condition in which the cells are sometimes sickle shaped. Two individuals with sickle-cell trait can produce children with all three phenotypes, as indicated in figure 18.15.

Among black Africans, the sickled cells seem to give protection against the malaria parasite, which uses red blood cells

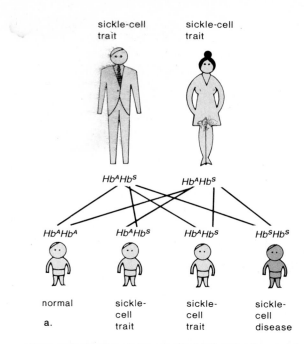

b.

Figure 18.15 *a.* Inheritance of sickle-cell disease. In this example, both parents have the sickle-cell trait. Therefore, each child has a 25% chance of having sickle-cell disease or of being perfectly normal and a 50% chance of having the sickle-cell trait. *b.* Sickled cells. Individuals with sickle-cell disease have sickled red blood cells that tend to clump, as illustrated here.

during its life cycle. Although infants with sickle-cell disease often die, those with sickle-cell trait are protected from malaria, especially from ages two to four. This means that in Africa, these children survive and grow to reproduce and to pass on the allele to their offspring. As many as 60% of blacks in malaria-infected regions of Africa have the allele. In the United States, about 10% of the black population carries the allele. A test can be done to detect the allele's presence; prenatal testing is also possible.

The red blood cells in persons with sickle-cell disease cannot pass through small blood vessels easily. The sickle-shaped cells either break down or they clog blood vessels, and the individual suffers from poor circulation, anemia, and sometimes internal hemorrhaging. Jaundice, episodic pain in the abdomen and joints, poor resistance to infection, and damage to internal organs are all symptoms of sickle-cell disease.

Persons with sickle-cell trait do not usually have any difficulties unless they undergo dehydration or mild oxygen deprivation. At such times, the cells become sickle shaped, clogging blood vessels and leading to pain and even death. A study of the occurrence of sudden deaths during army basic training showed that a person with sickle-cell trait was 40 times more likely to die compared to normal recruits. This has caused the Army Medical Corps to advise drill instructors to train recruits more gradually, to give them enough to drink, and to make allowances for heat and humidity when planning their workouts.

There are many exceptions to Mendel's laws. These include polygenic inheritance, multiple alleles, and degrees of dominance.

Do Practice Problems 4 located on page 373.

Sex-Linked Inheritance

Genes on the sex chromosomes are said to be **sex-linked.** As discussed previously (p. 348), the genes that determine the development of the sex organs are on the sex chromosomes. However, most of the genes on the sex chromosomes have nothing to do with sexual development and, instead, are concerned with other body traits. A few of these genes are on the Y chromosome, but most important ones discovered so far are only on the much larger X chromosome. Genes on the X chromosomes are said to be **X-linked.** The Y is blank for X-linked genes.

X-Linked Genetics Problems

Recall that when doing autosomal genetics problems, we represent the genotypes of males and females similarly, as shown in the following example for humans.

Key:
W = widow's peak
w = continuous hairline

Genotypes:
WW, Ww, or *ww*

However, as in the next example, when we set up the key for a sex-linked gene, males and females must be indicated by sex.

Key:
X^B = normal vision
X^b = color blindness

The possible genotypes in both males and females are

$X^B X^B$ = female with normal color vision
$X^B X^b$ = carrier female with normal color vision
$X^b X^b$ = female who is color blind
$X^B Y$ = male with normal vision
$X^b Y$ = male who is color blind

Note that the second genotype is a carrier female because although a female with this genotype appears normal, she is capable of passing on an allele for color blindness. Color-blind females are rare because they must receive the allele from both parents; color-blind males are more common since they need

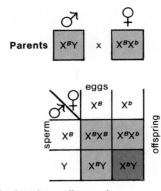

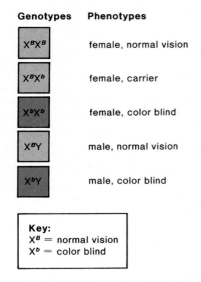

Results: females—all normal
males—1 normal: 1 color blind

Genotypes	Phenotypes
$X^B X^B$	female, normal vision
$X^B X^b$	female, carrier
$X^b X^b$	female, color blind
$X^B Y$	male, normal vision
$X^b Y$	male, color blind

Key:
X^B = normal vision
X^b = color blind

Figure 18.16   Cross involving X-linked genes. The male parent is normal and the female parent is a carrier; an allele for color blindness is located on one of her chromosomes. Therefore, each son stands a 50–50 chance of being color blind.

only one recessive allele in order to be color blind. The allele for color blindness has to be inherited from their mother because it is on the X chromosome; males only inherit the Y chromosome from their father.

Now, let us consider a particular cross. If a heterozygous woman reproduces with a man with normal vision, what are the chances of their having a color-blind daughter? A color-blind son?

Parents $X^B X^b \times X^B Y$

Inspection indicates that all daughters will have normal color vision because they all will receive an X^B from their father. The sons, however, have a 50% chance of being color blind, depending on whether they receive an X^B or an X^b from their mother. The inheritance of a Y chromosome from their father cannot offset the inheritance of an X^b from their mother.

Figure 18.16 illustrates the use of the Punnett square when doing X-linked problems. Notice that when indicating the results of a cross involving an X-linked gene, you give the phenotypic ratios for males and females separately.

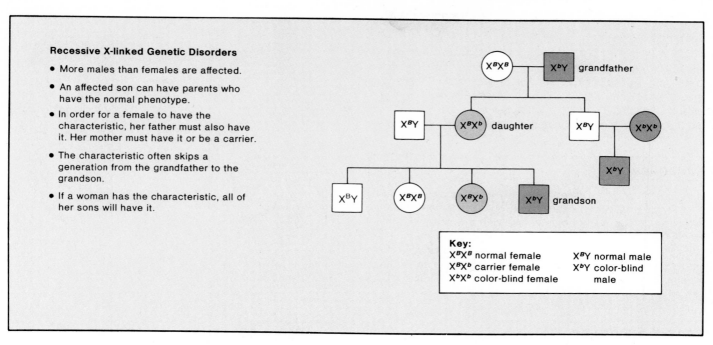

Recessive X-linked Genetic Disorders

- More males than females are affected.

- An affected son can have parents who have the normal phenotype.

- In order for a female to have the characteristic, her father must also have it. Her mother must have it or be a carrier.

- The characteristic often skips a generation from the grandfather to the grandson.

- If a woman has the characteristic, all of her sons will have it.

Key:
X^BX^B normal female X^BY normal male
X^BX^b carrier female X^bY color-blind
X^bX^b color-blind female   male

Figure 18.17   Pedigree chart for an X-linked recessive characteristic.

Figure 18.17 gives a pedigree chart for a recessive X-linked gene and lists ways to recognize this pattern of inheritance.

Recessive X-Linked Disorders

Color Blindness

In humans, there are three genes involved in distinguishing color because there are three different types of cones, the receptors for color vision (p. 245). Two of these are X-linked genes; one affects the green-sensitive cones, whereas the other affects the red-sensitive cones. About 6% of men in the United States are color blind due to a mutation involving green perception, and about 2% are color blind due to a mutation involving red perception.

Hemophilia

There are about 100,000 hemophiliacs in the United States. The most common type of hemophilia is hemophilia A, due to the absence or minimal presence of a particular clotting factor called factor VIII. *Hemophilia* is called the bleeder's disease because the affected person's blood is unable to clot. Although hemophiliacs do bleed externally after an injury, they also suffer from internal bleeding, particularly around joints. Hemorrhages can be checked with transfusions of fresh blood (or plasma) or concentrates of the clotting protein. Unfortunately, some hemophiliacs have contracted AIDS after using concentrated blood from donors, but this is less likely to occur if hemophiliacs use a purified form of the concentrate from donors who have been tested.

At the turn of the century, hemophilia was prevalent among the royal families of Europe. All of the affected males could trace their ancestry to Queen Victoria of England. Be-

cause none of Queen Victoria's forebearers or relatives was affected, it seems that the allele she carried arose by mutation either in Victoria or in one of her parents. Her carrier daughters, Alice and Beatrice, introduced the allele into the ruling houses of Russia and Spain. Alexis, the last heir to the Russian throne before the Russian Revolution, was a hemophiliac. The present British royal family has no hemophiliacs because Victoria's eldest son, King Edward VII, did not receive the allele and therefore could not pass it on to any of his descendants.

Muscular Dystrophy

Muscular dystrophy, as the name implies, is characterized by a wasting away of the muscles. The most common form, *Duchenne muscular dystrophy,* is X-linked and occurs in about one out of every 25,000 male births. Symptoms, such as waddling gait, toe walking, frequent falls, and difficulty in rising, may appear as soon as the child starts to walk. Muscle weakness intensifies until the individual is confined to a wheelchair. Death usually occurs during the teenage years; therefore, affected males are rarely fathers. The recessive allele remains in the population by passage from carrier mother to carrier daughter.

Recently, the gene for muscular dystrophy was isolated, and it was discovered that the absence of a protein, now called dystrophin, is the cause of the disorder. Much investigative work determined that dystrophin is involved in the release of calcium from the calcium-storage sacs (p. 228) in muscle fibers. The lack of dystrophin causes calcium to leak into the cell, which promotes the action of an enzyme that dissolves muscle fibers. When the body attempts to repair the tissue, the formation of fibrous tissue occurs, and this cuts off the blood supply so that more and more cells die.

A test now is available to detect carriers for Duchenne muscular dystrophy.

There are sex-linked genes on the X chromosome that have nothing to do with sex characteristics. Males have only one copy of these genes, and if they inherit a recessive allele, it is expressed.

Do Practice Problems 5 located on page 373.

Human Issue

Duchenne muscular dystrophy (DMD) is the most common lethal X-linked disorder, occurring in one out of every 3,500 male births. If you were a man in love with a woman with this disorder in her pedigree chart, would you (a) not marry the woman, (b) not have children if the woman tests positive for being a carrier, (c) use the egg selection technique described on page 323 and have only female children, or (d) leave it up to chance whether a boy with DMD is born?

Fragile-X Syndrome

Fragile-X syndrome is one of the most common genetic causes of mental retardation, second only to Down syndrome. It is called fragile-X syndrome because under the right laboratory conditions, a lesion can be seen in the X chromosome. Nevertheless, the condition is believed to be caused by a particular allele on the X chromosome.

The manner in which fragile-X syndrome is inherited does not follow the usual X-linked pattern of inheritance. For example, males with the defective allele can have normal intelligence, but their daughters can have fragile-X-positive sons. Also, investigators have found a number of girls with the fragile-X lesion who are subnormal in intelligence. It appears that the allele is turned off if inherited from the father but turned on if inherited from the mother. The hypothesis that the activity of an allele is dependent on the sex that passes the gene on is a new one in genetic inheritance.

Sex-Influenced Traits

Not all traits that we associate with the sex of the individual are due to sex-linked genes. Some are simply sex-influenced traits. Sex-influence traits are characteristics that often appear in one sex but only rarely appear in the other. It is believed that these traits are governed by genes that are turned on or off by hormones. For example, the secondary sex characteristics, such as the beard of a male and the developed breasts of a female, probably are controlled by the balance of hormones.

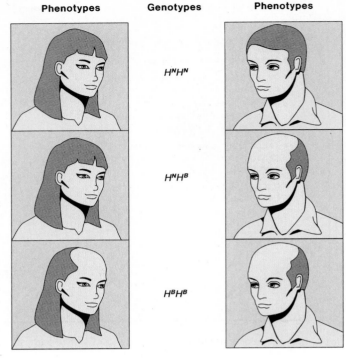

Phenotypes	Genotypes	Phenotypes
	$H^N H^N$	
	$H^N H^B$	
	$H^B H^B$	

H^N—normal hair growth
H^B—pattern baldness

Figure 18.18 Baldness is a sex-influenced characteristic. The presence of only one gene for baldness causes the condition in the male; the condition does not occur in the female unless she possesses both genes for baldness.

Baldness (fig. 18.18) is believed to be caused by the male sex hormone testosterone because males who take the hormone to increase masculinity begin to lose their hair. A more detailed explanation has been suggested by some investigators. It has been reasoned that due to the effect of hormones, males require only one allele for the trait to appear, whereas females require two alleles. In other words, the same allele acts as a dominant in males but as a recessive in females. This means that males born to a bald father and a mother with hair *at best* have a 50% chance of going bald. Females born to a bald father and a mother with hair *at worst* have a 25% chance of going bald.

Another sex-influenced trait of interest is the length of the index finger. In women, the index finger is at least equal to if not longer than the fourth finger. In males, the index finger is shorter than the fourth finger.

Summary

In keeping with Mendel's laws of inheritance, it is customary to use letters to indicate the genotype and gametes of individuals. Homozygous dominant is indicated by two capital letters, and homozygous recessive is indicated by two lowercase letters. Heterozygous is indicated by a capital letter and a lowercase letter. Contrary to the individual, gametes have one letter of each type, either capital or lowercase as appropriate. All possible combinations of letters can occur in the gametes for the dihybrid cross.

In doing an actual cross, it is assumed that all possible types of sperm fertilize all possible types of eggs. The results of some crosses may be determined by simple inspection, but certain others that commonly reoccur are given in table 18.2.

There are many exceptions to Mendel's laws, and these include polygenic inheritance (skin color), multiple alleles (ABO blood type), and incomplete dominance (curly hair).

Studies of human genetics have shown that there are many autosomal genetic disorders that can be explained on the basis of simple Mendelian inheritance. There are other disorders that are controlled by several genes (polygenic) or controlled by codominant alleles.

Some genes are sex-linked, most of which occur on the X chromosome while the Y is blank. More males than females have X-linked genetic disorders. Such disorders often skip a generation, passing from grandfather to grandson.

Practice Problems

Practice Problems I
1. For each of the following genotypes, give all possible gametes.
   a. WW
   b. WWSs
   c. Tt
   d. Ttgg
   e. AaBb
2. For each of the following, state whether a genotype or a gamete is represented.
   a. D
   b. Ll
   c. Pw
   d. LlGg

Practice Problems II
Using the information provided in figure 18.3, solve the following problems.
1. Both a man and a woman are heterozygous for freckles. What are the chances that their children will have freckles?
2. A woman is homozygous dominant for short fingers. Will any of her children have long fingers?
3. Both you and your sister or brother have attached earlobes, yet your parents have unattached ones. What are the genotypes of your parents?
4. A father has dimples, the mother does not have dimples; all the children have dimples. Dimples are dominant over no dimples. Give the probable genotype of all persons concerned.

Practice Problems III
Using the information in figure 18.3, solve these problems.
1. What is the genotype of the offspring if a man homozygous recessive for type of earlobes and homozygous dominant for type of hairline is married to a woman who is homozygous dominant for earlobes and homozygous recessive for hairline?
2. If the offspring of this cross marries someone of the same genotype, then what are the chances that this couple will have a child with a continuous hairline and attached earlobes?
3. A person who has dimples and freckles marries someone who does not. This couple produces a child who does not have dimples or freckles. What is the genotype of all persons concerned?

Practice Problems IV
1. What is the genotype of a person with straight hair? Could this individual ever have a child with curly hair?
2. What is the darkest child that could result from a mating between a light individual and a white individual?
3. What is the lightest child that could result from a mating between two medium-brown individuals?
4. From the following blood types, determine which baby belongs to which parents:

Mrs. Doe	Type A
Mr. Doe	Type A
Mrs. Jones	Type A
Mr. Jones	Type AB
Baby 1	Type O
Baby 2	Type B

5. Prove that a child does not have to have the blood type of either parent by indicating what blood types *might* be possible when a person with type A blood reproduces with a person with type B blood.

Practice Problems V
1. Both the mother and father of a hemophilic son appear to be normal. From whom did the son inherit the gene for hemophilia? What is the genotype of the mother, the father, and the son?
2. A woman is color blind. What are the chances that her sons will be color blind? If she is married to a man with normal vision, what are the chances that her daughters will be color blind? Will be carriers?
3. Both parents are right handed (R = right handed, r = left handed) and have normal vision. Their son is left handed and color blind. Give the genotype of all persons involved.
4. Both the husband and wife have normal vision. A woman has a color-blind daughter. What can you deduce about the girl's father?

Answers to Practice Problems

4. The husband is not the father.
3. RrXᴮXᵇ × RrXᴮY; rrXᵇY
2. 100%, None, 100%
1. His mother, XᴴXʰ, XᴴY, XʰY

Practice Problems 5

5. AB, O, A, B
4. Baby 1 = Doe; Baby 2 = Jones
3. White
2. Light
1. H'H, No

Practice Problems 4

3. DdFf × ddff; ddff
2. ¹/₁₆
1. Dihybrid

Practice Problems 3

4. DD × dd; Dd

3. Heterozygous
2. No
1. 75%

Practice Problems 2

2. a. gamete, b. genotype, c. gamete, d. genotype
ab
1. a. W, b. WS, Ws; c. T, t; d. Tg, tg; e. AB, Ab, aB,

Practice Problems 1

Study Questions

1. Explain why there is a pair of alleles for every trait except for sex-linked traits in males. (p. 357)
2. Relate Mendel's laws of inheritance to one-trait and two-trait problems. (pp. 357, 360)
3. What is the difference between genotype and phenotype? (p. 357)
4. What are the expected results from these crosses: heterozygous × heterozygous; heterozygous × recessive; dihybrid × dihybrid; dihybrid × double recessive? (p. 364)
5. What does the phrase "chance has no memory" mean? (p. 358)
6. Give four examples of exceptions to Mendel's laws. (pp. 366–368)
7. What is sex linkage? (p. 370) Give all possible genotypes for an X-linked trait, and discuss each. (p. 370)
8. How could you determine if a pedigree chart is depicting the inheritance of a dominant, recessive, or X-linked characteristic? (pp. 364, 371)

Objective Questions

1. Whereas an individual has two genes for every trait, the gametes have _____ gene for every trait.
2. The recessive allele for the dominant gene W is _____ .
3. Mary has a widow's peak and John has a continuous hairline. This would be a description of their _____ .
4. W = widow's peak and w = continuous hairline; therefore, only the phenotype _____ could be heterozygous.
5. Two heterozygotes, each having a widow's peak, already have a child with a continuous hairline. The next child has what chance of having a continuous hairline? _____
6. In a testcross, an individual having the dominant phenotype is crossed with an individual having the _____ phenotype.
7. How many letters are required to designate the genotype of a dihybrid individual? _____
8. If a dihybrid is crossed with a dihybrid, how many offspring out of sixteen are expected to have the dominant phenotype for both traits? _____
9. How many different phenotypes among the offspring are possible when a dihybrid is crossed with a dihybrid? _____
10. According to Mendel's law of independent assortment, a dihybrid can produce how many types of gametes having different combinations of genes? _____
11. Do sex-linked genes determine the sex of the individual? _____
12. If a male is color blind, he inherited the allele for color blindness from his _____ .
13. What is the genotype of a female who has a color-blind father but a homozygous normal mother? _____
14. In a pedigree chart, it is observed that although the children have a characteristic, neither parent has it. The characteristic must be inherited as a _____ gene.

Answers

1. one 2. w 3. phenotypes 4. widow's peak 5. 25% 6. recessive 7. four 8. nine 9. four 10. four 11. no 12. mother 13. $X^B X^b$ 14. recessive

Critical Thinking Questions

1. Mendel worked with garden peas, not humans. What is the significance of being able to apply Mendel's laws to humans?

2. Another investigator, Thomas H. Morgan, worked with fruit flies in the early 1900s. He performed this cross:

	females	males
P × P:	red-eyed ×	white-eyed
F_1:	red-eyed	
F_2:	3:1	

What data did Morgan need to show that the white eye in fruit flies is X-linked?

3. There are 4 ABO blood types determined by the A, B, or O genes. The O gene is recessive to the A and B genes and can only be expressed when homozygous (OO). Explain why the O blood type does not rapidly cease to exist when breeding occurs between persons who are blood type O and persons who are either blood type A, B, or AB.

Additional Genetics Problems

1. A woman heterozygous for polydactyly (dominant) is married to a normal man. What are the chances that their children will have six fingers and toes? (p. 357)
2. John cannot curl his tongue (recessive), but both his parents can curl their tongues. Give the genotypes of all persons involved. (p. 357)
3. Parents who do not have Tay-Sachs disease (recessive) produce a child who has Tay-Sachs. What are the chances that each child born to this couple will have Tay-Sachs? (p. 365)
4. A man with a widow's peak (dominant) who cannot curl his tongue (recessive) is married to a woman with a continuous hairline who can curl her tongue. They have a child who has a continuous hairline and cannot curl the tongue. Give the genotype of all persons involved. (p. 360)
5. Both Mr. and Mrs. Smith have freckles (dominant) and attached earlobes (recessive). Some of the children do not have freckles. What are the chances that the next child will have freckles and attached earlobes? (p. 360)
6. Mary has wavy hair (incomplete dominance) and marries a man with wavy hair. They have a child with straight hair. Give the genotype of all persons involved. (p. 369)
7. A man has type AB blood. What is his genotype? Could this man be the father of a child with type B blood? If so, what blood types could the child's mother have? (p. 368)

8. A woman with white skin has medium-brown parents. If this woman married a light man, what is the darkest skin color possible for their children? The lightest? (p. 367)
9. What is the genotype of a man who is color blind (X-linked recessive) and has a continuous hairline? If this man has children by a woman who is homozygous dominant for normal color vision and widow's peak, what will be the genotype and phenotype of the children? (p. 370)
10. Is the characteristic represented by the shaded individuals below inherited as a dominant, recessive, or X-linked recessive? (p. 364)

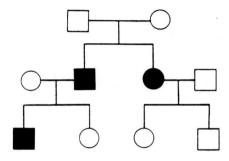

11. Fill in this pedigree chart to give the probable genotypes of the twins pictured in figure 18.12. (p. 367)

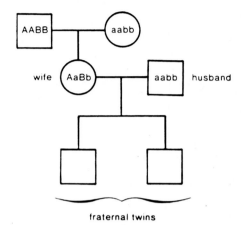

fraternal twins

Answers to Additional Genetics Problems

1. 50%
2. John = *tt*; parents = *Tt*
3. 25%
4. man = *Wwtt*; woman = *wwTT*; child = *wwtt*
5. 75%
6. Mary and husband = *HH'*; child = *H'H'*
7. AB: yes; A, B, O, AB
8. light; white
9. girls = X^BX^bWw; boys = X^bYWw both = normal vision and widow's peak
10. Autosomal recessive
11. *AaBb* and *aabb*

Selected Key Terms

allele (ah-lēl′)   an alternative form of a gene that occurs at a given chromosomal site (locus).   357

dihybrid (dī-hī′brid)   the offspring of parents who differ in two ways: shows the phenotype governed by the dominant alleles but carries the recessive alleles. 362

dominant allele (dom′ĭ-nant ah-lēl′)   form of a gene that expresses itself or a characteristic that is present even when the genotype is heterozygous.   357

genotype (jēn′o-tīp)   the genetic makeup of any individual.   357

heterozygous (het″er-ō-zī′gus)   having two different alleles (as *Aa*) for a given trait. 357

homozygous (hō″mo-zī′gus)   having identical alleles (as *AA* or *aa*) for a given trait; pure breeding.   357

monohybrid (mon″ō-hī′brid)   the offspring of parents who differ in one way only; shows the phenotype governed by the dominant allele, but carries the recessive allele.   358

pedigree chart (ped′i-grē chart)   a representation showing the pattern of inheritance of a condition among family members.   363

phenotype (fē′no-tīp)   the outward appearance of an organism caused by the genotype and environmental influences. 357

recessive allele (re-ses′iv ah-lēl′)   form of a gene that expresses itself or a characteristic that is present only when the genotype is homozygous.   357

sex-linked (seks′linkt)   alleles located on sex chromosomes.   370

testcross (test kros)   the crossing of a heterozygote with an organism homozygous recessive for the characteristic(s) in question in order to determine the genotype.   360

X-linked (eks′linkt)   an allele located on X chromosome.   370

Chapter Nineteen

DNA and

Biotechnology

Chapter Concepts

1 DNA is the genetic material, and therefore its structure and function constitute the molecular basis of inheritance.

2 DNA is able to replicate, mutate, and control the phenotype.

3 DNA controls the phenotype by controlling protein synthesis, a process that also requires the participation of RNA.

4 Molecular genetics, including genetic engineering, has expanded the capabilities of an industry now called biotechnology.

5 Modern-day biotechnology is expected to revolutionize medical care and agriculture. It also has many environmental applications.

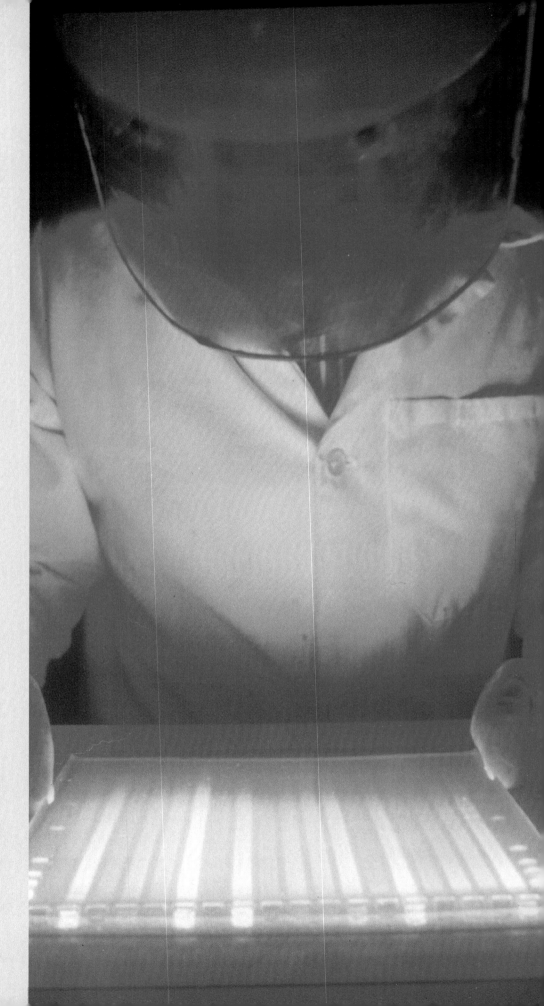

A laboratory technician is comparing several samples of "DNA fingerprints" under ultraviolet light. The DNA has been removed from the cells and treated with a special enzyme that cleaves the DNA into fragments. Each and every person has their own fragment length pattern, hence the term DNA fingerprint.

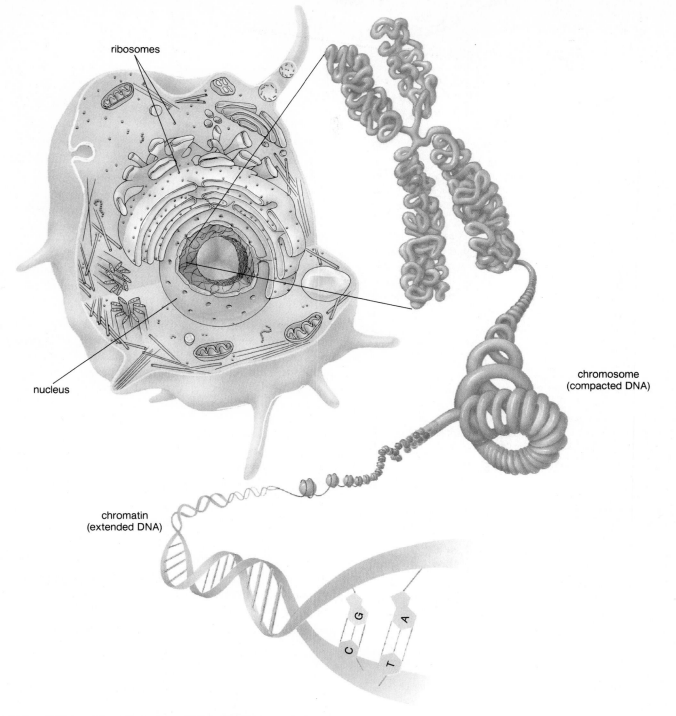

ribosomes

chromosome
(compacted DNA)

nucleus

chromatin
(extended DNA)

C
G
T
A

Figure 19.1   DNA location and structure. DNA is highly compacted in chromosomes, but it is extended as chromatin during interphase. It is during this time that DNA can be extracted from a cell and its structure studied.

Molecular genetics provides evidence of evolution. The same four types of nucleotides found in human DNA (fig. 19.1) make up the DNA of all organisms. Further, you can put a human gene into a bacterium, and the gene will perform its normal cellular function. It is not surprising, then, that you can also transfer genes between plants and animals. For certain, the very first cell or cells must have evolved into all the life forms we see about us and the evolutionary process must be dependent upon slight changes in the DNA. Only in recent times have we come to realize that in order to truly understand the evolution of plants and animals, we have to look inside the workings of a cell.

Nucleic Acids within the Cell

The chromosomes located within the nucleus of the cell (fig. 19.1) contain DNA, a nucleic acid having a structure we examined in chapter 1. Another nucleic acid, called RNA, is located within both the nucleus and the cytoplasm.

You will recall from the discussion on page 26 that the **DNA** structure is a double helix in which the two nucleotide strands are held together by hydrogen bonds between purine and pyrimidine bases. Thymine (T) is always paired with adenine (A), and guanine (G) is always paired with cytosine (C). This is called **complementary base pairing.** Besides one of these bases, each nucleotide in DNA contains a phosphate molecule and the sugar deoxyribose.

RNA is a single-stranded nucleic acid, and each polymer has a sequence of nucleotides that will also pair with a portion of DNA. However, the sugar within the nucleotides that make up an RNA strand is ribose—not deoxyribose. Also, the pyrimidine thymine does not appear in RNA; it is replaced by the pyrimidine uracil. Therefore, RNA contains these four bases: adenine (A), guanine (G), cytosine (C), and uracil (U). Table 19.1 summarizes the differences between DNA and RNA structure.

DNA has a structure like a twisted ladder: sugar-phosphate backbones make up the sides of the ladder; hydrogen-bonded bases make up the rungs of the ladder. The base A is always paired with the base T, and the base C is always paired with the base G. RNA differs from DNA in several respects (table 19.1).

DNA, the Hereditary Material

DNA, the hereditary material, has these four functions:

1. stores the genetic information that is needed by cells to carry on their routine activities;
2. replicates—makes copies of itself—that are passed on from cell to cell and from generation to generation;
3. controls the activities of the cell, thereby producing the phenotypic characteristics of the individual and the species;
4. undergoes mutations—permanent genetic changes passed on to the offspring—that ultimately account for the evolutionary history of life.

Replication

The double-stranded structure of DNA lends itself to replication because each strand can serve as a template for the formation of a complementary strand. A **template** is most often a mold used to produce a shape opposite to itself. In this case, the word *template* is appropriate because each new strand of DNA has a sequence of bases complementary to the bases of the old strand of DNA.

Replication requires the following steps (fig. 19.2):

1. The two strands that make up DNA become "unzipped" (i.e., the weak hydrogen bonds between the paired bases are broken).

Table 19.1 DNA Structure Compared to RNA Structure

	DNA	RNA
sugar	deoxyribose	ribose
bases	adenine, guanine, thymine, cytosine	adenine, guanine, uracil, cytosine
strands	double stranded with base pairing	single stranded
helix	yes	no

2. New complementary nucleotides, drawn from a "pool" of free nucleotides that are present in the cell, move into place by the process of complementary base pairing.
3. The adjacent nucleotides become joined through their sugar-phosphate components to form the new chain.
4. When the process is finished, two complete DNA molecules are present, identical to each other and to the original molecule.

This replication process is described as *semiconservative* because each double strand of DNA contains one old strand and one new strand. Although DNA replication can be easily explained, it is in actuality an extremely complicated process involving many steps and enzymes. The enzyme helicase assists the unwinding process, and the enzyme DNA polymerase joins the nucleotides together, for example. On occasion, errors are made that cause a change in the DNA and, in this way, a *mutation* can arise.

During replication, DNA becomes "unzipped" and a complementary strand forms opposite to each original strand. This is called semiconservative replication.

Genes and Enzymes

Many novel experiments and years of research allowed scientists to conclude that there is a relationship between genetic inheritance and the structure of enzymes and other proteins. For example, the metabolic pathway outlined in figure 19.3 was discovered in the early twentieth century. In this pathway, three genetic diseases are known. In the disorder called *phenylketonuria (PKU),* which was discussed on page 365, phenylpyruvate accumulates in the body and spills over into the urine because the enzyme needed to convert phenylalanine (a building block of protein) to tyrosine is missing. If the condition is not treated, the continued accumulation of phenylpyruvate can cause mental retardation. *Albinism* results because tyrosine cannot be converted to melanin, the natural pigment in human skin. The genetic disease *alkaptonuria* results if the enzyme needed to metabolize homogentisate is missing.

At first these conditions were simply called *inborn errors of metabolism,* and only later was it confirmed that the genetic fault lay in the absence of particular enzymes. This gave rise to the suggestion that genes in some way control the presence of enzymes in the individual. This was called the *one gene–one enzyme* theory.

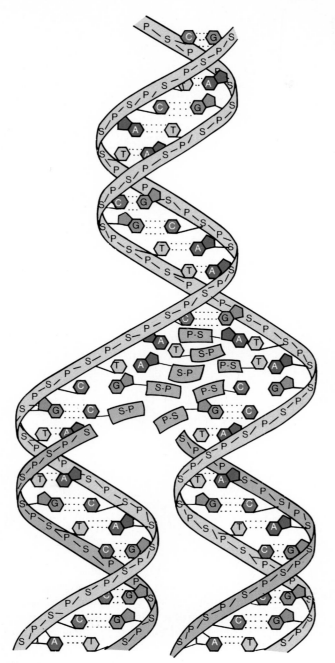

Region of parental DNA helix. (Both backbones are light.)

Region of replication. Parental DNA is unzipped and new nucleotides are pairing with those in parental strands.

Region of completed replication. Each double helix is composed of an old parental strand (light) and a new daughter strand (dark). Notice that each double helix is exactly like the other one and like the original parental strand.

Figure 19.2   DNA replication. Replication is called semiconservative because each new double helix is composed of an old parental strand and a new daughter strand.

Since enzymes are proteins, the concept was soon broadened to be the one gene–one protein theory. However, some proteins like hemoglobin have more than one type of polypeptide chain (fig. 6.3c). In persons with sickle cell disease, it is only the β polypeptide chain that has an altered sequence of amino acids compared to the normal β chain. Therefore, it may be more appropriate to state that a gene controls the sequence of amino acids in a polypeptide. Today we define a gene as a section of a DNA molecule that determines the sequence of amino acids in a single polypeptide chain of a protein.

It was a breakthrough to discover that there is a relationship between genetic inheritance and the primary structure of proteins in an individual.

It should be possible to cure genetic disorders by supplying needed enzymes. Researchers have recently reported, as described in the reading on page 381, that they can cure Gaucher's disease by supplying the patients with the enzyme they need to remain healthy.

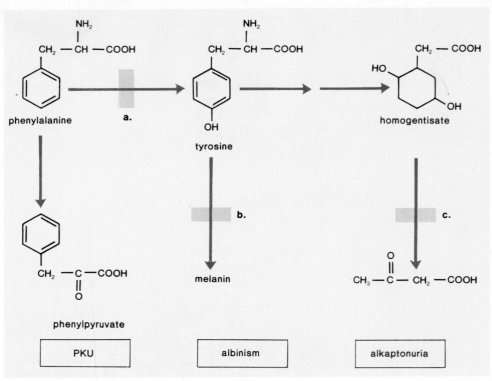

Figure 19.3   An albino (left) is unable to produce the pigment melanin. Metabolic pathway (right) by which phenylalanine is converted to other metabolites. *a.* If the enzyme that converts the phenylalanine to tyrosine is defective, phenylalanine is converted to phenylpyruvate instead, and the accumulation of this substance leads to PKU (phenylketonuria). *b.* If the enzyme that converts tyrosine to melanin is defective, albinism results. *c.* If homogentisate cannot be metabolized, alkaptonuria (the urine darkens) results.

Protein Synthesis

The fact that DNA controls the production of proteins at first may seem surprising when we consider that genes are located in the nucleus of higher cells but proteins are synthesized at the ribosomes in the cytoplasm. However, although DNA is found only in the nucleus (fig. 19.1), RNA exists in both the nucleus and the cytoplasm.

The central dogma of modern genetics is diagrammed in this manner:

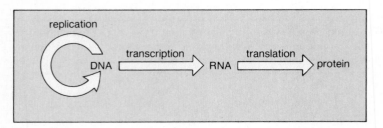

The diagram indicates that DNA not only serves as a template for its own replication, it is also a template for RNA formation. The arrow between DNA and RNA indicates that RNA is transcribed from the DNA template. **Transcription** is making an RNA molecule that is complementary to a portion of DNA. Following transcription, RNA moves into the cytoplasm. Photographs are available that show that radioactively labeled RNA moves from the nucleus to the cytoplasm, where protein synthesis occurs.

The arrow between RNA and protein in the diagram indicates that proteins are synthesized according to RNA's instructions. The specific type of RNA that has this job is called **messenger RNA (mRNA).** During **translation,** the information carried by RNA (in nucleotides) is used to produce the correct order of amino acids in a polypeptide. You can see that these terms are appropriate because transcribing a document means making a close copy of it, while translating a document means putting it in an entirely different language.

Code of Heredity

During transcription, DNA provides mRNA with a message that directs the order of amino acids during translation, when protein synthesis occurs. The message cannot be contained in the sugar-phosphate backbone because it is constant in every DNA

Effective Treatment for Genetic Disease

In one of medicine's rare triumphs over a genetic disorder, researchers said yesterday that they have dramatically reversed the symptoms of Gaucher's disease, a progressive, disabling and sometimes lethal condition that affects mainly Jews of Eastern European origin.

The treatment involves replacing an enzyme lacking in Gaucher's patients with a purified form obtained from human placentas.

A year after beginning treatment, "children who sat at desks and didn't participate in playground activities or sports now are doing them," said Norman W. Barton of the National Institutes of Health, where the work was carried out. "Adults are enjoying very full social schedules, in addition to working, rather than coming home and falling into bed and sleeping for 15 hours."

The success of the experimental treatment in a dozen patients, however, will pose problems. The enzyme must be taken for life at an estimated $100,000 to $200,000 a year, the researchers said, unless lower doses are found to maintain the patients. . . .

In a report on a yearlong study of the treatment, a federal researcher said that even the scientists were surprised at how quickly and effectively the enzyme reversed the Type 1 form of Gaucher's disease. The results are published in a paper by Barton, Dr. Roscoe Brady of the National Institutes of Health and Dr. S. H. Doppelt and Dr. Henry J. Mankin of Massachusetts General Hospital in Boston.

The feat represents a rare bright spot in the battle against genetic diseases, most of which still lack cures or effective treatments even when the defective gene has been identified.

Gaucher's disease is the most common inherited metabolic disease of East European Jews, said Barton, occurring in about one in every 400 births, though many non-Jews also are affected. Some who carry the pair of mutant genes that cause the condition have no apparent symptoms. Others suffer from enlarged spleens, progressive anemia and agonizingly painful "bone crises," Barton said.

As they grow older, patients suffer from easily broken bones and their bone marrow loses its ability to generate blood cells, resulting in sometimes lethal infections. The fundamental cause is the lack of the enzyme glucocerebrosidase, which the body needs to break down waste material from cells. As a result, tissues and organs become overloaded with fatty material. . . .

The ability to replace the missing enzyme was discovered only recently, after years of work to purify the enzyme and to determine the dose, Barton said. He said the placentas used in the manufacturing process by the technology firm Genzyme, Inc. are brought from France, which is one reason for the high cost. . . .

By Richard Saltus. *The Boston Globe* October 1990. Reprinted courtesy of *The Boston Globe.*

molecule. However, the order of the bases in DNA and mRNA can and does change. Therefore, it must be the bases that contain the message. The order of the bases in DNA must code for the order of the amino acids in a polypeptide. Can four bases provide enough combinations to code for 20 amino acids? It can if the code is a *triplet code;* each amino acid is dictated by a sequence of three bases.

Just as our alphabet forms words that can be arranged to provide information, so the bases of DNA form the triplets, whose sequences provide the information needed by an organism to develop, maintain itself, and reproduce. To understand the language of DNA, it was necessary to decipher the DNA code. To crack the code, artificial RNA was added to a medium containing bacterial ribosomes and a mixture of amino acids. Comparison of the bases in the RNA with the resulting polypeptide allowed investigators to decipher the code. Each three-letter unit of a messenger RNA is called a **codon.** All 64 codons have been determined (fig. 19.4). Each of 61 triplets correspond to a particular amino acid; the remaining three are stop codons that code for chain termination. The one codon that stands for the amino acid methionine is also a start codon that signals polypeptide initiation.

The Universal Genetic Code

Research indicates that the genetic code is essentially universal. The same codons stand for the same amino acids in all living things, including bacteria, plants, and animals. This illustrates the remarkable biochemical unity of living things and suggests that all living things have a common evolutionary ancestor.

DNA contains a code, and its message is passed to mRNA during transcription. Sixty-one of the 64 triplet codons stand for particular amino acids, and the other three codons are stop codons. During translation, the order of the codons in mRNA determines the order of the amino acids in a protein.

Transcription

During transcription, the DNA code is passed to mRNA. Therefore, the code is transcribed, or copied.

Messenger RNA

Following transcription (fig. 19.5), mRNA has a sequence of bases complementary to DNA; wherever A, T, G, or C is present

First Base	Second Base				Third Base
	U	C	A	G	
U	UUU phenylalanine	UCU serine	UAU tyrosine	UGU cysteine	U
	UUC phenylalanine	UCC serine	UAC tyrosine	UGC cysteine	C
	UUA leucine	UCA serine	UAA *stop*	UGA *stop*	A
	UUG leucine	UCG serine	UAG *stop*	UGG tryptophan	G
C	CUU leucine	CCU proline	CAU histidine	CGU arginine	U
	CUC leucine	CCC proline	CAC histidine	CGC arginine	C
	CUA leucine	CCA proline	CAA glutamine	CGA arginine	A
	CUG leucine	CCG proline	CAG glutamine	CGG arginine	G
A	AUU isoleucine	ACU threonine	AAU asparagine	AGU serine	U
	AUC isoleucine	ACC threonine	AAC asparagine	AGC serine	C
	AUA isoleucine	ACA threonine	AAA lysine	AGA arginine	A
	AUG (*start*) methionine	ACG threonine	AAG lysine	AGG arginine	G
G	GUU valine	GCU alanine	GAU aspartic acid	GGU glycine	U
	GUC valine	GCC alanine	GAC aspartic acid	GGC glycine	C
	GUA valine	GCA alanine	GAA glutamic acid	GGA glycine	A
	GUG valine	GCG alanine	GAG glutamic acid	GGG glycine	G

Figure 19.4   mRNA codons. These mRNA codons are complementary to the code in DNA. The DNA code is degenerate; most of the 20 amino acids have more than one codon. In this chart, notice that each of the codons comprise a letter from the first, second, and third bases. For example, find the square where C (for the first base) and A (for the second base) come together, and then look across to the right. You will notice the letters for the third base of the codons for histidine and glutamine.

in the DNA template, U, A, C, or G is incorporated into the mRNA molecule. A segment of the DNA helix unwinds and unzips, and complementary RNA nucleotides pair with DNA nucleotides of one strand. When these RNA nucleotides are joined by an enzyme called RNA polymerase, an mRNA molecule results. mRNA has a sequence of bases that are triplet codons complementary to the DNA triplet code.

After the mRNA strand is processed, it passes from the cell nucleus into the cytoplasm. There, it becomes associated with the ribosomes.

RNA Processing

Most genes in eukaryotes are interrupted by segments of DNA that are not part of the gene. These portions are called introns because they are *intra*gene segments. The other portions of the gene are called exons because they are ultimately *ex*pressed. A gene is expressed when a protein product results. When DNA is transcribed, the mRNA contains bases that are complementary to both exons and introns, but before the mRNA exits from the nucleus, it is *processed*—the nucleotides complementary to the introns are removed enzymatically.

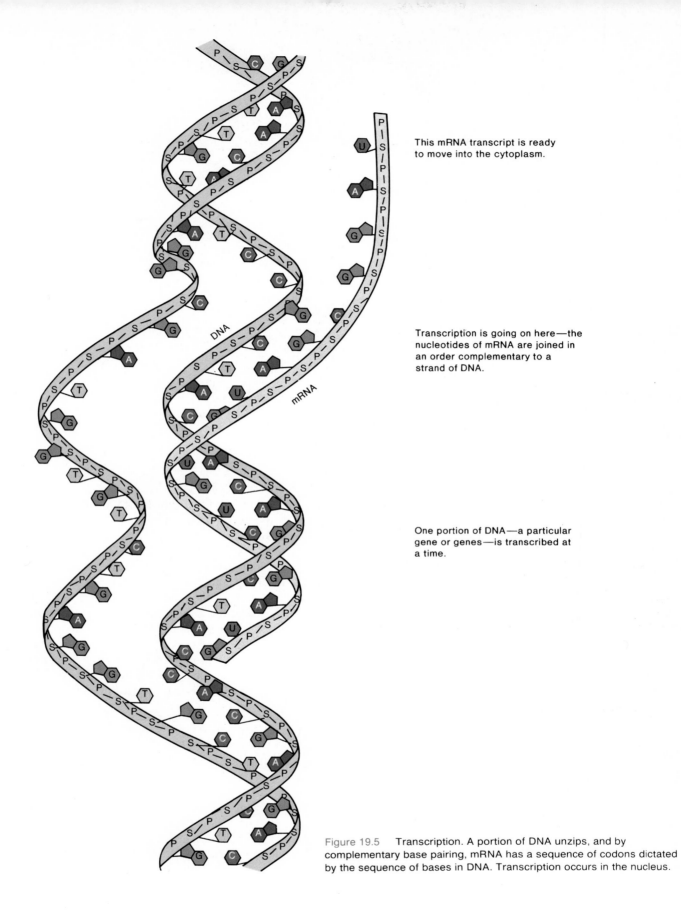

This mRNA transcript is ready to move into the cytoplasm.

Transcription is going on here—the nucleotides of mRNA are joined in an order complementary to a strand of DNA.

One portion of DNA—a particular gene or genes—is transcribed at a time.

DNA

mRNA

Figure 19.5   Transcription. A portion of DNA unzips, and by complementary base pairing, mRNA has a sequence of codons dictated by the sequence of bases in DNA. Transcription occurs in the nucleus.

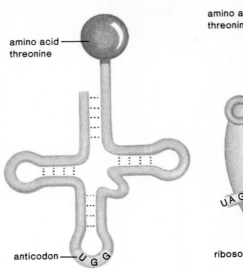

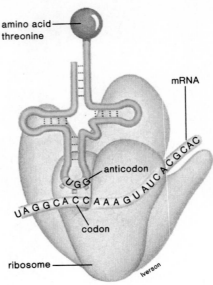

a. Transfer RNA (tRNA)

b. Transfer RNA (tRNA) at ribosome

Figure 19.6   Anticodon-codon base pairing. *a.* tRNA molecules have an amino acid attached to one end and an anticodon attached to the other end. If the anticodon is UGG, the amino acid is threonine (see text). *b.* The anticodon of a tRNA molecule is complementary to a codon. The pairing between codon and anticodon ensures that the sequence of amino acids in a polypeptide is that directed orginally by DNA.

There has been much speculation about the role of introns in the genes of eukaryotes. It is possible that introns allow crossing-over within a gene during meiosis. It is also possible that introns divide a gene into domains that can be joined in different combinations to give novel genes and protein products, a process that perhaps facilitates evolution.

During transcription, mRNA is made complementary to one of the DNA strands. It then contains a sequence of codons and moves into the cytoplasm, where it becomes associated with the ribosomes.

Translation

During translation, the sequence of codons in mRNA dictates the order of amino acids in a polypeptide. This is called translation because the sequence of bases in DNA is translated into a particular sequence of amino acids. Translation requires the involvement of several enzymes and two other types of RNA: ribosomal RNA and transfer RNA.

Ribosomal RNA

Ribosomal RNA (rRNA) makes up the ribosomes (see fig. 2.7), which are composed of two subunits, each with characteristic RNA and protein molecules. The rRNA molecules are transcribed from DNA in the region of the nucleolus. The proteins are manufactured in the cytoplasm but then migrate to the nucleolus, where the ribosomal subunits are assembled before they migrate to the cytoplasm. Ribosomes play an important role in coordinating protein synthesis.

Transfer RNA

Small molecules of **transfer RNA (tRNA)** bring the amino acids from the cytoplasm to the ribosomes located in the cytoplasm. A particular amino acid attaches to a tRNA at one end (fig. 19.6a). Attachment requires ATP energy, and the resulting bond is a high-energy bond represented by a wavy line. Therefore, the entire complex is designated as tRNA ∼ amino acid.

At the other end of each tRNA molecule, there is a specific **anticodon** complementary to an mRNA codon (fig. 19.6b). When the tRNA molecule comes to the ribosome, the anticodon pairs with a codon. Let us consider an example: For the codon ACC, what will be the tRNA molecule's anticodon, and what amino acid will be attached to the tRNA molecule? Inspection of figure 19.4 allows us to determine this:

codon	anticodon	amino acid
ACC	UGG	threonine

In this way, it is the order of the codons of the mRNA that determines the sequence of amino acids in the polypeptide.

During translation, tRNA molecules, each carrying a particular amino acid, travel to the mRNA, and through complementary base pairing between anticodon and codon, the tRNA molecules and therefore the amino acids in a polypeptide chain are sequenced in a predetermined order.

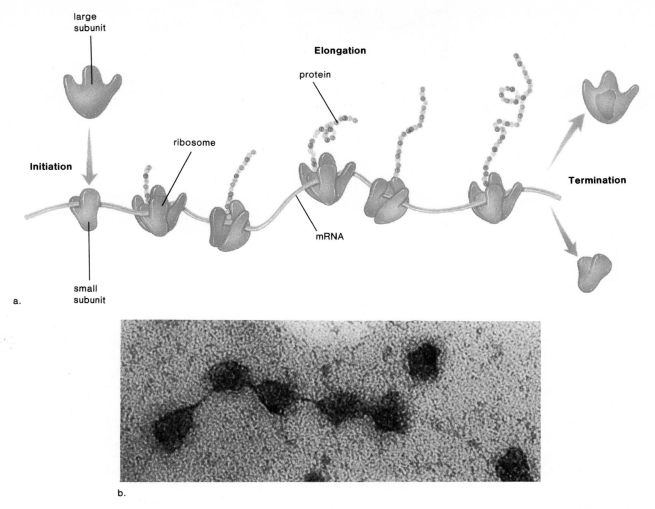

large
subunit

Elongation

protein

ribosome

Initiation

Termination

small
subunit

mRNA

a.

b.

Figure 19.7   Polysome structure. *a.* Several ribosomes move along an mRNA molecule at a time. They function independently of each other; therefore, several polypeptides can be made at the same time. *b.* Electron micrograph of a polysome.

Process of Translation

Protein synthesis requires three steps: initiation, elongation, and termination (fig. 19.7). During *initiation,* a ribosome binds to an mRNA molecule. Initiation always begins with a codon that stands for the amino acid methionine. First, the smaller ribosomal subunit binds to mRNA, and then the larger subunit joins to the smaller subunit, giving a complete ribosomal structure. *Elongation* occurs as the polypeptide chain grows in length. A ribosome is large enough to accommodate two tRNA molecules; the peptide chain attached to the tRNA molecule in the first position is transferred to the tRNA ~ amino acid complex in the second position. In this manner, the peptide chain grows and the primary structure of a protein comes about. The secondary and tertiary structures of a polypeptide appear after termination, as the amino acids, in a predetermined sequence within the polypeptide chain, interact with one another.

A polysome is several ribosomes moving along one mRNA at a time. Polysomes allow several polypeptides of the same type to be synthesized at once (fig. 19.7). *Termination* per ribosome occurs at a stop codon on the mRNA. The ribosome dissociates into its two subunits and falls off the mRNA molecule. Protein synthesis continues until the very last ribosome has come to the stop codon.

Protein synthesis requires the process of transcription and translation. Figure 19.8 gives a summary of protein synthesis.

Definition of a Gene

Classical (Mendelian) geneticists thought of a gene as a particle on a chromosome. To molecular geneticists, however, a gene is a sequence of DNA nucleotide bases that code for a product. Most often the gene product is a polypeptide chain. However, there are exceptions because rRNA and tRNA, for example, are gene products themselves. They are involved in protein synthesis, but they do not code for protein.

Other Features of DNA

Two features of DNA chemistry not previously discussed will be discussed here.

Name of Molecule	Special Significance	Definition	Name of Molecule	Special Significance	Definition
DNA	Code	Sequence of DNA bases in threes	rRNA	Ribosome	Site of protein synthesis
mRNA	Codon	Sequence of RNA bases complementary to DNA code	Amino acid	Building block for protein	Transported to ribosome by tRNA
tRNA	Anticodon	Sequence of 3 bases complementary to codon	Protein	Enzyme	Amino acids joined in a predetermined order

a.

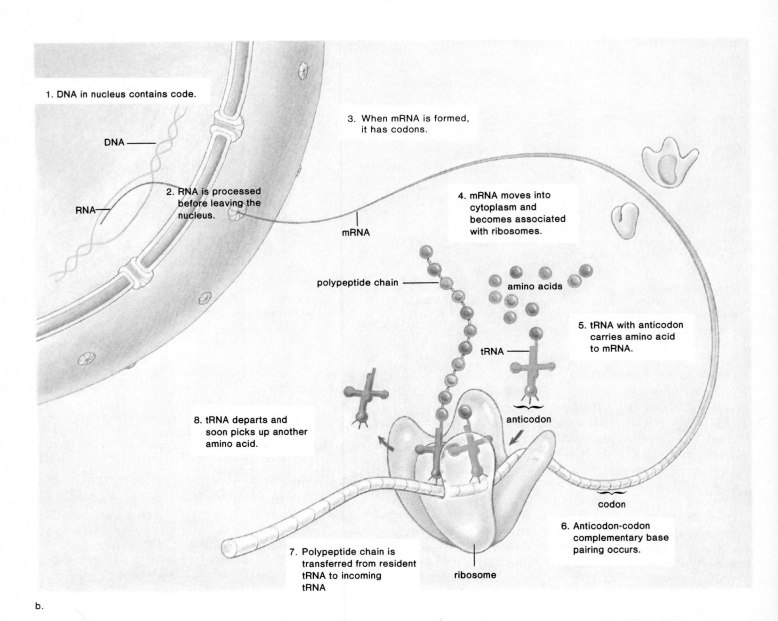

1. DNA in nucleus contains code.

DNA

RNA

2. RNA is processed before leaving the nucleus.

3. When mRNA is formed, it has codons.

mRNA

4. mRNA moves into cytoplasm and becomes associated with ribosomes.

polypeptide chain

amino acids

5. tRNA with anticodon carries amino acid to mRNA.

tRNA

anticodon

8. tRNA departs and soon picks up another amino acid.

codon

6. Anticodon-codon complementary base pairing occurs.

7. Polypeptide chain is transferred from resident tRNA to incoming tRNA

ribosome

b.

Figure 19.8   Summary of protein synthesis. *a*. List of participants in the process of protein synthesis. *b*. Diagram showing activities of participants. Transcription occurs in the nucleus, and translation occurs in the cytoplasm (blue). During translation, the codons borne by mRNA dictate the order of the amino acids in the polypeptide.

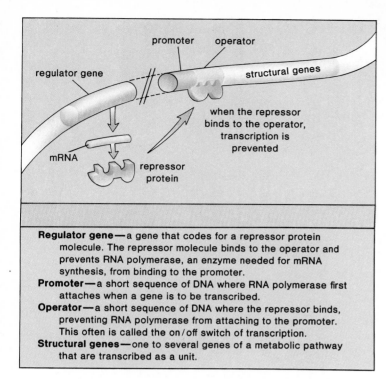

Regulator gene—a gene that codes for a repressor protein molecule. The repressor molecule binds to the operator and prevents RNA polymerase, an enzyme needed for mRNA synthesis, from binding to the promoter.

Promoter—a short sequence of DNA where RNA polymerase first attaches when a gene is to be transcribed.

Operator—a short sequence of DNA where the repressor binds, preventing RNA polymerase from attaching to the promoter. This often is called the on/off switch of transcription.

Structural genes—one to several genes of a metabolic pathway that are transcribed as a unit.

Figure 19.9   One model that explains the regulation of protein synthesis has the components listed here. With this model, the structural genes are normally turned off. However, when a specific inducer molecule is present, it combines with the repressor molecule, preventing it from attaching to the operator gene. Now transcription of the structural genes and protein synthesis occur. An inducer molecule could be the beginning substrate of a metabolic pathway—its very presence signifies that certain enzymes are needed to break it down.

Structural versus Regulatory Genes

Certain genes, called **regulatory genes,** control or influence whether or not other genes, called **structural genes,** are transcribed. It is the structural genes that code for metabolic enzymes and various cell components. One model by which regulatory genes may function is shown in figure 19.9. The model allows us to see that, in order to understand the phenotype, it is necessary to consider the entire genotype rather than each gene separately. In other words, the presence of a structural gene does not necessarily mean that it will be expressed (i.e., transcribed and translated) since it can be turned on or off by a regulatory gene.

The presence of regulatory genes is important to the occurrence of differentiation, the existence of specialized cells in the body. All cells have the same chromosomes and genes, yet some are intestinal cells, others are muscle cells, others are blood cells, etc. There is evidence that as development proceeds, certain genes get turned on and certain other genes get turned off so that specialization occurs.

Knowledge about regulatory genes is extremely important because genes can only be manipulated when we know how to turn them on and off. Also, mutations of regulatory genes probably account for some genetic diseases and/or the development of cancer.

Table 19.2   Types of Mutations

Mutation	Definition
chromosomal mutation	a rearrangement of chromosome parts, as described in figure 17.17, which may or may not result in a change of the phenotype
genetic mutation	a change in the genetic code for a gene or in the expression of the gene. Usually results in a change of the phenotype
germinal mutation	a mutation that manifests itself in the gametes so that it is passed on to offspring
somatic mutation	a mutation that occurs in the body cells and that very likely is not passed on to offspring

Structural genes code for proteins, and regulatory genes control the expression of structural genes.

Genetic Mutations

We have previously studied chromosomal mutations (fig. 17.17), but there are also genetic mutations. As you can see in table 19.2 a **genetic mutation** is any alteration in the code of a single gene or any change in its expression. Genetic mutations do not necessarily have a deleterious effect; some may have no effect at all and some may even have a beneficial effect.

Mutagenic Agents

Certain environmental influences like radiation, either UV light or X rays, and organic chemicals can cause genetic mutations to occur. Ultraviolet light most often causes two adjacent thymine bases to bond together. Such *thymine dimers* prevent replication of DNA and prevent mRNA transcription. Some chemicals can cause one type of base to be converted to a different type, but others are base analogues that are incorporated into DNA in place of a normal base. These, like AZT, which is used as a medication for AIDS patients (p. 301), also disrupt DNA replication and mRNA transcription.

Substitutions, Alterations, and Deletions of Bases

Mutations involving a change in the DNA sequence of bases are of three types (table 19.3). *Substitutions,* which involve a change in a single base, may have no effect at all if the new codon happens to stand for the same amino acid. Other base substitutions may lead to an amino acid substitution in the protein product. For example, a change from GAG to GUG would cause glutamic acid to be replaced by valine as it is in sickle-cell hemoglobin. This particular substitution causes a drastic deleterious effect in the phenotype. It's also possible that a base substitution might result in a new "stop" codon, and this would cause transcription to be terminated before the polypeptide is

Table 19.3 Gene Mutations

Base change	Example	Worse result
normal	TAC'GGC'ATG	
substitution	TAG'GGC'ATG	change in one amino acid or change to stop signal
deletion	ACG'GCA'TG	polypeptide altered completely
addition	ATA'CGG'CAT'G	polypeptide altered completely

fully formed. This type of base change is being investigated as a possible cause for hemophilia, the sex-linked blood clotting disorder (p. 371).

Additions and *deletions* of bases are expected to result in profound alterations in the DNA code. If the altered allele codes for an enzyme, the enzyme would most likely be nonfunctional because the sequence of amino acids would be so greatly affected.

Transposons

Transposons are specific DNA sequences that have the remarkable ability to move within and out of chromosomes. Their movement to a new location sometimes alters neighboring genes, particularly by increasing or decreasing their expression. Although "movable elements" in corn were described 40 years ago, their significance was only recently realized. So-called "jumping genes" have now been discovered in bacteria, fruit flies, and humans, and it is likely that all organisms have such elements.

Biotechnology

Biotechnology, the use of a natural biological system to produce a product or to achieve an end desired by human beings, is not new. Plants and animals have been bred to give a particular phenotype since the dawn of civilization. Also, the biochemical capabilities of microorganisms have been exploited for a very long time. For example, the baking of bread and the production of wine are dependent on yeast cells to carry out fermentation reactions.

Today, however, biotechnology is first and foremost an industrial process (fig. 19.10) that provides products because we are able to genetically engineer bacteria to produce human proteins or other proteins of interest. Biotechnology is also being investigated as a way to alter the genotype and, subsequently, the phenotype of plants and animals. It is also hoped that eventually it may be possible to carry out gene therapy in human beings to correct inherited disorders.

Biotechnology Techniques

Genetically engineered, or simply bioengineered, bacteria contain a foreign gene, such as a human gene, and therefore are capable of producing a human protein (fig. 19.11). Often a human gene is carried into a bacterium as a part of recombinant DNA.

a.

b.

c.

d.

Figure 19.10 Biotechnology is an industrial endeavor. *a.* Laboratory procedures must be adapted to mass-produce the product. *b.* Microbes are grown in huge tanks called fermenters because they were first used for yeast fermentation to produce wine. *c.* The product is purified, and *d.* packaged.

Recombinant DNA

Recombinant DNA contains DNA from two or more different sources, such as bacterial and human DNA. Plasmids are frequently used as **vectors** to carry recombinant DNA into bacteria. This is logical because plasmids are taken from bacteria in the first place. **Plasmids** are small accessory rings of DNA that carry genes not present in the bacterial chromosome.

The introduction of a human gene into a plasmid to give recombinant DNA requires two different types of enzymes. The

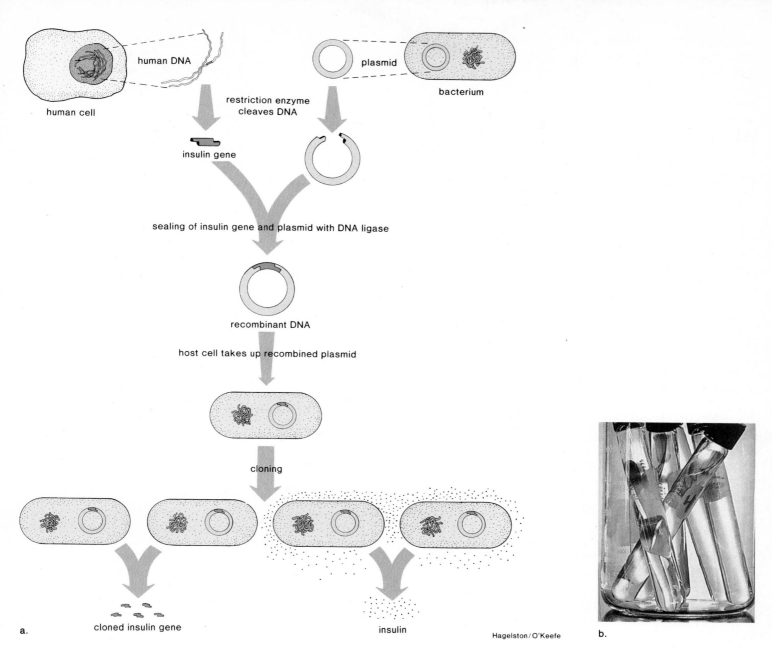

a.

b.

Hagelston/O'Keefe

Figure 19.11   *a.* Methodology for cloning a human gene. Human DNA and plasmid DNA are cleaved by the same type restriction endonuclease enzyme and spliced together using the enzyme DNA ligase. After a bacterium takes up the plasmid, gene cloning occurs and the product is produced. The investigator can retrieve either the genes or the product for further analysis or use. *b.* Test tubes containing restriction endonuclease enzyme.

first type of enzyme is a **restriction endonuclease enzyme.** Restriction enzymes have the capability of cutting a molecule of DNA into discrete pieces. These enzymes occur naturally in bacteria. If present, they protect bacteria from infection by cutting up viral DNA so that the viruses are unable to reproduce. They are called restriction enzymes because they restrict the growth of viruses. Each type of restriction enzyme—and hundreds are now known—cleaves DNA at specific locations called restriction sites. For example, there is a restriction enzyme that always cleaves double-stranded DNA wherever it has this sequence of bases:

. . . . .G A A T T C. . . . .
. . . . .C T T A A G. . . . .

Furthermore, this enzyme always cleaves each strand between a G and an A in this manner:

. . . . .G   A A T T C. . . . .
. . . . .C T T A A   G. . . . .

Notice that there is now a gap into which a piece of foreign DNA could be placed. It is only necessary that the foreign DNA piece end with bases complementary to those exposed by using the restriction enzyme. This is assured by cleaving the foreign DNA with the same type restriction enzyme. It must be the same type of restriction enzyme because different enzymes recognize and cleave different nucleotide sequences in DNA. The single-stranded ends that result from restriction enzyme use are called "sticky ends" because they facilitate the insertion of foreign DNA into the vector DNA.

The second enzyme needed is **DNA ligase,** an enzyme present in all cells, that ordinarily seals any breaks in a DNA molecule. Genetic engineers use this enzyme to seal the foreign piece of DNA into a plasmid being used as a vector. Gene splicing is now complete and a **recombinant DNA** molecule has been prepared. A recombinant DNA molecule is one that contains DNA from two different sources.

Human Issue

When the recombinant DNA technique was first devised, scientists were concerned about creating mutant bacteria that could possibly escape into the environment and cause harm. However, nothing untoward has yet developed because of DNA-splicing techniques. Indeed, a number of benefits have resulted. Still, concerns of potential hazards persist, and the governmental approval process for genetically engineered organisms and their products can take years. Do you support a cautious and restrictive approach or a free and open approach? What type of approval process would you suggest? By whom?

Bacteria will take up recombined plasmids and after a plasmid has entered, the plasmid will be copied whenever the bacterium reproduces. Eventually, there are many copies of the plasmid and, therefore, many copies of the human gene such as the gene that codes for insulin, a protein that is a hormone which helps regulate sugar metabolism in humans (p. 266).

At this point, the human gene has been **cloned** (i.e., there are many copies of it (fig. 19.11) and it is possible to remove these copies from the bacteria, if desired). Also, the bacteria are producing insulin and this product can be collected, purified, and sold.

Other Biotechnology Techniques

Aside from recombinant DNA, other techniques are common in biotechnology, including the following:

1. Gene sequencing, or determining the order of nucleotides in a gene. There are automated DNA sequencers that make use of computers to sequence genes at a fairly rapid rate.
2. Manufacture of a gene; that is, the nucleotides are joined in the correct sequence, or if desired, a mutated gene is prepared by altering the sequence.
3. Insertion of a gene directly into a host cell without using a plasmid. Animal cells, in particular, do not take up plasmids but DNA can be microinjected into them.
4. Insertion of the regulator—structural gene complex into a vector.

5. Placement of the isolated or constructed gene linked to appropriate regulatory regions in another cell where it undergoes replication and directs protein synthesis.

Biotechnology Products

Table 19.4 shows at a glance some of the types of biotechnology products that are now available. Monoclonal antibodies (p. 145) are not now but may soon be produced by recombinant DNA technology.

Hormones and Similar Types of Proteins

One impressive advantage of biotechnology is it allows mass production of proteins that are very difficult to obtain otherwise. For example, insulin previously was extracted from the pancreas glands of slaughtered cattle and pigs; it was expensive and sometimes caused allergic reactions in recipients. Human growth hormone previously was extracted from the pituitary gland of cadavers, and it took 50 glands to obtain enough for one dose. Few of us knew of tPA (tissue plasminogen activator), a protein present in the body in minute amounts that activates an enzyme to dissolve blood clots. Now tPA is a biotechnology product that is used to treat heart attack victims. Erythropoietin was not available at all and now it can be used to prevent anemia in patients with kidney failure receiving hemodialysis, for example.

Human Issue

Now that human growth hormone (HGH) is readily available, it can be more readily administered. In the past, the drug was only given to youngsters who had a severe deficiency of the hormone and were abnormally short. Now some parents want their children to take the drug so that they will be more likely to succeed in business or become an atheletic star. And adults want to take the drug because there is evidence to suggest that it causes the adult body to lose fat and gain muscle. Should HGH be prescribed to anyone who wishes to take it or should there be restrictions on its use? If there should be restrictions, what should they be?

Other biotechnology products that are expected soon are clotting factor VIII to treat hemophilia; human lung surfactant for premature infants with respiratory distress syndrome; and atrial natriuretic factor to treat hypertension. Factor VIII and lung surfactant have been available from other sources, but not as a biotechnology product.

Hormones for use in animals are also biotechnology products. Cows given bovine growth hormone (bGH) produce 25% more milk than usual, which should make it possible for dairy farmers to maintain fewer cows and to cut down on overhead expenses.

DNA Probes

A **DNA probe** is a piece of single-stranded DNA, often radioactive, that can be used to detect the presence of a certain allele in a cell or body fluid. Today, a probe can be prepared by using

Table 19.4 Representative Biotechnology Products

Hormones and similar types of proteins	DNA probes	Vaccines
Treatment of humans	*Diagnostics in humans*	*Use in humans*
insulin growth hormone tPA (tissue plasminogen activator) erythropoietin interleukin-2 clotting factor VIII* human lung surfactant* atrial natriuretic factor*	various diseases (e.g., sexually transmitted diseases) inherited disorders (e.g., cystic fibrosis) various cancers (e.g., chronic myelogenous leukemia) detection of criminals paternity suits	AIDS* herpes* (oral and genital) hepatitis A,* B, and C* malaria*

*Expected in the near future. These may be available now, but presently they are not made by recombinant DNA technology.

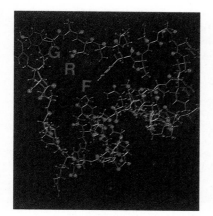

a.

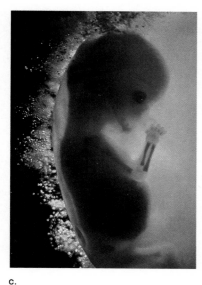

c.

Figure 19.12   Possible biotechnology scenario. *a.* A protein is received from a cell and the amino acid sequence is determined. (This hormone is called growth hormone-releasing factor.) From this, the sequence of nucleotides in DNA can be decoded. *b.* The DNA synthesizer can be used to string nucleotides together in the correct order. *c.* A small section of the gene can be used as a DNA probe to test fetal cells for a defect in the gene. The manufactured gene could also be placed in a bacterium to produce more of the protein, which could then be used for treatment of the defect.

a DNA synthesizer that links nucleotides together in the same order as a portion of a gene. The probe will bind by complementary base pairing to a particular DNA sequence. In the process called DNA fingerprinting, several probes are used to compare the DNA in hair or body fluid taken from the scene of a crime with that of DNA obtained from blood cells of a suspected criminal. If the data match, the criminal is caught just as if he or she had left fingerprints behind. DNA probes can

also be used in court cases to identify the actual parents, either mother or father, of an individual.

In medicine, DNA probes are used to diagnose an infection by indicating if the gene of an infectious organism is present. When available, a DNA probe can tell us whether a gene coding for a hereditary defect might be present (fig. 19.12). When a couple is using in vitro fertilization (p. 293), a probe can be used to test for a genetic disorder even before an embryo is placed

Polymerase Chain Reaction

The polymerase chain reaction (PCR) is a way to make multiple copies of a single gene, or any specific piece of DNA, in a test tube. Further, the process is very specific—the targeted DNA sequence can be less than one part in a million of the total DNA sample! This means that a single gene among all the human genes can be amplified (copied) using PCR.

The process takes its name from DNA polymerase, the enzyme that carries out DNA replication in a cell. The reaction is a chain reaction because DNA polymerase is allowed to carry out replication over and over again until there are a million or more copies of the targeted DNA. The polymerase chain reaction will not replace DNA cloning. Cloning provides many more copies of a gene than this, and it still will be used whenever a large quantity of a gene or a protein product is needed.

Before carrying out PCR, it is necessary to have available *primers*—sequences of about 20 bases that are complementary to the bases on either side of the "target DNA." The primers are needed because DNA polymerase does not start the replication process—it continues or extends the process. After the primers bind complimentarily to the DNA strand, DNA polymerase copies the target DNA and only the target DNA between the primers. Therefore, PCR is very specific.

PCR has been in use for several years, but a recent advance has been the introduction of automated PCR machines allowing almost any laboratory to carry out the procedure.

Automation became possible after a thermostable DNA polymerase was extracted from a bacterium. Using this enzyme means that there is no need to add more DNA polymerase each time a high temperature is used to separate double stranded DNA so that replication can occur once again.

The accompanying figure shows that after PCR amplication, it becomes a lot easier to use a probe to detect that target DNA is present. (The binding of the probe is detectable because the probe is labeled either radioactively, or with a fluorescent dye.) Therefore, because of the polymerase chain reaction, DNA probes are increasingly used for all sorts of purposes, including those discussed in the following paragraphs.

Genetic Disorders

PCR analysis, PCR amplification followed by the use of a DNA probe, can be used to determine if a mutant gene responsible for a genetic disorder is present in a single fetal or embryonic cell. Therefore, following in vitro fertilization, PCR analysis can be used to ensure the implantation of only normal embryos—ones that lack the faulty gene.

Infectious Diseases

Blood tests currently used to diagnose human illness often detect the presence of antibodies to the pathogen rather than the pathogen itself. For example, the blood test for AIDS does not detect HIV but confirms the presence of antibodies to HIV in the blood. PCR analysis can be used to detect the presence of the pathogen before the immune system has even begun an antibody response. It then becomes possible to begin treatment of a disease like AIDS earlier than before.

Cancer Diagnosis

It has long been the custom for physicians to save the tissues of persons who have died from various illnesses. Some of these are cancer patients. PCR is making it possible to analyze the DNA of deceased patients and of living patients who currently have a particular type of cancer. Using this

---

in the uterus. This ensures that a couple will only have children that are free of a particular genetic disorder.

As discussed later in this chapter, DNA probes are also being used to help investigators map the human chromosomes. More and more uses for DNA probes are being found because the polymerase chain reaction allows "targeted" pieces of DNA to be amplified before the probe is applied. The reading on this page gives more information about the polymerase chain reaction.

Vaccines

Vaccines are used to make people immune so that they do not become ill when exposed to an infectious organism. This comes about because a vaccine causes the build up of antibodies that protect the individual from the infectious organism. In the past, vaccines were made from treated bacteria or viruses. But bacteria and viruses have surface proteins, and a gene for just one

of these can be placed in a plasmid. When bioengineered bacteria produce many copies of the surface protein, these copies can be used as a vaccine. Right now, the only vaccine produced through biotechnology is hepatitis B, but others such as a vaccine for malaria and another for AIDS are expected.

Transgenic Organisms

Free-living organisms in the environment that have had a foreign gene inserted into them are called **transgenic organisms.**

Transgenic Bacteria

Naturally occurring bacteria have been bioengineered to perform a service in the environment. For example, in agriculture, engineered bacteria can be used to promote the health of plants. Certain bacteria that normally live on plants have been changed from frost-plus to frost-minus bacteria. Whereas before they promoted frost damage of plants, now they prevent it. Also, a

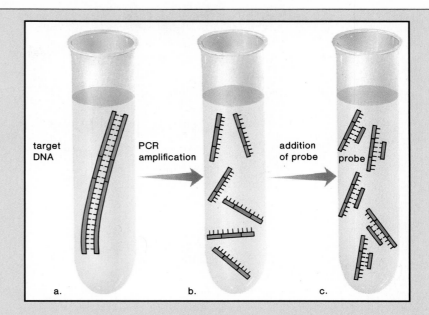

PCR analysis. *a.* DNA is removed from a cell and placed in a test tube along with appropriate primers, DNA polymerase, and a supply of nucleotides. *b.* Following PCR amplication, many copies of target DNA (red) are present. *c.* Binding of a labeled DNA probe (blue) allows the scientist to determine that a particular DNA segment was indeed present in the original sample.

method, for example, investigators found that 90% of pancreatic tumors contain a particular mutated gene.

Human Genome Project

Investigators are beginning to use PCR to help them map the human chromosomes. The DNA of a single sperm cell can be analyzed to detect the presence of genetic markers, even ones quite close together. PCR also allows scientists to determine whether a particular DNA sequence in one chromosomal fragment is also found in another fragment so they can be placed in the correct order.

Evolutionary Relations of Organisms

Scientists used PCR to amplify DNA segments from a quagga, an extinct zebralike animal whose only remains consisted of bits of dried skin. This allowed them to determine that the quagga was a zebra rather than a horse. In other studies, DNA sequences from a 7,000-year-old mummified human brain and from a 17 to 20 million-year-old plant fossil were analyzed. It is expected that PCR will contribute greatly to information about evolutionary relationships.

bacterium that normally colonizes the roots of corn plants now has been given genes (from another bacterium) that code for an insect-killing toxin.

Bacteria can be used for *bioremediation* which is a new term that means pollution clean-up. There are naturally occurring bacteria that can degrade most any type of chemical or material. For example, bacteria recently were used to help clean up the beaches of Alaska after a massive oil spill. These were naturally occurring bacteria that eat oil, but there are genetically engineered bacteria that could have done a better job (fig. 19.13).

Many major mining companies already use bacteria to obtain various metals. Genetic engineering may enhance the ability of bacteria to extract copper, uranium, and gold from low-grade sources. At least two mining companies plan to test genetically engineered organisms with enhanced bioleaching capabilities.

Ecological Considerations There are those who are very concerned about the deliberate release of genetically engineered microbes (GEMs) into the environment. Ecologists point out that these bacteria might displace those that normally reside in an ecosystem, and the effects could be deleterious. Others rely on past experience with GEMs, primarily in the laboratory, to suggest that these fears are unfounded. Tools now are available to detect, to measure, and even to disable cell activity in the natural environment. It is hoped that these eventually will allow GEMs to play a significant role in agriculture and environmental protection.

Transgenic Plants

In one of the first experiments to show that transgenic plants are possible, a gene for the enzyme luciferase was transplanted from a firefly into a tobacco plant. Whenever the plant was sprayed with luciferin, it glowed, proving that the inserted gene was indeed present and active (fig. 19.14). Plants, in particular,

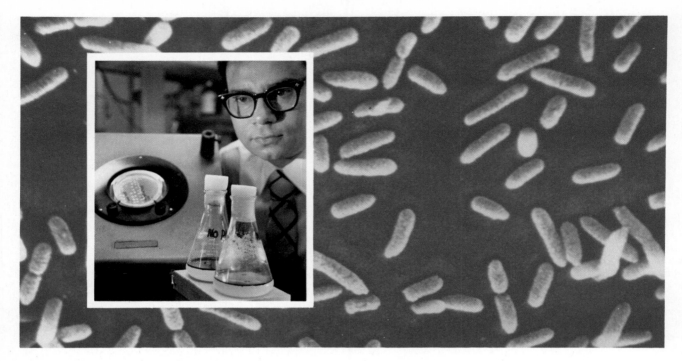

Figure 19.13 Oil-eating bacteria have been engineered and patented by investigator Dr. Chakrabarty. In the inset, the flask toward the rear contains oil and no bacteria; the flask toward the front contains the bacteria and is almost clear of oil. Now that engineered organisms (e.g., bacteria and plants) can be patented, there is even more impetus to create them.

Figure 19.14 This transgenic plant glows when sprayed with luciferin (a chemical that emits light) because its cells contain the protein luciferase, a firefly enzyme that breaks down the luciferin.

lend themselves to genetic manipulation because it is possible to grow plant cells in tissue culture, and each cell can be stimulated to produce an entire plant (fig. 19.15).

Unlike animal cells, plant cells have a cell wall and this must be removed to give "naked" cells called protoplasts before they will take up a plasmid. The only possible plasmid for bioengineering plant cells, the Ti plasmid, is taken up by only certain plant cells. Unfortunately the plants of agricultural significance do not take up the Ti plasmid and new techniques have been developed to introduce foreign DNA into plant cells. For example, it is possible to irradiate protoplasts with a laser beam while they are suspended in a liquid containing foreign DNA. Laser beams make tiny self-sealing holes in the cell membrane through which genetic material can enter. Presently, about 50 types of genetically engineered plants that resist either insects, viruses, or herbicides now have entered small-scale field trials. The major crops that can be improved in this way are soybean, cotton, alfalfa, and rice. However, even genetically engineered corn is expected to reach the marketplace by the year 2000.

There are ecological considerations about these bioengineered plants also. Some are concerned that if herbicide resistant plants are available, farmers will tend to use more herbicide and the environment will be degraded. If plants are resistant to insects because they produce a toxin, perhaps resistant insects will evolve.

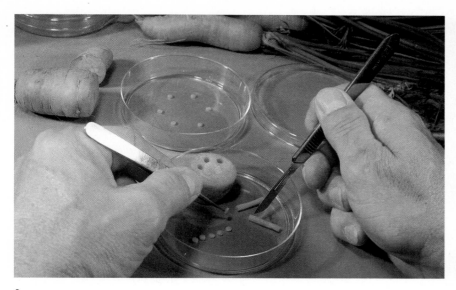

a.

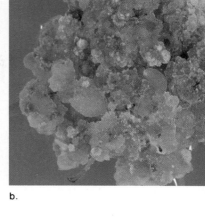

b.

c.

d.

Figure 19.15   Cloning of entire plants from tissue cells. *a.* Sections of carrot root are cored, and thin slices are placed in a nutrient medium. *b.* After a few days, the cells form a callus, a lump of undifferentiated cells. *c.* After several weeks, the callus begins sprouting cloned carrot plants. *d.* Eventually, the carrot plants can be moved from culture medium to pots.

On the other hand, it is hoped that one day even more agriculturally significant plants will be produced by bioengineering. Plants are needed which:

are more nutritious and in which the seed contains all the amino acids required by humans. Protein-enhanced beans, corn, soybeans, and wheat are now being developed.

require less fertilizer and which are able to make use of nitrogen from the atmosphere. (Plants that do not take nitrogen from the soil only). Certain bacteria, however, can make use of atmospheric nitrogen. If the required genes of bacteria were cloned, perhaps they could be transferred to plants.

have increased ability to grow under unfavorable environmental conditions such as lack of water or in salty soil. Less irrigation of crops would then be needed. First, it will be necessary to identify which genes in plants might allow them to do this.

Transgenic Animals

Genetic engineering of animals has begun. Animal cells will not take up plasmids, but it is possible to microinject foreign genes into eggs before they are fertilized. The most common procedure attempted today is to microinject bovine growth hormone (bGH) into the eggs of various animals. It is hoped that the gene will establish itself and be transmitted to all the cells of the developing organism and even be passed along to the next generation of offspring. This procedure has been used in fishes, chickens, cows, pigs, rabbits, and sheep in the hope of producing bigger varieties.

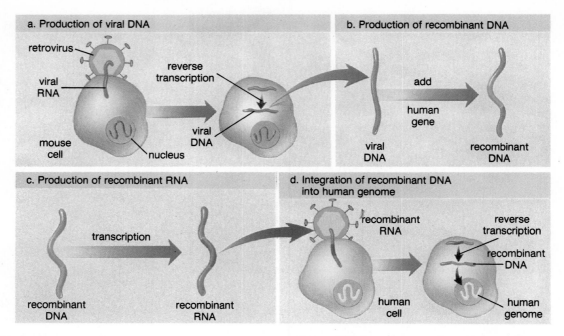

a. Production of viral DNA

retrovirus

viral RNA

reverse transcription

mouse cell

nucleus

viral DNA

b. Production of recombinant DNA

add human gene

viral DNA

recombinant DNA

c. Production of recombinant RNA

transcription

recombinant DNA

recombinant RNA

d. Integration of recombinant DNA into human genome

recombinant RNA

reverse transcription

recombinant DNA

human cell

human genome

Figure 19.16 Retroviruses as vectors. *a*. A retrovirus has an RNA chromosome. When this chromosome enters a cell, reverse transcription occurs and viral DNA is produced. *b*. A foreign human gene can now be inserted into viral DNA. *c*. Transcription inside a host cell produces an RNA copy of the recombinant DNA. *d*. A virus now is used to carry the recombinant RNA into a human host cell. Following reverse transcription, recombinant DNA is inserted into the human genome.

In another experiment, the sheep gene that codes for the milk protein betalactoglobulin (BLG) was microinjected into mice eggs. (Mice normally produce a milk that has no BLG at all). Of 46 offspring successfully weaned, 16 carried the BLG sequence, and the females among them later produced BLG-rich milk. Some of these females passed on the BLG gene to their offspring.

Gene Therapy in Humans

Investigators are striving to use biotechnology for **gene therapy,** the use of a transplanted gene to cure or to treat human genetic disorders and to treat other human ills as well.

Human Issue

Our new knowledge of the workings of DNA have pushed us into a new era—one in which it may even be possible to cure genetic diseases. For example, we may develop the capability to splice genes that correct specific genetic defects into human zygotes. Exploring and developing this capability would require experimentation with human embryos and artificial implantation into a woman's uterus. Does the prospect of children who are free of significant genetic diseases justify extensive experimentation with human embryos and the risk of creating children who are the result of failed genetic experiments? Who should answer this question?

Bioengineering Bone Marrow Stem Cells

Some genetic disorders are caused by the inheritance of a faulty code for a particular protein that functions in blood cells. Recall that red bone marrow contains stem cells that give rise to the many types of blood cells in the body. Some investigators believe that it will be possible to introduce a normal functioning gene into stem cells and, thereafter, all types of blood cells will produce functioning enzyme. The virus to be used is a retrovirus which uses RNA instead of DNA for its genes. Retroviruses you will recall (p. 299) are unique because they have an enzyme that on occasion transcribes RNA to DNA, which then becomes part of the host cell's DNA. When RNA viruses are used for gene therapy, they have been equipped with recombinant RNA (fig. 19.16). After the recombinant RNA enters a human cell, such as a bone marrow stem cell, reverse transcription occurs and recombinant DNA enters a human chromosome.

Bioengineering Other Types of Cells

Using the same method described in figure 19.16, a retrovirus can also be used to bioengineer human lymphocytes (a type of white blood cell) directly. For example, during a clinical trial, children with severe combined immune deficiency syndrome (SCID)[1], a genetic disorder, will have the gene they need introduced into their lymphocytes. The enzyme that is lacking is needed for the maturation of T and B cells; as a precaution, all participants in the trial will receive the enzyme directly.

[1]SCID is often called the "bubble-baby" disease after David, a young person who lived under a plastic dome to protect himself from infection.

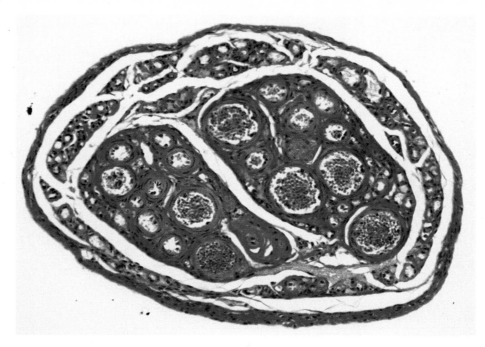

Figure 19.17   This organoid was made by treating Gore-Tex fibers with a growth factor that stimulates blood vessel formation. The organoid could contain bioengineered cells carrying a normal gene to make up for a defective gene inherited by the individual.

Cells that normally line blood cells (endothelial cells) are easily obtained and cultured in the laboratory, and they can be bioengineered to make a particular enzyme. Whereas bone marrow stem cells and lymphocytes can be withdrawn and returned to the body by injection, endothelial cells need a special delivery system. One novel idea is to put the cells in organoids, artificial organs that can be implanted in the abdominal cavity. Organoids have been made by coating angel-hair Gore-Tex fibers with a substance called collagen and adding a growth factor for blood vessels (fig. 19.17). Once in the body, the endothelial cells begin to line the newly developing blood vessels which spread out to adjoining organs.

Mapping the Human Chromosomes

If investigators knew the order and precise location of the genes on the human chromosomes, it would facilitate laboratory research and medical diagnosis and treatment. Several methods have been used to attempt mapping the human chromosomes and we will discuss some of these.

Human-Mouse Cell Data

Human and mouse cells are mixed together in a laboratory dish and, in the presence of a fusing agent, they will fuse (fig. 19.18). As the cells grow and divide, some of the human chromosomes are lost, and eventually the daughter cells contain only a few human chromosomes, each of which can be recognized by its distinctive banding pattern. Analysis of the proteins made by the various human mouse cells enables scientists to determine which genes are to be associated with which human chromosome.

Sometimes it is possible to obtain a human-mouse cell that contains only one human chromosome or even a portion of a chromosome. This technique has been very helpful to those researchers who have been studying the genes that are located on number-21 chromosome (figure 17.18).

Genetic Marker Data

A genetic marker is a place on the chromosome where the sequence of bases differs from one person to another. Genetic markers can be discovered by using restriction endonuclease enzymes because, as mentioned previously (p. 389), each one always cleaves DNA at a specific sequences of bases. If an individual has this sequence of bases at a particular location, then the restriction enzyme is able to cleave the chromosome at this location. If an individual lacks this sequence of bases, then the restriction enzyme is unable to cleave the chromosome at this location. Scientists say that "restriction fragment length polymorphisms (RFLPs)" can be observed. Polymorphisms mean many changes in structure; in this case, polymorphisms exist in the structure of the chromosomes taken from different individuals.

Genes can be assigned a location on a chromosome according to their relative relationship to genetic markers. Genetic markers can also be used to tell if one has a genetic disorder when a particular marker is always inherited with a particular genetic disorder. The test for the genetic disorder consists of testing an individual for the presence of the marker, instead of for the specific faulty allele (fig. 19.19). For a marker to be dependable, it should be inherited with the defective allele at least 98% of the time.

The current tests of sickle-cell disease, Huntington disease, and Duchenne muscular dystrophy are all based on the presence of a marker.

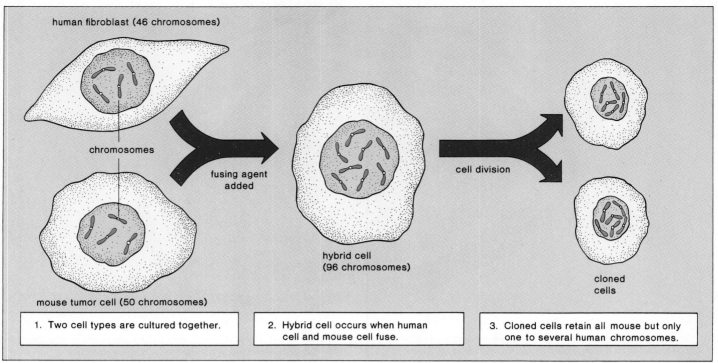

human fibroblast (46 chromosomes)

chromosomes

fusing agent added

hybrid cell (96 chromosomes)

cell division

cloned cells

mouse tumor cell (50 chromosomes)

1. Two cell types are cultured together.

2. Hybrid cell occurs when human cell and mouse cell fuse.

3. Cloned cells retain all mouse but only one to several human chromosomes.

O'Keefe

Figure 19.18   In the presence of a fusing agent, human fibroblast cells sometimes join with mouse tumor cells to give hybrid cells having nuclei that contain both types of chromosomes. Subsequent cell division of the hybrid cell produces clones that have lost most of their human chromosomes, allowing the investigator to study these chromosomes separate from all other human chromosomes.

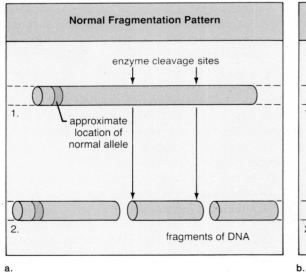
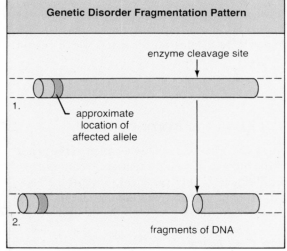

Normal Fragmentation Pattern

enzyme cleavage sites

1.

approximate location of normal allele

2.   fragments of DNA

a.

Genetic Disorder Fragmentation Pattern

enzyme cleavage site

1.

approximate location of affected allele

2.   fragments of DNA

b.

Figure 19.19   Use of a genetic marker to detect the approximate location of a gene or to test for a genetic disorder when the exact location of the gene is unknown. *a.* DNA from the normal individual has certain restriction enzyme cleavage sites near the gene in question. *b.* DNA from another individual lacks one of the cleavage sites, and this loss indicates that they almost certainly have the genetic disorder because experience has shown that this genetic marker is almost always present when an individual has the disorder. In other cases, a gain in a cleavage site is the genetic marker.

Human Genome Project

The goal of the *Human Genome Project* is to identify the location of the approximately 100,000 human genes on all the chromosomes. In order to create this **genetic map,** genetic markers will be used to index the chromosomes. Known and newly discovered genes will be assigned locations between the markers.

Eventually, the project also wants to determine the sequence of the 3 billion bases in the human genome. In order to create this **physical map,** researchers will use laboratory procedures to determine the sequence of the DNA bases (fig. 19.20).

Recently it's been discovered that genetic markers contain unique stretches of DNA called sequence-tagged sites, or STS's. Hopefully, these can be used to create links between the genetic map and the physical map.

The human genome project will require millions of dollars and many years to complete. Just how successful and useful it will be cannot yet be determined.

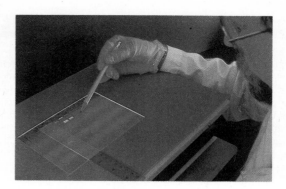

Figure 19.20   Gel electrophoresis is used to sequence DNA. The DNA is cleaved into fragments that vary in length by one nucleotide. Each fragment ends with an A, T, C, or G, marked by one of four flourescent dyes. The fragments are subjected to gel electrophoresis, where they migrate on a gel in an electric field. Shorter fragments migrate further than longer fragments. Under proper lighting, the investigator (or a computer) can simply read off the sequence of the nucleotides.

Summary

DNA is the genetic material; it can store information, replicate, and mutate. During replication, DNA becomes "unzipped" and then a complementary strand forms opposite to each orginial strand. DNA directs protein synthesis. During transcription, mRNA is made complementary to one of the DNA strands. It then contains codons and moves to the cytoplasm where it becomes associated with the ribosomes. During translation, tRNA molecules attached to their own particular amino acid travel to the mRNA, and through complementary base pairing, the tRNAs and, therefore, the amino acids in a polypeptide chain are sequenced in a predetermined way. Aside from structural genes, there are also regulatory genes that control the expression of structural genes.

Genes are subject to mutations that may be substitutions, additions, or deletions. Transposons are movable elements that cause mutations. Mutagenic agents also promote genetic mutations.

Biotechnology is being expanded greatly by DNA technology. To achieve a genetically engineered cell, a vector such as a plasmid is used to carry a foreign gene into a cell. First, a restriction endonuclease enzyme is used to cleave plasmid DNA and foreign DNA. Then, the vector is sealed by DNA ligase. When the plasmid replicates, the foreign gene is cloned. The protein produced by the foreign gene also can be collected. This is the basis for the production of biotechnology products, such as hormones, DNA probes, and vaccines.

Transgenic organisms also have been made. Plants lend themselves to genetic manipulation because whole plants will grow from cultured cells. Plants resistant to pests and herbicides that contain complete protein soon will be available commercially. The possibility of developing improved livestock through biotechnology also exists.

Human gene therapy is being investigated. Researchers are currently interested in using lymphocytes and endothelial cells as delivery systems for genetic drugs. The United States government is committed to mapping and sequencing the entire human genome. It is hoped this information will assist human gene therapy.

Study Questions

1. Compare and contrast the structure of RNA with that of DNA. (p. 378)
2. What are the four functions of the hereditary material? (p. 378)
3. Explain how DNA replicates. Why is this replication called semiconservative? (p. 378)
4. If the code is TTA'TGC'TCC'TAA, what are the codons and what is the sequence of amino acids? Show how a deletion or duplication could affect the code. (p. 382)
5. List the steps involved in protein synthesis, mentioning the process of transcription and translation and the roles of DNA, mRNA, rRNA, and tRNA. (p. 386)

6. You are a scientist who has decided to "clone a gene." Tell precisely how you would proceed. (p. 389)
7. Name three categories of biotechnological products that are now available, and discuss their advantages. (p. 390)
8. Naturally occurrng bacteria have been bioengineered to perform what services? What are the ecological concerns regarding their release into the environment? (p. 392)

9. Why are plants good candidates for genetic engineering, and what are the problems involved in using this method to achieve agriculturally significant plants? (p. 394)
10. Describe the current situation in regard to transgenic animals. (p. 395)
11. What types of problems have to be solved before gene therapy in humans becomes a reality? (p. 396)
12. What is the human genome project, and why might the project be useful? (p. 399)

Objective Questions

1. Replication of DNA is semiconservative, meaning that each new double helix is composed of an _____ strand and a _____ strand.
2. The DNA code is a _____ code, meaning that every three bases stands for an _____ .
3. The three types of RNA that are necessary to protein synthesis are _____ , _____ , _____ , and _____ .
4. Which of these types of RNA carries amino acids to the ribosomes? _____

5. The sequence of mRNA codons dictates the sequence of amino acids in a protein. This step in protein synthesis is called _____ .
6. The two types of enzymes needed to make recombinant DNA are _____ and _____ .
7. DNA probes seek out and bind to _____ .
8. Plants that have been cultured from a protoplast that contains a foreign gene are called _____ plants.
9. _____ , not plasmids, are used commonly as vectors in research with human cells.

10. The _____ project, backed by the United States government, is expected to result in complete mapping of the human chromosomes.

Answers

Critical Thinking Questions

1. Drawing from your knowledge of the reproductive process, support the suggestion that we are nothing but transport vehicles or hosts for DNA. What arguments can you think of to counter this suggestion?

2. Both mRNA and tRNA are transcribed from DNA in the nucleus, but they have different functions in the cytoplasm. How might they be "guided" to take up their respective functions?

3. How does the success of genetic engineering techniques lend support to the belief that organisms are chemical and physical machines?

Selected Key Terms

anticodon (an''ti-ko'don) a "triplet" of three nucleotides in transfer RNA that pairs with a complementary triplet (codon) in messenger RNA. 384

cloned (klond) genes from an external source that have been reproduced by bacteria. 390

codon (ko'don) a "triplet" of three nucleotides in messenger RNA that directs the placement of a particular amino acid into a polypeptide chain. 381

complementary base pairing (kom''ple-men'ta-re bas par'ing) pairing of bases between nucleic acid strands; adenine is always paired with either thymine (DNA) or uracil (RNA) and cytosine is always paired with guanine. 378

gene therapy (jen ther-ah-pe) use of transplanted genes to treat a human disorder. 396

messenger RNA (mes'en-jer) mRNA; a nucleic acid (ribonucleic acid) complementary to genetic DNA and bearing a message to direct cell protein synthesis at the ribosome. 380

plasmid (plaz'mid) a circular DNA segment that is present in bacterial cells, but is not part of the bacterial chromosome. 388

recombinant DNA (re-kom'bi-nant) DNA having genes from two different organisms often produced in the laboratory by introducing foreign genes into a bacterial plasmid. 390

regulatory genes (reg'u-lah-tor''e jenz) genes that code for proteins involved in regulating the activity of structural genes. 387

replication (rep''li-ka'shun) the duplication of DNA; occurs during interphase of the cell cycle. 378

ribosomal RNA (ri'bo-som''al) rRNA; RNA occurring in ribosomes, structures involved in protein synthesis. 384

structural genes (struk'tur-al jenz) genes that direct the synthesis of enzymes and also structural proteins in the cell. 387

template (tem'plat) a pattern that serves as a mold for the production of an oppositely shaped structure; one strand of DNA is a template for the complementary strand. 378

transcription (trans-krip'shun) the process that results in the production of a strand of mRNA (also, tRNA, rRNA) that is complementary to a segment of DNA. 380

transfer RNA (trans'fer) tRNA; molecule of RNA that carries an amino acid to a ribosome engaged in the process of protein synthesis. 384

translation (trans-la'shun) the process involving mRNA, ribosomes, and tRNA molecules that results in a synthesis of a polypeptide having an amino acid sequence dictated by the sequence of codons in mRNA. 380

vector (vek'-tor) a carrier, such as a plasmid or virus, for recombinant DNA that introduces a foreign gene into a host cell. 388

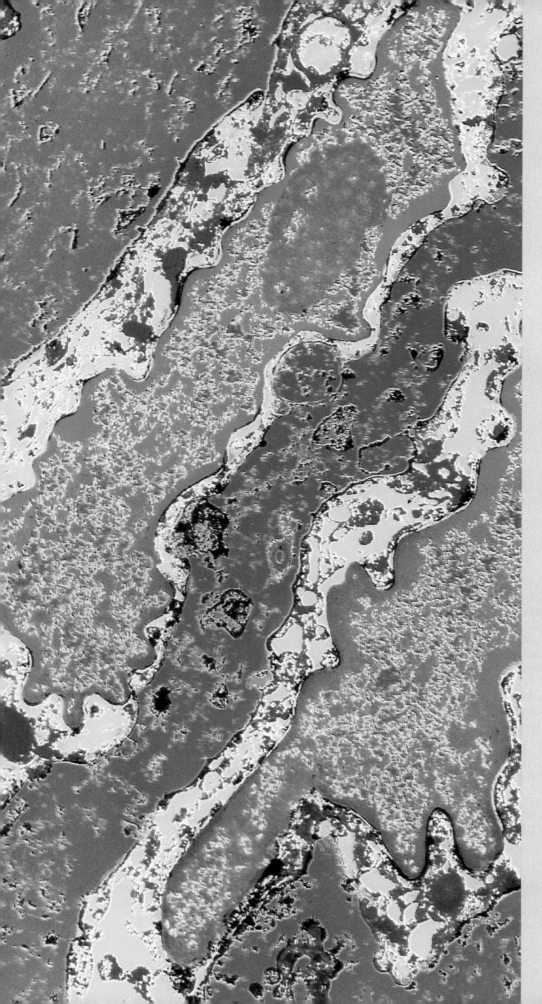

Chapter Twenty

Cancer

Chapter Concepts

1 Cancer cells are transformed; they have certain specific characteristics that distinguish them from normal cells.

2 Mutations brought on by environmental factors such as certain chemicals and radiation sometimes cause the development of cancer.

3 Cancer cells become tumors in various parts of the body; some of these are more common than others.

4 Treatment of cancer includes some old strategies (surgery, radiation, and chemotherapy) and some new strategies (immunotherapy).

5 Various suggestions are available for avoiding cancer, and there is a recent emphasis on dietary guidelines.

Colored transmission electron micrograph of nuclei from cancer cells. A cytologist looking at these deformed nuclei would know they are from cancer cells of a particular type.

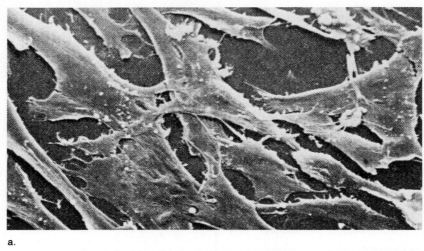

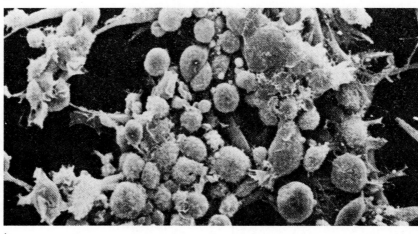

Figure 20.1   *a.* Normal fibroblasts are flat and extended. *b.* After being infected with a cancer-causing virus, the cells become round and cluster in piles.

Cancer is not a new disease; fossil bones of dinosaurs and early humans sometimes show evidence of the disease. Its prominence among modern humans is due to the fact that many humans now live long enough to get cancer. Knowledge about and treatment of cancer, like biotechnology, is a perfect example of how basic research can be intertwined with practical application. It is only in the past 50 years that generalized hypotheses about cancer have led to specific findings that now help shape the treatment that people receive. This chapter explores what we know about cancer.

Characteristics of Cancer Cells

Although cancer cells (fig. 20.1) are derived from normal cells, they are transformed. Because of this **transformation,** they have characteristics that distinguish them from normal cells, and they pass these characteristics on to daughter cells when they divide.

Cancer Cells Lack Differentiation

Cancer cells are nonspecialized and do not contribute to the functioning of an organ. A cancer cell does not look like a dif-

ferentiated epithelial, muscular, nervous, or connective tissue cell; therefore, a histologist can pick them out when looking at a slide of a particular organ. Because cancer cells are nondifferentiated, they can undergo the cell cycle repeatedly, and in this way, they are immortal.

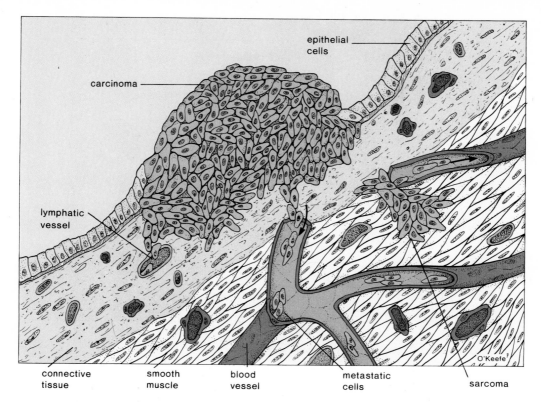

Figure 20.2   In the body, cancer cells form a tumor, a disorganized mass of cells undergoing uncontrolled growth. As a tumor grows, it invades underlying tissues; some of the cells leave the primary tumor and move through layers of tissue into blood or lymphatic vessels. After traveling through these vessels, the cells start new tumors elsewhere in the body. A carcinoma is a cancer that begins in epithelial tissue; a sarcoma is a cancer that begins in connective tissue.

Cancer Cells Grow Unchecked

Cancer cells undergo uncontrolled and disorganized growth. In tissue culture, cancer cells lack the contact inhibition exhibited by normal cells. Normal cells in tissue culture grow in only one layer because they adhere to the glass and stop dividing once they make contact with their neighbors. Cancer cells have lost all restraint and grow in multiple layers most likely because of cell surface changes. In the body, a cancer cell divides to form a growth, or **tumor,** that invades and destroys neighboring tissue (fig. 20.2). This new growth, termed *neoplasia,* is made up of cells that are disorganized, a condition termed *anaplasia.* To support their growth, cancer cells release a growth factor that causes neighboring blood vessels to branch into the cancerous tissue. This phenomenon has been termed vascularization, and some modes of treatment are aimed at preventing vascularization from occurring.

Cancer Cells Metastasize

Cancer cells detach from the tumor and spread around the body. This is called **metastasis.** To accomplish this, cancer cells often must make their way across a basement membrane (p. 49) and into a blood vessel or lymphatic vessel (fig. 20.2). It has been discovered that cancer cells have receptors that allow them to adhere to basement membranes; they also produce proteolytic enzymes that degrade the membrane and allow them to invade the underlying tissues. Cancer cells also tend to be motile. This may be associated with a disorganized internal cytoskeleton and their lack of intact microfilament bundles. After traveling through the blood of lymph, cancer cells start new tumors elsewhere in the body.

If the original tumor is found before metastasis has occurred, the chances of a cure are greatly increased. This is the rationale for early detection of cancer. In fact, sometimes a person is not even considered to have cancer unless metastasis has occurred. Benign tumors are those that remain in one place, malignant tumors are those that metastasize.

Cancer cells have been transformed; they lack differentiation, grow uncontrollably, and metastasize. These characteristics distinguish cancer cells from normal cells.

Causes of Cancer

Carcinogenesis (the development of cancer) is often described as a two-step process involving (1) initiation and (2) promotion. Although this is most likely an oversimplified view of carcinogenesis, it still allows us to understand that the development of

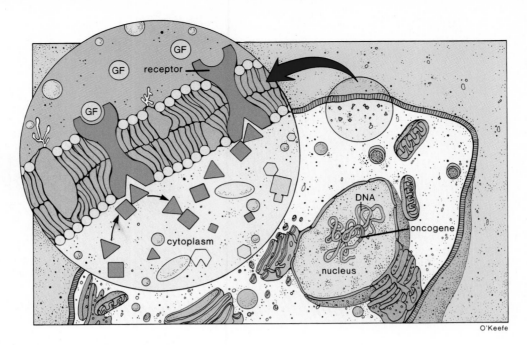

Figure 20.3 Growth of cells involves many elements that begin at the cell membrane. Receptor at left receives growth factor (GF), and this initiates chemical reaction in cytoplasm. Receptor at right is malfunctioning since reaction takes place without GF being in place. Note oncogene in nucleus.

cancer requires not just one event, but more than one event and, indeed, most likely several events.

Proto-Oncogenes and Oncogenes

It is now clear that initiation of carcinogenesis requires a mutation in a normal gene involved in the regulation of cell growth. Each cell contains many **proto-oncogenes** so called because a mutation can cause them to become an **oncogene,** or a cancer-causing gene. The cell membrane contains receptors that respond to growth factors (fig. 20.3). When a growth factor binds to its receptor, the receptor interacts with a series of signaling proteins which, in turn, communicate with the cell's DNA. An oncogene can be a gene that codes for one of the crucial elements along this pathway—growth factor, receptor, or signaling protein—because a change affecting any of these points can cause a cell to grow out of control.

Investigators have shown that an oncogene that causes both lung cancer and bladder cancer differs from a normal gene by a change in only one nucleotide. However, almost any type of mutation can convert a proto-oncogene into an oncogene. For instance, in addition to a gene mutation, a chromosome rearrangement (fig. 17.17a) may place a normally dormant structural gene next to an active regulatory gene. If this structural gene is a proto-oncogene, it may now become an oncogene.

Studies of a rare childhood cancer of the eye called retinoblastoma have revealed that there are tumor-suppressor genes (anti-oncogenes). Retinoblastoma is not caused by an oncogene that turns on growth, but by the lack of *tumor-suppressor genes* that switch off growth. If mutations occur in both copies of the retinoblastoma gene within a single cell, the cell grows uncontrollably. Recently, investigators have outlined several steps that are required for colon cancer to appear. Among these are the change of proto-oncogene to an oncogene and then the loss of tumor suppressor genes from chromosomes 5, 18, and 17.

Heredity

Particular types of cancer seem to run in some families. For example, breast cancer may be seen among female members; a tendency toward melanoma (skin cancer) may take place in both sexes or prostate cancer may occur among the male members of a family. The existence of oncogenes can explain these findings: mutated genes are passed from parent to child and among these mutations there certainly could be an oncogene. Cancer itself would not develop until even more mutations occurred among proto-oncogenes or suppressor genes.

Researchers have recently found a tumor-suppressor gene they call p53. This gene is responsible for the Li-Fraumeni-syndrome, a syndrome characterized by a high rate of lethal cancers among family members. When an individual inherits only one normal allele, that individual has a 50% chance of developing cancer by age 30 because of the high chance that the normal allele will mutate.

a.

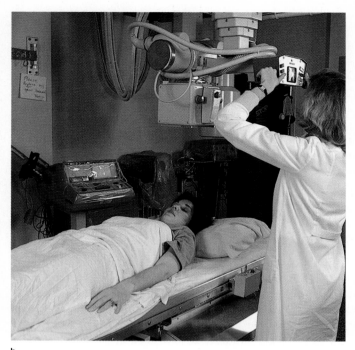

b.

Figure 20.4 *a.* Chemical mutagens include pesticides, manufacturing chemicals, and hallucinogenic drugs. Suspected mutagens are often tested in bacteria, fruit flies, and mice. If cancer develops in the mice, then the government either allows the product to be sold with a warning label or it bans the sale of the product entirely. *b.* Patient undergoing X-ray diagnosis. While many people are concerned about radiation from the nuclear power industry, medical diagnostic radiation actually accounts for at least 90 percent of human exposure to radiation. Everyone should be aware of this potential danger and have X rays only when necessary. All nonvisible short wavelengths of radiation are believed to be mutagenic, but the greater the amount, the greater the risk.

Carcinogens

A **carcinogen** is an environmental agent that can initiate or promote cancer. Well-known carcinogens are (1) certain chemicals, (2) radiations (fig. 20.4), and (3) viruses.

Chemical Carcinogens

Thousands of chemicals are capable of causing a mutation that can lead to cancer. Tobacco smoke and charcoal-broiled meats contain some of the same hydrocarbons that can initiate or promote cancer. Animal testing of certain food additives such as red dye #2 and the synthetic sweetener saccharin has shown that these substances are cancer-causing in high doses. Industrial chemicals such as benzene and carbon tetrachloride and industrial materials such as vinyl chloride and asbestos fibers are associated with the development of cancer. A recent problem has been the protection of underground water supplies from cancer-causing industrial wastes.

Pesticides and herbicides are dangerous not only to pests and plants, but also to our own health (fig. 20.4a) because they contain organic chemicals that can cause cancer.

Radiation

Human beings are exposed to natural radiation and man-made radiation. Ultraviolet radiation in sunlight and tanning lamps is responsible for the dramatic increases seen in skin cancer the past several years. The public is also now aware that naturally occurring radioactive radon gas can cause cancer, particularly if concentrated inside homes. In the very rare house with an extremely high level of exposure, the risk of developing lung cancer is thought to be the equivalent of smoking a pack of cigarettes a day.

Most people are familiar with the damaging effects of radioactive emissions from nuclear power plants or bombs. However diagnostic X rays account for most of our exposure to man-made sources of radiation (fig. 20.4b). If a fetus in the womb is exposed to X rays, the development of leukemia is a possibility.

Viruses

Both DNA and RNA viruses (fig. 15.3) can contribute to development of cancer when they bring oncogenes into the cell. We now know that any oncogene in a virus is actually a human gene that was picked up earlier, by chance, when the virus infected a previous human cell. The gene could have mutated before or after it was picked up, but, in any case, when the virus infects a new cell, the human oncogene is passed on and transcribed leading to the development of cancer.

Three DNA viruses have been linked to common cancers in humans: hepatitis B virus to liver cancer, Epstein-Barr virus to lymphomas and nasopharyngeal cancer, and certain strains of human papilloma virus to cervical and other genital cancers.

Retroviruses, in particular, are known to cause cancers in animals. In humans, the retrovirus HTLV (human T-cell leukemia/lymphotropic virus) has been shown to cause adult T-cell leukemia. You would expect from our previous study of retroviruses (fig. 15.3b) that a retrovirus causes the oncogene to become integrated into host DNA (fig. 20.5).

Immune System Failure

Most likely cancer cells do have altered cell membrane antigens that should normally subject them to attack by T cells. Whenever cancer develops, it would seem that cell-mediated immunity carried out by T cells has not been activated. Whether cancer cells are able to protect themselves in some way from attack or whether immunity is weakened in cancer patients is not known (fig. 20.6).

When cancer develops, proto-oncogenes may have mutated to oncogenes. Certain chemicals and radiation can bring about this mutation. Or an oncogene can be inherited from a parent or brought into a cell by a virus. Most likely more than one genetic change is required before cancer develops. Also, the immune system has failed to kill off the transformed cells.

Types of Cancer

Cancers are classified according to the type of tissue from which they arise. **Carcinomas** are cancers of the epithelial tissues (p. 49), and adenocarcinomas are cancers of glandular epithelial cells. Since epithelial tissue covers the surface of the body and lines internal cavities, carcinomas include cancer of the skin, breast, liver, pancreas, intestines, lung, prostate, and thyroid. These cancers may be spread to other body parts by the lymphatic system. **Sarcomas** are cancers that arise in connective tissue (p. 54), such as muscle, bone, and fibrous connective tissue. These cancers may be spread to other body parts by the bloodstream. **Leukemias** are cancers of the blood, and **lymphomas** are tumors of lymphoid tissue.

Of the Skin

Skin cancer (p. 57) is the most common form of cancer but does not cause the most deaths (table 20.1).

There are three types of skin cancer. The most common and least dangerous is *basal-cell cancer,* which develops in the deepest layer of the epidermis (fig. 3.13), usually where oil gland distribution is greatest. Often, the cancer may look like a pimple, but then it forms a gray border and ulcerates. *Squamous-cell*

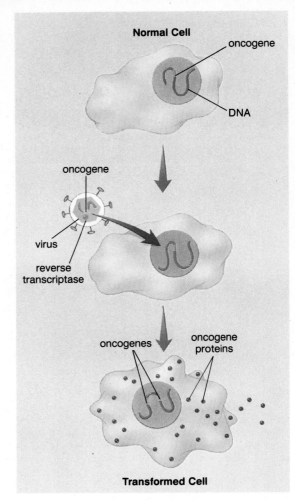

Figure 20.5 Transformation of normal cells into cancerous cells by a retroviral infection. The normal cell contains one oncogene but is not cancerous. The retrovirus also contains an oncogene, which it passes to the cell. Now the cell is transformed and becomes cancerous because it produces a protein, most likely a growth factor, that causes the cell to become abnormal.

cancer occurs in the upper layers of the epidermis. It is generally characterized by extremely dry, rough, scaly patches but can vary greatly in appearance. The third type of skin cancer is more rare but much more virulent. Malignant *melanomas* usually appear as dark brown or black patches like moles. They may also develop from moles.

Skin cancers are usually surgically removed, sometimes by electrosurgery, the use of an electric current to destroy the cancerous cells. Cryosurgery destroys the cells by freezing them. Following their removal, the person is free of cancer but will most likely have a scar.

The best protection against skin cancer is to avoid the sun, especially in the middle of the day. Also the use of sunscreens that contain a chemical to absorb UV rays is recommended.

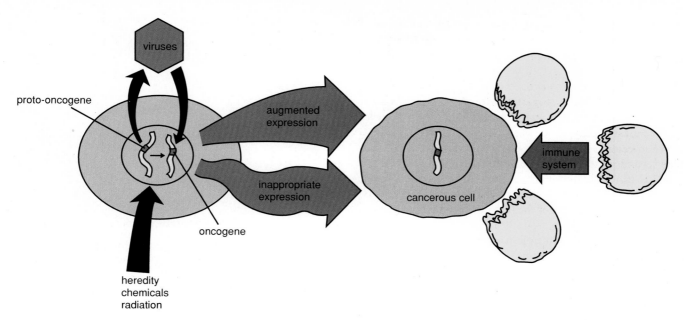

Figure 20.6 Summary of the development of cancer. A virus can pass an oncogene to a cell. A normal gene, called a proto-oncogene, can become an oncogene due to a mutation caused by a chemical or by radiation. The oncogene either expresses itself to a greater degree than normal or expresses itself inappropriately. Thereafter, the cell becomes cancerous. Cancer cells are usually destroyed by the immune system, and the individual only develops cancer when the immune system fails to perform this function.

Table 20.1 Cancer Deaths

Males	Females
lung 35%	lung 20%
prostate 11%	breast 18%
colon and rectum 11%	colon and rectum 14%
leukemia and lymphomas 9%	leukemia and lymphomas 9%
bladder 5%	ovary 5%
pancreas 5%	pancreas 5%

Of the Male Reproductive Organs

Cancer of the penis almost always involves the foreskin or glans penis and is rare in circumcised men. The tumor begins on the inner surface of the foreskin and metastases occur by way of the lymph nodes in the groin. Tumors of the scrotal skin are rare and most often due to exposure to a carcinogen at work. For example, chimney sweeps get scrotal skin cancer due to exposure to soot. Testicular tumors are not common, but when they occur, they are usually malignant and metastasis to the liver and lung is seen.

The risk of cancer of the prostate is not great until a male is 60 years old; then the risk increases with every decade. It is

the second most common cause of death from cancer in males, the first being lung cancer. If treatment begins while tumors remain small and confined to the area, 84% of patients survive prostate cancer. Presently, the use of an ultrasound probe by way of the rectum has doubled the detection rate. Some experts believe that ultrasound in men might become the equivalent of mammograms in women.

Unfortunately, impotency is a side effect in at least 25% of patients after treatment of prostate cancer by surgery or radiation. In addition, hormone therapy for metastasis increases the chance of heart attack and stroke.

Of the Female Reproductive Organs

Cancer of the breast is the most common cancer in females. Most cases of breast cancer occur beyond the age of 40 in women who have at least two close relatives who have also had breast cancer. It is possible that diet is a factor because breast cancer is rare among the Japanese, who eat less meat and fat. Early detection is extremely advisable, and all women should personally examine their breasts once a month in a manner recommended by the medical profession. Secretions from and changes in the appearance of the nipple should be noted, and all lumps, especially those that are hard and rough, should be reported to a physician immediately. Some doctors recommend that mammography, examination of the breast by low level X ray, be routinely done in older women. The X-ray image is called a mammogram.

For years, the surgical treatment for breast cancer has been a radical mastectomy, a procedure that involves total removal of the affected breast together with the axillary lymph nodes and muscles of the chest wall. There are those who believe that a simple mastectomy, which is much less disfiguring, is just as effective in certain cases.

The second most common cancer of the female reproductive organs is cancer of the endometrium. It is more common during or after menopause. Cancer of the cervix (fig. 14.7) is the third most common cancer of the female reproductive system. It is most frequent in women from 30 to 50 years of age. Sexually active women having multiple partners are at greater risk and so are those who have a genital warts infection (p. 302). Cancer of the cervix can be detected by a **Pap test,** a microscopic examination of the cervical cells. All women are encouraged to have a Pap test every year. Cancer of the uterus and cervix often requires a hysterectomy in which these organs are removed.

Five percent of all female cancer deaths are caused by ovarian cancer. Most malignant ovarian tumors begin in surface epithelium but may eventually involve all ovarian cell types. If detected early, ovarian tumors can be treated successfully.

Of the Respiratory Tract

Cancerous growths in the mouth make up one in twenty of all cases of human cancers. Malignant growths on the lips are a frequent finding in areas of the world where strong sunlight is usual. The tongue is another site of cancer, often starting as an innocuous-looking area of whitening of the normally pink surface. A frequent cause of tongue and mouth cancer is tobacco smoking, especially in pipes and cigars.

Lung cancer (p. 164) has been on the increase since the turn of the century. Recently, however, there has been a decline in the number of men with lung cancer but a large increase in the number of women with lung cancer so that lung cancer now surpasses breast cancer as a cause of death (table 20.1). Coal miners and radioactive ore miners inhale fumes that can lead to lung cancer, and persons who manufacture products containing asbestos are subject to a tumor of the pleura (fig. 8.7). Only about 1% of lung cancer patients have alveolar cell carcinoma. The rest have cancer of the bronchus (fig. 20.7). Other types of bronchial cancer are oat (small) cell cancer, large-cell cancer, and adenocarcinoma that arises in the glandular cells. For the ordinary individual, smoking is the most frequent cause of lung cancer. Coughing, spitting blood, and loss of weight are symptoms of lung cancer. In some cases, there are no symptoms until a shadow is noticed on a routine chest X ray. Usually, lung cancer is initially treated by an attempted removal of the cancerous tissue.

Figure 20.7   Cancer cells in the bronchi, respiratory tubes that conduct air to the lungs. The normal cells are pink and have cilia that sweep impurities up into the throat. The cancerous cells are green and lack cilia. If an individual stops smoking, there is a good chance the tissue will become normal again.

Of the Digestive Tract and Accessory Organs

Cancer of the esophagus, stomach, colon, liver, and pancreas are common enough to be listed among the top ten most frequently diagnosed types of cancer. We mentioned before that diet has been linked to cancer of the breast, and it is certainly not surprising that the same would be true for the digestive organs. Since 1900 the incidence of colon cancer has risen, while that of stomach cancer has fallen in the United States. The average diet has also changed during this time period. The intake of fats has risen by 40% while that of pickled and smoked food has decreased dramatically.

Cancers of the esophagus and stomach are usually carcinomas. Cancers of the esophagus metastasize early, and surgery may not be helpful. Surgery is helpful in about 10% of patients with cancer of the stomach. Persistent indigestion is sometimes a warning sign for cancer of the stomach. Cancers of the colon and rectum often arise from precancerous polyps (protruding growths from the mucous membrane). These are usually adenocarcinomas that can be detected by a sigmoidoscope and subsequently surgically removed. Cancers of the liver may occasionally be successfully treated if identified in time to remove all of the tumor surgically. On the other hand, the average survival for persons with cancer of the pancreas is relatively short even if treatment is attempted.

Of the Blood and Lymphatic Organs

Leukemias are a group of cancers of the blood-forming tissues, which usually result in an overproduction of abnormal white blood cells. Two common forms of leukemia are acute lymphocytic leukemia with malignant lymphocytes, which affects mostly young children, and AML (acute myelocytic leukemia) with malignant granular leukocytes, which strikes middle-age adults. In chronic leukemia, the cells are fairly normal and the disease progresses slowly. In acute leukemia, the cells may be immature and poorly differentiated, and the disease progresses rapidly. As the cells accumulate in the blood and lymph, the patient becomes anemic, bleeds easily, and is susceptible to infections. In the end, the leukemic cells invade the body's vital organs, impeding their functions and leading to death. A combination of chemotherapy and radiation treatments has proven helpful to patients with leukemia.

Lymphomas are cancers of the lymphoid organs such as lymph nodes, spleen, and the thymus gland. Hodgkin's disease, which is characterized by the presence of two kinds of abnormal lymph cells, is also classified as a lymphoma. Lymphomas can spread from one lymph area to another by way of the lymphatic system.

Cancers can be categorized according to the type of tissue and organ affected.

Detection, Treatment, and Prevention

Individuals should be aware of the seven danger signals for cancer (table 20.2) and inform their doctors when they notice any one of these. Thereafter, cancer can be detected by physical examination assisted by various means to view the internal organs. For example, ultrasound (high-frequency sound waves) and CAT (computer-assisted axial tomography) scanner, which uses X rays along with a computer to produce cross-sectional pictures of the body, are relatively new methods used to detect tumors. A new idea is to develop blood tests for cancer because some tumors can be associated with a particular chemical in the blood. For example, investigators believe that prostate specific antigen (PSA) is a substance that increases in the blood according to the size of a prostate tumor.

Surgery, Radiation, and/or Chemotherapy

Tumors can be surgically removed, but there is always the danger that malignant tumors have already metastasized. Therefore, when a growth is malignant, surgery is often preceded by or followed by radiation therapy. Radiation destroys the more rapidly dividing cancer cells, but causes less damage to the more slowly dividing normal cells. The use of radioactive protons is preferred over X ray whenever a hospital provides this form of treatment. Proton beams can be aimed at the tumor like an automatic rifle hitting the bull's eye of a target.

Chemotherapy is the use of drugs to kill cancer cells. It is another means of curing cancer and/or catching any stray cells left behind when a malignant growth is removed surgically.

Table 20.2 Danger Signals for Cancer

1. Unusual bleeding or discharge
2. A lump or thickening in the breast or elsewhere
3. A sore that does not heal
4. Change in bowel or bladder habits
5. Persistent hoarseness or cough
6. Persistent indigestion or difficulty in swallowing
7. Change in a wart or mole

Whenever possible, chemotherapy is specifically designed for the particular cancer cell. For example, in Allen's lymphoma/leukemia, it is known that a small portion of a chromosome 9 is missing and, therefore, DNA metabolism differs in the cancerous cells compared to normal cells. Specific chemotherapy for this cancer provides the patient with a drug designed to exploit this metabolic difference in order to destroy the cancerous cells.

Tailoring treatment to the particular cancer is a new trend. Breast cancers can be divided into those that are estrogen sensitive and those that are not. This is detected by testing excised cancer cells for the presence of estrogen receptors in the cell membrane. If many estrogen receptors are present, the patient is given a drug that blocks estrogen binding. Otherwise, the patient is given standard chemotherapy.

Cancer cells sometimes become resistant to chemotherapy (even when several drugs are used in combination) through the normal natural selection process discussed on page 420. It has been discovered that the cell membrane in resistant cells contains a carrier that pumps toxic chemicals out of the cell. Researchers are testing drugs known to poison the pump in an effort to restore sensitivity to chemotherapy.

Immunotherapy

Immunotherapy is the use of any immune system component such as antibodies, cytotoxic T cells, or lymphokines to promote the health of the body, such as when cancer is cured. Recently, monoclonal antibodies produced by the technique described in figure 7.14 have been used to carry a drug directly to individual cancer cells. In that way, only the tumor cells are exposed to the drug.

As mentioned previously, the very development of cancer signifies that the immune system is not working properly on its own. Specifically, it appears that cell-mediated immunity is lacking. If this is correct, then the use of lymphokines (p. 143) might awaken T cells and lead to the destruction of the cancer. Interferon was the first type of lymphokine therapy to be investigated. Unfortunately, the results, in general, have been disappointing; however, interferon has been found to be effective in up to 90% of patients with a type of leukemia known as hairy-cell leukemia (because of the hairy appearance of the malignant cells).

1. *Avoid obesity.* Women who are obese have a 55% greater risk, and men have a 33% greater risk of cancer than those of normal weight.
2. *Lower total fat intake.* A high-fat intake has been linked to development of breast, colon, and prostate cancer.
3. *Eat more high-fiber foods* such as those in whole-grain cereals, fruits, and vegetables. Studies have indicated that a high-fiber diet is protective against colon cancer, a frequent cause of cancer deaths (table 20.1). It is worth noting that foods high in fiber tend to be low in fat!
4. *Increase consumption of foods that are rich in vitamin A and C.* Beta-carotene, a precursor of vitamin A, is found in dark green leafy vegetables, carrots, and various fruits. Vitamin C is present in citrus fruits. These vitamins are called antioxidants because in cells they prevent the formation of free radicals (organic ions that have a nonpaired electron) that can possibly damage DNA. Vitamin C also prevents the conversion of nitrates and nitrites into carcinogenic nitrosamines in the digestive tract. Processed foods may have had nitrates and nitrites added as preservatives.
5. *Cut down on consumption of salt-cured, smoked, and nitrite-cured foods.* Salt-cured or pickled foods may increase the risk of stomach and esophageal cancer. Smoked foods like ham and sausage contain chemical carcinogens similar to those in tobacco smoke. Nitrites are sometimes added to processed meats (e.g., hot dogs and cold-cuts) and other foods to protect them from spoilage; as mentioned previously, nitrites are converted to nitrosamines in the digestive tract.
6. *Include vegetables of the cabbage family in the diet.* Aside from cabbage, broccoli, brussels sprouts, kohlrabi, and cauliflower are in the cabbage family. These vegetables may reduce the risk of gastrointestinal and respiratory tract cancers.
7. *Be moderate in the consumption of alcohol.* People who drink and smoke are at an unusually high risk of cancer of the mouth, larynx, and esophagus.

The interleukins (p. 144) are also being clinically tested. In one technique being used, cytotoxic T cells are withdrawn from the patient and activated by culturing them in the presence of an interleukin. The cells are then reinjected into the patient, who is also given doses of interleukin to maintain the killer activity of the T cells. This type of therapy shows some promise, but is still regarded as experimental. In the near future, there are plans to use genetically engineered cytotoxic T cells to produce tumor necrosis factor, a lymphokine that acts against tumors.

New methods are available for the treatment of cancer. Radiation and chemotherapy are being improved and immunotherapy is being developed.

Survival Rates

When cancer patients live five years beyond the time of diagnosis and treatment, it is generally considered that they are cured. Presently, less than 50% of all cancer patients survive the five-year period, but there has been an increase in the survival rate for certain cancers. For localized breast cancer, the survival rate has risen from 78% in the 1940s to 96% today. The survival rate for prostate cancer has steadily improved over the past 20 years. The five-year survival rate for prostate cancer has risen from 50% to 68% for whites and from 35% to 58% for blacks. There has been a recent dramatic increase in survival rate for children with acute lymphocytic leukemia. In some medical centers, the overall children's survival rate has risen to 75%.

Prevention of Cancer

There is clear evidence that the risk of certain types of cancer can be reduced by adopting certain behaviors. In general, the avoidance of carcinogenic chemicals and excessive radiation is helpful. Specifically, avoiding excessive sunlight reduces the risk of skin cancer, for example, and not smoking cigarettes and cigars reduces the risk of lung cancer and of other types of cancer as well.

Recently, there has been an emphasis on dietary considerations that help prevent cancer (fig. 20.8). Suggested guidelines are given in the accompanying reading. These are based on recommendations by the American Cancer Society and other organizations.

Figure 20.8   There are some data to suggest that the diet can influence the development of cancer. Fresh fruits, especially those high in vitamin A and C, and vegetables, especially those in the cabbage family, are believed to reduce the risk of cancer.

Human Issue

The National Cancer Institute believes that a 50% reduction in cancer incidence and mortality is possible by the year 2000 if the following measures are taken: (1) reduce percentage of adults who smoke to 15% or less, (2) reduce average consumption of fat to 25% or less of total calories, (3) double the average consumption of fiber, (4) increase to 80% the percentage of women ages 50–70 who have a regular physical breast exam and mammography, (5) increase to 90% the percentage of women ages 40–70 who have a Pap test every three years, and (6) increase the percentage of people who are receiving the best cancer treatments available.

Do you feel that the prevention of cancer is a social matter or a private matter? Should governmental agencies try to lower the statistics of cancer or should the individual assume responsibility for not getting cancer? If a governmental agency is responsible, what exactly should they do? If the individual is responsible, do they have an obligation to assume a healthy life-style that reduces the risk of cancer?

Summary

Cancer cells are transformed and differ from normal cells. They lack differentiation, grow unchecked, and metastasize. In cancer cells, proto-oncogenes have become oncogenes. This can be brought about by carcinogenic chemicals and radiation, or an oncogene can be inherited from a parent or brought into a cell by a virus. Most likely, more than one genetic change is required before cancer develops. Also, the immune system has failed to kill off the transformed cells.

Cancers are classified according to the type of tissues from which they arise: carcinomas are cancers of the epithelial tissues, sarcomas of connective tissue, leukemias of the blood tissues, and lymphomas of lymphoid tissue. Tumors can occur in any organ, but certain ones are more common than others.

Treatment has long centered on surgery, radiation, and chemotherapy, but these

treatments are being refined and modernized. Immunotherapy is a new type of treatment involving the use of monoclonal antibodies; treated, or even genetically engineered, cytotoxic T cells; and lymphokines. Human behaviors such as using sunscreens and not smoking and drinking can help avoid cancers. The new stress is on dietary guidelines to prevent the development of cancer.

Study Questions

1. Discuss three characteristics of cancer cells. (p. 402)
2. **What is a proto-oncogene, an oncogene? (p. 404)**
3. What is the function of proto-oncogenes in a cell? (p. 404)
4. Name two types of environmental carcinogens. How can an oncogene be brought into a cell? (p. 405)

5. Why does development of cancer signify a loss of immunity to cancer? (p. 406)
6. What are the most common causes of cancer deaths in men and in women? Discuss significant features of each type of cancer. (p. 407)
7. How can cancer be detected? (p. 409)

8. What kinds of treatment are available to cure cancer? Discuss the rationale for each. (p. 409)
9. Give a suggestion for avoiding (a) skin cancer, (b) lung cancer, (c) colon cancer, (d) breast cancer, and (e) gastrointestinal cancer. (p. 410)
10. What two vitamins, in particular, may help guard against cancer? (p. 410)

Objective Questions

1. Cancer cells have been _____ ; they have characteristics that distinguish them from normal cells.
2. Cancer cells lack _____ ; they are nonspecialized and do not contribute to the function of an organ.
3. The spread of cancer from the place of origin is called _____ .
4. During the initiation of cancer, a _____ becomes an _____ .

5. An environmental agent that can initiate or promote cancer is called a _____ .
6. You can inherit an oncogene, but one can also be carried into a cell by a _____ .
7. Smoking cigarettes is associated with _____ cancer and ultraviolet radiation is associated with _____ cancer.
8. The papilloma virus is associated with _____ cancer.

9. Increasing dietary fiber is believed to be protective against _____ cancer and decreasing fat intake is protective against _____ cancer.

Answers

1. transformed 2. differentiation 3. metastasis
4. proto-oncogene, oncogene 5. carcinogen
6. virus 7. lung, skin 8. cervical 9. colon; breast, colon, and prostate

Critical Thinking Questions

1. A number of retroviruses carry oncogenes in animals. Oncogenes are not viral in origin; they are human in origin. Explain.
2. The use of antibiotics is based on the fact that the metabolism of human cells is different from bacterial cells; therefore, the antibiotic kills only the bacterial cells. Chemotherapy is based on the fact that the metabolism of normal human cells is different from cancer cells. Explain.
3. Unsaturated fatty acids have a tendency to form free radicals (organic ions that have a nonpaired electron) which may damage DNA. With this knowledge in mind, hypothesize why a low-fat diet reduces the risk of breast, colon, and prostate cancer.

Selected Key Terms

carcinogen (kar-sin'o-jen)   an environmental agent that either initiates or promotes the development of cancer.   405

carcinoma (kar''si-no'mah)   a cancer of the epithelial tissue such as that lining the gastrointestinal or respiratory tracts.   406

chemotherapy (ke''mo-ther'ah-pe)   the use of a drug to selectively kill off cancer cells as opposed to normal cells.   409

immunotherapy (i-mu''no-ther'ah-pe)   the use of any immune system component such as antibodies, cytotoxic T cells, or lymphokines to promote the health of the body such as curing cancer.   409

leukemia (loo-ke'me-ah)   cancer of the blood-forming tissues leading to the overproduction of abnormal white blood cells.   406

lymphoma (lim-fo'mah)   cancer of the lymphoid organs such as lymph nodes, spleen, and thymus gland.   406

metastasis (me-tas'tah-sis)   the spread of cancer from the place of origin throughout the body caused by the ability of cancer cells to migrate and invade tissues.   403

oncogene (ong'ko-jen)   a cancer-causing gene that codes for a growth factor, or a growth factor receptor in the cell membrane, or a signaling protein between a growth factor receptor and nucleus of the cell.   404

Pap test (pap test)   a microscopic examination of cervical cells for the purpose of detecting cervical cancer.   408

proto-oncogene (pro''to-ong'ko-jen)   a normal gene that becomes an oncogene through mutation.   404

sarcoma (sar-ko'mah)   a cancer that arises in connective tissue, such as muscle, bone, and fibrous connective tissue.   406

transformation (trans''for-ma'shun)   the change of a normal cell into a cancer cell; transformed cells lack differentiation, grow uncontrollably, and metastasize.   402

tumor (too'mor)   growth containing cells derived from a single mutated cell that has repeatedly undergone cell division; benign tumors remain at the site of origin and malignant tumors metastasize.   403

Further Readings for Part V

Antebi, E., and D. Fishlock. 1986. *Biotechnology: Strategies for life*. Cambridge, Mass.: The MIT Press.

Bishop, J. M. March 1982. Oncogenes. *Scientific American*.

Chambon, P. May 1981. Split genes. *Scientific American*.

Chilton, M. June 1983. A vector for introducing new genes into plants. *Scientific American*.

Darnell, J. E. October 1983. The processing of RNA. *Scientific American*.

Dickerson, R. E. December 1983. The DNA helix and how it is read. *Scientific American*.

Drlica, K. 1984. *Understanding DNA and gene cloning*. New York: John Wiley and Sons.

Elkington, J. 1985. *The gene factory: Inside the science and business of biotechnology*. New York: Carroll and Graf Publishers.

Feldman, M., and L. Eisenback. November 1988. What makes a tumor cell metastatic? *Scientific American*.

Glover, D. M. 1984. *Gene cloning: The mechanics of DNA manipulation*. New York: Chapman and Hall.

Guttmacher, Alan F. 1986. *Pregnancy, birth, and family planning*. New York: New American Library.

Kartner, Norbert, and Victor Ling. March 1989. Multidrug resistance in cancer. *Scientific American*.

Kieffer, G. H. 1987. *Biotechnology, genetic engineering, and society*. Reston, Va.: National Association of Biology Teachers.

Mullis, Kary B. April 1990. The unusual origin of the polymerase chain reaction. *Scientific American*.

Nathans, Jeremy. February 1989. The genes for color vision. *Scientific American*.

Neufeld, Peter J., and Neville Colman. May 1990. When science takes the witness stand. *Scientific American*.

Patterson, D. August 1987. The causes of Down syndrome. *Scientific American*.

Ptashne, M. January 1989. How gene activators work. *Scientific American*.

Rosenberg, Steven A. May 1990. Adoptive immunotherapy for cancer. *Scientific American*.

Ross, Jeffrey. April 1989. The turnover of messenger RNA. *Scientific American*.

Sapienza, Carmen. October 1990. Parental imprinting of genes. *Scientific American*.

Shepard, J. F. May 1982. The regeneration of potato plants from leaf-cell protoplasts. *Scientific American*.

Tompkins, J. S., and C. Rieser. June 1986. Special report: Biotechnology. *Science Digest*.

Torrey, J. G. July–August 1985. The development of plant biotechnology. *American Scientist*.

Verma, Inder M. November 1990. Gene therapy. *Scientific American*.

Weinberg, R. A. September 1988. Finding the anti-oncogene. *Scientific American*.

White, R., and J. Lalouel. February 1988. Chromosome mapping with DNA markers. *Scientific American*.

Part Six

Human Evolution and Ecology

Evidences for the theory of evolution are drawn from many areas of biology. Charles Darwin was the first to recognize these evidences formally, and he suggested that those organisms best suited to the environment are the ones that survive and reproduce most successfully. Evolution causes life to have a history, and it is possible to trace the ancestry of humans even to the first cell or cells.

The diverse forms of life that are now present on earth live within ecosystems, where energy flows and chemicals cycle. Because population sizes remain constant, natural ecosystems tend to require the same amount of energy and chemicals each year.

Humans have created their own ecosystems, consisting of the country and city, which differ from natural ecosystems in that the populations constantly increase in size and ever greater amounts of energy and raw materials are needed each year. Since 1850, the human population has expanded so rapidly that some doubt there will be sufficient energy and food to permit the same degree of growth in the future. The human ecosystem depends on the natural ecosystems not only because they absorb pollutants but also because they are inherently stable. Every possible step should be taken to protect natural ecosystems to help ensure the continuance of the human ecosystem.

Chapter Twenty-One

Evolution

Chapter Concepts

1 The fossil record and comparative anatomy, embryology, and biochemistry all provide evidences of evolution.

2 Charles Darwin formulated a mechanism for an evolutionary process that results in adaptation to the environment.

3 Humans are primates, and many of their physical traits are the result of their ancestors' adaptations to living in trees.

4 Tool use, walking erect, and intelligence are all important advances made during human evolution.

5 All human races are classified as *Homo sapiens.*

Human remains appear very late in the fossil record. In fact, if the history of the earth is compared to a 24-hour period, humans do not appear until a minute before midnight. (See figure 21.3.)

Figure 21.1 Artificial selection in dogs. All the different types of dogs like *a.* Shetland sheepdog, *b.* dalmation, *c.* beagle, and *d.* bulldog are descended from the original domesticated dog. Humans have brought about the different types by controlling how the dogs breed.

What living things are able to shape the course of evolution? Human beings, of course. When human beings alter the environment, they are influencing which organisms will survive and which will die out in a particular locale. More directly humans carry out breeding programs to select which plants or animals will reproduce more than others. The end result can be specific, discrete types (fig. 21.1); all of the different forms of domesticated dogs belong to the same species—*Canis familiaris*—and selection has brought about the many varieties. It's obvious to us that genes control the characteristics we wish to see selected. One of the great accomplishments of twentieth-century science is to give the concept of evolution a genetic basis.

Evidences for Evolution

Fossil Record

Our knowledge of the history of life is based primarily on the fossil record. **Fossils** are the remains or evidence of some organism that lived long ago. Preserved fossils are most often found in sedimentary rock. Weathering produces sediments that are carried by streams and rivers into the oceans and other large bodies of water. There they slowly settle and are converted into sedimentary rock. Later sedimentary rocks are uplifted from below sea level to form new land. Now researchers are freely able to search for any fossil remains trapped in the rocks (fig. 21.2). The oldest fossils found are of prokaryotic cells dated some 3.5 billion years ago. Thereafter the fossils get more and more complex. For example, among animals simple eukaryotic forms were followed by invertebrates, and these were followed by vertebrates, until humans evolved only about one million years ago. This is clear evidence that there has been a history of life (fig. 21.3) based on evolutionary events. The later-evolving animals are descended from the earlier ones. The recent controversies among evolutionists are not about whether evolution occurred but simply about the manner in which it may have occurred (see reading on page 418).

Comparative Anatomy

A comparative study of the anatomy of groups of organisms has shown that each has a *unity of plan*. For example, all vertebrate animals have essentially the same type of skeleton. Unity of plans allows organisms to be classified into various groups. Organisms most similar to one another are placed in the same **species,** similar species are placed in a **genus,** similar genera in a family; thus, we proceed from **family** to **order** to **class** to **phylum** to **kingdom.** The classification of any particular organism indicates to which kingdom, phylum, class, order, family, genus, and species the organism belongs. According to the **binomial system** of naming organisms, each organism is given a two-part name, which consists of the genus and species to which it belongs (table 21.1). Thus, for example, a human is *Homo sapiens*[1] and the domesticated cat is *Felis domestica.* **Taxonomy** is the branch of biology that is concerned with classification, and biologists who specialize in classifying organisms are called taxonomists.

Human Issue

That organisms have arisen by evolution is supported by investigations using accepted scientific methodology, including hypothesis testing, experimentation, observation, measurement, and predictions from scientific models. Moreover, ideas about evolution—as in all fields of science—are continually subjected to the critical scrutiny of the scientific community via scientific papers and debate. The creation story does not rely on such protocol, and is therefore considered unscientific by most scientists. Do you think science teachers should be required to teach the creation story? Do you think what is or is not science should be determined by a local school board? To what extent do you think the public should be involved in what is taught as science in public schools?

A unity of plan is explainable by descent from a **common ancestor.** Species that share a recent common ancestor will share a large number of the same genes and therefore will be quite similar to each other and to this ancestor. Species that share a

Figure 21.2   Field researchers examine rocks near North Pole, Australia, for evidence of fossils. At this location, researchers have found fossils from the earliest form of life.

Table 21.1   The Classification of Modern Humans

kingdom	Animalia (animals)
phylum	Chordata (chordates)
class	Mammalia (mammals)
order	Primates (primates)
suborder	Anthropoidea (anthropoids)
superfamily	Hominoidea (hominoids)
family	Hominidae (hominids)
genus	*Homo* (humans)
species	*sapiens* (modern humans)

more distant common ancestor will have fewer genes in common and will be less similar to each other and to this ancestor because differences arise as organisms continue on their own evolutionary pathways. This principle allows biologists to construct **evolutionary trees,** diagrams that tell how various organisms are believed to be related to one another. All evolutionary trees have a branch-like pattern (fig. 21.4), indicating that evolution does not proceed in a single steplike manner; rather, evolution proceeds by way of common ancestors that often give rise to two different groups of organisms. For example, reptiles are believed to have produced both birds and mammals. Also, all organisms continue to evolve. For example reptiles are still evolving even though they gave rise to birds and mammals.

[1]Varieties of the same species are also given a subspecies designation. Neanderthals are *Homo sapiens neanderthalensis* and modern humans are *Homo sapiens sapiens.*

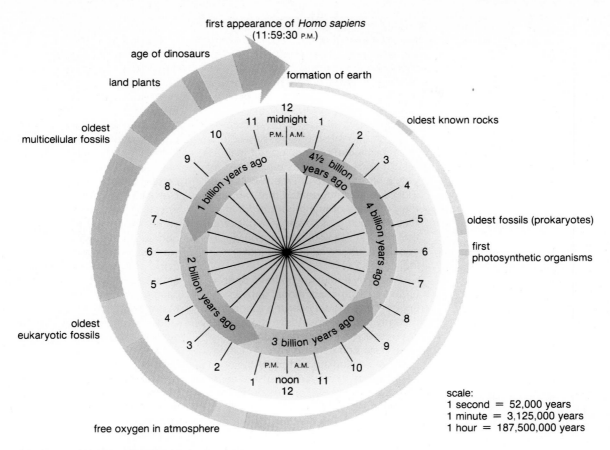

Figure 21.3   The outer ring of this diagram shows the history of the earth as it would be measured on a 24-hour time scale. (The inner ring shows the actual years starting at 4½ billion years ago.) A very large portion of life's history is devoted to the evolution of single-celled organisms. The first multicellular organisms do not appear until just after 8 P.M. and humans are not on the scene until less than a minute before midnight.

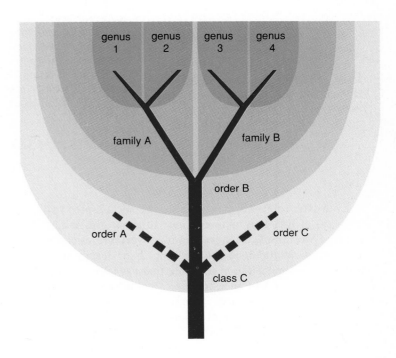

Figure 21.4   Evolutionary trees can show the relationship between different taxonomic categories. The organisms in family A and genera 1 and 2 are closely related as are those in family B and genera 3 and 4. A common ancestor can be found wherever a branch point occurs in the tree.

New Questions from the Fossil Record

There is no doubt that the fossil record supplies data to support the theory of evolution. The record shows that life began with simple organisms and advanced to the complex organisms in existence today. Also, direct ancestors for every living thing can be found in the fossil record.

A few years ago, Stephen J. Gould and Niles Eldredge, two paleontologists, wondered if the fossil record also gives us specific information regarding the tempo (rate) of evolutionary change. Based primarily on experimental genetic data in the laboratory, most biologists believe that a single species changes slowly—over thousands or even millions of years—into other species. This so-called *gradualistic model* of evolutionary change implies that the fossil record contains a plentiful supply of *intermediate forms,* which are fossil remains of life having characteristics of two different groups. Some intermediate forms have been found, but not an overwhelming number.

In contrast to the gradualistic model, Gould and Eldredge, upon close examination of the fossil record, proposed a *punctuated equilibrium model* of evolutionary change. The record shows that each species tends to remain the same for hundreds of thousands of millions of years. Then, relatively suddenly, new species appear. In other words, an *equilibrium* phase (long periods without change) is *punctuated* by a rapid burst of change. According to this model there would be few intermediate forms in the fossil record because such forms existed for only a short period of time. Gould and Eldredge also suggested that, without the presence of any other pertinent factors, such rapid change might require that natural selection select entire species for survival and not just individuals.

Thus far, evolutionists have been unable to supply additional data to support one model over the other. Very recently and unexpectedly, though, they have received some help from geologists and astronomers. For many years, evolutionists and other biologists have tried to explain the mass extinction of the dinosaurs. Dinosaurs dominated the earth for 135 million years, but suddenly they and much of the rest of life on earth vanished. In 1979, a Berkeley geologist, Walter Alvarez, and his colleagues found that Cretaceous clay contained an abnormally high level of iridium. This could have been caused by a worldwide fallout of radioactive material created by an asteroid impact at the time of the dinosaur disappearance. In 1984, paleontologists David Raup and John Sepkoski of the University of Chicago discovered that the dinosaurs are not alone; rather, the fossil record of marine animals shows that mass extinctions have occurred every 26 million years or so.

Surprisingly, astronomers, taking a hint from Alvarez, can give an explanation for these occurrences. Our solar system is in the Milky Way, a starry galaxy that is 100,000 light years[1] in diameter and 1,500 or so light years thick. Our sun moves up and down as it orbits in the Milky Way. Scientists predict that this vertical movement will cause our solar system to approach certain other members of the Milky Way every 26–33 million years, producing an unstable situation that could lead to comet bombardment of the earth. This bombardment can be likened to a worldwide atomic bomb explosion. A cloud of dust will mushroom into the atmosphere and shade out the sun, thus causing the plants to freeze and die. Mass extinction of animals will follow. Once the cosmic winter is over, plant seeds will germinate. This new growth will serve as food for the few remaining animals. Just as Darwin's finches rapidly evolved on the Galápagos Islands because of the lack of competition, these animals will also undergo rapid evolution.

This scenario suggests that the punctuated equilibrium model of evolutionary change does have merit. The equilibrium phase, during which evolutionary change is extremely slow, would occur between times of mass extinction, and the punctuated phase would occur immediately after mass extinctions. It is only necessary to return to the fossil record to see if this hypothesis holds.

[1]One light year, the distance light travels in a year, is about 6 trillion miles.

Even after related organisms have become adapted to different ways of life, they may continue to show similarities of structure. For example, the forelimb of all vertebrates contains the same fundamental bone structure (fig. 21.5) despite specific specializations. Similarities in structure that have arisen through descent from a common ancestor are called **homologous structures.** Homologous structures indicate that organisms are related. Sometimes two groups of organisms have structures that function similarly but are constructed differently. In contrast to homologous structures, **analogous structures,** such as an insect wing and bird wing, have similar functions but differ in anatomy, and therefore we know they evolved independently of one another.

Comparative Embryology

Some groups of organisms share the same type of embryonic stages. As would be expected if all vertebrates are related, their embryonic stages are similar. During development, a human embryo at one point has gill clefts—even though it will never

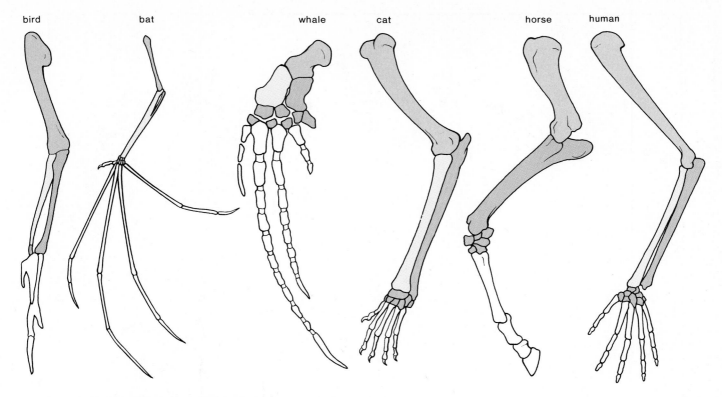

bird   bat   whale   cat   horse   human

Figure 21.5 Homologous structures. The bones are color coded so that you may note the similarity in the bones of the forelimbs of these vertebrates. This similarity is to be expected since all vertebrates trace their ancestry to a common ancestor.

breathe by means of gills as do fishes—and a rudimentary tail—even though it will never have a long tail as do some four-legged vertebrates. In this way, embryological observations indicate evolutionary relationships.

Vestigial Structures

An organism may have structures that are underdeveloped and seemingly useless, and yet they are fully developed and functional in related organisms. These **vestigial** structures are understandable when we realize that related organisms have shared a common evolutionary history. As examples of vestigial structures, consider that humans have the remnants of a nictitating membrane in the inner corner of the eye. In reptiles, birds, and some other mammals the nictitating membrane can be drawn over the eye like a third eyelid. The hair covering in many mammals is a thick fur, but humans usually have only the remnants of body hair.

Comparative Biochemistry

Almost all living organisms use the same basic biochemical molecules, including DNA, ATP, and many identical or nearly identical enzymes. For example, the metabolic enzymes involved in cellular respiration and in the synthesis of cellular macromolecules are the same in all organisms. It would seem, then,

that the molecules unique to living things appeared very early in the evolution of life and have been passed on ever since.

Analyses of amino acid sequences in certain proteins like hemoglobin and cytochrome C and analyses of DNA nucleotide differences are used to determine how distantly related animals are. It is assumed that the number of differences will reflect how long ago the two species shared a common ancestor. Investigators have been gratified to find that evolutionary trees based on biochemical data are quite similar to those based on anatomical data. Whenever the same conclusions are drawn from independent data, they substantiate scientific theory—in this case organic evolution—even more than usual.

Biogeography

It is observed that similar but geographically separate environments have different plants and animals that are similarly adapted. For example, figure 21.6 compares several North American animals with African animals that are adapted to living in a grassland environment. First, you will notice that although each group of animals could live in the other's biogeographic region, they do not. Why? They do not because geographic separation made it impossible for a common ancestor to produce descendants for both regions. On the other hand, notice that the same type of adaptations are seen in both groups of animals. This phenomenon, called **convergent evolution,** sup-

North American

African

Figure 21.6   Parallel evolution. The grasslands in Africa and America have similarly adapted but different animals. Bison and pronghorn in North America and zebra and springbok in Africa are running animals that feed on grass; coyote and bobcat in North America and lion and cheetah in Africa are meat eaters that feed on the running animals. This shows that similarly adapted animals evolve in similar environments.

ports the belief that the evolutionary process causes organisms to be adapted to their environments.

Evolution explains the history and diversity of life. Evidences for evolution can be taken from the fossil record and comparative anatomy, embryology, and biochemistry. Also, vestigial structures and biogeography support the occurrence of evolution.

Mechanisms of Evolution

Based on similar types of evidence, such as we have just outlined, Charles Darwin formulated a theory of **natural selection** of evolution around 1860. The theory of natural selection is nonteleological. A **teleological** mechanism is one in which the end result has been determined from the beginning. A nonteleological mechanism is one in which the end result cannot be determined. Do humans breathe air in order to live on land (teleological), or have circumstances brought about this ability over time (nonteleological)?

The following are critical to understanding **natural selection.**

Variations

Individual members of a species vary in their physical characteristics (fig. 21.7). Such variations can be passed on from generation to generation. Darwin was never able to determine the cause of variations, nor how they are passed on. Today, we realize that genes determine the appearance of an organism and that mutations (permanent genetic changes) can cause new variations to arise.

Struggle for Existence

In Darwin's time, a socioeconomist, Thomas Malthus, had stressed the reproductive potential of human beings. He proposed that death and famine were inevitable because the human population tended to increase faster than the supply of food. Darwin applied this concept to all organisms and saw that members of any plant or animal population must compete with one another for available resources. Competition must of necessity occur because reproduction produces more members of a population than can be sustained. Darwin calculated the reproduc-

Figure 21.7   Variations among individuals. It is easy for us to note that humans vary one from the other, but the same is true for populations of any type organism.

tive potential of elephants. Assuming a life span of about 100 years and a breeding span of from thirty to ninety years, a single female will probably bear no fewer than six young. If all these young survived and continued to reproduce at the same rate, after only 750 years the descendants of a single pair of elephants would number about 19 million! Such reproductive potential necessitates a *struggle for existence,* and only certain organisms will survive and reproduce.

Survival of the Fittest

Darwin noted that in *artificial selection* humans choose which plants or animals will reproduce. This selection process brings out certain traits. For instance, there are many varieties of pigeons, each of which is derived from the wild rock dove. In a similar way, several varieties of vegetables can be traced to a single type.

In nature, it is the environment that selects those members of the population that will reproduce to a greater degree than other members. In contrast to artificial selection, the natural selection process is not teleological. Selection occurs only because certain members of a population happen to have a variation that makes them more suited to the environment. For example, any variation that increases the speed of a hoofed animal will help it escape predators and live longer; a variation that reduces water loss will help a desert plant survive; and one that increases the sense of smell will help a wild dog find its prey. Therefore, we would expect organisms with these traits to eventually reproduce to a greater extent.

Whereas Darwin emphasized that only certain organisms survived to reproduce, modern evolutionists emphasize that competition results in unequal reproduction. If certain organisms can acquire a greater share of available resources and if they have the ability to reproduce, then their chances of reproduction are greater than those of their cohorts.

Adaptation

Natural selection brings about **adaptation**—ultimately a species becomes adapted or suited to the environment. The process is slow, but each subsequent generation will include more individuals that are adapted than the previous generation. Eventually a species may become so adapted to a particular environment that it is unable to adapt to a new and changing environment. Extinction of this species may now occur (fig. 21.8).

The following listing summarizes the theory of evolution as developed by Darwin.

1. There are inheritable variations among the members of a population.
2. Many more individuals are produced each generation than can survive and reproduce.
3. Individuals with adaptive characteristics are more likely to be selected to reproduce by the environment.
4. Gradually, over long periods of time, a population can become well adapted to a particular environment.
5. The end result of organic evolution is many different species, each adapted to specific environments.

Figure 21.8   Out on an evolutionary limb? Some believe the saber-toothed tiger became extinct because its canine teeth became too long to use. But its real problem seems to have been a lack of speed. When its large, slow-moving prey became extinct, it could not adjust to catching smaller, more speedy prey.

Table 21.2   Origin of the First Cell

Primitive Earth
    Cooling
Gases
    Energy Capture
Small Molecules: Amino Acids, Glucose, Nucleotides
    Polymerization
Macromolecules: Proteins, Carbohydrates
    Metabolism
Protocell: Heterotrophic Fermenter
    Hereditary Material
Cell: Reproduction
    Organic Evolution
Autotrophic Cell: Gives Off Oxygen
    Aerobic Respiration
Animal-Like Heterotroph

Organic Evolution

Figure 21.3 outlines the evolutionary history of life on earth. Organic evolution began with the evolution of the first cell or cells. Scientists have pieced together a reasonable hypothesis (table 21.2) concerning the evolution of the first cell(s), which they call the origin of life.

Origin of Life

The sun and the planets probably formed from aggregates of dust particles and debris about 4.6 billion years ago. Intense heat produced by gravitational energy and radioactivity of some atoms caused the earth to become stratified into a core, mantle, and crust. Heavier atoms of iron and nickel became the molten liquid core, and dense silicate minerals became the semiliquid mantle. The unstable mantle caused the thin crust to move continually, so there were no stable land masses during the time life was evolving.

Primitive Atmosphere

The *primitive atmosphere* was not the same as today's atmosphere (fig. 21.9). It is now thought that the primitive atmosphere was produced after the earth formed by outgassing from the interior, particularly by volcanic action. The atmosphere would have consisted mostly of water vapor (H_2O), nitrogen gas (N_2), and carbon dioxide (CO_2), with only small amounts of hydrogen (H_2) and carbon monoxide (CO). The primitive atmosphere with little, if any, free oxygen was a reducing atmosphere as opposed to the oxidizing atmosphere of today. This was fortuitous because oxygen (O_2) attaches to organic molecules, preventing them from joining together to form larger molecules.

Chemical Evolution

Water, present at first as vapor in the atmosphere, formed dense, thick clouds, but cooling eventually caused the vapor to condense to liquid, and rain began to fall. This rain was in such quantity that it produced the oceans of the world (fig. 21.9a). The gases, dissolved in the rain, were carried down into the newly forming oceans.

The remaining steps shown in table 21.2 took place in the sea, where life arose. The dissolved gases, although relatively inert, are believed to have reacted together to form simple organic compounds when they were exposed to the strong outside *energy sources* present on the primitive earth (fig. 21.9b). These energy sources included heat from volcanoes and meteorites, radioactivity from the earth's crust, powerful electric discharges in lightning, and solar radiation, especially ultraviolet radiation. In a classic experiment, Stanley Miller showed that an atmosphere containing methane and ammonia could have produced organic molecules. These gases were dissolved in water and circulated in a closed container past an electric spark. After a week's run, he analyzed the contents of the reaction mixture and found, among other organic compounds, amino acids and nucleotides. Other investigators have achieved the same results by utilizing carbon monoxide and nitrogen gas dissolved in water. On the basis of these experiments, it is surmised that the primitive oceans were most likely a thick organic soup.

The next step was the condensation of small organic molecules to produce the macromolecules characteristic of living things: polynucleotides, polypeptides, and polysaccharides. It is possible that macromolecules could have formed in the ocean, but it is more likely that small organic molecules were washed ashore, where they adhered to clay particles and were exposed to dry heat that encouraged polymerization. S. W. Fox has heated mixtures of amino acids at 130°–180° to form amino acid polymers that he calls proteinoids (fig. 21.9c). When proteinoids are placed in water, they form microspheres, structures that have some cell-like properties. They are, so to speak, the beginnings of cells, or **protocells.**

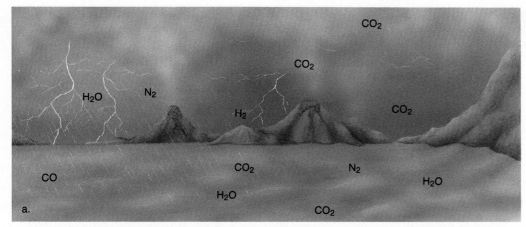

a. The primitive atmosphere contained gases, including water vapor that escaped from volcanoes; as the latter cooled, some gases were washed into the ocean by rain.

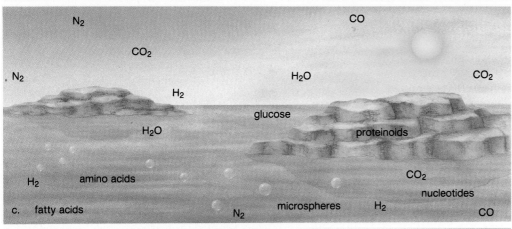

b. The availability of energy from volcanic eruption (shown here) and lightning allowed gases to form simple organic molecules.

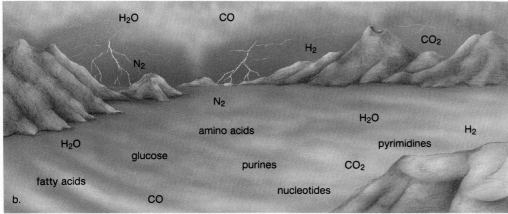

c. Amino acids splashed up onto rocky coasts could have polymerized into polypeptides (proteinoids) that would have become microspheres when they reentered the water.

d. Eventually various types of prokaryotes and then eukaryotes evolved. Some of the prokaryotes were oxygen-producing photosynthesizers. The presence of oxygen in the atmosphere was needed for aerobic respiration to evolve.

Figure 21.9   A model for the origin of life.

In contrast, Oparin showed that mixtures of macromolecules could join together in water to form **coacervate droplets.** Both microspheres and coacervates can take up molecules from the environment to form a lipid-protein film like a simple cell membrane. When they take up enzymes, they carry on some metabolic functions. Fox believes that after protocell formation, proteins and nucleic acids could have gradually evolved and increased in complexity until a true cell capable of reproduction came into being.

Some biologists disagree with this hypothesis. They believe chemical evolution may have proceeded by the initial development of nucleic acid genetic material rather than protein coacervates or microspheres. The most primitive nucleic acids might have reproduced themselves and subsequently acquired the ability to direct the synthesis of peptides. Regardless, the first true cell had to perform two functions: metabolism and reproduction.

A chemical evolution led to the protocell, which was a heterotroph that carried on anaerobic respiration. The protocell became a true cell once it could reproduce. Thereafter organic evolution began.

Organic Evolution Begins

The first primitive cell(s) possessing rudimentary enzymes, genes, and a selectively permeable membrane must have been a heterotroph living off preformed organic molecules in the primitive ocean, and it must have carried on anaerobic respiration since there was no oxygen (O_2) in the atmosphere. It was, then, a *heterotrophic fermenter.*

Once the preformed organic molecules were depleted, organic evolution would have favored any cell capable of making its own food. The first autotrophs probably lacked a light-absorbing pigment and, therefore, were not photosynthesizers. They could have been chemosynthetic organisms that extracted electrons from inorganic molecules (other than water) and used these electrons to generate ATP by electron transport. Or even if they did possess a pigment able to capture light energy, in order to reduce carbon dioxide, they probably did not at first use water as a hydrogen source and, thus, did not give off oxygen. The first autotroph to use water would have been selected because water is such a plentiful molecule. Now oxygen would have been given off as a by-product of photosynthesis (fig. 21.9d).

The presence of free oxygen (O_2) changed the character of the atmosphere; it became an oxidizing atmosphere instead of a reducing atmosphere. Abiotic synthesis of organic molecules was no longer possible because any organic molecules that happened to form would have been broken down by oxidation. As oxygen levels increased, cells capable of aerobic respiration evolved. This is the type of respiration used by the vast majority of organisms today.

The buildup of oxygen in the atmosphere caused the development of an ozone (O_3) layer. This filters out ultraviolet rays, shielding the earth from dangerous radiation. Prior to this time, organisms probably lived only deep in the oceans where they were not exposed to the intense radiation striking the earth's surface. Now life could safely spread to shallow waters and eventually move onto the land.

Plants were probably the first living organisms on land. This is reasonable because animals are dependent upon plants as a food source. The first vertebrates (animals with backbones) to dominate the land were the amphibians, whose name implies that their life cycle has two phases—one spent in the water and one spent on land. Frogs are amphibians and during their tadpole stage, they metamorphose into an adult frog. Next came the reptiles that have no need to return to water for reproductive purposes because the female lays a shelled egg. Within the egg are the extraembryonic membranes (fig. 16.7) that service all the needs of the developing organism. Both birds and mammals evolved from the reptiles. We are mammals (table 21.1), animals characterized by the presence of hair and mammary glands. Even so, the brain is most highly evolved among mammals and reaches its largest size and complexity in the primates, the order to which humans belong.

The evolution of humans did not begin until about 5 million years ago; therefore, we have been evolving only 0.1% of the total history of the earth.

Organic evolution involves the evolution of all life forms, including vertebrates. The first land vertebrates were the amphibians, which gave rise to the reptiles, from which the birds and mammals arose. Humans are primates, a type of mammal.

Human Issue

Which position do you support? *a.* It should be illegal to teach human evolution, and the concept should be removed from all biology textbooks. *b.* By law, biology textbooks should have to include, along with the concept of evolution, the biblical account of human origins. *c.* Since the biblical account is not a part of scientific knowledge, biology textbooks need not discuss it. *d.* It should be made illegal to include the biblical account of human origins in biology textbooks.

Human Evolution

Humans are mammals in the order Primates (table 21.1). **Primates** were originally adapted to an arboreal life. Long and freely movable arms, legs, fingers, and toes allowed them to reach out and grasp an adjoining tree limb. The opposable thumb and toe, meaning that the thumb and toe could touch each of the other digits, were also helpful. Nails replaced claws; this meant that primates could also easily let go of tree limbs.

The brain became well developed, especially the cerebral cortex and frontal lobes, the highest portions of the brain. Also, the centers for vision and muscle coordination were enlarged. The face became flat so that the eyes were directed forward, allowing the two fields of vision to overlap. The resulting stereoscopic (three-dimensional) vision enabled the brain to determine depth. Color vision aided the ability to find fruit or prey.

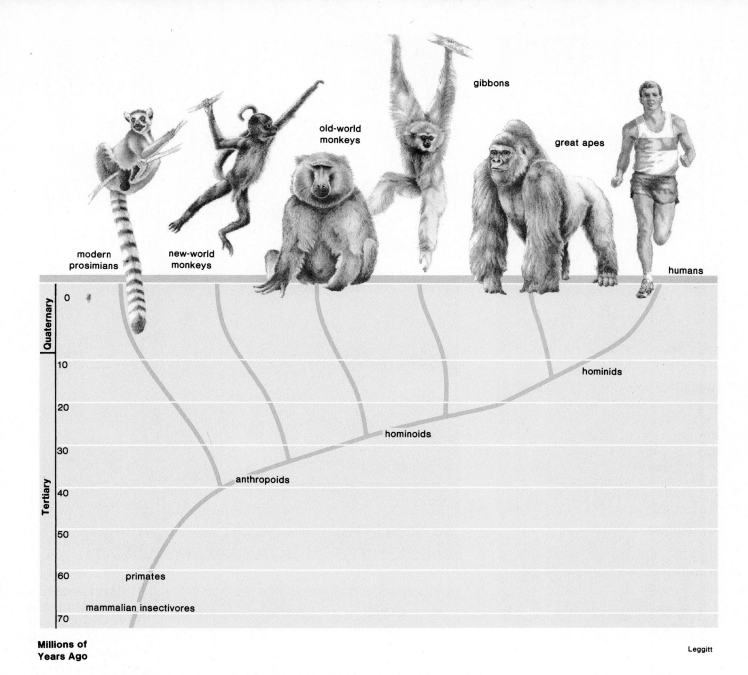

modern prosimians

new-world monkeys

old-world monkeys

gibbons

great apes

humans

Quaternary

Tertiary

0

10

20

30

40

50

60

70

hominids

hominoids

anthropoids

primates

mammalian insectivores

Millions of Years Ago

Leggitt

Figure 21.10   Primate evolution. There are at least 4 lines of evolution among the primates, which arose from mammalian insectivores. Prosimians include lemurs and tarsiers; monkeys include new-world and old-world monkeys; apes include gibbons and great apes (orangutan, gorilla, and chimpanzee); and humans.

One birth at a time became the norm; it would have been difficult to care for several offspring, as large as primates, in trees. The period of postnatal maturation was prolonged, giving the immature young an adequate length of time to learn behavior patterns.

Prosimians

The first primates were **prosimians,** a term meaning "premonkeys." The prosimians are represented today by several types of

animals, including the lemurs (fig. 21.10), which have a squirrel-like appearance, and the tarsiers, curious mouse-sized creatures with enormous eyes suitable for their nocturnal way of life. Tarsiers have a flattened face, and their digits terminate in nails.

Anthropoids

Monkeys, along with apes and humans, are **anthropoids.** Monkeys evolved from the prosimians about 38 million years ago, when the weather was warm and vegetation was like that of a

a.

c.

b.

Figure 21.11 Ape diversity. *a.* Of the apes, gibbons are the most distantly related to humans. They dislike coming down from trees, even at watering holes. They extend a long arm into the water and then drink collected moisture from the back of the hand. *b.* Orangutans are solitary except when they reproduce. Their name means "forest man"; early Malayans believed that orangutans were intelligent and could speak but did not because they were afraid of being put to work. *c.* Gorillas are terrestrial and live in groups in which a silver-backed male is always dominant. *d.* Of the apes, chimpanzees seem the most humanlike.

d.

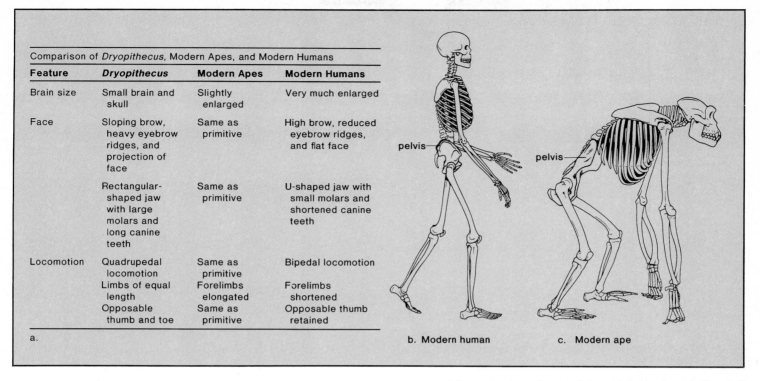

Comparison of *Dryopithecus*, Modern Apes, and Modern Humans

Feature	*Dryopithecus*	Modern Apes	Modern Humans
Brain size	Small brain and skull	Slightly enlarged	Very much enlarged
Face	Sloping brow, heavy eyebrow ridges, and projection of face	Same as primitive	High brow, reduced eyebrow ridges, and flat face
	Rectangular-shaped jaw with large molars and long canine teeth	Same as primitive	U-shaped jaw with small molars and shortened canine teeth
Locomotion	Quadrupedal locomotion	Same as primitive	Bipedal locomotion
	Limbs of equal length	Forelimbs elongated	Forelimbs shortened
	Opposable thumb and toe	Same as primitive	Opposable thumb retained

a.

b. Modern human   c. Modern ape

Figure 21.12   Evolution of anatomical differences between apes and humans. *a.* As outlined in the table, the dryopithecines had primitive characteristics that could have led to both modern-day apes and humans. Compare the modern human skeleton, *b.* to the modern ape (gorilla) skeleton *c.* Humans are bipedal, while modern-day apes are knuckle walkers. In the ape, the pelvis is very long and tilts forward, whereas in humans, it is short and upright. The ape shoulder girdle is more massive than that of humans, and the head hangs forward

because of the angle of attachment of the vertebral column to the skull. In apes, the foramen magnum (hole in skull through which the spinal cord passes) is well to the rear of the skull; in humans it is almost directly in the bottom center of the skull. Also, in humans, the spine has an S-shaped curve that allows better weight distribution and improved balance when the body is upright.

tropical rain forest. There are two types of monkeys: the New World monkeys, which have long prehensile (grasping) tails and flat noses, and the Old World monkeys, which lack such tails and have protruding noses. Two of the well-known New World monkeys are the spider monkey and the capuchin, the "organ grinder's monkey." Some of the better-known Old World monkeys are now ground dwellers, such as the baboon and the rhesus monkey, which has been used in medical research.

Apes and humans, the other anthropoids, evolved later.

Primates evolved from shrewlike mammals and became adapted to living in trees, as exemplified by skeletal features, good vision, and reproductive habits. Primates are represented by prosimians, monkeys, apes, and humans. Monkeys, apes, and humans are anthropoids.

Hominoids

Humans are more closely related to apes (fig. 21.11) than to monkeys. **Hominoids** include apes and humans. There are four types of apes: the **gibbon,** the **orangutan,** the **gorilla,** and the **chimpanzee.** The gibbon is the smallest of the apes, with a body weight ranging from 5 to 10 kg. Gibbons have extremely long arms that are specialized for swinging between tree limbs. The orangutan is large (75 kg), but nevertheless spends a great deal of time in trees. In contrast, the gorilla, the largest of the apes

(185 kg), spends most of its time on the ground. Chimpanzees, which are at home both in the trees and on the ground, are the most humanlike of the apes in appearance and are used sometimes in psychological experiments.

Hominoid Ancestor   About 25 million years ago, the weather began to turn cooler and drier. In places, the tropical forests gave way to grasslands. At this time, apes became abundant and widely distributed in Africa, Europe, and Asia. Among these, members of the genus ***Dryopithecus*** are of particular interest because they are thought to be a possible hominoid ancestor—the last common ancestor to all other apes and humans. The dryopithecines were forest dwellers that probably spent most of the time in trees. The bones of their feet, however, indicate that if they did sometimes walk on the ground, they walked on all four limbs, using the knuckles of their hands to support part of their weight. Such "knuckle walking" is retained by modern apes but not by humans, who stand erect (fig. 21.12).

The skull of the dryopithecines had a sloping brow, heavy eyebrow ridges (called supraorbital ridges), and jaws that projected forward. These, too, are features retained by apes but not by humans. In contrast to apes, humans have a high brow and lack the supraorbital ridges. The human face is flat, and there is no snout, and the canine teeth are comparable in size to the other human teeth.

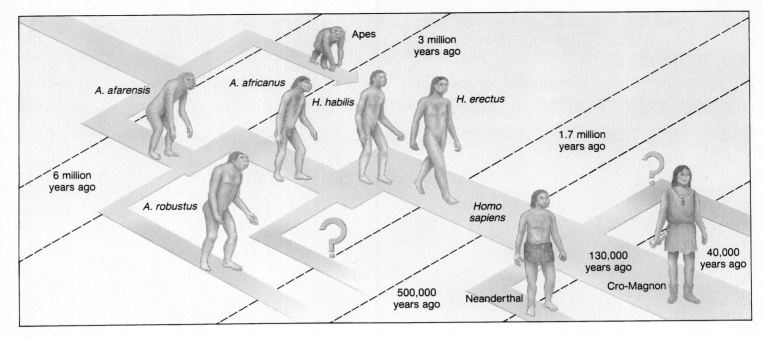

Figure 21.13   Hominid evolutionary tree. Adaptive radiation is evident because this tree indicates that various hominids coexisted. This is one possible tree; others have been suggested.

Hominids

It is not known when the human lineage split from that of the apes, but biochemical evidence can be used to suggest a date. Biochemists have compared the sequence of DNA bases in modern apes and humans. Based on the finding that there is only a 2% difference between the sequence of DNA bases in humans and that in chimpanzees, they believe that the split between these apes and humans occurred only 5–10 million years ago.

Fossil evidence indicates that hominids evolved in Africa, but we do not know what the very first **hominids,** humans and fossils in the human lineage, were like nor whether they lived on trees or walked on the ground. The cooling trend that began earlier continued. If the first hominids occasionally did come down out of the trees, perhaps they already had an upright bipedal posture (walking on two limbs). It is possible that bipedal walking started through balancing erect on tree branches, as gibbons sometimes do. A bipedal posture would have enhanced survival on the savanna because it would have allowed hominids to move efficiently between trees.

The first hominid (ancestor leading to humans) arose at a time when a change in weather reduced the size of the African forests, making it advantageous to move on the ground.

New ideas regarding speciation have been used to interpret the hominid fossil record. Although at one time scientists attempted to place each hominid fossil in a straight line from the most primitive to the most advanced, it now is reasoned that several hominid species existed at the same time. In other words, hominids underwent adaptive radiation. Figure 21.13 indicates a possible evolutionary tree for known hominids.

Australopithecines

The fossils classified in the genus *Australopithecus* (Southern Apeman) were found in southern Africa. These fossils date from about 4 million to 1.5 million years ago. Three species of australopithecines have been identified; *A. afarensis, A. africanus, and A. robustus.* Because of the small brain, the large canine teeth, and the protruding face, it is believed that *A. afarensis,* also called Lucy (from the song "Lucy in the Sky with Diamonds") is the most primitive of the three fossils.

The australopithecines (fig. 21.14) were about 1.5 m tall, with a brain that ranged in size from 300 to 500 cc (cubic centimeters). Their overall anatomy forms a connecting link between apes and humans. For example, they had large, long-fingered hands retaining an ability to easily grasp tree limbs. Their arms were shorter than apes but their legs were only a little longer compared to their size. The foot and hipbones indicate that they were able to walk erect. Notice that this means that an enlarged brain was *not* needed for the evolution of bipedal locomotion. However, their stride was short and, therefore, walking and running would have been slowed. Almost certainly they did not talk or use fire. They probably could have driven off large carnivores by screaming, charging, and throwing objects as apes do today. Also like today's terrestrial apes, they probably slept in trees or on cliffs.

The hominid *A. robustus,* as its name implies, was larger than *A. africanus.* Certain anatomical differences may be as-

Figure 21.14 *Australopithecus africanus.* While at one time these hominids were thought to have been hunters, a study of cut marks left on animal bones suggests that they most likely were scavengers that sought and confiscated the kill of coexisting predators and scavengers.

sociated with diet. The massive cranium and face and the large cheek teeth of *A. robustus* indicate a vegetarian diet. By comparison, the gracile facial features of *A. africanus* indicate a more varied diet, possibly including meat, because facial bones—notably eyebrow ridges—facial muscles, and teeth need not be as large in meat eaters. It now is believed that the australopithecines were not hunters but scavenged the meat that they ate.

The earliest stone tools were found in East Africa and date from about 2 million years ago. Some of these very crude stone tools were found among the remains of *A. robustus.* Perhaps this australopithecine was a toolmaker just as *Homo habilis* people were. Supporting this contention is the broadness of the thumb, which probably could flex at the terminal knuckle. Even so, the manner in which the teeth developed and the final thickness of the enamel suggest that the hominid *A. robustus* is on a side branch of the human evolutionary tree and became extinct. Only the hominid *A. africanus* is believed to have possibly evolved into *Homo habilis.*

One possible human evolutionary tree shows the bipedal *Australopithecus afarensis* as a common ancestor to two other members of this genus. *A. africanus* may be in the mainstream of human evolutionary history, but *A. robustus* is believed to have become extinct.

Humans

Humans are distinguishable by certain traits. Among these are a bipedal gait, a large brain, the making of tools, the eating of meat and green plants (omnivorous), the pair-bonding between male and female, and the ability to create language, art, and music.

Paleontologists disagree on which fossils belong to the human genus. Some contend that even the three australopithecine fossils should be included in the genus, and others want to exclude *Homo habilis,* the first fossil given the designation *Homo.*

Homo Habilis

Homo habilis is dated at about 3 million years ago, making these people contemporary with the hominid *A. robustus.* The skeletal remains indicate that there may have been two species of *Homo habilis* people who evolved from *A. africanus. Homo habilis* means "handyman"; the fossil was given this name because it was found with stone tools. At the time this fossil was named, it was thought that *Homo habilis* people were the first to make tools. Now, however, it appears the hominid *A. robustus* may have made tools also.

The skull of *Homo habilis* has a brain capacity of 600–800 cc, while modern humans have a brain size of 1,360 cc. At

Figure 21.15 *Homo erectus* people. These people had the use of fire, made stone tools, and may have been big-game hunters.

one time, it was thought that only people with a brain size of at least 1,000 cc could make tools. Because *Homo habilis* people had a smaller brain than this and yet made tools, are we to think that the making of tools was necessary to the evolution of an enlarged brain? This seems to be an unnecessary question. Increased brain capacity, no matter how slight, would have permitted better toolmaking. This combination would have been selected because toolmaking would have fostered survival in a grassland habitat. Therefore, as the brain became increasingly larger, tool use became more sophisticated.

Homo habilis, people of at least 2 million years ago, may not have been highly intelligent but did make tools. Intelligence and the making of tools probably evolved together.

Homo erectus

Homo erectus people (fig. 21.15) evolved in Africa about 1.5 million years ago and then later spread throughout Eurasia. Their fossils have an average brain size of 1,000 cc, but the shape of the skull indicates that the areas of the brain necessary for memory, intellect, and language probably were not well developed.

Homo erectus people had a posture and locomotion (i.e., a striding gait, fig. 21.16) similar to that seen in modern humans. They made tools of a better quality than the tools of predecessors—these tools had symmetrical bifaces. *Homo erectus* people used fire and ate meat. How much meat they ate is questionable, and some researchers do not believe that they were big-game hunters. However, even though concrete evidence is lacking, circumstantial evidence suggests that *Homo erectus* people could hunt big game:

Social cooperation was present. Old and diseased skeletons were found, indicating that frail and sick people were allowed to live beyond the time of usefulness. Because humans are relatively small, they need to cooperate in order to kill large animals.

They had sharp cutting devices. The stone tools found with fossil remains are of a better quality than those found previously.

They had a wide geographic range. Hunters need to spread out in order to find sufficient game. *Homo erectus* people spread out into Europe and Asia.

The newborn's brain was small. The adult pelvis of *Homo erectus* is too narrow to allow a large-brained baby to

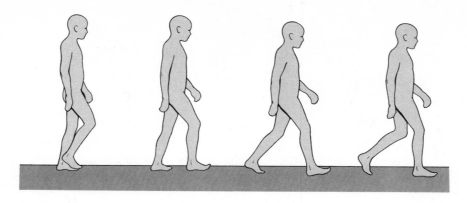

Figure 21.16   The striding gait of modern humans. Each limb alternately goes through a stance phase and a swing phase. In the first drawing, the left limb is in the stationary stance phase and the right limb is beginning the swing phase. First, the knee is bent and extended forward. Then, as the knee straightens, the heel and the ball of the foot touch the ground. In the last 2 drawings, the left limb has entered the swing phase and the right limb is entering the stance phase.

pass through. The brain of the newborn most likely increased in size after birth, which could only happen if the lactating mother received good nutrition (such as hunting might provide).

Homo erectus people had a large brain, walked like we do, and used fire. They even may have been big-game hunters.

Homo sapiens

Many years (about 1.5 million) separate people from *Homo erectus* the Neanderthal people. During these years, *Homo sapiens* people must have evolved. This early version of our own species probably was not radically different from *Homo erectus,* however.

There is evidence that by 100,000 years ago, there were at least three distinct populations of humans. The Neanderthal people lived in Europe and western Asia; in Africa, there were people increasingly like modern humans in anatomy; and in eastern Asia, there were people unlike either of these other two populations. We only have detailed information about the Neanderthal and Cro-Magnon peoples.

Neanderthals

The **Neanderthal** people (*Homo sapiens neanderthalis*) take their name from Germany's Neander Valley, where one of the first Neanderthal skeletons was discovered. Their earliest fossils are 130,000 years old, and their latest fossils are only 32,000 years old. Neanderthals evolved and lived in Eurasia during the last Ice Age.

The Neanderthals (fig. 21.17) had massive eyebrow ridges, and the nose, the jaws, and the teeth protruded far forward. The forehead was low and sloping, and the lower jaw sloped back without a chin. Surprisingly, the Neanderthal brain was, on the average, slightly larger than that of modern humans. The brain was 1,400 cc, whereas that of most modern humans is 1,360 cc.

The Neanderthals were heavily muscled, especially in the shoulders and the neck. The bones of the limbs were shorter and thicker than those of modern humans. It is hypothesized that a larger brain than that of modern humans was required for control of the extra musculature.

Were the Neanderthals adapted to a cold climate? Did they have a modern appearance? These questions have been answered variously. Recently, it was fashionable to stress how the Neanderthals were like us in appearance; currently, researchers once again are emphasizing their primitive features—the low forehead, the lack of a chin, and the squat skeleton with massive musculature, for example. Nevertheless, the Neanderthals seem to have been more culturally advanced than *Homo erectus* people. Although most remains were found in caves, there is evidence that those who lived in the open built houses. They buried their dead with flowers—an expression of grief perhaps—and with tools, as if they thought these would be needed in a future life. Some researchers even have suggested that Neanderthals had a religion and that because great piles of bear skulls were found in their caves, the bear played a role in this religion.

The Neanderthals were more massive than modern humans, and they had a larger brain. Perhaps they were adapted to a cold climate because they lived in Eurasia during the last Ice Age.

Cro-Magnon

Everyone agrees that **Cro-Magnon** (*Homo sapiens sapiens*), a fossil named for Cro-Magnon, France, where these remains were first found, has a modern appearance. There is good evidence now that modern humans evolved in Africa some 60,000 years ago, and they then spread into Europe and Asia. It seems unlikely, then, that the Neanderthals, which have not been found in Africa, evolved into Cro-Magnon people. While some interbreeding may have taken place between these two types of people after Cro-Magnon arrived in Europe, most likely Cro-Magnons simply replaced the Neanderthals, who died out.

The Cro-Magnons had an advanced form of stone technology that included the making of compound tools; stone flakes

Figure 21.17 The Neanderthals. The nose and the mouth of these people protruded from the face, and the muscles were massive. They made stone tools and perhaps were hunters.

were fitted to a wooden handle. They were the first to throw spears, enabling them to kill animals from a distance. They also were the first to make knifelike blades. They were such accomplished hunters that some researchers believe they were responsible for the extinction of many larger mammals, such as the giant sloth, the mammoth, the saber-toothed tiger, and the giant ox during the Upper Pleistocene Epoch.

Because language would have facilitated their ability to hunt such large animals, it is quite possible that meaningful speech began at this time. There are very few DNA base sequence differences between modern humans and modern apes. Is it possible that these different sequences are in genes controlling the ability to speak? Perhaps the ability to speak led to language and the development of modern culture. Some believe so. It is our culture and way of life that separate us from the apes. Others point out that cooperative hunting requires language; therefore, this would have led to socialization and the advancement of culture. Humans are believed to have lived in small groups, the men going out to hunt by day, while the women remained at home with the children. It is possible that a hunting way of life shaped the behavior patterns that still are seen today.

Cro-Magnon people lived during the Reindeer Age, a time when great reindeer herds spread across Europe. They used every part of the reindeer, including the bones and the antlers, from

which they sculpted many small figurines. They also painted beautiful drawings of animals on cave walls in Spain and France (fig. 21.18). Perhaps they also had a religion, and these artistic endeavors were an important part of their form of worship.

If Cro-Magnon people did cause the extinction of many types of animals, it may account for the transition from a hunting existence to an agricultural existence about 12,000–15,000 years ago. This agricultural period extended from that time to about 200 years ago, when the Industrial Revolution began. At this time, many people began to live in cities, in large part divorced from nature and endowed with the philosophy of exploitation and control of nature. Only recently have we begun to realize that the human population, like all other organisms with which we share an evolutionary history, should work with, rather than against, nature.

Cro-Magnons were modern in appearance. They were expert hunters and developed an advanced culture. They could speak and had a language, which may account for both their hunting ability and their culture.

Human Races   All human races of today are also classified as *Homo sapiens sapiens*. This is consistent with the biological

Figure 21.18   Cro-Magnon people painting on cave walls. Some of these paintings can still be observed today. Of all the animals, only humans developed a culture that includes technology and the arts.

definition of species because it is possible for all types of humans to interbreed and bear fertile offspring. The close relationship between the races is supported by biochemical data showing that differences in amino acid sequence between two individuals of the same race are as great as those between two individuals of different races.

It is generally accepted that racial differences developed as adaptations to climate. Although it might seem as if dark skin is a protection against the hot rays of the sun, it has been suggested that it is actually a protection against ultraviolet ray absorption. Dark-skinned persons living in southern regions and white-skinned persons living in northern regions absorb the same

amount of radiation. (Some absorption is required for vitamin D production.)

Differences in body shape represent adaptations to temperature. A squat body with shortened limbs and nose retains more heat than an elongated body with longer limbs and nose. Also, the "almond" eyes, flattened nose and forehead, and broad cheeks of the Oriental are believed to be adaptations to the extremely cold weather of the Ice Age.

Although it has always seemed to some that physical differences might warrant assigning human races to different species, this contention is not borne out by the biochemical data mentioned previously.

Summary

Evolution explains the history and diversity of life. Evidences for evolution can be taken from the fossil record and comparative anatomy, embryology, and biochemistry. Also, vestigial structures and biogeography support the occurrence of evolution.

Darwin not only presented evidences in support of organic evolution, he showed that evolution was guided by natural selection. Due to reproductive potential, there is a struggle for existence between members of the same species. Those members that possess variations more suited to the environment will most likely acquire more resources and so have more offspring than other members. Because of this natural selection process, there is a gradual change in species composition, which leads to adaptation to the environment.

Humans are primates that evolved from shrewlike insectivores and became adapted to living in trees. The first primates were prosimians, followed by the monkeys, apes, and hominids. The latter three are all anthropoids, the latter two are hominoids, and the last contains the australopithecines and humans.

Humans and apes share a common ancestor which may possibly have been *Dryopithecus,* a fossil commonly regarded as the first hominid. The first hominid lived at a time when a change in weather made it advantageous to dwell on the ground.

One possible hominid evolutionary tree shows *Australopithecus afarensis* as a common ancestor to two other species of this genus and *Homo habilis,* the first fossil to be

placed in the genus *Homo. Homo habilis* peoples may not have been highly intelligent but they did make tools. Intelligence and making of tools probably evolved together.

Homo erectus had a large brain and walked with a striding gait. These people used fire and probably were the first true hunters. *Homo sapiens neanderthalensis* was not as primitive as formerly thought. The Neanderthals are believed to have interbred with or to have been replaced by Cro-Magnon, the first *Homo sapiens sapiens.* Cro-Magnon was an expert hunter. Hunting promoted language and socialization. In a relatively short time, humans developed an advanced culture that has tended to separate them from other organisms in the biosphere. All human races belong to the same species.

Study Questions

1. Show that the fossil record, comparative anatomy, comparative embryology, vestigial structures, comparative biochemistry, and biogeography all give evidence that evolution occurred. (pp. 415–419)
2. What are five aspects to Darwin's mechanism for evolution? (p. 421)
3. Describe the events which led up to the origin of the first cell(s). (p. 422)
4. Name several primate characteristics still retained by humans. (p. 424)
5. Draw an evolutionary tree that includes all primates. (p. 425)
6. What animals mentioned in this chapter, whether living or extinct, are anthropoids? Hominoids? Hominids? Humans? (pp. 425–429)
7. How might adaptations to a grassland habitat have influenced the evolution of humans? (p. 428)
8. Draw a hominid evolutionary tree. (p. 428)
9. Which humans were tool users? Walked erect? Had a striding gait? Used fire? Drew pictures? (p. 429)
10. What evidence do we have that all races of humans belong to the same species? Name several races of humans. (p. 432)

Objective Questions

Match the phrases in questions 1–4 with those in this key:
   a. biogeography
   b. fossil record
   c. biochemistry
   d. anatomy

1. species change over time
2. forms of life are variously distributed
3. a group of related species has a unity of plan
4. same types of molecules are found in all living things

5. Darwin's theory of natural selection is _____ , meaning that the organism is unable to predetermine how it shall evolve.
6. Evolutionary success is judged by _____ success, or the number of an organism's offspring.
7. A _____ evolution is believed to have produced the first cell(s).
8. Anthropoids include _____ , _____ , _____ .

9. The australopithecines could probably walk _____ , but had a _____ brain.
10. The two varieties of *Homo sapiens* from the fossil record are _____ and _____ .

Answers

1. b 2. a 3. d 4. c 5. nonteleological 6. reproductive 7. chemical 8. monkeys, apes, humans 9. erect, small 10. Neanderthal, Cro-Magnon

Critical Thinking Questions

1. The evidences for evolution are not convincing to everyone. What types of problems are involved in producing convincing evidence?
2. Do the mechanisms of evolution discussed on pages 000–000 apply to human beings today?
3. Why would you not expect the "origin of life" to occur today?

Selected Key Terms

analogous structure (ah-nal'o-gus struk'chur) similar in function but not in structure; particularly in reference to similar adaptations. 418

anthropoids (an'thro-poidz) higher primates, including only monkeys, apes, and humans. 425

Australopithecus (aw''strah-lo-pith'e-cus) the first generally recognized hominid. 428

common ancestor (ko'mun an'ses-tor) an ancestor to two or more branches of evolution. 416

Cro-Magnon (kro-mag'non) the common name for the first fossils to be accepted as representative of modern humans. 431

Dryopithecus (dri''o-pith'e-cus) a genus of extinct apes that may have included or resembled a common ancestor to both apes and humans. 427

evolutionary trees (ev''o-lu''shun-ar-e trēz) diagrams describing the evolutionary relationship of groups of organisms. 416

hominid (hom'ĭ-nid) member of a family of upright, bipedal primates that includes australopithecines and modern humans. 428

hominoid (hom'ĭ-noid) member of a superfamily that contains humans and the great apes and humans. 427

Homo erectus (ho'mo ē-rek'tus) the earliest nondisputed species of humans, named for their erect posture that allowed them to have a striding gait. 430

Homo habilis (ho'mo hah'bĭ-lis) an extinct species that may include the earliest humans, having a small brain but quality tools. 429

homologous structures (ho-mol'o-gus struk'churz) similar in structure but not necessarily function; homologous structures in animals share a common ancestry. 418

Neanderthal (ne-an'der-thawl) the common name for an extinct subspecies of humans whose remains are found in Europe and Asia. 431

primates (pri'māts) animals that belong to the order Primates, the order of mammals that includes prosimians, monkeys, apes, and humans. 424

prosimians (pro-sim'e-anz) primitive primates such as lemurs, tarsiers, and tree shrews. 425

taxonomy (tak-son'o-me) the science of naming and classifying organisms. 416

vestigial (ves-tij'e-al) the remains of a structure that was functional in some ancestor but is no longer functional in the organism in question. 419

Chapter Twenty-Two

Ecosystems

Chapter Concepts

1 Ecosystems are units of the biosphere in which populations interact with each other and with the physical environment.

2 Natural ecosystems, which use solar energy efficiently and utilize chemicals that cycle, produce little pollution and waste.

3 The human ecosystem which comprises both country and city utilizes fossil fuel energy inefficiently and uses material resources that do not cycle. Therefore there is much pollution and waste.

The flow of energy and the cycling of materials characterizes ecosystems. The fish provides nutrient molecules to the brown bear. The bear's cells will use the nutrient molecules as a source of energy and as chemical molecules. The chemicals will eventually be passed on to other organisms but the energy will be dissipated.

Ecologists like to use the catchy phrase "everything is connected to everything else." By this, they mean that you cannot affect one part of the environment without affecting another part. For example, would you suppose that trees have anything to do with the daily environmental temperature? Well, they do and in perhaps an unexpected way. The accumulation of carbon dioxide (and other gases) in the air acts like the glass of a greenhouse and allows solar heat to be trapped near the surface of the earth. Environmentalists predict a gradual rise in the earth's daily temperature, especially if we continue to do away with the world's tropical rain forests. They act like a giant sponge that absorbs carbon dioxide, slowing down the greenhouse effect caused by all the carbon dioxide we pour into the environment from the burning of fossil fuels.

Characteristics of Ecosystems

When the earth was first formed, the outer crust was covered by ocean and barren land. Over time, plants colonized the land so that eventually it supported many complex communities of living things. A community is made up of all the populations in a particular area (table 22.1).

Succession

Succession occurs when a sequence of communities replaces one another in an orderly and predictable way, until finally there is a climax community, a mix of plants and animals typical of that area that remains stable year after year. For example, in the United States not too long ago, a deciduous forest was typical of the East, a prairie was common to the Midwest, and a semidesert covered the Southwest (fig. 22.1).

Primary succession is a series of sequential events by which bare rock becomes capable of sustaining many organisms. *Secondary succession* also is observed when a climax community that has been disturbed once again takes on its former state. Figure 22.2 shows a possible secondary succession for abandoned farmland in the eastern United States.

The process of succession leads to a climax community, which contains the plants and animals characteristic of the area. The populations within a community interact with themselves and with the physical environment, forming an ecosystem.

Ecosystem Composition

Each population in an ecosystem has a habitat and a niche. The **habitat** of an organism is its place of residence (that is, where it can be found) such as under a log or at the bottom of the pond. The **niche** of an organism is its profession or total role in the community. A description of an organism's niche (table 22.2) includes its interactions with the physical environment and with the other organisms in the community. One important aspect of niche is how the organism acquires its food.

Producers are autotrophic organisms with the ability to carry on photosynthesis and to make food for themselves (and indirectly for the other populations as well). In terrestrial eco-

Table 22.1 Ecological Terms

Term	Definition
ecology	study of the interactions of organisms with each other and with the physical environment
population	all the members of the same species that inhabit a particular area
community	all the populations that are found in a particular area
ecosystem	a community and its physical environment; has living (biotic) components and nonliving (abiotic) components
biosphere	the portion of the surface of the earth (air, water, and land) where living things exist

systems, the producers are predominantly green plants, while in freshwater and saltwater ecosystems, the dominant producers are various species of algae.

Consumers are heterotrophic organisms that use preformed food. It is possible to distinguish four types of consumers, depending on their food source. **Herbivores** feed directly on green plants; they are termed primary consumers. **Carnivores** feed only on other animals and are, therefore, secondary or tertiary consumers. **Omnivores** feed on both plants and animals. Therefore, a caterpillar feeding on a leaf is a herbivore; a green heron feeding on a fish is a carnivore; a human eating both leafy green vegetables and beef is an omnivore. **Decomposers** are organisms of decay, such as bacteria and fungi, that break down **detritus,** nonliving organic matter, to inorganic matter, which can be used again by producers. In this way, the same chemical elements can be used over and over again in an ecosystem.

When we diagram the components of an ecosystem, as in figure 22.3, it is possible to illustrate that every ecosystem is characterized by two fundamental phenomena: energy flow and chemical cycling. Energy flow begins when producers absorb solar energy, and chemical cycling begins when producers take in inorganic nutrients from the physical environment. Thereafter, producers make food for themselves and indirectly for the other populations of the ecosystem. Energy flow occurs because all the energy content of organic food eventually is lost to the environment as heat. Therefore, most ecosystems cannot exist without a continual supply of solar energy. However, the original inorganic elements are cycled back to the producers, and no new input is required.

Within an ecosystem, energy flows and chemicals cycle.

Energy Flow

Energy flows through an ecosystem because when one form of energy is transformed into another form, there is always a loss of some usable energy as heat. For example, the conversion of energy in one molecule of glucose to 38 molecules of ATP represents only 50% of the available energy in a glucose molecule.

a.

c.

b.

Figure 22.1   Three major biomes in the United States, each containing its own mix of plants and animals. Temperature and amount of rainfall largely determine which type biome will be found where. *a.* A deciduous forest is typical of eastern United States. *b.* A prairie biome is found in the Midwest. *c.* A semidesert is located in the Southwest.

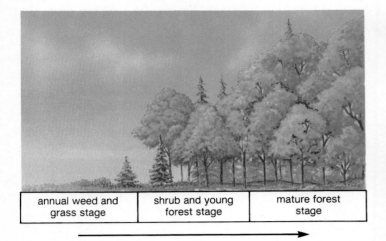

Figure 22.2   Secondary succession. This drawing shows a possible sequence of events by which abandoned farmland becomes a climax community again. During secondary succession in the Northeast, grass and weeds are followed by shrubs and trees and finally by a mature climax forest.

| annual weed and grass stage | shrub and young forest stage | mature forest stage |

Table 22.2   Aspects of Niche

Plants	Animals
season of year for growth and reproduction	time of day for feeding and season of year for reproduction
sunlight, water, and soil requirements	habitat and food requirements
relationships with other organisms	relationships with other organisms
effect on abiotic environment	effect on abiotic environment

The rest is lost as heat. This means that as one population feeds on another and as decomposers work on detritus, all of the captured solar energy that was converted to chemical-bond energy by algae and plants is returned to the atmosphere as heat. Therefore, energy flows through an ecosystem and does not cycle.

Food Chains and Food Webs

Energy flows through an ecosystem as the individuals of one population feed on those of another. A **food chain** indicates who eats whom in an ecosystem. Figure 22.4 depicts examples of a terrestrial food chain and an aquatic food chain. It is important to realize that each represents just one path of energy flow through an ecosystem. Natural ecosystems have numerous food chains, each linked to others to form a complex **food web.** For example, figure 22.5 shows a deciduous forest ecosystem in which plants are eaten by a variety of insects, and, in turn, these are eaten by several different birds, while any one of the latter may be eaten by a larger bird, such as a hawk. Therefore, energy flow is better described in terms of **trophic** (feeding) **levels,** each one further removed from the producer population, the first (photosynthetic) trophic level. All animals acting as primary

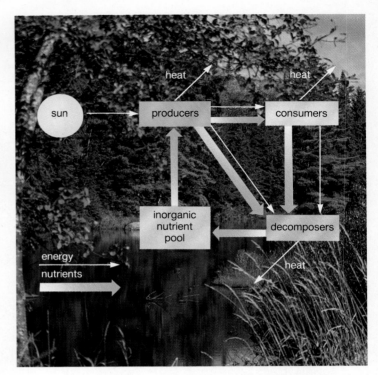

Figure 22.3   Ecosystem composition. The diagram illustrates energy flow and chemical cycling through an ecosystem. Energy does not cycle because all the energy derived from the sun eventually dissipates as heat.

consumers are part of a second trophic level, and all animals acting as secondary consumers are part of the third level, and so on.

The populations in an ecosystem form food chains in which the producers produce food for the other populations, which are consumers. While it is convenient to study food chains, the populations in an ecosystem actually form a food web, in which food chains join with and overlap one another.

One of the food chains depicted in figure 22.4 is part of the forest food web shown in figure 22.5 and the other food chain is part of the aquatic food web from the freshwater pond ecosystem shown in figure 22.6. Both of these food chains are called *grazing food chains* because the primary consumer feeds on a photosynthesizer. In some ecosystems (forests, rivers, and marshes), the primary consumer feeds mostly on detritus (dead organisms). The *detritus food chain* accounts for more energy flow than the grazing food chain whenever most organisms die without having been eaten. In the forest, an example of a detritus food chain is

detritus ———→ soil bacteria ———→ earthworms

A detritus food chain often is connected to a grazing food chain, as when earthworms are eaten by a robin. Eventually, however, as dead organisms decompose, all the solar energy that was taken up by the producer populations is dissipated as heat. Therefore, energy does not cycle.

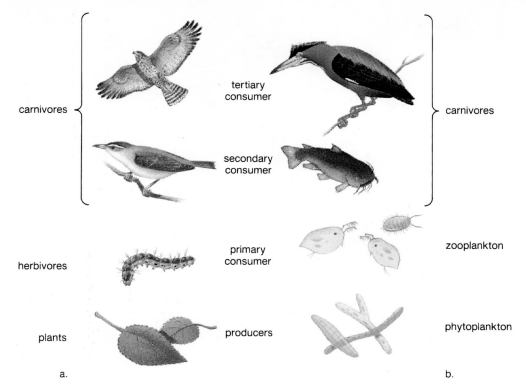

carnivores {
tertiary consumer
secondary consumer
} carnivores

herbivores — primary consumer — zooplankton

plants — producers — phytoplankton

a.

b.

Figure 22.4   Examples of food chains. *a.* Terrestrial. *b.* Aquatic.

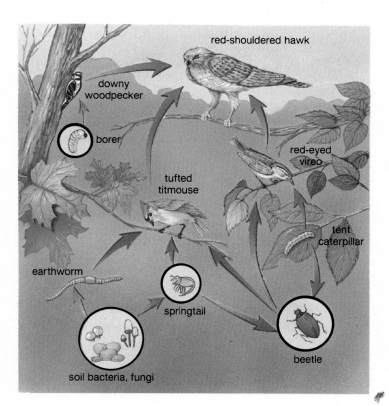

red-shouldered hawk

downy woodpecker

borer

red-eyed vireo

tufted titmouse

tent caterpillar

earthworm

springtail

beetle

soil bacteria, fungi

Figure 22.5   A deciduous forest ecosystem. The arrows indicate the flow of energy in a food web.

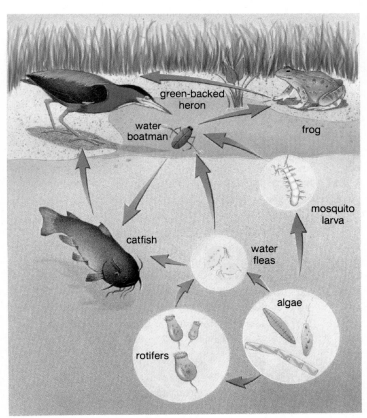

green-backed heron

water boatman

frog

catfish

water fleas

mosquito larva

algae

rotifers

Figure 22.6   A freshwater pond ecosystem. The arrows indicate the flow of energy in a food web.

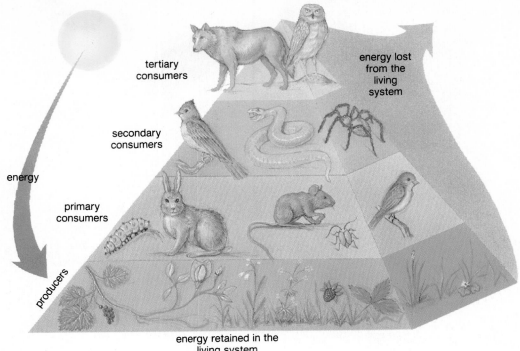

Figure 22.7 Pyramid of energy. At each step in the pyramid, an appreciable portion of energy originally trapped by the producer is dissipated as heat. Accordingly, organisms in each trophic level pass on less energy than they received.

Ecological Pyramids

The trophic structure of an ecosystem can be summarized in the form of an **ecological pyramid.** The base of the pyramid represents the producer trophic level, and the apex is the highest-level consumer, called the top predator. The other consumer trophic levels are in between the producer and the top predator levels. There are three kinds of pyramids. One is a *pyramid of numbers,* based on the number of organisms at each trophic level. A second is the *pyramid of biomass.* Biomass is the weight of living material at some particular time. To calculate the biomass for each trophic level, an average weight for the organisms at each level is determined and then the number of organisms at each level is estimated. Multiplying the average weight by the estimated number gives the approximate biomass for each trophic level. A third pyramid, the *pyramid of energy* (fig. 22.7), illustrates that each succeeding trophic level is smaller than the previous level. Less energy is found in each succeeding trophic level for the following reasons.

1. Of the food available, only a certain amount is captured and eaten by the next trophic level.
2. Some of the food that is eaten cannot be digested and exits the digestive tract as waste.
3. Only a portion of the food that is digested becomes part of the organism's body. The rest is used as a source of energy.

In regard to the last point, we have to realize that a significant portion of food molecules is used as an energy source for ATP buildup in mitochondria. This ATP is needed to build the proteins, carbohydrates, and lipids that compose the body. ATP also is needed for such activities as muscle contraction and nerve conduction. Figure 22.7 indicates the amount of energy in one trophic level that is unavailable to the next trophic level.

The energy considerations associated with ecological pyramids have implications for the human population. It generally is stated that only about 10% of the energy available at a particular trophic level is incorporated into the tissues at the next level. This being the case, it can be estimated that 100 kg of grain could, if consumed directly, result in 10 human kg; however, if fed to cattle, the 100 kg of grain would result in only 1 human kg. Therefore, a larger human population can be sustained by eating grain than by eating grain-fed animals. Humans generally need some meat in their diet, however, because this is the most common source of the essential amino acids, as discussed in chapter 4.

In a food web, each successive trophic level has less total energy content. This is because some energy is lost to the environment as heat when energy is transferred from one trophic level to the next.

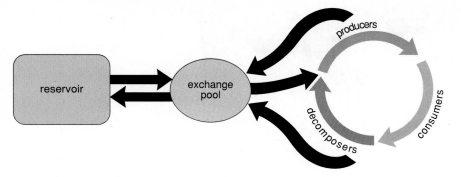

Figure 22.8 Components of a chemical cycle. The reservoir stores the chemical and the exchange pool makes it available to producers. The chemical then cycles through food chains. Decomposition returns the chemical to the exchange pool once again if it has not already returned by another process.

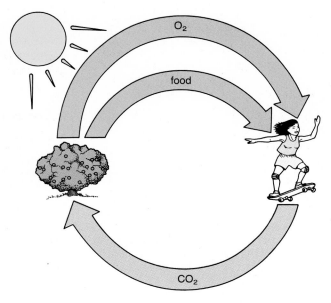

Figure 22.9 Relationship between photosynthesis and respiration. Animals are dependent on plants for a supply of oxygen (O_2) and organic food and plants are dependent on animals for a supply of carbon dioxide (CO_2).

Chemical Cycling

In contrast to energy, inorganic nutrients do cycle through large natural ecosystems. Because there is minimal input from the outside, the various elements essential for life are used over and over. For each element, the cycling process (fig. 22.8) involves (1) a reservoir—that portion of the earth that acts as a storehouse for the element; (2) an exchange pool—that portion of the environment from which the producers take their nutrients; and (3) the biotic community—through which elements move along food chains to and from the exchange pool.

The Carbon Cycle

The relationship between photosynthesis and respiration should be kept in mind when discussing the carbon cycle. Recall that for simplicity's sake, this equation in the forward direction represents respiration, and in the other direction, it is used to represent photosynthesis:

$$C_2H_{12}O_6 + 6O_2 \rightleftharpoons 6CO_2 + 6H_2O$$

The equation tells us that respiration releases carbon dioxide (CO_2), the molecule needed for photosynthesis. However, photosynthesis releases oxygen (O_2), the molecule needed for respiration. From figure 22.9, it is obvious that animals are dependent on green organisms, not only to produce organic food and energy but also to supply the biosphere with oxygen.

In the carbon cycle, organisms in both terrestrial and aquatic ecosystems (fig. 22.10) exchange carbon dioxide with the atmosphere. On land, plants take up carbon dioxide from the air, and through photosynthesis, they incorporate carbon into food that is used by autotrophs and heterotrophs alike. When organisms respire, a portion of this carbon is returned to the atmosphere as carbon dioxide.

In aquatic ecosystems, the exchange of carbon dioxide with the atmosphere is indirect. Carbon dioxide from the air combines with water to give bicarbonate (HCO_3^-), a source of carbon for algae that produce food for themselves and for heterotrophs. Similarly, when aquatic organisms respire, the carbon dioxide they give off becomes bicarbonate. The amount of bicarbonate in the water is in equilibrium with the amount of carbon dioxide in the air.

Carbon Reservoirs

Living and dead organisms contain organic carbon and serve as one of the reservoirs for the carbon cycle. The world's biota, particularly trees, contain 800 billion metric tons of organic carbon, and an additional 1,000–3,000 billion metric tons are estimated to be held in the remains of plants and animals in the

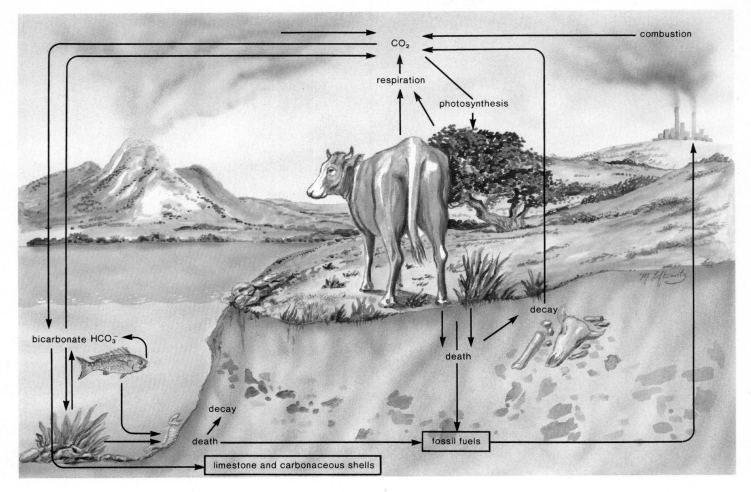

Figure 22.10   Carbon cycle. Photosynthesizers take carbon dioxide or (CO_2) from the air or bicarbonate (HCO_3^-) from the water. They and all other organisms return carbon dioxide to the environment. The carbon dioxide level is also increased when volcanoes erupt and fossil fuels are burned. Presently, the oceans are a primary reservoir for carbon in the form of limestone and carbonaceous shells.

soil. Before decomposition can occur, some of these remains are subjected to physical processes that transform them into coal, oil, and natural gas. We call these materials the fossil fuels. Most of the fossil fuels were formed during the Carboniferous Period, 280–350 million years ago, when an exceptionally large amount of organic matter was buried before decomposing. Another reservoir is the inorganic carbonate that accumulates in limestone and in carbonaceous shells. The oceans abound in organisms, some microscopic, that are composed of carbonaceous shells and accumulate in ocean bottom sediments. Limestone is formed from these sediments by geological transformation.

Human Influence on the Carbon Cycle

The activities of human beings have increased the amount of carbon dioxide (CO_2), and other gases in the atmosphere. Data from monitoring stations record an increase of 20 ppm (parts per million) in carbon dioxide in only 22 years. (This is equivalent to 42 billion metric tons of carbon.) This buildup is attributed primarily to the burning of fossil fuels and the destruction of the world's forests. When we do away with for-

ests, we reduce a reservoir that takes up excess carbon dioxide. At this time, the oceans are believed to be taking up most of the excess carbon dioxide; the burning of fossil fuels in the last 22 years has probably released 78 billion metric tons of carbon, yet the atmosphere registers an increase of "only" 42 billion metric tons.

Human Issue

Today there are two major biomes that are in danger of being overexploited by humans—the tropical rain forests, such as the Amazon basin of South America, and the tundra in Alaska. Wildlife is rich in both of these places, but the diversity of life in the rain forest is beyond imagination. Therefore exploitation of the rain forests will cause mass extinctions—many different species will be no more. Should we be concerned about what happens in places so distant from us? Can you think of any reasons why it matters to you that pollution might be gaining a foothold in formerly pristine environments or that species of plants and animals are becoming extinct?

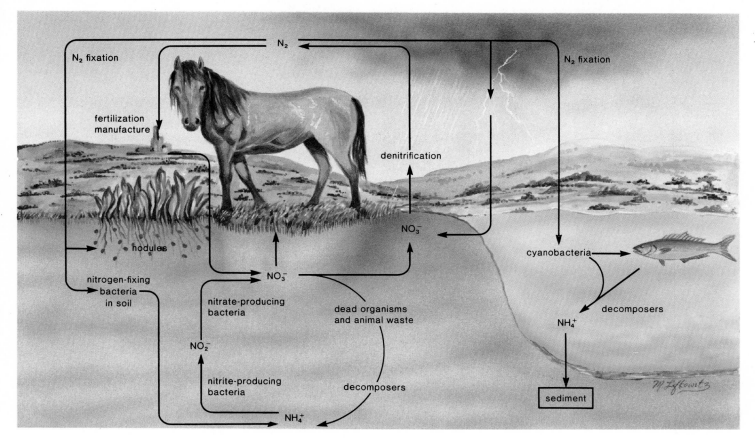

Figure 22.11 Nitrogen cycle. Several types of bacteria are at work: nitrogen-fixing bacteria reduce nitrogen gas (N_2); nitrifying bacteria, which include both nitrite-producing and nitrate-producing bacteria, convert ammonium (NH^+_4) to nitrate; and the denitrifying bacteria convert nitrate back to nitrogen gas. Humans contribute to the cycle by using nitrogen gas to produce nitrate for fertilizers.

As discussed previously and as mentioned on page 465, there is much concern that an increased amount of carbon dioxide (and other gases) in the atmosphere is causing global warming. These gases allow the sun's rays to pass through, but they absorb and reradiate heat back to the earth, a phenomenon called the *greenhouse effect*.

In the carbon cycle, carbon dioxide is removed from the atmosphere by photosynthesis but is returned by respiration. Living things and dead matter are carbon reservoirs. The oceans, because they abound with carbonaceous shells and accumulate limestone, are also major carbon reservoirs.

The Nitrogen Cycle

Nitrogen is an abundant element in the atmosphere. Nitrogen gas (N_2) makes up about 78% of the atmosphere by volume, yet nitrogen deficiency commonly limits plant growth. Plants cannot incorporate nitrogen gas into organic compounds and, therefore, depend on various types of bacteria to make nitrogen available to them (fig. 22.11).

Nitrogen Fixation

The reduction of nitrogen gas (N_2) prior to the incorporation of nitrogen into organic compounds is called **nitrogen fixation.**

For example, some cyanobacteria in aquatic ecosystems and some free-living bacteria in the soil reduce nitrogen gas to ammonium (NH_4^+), which then is subject to processes leading to organic compounds. Other *nitrogen-fixing bacteria* infect and live in nodules on the roots of legumes (fig. 22.11). They fix atmospheric nitrogen gas and incorporate nitrogen into organic compounds that not only they themselves use, but the host plant as well.

Nitrogen fixation also occurs after plants take up nitrates (NO_3^-) from the soil. Plants use nitrates as their source of nitrogen when they make amino acids and nucleic acids, for example.

Nitrification and Denitrification

Nitrogen gas (N_2) is converted to nitrate (NO_3^-) in the atmosphere when cosmic radiation, meteor trails, and lightning provide the high energy needed for nitrogen to react with oxygen. Also, humans make a most significant contribution to the nitrogen cycle when they convert nitrogen gas to nitrate for use in fertilizers.

Ammonium (NH_4^+) in the soil also is converted to nitrate by bacteria in a process called nitrification. Ammonium gets into the soil in two ways. As mentioned previously, some free-living bacteria in the soil reduce nitrogen gas to ammonium. Decomposers also produce ammonium when they decompose waste and dead organic matter. Other types of bacteria carry

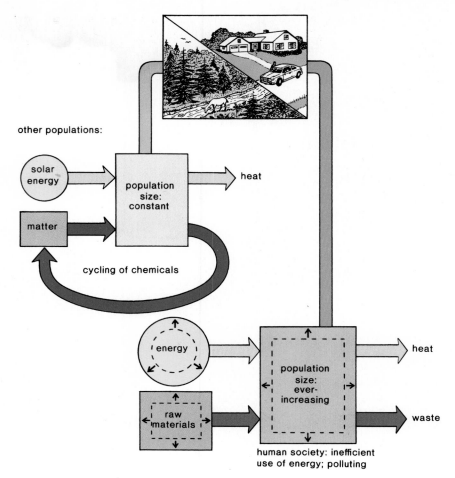

other populations:

solar
energy

population
size:
constant

heat

matter

cycling of chemicals

energy

population
size:
ever-
increasing

heat

raw
materials

waste

human society: inefficient
use of energy; polluting

Figure 22.12   Human ecosystem versus a natural ecosystem. In natural ecosystems, population sizes remain about the same year after year, materials cycle, and energy is used efficiently. In the human ecosystem, the population size consistently increases, resulting in much pollution because of inadequate cycling of materials and inefficient use of supplemental energy.

out nitrification. First, nitrite-producing bacteria convert ammonium to nitrite (NO_2^-), and then nitrate-producing bacteria convert nitrite to nitrate. These two groups of bacteria are called the *nitrifying bacteria*. Notice that there is a subcycle in the nitrogen cycle that involves only nitrates, ammonium, and nitrites. This subcycle does not depend on the presence of nitrogen gas at all (fig. 22.11).

Denitrification is the conversion of nitrate to nitrogen gas. There are *denitrifying bacteria* in both aquatic and terrestrial ecosystems. Denitrification counterbalances nitrogen fixation but not completely. There is more nitrogen fixation, especially due to fertilizer production.

In the nitrogen cycle: nitrogen-fixing bacteria (in nodules and in the soil) reduce nitrogen gas, and, thereafter, nitrogen can be incorporated into organic compounds; nitrifying bacteria convert ammonium to nitrate; denitrifying bacteria convert nitrate back to nitrogen gas.

The Human Ecosystem

Mature natural ecosystems tend to be stable and to exhibit the characteristics listed in table 22.3.

Table 22.3   Ecosystems

Natural	Human
independent	dependent
cyclical (except energy)	noncyclical
nonpolluting	polluting
renewable solar energy	nonrenewable fossil fuel energy
conserves resources	uses up resources

Pollution, defined as any undesirable change in the environment that can be harmful to humans and other life, and excessive waste do not occur normally. The human ecosystem that replaces natural ecosystems is quite different, however.

Human beings have replaced natural ecosystems with one of their own making, as is depicted in figure 22.12. This ecosystem essentially has two parts: the *country,* where agriculture and animal husbandry are found, and the *city,* where most people live and where industry is carried on. This representation of the human ecosystem, although simplified, allows us to see that the system requires two major inputs: *fuel energy* and *raw materials* (e.g., metals, wood, synthetic materials). The use of these necessarily results in *pollution* and *waste* as outputs.

Figure 22.13   The country. Crops are tended with heavy farming equipment that operates on fossil fuel. High yields are dependent upon a generous supply of fertilizers, pesticides, herbicides, and water.

Human Issue

Modern farming activities have had devastating effects on the environment. For example, cultivated fields do not retain the excellent soil-holding capabilities provided by natural vegetation. Soil erosion from these fields creates the murky, turbid waters characteristic of many of our rivers and streams. Water runoff from agricultural fields into streams carries with it fertilizers, leading to eutrophication of lakes and rivers (see p. 461). Therefore, in destroying the original natural vegetation of terrestrial ecosystems, modern agricultural practices have severely disrupted aquatic ecosystems. Who is responsible for preventing the profound ecological impacts exerted by modern agriculture? Is it appropriate to blame farmers?

The Country

Modern United States agriculture produces exceptionally high yields per acre, but this bounty is dependent on a combination of the five variables given here:

1. **Planting of a few genetic varieties.** The majority of farmers specialize in growing one of these. Wheat farmers plant the same type of wheat, and corn farmers plant the same type of corn (fig. 22.13). This so-called monoculture agriculture is subject to attack by a single type of parasite. For example, a single parasitic mold reduced the 1970 corn crop by 15%, and the results could have been much worse because 80% of the nation's corn acreage was susceptible.

2. **Heavy use of fertilizers, pesticides, and herbicides.** *Fertilizer* production requires a large energy input, and fertilizer runoff contributes to water pollution. *Pesticides* reduce soil fertility because they kill off beneficial soil organisms as well as pests, and some pesticides, for example alar, have been accused of increasing the long-term risk of cancer, particularly in children. *Herbicides,* especially those containing the contaminant dioxin, have been charged with causing adverse reproductive effects and cancer.

3. **Generous irrigation.** River waters sometimes are redirected for the purpose of irrigation, in which case "used water" returns to the river carrying a heavy concentration of salt. The salt content of the Rio Grande River in the Southwest is so high that the government has built a treatment plant to remove the salt. Water also is taken sometimes from aquifers (underground rivers), whose water content can be so reduced that it becomes too expensive to pump out more water. Farmers in Texas already are facing this situation.

4. **Excessive fuel consumption.** Energy is consumed on the farm for many purposes. Irrigation pumps already have been mentioned, but large farming machines also are used to spread fertilizers, pesticides, and herbicides and to sow and to harvest the crops. It is not incorrect to suggest that modern farming methods transform fossil fuel energy into food energy.

Supplemental fossil fuel energy also contributes to animal husbandry yields. At least 50% of all cattle are kept in *feedlots,* where they are fed grain. Chickens are raised in a completely artificial environment, where the climate is controlled and each bird has its own cage to which food is delivered on a conveyor belt. Animals raised under these conditions often have antibiotics and hormones added to their feed to increase yield.

5. **Loss of land quality.** Evaporation of excess water on irrigated lands can result in a residue of salt. This process, termed salinization, makes the land unsuitable for the growth of crops. Between 25% and 35% of the irrigated western croplands are thought to have excessive salinity. Soil erosion is also a serious problem. It is said that we are *mining the soil* because farmers are not taking measures to prevent the loss of topsoil. The Department of Agriculture estimates that erosion is causing a steady drop in the productivity of land equivalent to the loss of 1.25 million acres per year. Even more fertilizers, pesticides, and energy supplements will be required to maintain yield.

Organic Farming

Some farmers have given up this modern means of farming and instead have adopted organic farming methods. This means that they do not use applications of fertilizers, pesticides, or herbicides. They use cultivation of row crops to control weeds, crop rotation to combat major pests, and the growth of legumes to supply nitrogen fertility to the soil. Some farmers use natural predators and parasites instead of pesticides to control insects (fig. 22.14). Most of these farmers switched farming methods because they were concerned about the health of their family and livestock and had found that the chemicals were sometimes ineffective.

A study of about 40 farms showed that organic farming for the most part was just as profitable as conventional farming. Crop yields were lower, but so were operating costs. Organic farms required about two-fifths as much fossil energy to produce one dollar's worth of crop. The method of plowing and the utilization of crop rotation resulted in one-third less soil erosion.

Figure 22.14 Biological control of pests. Here, ladybugs are being used to control the cottony-cushion scale insect on citrus trees. If a pesticide were used, the ladybugs would be killed.

The researchers concluded it would be well to determine how far farmers can move in the direction of reduced agricultural chemical use and still maintain the quality of the product. They noted that a modest application of fertilizer would have improved the protein content of the crop.

If biotechnology techniques eventually are able to endow plants with an innate resistance to pests and the ability to fix nitrogen, organic farming most likely will be practiced by most farmers.

The City

The city (fig. 22.15) is dependent on the country to meet its needs. For example, each person in the city requires several acres of land for food production. Overcrowding in cities does not mean that less land is needed; each person still requires a certain amount of land to ensure survival. Unfortunately, however, as the population increases, the suburbs and the cities tend to encroach on agricultural areas and rangeland.

The city houses workers for both commercial businesses and industrial plants. Solar and other renewable types of energy rarely are used; cities currently rely mainly on fossil fuel in the

Figure 22.15   The city. The city is dependent upon the country to supply it with food and other resources. The larger the city, the more resources required to support it.

form of oil, gas, electricity, and gasoline. The city does not conserve resources. An office building, with continuously burning lights and windows that cannot be opened, is an example of energy waste. Another example is people who drive cars long distances instead of carpooling or taking public transportation and who drive short distances instead of walking or bicycling. Materials are not recycled, and products are designed for rapid replacement.

The burning of fossil fuels for transportation, commercial needs, and industrial processes causes air and water pollution (p. 460). This pollution is compounded by the chemical and solid waste pollution that results from the manufacture of many products. Consider that any product used by the average consumer (house, car, washing machine) causes pollution and waste, both during its production and when it is disposed. Humans themselves produce much sewage that is discharged into bodies of water, often after only minimal treatment.

The Solution

Table 22.3 lists the characteristics of the human ecosystem as it now exists. Just as the city is not self-sufficient and depends on the country to supply it with food, so the whole human ecosystem is dependent on the natural ecosystems to provide resources and to absorb waste. Fuel combustion by-products, sewage, fertilizers, pesticides, and solid wastes all are added to natural ecosystems in the hope that these systems will cleanse the biosphere of these pollutants. But we have replaced natural ecosystems with our human ecosystem and have exploited natural ecosystems for resources, adding even more pollutants, to the extent that the remaining natural ecosystems have become overloaded.

Natural ecosystems have been destroyed and overtaxed because the human ecosystem is noncyclical and because an ever-increasing number of people want to maintain a standard of living that requires many goods and services. But we can call a

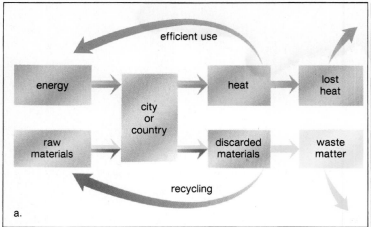
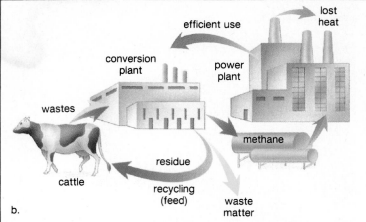

Figure 22.16   Modified human ecosystem. *a.* In order to cut down on lost heat and waste matter, heat could be used more efficiently and discarded materials could be recycled. *b.* For example, instead of allowing cattle waste to enter a water supply, the cattle waste could be sent to a conversion plant that produces methane gas. (The remaining residue could be converted into feed for cattle.) Excess heat, which arises from the burning of methane gas to produce electricity, could be cycled back to the conversion plant.

halt to this spiraling process if we achieve zero population growth and if we conserve energy and raw materials. Conservation can be achieved in three ways: (1) wise use of only what is actually needed; (2) recycling of nonfuel materials, such as iron, copper, lead, and aluminum; and (3) use of renewable energy resources and development of more efficient ways to utilize all forms of energy. Figure 22.16*a* presents a diagrammatic representation of what is needed to maintain the delicate balances of the human and natural ecosystems. As a practical example, consider a plant that was built in Lamar, Colorado, which produces methane from feedlot animals' wastes.

Summary

The process of succession from either bare rock or disturbed land results in climax communities. An ecosystem is a community of organisms plus the physical environment. Each population in an ecosystem has a habitat and a niche. Some populations are producers and some are consumers. Consumers can be herbivores, carnivores, omnivores, or decomposers.

Energy flow and chemical cycling are important aspects of ecosystems. Food chains are paths of energy flow through an ecosystem. Grazing food chains always begin with a producer population that is capable of producing organic food, followed by a series of consumer populations. Detritus food chains begin with dead organic matter that is consumed by decomposers. Eventually, all members of a food chain die and decompose. Then the very same chemical elements again are made available to the producer population, although the energy has been dissipated as heat. Therefore, energy does not cycle through an ecosystem.

The food chains form an intricate food web in which there are various trophic (feeding) levels. All producers are on the first level, all primary consumers are on the second level, and so forth. To illustrate that energy does not cycle, it is customary to arrange the various trophic levels in the form of an energy pyramid. A pyramid results because each trophic level contains less energy than the previous level.

Each chemical cycle involves a reservoir, where the element is stored; an exchange pool, from which the populations take and return nutrients; and the populations themselves. In the carbon cycle, the reservoir is organic matter, carbonaceous shells, and limestone. The exchange pool is the atmosphere; photosynthesis removes carbon dioxide (CO_2), and respiration and combustion add carbon dioxide. In the nitrogen cycle, the reservoir is the atmosphere, but nitrogen gas (N_2) must be converted to nitrate (NO_3^-) for use by producers. Nitrogen-fixing bacteria, particularly in root nodules, make organic nitrogen available to plants. Other bacteria active in the nitrogen cycle are the nitrifying bacteria, which convert ammonium to nitrate, and denitrifying bacteria, which convert nitrate to nitrogen gas again.

In mature natural ecosystems, the populations usually remain the same size and need the same amount of energy each year. Additional material inputs are minimal because matter cycles. In the human ecosystem, the population size constantly increases; more energy is needed each year, and additional material inputs also increase. In the country, farmers plant only certain high-yield varieties of plants, which require such supplements as fertilizers, pesticides, and water. In the city, the populace is wasteful of energy and materials. Therefore, there is much pollution. It would be beneficial for us and future generations to find ways to use excess heat and to recycle materials.

Study Questions

1. What is succession, and how does it result in a climax community? (p. 436)
2. Define habitat and niche. (p. 436)
3. Name four different types of consumers found in natural ecosystems. (p. 436)
4. What is the difference between a food chain and a food web? Define a trophic level. (p. 438)

5. Give an example of a grazing food chain and a detritus food chain for a terrestrial and for an aquatic ecosystem. (p. 439)
6. Draw an energy pyramid, and explain why such a pyramid can be used to verify that energy does not cycle. (p. 440)
7. What are the reservoir and the exchange pool of a chemical cycle? (p. 441)

8. Describe the carbon cycle. How do humans contribute to this cycle? (p. 441)
9. Describe the nitrogen cycle. How do humans contribute to this cycle? (p. 443)
10. Contrast the characteristics of mature natural ecosystems with those of the human ecosystem. (p. 444)

Objective Questions

1. Chemicals cycle through the populations of an ecosystem, but energy is said to _____ because all of it eventually is dissipated as heat.
2. When organisms die and decay, chemical elements are made available to _____ populations once again.
3. Organisms that feed on plants are called _____ .
4. A pyramid of energy illustrates that there is a loss of energy from one _____ level to the next.
5. There is a loss of energy because one form of energy can never be _____ completely into another form.
6. Forests are a(n) _____ for carbon in the carbon cycle.
7. In the carbon cycle, when organisms _____ , carbon dioxide (CO_2) is returned to the exchange pool.
8. Humans make a significant contribution to the nitrogen cycle when they convert nitrogen gas (N_2) to _____ for use in fertilizers.

9. During the process of denitrification, nitrate is converted to _____ .
10. Natural ecosystems utilize the same amount of energy per year, but the human ecosystem utilizes a(n) _____ .

Label this diagram.
See figure 22.3 (p. 438) in text.

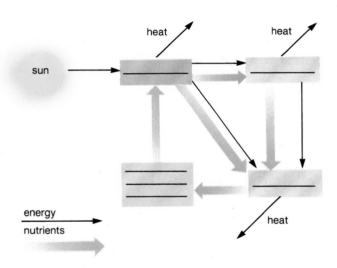

Critical Thinking Questions

1. Argue against the suggestion that natural ecosystems are stable.

2. Sometimes we are under the impression that all bacteria are harmful. What evidence can you give from this chapter to counter this belief?

3. Explain how humans and animals are an essential part of the carbon cycle.

Selected Key Terms

consumer (kon-su′mer)   a member of a population that feeds on members of other populations in an ecosystem.   436

decomposer (de-kom-po′zer)   organism of decay (fungi and bacteria) in an ecosystem.   436

denitrification (de-ni′′tri-fi-ka′shun)   the process of converting nitrate to nitrogen gas; is a part of the nitrogen cycle.   444

ecological pyramid (e′′ko-log′i-kal pir′ah-mid)   pictorial graph representing the biomass, organism number, or energy content of each trophic level in a food web from the producer to the final consumer populations.   440

ecology (e-kol′o-je)   the study of the relationship of organisms between each other and the physical environment.   436

ecosystem (ek′′o-sis′tem)   a biological community, together with the associated abiotic environment.   436

food chain (food chān)   a succession of organisms in an ecosystem that are linked by an energy flow and the order of who eats whom.   438

food web (food web)   the complete set of food links between populations in a community.   438

habitat (hab′i-tat)   the natural abode of an animal or plant species.   436

niche (nich)   total description of an organism's functional role in an ecosystem, from activities to reproduction.   436

nitrogen fixation (ni′tro-jen fik-sa′shun)   a process whereby nitrogen gas is reduced prior to the incorporation of nitrogen into organic compounds.   443

pollution (po-lu′shun)   detrimental alteration of the normal constituents of air, land, and water due to human activities.   444

producer (pro-du′ser)   organism that produces food and is capable of synthesizing organic compounds from inorganic constituents of the environment; usually the green plants and algae in an ecosystem.   436

succession (suk-se′shun)   a series of ecological stages by which the community in a particular area gradually changes until there is a climax community that can maintain itself.   436

trophic level (trof′ik lev′el)   feeding level of one or more populations in a food web.   438

Chapter
Twenty-Three

Population

Concerns

Chapter Concepts

1 A population undergoing exponential growth has an ever-greater increase in numbers and a shorter doubling time, and it may outstrip the carrying capacity of the environment.

2 The world is divided into the developed countries and the less-developed countries; mainly the less-developed countries presently are undergoing exponential population growth.

3 Human activities cause land, water, and air pollution and threaten the integrity of the biosphere.

4 A sustainable world is possible if economic growth is accompanied by ecological preservation.

Industry and ecology are often in conflict. We desire the products of industry but know that pollution will make the world uninhabitable. To balance these concerns properly is the challenge that our leaders and all humans face.

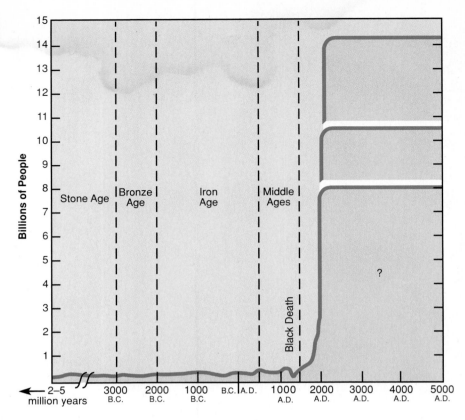

Figure 23.1 Growth curve for human population. The human population is now undergoing rapid exponential growth. It is predicted that the growth rate will decline and that the population size will level at 8, 10.5, or 14.2 billion, depending upon the speed with which the growth rate declines.

Imagine a watering hole that can accommodate one hundred rabbits. If, at first, there are only two rabbits and each pair of rabbits only produces four rabbits, how many doublings could there be without overtaxing the watering hole? You're correct if you say four and incorrect if you say five. The problem is that you can't just consider the newly arrived rabbits—you have to add in the number of rabbits already there.

$$2 - 4 - 8 - 16 - 32 - 64 - 128$$

Also, notice that when it is time to stop, there are only 62 rabbits (30 + 32). That's one of the unusual things about population growth—at one point in time it seems as if there is plenty of room and then boom! There's not enough room all of a sudden.

Human Population Growth

The human growth curve is an *exponential curve* (fig. 23.1). In the beginning, growth of the human population was relatively slow, but as more reproducing individuals were added, growth increased, until the curve began to slope steeply upward. It is apparent from the position of 1990 on the growth curve in figure 23.1 that growth is now quite rapid. The world population increases at least the equivalent of a medium-sized city every day (200,000), and the combined populations of the United Kingdom, Norway, Ireland, Iceland, Finland, and Denmark

every year. These startling figures are a reflection of the fact that a very large world population is undergoing exponential growth.

Mathematically speaking, **exponential growth,** or geometric increase, occurs in the same manner as compound interest; that is, the percentage increase is added to the principal before the next increase is calculated. Referring specifically to populations, consider the hypothetical population sizes in table 23.1. This table illustrates the circumstances of world population growth at the moment: the percentage increase has decreased and yet the size of the population grows by a greater amount each year. The increase in size is dramatically large because the world population is very large.

In our hypothetical examples (table 23.1), an initial increase of 2% added to the original population size followed by a 1.99% increase results in the third-generation size listed in the last column. Note the following observations:

1. In each instance, the second generation has a larger increase than the first generation because the second generation's population is larger than the first.
2. Because of exponential growth, the lower percentage increase (i.e., 1.99% compared to 2%) still brings about larger population growth.

Table 23.1 Exponential Growth of Hypothetical Populations

Population size	Percentage increase	Actual increase in numbers	Population size	Percentage increase	Actual increase in numbers	Population size
500,000,000	2.00	10,000,000	510,000,000	1.99	10,149,000	520,149,000
3,000,000,000	2.00	60,000,000	3,060,000,000	1.99	60,894,000	3,120,894,000
5,000,000,000	2.00	100,000,000	5,100,000,000	1.99	101,490,000	5,201,490,000

3. The larger the population, the larger the increase for each generation.

The percentage increase is termed the **growth rate,** which is calculated per year.

The Growth Rate

The growth rate of a population is determined by considering the difference between the number of persons born (birthrate, or natality) and the number of persons who die per year (death rate, or mortality). It is customary to record these rates per 1,000 persons. For example, the USSR at the present time has a birthrate of 20 per 1,000 per year, but it has a death rate of 10 per 1,000 per year. This means that Russia's population growth, or simply its growth rate, is

$$\frac{20 - 10}{1,000} = \frac{10}{1,000} = \frac{1.0}{100} = 1.0\%$$

Notice that while birthrate and death rate are expressed in terms of 1,000 persons, the growth rate is expressed per 100 persons, or as a percentage.

After 1750, the world population growth rate steadily increased until it peaked at 2% in 1965, but it has fallen slightly since then to 1.7%. Yet, there is an ever-greater increase in the world population each year because of exponential growth. The explosive potential of the present world population can be appreciated by considering the doubling time.

The Doubling Time

Table 23.2 shows that the **doubling time** (d) for a population can be calculated by dividing 70 by the growth rate (gr):

$$d = \frac{70}{gr}$$

d = Doubling time
gr = Growth rate
70 = Demographic constant

If the present world growth rate of 1.8% continues, the world population will double in 39 years.

$$d = \frac{70}{1.7} = 39 \text{ years}$$

This means that in 39 years, the world would need double the amount of food, jobs, water, energy, and so on to maintain the same standard of living.

It is of grave concern to many individuals that the amount of time needed to add each additional billion persons to the world population has taken less and less time (table 23.3). The world

Table 23.2 Relationship between Growth Rate and the Doubling Time of a Population

Growth rate (%)	Doubling time (years)
0.25	280
0.50	140
1.00	70
2.00	35
3.00	23

Table 23.3 World Population Increase

Billions of people	Time needed (years)	Year of increase
first	2–5 million	1800
second	130	1930
third	30	1960
fourth	15	1975
fifth	12	1987
sixth (projected)	11	1998

Source: Elaine M. Murphy, *World Population: Toward the Next Century,* Washington, DC: Population Reference Bureau. November 1981, page 3.

reached its first billion around 1800—some 2 million years after the evolution of humans. Adding the second billion took about 130 years, the third billion took about 30 years, and the fourth billion took only about 15 years. However, if the growth rate continues to decline, this trend would reverse itself, and eventually there would be zero population growth. Then population size would remain steady. Therefore, figure 23.1 shows three possible logistic curves: the population may level off at 8, 10.5, or 14.2 billion, depending on the speed with which the growth rate declines.

The Carrying Capacity

Examining the growth curves for nonhuman populations reveals that the populations tend to level off at a certain size. For example, figure 23.2 gives the actual data for the growth of a fruit fly population reared in a culture bottle. At the beginning, the fruit flies were adjusting to their new environment and growth was slow. Then, because food and space were plentiful, they began to multiply rapidly. Notice that the curve begins to rise dramatically just as the human population curve does now. At this time, it can be said that the population is demonstrating its **biotic potential.** Biotic potential is the maximum growth rate

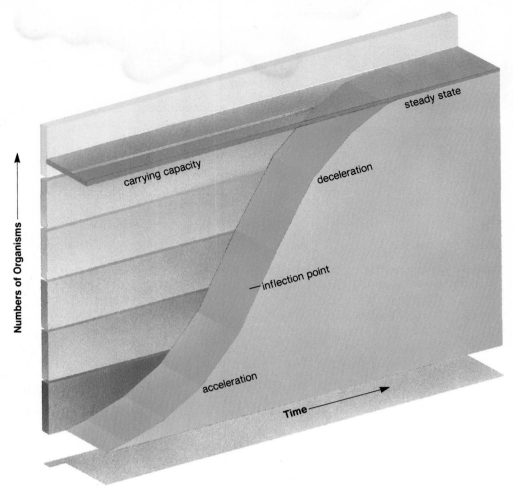

Figure 23.2   The logistic or S-shaped growth curve. The population, such as fruit flies in a culture bottle, initially grows exponentially but begins to slow down as resources become limited. Population size remains stable when the carrying capacity of the environment is reached.

under ideal conditions. Biotic potential usually is not demonstrated for long because of an opposing force called **environmental resistance.** Environmental resistance includes all the factors that cause early death of organisms and therefore prevents the population from producing as many offspring as it might otherwise do. As far as the fruit flies are concerned, we can speculate that environmental resistance included the limiting factors of food and space. The waste given off by the fruit flies also may have contributed to keeping the population size down. When environmental resistance sets in, biotic potential no longer is possible, and the slope of the growth curve begins to decline. This is called the inflection point of the curve.

The eventual size of any population represents a compromise between the biotic potential and the environmental resistance. This compromise occurs at the **carrying capacity** of the environment. The carrying capacity is the maximum population that the environment can support—for an indefinite period.

The carrying capacity of the earth for humans is not certain. Some authorities think the earth potentially is capable of supporting 50–100 billion people. Others think we already have more humans than the earth can support adequately.

Human Population Growth

The human population has undergone three periods of exponential growth. *Toolmaking* may have been the first technological advance that allowed the human population to enter a period of exponential growth. *Cultivation of plants* and *animal husbandry* may have allowed a second period of growth. The *Industrial Revolution,* which occurred about 1850, promoted the third phase. At first, only certain countries of the world became industrialized, and these countries now are called the *more-developed countries.*

More-Developed Countries

The Industrial Revolution, which also was accompanied by a medical revolution, took place in the Western world. In addition to European and North American countries, Russia and Japan also became industrialized. Collectively, these countries often are referred to as the *more-developed countries.* The developed countries doubled their size between 1850 and 1950 (fig. 23.3), largely due to a decline in the death rate. This decline is at-

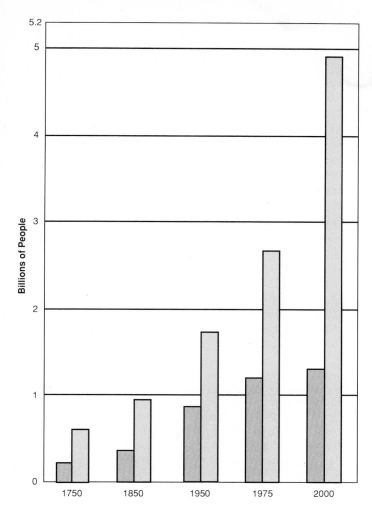

Figure 23.3 Size of human population in more-developed versus less-developed countries. The population size of the more-developed countries (blue) increased between 1850 and 1950, but it is expected to increase little between 1975 and 2000. In contrast, the population size of the less-developed countries (yellow) increased in the past and is also expected to increase dramatically in the future.

From "The Populations of the Underdeveloped Countries" by Paul Demeny. Copyright © 1974 by Scientific American, Inc. All rights reserved.

tributed to the influence of modern medicine and improved socioeconomic conditions. Industrialization raised personal incomes, and better housing permitted improved hygiene and sanitation. Numerous infectious diseases, such as cholera, typhus, and diphtheria, were brought under control.

The decline in the death rate in the more-developed countries was followed shortly by a decline in the birthrate. Between 1950 and 1975, populations in the more-developed countries showed only modest growth (fig. 23.3) because the growth rate fell from an average of 1.1% in 1950 to 0.8% in 1975.

The Demographic Transition

Overall, the growth rate in more-developed countries has gone through three phases (table 23.4). In phase I, prior to 1850, the growth rate was low because a high death rate canceled out the effects of a high birthrate; in phase II, the growth rate was high

Table 23.4 Analysis of Annual Growth Rates in More-Developed Countries

Phase	Birthrate	Death rate	Annual growth rate
I	high	high	low
II	high	low	high
III	low	low	low

because of a lowered death rate; and in phase III, the growth rate was again low because the birthrate had declined. These phases now are known as the **demographic transition.** In seeking a reason for the transition, it has been suggested that as industrialization occurred, the population became concentrated in the cities. Urbanization may have contributed to the decline in the growth rate because in the city, children were no longer the boon they were in the country. Instead of contributing to the yearly income of the family, they represented a severe drain on its resources. It also could be that urban living made people acutely aware of the problems of crowding, and for this reason, the birthrate declined. Also, some investigators believe that there was a direct relationship between improvement in socioeconomic conditions and the birthrate. They point out that as the developed nations became wealthier, as infant mortality was reduced, and as educational levels increased, the birthrate declined.

Regardless of the reasons for the demographic transition, it caused the rate of growth to decline in the more-developed countries. The growth rate for the more-developed countries is now about 0.6%, and their overall population size is about one-third that of the less-developed countries. A few more-developed countries—Austria, Denmark, Germany, Hungary, Sweden—are not growing or actually are losing population.

At this time, the United States has a population of over 249 million and still is growing. It is projected that the population could level off at about 300 million around 2080 and then start to decline. This projection assumes that at that time, the average number of children per woman will be 1.8, life expectancy will be age 81, and immigration will have declined to 500,000 newcomers a year. If immigration was not occurring, the United States population would decline to 220 million by 2080.

Human Issue

The more-developed and less-developed countries often have economic ties. For example, United States industries open factories in less-developed countries in order to take advantage of cheap labor, and the United States imports many raw materials from the less-developed countries.

Do you think that these economic ties benefit less-developed countries by supplying them with jobs and money, or are these countries being unfairly exploited? To what degree and in what way should more developed countries help less-developed countries?

Less-Developed Countries

Countries such as those in Latin America, Asia, and Africa collectively are called the *less-developed countries* because they either are nonindustrialized or newly industrialized. (Some researchers are in favor of placing the nonindustrialized countries in a third category, called the least-developed countries.) The mortality rate of the less-developed countries began to decline steeply following World War II. This decline was prompted not by socioeconomic development, but by the importation of modern medicine from the developed countries. Various illnesses were brought under control due to immunization, use of antibiotics, improved sanitation, and use of insecticides. Although the death rate declined, the birthrate did not decline to the same extent; therefore, the populations of the less-developed countries began—and still are today—increasingly dramatically (fig. 23.3). The less-developed countries were unable to cope adequately with such rapid population expansion. Today, many people in these countries are underfed, ill housed, unschooled, and living in abject poverty. Many of these poor have fled to the cities, where they live in makeshift shanties on the outskirts.

The growth rate of the less-developed countries as a whole finally peaked at 2.4% during 1960–1965. Since that time, the decrease in the death rate has slowed, and the birthrate is falling slowly. It is hoped that the overall growth rate will decline to 1.8% by the end of the century. At that time, about two-thirds of the world population will be in the less-developed countries.

Some investigators believe that the demographic transition occurs when a less-developed country begins to enjoy the benefits of economic development. Yet, during the 1980s, population outstripped economic growth in most less-developed countries, and two-thirds experienced a fall in per capita income. The least-developed countries, located largely in Africa, do not show any evidence yet of the demographic transition, and in these countries, the environmental resources are being depleted. Just 40 years ago, Ethiopia, for example, had a 30% forest cover; 12 years ago, it was down to 4%; and today, it may be 1%. Under such circumstances, the economic growth needed to check the birthrate may not occur.

Others point out that the countries with the greatest decline in the birthrate are those with the best family-planning programs. From this, it can be argued that such programs indeed can help to bring about a stable population size in the less-developed countries. Nevertheless, it has been found that certain socioeconomic factors also have contributed to a decline in the developing countries' growth rate. A relatively high gross national product (GNP), urbanization, low infant mortality, increased life expectancy, literacy, and education all had a dampening effect on the growth rate.

Age Structure Comparison

Laypeople are sometimes under the impression that if each couple had two children, zero population growth would take place immediately. However, **replacement reproduction,** as it is called, still would cause most countries today to continue growth due to the age structure of the population. If more young women are entering the reproductive years than there are older women leaving them behind, then replacement reproduction will give a positive growth rate.

Reproduction is at or below replacement level in some 20 more developed countries, including the United States. Even so, some of these countries will continue to grow modestly, in part because there was a baby boom after World War II. Young women born in these baby boom years are now in their reproductive years, and even if each one has less than two children, the population still will grow. Also, keep in mind that even the smallest of growth rates can add a considerable number of individuals to a large country. For example, a growth rate of 0.7% added about 1.5 million people to the United States population in 1990.

Many more developed countries have a stabilized age structure diagram (fig. 23.4), but most less-developed countries have a youthful profile—a large proportion of the population is below the age of 15. Since there are so many young women entering the reproductive years, the population still will expand greatly even after replacement reproduction is attained. The more quickly replacement reproduction is achieved, however, the sooner zero population growth will result.

The Human Population and Pollution

As the human population increases in size, more energy and materials are consumed. Because the human population does not use energy efficiently and does not recycle materials (see fig. 22.17), pollutants are added to all parts of the biosphere—land, water, and air.

Land Degradation

The land has been degraded in many ways. Here, we will discuss only those of the greatest concern.

Soil Erosion and Desertification

Soil erosion causes the productivity of agricultural lands to decline. It occurs when wind blows and rain washes away the topsoil to leave the land exposed without adequate cover. The United States Department of Agriculture estimates that erosion is causing a steady drop in the productivity of farmland equivalent to the loss of 1.25 million acres per year. To maintain the productivity of eroding land, more fertilizers, more pesticides, and more energy must be used.

One answer to the problem of erosion is to adopt soil conservation measures. For example, farmers could use strip-cropping and contour farming (fig. 23.5).

Desertification leads to the transformation of marginal lands to desert conditions. Desertification has been particularly evident along the southern edge of the Sahara Desert in Africa, where it is estimated that 350,000 square miles of once-productive grazing land has become desert in the last 50 years. However, desertification also occurs in this country. The United States Bureau of Land Management, which opens up federal

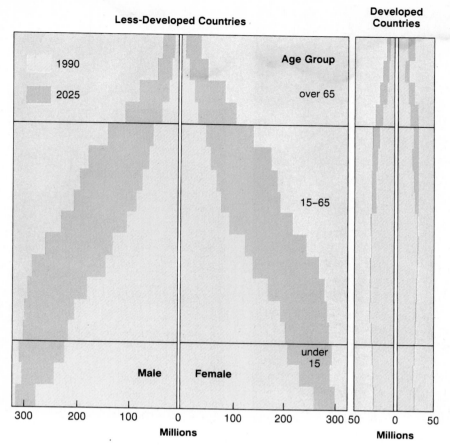

Less-Developed Countries

Developed Countries

1990

2025

Age Group

over 65

15–65

Male   Female

under 15

300   200   100   0   100   200   300   50   0   50

Millions   **Millions**

Figure 23.4 Age structure diagram for less-developed and more-developed countries. The less-developed countries contain 75 percent of the world population versus 25 percent in the more-developed countries. The less-developed countries will continue to expand for many years because of their youthful profile, but the more-developed countries will approach stabilization.

Figure 23.5 Contour farming. Crops are planted according to the lay of the land to reduce soil erosion. This farmer has planted alfalfa in between strips of corn to replenish the nitrogen content of the soil. Alfalfa, a legume, has root nodules that contain nitrogen-fixing bacteria.

lands for grazing, reports that much of the rangeland it manages is in poor or bad condition, with much of its topsoil gone and with greatly reduced ability to support forage plants.

Tropical Rain Forest Destruction

In developed countries, much of the hardwood forests long ago were converted to rapidly growing softwood forest plantations to provide humans with a source of wood. However, there are still virgin rain forests (fig. 23.6a) in Southeast Asia and Oceania, Central and South America, and Africa. These forests are severely threatened by human exploitation. The people living in developed countries want all sorts of things made from beautiful and costly tropical woods. These desires have created a market for such lumber. Most of the loss due to logging is in Southeast Asia, and in Malaysia and the Philippines, for example, the best commerical forests already are gone. Indonesia is cutting heavily to feed its new wood-exporting business, and Brazil is racing to catch up. Much of this wood goes to Japan, the United States, and Europe.

Another reason that tropical rain forests are undergoing destruction is a result of the needs of the people who live there.

a.

c.

b.

Figure 23.6   Tropical rain forest sustainability. *a.* In its natural state, a tropical rain forest is immensely rich in vegetation and breathtakingly beautiful. *b.* Slash-and-burn agriculture is the first step toward destruction of a portion of the rain forest. With this method, agriculture cannot be sustained for more than a few years because the soil is infertile and does not hold moisture. *c.* If the trees are not cut down, rubber tappers can earn a sustainable living by tapping the same trees year after year.

For example, in Brazil, there are large numbers of persons who have no means to support their family. To ease social unrest, the government allows citizens to own any land they clear in the Amazon Forest (occurs along the Amazon River). In tropical rain forests, it is customary to practice *slash-and-burn agriculture,* in which trees are cut down and burned to provide space to raise crops. Unfortunately, however, the land is fertile for only a few years. In tropical rain forests, the inorganic nutrients immediately cycle back to producers and do not accumulate in the soil. Therefore, despite the massive amount of growth above ground, the soil itself is nutrient-poor. Once the cleared land is unable to sustain crops, the farmer moves on to another part of the rain forest to slash and burn again.

Cattle ranchers are the greatest beneficiaries of deforestation, and increased ranching is therefore another reason for tropical rain forest destruction. The ranchers in Brazil are so aggressive that they actually force colonists to sell them newly cleared land at gunpoint. The cattle that are raised provide beef for export, and much of it is bought by U.S. fast-food chains. Because the imported beef is cheaper, hamburgers can be sold for 5 cents less than if the beef was purchased in the United States. The destruction of tropical forests on this account has been termed the "hamburger connection."

A newly begun pig-iron industry also indirectly results in further exploitation of the rain forest. The pig iron must be processed before it is exported, and smelting the pig iron requires the use of charcoal. The largest pig-iron company acknowledges having paid for construction of 1,500 small makeshift ovens used by peasants who burn trees from the rain forests to produce the charcoal.

There are currently three primary reasons for tropical rain forest destruction: (1) logging to provide hardwoods for export, (2) slash-and-burn agriculture, and (3) cattle ranching. Industrialization in countries having extensive tropical rain forests no doubt will become an important future reason also.

The Loss of Biological Diversity Tropical rain forests are much more biologically diverse than temperate forests. For example, temperate forests across the entire United States contain about 400 tree species. In the rain forest, a typical four-square-mile area holds as many as 750 types of trees. The fresh waters of South America are inhabited by an estimated 5,000 fish species; on the eastern slopes of the Andes, there are 80 or more species of frogs and toads; and in Ecuador, there are more than 1,200 species of birds—roughly twice as many as those inhabiting all of the United States and Canada. Therefore, a very serious side effect of deforestation in tropical countries is the loss of biological diversity.

Altogether, about half of the world's species are believed to live in tropical forests. A National Academy of Sciences study estimated that a million species of plants and animals are in danger of disappearing within 20 years as a result of deforestation in tropical countries. Many of these life forms never have been studied, and yet many possibly could be useful. At present, our entire domesticated crop production around the world relies on fewer than 30 species of plants and animals that have been

domesticated during the last 10,000 years! It is quite possible that many additional species of wild plants and animals now living in tropical forests could be domesticated. As matters now stand, the clearing of tropical forests very likely will prevent humans from ever having the opportunity to utilize more than a tiny fraction of the earth's biological diversity.

While it may seem like an either-or situation—either biological diversity or human survival—there is a growing recognition that this is not the case. Natives who harvest rubber from the trees (fig. 23.6c) can have a sustainable income from the same trees year after year. A recent study calculated the market value of rubber and exotic produce, like the aguaje palm fruit, that can be harvested continually from the Amazon forest. It concluded that selling these products would yield more than twice the income of either lumbering or cattle ranching.

There is much worldwide concern about the loss of biological diversity due to the destruction of tropical rain forests. The myriad of plants and animals that live there possibly could benefit human beings.

Human Issue

Do you think humans have a right to exploit the environment in any way they see fit, or do you think natural areas should be protected? Should they be protected for the sake of the plants and animals living there and/or for the sake of future generations of humans? Where do you stand on the necessity to protect the natural environment?

Waste Disposal

Every year, the United States population discards billions of metric tons of solid wastes, much of it on land. Solid wastes include not only household trash (fig. 23.7) but also sewage sludge, agricultural residues, mining refuse, and industrial wastes. Some of these solid wastes contain substances that cause human illness and sometimes even death; they are called **hazardous wastes.**

Hazardous wastes, such as heavy metals, chlorinated hydrocarbons, or organochlorides, and nuclear wastes, are subject to biological magnification (fig. 23.8). Since decomposers are unable to break down these wastes, they remain in dead organic matter. Further, the wastes are retained in the body of living organisms rather than excreted. Therefore, hazardous wastes will become more concentrated as they pass along a food chain. Notice in figure 23.8 that the number of dots representing DDT become more concentrated as it passes from producer to tertiary consumer. Biological magnification is most apt to occur in aquatic food chains; there are more links in aquatic food chains than there are in terrestrial food chains. Humans are the final consumers in both types of food chains, and in some areas, human milk contains detectable amounts of DDT and PCBs, which are organochlorides.

The dumping of hazardous wastes directly endangers public health. Chemical wastes buried over a quarter of a century ago in Love Canal, near Niagara Falls, have seriously

Figure 23.7   Dump sites not only cause land and air pollution, they also allow chemicals to enter groundwater. The groundwater may later become part of human drinking water.

damaged the health of some residents there. Similarly, the town of Times Beach, Missouri, is abandoned because workers spread an organochloride (dioxin)-laced oil on the city streets, leading to a myriad of illnesses among its citizens. In other places, such as in Holbrook, Massachusetts, manufacturers have left thousands of drums in abandoned or uncontrolled sites, where toxic chemicals are oozing out into the ground and are contaminating the water supply. Illnesses, especially forms of cancer, are quite common not only in Holbrook but also in adjoining towns.

Water Pollution

Pollution of surface water, groundwater, and the oceans is of major concern today.

Surface Water Pollution

All sorts of pollutants from various sources enter surface waters, as depicted in figure 23.9. Sewage treatment plants can be built to help degrade organic wastes, which otherwise can cause oxygen depletion in lakes and rivers. As the oxygen level de-

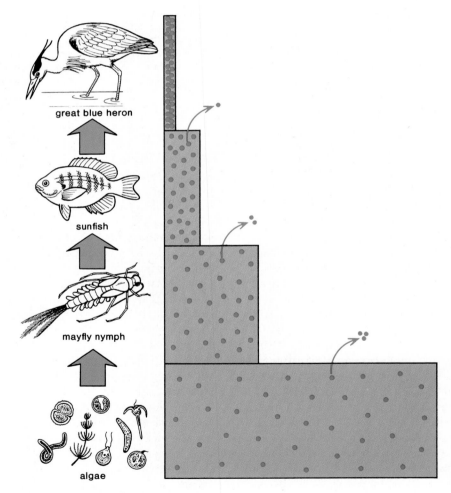

great blue heron

sunfish

mayfly nymph

algae

Figure 23.8   Biological magnification. A poison (*dots*), such as DDT, that is minimally excreted (*arrows*) becomes maximally concentrated as it passes along a food chain due to the reduced size of the trophic levels.

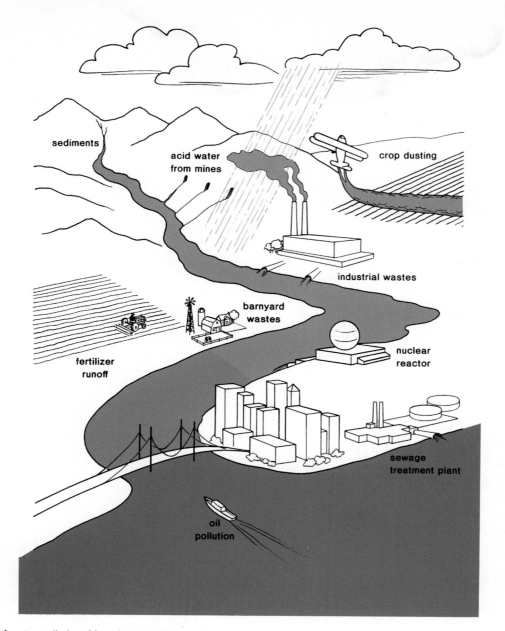

Figure 23.9   Sources of water pollution. Many bodies of water are
dying due to the introduction of sediments and surplus nutrients.
Source: Adapted from U.S. Environmental Protection Agency, Office of Water Supply and Solid
Waste Management Programs, "Waste Disposal Practices and Their Effects on Ground Water"
Executive Summary. (Washington, D.C., U.S. Government Printing Office, 1977).

creases, the diversity of life is greatly reduced. Also, human feces can contain pathogenic microorganisms that cause cholera, typhoid fever, and dysentery. In less-developed countries, where the population is growing and where waste treatment is practically nonexistent, many children die each year from these diseases.

Typically, sewage treatment plants use bacteria to break down organic matter to inorganic nutrients, like nitrates and phosphates, which then enter surface waters. These types of nutrients, which also can enter waters by fertilizer runoff and soil erosion, lead to **cultural eutrophication,** an acceleration of the natural process by which bodies of water fill in and disappear.

First, the nutrients cause overgrowth of algae. Then, when the algae die, oxygen is used up by the decomposers, and the water's capacity to support life is reduced. Massive fish kills are sometimes the result of cultural eutrophication.

Industrial wastes can include heavy metals and organochlorides, such as pesticides. These materials are not degraded readily under natural conditions nor in conventional sewage treatment plants. Sometimes, they accumulate in the bottom mud of deltas, and estuaries of highly polluted rivers cause environmental problems if they are disturbed. Industrial pollution is being addressed in many industrialized countries but usually has low priority in less-developed countries.

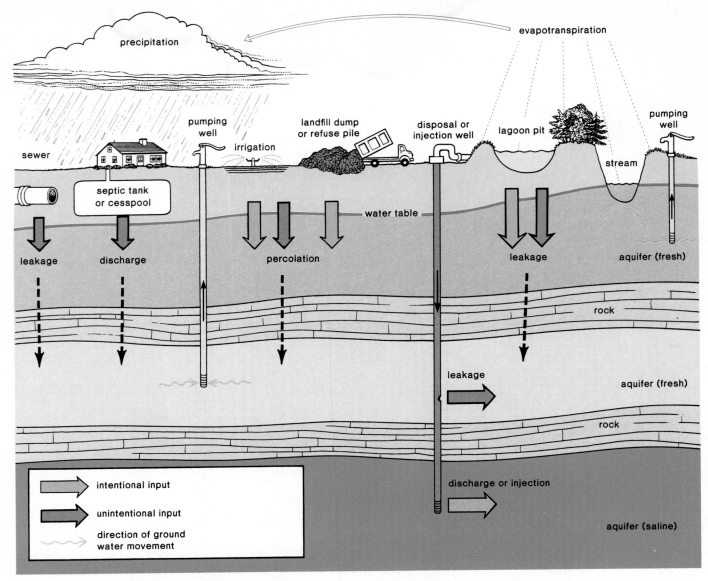

precipitation

evapotranspiration

sewer

pumping well

irrigation

landfill dump or refuse pile

disposal or injection well

lagoon pit

pumping well

septic tank or cesspool

water table

stream

leakage

discharge

percolation

leakage

aquifer (fresh)

rock

leakage

aquifer (fresh)

rock

discharge or injection

aquifer (saline)

intentional input

unintentional input

direction of ground water movement

Figure 23.10   Sources of groundwater pollution. Discontinuance of these means of industrial waste disposal has been difficult to achieve because citizens do not want to have waste disposal plants located near them.

Source: Adapted from U.S. Environmental Protection Agency, Office of Water Supply and Solid Waste Management Programs, "Waste Disposal Practices and Their Effects on Ground Water" *Executive Summary*, (Washington, D.C., U.S. Government Printing Office, 1977).

Some pollutants enter bodies of water from the atmosphere. Acid deposition has caused many lakes to become sterile in the industrialized world because acid leaches aluminum and iron out of the soil. A high concentration of these ions kills fishes and other forms of aquatic life. Adding lime is sometimes helpful against acidification of a lake.

Groundwater Pollution

Figure 23.10 shows the ways—both intentionally and unintentionally—that pollutants can reach underground rivers called aquifers. In areas of intensive animal farming or where there are many septic tanks, ammonium (NH_4^+) released from animal and human waste is converted by soil bacteria to soluble nitrate that moves down through the soil (percolates) into underground water supplies. Between 5% and 10% of all wells examined in the United States have nitrate levels higher than the recommended maximum.

Industry also pollutes aquifers. Previously, industry was accustomed to running wastewater into a pit. The pollutants then could seep into the ground. Wastewater and chemical wastes also have been injected into deep wells from which the pollutants constantly discharge. Both of these customs have been or are in the process of being phased out. However, it is very difficult for industry to find other ways to dispose of wastes. More adequately managed and controlled waste treatment plants are needed, but because citizens do not wish to live near such plants, towns often are successful in preventing their construction.

Oceanic Pollution

Coastal regions are not only the immediate receptors for local pollutants, they are also the final receptors for pollutants carried by rivers that empty at the coast. Waste dumping also occurs at sea, and ocean currents sometimes transport both trash and pollutants back to shore. Examples are the nonbiodegradable plastic bottles, pellets, and containers that now commonly litter beaches and the ocean's surfaces (fig. 23.11). Some of these, such as the plastic that holds a six-pack of beer, cause the death of birds, fishes, and marine mammals that mistake them for food and get entangled in them.

Offshore mining and shipping add pollutants to the oceans. Some 5 million metric tons of oil a year, or more than one gram per 100 square meters of the oceans' surfaces, end up in the oceans. Large oil spills kill plankton, fish larvae, and shellfishes, as well as birds and marine mammals. The largest spill may have occurred on March 24, 1989, when the tanker *Exxon Valdez* struck a reef in Alaska's Prince William Sound and leaked 44 million liters of crude oil. During the war with Iraq, 120 million liters were released from onshore storage tanks into the Persian Gulf—an event that was called environmental terrorism. Although petroleum is biodegradable, the process takes a long time because the low-nutrient content of seawater does not support a large bacterial population. Once the oil washes up onto beaches, it takes many hours of work and millions of dollars to clean it up.

Adequate sewage treatment and waste disposal are necessary to prevent the pollution of rivers and oceans. New methods also are needed to prevent pollution of underground water supplies.

Air Pollution

The atmosphere has two layers, the stratosphere and the troposphere. The stratosphere is a layer that lies 15–50 km above the surface of the earth. Here, the energy of the sun splits oxygen (O_2) molecules. These individual oxygen (O) atoms then combine with molecular oxygen to give ozone (O_3). This ozone layer is called a shield because it absorbs the ultraviolet rays of the sun, preventing them from striking the earth. If these rays did penetrate the atmosphere, life on earth would not be possible because living things cannot tolerate heavy doses of ultraviolet radiation. The troposphere is the atmospheric layer closest to the earth's surface; it ordinarily contains the gases nitrogen (N_2)—78%, oxygen (O_2)—21%, and carbon dioxide (CO_2)—0.3%.

Four major concerns—photochemical smog, acid deposition, global warming and the destruction of the ozone shield—are associated with the air pollutants listed in figure 23.12. You can see that fossil fuel burning and vehicle exhaust are primary sources of gases associated with air pollution. These two sources are related because gasoline is derived from petroleum, a fossil fuel.

Figure 23.11   Oceanic pollution. The possibility of oceanic pollution is brought home to the public when medical wastes are washed back onto beaches.

Photochemical Smog

Photochemical smog contains two air pollutants—nitrogen oxides (NO_x) and hydrocarbons (HC)—that react with one another in the presence of sunlight to produce ozone (O_3) and PAN (peroxylacetyl nitrate). Both nitrogen oxides and hydrocarbons come from fossil fuel combustion, but additional hydrocarbons come from various other sources as well, including industrial solvents.

Ozone and PAN commonly are referred to as oxidants. Breathing ozone affects the respiratory and nervous systems, resulting in respiratory distress, headache, and exhaustion. These symptoms are particularly apt to appear in young people; therefore, in Los Angeles, where ozone levels are often high, schoolchildren must remain inside the school building whenever the ozone level reaches 0.35 ppm (parts per million by weight). Ozone is especially damaging to plants, resulting in leaf mottling and reduced growth (fig. 23.13).

Normally, warm air near the ground is able to escape into the atmosphere. Sometimes, however, air pollutants, including smog and soot, are trapped near the earth due to a long-lasting thermal inversion. During a *thermal inversion,* cold air is at ground level beneath a layer of warm, stagnant air above. This often occurs at sunset, but turbulence usually mixes these layers during the day. Some areas surrounded by hills are particularly susceptible to the effects of a temperature inversion because the air tends to stagnate and there is little turbulent mixing.

Acid Deposition

The coal and oil burned by power plants releases sulfur dioxide (SO_2), and automobile exhaust contains nitrogen oxides (NO_x); both of these are converted to acids when they combine with water vapor in the atmosphere, a reaction that is promoted by ozone in smog. These acids return to earth as either wet deposition (acid rain or snow) or dry deposition (sulfate and nitrate salts).

Acid deposition now is associated with dead or dying lakes and forests, particularly in North America and Europe. Acid deposition also corrodes marble, metal, and stonework, an effect that is noticeable in cities. It also can degrade our water supply

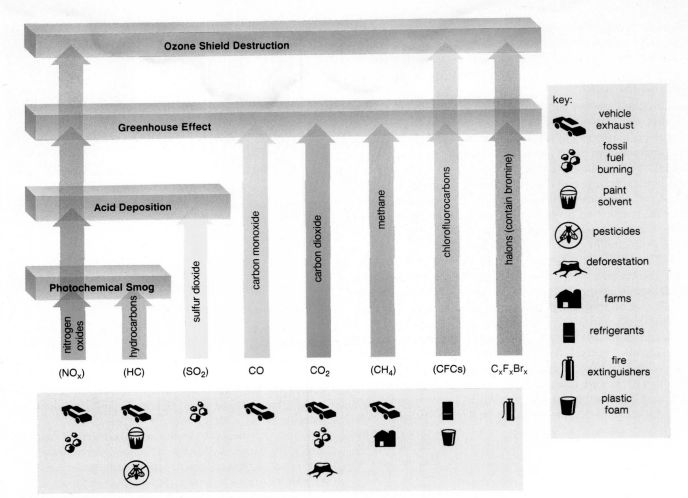

Figure 23.12   Air pollutants. These are the gases, along with their sources, that contribute to 4 environmental effects of major concern: photochemical smog, acid deposition, the greenhouse effect, and ozone shield destruction. An examination of the sources of these gases shows that vehicle exhaust and fossil fuel burning are the chief contributors.

Figure 23.13   Effect of ozone on plants. The milkweed in *a*. was exposed to ozone and appears unhealthy; the milkweed in *b*. was grown in an enclosure with filtered air and appears healthy.

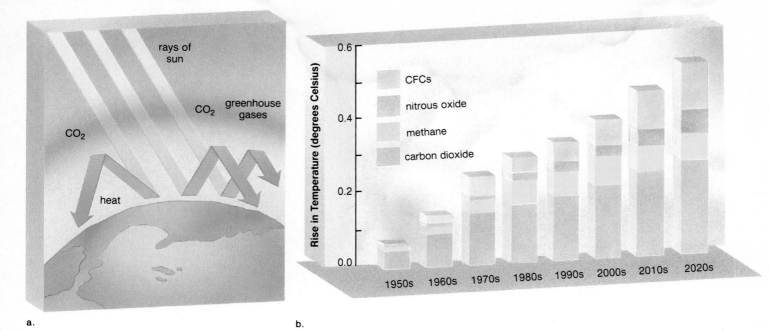

a.   **b.**

Figure 23.14   Global warming. *a.* The greenhouse effect is caused by the accumulation of certain gases in the atmosphere, such as carbon dioxide (CO₂). These gases allow the rays of the sun to pass through, but the gases absorb and reradiate heat back to the earth. *b.* The greenhouse gases. This graph shows the fraction of warming caused by the carbon dioxide, CFCs, methane, and nitrous oxide from 1950 to 2020. There was no accumulation of CFCs in the 1950s because they were not, as yet, being manufactured to any degree. By 2020, carbon dioxide and all of the other gases taken together will each contribute about 50 percent to the projected global warming.

by leaching heavy metals from the soil into drinking-water supplies. It may even be contributing to the incidence of illness as discussed on page 16.

Global Warming

It is predicted that certain air pollutants will cause the temperature of the earth to rise, perhaps by as much as 0.8°C per decade (1.36° F/decade). These gases let the sun's rays pass through, but they absorb and reradiate their heat back toward the earth (fig. 23.14*a*). This is called the **greenhouse effect** because the glass of a greenhouse allows sunlight to pass through, then traps the resulting heat inside the structure. The pollutants that cause global warming include the following:

carbon dioxide (CO₂) (from fossil fuel and wood burning)
nitrous oxide (NO₂) (primarily from fertilizer use and animal wastes)
methane (from biogas, bacterial decomposition, particularly in the guts of animals, sediments, and flooded rice paddies)
CFCs (from Freon) and halons (from fire extinguishers)

It so happens that together, nitrous oxide, methane, and CFCs cause a more severe greenhouse effect than carbon dioxide, and their combined effect is expected to equal the effect from carbon dioxide by the year 2020 (fig. 23.14*b*). Just now, carbon dioxide levels are 25% higher than they were in 1860. Most of this excess is due to industrialization, but tropical *deforestation*

is now a major contributor as well. Burning one acre of primary forest puts 200,000 kg of carbon dioxide into the air; moreover, the trees are no longer available to act as a sink to take up carbon dioxide during photosynthesis.

It appears that global warming has already begun: the four hottest years on record have occurred during the 1980s; there has been greater warming in the winters than summers; there has been greater warming at high altitudes than near the equator; the stratosphere is cooler and the lower atmosphere is warmer than formerly. All of these effects had been predicted by computer models of the greenhouse effect.

The ecological effects of global warming are expected to be severe. First of all, the sea level will rise—melting of the polar ice caps will add more water to the sea and water expands when it heats as well. This will cause flooding of many coastal regions and the possible loss of many cities, like New York, Boston, Miami, and Galveston in the United States. Coastal ecosystems, such as marshes, swamps, and bayous, would normally move inland to higher ground as the sea level rises, but many of these ecosystems are blocked in by artificial structures and may be unable to move inland. If so, the loss of fertility will be immense.

There also may be food loss because of regional changes in climate. Not only will there be greater heat, there will be drought in the midwestern United States, and the suitable climate for growing wheat and corn will shift to Canada, where the soil is not as suitable.

Drastic measures are recommended to hold the quantity of greenhouse gases to their present level. A 50% decrease in consumption of fossil fuels is recommended, and we must find more efficient ways to acquire energy from cleaner fuels, such as natural gas. We should use alternative energy sources, such as solar and geothermal energy and even perhaps nuclear power, more aggressively.

Not only should tropical rain forest deforestation be halted, extensive reforesting all over the globe should take place. These forests could be used as a source of wood for fuel, and at the same time, the carbon dioxide given off by the burning of the fuel would be absorbed by the new trees coming along.

Manufacture and use of CFCs should be eliminated completely. The United States and the European countries already have agreed to reduce CFC production by 85% as soon as possible and to try to ban the chemicals altogether by the end of the century.

Ozone Shield Destruction

Chlorofluorocarbons, or CFCs, of which the most important is Freon, are heat-transfer agents used in refrigerators and air conditioners. They also are used as foaming agents in such products as styrofoam cups and egg cartons. Formerly, they were used as propellants in spray cans, but this application now is banned in the United States. Halons are anti-fire agents used in fire extinguishers.

It was known that CFCs would drift up into the stratosphere, but it was believed that they would be nonreactive there during their 150-year life span. However, it now is apparent that when the temperature drops, these compounds react chemically on frozen particle surfaces within stratospheric clouds and release active chlorine that then can react with the ozone in the **ozone shield.** (Similarly, halons release bromine, which reacts with ozone.) Once freed, a single atom of chlorine destroys about 100,000 molecules of ozone before finally settling to the earth's surface as chloride years later. Measurements suggest that 3% and perhaps up to 5% of the global ozone layer already has been destroyed by CFCs and halons. A more drastic effect has been found in the Antarctic and the Arctic. At the South Pole, up to 50% of the ozone is destroyed each spring over an area the size of North America (fig. 23.15). There is a similar reduction, called an ozone hole, but not as severe, over the Arctic as well.

Depletion of the ozone layer will allow more ultraviolet rays to enter the troposphere. The incidence of human cancer, especially skin cancer, can be expected to increase, and plants and animals living in the top microlayer of the oceans will begin to die. Increased ultraviolet radiation also will hasten the rate at which smog is formed. There has been an international agreement to halve CFC emissions by 1998 and freeze halon emissions by 1992.

Outdoor air pollutants are involved in causing four major environmental effects: photochemical smog, acid deposition, global warming, and ozone shield destruction. Each pollutant may be involved in more than one of these.

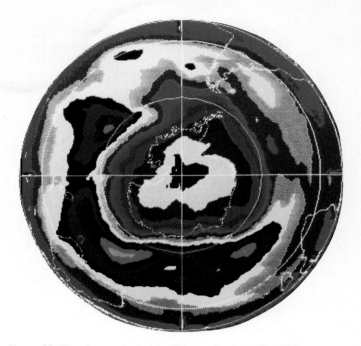

Figure 23.15   Ozone depletion. This has been confirmed by observations of a drastic decrease in ozone at the South Pole of Antarctica each spring during the late 1970s and the 1980s. The ozone hole appears as a large white area at the center of this picture, a picture taken by the total ozone mapping spectrometer aboard NASA's Nimbus 7 satellite on 5 October 1987, when the ozone loss reached nearly 60%.

While each of these environmental effects is bad enough when considered separately, they actually feed on one another, making the total effect much worse than is predicted for the total of each separately.

A Sustainable World

Economic growth often is accompanied by environmental degradation, and there is great concern that as the less-developed countries become more developed, environmental degradation will increase to the point that the human population will outstrip the carrying capacity of the planet. Without economic growth, however, the demographic transition may not occur, and the sheer number of people in the less-developed countries will bring about environmental degradation to such a degree that the effects will be felt worldwide, not just in the immediate area.

The answer to this dilemma is economic growth without the side effect of environmental degradation. This is called sustainable growth. Certain developed countries, such as the Scandinavian countries and the United States to a degree, are beginning to learn to protect the environment. Energy consumption is decreasing even as economic growth continues. Industries are beginning to recycle their waste to prevent

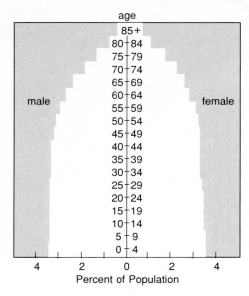

Figure 23.16   Age and sex structure of a hypothetical human population that remains the same size each year.
Source: United States Bureau of the Census.

Figure 23.17   In the steady state, environmental preservation will be an important consideration.

environmental pollution. Citizens are learning to recycle their trash. More should be done, and ecologically sound practices must be exported to the less-developed countries very quickly. Only sustainable economic development will ensure the continuance of the world's human population.

Once zero population growth has been achieved, then we can begin to consider the possibility of a steady state in which population and resource consumption remain constant.

The Steady State

A stable population size would be a new experience for humans. Figure 23.16 shows the age structure for a hypothetical stable population. Such a population has many benefits, as discussed here:

1. Over 40% of the people would be fairly youthful, with only about 15% in the senior citizen category. There would be proportionately fewer children and teenagers than in a rapidly expanding population.

2. The quality of life for children might increase substantially since fewer unwanted babies might be born and the opportunity would exist for children to receive the loving attention each needs.

3. There might be increased employment opportunities for women and a generally less competitive workplace

because newly qualified workers would enter the job market at a more moderate rate.

4. Creativity need not be impaired. A study of Nobel Prize winners showed that the average age at which the prize-winning work was done was over 30.

Environmental preservation would be the most important consideration in a steady-state world. Renewable energy sources, such as solar energy, would play a greater role in meeting energy needs. Pollution would be minimized. Ecological diversity would be maintained, and overexploitation would cease (fig. 23.17). Ecological principles would serve as guidelines for specialists in all fields, creating a unified approach to the environment. In a steady-state world, all people would strive to be aware of the environmental consequences of their actions, consciously working toward achieving balance in the ecosystems of their planet.

In a *steady-state society,* there would be no yearly increase in population nor resource consumption. It is forecast that under these circumstances, the quality of life would improve.

What would our culture be like if we had steady-state manufacturing and a steady-state population? Perhaps it would be greatly improved. Certainly, there are no limits to growth in knowledge, education, art, music, scientific research, human rights, justice, and cooperative human interactions. In a steady-state world, the general sense of fearful competition among peoples might diminish, allowing human compassion and creativity to prosper as never before.

Summary

The human population is expanding exponentially, and even though the growth rate has declined, there is a large increase each year—the doubling time is now about 40 years. The developed countries underwent a demographic transition between 1950 and 1975, but the less-developed countries just now are undergoing demographic transition. In these countries, where the average age is less than 15, it will be many years before reproduction replacement will mean zero population growth.

On land, soil erosion sometimes reduces soil quality and leads to desertification. The tropical rain forests are being reduced in size, and the loss of biological diversity will be immense. Solid wastes, including hazardous wastes, are deposited on land. These are not biodegradable and are subject to biological magnification.

Surface waters, groundwater, and the oceans all are being polluted. Organic materials can be broken down in sewage treatment plants, but the nutrients made available to algae can lead to cultural eutrophication. It is difficult to rid surface waters and particularly groundwater of hazardous wastes. The oceans are the final recipients of all the pollutants that enter water. In addition, some materials are dumped purposefully or accidentally directly into the oceans.

In the air, hydrocarbons and nitrogen oxides (NO_x) react to form smog, which contains ozone and PAN. Sulfer dioxide (SO_2) and nitrogen oxides react with water vapor to form acids that contribute to acid deposition. Ozone destruction is associated with CFCs and halons, which rise into the stratosphere and react with frozen particles within clouds to release chlorine. Chlorine causes ozone to break down. Several gases—carbon dioxide (CO_2), nitrous oxide (NO_2), methane, and CFCs—are called the greenhouse gases because they trap heat.

Economic growth is required particularly in the less-developed countries. All countries of the world should try to have sustainable growth, and this necessitates preservation of the environment. Eventually, it may be possible to have a steady state in which neither population nor resource consumption increases.

Study Questions

1. Define exponential growth. (p. 452) Draw a growth curve to represent exponential growth, and explain why a curve representing population growth usually levels off. (p. 453)
2. Calculate the growth rate and the doubling time for a population in which the birthrate is 20 per 1,000 and the death rate is 2 per 1,000. (p. 453)
3. Define demographic transition. When did the developed countries undergo demographic transition? When did the less-developed countries undergo demographic transition? (p. 455)
4. Give at least three differences between the more-developed countries and the less-developed countries. (pp. 454–456)
5. Give two reasons why the quality of the land is being degraded today. What is desertification? (p. 456)
6. Give three reasons why tropical rain forests are being destroyed. What is another possible reason in the future? (p. 457)
7. What is the primary ecological concern associated with the destruction of rain forests? (p. 459)
8. What are the three types of hazardous wastes that contribute to pollution on land? (p. 459)
9. What are several ways in which underground water supplies can be polluted? (p. 462)
10. What substances contribute to air pollution? What are their sources? Which ones are associated with photochemical smog, acid deposition, depletion of the ozone shield, or the greenhouse effect? (pp. 463–466)

Objective Questions

1. After a country has undergone the demographic transition, the death rate and the birthrate both are _____ (high or low).
2. If a country has a pyramid-shaped age structure diagram, most individuals are _____ (prereproductive, reproductive, or postreproductive).
3. Less-developed countries are not as _____ as developed countries.
4. When a population is undergoing exponential growth, the increase in number of people each year is _____ than the year before. (higher or lower)
5. The chemical best associated with ozone depletion is _____ .
6. The gas best associated with the greenhouse effect is _____ .
7. Sewage is biodegradable, but the nutrients released can lead to _____ of surface water.
8. Pesticides and radioactive wastes both are subject to biological _____ .

Answers

1. low 2. prereproductive 3. industrialized 4. higher 5. CFCs 6. carbon dioxide (CO_2) 7. cultural eutrophication 8. magnification

Critical Thinking Questions

1. Can you think of any reasons why it might be inappropriate to imply that a growth curve such as that in figure 23.2 can be applied to the human population?
2. The Mexican government at one time encouraged large families because it was believed that the greater the number of people, the greater the work force, and the greater the prosperity. What's wrong with this type of thinking?
3. Is sending food the best answer to preventing a famine? If we really wanted to help the starving people of the world, what would we be doing?

Selected Key Terms

acid deposition (as'id dep''o-zish'un)   the return to earth as rain or snow of the sulfate or nitrate salts of acids produced by commerical and industrial activities on earth.   463

biotic potential (bi-ot'ik po-ten'shal)   the maximum population growth rate under ideal conditions.   453

carrying capacity (kar'e-ing kah-pas'i-te)   the largest number of organisms of a particular species that can be maintained indefinitely in an ecosystem.   454

cultural eutrophication (kul'tu-ral u''tro-fi-ka'shun)   enrichment of a body of water, causing excessive growth of producers and then death of these and other inhabitants.   461

demographic transition (dem-o-graf'ik tran-zi'shun)   the change from a high birthrate to a low birthrate so that the growth rate is lowered.   455

doubling time (dŭ'b'l-ing tīm)   the number of years it takes for a population to double in size.   453

environmental resistance (en-vi''ron-men'tal re-zis'tans)   sum total of factors in the environment that limit the numerical increase of a population in a particular region.   454

exponential growth (eks''po-nen'shal grōth)   growth, particularly of a population, in which the increase occurs in the same manner as compound interest.   452

greenhouse effect (grēn'hows ĕ-fekt)   buildup in the atmosphere of CO_2 and other gases that retain and reradiate heat, causing global warming.   465

hazardous waste (haz'er-dus wāst)   waste containing chemicals hazardous to life.   459

photochemical smog (fo''to-kem'ĭ-kal smog)   air pollution that contains nitrogen oxides (NO_x) and hydrocarbons that react to produce ozone and peroxylacetyl nitrate (PAN).   463

ozone shield (o'zōn shēld)   a layer of O_3 present in the upper atmosphere that protects the earth from damaging ultraviolet light. Nearer the earth, ozone is a pollutant.   466

replacement reproduction (re-plās'ment re''pro-duk'shun)   a population whose average number of children is one per person.   456

Further Readings for Part VI

Begon, J., J. Harper, and C. Townsend. 1986. *Ecology: Individuals, populations, and communities*. Sunderland, Mass.: Sinauer, Associates.

Berner, Robert A., and Antonio C. Lasaga. March 1989. Modeling the geochemical carbon cycle. *Scientific American*.

Brown, L. R., et al. 1990. *State of the world: 1990*. New York: W. W. Norton and Co.

Bushbacher, R. J. January 1986. Tropical deforestation and pasture development. *BioScience*.

Earthworks Group. 1989. *50 Simple things you can do to save the earth*. Berkeley, CA: Earthworks Press.

Environment. All issues of this journal contain articles covering modern ecological problems.

Hamakawa, Y. April 1987. Photovoltaic power. *Scientific American*.

Houghton, Richard A., and George M. Woodwell. April 1989. Global climatic change. *Scientific American*.

Jones, Phillip D., and Tom M. L. Wigley. August 1990. Global warming trends. *Scientific American*.

Miller, J. T. 1988. *Living in the environment*. 5th ed. Belmont, Calif.: Wadsworth.

Mohnen, V. A. August 1988. The challenge of acid rain. *Scientific American*.

Nebel, B. J. 1987. *Environmental science: The way the world works*. 2d ed. Englewood Cliffs, N.J.: Prentice-Hall.

Newell, Reginald E., Henry G. Reichle, Jr., and Wolfgang Seiler. October 1989. Carbon monoxide and the burning earth. *Scientific American*.

O'Leary, P. R., P. W. Walsh, and R. H. Ham. December 1988. Managing solid waste. *Scientific American*.

Power, J. F., and R. F. Follett. March 1987. Monoculture. *Scientific American*.

Reganold, John P., Robert I. Papendik, and James F. Parr. June 1990. Sustainable agriculture. *Scientific American*.

Repetto, Robert. April 1990. Deforestation in the tropics. *Scientific American*.

Ricklefs, R. E. 1986. *Ecology*. 3rd ed. New York: Chiron Press.

Schneider, S. H. May 1987. Climate modeling. *Scientific American*.

Scientific American. September 1989. Managing planet earth. Special issue.

Shaw, R. W. August 1987. Air pollution by particles. *Scientific American*.

Smith, R. L. 1985. *Ecology and field biology*. 3rd ed. New York: Harper & Row, Publishers.

White, Robert M. July 1990. The great climate debate. *Scientific American*.

Appendix A

Table of Chemical Elements

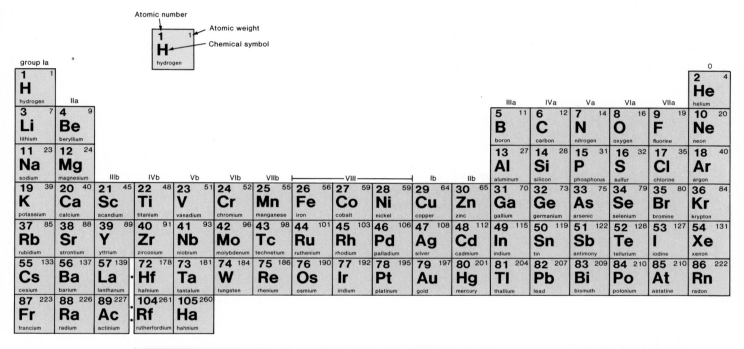

Appendix B

The Metric System

The Metric System

Standard metric units		Abbreviation
standard unit of mass	gram	g
standard unit of length	meter	m
standard unit of volume	liter	l
Common prefixes		**Examples**
kilo	1,000	a kilogram is 1,000 grams
centi	0.01	a centimeter is 0.01 meter
milli	0.001	a milliliter is 0.001 liter
micro (μ)	one-millionth	a micrometer is 0.000001 (one-millionth) of a meter
nano (n)	one-billionth	a nanogram is 10^{-9} (one-billionth) of a gram
pico (p)	one-trillionth	a picogram is 10^{-12} (one-trillionth) of a gram

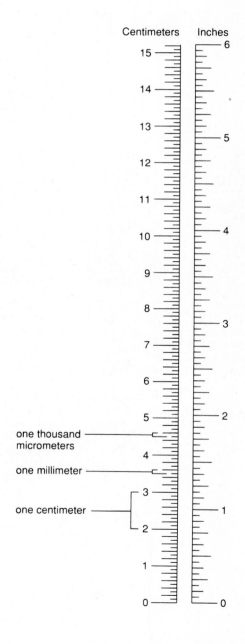

one thousand micrometers

one millimeter

one centimeter

Units of Length

Unit	Abbreviation	Equivalent
meter	m	approximately 39 in
centimeter	cm	10^{-2} m
millimeter	mm	10^{-3} m
micrometer	μm	10^{-6} m
nanometer	nm	10^{-9} m
angstrom	Å	10^{-10} m

Length conversions

1 in = 2.5 cm	1 mm = 0.039 in
1 ft = 30 cm	1 cm = 0.39 in
1 yd = 0.9 m	1 m = 39 in
1 mi = 1.6 km	1 m = 1,094 yd
	1 km = 0.6 mi

To convert	Multiply by	To obtain
inches	2.54	centimeters
feet	30	centimeters
centimeters	0.39	inches
millimeters	0.039	inches

Think Metric

Length

1. The speed of a car is 60 miles/hr or 100 km/hr.
2. A man who is 6 feet tall is 180 cm.
3. A 6-inch ruler is 15 cm.
4. One yard is almost a meter (0.9 m).

Units of Volume

Unit	Abbreviation	Equivalent
liter	l	approximately 1.06 qt
milliliter	ml	10^{-3} l (1 ml = 1 cm^3 = 1 cc)
microliter	μl	10^{-6} l

Volume conversions

1 tsp = 5 ml	1 pt = 0.47 l	1 ml = 0.03 fl oz
1 tbsp = 15 ml	1 qt = 0.95 l	1 l = 2.1 pt
1 fl oz = 30 ml	1 gal = 3.8 l	1 l = 1.06 qt
1 cup = 0.24 l		1 l = 0.26 gal

To convert	Multiply by	To obtain
fluid ounces	30	milliliters
quarts	0.95	liters
milliliters	0.03	fluid ounces
liters	1.06	quarts

Think Metric

Volume

1. One can of beer (12 oz) contains 360 ml.
2. The average human body contains between 10 and 12 pints of blood or between 4.7 and 5.6 liters.
3. One cubic foot of water (7.48 gallons) is 28.426 liters.
4. If a gallon of unleaded gasoline costs $1.00, a liter costs 38¢.

Units of Weight

Unit	Abbreviation	Equivalent
kilogram	kg	10^3 g (approximately 2.2 lb)
gram	g	approximately 0.035 oz
milligram	mg	10^{-3} g
microgram	μg	10^{-6} g
nanogram	ng	10^{-9} g
picogram	pg	10^{-12} g

Weight conversions

1 oz = 28.3 g
1 lb = 453.6 g   1 g = 0.035 oz
1 lb = 0.45 kg   1 kg = 2.2 lb

To convert	Multiply by	To obtain
ounces	28.3	grams
pounds	453.6	grams
pounds	0.45	kilograms
grams	0.035	ounces
kilograms	2.2	pounds

Think Metric

Weight

1. One pound of hamburger is 448 grams.
2. The average human male brain weighs 1.4 kg (3 lb 1.7 oz).
3. A person who weighs 154 lbs weighs 70 kg.
4. Lucia Zarate weighed 5.85 kg (13 lbs) at age 20.

Units of Temperature

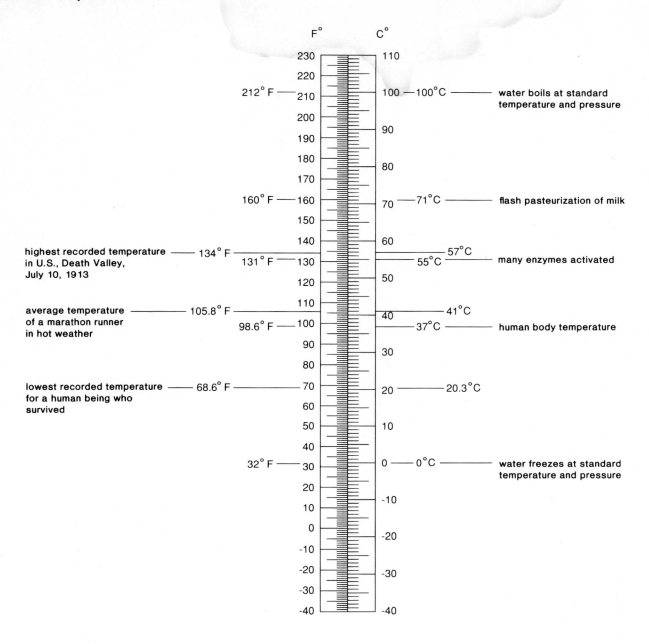

To convert temperature scales:

Fahrenheit to Centigrade $°C = \frac{5}{9} (°F - 32)$

Centigrade to Fahrenheit $°F = \frac{9}{5} (°C) + 32$

Glossary

A

accommodation (ah-kom″o-da′shun)   lens adjustment in order to see close objects *243*

acetylcholine (as″e-til-ko′len)   ACh; a neurotransmitter substance secreted at the ends of many neurons; responsible for the transmission of a nerve impulse across a synaptic cleft *194*

acetylcholinesterase (as″e-til-ko″lin-es′ter-as)   AChE; an enzyme that breaks down acetylcholine *194*

acid (as′id)   a solution in which pH is less than 7; a substance that contributes or liberates hydrogen ions in a solution *17*

acid deposition (as′id de′po-si″shun)   an accumulation of acids from the atmosphere into lakes and forests, particularly in North America and Europe *463*

acromegaly (ak″ro-meg′ah-le)   condition resulting from an increase in growth hormone production after adult height has been achieved *260*

acrosome (ak′ro-som)   covering on the tip of a sperm cell's nucleus believed to contain enzymes necessary for fertilization *276*

ACTH (adrenocorticotropic hormone)   hormone secreted by the anterior lobe of the pituitary gland that stimulates activity in the adrenal cortex *261*

actin (ak′tin)   one of the two major proteins of muscle; makes up thin filaments in myofibrils of muscle cells. *See also* myosin *230*

action potential (ak′shun po-ten′shal)   the change in potential propagated along the membrane of a neuron; the nerve impulse *192*

active site (ak′tiv sit)   the region on the surface of an enzyme where the substrate binds and where the reaction occurs *41*

active transport (ak′tiv trans′port)   transfer of a substance into or out of a cell against a concentration gradient by a process that requires a carrier and the expenditure of energy *35*

adaptation (ad″ap-ta′shun)   the fitness of an organism for its environment, including the process by which it becomes fit, in order that it may survive and reproduce; also the adjustment of sense receptors to a stimulus so that the stimulus no longer excites them *421*

adenosine triphosphate (ah-den′o-sen tri-fos′fat)   *See* ATP.

ADH (antidiuretic hormone)   hormone released from the posterior lobe of the pituitary gland that enhances the conservation of water by the kidneys *180, 258*

adrenocorticotropic hormone (ah-dre″no-kor″te-ko-trop′ik hor′mon)   *see* ACTH

aerobic cellular respiration (a″er-ob′ik sel′u-lar res″pi-ra′-shun)   the reactions of glycolysis, Krebs cycle, and electron transport system that provide energy for ATP production *39*

agglutination (ah-gloo″ti-na′shun)   clumping of cells, particularly in reference to red blood cells involved in an antigen-antibody reaction *127*

aging (aj′ing)   the degenerative biological changes that are observed as a human grows older *329*

agranular leukocytes (ah-gran′u-lar loo′ko-sitz)   white blood cells that do not contain distinctive granules *124*

AIDS (adz)   acquired immunodeficiency syndrome, a disease caused by HIV and transmitted via body fluids; characterized by failure of the immune system *299*

albumin (al-bu′min)   plasma protein of the blood with transport and osmotic functions *115*

aldosterone (al″do-ster′on)   a hormone secreted by the adrenal cortex that functions in regulating sodium and potassium concentrations of the blood *181, 265*

allantois (ah-lan′ to-is)   one of the extraembryonic membranes; in reptiles and birds a pouch serving as a repository for nitrogeneous waste; in mammals a source of blood vessels to and from the placenta *318*

allele (ah-lel′)   an alternative form of a gene that occurs at a given chromosome site (locus) *357*

all-or-none law (awl′ or nun′ law)   states that muscle fibers either contract maximally or not at all, and neurons either conduct a nerve impulse completely or not at all *224*

alveoli (al-ve′o-li)   saclike structures that are the air sacs of a lung *154*

amino acid (ah-me′no as′id)   a unit of protein that takes its name from the fact that it contains an amino group (NH_2) and an acid group (COOH) *20*

ammonia (ah-mo′ne-ah) NH_3, a nitrogenous waste product resulting from deamination of amino acids *169*

amnion (am′ne-on)   an extraembryonic membrane; a sac around the embryo containing fluid *313*

ampulla (am-pul′lah)   base of a semicircular canal in the inner ear *248*

amylase (am′i-las)   a starch-digesting enzyme secreted by salivary glands and the pancreas *78*

analogous structure (ah-nal′o-gus struk′chur)   similar in function but not in structure, particularly in reference to similar adaptations *418*

anaphase (an′ah-faz)   stage in mitosis during which chromatids separate, forming chromosomes *342*

anaphase I (an′ah-faz wun)   phase of meiosis I during which homologous chromosomes separate and move to the poles of the spindle *344*

anaphase II (an′ah-faz too)   phase of meiosis II during which the centromeres divide and sister chromatids separate and move toward the poles *344*

anthropoids (an′thro-poidz)   higher primates, including only monkeys, apes, and humans *425*

antibiotic (an″ti-bi-ot′ik)   a drug that has the ability to kill microorganisms without doing harm to a human patient *305*

antibody (an′ti-bod″e)   a protein produced in response to the presence of some foreign substance in the blood or tissues *124*

anticodon (an″ti-ko′don)   a "triplet" of 3 nucleotides in transfer RNA that pairs with a complementary triplet (codon) in messenger RNA *384*

antidiuretic hormone (an''tĭ-dī''u-ret'ik hor' mōn) *see* ADH

antigen (an'tĭ-jen) a foreign substance, usually a protein, that stimulates the immune system to produce antibodies 125

anus (a'nus) outlet of the digestive tube 75

anvil (an'vil) also called the incus; the middle one of the auditory ossicles, which amplify the sound waves in the middle ear 247

aorta (a-or'tah) major systemic artery that receives blood from the left ventricle 101

appendicular skeleton (ap''en-dik'u-lar skel'ĕ-ton) portion of the skeleton forming the upper extremities, pectoral girdle, lower extremities, and pelvic girdle 218

appendix (ah-pen'diks) a small, tubular appendage that extends outward from the cecum of the large intestine 75

aqueous humor (a'kwe-us hu'mor) watery fluid that fills the space between the cornea and lens of the eye 240

arterial duct (ar-te're-al dukt) ductus arteriosus; fetal connection between the pulmonary artery and the aorta 325

arteriole (ar-te're-ōl) a branch from an artery that leads into a capillary 94

artery (ar'ter-e) a vessel that takes blood away from the heart; characteristically possessing thick elastic walls 94

artificial insemination (ar''tĭ-fish'al in-sem'' ĭ-na'shun) introduction of semen into the vagina or uterus by artificial means 292

aster (as'ter) short rays of microtubules that appear at the ends of the spindle apparatus in animal cells during cell division 341

ATP (adenosine triphosphate) a compound containing adenine, ribose, and three phosphates, two of which are high-energy phosphates. It is the "common currency" of energy for most cellular processes 26

atom (at'om) smallest unit of matter nondivisible by chemical means 13

atria (a'tre-ah) chambers; particularly the upper chambers of the heart that lie above the ventricles (*sing.* atrium) 96

atrioventricular (a''tre-o-ven-trik'u-lar) a structure in the heart that pertains to both the atria and ventricles; for example, an atrioventricular valve is located between an atrium and a ventricle 96

atrioventricular node (a''tre-o-ven-trik'u-lar nōd) *see* AV node 98

auditory canal (aw'dĭ-to''re kah-nal') a tube in the outer ear that leads to the tympanic membrane 247

auditory nerve (aw'dĭ-to''re nurv) a nerve sending the signal for sound from the inner ear to the temporal lobe of the brain 250

Australopithecus (aw''strah-lo-pith'e-kus) the first recognized hominid genus 428

autonomic nervous system (aw''to-nom'ik ner'vus sis'tem) a branch of the peripheral nervous system administering motor control over internal organs 199

autosome (aw'to-sōm) a chromosome other than a sex chromosome 337

AV node (a-ve nōd) a small region of neuromuscular tissue located near the septum of the heart that transmits impulses from the SA node to the ventricular walls 98

axial skeleton (ak'se-al skel'ĕ-ton) portion of the skeleton that supports and protects the organs of the head, neck, and trunk 217

axon (ak'son) process of a neuron that conducts nerve impulses away from the cell body 189

B

bacteria (bak-te're ah) prokaryotes that lack the organelles of eukaryotic cells 303

basal bodies (ba'sal bod'ez) short cylinders having a circular arrangement of 9 microtubule triplets (9 + 0 pattern) located within the cytoplasm at the bases of cilia and flagella 41

base (bās) a solution in which pH is more than 7; a substance that contributes or liberates hydroxide ions in a solution; alkaline; opposite of acidic. Also, in genetics the chemicals adenine, guanine, cytosine, thymine, and uracil that are found in DNA and RNA 17

bile (bīl) a secretion of the liver that is temporarily stored in the gallbladder before being released into the small intestine where it emulsifies fat 76

binary fission (bi'na-re fish'un) reproduction by simple cell division that does not involve a mitotic spindle 304

binomial system (bi-no'me-al sis'tem) the assignment of two names to each organism, the first of which designates the genus and the second of which designates the species 416

biotechnology (bi''o-tek-nol'-o-je) the use of biological organisms to mass-produce a product of commercial interest or to achieve a desired end 388

biotic potential (bi-ot'ik po-ten'shal) the maximum population growth rate under ideal conditions 453

blastocyst (blas'to-sist) an early stage of embryonic development that consists of a hollow ball of cells 312

blind spot (blīnd spot) area where the optic nerve passes through the retina and where vision is not possible due to the lack of rods and cones 241

blood (blud) connective tissue composed of cells separated by plasma; transports substances in the cardiovascular system 55

blood pressure (blud presh'ur) the pressure of the blood against a blood vessel 104

B lymphocyte (bē lim'fo-sīt) a lymphocyte that matures in the bone marrow and differentiates into antibody-producing plasma cells when stimulated by the presence of a specific antigen 135

bone (bōn) connective tissue having a hard matrix of calcium salts deposited around protein fibers 55

Bowman's capsule (bo'manz kap'sūl) a double-walled cup that surrounds the glomerulus at the beginning of the kidney tubule 174

bradykinin (brad''e-ki'nin) a substance found in damaged tissue that initiates nerve impulses resulting in the sensation of pain 126

breathing (brēth'ing) entrance and exit of air into and out of the lungs 150

bronchi (brong'ki) the two major divisions of the trachea leading to the lungs 154

bronchiole (brong'ke-ōl) the smaller air passages in the lungs 154

buffer (buf'er) a substance or compound that prevents large changes in the pH of a solution 17

C

calcitonin (kal''si-to'nin) hormone secreted by the thyroid gland that helps to regulate the level of blood calcium 263

capillaries (kap'ĭ-lar''es) microscopic vessels located in the tissues connecting arterioles to venules through the thin walls of which molecules either exit or enter the blood 94

carbohydrate (kar''bo-hi'drāt) organic compounds with the general formula $(CH_2O)_n$ including sugars and glycogen 20

carcinogen (kar-sin'o-jen) an environmental agent that either initiates or promotes the development of cancer 405

carcinoma (kar''si-no'mah) a cancer of the epithelial tissue such as that lining the gastrointestinal or respiratory tracts 406

cardiac muscle (kar'de-ak mus'el) heart muscle (myocardium) consisting of striated muscle cells that interlock 56

carnivore (kar'ni-vōr) an animal that feeds only on animals 436

carpal (kar'pal) a bone that is located in the human wrist 219

carrier (kar'e-er) a molecule that combines with a substance and actively transports it through the cell membrane; an individual that transmits an infectious or genetic disease 34, 364

carrying capacity (kar'e-ing kah-pas'ĭ-te) the largest number of organisms of a particular species that can be maintained indefinitely in an ecosystem 454

cartilage (kar'ti-lij) a connective tissue, usually part of the skeleton, that is composed of cells in a flexible matrix 54

cell body (sel bod'e) portion of a nerve cell that includes a cytoplasmic mass and a nucleus and from which the nerve fibers extend 189

cell cycle (sel si'kl)   a cyclical series of phases that includes cellular events before, during, and after mitosis *342*

cell membrane (sel mem'brān)   a membrane that surrounds the cytoplasm of cells and regulates the passage of molecules into and out of the cell *33*

central canal (sen'tral kah-nal')   tube within the spinal cord that is continuous with the ventricles of the brain and contains cerebrospinal fluid *201*

central nervous system (sen'tral ner'vus sis'tem)   CNS; the brain and spinal cord *201*

centriole (sen'tre-ōl)   a short, cylindrical organelle that contains microtubules in a 9 + 0 pattern and is associated with the formation of the spindle during cell division *39*

centromere (sen'tro-mēr)   a region of attachment for a chromosome to a spindle fiber that is generally seen as a constricted area *338*

cerebellum (ser''ē-bel'um)   the part of the brain that controls muscular coordination *203*

cerebral hemisphere (ser'ē-bral hem'i-sfēr)   one of the large, paired structures that together constitute the cerebrum of the brain *203*

cerebrospinal fluid (ser''ē-bro-spi'nal floo'id)   fluid present in ventricles of brain and in central canal of spinal cord *201*

cerebrum (ser'ē-brum)   the main portion of the brain that is responsible for consciousness *203*

chemotherapy (ke''mo-ther'ah-pe)   the use of a drug to selectively kill off cancer cells as opposed to normal cells *409*

chimpanzee (chim-pan'ze)   a small ape that is closely related to humans and is frequently used in psychological studies *427*

chlamydia (klah-mid'e-ah)   an organism that causes a sexually transmitted disease particularly characterized by urethritis *306*

chorion (ko're-on)   an extraembryonic membrane that forms an outer covering around the embryo and contributes to the formation of the placenta *312*

chorionic villi (ko''re-on'ik vil'i)   treelike extensions of the chorion of the embryo, projecting into the maternal tissues *318*

choroid (ko'roid)   the vascular, pigmented middle layer of the wall of the eye *240*

chromatids (kro'mah-tidz)   the two identical parts of a chromosome following replication of DNA *338*

chromatin (kro'mah-tin)   threadlike network in the nucleus that is made up of DNA and proteins *35*

chromosomes (kro'mo-sōmz)   rod-shaped bodies in the nucleus, particularly during cell division, that contain the hereditary units or genes *35*

cilia (sil'e-ah)   hairlike projections that are used for locomotion by many unicellular organisms and that have various purposes in higher organisms *41*

ciliary muscle (sil'e-er''e mus'el)   a muscle that controls the curvature of the lens of the eye *240*

circumcision (ser''kum-sizh'un)   removal of the foreskin of the penis *278*

class (klas)   in taxonomy, the category below phylum and above order *416*

clavicle (klav'i-k'l)   a slender, rodlike bone located at the base of the neck that runs between the sternum and the shoulders *218*

clone (klōn)   sexually produced organisms having the same genetic makeup; also DNA fragments from an external source that have been reproduced by *E. coli* *390*

clotting (klot'ing)   agglutination of red blood cells in which they clump together due to an antigen-antibody reaction *121*

coacervate (ko-as'er-vāt)   a mixture of polymers that may have preceded the origination of the first cell or cells *424*

cochlea (kok'le-ah)   that portion of the inner ear that resembles a snail's shell and contains the organ of Corti, the sense organ for hearing *248*

cochlear canal (kok'le-ar kah-nal')   canal within the cochlea bearing small hair cells that function as hearing receptors *248*

codon (ko'don)   a "triplet" of three nucleotides in messenger RNA that directs the placement of a particular amino acid into a polypeptide chain *381*

coenzyme (ko-en'zīm)   a nonprotein molecule that aids the action of an enzyme, to which it is loosely bound *41*

collecting duct (kŏ-lekt'ing dukt)   a tube that receives urine from several distal convoluted tubules *176*

colon (ko'lon)   the section of large intestine between the cecum and the rectum *75*

columnar epithelium (ko-lum'nar ep''i-the-le-um)   pillar-shaped cells usually having the nuclei near the bottom of each cell and found lining the digestive tract, for example *49*

common ancestor (kom'un an'ses-ter)   a predecessor in evolution that leads to two or more lines of descent *416*

compact bone (kom-pakt' bōn)   hard bone consisting of Haversian systems cemented together *55, 216*

complement system (kom'plē-ment sis'tem)   a group of proteins in plasma that are involved in both general and specific immune effects *135*

complementary base pairing (kom''plē men''tă-re bās pār'ing)   pairing of bases found in DNA and RNA; adenine is always paired with either thymine (DNA) or uracil (RNA) and cytosine is always paired with guanine *378*

compound (kom'pownd)   Two or more atoms of different elements that are chemically combined *14*

cones (kōnz)   bright-light receptors in the retina of the eye that detect color and provide visual acuity *241*

connective tissue (kŏ-nek'tiv tish'u)   a type of tissue, characterized by cells separated by a matrix, that often contains fibers *54*

consumers (kon-su'merz)   organisms of one population that feed on members of other populations in an ecosystem *436*

coronary artery (kor'ŏ-na-re ar'ter-e)   an artery that supplies blood to the wall of the heart *101*

corpus callosum (kor'pus kah-lo'sum)   a mass of white matter within the brain, composed of nerve fibers connecting the right and left cerebral hemispheres *205*

corpus luteum (kor'pus lut'e-um)   a body, yellow in color, that forms in the ovary from a follicle that has discharged its egg *283*

cortex (kor'teks)   the outer layer of an organ such as the cortex of the cerebrum, adrenal gland, or kidney *174*

cortisol (kor'ti-sol)   a glucocorticoid secreted by the adrenal cortex *264*

covalent bond (ko-va'lent bond)   chemical bond created by the sharing of electrons between atoms *15*

Cowper's glands (kow'perz'glandz)   two small structures located below the prostrate gland in males *278*

cranial nerve (kra'ne-al nerv)   nerve that arises from the brain *197*

creatine phosphate (kre'ah-tin fos'fāt)   a compound unique to muscles that contains a high-energy phosphate bond *171, 230*

creatinine (kre-at'ĭ-nin)   excretion product from creatine phosphate breakdown *171*

cretinism (kre'tin-izm)   a condition resulting from a lack of thyroid hormone in an infant *262*

cri du chat (kre-du-shah)   condition created by a chromosome number-5 deletion, characterized by an infant cry sounding like a cat's meow and having various physical abnormalities *352*

Cro-Magnon (kro-mag'non)   the common name for the first fossils to be accepted as representative of modern humans *431*

crossing over (kros'ing o'ver)   the exchange of corresponding segments of genetic material between nonsister chromatids of homologous chromosomes during meiosis *344*

cuboidal epithelium (ku-boi'dal ep''i-the'le-um)   cube-shaped cells found lining, for example, the kidney tubules *49*

cultural eutrophication (kul'tu-ral u''tro-fi-ka'shun)   enrichment of a body of water causing excessive growth of producers and then death of these and other inhabitants *461*

cystic fibrosis (sis'tik fi-bro'sis)   a lethal genetic disease involving problems with the functions of the mucous membranes in the respiratory and digestive tracts *365*

cytokinesis (si''to-ki-ne'sis)   division of the cytoplasm of a cell *342*

cytoplasm (si'to-plazm)   the ground substance of cells located between the nucleus and the cell membrane *33*

cytoskeleton (si'to-skel'ĕ-ton)   filamentous protein structures found throughout the cytoplasm that help maintain the shape of the cell *39*

cytotoxic T cell (si''to-tok'sik te sel)   T lymphocyte that attacks cells bearing foreign antigens *138*

D

deamination (de-am''i-na'shun)   removal of an amino group (—NH₂) from an amino acid or other organic compound *77*

decomposer (de-kom-po'zerz)   organisms of decay (fungi and bacteria) in an ecosystem *436*

deletion (de-le'shun)   a chromosome mutation that results in loss of a portion of the chromosome *349*

demographic transition (dem-o-graf'ik tran-zi'shun)   the change from a high birthrate to a low birthrate so that the growth rate is lowered *455*

dendrite (den'drīt)   process of a neuron, typically branched, that conducts nerve impulses toward the cell body *189*

denitrification (de-ni''tri-fi-ka'shun)   the process of converting nitrate to nitrogen gas; is a part of the nitrogen cycle *444*

dermis (der'mis)   the thick skin layer that lies beneath the epidermis *59*

detritus (de-tri'tus)   nonliving organic matter *436*

diabetes insipidus (di''ah-be'tēz in-sip'ĭ-dus)   condition characterized by an abnormally large production of urine due to a deficiency of antidiuretic hormone *258*

diabetes mellitus (di''ah-be'tez mĕ-li'tus)   condition characterized by a high blood glucose level and the appearance of glucose in the urine due to a deficiency of insulin *266*

diaphragm (di'ah-fram)   a sheet of muscle that separates the chest cavity from the abdominal cavity. Also, a birth control device inserted in front of the cervix in females *155*

diastole (di-as'to-le)   relaxation of heart chambers *98*

diastolic blood pressure (di-a-stol'ik blud presh'ur)   arterial blood pressure during the diastolic phase of the cardiac cycle *104*

differentiation (dif'er-en''she-a'shun)   the process and developmental stages by which a cell becomes specialized for a particular function *312*

diffusion (di-fu'zhun)   the movement of molecules from an area of greater concentration to an area of lesser concentration *34*

digit (dij'it)   a finger or toe *219*

dihybrid (di-hi'brid)   the offspring of parents who differ in two ways; shows the phenotype governed by the dominant alleles, but carries the recessive alleles *362*

diploid (dip'loid)   the 2N number of chromosomes; twice the number of chromosomes found in gametes *338*

disaccharide (di-sak'ah-rīd)   a sugar such as maltose that contains two units of monosaccharide *21*

distal convoluted tubule (dis'tal kon'vo-lūt-ed too'būl)   highly coiled region of a nephron that is distant from Bowman's capsule because it occurs after the loop of Henle *176*

DNA (deoxyribonucleic acid)   a nucleic acid, found especially in the nucleus and containing a triplet genetic code *24, 378*

DNA ligase (de'en-ā lig'ās)   a bacterial enzyme that seals breaks in the DNA molecule after cleavage, for example, by a restriction enzyme *390*

DNA probe (de'en-ā prōb)   a single strand of DNA used to locate a complementary DNA strand in a cell or body fluid by its binding activity *390*

dominant allele (dom'i-nant ah-lēl')   hereditary factor that expresses itself even when there is only one copy in the genotype *357*

dorsal root ganglion (dor'sal rōot gang'gle-on)   mass of sensory neuron cell bodies located in the dorsal root of a spinal nerve *198*

double helix (dŭ'b'l he'liks)   a double spiral often used to describe the three-dimensional shape of DNA *26*

doubling time (dŭ'b'ling tīm)   the number of years it takes for a population to double in size *453*

Down syndrome (down sin'drōm)   human congenital disorder associated with an extra number-23 chromosome *351*

Dryopithecus (dri''o-pith'e-kus)   extinct genus of apes that may have included or resembled a common ancestor to both apes and humans *427*

duodenum (du''o-de'num)   the first portion of the small intestine into which ducts from the gallbladder and pancreas enter *74*

duplication (dü pli' kā shən)   a chromosome mutation in which a chromosome contains two groups of identical genes; replication of DNA *349*

dyad (di'ad)   a chromosome having two chromatids held together at a centromere *343*

E

ecological pyramid (e''ko-log'i-kal pir'ah-mid)   a pictorial representation of the trophic structure of an ecosystem with producers at the base and the highest level consumers at the top *440*

ecology (e-kol'o-je)   the study of the relationship of organisms between themselves and the physical environment *436*

ecosystem (ek''o-sis'tem)   a biological community together with the associated abiotic environment *436*

ectopic pregnancy (ek-top'ik preg'nan-se)   implantation and development of a fertilized egg outside the uterus *315*

edema (ĕ-de'mah)   swelling due to tissue fluid accumulation in the intercellular spaces *132*

effector (ef-fek'tor)   a structure that allows a response to environmental stimuli such as the muscles and glands *189*

elastic cartilage (e-las'tik kar'ti-lij)   cartilage composed of elastic fibers that allows for greater flexibility *55*

electrocardiogram (e-lek''tro-kar'de-o-gram'')   ECG or EKG; a recording of the electrical activity associated with the heartbeat *100*

electroencephalogram (e-lek''tro-en-sef'ah-lo-gram'')   EEG; a graphic recording of the brain's electrical activity *205*

electron (e-lek'tron)   a subatomic particle that has almost no weight and carries a negative charge; travels in an orbital, called a shell, about the nucleus of an atom *13*

element (el'ĕ-ment)   the simplest of substances consisting of only one type of atom; i.e., carbon, hydrogen, oxygen *13*

embryo (em'bre-o)   the developing organism, particularly during the early stages *312*

endocrine gland (en'do-krin gland)   a gland that secretes hormones directly into the blood or body fluids *49, 255*

endometrium (en-do-me'tre-um)   the lining of the uterus that becomes thickened and vascular during the uterine cycle *283*

endoplasmic reticulum (ER) (en-do-plaz'mic re-tik'u-lum)   a complex system of tubules, vesicles, and sacs in cells; sometimes having attached ribosomes. *36*

endospore (en'do-spōr)   a resistant body formed by bacteria when environmental conditions worsen *304*

environmental resistance (en-vi''ron-men'tal re-zis'tans)   sum total of factors in the environment that limit the numerical increase of a population in a particular region *454*

enzyme (en'zīm)   a protein catalyst that speeds up a specific reaction or a specific type of reaction *19*

epidermis (ep''i-der'mis)   the outer skin layer composed of stratified squamous epithelium *58*

epididymis (ep''i-did'ĭ mis)   coiled tubules next to the testes where sperm mature and may be stored for a short time *278*

epiglottis (ep''i-glot'is)   a structure that covers the glottis during the process of swallowing *72*

epithelial tissue (ep''i-the'le-al tish'u)   a type of tissue that lines cavities and covers the external surface of the body *49*

erection (ě-rek′shun)   referring to the penis when it is turgid and erect as opposed to flaccid and lacking turgidity *278*

erythrocyte (ě-rith′ro-sīt)   non-nucleated, hemoglobin containing blood cells capable of carrying oxygen; the red blood cell *55*

esophagus (ě-sof′ah-gus)   a tube that transports food from the mouth to the stomach *72*

essential amino acid (e-sen′shal ah-me′no as′id)   amino acid required for health that cannot be synthesized by body cells *79*

eustachian tube (u-sta′ke-an tūb)   an air tube that connects the pharynx to the middle ear *248*

evolutionary tree (ev″o-lu′shun-ar-e tre)   a diagram describing the phylogenetic relationship of groups of organisms *416*

exocrine gland (ek′so-krin gland)   secreting externally; particular glands with ducts whose secretions are deposited into cavities, such as salivary glands *49*

exophthalmic goiter (ek″sof-thal′mik goi′ter)   an enlargement of the thyroid gland accompanied by an abnormal protrusion of the eyes *263*

expiration (eks″pi-ra′shun)   process of expelling air from the lungs; exhalation *150*

exponential growth (eks″po-nen′shal grōth)   growth, particularly of a population, in which the total number increases in the same manner as compound interest *452*

external respiration (eks-ter′nal res″pi-ra′shun)   exchange of oxygen and carbon dioxide between air and blood in the alveoli of the lungs *150*

extraembyronic membranes (eks″trah-em″bre-on′ik mem′brānz)   membranes that are not a part of the embryo but are necessary to the continued existence and health of the embryo *312*

F

facilitated transport (fah-sil′ĭ-tāt-ed trans′port)   passive transfer of a substance into or out of a cell along a concentration gradient by a process that requires a carrier *35*

family (fam′i-le)   a rank in taxonomic classification above genus and below order *416*

fatigue (fah-tēg′)   muscle relaxation in the presence of stimulation due to energy reserve depletion *225*

fatty acid (fat′e as′id)   an organic molecule having a long chain of carbon atoms and ending in an acidic group *22*

femur (fe′mur)   the thighbone found in the upper leg *219*

fermentation (fer″men-ta′shun)   anaerobic breakdown of carbohydrates that results in end products such as lactic acid *43*

fetus (fe′tus)   human development in its later stages following the embryonic stages *312*

fibrin (fi′brin)   insoluble, fibrous protein formed from fibrinogen during blood clotting *122*

fibrinogen (fi-brin′o-jen)   plasma protein that is converted into fibrin threads during blood coagulation *122*

fibrocartilage (fi″bro-kar′tĭ-lij)   cartilage with a matrix of strong collagenous fibers *55*

fibrous connective tissue (fi′brus ko-nek′tiv tish′u)   tissue composed mainly of closely packed collagenous fibers and found in tendons and ligaments *54*

fibula (fib′u-lah)   a long slender bone located on the lateral side of the tibia *220*

fimbriae (fim′bre-e″)   fingerlike extensions from the oviduct near the ovary *283*

flagella (flah-jel′ah)   slender, long processes used for locomotion by, for example, sperm *41*

focusing (fo′kus-ing)   manner by which light rays are bent by the cornea and lens, creating an image on the retina *242*

follicle (fol′i-kl)   a structure in the ovary that produces the egg and particularly the female sex hormone, estrogen *282*

follicle-stimulating hormone (fol′i-kl stim′u-la″ting hor′mon)   see FSH

food chain (food chān)   a series of organisms in a feeding chain, starting with a producer and ranging through succeeding levels of consumers *438*

food web (food web)   the complete set of food links between populations in a community *438*

foramen magnum (fo-ra′men mag′num)   opening in the occipital bone of the skull through which the spinal cord passes *217*

foreskin (for′skin)   skin covering the glans penis in uncircumcised males *278*

formed element (form′d el′ě-ment)   those elements that make up the solid portion of blood; they are either cells (erythrocytes and leukocytes) or derived from a cell (platelets). *114*

fossils (fos′lz)   any remains of an organism that have been preserved in the earth's crust *415*

fovea centralis (fo′ve-ah sen-tra′lis)   region of the retina consisting of densely packed cones, that is responsible for the greatest visual acuity *241*

frontal lobe (frun′tal lōb)   area of the cerebrum responsible for voluntary movements and higher intellectual processes *203*

FSH (follicle-stimulating hormone)   hormone secreted by the anterior pituitary gland that stimulates the development of an ovarian follicle in a female or the production of sperm cells in a male *280*

fungus (fung′ gus)   an organism, usually composed of strands called hyphae, that lives chiefly on decaying matter; e.g., mushroom and mold *309*

G

gallbladder (gawl′blad-er)   saclike organ associated with the liver that stores and concentrates bile *76*

gallstones (gawl′stonz)   precipitated crystals of cholesterol or calcium carbonate formed from bile within the gallbladder or bile duct *77*

gamete (gam′ēt)   a reproductive cell that joins with another in fertilization to form a zygote; egg or sperm *338*

ganglion (gang′gle-on)   a collection of neuron cell bodies outside the central nervous system *196*

gastric gland (gas′trik gland)   gland within the stomach wall that secretes gastric juice *73*

gene therapy (jēn ther′ah-pe)   the use of biotechnology to treat genetic disorders and illnesses *396*

genetic mutation (jě-net′ik mu-ta′shun)   a permanent change in the DNA code such that there is an observed change in the phenotype of the individual *387*

genital herpes (jen′i-tal her′pēz)   a sexually transmitted disease characterized by open sores on the external genitalia *301*

genotype (je′no-tīp)   the genetic makeup of any individual *357*

genus (je′nus)   a rank in taxonomic classification above species and below family *416*

germ layers (jerm la′ers)   primary tissues of an embryo (ectoderm, mesoderm, endoderm) that give rise to the major tissue systems of the adult animal *315*

gerontology (jer″on-tol′o-je)   the study of older individuals, including the biological degenerative changes associated with the elderly *329*

gibbon (gib′on)   the smallest ape; well known for its arm-swinging form of locomotion *427*

glial cells (gli′al selz)   cells that support, protect, and nourish neurons within the brain and spinal cord *57*

globin (glo′bin)   the protein portion of a hemoglobin molecule *116*

glomerular filtrate (glo-mer′u-lar fil′trāt)   the filtered portion of blood that is contained within Bowman's capsule *177*

glomerulus (glo-mer′u-lus)   a cluster; for example, the cluster of capillaries surrounded by Bowman's capsule in a kidney tubule *176*

glottis (glot′is)   slitlike opening in the larynx between the vocal cords *72*

glucagon (gloo′kah-gon)   hormone secreted by the pancreatic islets of Langerhans that causes the release of glucose from glycogen *266*

glucose (gloo′kōs)   the most common six-carbon sugar *20*

glycerol (glis′er-ol)   an organic compound that serves as a building block for fat molecules *22*

glycogen (gli'ko-jen) a polysaccharide that is the storage compound for sugar in the liver *21*

Golgi apparatus (gol'ge ap''ah-ra'tus) an organelle that consists of a stack of saccules and functions in the packaging and secretion of cellular products *38*

gonad (go'nad) an organ that produces sex cells; the ovary, which produces eggs, and the testis, which produces sperm *338*

gonadotropic hormone (go-nad''o-trop'ik hor'mōn) a type of hormone that regulates the activity of the ovaries and testes; principally FSH and LH (ICSH) *261*

gonorrhea (gon''o-re'ah) contagious sexually transmitted disease caused by bacteria and leading to inflammation of the urogenital tract *305*

gorilla (go-ril'ah) the largest of the great apes, which are as closely related to humans as to other apes *427*

Graafian follicle (graf'e-an fol'li-k'l) mature follicle within the ovaries that houses a developing egg *282*

granular leukocytes (gran'u-lar loo'ko-sitz) white blood cells that contain distinctive granules *124*

greenhouse effect (grēn'hows ē-fekt') buildup in the atmosphere of carbon dioxide and other gases that retain and reradiate heat causing global warming *465*

growth hormone (grōth hor'mōn) a hormone released by the anterior lobe of the pituitary gland that promotes the growth of the organism; GH or somatotropin *259*

growth rate (grōth rāt) percentage of increase or decrease in the size of a population *453*

gummas (gum'ahz) large unpleasant sores that sometimes occur during the tertiary stage of syphilis *307*

H

habitat (hab'i-tat) the natural abode of an animal or plant species *436*

hammer (ham'er) also called the malleus; the first of the auditory ossicles that amplify the sound waves in the middle ear *247*

haploid (hap'loid) the N number of chromosomes; half the diploid number; the number characteristic of gametes that contain only one set of chromosomes *338*

hard palate (hard pal'at) anterior portion of the roof of the mouth that contains several bones *71*

hazardous wastes (haz'ard-us wāsts) wastes containing chemicals dangerous to life *459*

HCG (human chorionic gonadotropic) a gonadotropic hormone produced by the chorion that functions to maintain the uterine lining *286*

heart (hart) muscular organ located in thoracic cavity responsible for maintenance of blood circulation *96*

helper T cells (help'er T selz) T lymphocytes that stimulate certain other T and B lymphocytes to perform their respective functions; T$_H$ cells *138*

heme (hēm) the iron-containing portion of a hemoglobin molecule *116*

hemoglobin (he'mo-glo''bin) a red, iron-containing pigment in blood that combines with and transports oxygen *116*

hepatic portal vein (hě-pat'ik por'tal vān) vein leading to the liver formed by the merging of blood vessels from the villi of the small intestine *76*

herbivore (her'bi vōr) an animal that feeds directly on plants *436*

herpes simplex virus (her'pēz sim'plex vi'rus) a virus of which type I causes cold sores and type II causes genital herpes *301*

heterozygous (het''er-o-zi'gus) having two different alleles (as *Aa*) for a given trait *357*

histamine (his'tah-min) substance produced by basophil-derived mast cells in connective tissue that causes capillaries to dilate and release immune and other substances *126*

HIV viruses responsible for AIDS; human immunodeficiency virus *299*

homeostasis (ho''me-o-sta'sis) the constancy of conditions, particularly the environment of the body's cells; constant temperature, blood pressure, pH, and other body conditions *62*

hominids (hom'i-nidz) members of a family containing upright, bipedal primates including modern humans *428*

hominoids (hom'i-noidz) members of a superfamily containing humans and the great apes *427*

Homo erectus (ho'mo ē-rek'tus) the earliest nondisputed species of humans, named for their erect posture that allowed them to have a striding gait *430*

Homo habilis (ho'mo hah'bi-lis) an extinct species that may include the earliest humans, having a small brain but quality tools *429*

homologous (ho-mol'o-gus) similarly constructed; homologous chromosomes have the same shape and contain genes for the same traits; homologous structures in animals share a common ancestry *337, 418*

homozygous (ho''mo-zi'gus) having identical alleles (as *AA* or *aa*) for a given trait; pure breeding *357*

hormone (hor'mōn) a chemical messenger produced in small amounts in one body region that is transported to another body region *255*

host (hōst) an organism on or in which another organism lives *298*

human chorionic gonadotropic hormone (hu'man ko''re-on'ik go-nad''o-trop'ik hor'mōn) *see* HCG

humerus (hu'mer-us) a heavy bone that extends from the scapula to the elbow *219*

Huntington disease (hunt'ing-tun dĭ zēz) a fatal genetic disease marked by neurological disturbances and failure of brain regions *366*

hyaline cartilage (hi'ah-līn kar'tĭ-lij) cartilage composed of very fine collagenous fibers and a matrix of a clear, milk glass appearance *55*

hydrogen bond (hi'dro-jen bond) a weak attraction between a hydrogen atom carrying a partial positive charge and an atom of another molecule carrying a partial negative charge *17*

hydrolysis (hi-drol'i-sis) the splitting of a bond within a larger molecule by addition of water *19*

hydrolytic enzyme (hi-dro-lit'ik en'zīm) describes an enzyme in which the substrate is broken down by the addition of water *78*

hypertonic solution (hi''per-ton'ik so-lu'shun) one that has a greater concentration of solute, a lesser concentration of water than the cell *35*

hypothalamus (hi''po-thal'ah-mus) a region of the brain; the floor of the third ventricle that helps maintain homeostasis *203*

hypotonic solution (hi''po-ton'ik so-lu'shun) one that has a lesser concentration of solute, a greater concentration of water than the cell *35*

I

immunotherapy (i-mu''no ther'ah-pe) the use of any immune system component such as antibodies, cytotoxic T cells, or interleukins to promote the health of the body, such as curing cancer *409*

implantation (im''plan-ta'shun) the attachment and penetration of the embryo to the lining (endometrium) of the uterus *286*

impotency (im'po-ten''se) failure of the penis to achieve erection *280*

induction (in-duk'shun) in development, the ability of one body part to influence the development of another part *317*

inflammatory reaction (in-flam'ah-to''re re-ak'shun) a tissue response to injury that is characterized by dilation of blood vessels and an accumulation of fluid in the affected region *126*

inner cell mass (in'er sel mas) the portion of a blastocyst that will develop into the embryo and fetus *312*

innervate (in'er-vāt) to activate an organ, muscle, or gland by motor neuron stimulation *189*

innominate (i-nom'ĭ-nāt) one of two hipbones that form the pelvis *219*

insertion (in-ser'shun) the end of a muscle that is attached to a movable part *222*

inspiration (in''spi-ra'shun) the act of breathing in; inhalation *150*

insulin (in′su-lin) a hormone produced by the pancreas that regulates carbohydrate storage 266

interferon (in″ter-fēr′on) a protein formed by a cell infected with a virus that can increase the resistance of other cells to the virus 135

internal respiration (in-ter′nal res″pi-ra′shun) exchange between blood and tissue fluid 150

interneuron (in″ter-nu′ron) a neuron that is found within the central nervous system and takes nerve impulses from one portion of the system to another 189

interstitial cells (in″ter-stish′al selz) hormone-secreting cells located between the seminiferous tubules of the testes 278

inversion (in-ver′zhun) a chromosome mutation caused by a 180 degree turn-around by the fragment of a chromosome 349

in vitro fertilization IVF (in ve′tro fer′tĭ-lĭ-za′shun) union of egg and sperm in laboratory glassware. The resulting embryo may be introduced into a prepared uterus where it develops naturally 293

ion (i′on) an atom or group of atoms carrying a positive or negative charge 15

ionic bond (i-on′ik bond) a chemical attraction between a positive and negative ion 15

iris (ī′ris) a muscular ring that surrounds the pupil and regulates the passage of light through this opening 240

islets of Langerhans (i′lets uv lahng′er-hanz) distinctive groups of cells within the pancreas that secrete insulin and glucagon 266

isotonic solution (i′so-ton′ik so-lu′shun) one that contains the same concentration of solutes and water as does the cell 35

isotope (i′so-tōp) atoms with the same number of protons and electrons but differing in the number of neutrons and, therefore, in weight 14

K

karyotype (kar′e-o-tīp) the arrangement of all the chromosomes by homologous pairs in a fixed order 337

kidneys (kid′nēz) organs in the urinary system that concentrate and excrete urine 173

kingdom (king′dum) the largest taxonomic category into which organisms are placed: Monera, Protista, Fungi, Plants, and Animals 416

Klinefelter syndrome (klīn′fel-ter sin′drōm) a condition caused by the inheritance of XXY chromosomes 353

Krebs cycle (krebz si′kl) a series of reactions found within mitochondria that give off carbon dioxide. Also called the citric acid cycle because the reactions begin and end with citric acid 42

L

labia (la′be-ah) fleshy borders or liplike folds of skin, as in the labia majora and labia minora of the female genitalia 283–284

lacteal (lak′te-al) a lymph vessel in a villus of the intestinal wall of mammals 74

lacuna (lah-ku′nah) a small pit or hollow cavity, as in bone or cartilage, where a cell or cells are located 54

lanugo (lah-nu′go) short, fine hair that is present during the later portion of fetal development 324

large intestine (′lärj in-tes′tin) the last major portion of the gut consisting of the cecum, colon, rectum, and anal canal 75

larynx (lar′ingks) structure that contains the vocal cords; voice box 151

lens (lenz) a clear, membranelike structure found in the eye behind the iris. The lens brings objects into focus 240

leukemia (loo-ke′me-ah) cancer of the blood-forming tissues leading to the overproduction of abnormal white blood cells 406

leukocyte (lu′ko-sīt) refers to several types of colorless, nucleated blood cells that, among other functions, resist infection; white blood cells 55

LH (luteinizing hormone) hormone produced by the anterior pituitary gland that stimulates the development of the corpus luteum in females and the production of testosterone in males 280

ligament (lig′ah-ment) a strong connective tissue that joins bone to bone 54, 220

limbic system (lim′bik sis′tem) an area of the forebrain implicated in visceral functioning and emotional responses; involves many different centers of the brain 206

lipase (li′pās) an enzyme secreted by the pancreas that digests or breaks down fats 78

lipid (lip′id) a group of organic compounds that are insoluble in water; notably fats, oils, and steroids 22

liver (liv′er) a large organ in the abdominal cavity that has many functions vital to continued existence such as production of blood proteins and detoxification of harmful substances 76

loop of Henle (lōōp uv hen′le) that part of a nephron that lies between the proximal convoluted tubule and the distal convoluted tubule; consists of a descending portion and ascending portion; concerned with reabsorption of water 176

loose connective tissue (lōōs kŏ-nek′tive tish′u) tissue composed mainly of fibroblasts that are separated by collagen and elastin fibers and found beneath epithelium 54

lumen (lu′men) the cavity inside any tubular structure, such as the lumen of the gut 72

luteinizing hormone (lu′te-in-īz″ing hor′mon) see LH

lymph (limf) fluid having the same composition as tissue fluid and carried in lymph vessels 132

lymphatic system (lim-fat′ik sis′tem) vascular system that takes up excess tissue fluid and transports it to the bloodstream 132

lymphokine (lim′fo-kīn) chemicals, secreted by T cells, that stimulate immune cells 140

lymphoma (lim-fo′mah) cancer of the lymphoid organs such as lymph nodes, spleen, and thymus gland 406

lysosome (li′so-sōm) an organelle in which digestion takes place due to the action of powerful hydrolytic enzymes 38

M

macromolecule (mak′ro mäl i kyül) a large molecule that is formed by the joining together of unit molecules; also called a polymer; for example, protein, fat, carbohydrate, and nucleic acid 19

macrophage (mak′ro-fāj) an enlarged monocyte that ingests foreign material and cellular debris 126

major histocompatibility protein (ma′jor his″to-kom-pat″ĭ-bil′ĭ-te pro′te-in) see MHC protein

matrix (ma′triks) the secreted basic material or medium of biological structures, such as the matrix of cartilage or bone 54

medulla (mĕ-dul′ah) the inner portion of an organ; for example, the adrenal medulla 174

medulla oblongata (mĕ-dul′ah ob″long-gah′tah) the lowest portion of the brain that is concerned with the control of internal organs 202

meiosis (mi-o′sis) type of cell division that occurs during the production of gametes by means of which four daughter cells receive the haploid number of chromosomes 343

meiosis I (mi-o′sis wun) that portion of meiosis during which homologous chromosomes come together and then later separate 343

meiosis II (mi-o′sis tōō) that portion of meiosis during which sister chromatids separate, resulting in four haploid daughter cells 343

memory B cells (mem′o-re bee selz) cells derived from B and T lymphocytes that remain within the body for some time and account for the presence of long-lasting active immunity 137

memory T cells (mem′o-re te sels) persistent population of cells capable of secreting lymphokines to stimulate macrophages and B cells 140

meninges (me-nin′jez) protective membranous coverings about the central nervous system 201

meniscus (mĕ-nis′kus)   a piece of fibrocartilage that separates the surfaces of bones in the knee   *220*

menopause (men′o-pawz)   termination of the uterine cycle in older women   *287*

menstruation (men″stroo-a′shun)   loss of blood and tissue from the uterus at the end of a female uterine cycle   *285*

messenger RNA (mes″n-jer)   mRNA; a nucleic acid complementary to genetic DNA and bearing codons to direct cell protein synthesis at the ribosome   *380*

metabolism (mĕ-tab′o-lizm)   all of the chemical changes that occur within cells considered together   *41*

metacarpal (met″ah-kar′pal)   a bone found in the palm of the hand   *219*

metafemale (met′ah-fe′māl)   a female who has three X chromosomes   *353*

metaphase (met′ah-fāz)   stage in mitosis during which chromosomes are at the equator of the spindle   *342*

metaphase I (met″ah-fāz wun)   phase of meiosis I during which tetrads are at the equator of the spindle   *344*

metastasis (mĕ-tas′tah-sis)   the spread of cancer from the place of origin throughout the body caused by the ability of cancer cells to migrate and invade tissues   *403*

metatarsal (met″ah-tar′sal)   a bone found in the foot between the ankle and the toes   *220*

MHC protein (major histocompatibility protein);   a surface molecule that serves as an identifying marker   *140*

microfilament (mi″kro-fil′ah-ment)   an extremely thin fiber found within the cytoplasm that is involved in the maintenance of cell shape and movement of cell contents   *39*

microtubule (mi″kro-tu′būl)   an organelle composed of thirteen rows of globular proteins; found in multiple units in several other cell organelles such as the centriole, cilia, and flagella   *39*

middle ear (mid′l ēr)   portion of the ear consisting of the tympanic membrane, the oval and round windows, and the ossicles, where sound is amplified   *247*

mineral (min′er-al)   an inorganic, homogeneous substance   *85*

mitochondrion (mi″to-kon′dre-on)   an organelle in which aerobic cellular respiration produces the energy molecule, ATP   *38*

mitosis (mi-to′sis)   cell division by means of which two daughter cells receive the exact chromosome and genetic makeup of the mother cell; occurs during growth and repair   *339*

mixed nerve (mikst nerv)   nerve that contains both the long dendrites of sensory neurons and the long axons of motor neurons   *197*

molecule (mol′ĕ-kūl)   smallest unit of a chemical compound that still has the properties of that compound; a molecule

can also form when two or more atoms of the same element react with one another by covalent bonding   *14*

monoclonal antibodies (mon″o-klōn′al an′ti-bod″ēz)   antibodies of one type that are produced by cells derived from a lymphocyte that has fused with a cancer cell   *144*

monoculture agriculture (mon′o-kul′tur ag′ri-kul″tūr)   planting of a single type of crop that is subject to destruction by a single type of pest or parasite   *445*

monohybrid (mon″o-hi′brid)   the offspring of parents who differ in one way only; shows the phenotype of the dominant allele, but carries the recessive allele   *358*

monosaccharide (mon″o-sak′ah-rīd)   a simple sugar; a carbohydrate that cannot be decomposed by hydrolysis   *20*

morphogenesis (mor″fo-jen′ĕ-sis)   the movement of cells and tissues to establish the shape and structure of an organism   *312*

morula (mor′u-lah)   an early stage in development in which the embryo consists of a mass of cells, often spherical   *312*

motor nerve (mo′tor nerv)   nerve containing only the long axons of motor neurons   *196*

motor neuron (mo′tor nu′ron)   a neuron that takes nerve impulses from the central nervous system to an effector   *189*

muscle action potential (mus′el ak′shun po-ten′shal)   an electrochemical change due to increased sarcolemma permeability that is propagated down the T system and results in muscle contraction   *231*

muscle spindle (mus′el spin′dul)   modified skeletal muscle fiber that can respond to changes in muscle length   *225*

muscular tissue (mus′ku-lar tish′u)   a type of tissue that contains cells capable of contracting; skeletal muscles are attached to the skeleton, smooth muscle is found within walls of internal organs, and cardiac muscle comprises the heart   *56*

mutation (mu-ta′shun)   a chromosomal or genetic change that is inherited either by daughter cells following mitosis or by an organism following reproduction   *349*

myelin sheath (mi′ĕ-lin shēth)   the fatty cell membranes that cover long neuron fibers and give them a white, glistening appearance   *190*

myocardium (mi″o-kar′de-um)   heart (cardiac) muscle consisting of striated muscle cells that interlock   *96*

myofibrils (mi″o-fi′brilz)   the contractile portions of muscle fibers   *228*

myogram (mi′o-gram)   a recording of a muscular contraction   *223*

myosin (mi′o-sin)   one of two major proteins of muscle; the thick filament in myofibrils made of protein and capable of breaking down ATP   *230*

myxedema (mik″sĕ-de′mah)   a condition resulting from a deficiency of thyroid hormone in an adult   *262*

N

natural selection (nat′u-ral sĕ-lek′shun)   the process by which better adapted organisms are favored to reproduce to a greater degree and pass on their genes to the next generation   *420*

NE   *see* norepinephrine

Neanderthal (ne-an′der-thawl)   the common name for an extinct subspecies of humans whose remains are found in Europe, Asia, and Africa   *431*

negative feedback (neg′ah-tiv fēd′bak)   a homeostatic control system that responds to an imbalance and is inhibited by that response   *63*

nephron (nef′ron)   the anatomical and functional unit of the vertebrate kidney; kidney tubule   *174*

nerve (nerv)   a bundle of long nerve fibers that run to and/or from the central nervous system   *57*

nerve impulse (nerv im′puls)   an electrochemical change due to increased neurolemma permeability that is propagated along a neuron from the dendrite to the axon following excitation   *190*

neurilemma (nu″ri-lem′ah)   a thin, membranous covering surrounding the myelin of a nerve fiber   *190*

neurofibromatosis (nu″ro-fi-brō-mah-tō′sis)   a genetic disease marked by development of neurofibromas under the skin and muscles   *366*

neuromuscular junction (no″ro-mus′ku-lar junk′shun)   the point of contact between a nerve cell and a muscle cell   *231*

neuron (nu′ron)   nerve cell that characteristically has three parts: dendrite, cell body, axon   *56*

neurotransmitter substance (nu″ro-trans mit′er sub′stans)   a chemical made at the ends of axons that is responsible for transmission across a synapse   *194*

neutron (nu′tron)   a subatomic particle that has a weight of one atomic mass unit, carries no charge, and is found in the nucleus of an atom   *13*

niche (nich)   the functional role and position of an organism in the ecosystem   *436*

nitrogen fixation (ni′tro-jen fik-sa′shun)   a process whereby free atmospheric nitrogen is converted into compounds, such as ammonia and nitrates, usually by soil bacteria   *443*

nondisjunction (non″dis-junk′shun)   the failure of homologous chromosomes or sister chromatids to separate during the formation of gametes   *351*

nongonococcal urethritis (non″gon-o-kok′al u″rĕ-thri′tis)   NGU; an infection of the urinary tract by an organism other than *N. gonorrheae*   *307*

norepinephrine (nor″ep′i-nef′ron)   NE; excitatory neurotransmitter active in the peripheral and central nervous systems   *194, 264*

notochord (no′to-kord) dorsal supporting rod that exists only during embryonic development and is replaced by the vertebral column *317*

nuclear envelope (nu′kle-ar en′vĕ-lōp) double-layered membrane enclosing the nucleus *35*

nucleic acid (nu-kle′ik as′id) a large organic molecule made up of nucleotides joined together; for example, DNA and RNA *24*

nucleolus (nu-kle′o-lus) an organelle found inside the nucleus; composed largely of RNA for ribosome formation; (*pl:* nucleoli) *36*

nucleotide (nu′kle-o-tīd) a molecule consisting of three subunits: phosphoric acid, a five-carbon sugar, and a nitrogenous base; a building block of a nucleic acid *24*

nucleus (nu′kle-us) a large organelle containing the chromosomes and acting as a control center for the cell; also the center of an atom *35*

O

obesity (o-bēs′i-te) an excessive accumulation of adipose tissue; usually the condition of exceeding the desirable weight by more than 20% *88*

occipital lobe (ok-sip′i-tal lōb) area of the cerebrum responsible for vision, visual images, and other sensory experiences *203*

olfactory receptor (ol-fak′to-re re-sep′tor) sense organ whose stimulation results in the sensation of smelling *239*

omnivore (om′ni-vor) an animal that feeds on both plants and animals *436*

oncogene (ong′ko-jēn) a cancer-causing gene that codes for a growth factor, or a growth factor receptor in the cell membrane, or a signaling protein between receptor and nucleus of the cell *404*

oogenesis (o′′o-jen′ĕ-sis) production of the egg in females by the process of meiosis and maturation *346*

optic nerve (op′tik nerv) nerve composed of the ganglion cell fibers that form the innermost layer of the retina *241*

orangutan (o′rang′oo-tan′′) one of the great apes; large with long red hair *427*

order (or′der) in taxonomy, the category below class and above family *416*

organ (or′gan) a structure composed of two or more tissues functioning as a unit *57*

organelle (or′′gah-nel′) specialized structures within cells, such as the nucleus, mitochondria, endoplasmic reticulum *35*

organ of Corti (or′gan uv kor′ti) the organ that contains the hearing receptors in the inner ear *248*

orgasm (or′gazm) physical and emotional climax during sexual intercourse; results in male ejaculation *280*

origin (or′i-jin) end of a muscle that is attached to a relatively immovable part *222*

osmosis (oz-mo′sis) the movement of water from an area of greater concentration of water to an area of lesser concentration of water across a semipermeable membrane *34*

osmotic pressure (oz-mot′ik presh′ur) pressure generated by the osmotic flow of water *35*

osteoblast (os′te-o-blast′′) a bone-forming cell *216*

osteoclast (os′te-o-klast′′) a cell that causes the erosion of bone *216*

osteocyte (os′te-o-sīt′′) a mature bone cell *217*

otoliths (o′to-liths) granules that lie above and whose movement stimulates ciliated cells in the utricle and saccule *248*

outer ear (out′er ēr) portion of the ear consisting of the pinna and the auditory canal *247*

oval opening (o′val o′pen-ing) foramen ovale; an opening between the 2 atria in the fetal heart *325*

oval window (o′val win′do) membrane-covered opening between the stapes and the inner ear *247*

ovarian cycle (o-va′re-an si′kl) monthly occuring changes in the ovary that affect the level of sex hormones in the blood *284*

ovaries (o′var-ez) female gonads, the organs that produce eggs and estrogen and progesterone *282*

ovulation (o′′vu-la′shun) the discharge of a mature egg from the follicle within the ovary *283*

oxidation (ok′′si-da′shun) the loss of electrons (inorganic) or the removal of hydrogen atoms (organic) *17*

oxygen debt (ok′si-jen det) the amount of oxygen needed to metabolize lactic acid that accumulates during vigorous exercise *230*

oxytocin (ok′′se-to′sin) hormone released by the posterior pituitary that causes contraction of the uterus and milk letdown *259*

ozone shield (o′zōn shēld) a layer of O_3 present in the upper atmosphere that protects the earth from damaging ultraviolet light. Nearer the earth, ozone is a pollutant *466*

P

pacemaker (pās′māk-er) a small region of neuromuscular tissue that initiates the heartbeat; also SA node *98*

pancreas (pan′kre-as) an elongated, flattened organ in the abdominal cavity that secretes digestive enzymes into the duodenum and produces hormones, notably insulin *76, 266*

pancreatic amylase (pan′′kre-at′ik am′i-lās) enzyme that digests starch to maltose *78*

Pap test (pap test) sampling of the cells from the cervix that is examined to determine if a woman has cancer of the cervix *283, 408*

parasite (par′ah-sīt) an organism that resides externally on or internally within another organism and does harm to this organism *298*

parasympathetic nervous system (par′′ah-sim′′pah-thet′ik ner′vus sis′tem) a portion of the autonomic nervous system that usually promotes those activities associated with a normal state *201*

parathyroid hormone (par′′ah-thi′roid hor′mōn) (PTH); a hormone secreted by the parathyroid glands that affect the level of calcium and phosphate in the blood *263*

parietal lobe (pah-ri′ĕ-tal lōb) area of the cerebrum responsible for sensations involving temperature, touch, pressure, and pain, as well as speech *203*

parturition (par′′tu-rish′un) passageway and delivery of a newborn organism through the terminal portion of the female reproductive tract *327*

pectoral girdle (pek′tor-al ger′dl) portion of the skeleton that provides support and attachment for the arms *218*

pedigree chart (ped′i-gre ′chärt) record of the inheritance pattern for a given trait in a particular family *363*

pelvic girdle (pel′vik ger′dl) portion of the skeleton to which the legs are attached *219*

pelvic inflammatory disease (pel′vik in-flam′ah-to′′re di-zēz′) PID; a disease state of the reproductive organs caused by an organism that is sexually transmitted *306*

pelvis (pel′vis) a bony ring formed by the innominate bones. Also a hollow chamber in the kidney that lies inside the medulla and receives freshly prepared urine from the collecting ducts *174*

penis (pe′nis) male copulatory organ *278*

pepsin (pep′sin) a protein-digesting enzyme secreted by gastric glands *78*

peptide bond (pep′tīd bond) the bond that joins two amino acids *20*

periodontitis (per′′e-o-don-ti′tis) inflammation of the gums *71*

peripheral nervous system (pĕ-rif′ er-al ner′vus sis′tem) PNS; nerves and ganglia that lie outside the central nervous system *196*

peristalsis (per′′i-stal′sis) a rhythmic contraction that serves to move the contents along in tubular organs such as the digestive tract *72*

peritubular capillary (per′′i-tu′bu-lar kap′i-lar′′e) capillary that surrounds a nephron and functions in reabsorption during urine formation *176*

peroxisome (pĕ-roks′i-sōm) an organelle involved in oxidation of molecules and other metabolic reactions *37*

pH (pe āch) a measure of the hydrogen ion concentration; any pH below 7 is acid and any pH above 7 is basic *17*

phagocytosis (fag′′o-si-to′sis) the taking in of bacteria and/or debris by engulfing *126*

phalanges (fah-lan'jēz)   bones of the finger and thumb 219

pharynx (far'ingks)   a common passageway (throat) for both food intake and air movement 71

phenotype (fe'no-tīp)   the outward appearance of an organism caused by the genotype and environmental influences 357

phenylketonuria (fen''il-ke''to-nu'rē-ah)   PKU; a genetic disease stemming from the lack of an enzyme to metabolize the amino acid phenylalanine 365

photochemical smog (fo''to-kem'i-kal smog)   a combination of nitrogen oxides and hydrocarbons reacting in sunlight to produce ozone and PAN 463

phylum (fi'lum)   a taxonomic category applied to animals that follows kingdom and lies above class 416

physiograph (fiz'e-o-graf)   instrument used to record a myogram 223

PID   see pelvic inflammatory disease

pineal gland (pin'ē-al gland)   a gland either at the skin surface (fish, amphibians) or in the third ventricle of the brain, producing melatonin 268

pinna (pin'nah)   outer, funnellike structure of the ear that picks up sound waves 247

pituitary gland (pi-too'i-tār''e gland)   anterior portion produces six types of hormones and is controlled by hypothalamic-releasing and release-inhibiting hormones; posterior portion is connnected by a stalk to the hypothalamus 256

placenta (plah-sen'tah)   a structure formed from the chorion and uterine tissue through which nutrient and waste exchange occur for the embryo and later the fetus 287

plasma cell (plaz'mah sel)   a cell derived from a B cell lymphocyte that is specialized to mass-produce antibodies 137

plasmid (plaz'mid)   a circular DNA segment that is present in bacterial cells but is not part of the bacterial chromosome 388

platelet (plāt'let)   cell-like disks formed from fragmentation of megakaryocytes that initiate blood clotting in the blood 122

pleural membranes (ploor'al mem'brānz)   serous membranes that enclose the lungs 155

polar bodies (po'lar bod'ēz)   nonfunctioning daughter cells that have little cytoplasm and are formed during oogenesis 346

pollution (pō-lu'shun)   contamination of air, water, or soil with undesirable amounts of material or heat. The material can be an overabundance of a natural substance or a small amount of a toxic substance 444

polypetide (pol''e-pep'tīd)   a molecule composed of many amino acids linked together by peptide bonds 20

polysaccharide (pol''e-sak'ah-rīd)   a macromolecule composed of many units of sugar 21

postganglionic axon (pōst''gang-gle-on'ik ak'son)   in the autonomic nervous system, the axon that leaves, rather than goes to, a ganglion 199

postsynaptic membrane (pōst''si-nap'tik mem'brān)   a membrane that is part of a synapse and receives a neurotransmitter substance 194

preganglionic axon (pre''gang-gle-on'ik ak'son)   in the autonomic nervous system, the axon that goes to, rather than leaves, a ganglion 199

pressure filtration (presh'ur fil-tra'shun)   process by which small molecules leave a capillary due to blood pressure 177

presynaptic membrane (pre''si-nap'tik mem'brān)   a membrane that is part of a synapse and releases a neurotransmitter substance 194

primates (pri'māts)   animals that belong to the order Primates, the order of mammals that includes prosimians, monkeys, apes, and humans 424

producers (pro-du'serz)   organisms that produce food and are capable of synthesizing organic compounds from inorganic constituents of the environment; usually the green plants and algae in an ecosystem 436

prolactin (pro-lak'tin)   a hormone secreted by the anterior pituitary that stimulates the production of milk from the mammary glands 260

prophase (pro'fāz)   early stage in mitosis during which chromosomes appear 341

prophase I (pro'fāz wun)   phase of meiosis I during which the spindle appears and the nuclear envelope and nucleolus disappear 344

proprioceptor (pro''pre-o-sep'tor)   sensory receptor that assists the brain in knowing the position of the limbs 238

prosimians (pro-sim'e-anz)   primitive primates such as lemurs, tarsiers, and tree shrews 425

prostate gland (pros'tāt gland)   a gland in males that is located about the urethra at the base of the bladder; produces most of the seminal fluid 278

protein (pro'te-in)   a macromolecule composed of one or several long polypeptides 19

prothrombin (pro-throm'bin)   plasma protein made by liver that must be present in blood before clotting can occur 122

protocell (pro'to-sel)   a structure that precedes the evolution of the true cell in the history of life 422

proton (pro'ton)   a subatomic particle found in the nucleus of an atom that has a weight of one atomic mass unit and carries a positive charge; a hydrogen ion 13

proto-oncogene (pro''to-ong'ko-jēn)   a normal gene that becomes an oncogene through mutation 404

protozoan (pro''to-zo'an)   unicellular, generally colorless and motile microorganism 308

proximal convoluted tubule (prok'si-mal kon'vo-lūt-ed too'būl)   highly coiled region of a nephron near Bowman's capsule 174

pseudostratified (su''do-strat'i-fīd)   the appearance of layering in some epithelial cells when actually each cell touches a baseline and true layers do not exist 49

pulmonary circuit (pul'mo-ner''e ser-ket)   the blood vessels that take deoxygenated blood to and oxygenated blood away from the lungs 101

pulse (puls)   vibration felt in arterial walls due to expansion of the aorta following ventricle contraction 104

Punnett square (pun'et skwar)   a gridlike device that enables one to calculate the results of simple genetic crosses by lining gametic genotypes of two parents on the outside margin and their recombination in boxes inside the grid 348

pupil (pu'pil)   an opening in the center of the iris of the eye 240

purines (pu'rinz)   nitrogenous bases found in DNA and RNA that have two interlocking rings, as in adenine and guanine 24

pus (pus)   thick yellowish fluid composed of dead phagocytes, dead tissue, and bacteria 127

pyrimidines (pi-rim'i-dinz)   nitrogenous bases found in DNA and RNA that have just one ring, as in cytosine and thymine 24

R

radius (ra'de us)   an elongated bone located on the thumb side of the lower arm 219

receptor (re-sep'tor)   a sense organ specialized to receive information from the environment. Also a structure found in the membrane of cells that combines with a specific chemical in a lock-and-key manner 189, 237

recessive allele (re-ses'iv ah-lēl')   hereditary factor that only expresses itself when two copies are present in the genotype 357

recombinant DNA (re-kom'bi-nant de'en-ā)   DNA having genes from two different organisms, often produced in the laboratory by introducing foreign genes into a bacterial plasmid 390

red marrow (red mar'o)   blood-cell-forming tissue located in spaces within certain bones 216

reduction (re-duk'shun)   the gain of electrons (inorganic); the addition of hydrogen atoms (organic) 17

reflex (re'fleks)   an inborn autonomic response to a stimulus that is dependent on the existence of fixed neural pathways 198

regulatory gene (reg'u-lah-tor''e jēn)   genes that regulate the activity of structural genes 387

REM sleep (rapid eye movement)   a stage in sleep that is characterized by eye movements and dreaming   205

replacement reproduction (re-plās'ment re''pro-duk'shun)   replacement of only one's self in the population, as when a couple has a maximum of two children   456

replication (rep''li-ka'shun)   the duplication of DNA; occurs when the cell is not dividing   340, 378

residual volume (re-zid'u-al vol'um)   the amount of air (about 1,200 cc) that remains in the lungs after the most forceful expiration   157

respiratory chain (re-spi'rah-to''re chan)   a series of carriers within the inner mitochondrial membrane that pass electrons one to the other from a higher level to a lower energy level; the energy released is used to build ATP; also called the electron transport system; the cytochrome system   43

resting potential (rest'ing po-ten'shal)   the voltage recorded from inside a neuron when it is not conducting nerve impulses   191

restriction endonuclease enzyme (re-strik'shun en''do-nu'kle-ās en'zīm)   a type of enzyme that cuts DNA at a location that contains a certain sequence of bases; hundreds of such enzymes exist   389

retina (ret'i-nah)   the innermost layer of the eyeball that contains the rods and cones   241

Rh factor (ar'ach fak'tor)   a type of antigen on the red blood cells   127, 299

rhodopsin (ro-dop'sin)   visual purple, a pigment found in the rods of one type of receptor in the retina of the eye   245

ribonucleic acid (ri''bo-nu-kle'ik as'id)   see RNA

ribosomal RNA (ri'bo-sōm''al)   rRNA; RNA found in ribosomes   384

ribosomes (ri'bo-sōmz)   minute particles, found attached to endoplasmic reticulum or loose in the cytoplasm, that are the site of protein synthesis   37

ribs (ribz)   bones hinged to the vertebral column and sternum that along with muscle define the top and sides of the chest cavity   155, 218

RNA (ribonucleic acid)   a nucleic acid important in the synthesis of proteins that contains the sugar ribose; the bases uracil, adenine, guanine, cytosine; and phosphoric acid   24, 378

rods (rodz)   dim-light receptors in the retina of the eye that detect motion but no color   241

rough endoplasmic reticulum (ruf en''do-plaz'mik rē-tik''u-lum)   RER, endoplasmic reticulum having attached ribosomes   36

round window (rownd win'do)   a membrane-covered opening between the inner ear and the middle ear   247

S

saccule (sak'ūl)   a saclike cavity that makes up part of the membranous labyrinth of the inner ear; contains receptors for static equilibrium   248

salivary amylase (sal'i-ver-e am'i-lās)   an enzyme in the saliva that initiates the digestion of starch   78

salivary gland (sal'i-ver-e gland)   a gland that makes saliva and secretes it into the mouth   71

SA node (es'ā nōd)   sinoatrial node; a small region of neuromuscular tissue that initiates the heartbeat. Also called the pacemaker   98

saprophyte (sap'ro-fīt)   a heterotrophic organism such as bacteria and fungi that externally breaks down dead organic matter before absorbing the products   304

sarcolemma (sar''ko-lem'ah)   the membrane that surrounds striated muscle cells   228

sarcoma (sar-ko'mah)   a cancer that arises in connective tissue, such as muscle, bone, and fibrous connective tissue   406

sarcomere (sar'ko-mēr)   the structural and functional unit of a myofibril   228

sarcoplasmic reticulum (sar''ko-plaz'mik re-tik'u-lum)   membranous network of channels and tubules within a muscle fiber corresponding to the endoplasmic reticulum of other cells   228

scapula (skap'u-lah)   a broad somewhat triangular bone located on either side of the back   218

sclera (skle'rah)   white fibrous outer layer of the eyeball   240

scrotum (skrōt-em)   the sacs that contain the testes   275

selectively permeable (sė-lek'tiv-le per'me-ah-b'l)   indicates condition in the living cell membrane in which permeability is a regulated process   34

selective reabsorption (sė-lek'tiv re''ab-sorp'shun)   one of the processes involved in the formation of urine; involves the greater reabsorption of nutrient molecules compared to waste molecules from the contents of the nephron into the blood   177

semicircular canal (sem''e-ser'ku-lar kah-nal')   tubular structures within the inner ear that contain the receptors responsible for the sense of dynamic equilibrium   248

semilunar valves (sem''e-lu'nar val'vz)   valves resembling half-moons located between the ventricles and their attached vessels   96, 248

seminal fluid (sem'i-nal floo'id)   a fluid produced by various glands situated along the male reproductive tract   278

seminal vesicle (sem'i-nal ves'i-k'l)   a convoluted saclike structure attached to the vas deferens near the base of the bladder in males   278

seminiferous tubules (sem''i-nif'er-us too'bulz)   highly coiled ducts within the male testes that produce and transport sperm   275

sensory nerve (sen'so-re nerv)   nerve containing only sensory neuron dendrites   196

sensory neuron (sen'so-re nu'ron)   a neuron that takes the nerve impulse to the central nervous system; afferent neuron   189

septum (sep'tum)   partition or wall such as the septum in the heart, which divides the right half from the left half   96

serum (se'rum)   light-yellow liquid left after clotting of the blood   122

sex chromosome (seks kro'mo-sōm)   a chromosome responsible for the development of characteristics associated with maleness or femaleness; an *X* or *Y* chromosome   337

sex linked (seks lingkt)   alleles located on sex chromosomes   370

sickle-cell disease (sik'l sel di-zēz')   a genetic disorder due to the homozygous genotype of the sickle-cell allele; producing sickle-shaped cells and loss of oxygen-carrying power in the blood   369

sickle-cell trait (sik'l sel trāt)   characteristics due to the heterozygous genotype of the sickle-cell allele; producing sickle-shaped cells in the blood under low oxygen conditions   369

simple goiter (sim'p'l goi'ter)   condition in which an enlarged thyroid produces low levels of thyroxin   261

sinus (si'nus)   a cavity, as the sinuses in the human skull   217

skeletal muscle (skel'ė-tal mus'el)   the contractile tissue that comprises the muscles attached to the skeleton; also called striated muscle   56

sliding filament theory (slīd'ing fil'ah-ment the'o-re)   the movement of actin in relation to myosin in explaining the mechanics of muscle contraction   230

small intestine (smal in-tes'tin)   long, tubelike chamber of the digestive tract between the stomach and large intestine   74

smooth endoplasmic reticulum (smooth en''do-plaz'mik rē-tik'u-lum)   SER, endoplasmic reticulum without attached ribosomes   36

smooth muscle (smooth mus'el)   the contractile tissue that comprises the muscles found in the walls of internal organs   56

soft palate (soft pal'at)   entirely muscular posterior portion of the roof of the mouth   71

somatic nervous system (so-mat'ik ner'vus sis'tem)   that portion of the PNS containing motor neurons that control skeletal muscles   198

species (spe'shēz)   a group of similarly constructed organisms that are capable of interbreeding and producing fertile offspring; the taxonomic category below genus 416

spermatogenesis (sper''mah-to-jen'ē-sis)   production of sperm in males by the process of meiosis and maturation 346

sphincter (sfingk'ter)   a muscle that surrounds a tube and closes or opens the tube by contracting and relaxing 73

spinal nerve (spi'nal nerv)   a nerve that arises from the spinal cord 197

spindle fibers (spin'd'l fi'berz)   microtubule bundles in cells that are involved in the movement of chromosomes during mitosis and meiosis 341

spongy bone (spun'je bōn)   bone found at the ends of long bones that consists of bars and plates separated by irregular spaces 55, 216

squamous epithelium (skwa'mus ep''i-the'le-um)   flat cells found lining the lungs and blood vessels, for example 49

stereoscopic vision (ste''re-o-skop'ik vizh'un)   the product of two eyes and both cerebral hemispheres functioning together so that depth perception results 243

sterilization (ster''i-li-za'shun)   the inability to reproduce; a surgical procedure eliminating reproductive capability; the absence of living organsims due to exposure to environmental conditions that are unfavorable to sustain life 288, 304

sternum (ster'num)   the breastbone to which the ribs are ventrally attached 218

steroid (ste'roid)   lipid-soluble, biologically active molecules having four interlocking rings; examples are cholesterol, progesterone, testosterone 24

stirrup (stir-ep)   the innermost of the ossicles of the middle ear, which fits against the oval window; also called the stapes 247

stomach (stum'ak)   a muscular sac which mixes food with gastric juices to form chyme which enters the small intestine 73

stratified (strat'i-fīd)   layered, as in stratified epithelium, which contains several layers of cells 49

stretch receptors (strech re-sep'torz)   muscle fibers that upon stimulation cause muscle spindles to increase the rate at which they send impulses to the CNS 238

striated (stri'āt-ed)   having bands; cardiac and skeletal muscle are striated with bands of light and dark 56

structural gene (struk'tūr-al jēn)   gene that determines protein structure 387

subcutaneous layer (sub''ku-ta'ne-us)   a tissue layer found in skin that lies just beneath the dermis and tends to contain fat cells 59

succession (suk-sĕ'shun)   a series of ecological stages by which the community in a particular area gradually changes until there is a stable community that can maintain itself 436

summation (sum-ma'shun)   ever greater contraction of a muscle due to constant stimulation that does not allow complete relaxation to occur 195, 224

suppressor T cell (sŭ-pres'or te sel)   T lymphocyte that suppresses certain other T and B lymphocytes from continuing to divide and perform their respective function, T_s cells 140

sympathetic nervous system (sim''pah-thet'ik ner'vus sis'tem)   that part of the autonomic nervous system that generally causes effects associated with emergency situations 201

synapse (sin'aps)   the region between two nerve cells where the nerve impulse is transmitted from one to the other; usually from axon to dendrite 194

synapsis (si-nap'sis)   the attracting and pairing of homologous chromosomes during meiosis 343

synaptic cleft (si-nap'tik kleft)   small gap between presynaptic and postsynaptic membranes of a synapse 194

synaptic ending (si-nap'tik end'ing)   the knob at the end of an axon in a synapse 194

syndrome (sin'drōm)   a group of symptoms that together characterize a disease condition 351

synovial joint (si-no've-al joint)   a freely movable joint 220

synthesis (sin'the-sis)   to build up, such as the combining together of two small molecules to form a larger molecule 19

syphilis (sif'i-lis)   chronic, contagious sexually transmitted disease caused by a bacterium that is a spirochete 307

systemic circuit (sis-tem'ik ser'kit)   that part of the circulatory system that serves body parts other than the gas-exchanging surfaces in the lungs 101

systole (sis'to-le)   contraction of the heart chambers, particularly the left ventricle 98

systolic blood pressure (sis-tol'ik blud presh'ur)   arterial blood pressure during the systolic phase of the cardiac cycle 104

T

tarsal (tahr'sal)   a bone of the ankle in humans 220

taste bud (tāst bud)   organ containing the receptors associated with the sense of taste 239

taxonomy (tak-son'o-me)   the science of naming and classifying organisms 416

Tay-Sachs disease (tā saks' di zēz)   a lysosomal storage disease that is inherited and causes neurological impairment and death 365

tectorial membrane (tek-to're-al mem'brān)   a membrane in the organ of Corti that lies above and makes contact with the receptor cells for hearing 248

teleological (te''le-o-loj'i-kal)   assuming a process that is directed toward a final goal or purpose 420

telophase (tel'o-fāz)   stage of mitosis during which the diploid number of chromosomes are located at each pole 342

telophase I (tel'o-fāz wun)   phase of meiosis I during which the nuclear envelope and nucleolus reappear as the spindle disappears 344

template (tem'plat)   a pattern that serves as a mold for the production of an oppositely shaped structure; one strand of DNA is a template for the complementary strand 378

temporal lobe (tem'po-ral lōb)   area of the cerebrum responsible for hearing and smelling, the interpretation of sensory experience and memory 203

tendon (ten'don)   a tissue that connects muscle to bone 54, 222

testcross (test kros)   a genetic mating in which a possible heterozygote is crossed with an organism homozygous recessive for the characteristic(s) in question in order to determine the genotype 360

testes (tes'tez)   the male gonads, the organs that produce sperm and testosterone 275

testosterone (tes-tos'te-rōn)   the most potent of the androgens 281

tetanus (tet'ah-nus)   sustained muscle contraction without relaxation 224

tetany (tet'ah-ne)   severe twitching caused by involuntary contraction of the skeletal muscles due to a lack of calcium 263

tetrad (tet'rad)   a set of four chromatids resulting from the pairing of homologous chromosomes during meiosis 343

thalamus (thal'ah-mus)   a mass of gray matter located at the base of the cerebrum in the wall of the third ventricle; receives sensory information and selectively passes it to the cerebrum 203

thrombin (throm'bin)   the enzyme derived from prothrombin that converts fibrinogen to fibrin threads during blood clotting 122

thymus (thi'mus)   a lymphatic organ that lies in the neck and chest area and is absolutely necessary to the development of immunity 268

thyroid stimulating hormone (thi'roid stim'u-lāt''ing hor'mon)   see TSH

thyroxin (thi-rok'sin)   the hormone produced by the thyroid that speeds up the metabolic rate 262

tibia (tib'e-ah)   the shinbone found in the lower leg 220

tidal volume (tīd'al vol'um)   amount of air that enters the lungs during a normal, quiet inspiration 157

T lymphocyte (lim'fo-sīt)   a lymphocyte that matures in the thymus and occurs in four varieties, one of which kills antigen-bearing cells outright 135

tone (tōn)   the continuous partial contraction of muscle; also the quality of a sound   225

trachea (tra'ke-ah)   the windpipe that serves as a passageway for air   154

transcription (trans-krip'shun)   the process that results in the production of a strand of mRNA that is complementary to a segment of DNA   380

transfer RNA (trans'fer)   tRNA; molecule of RNA that carries an amino acid to a ribosome in the process of protein synthesis   384

transgenic organism (trans-jen'ik or'gah-nizm)   plant or animal that has received a foreign gene, thereby changing its phenotype   392

transformation (trans''for-ma'-shun)   the change of a normal cell into a cancer cell; transformed cells lack differentiation, grow uncontrollably, and metastasize   402

translation (trans-la'shun)   the process involving mRNA, ribosomes, and tRNA that results in a synthesis of a polypeptide having an amino acid sequence dictated by the sequence of codons in mRNA   380

translocation (trans''lo-ka'shun)   chromosome mutation due to the attachment of a broken chromosome fragment to a nonhomologous chromosome   349

transposon (trans-po'zon)   motile genetic elements called jumping genes that can move between chromosomes and thereby bring about genetic mutations   388

trophic level (tro'fic levl)   a categorization of species in a food web according to their feeding relationships from the first level autotrophs through succeeding levels of herbivores and carnivores   438

trophoblast (trof'o-blast)   an outer layer of cells that surrounds the human embryo and, when thickened by a layer of mesoderm, becomes the chorion, an extraembryonic membrane   312

trypsin (trip'sin)   a protein-digesting enzyme secreted by the pancreas   78

TSH (thyroid-stimulating hormone)   hormone that causes the thyroid to produce thyroxin   261

tubal ligation (tu'bal li-ga'shun)   cutting of the oviducts in females to cause sterilization   288

tubular excretion (too'bu-lar eks-kre'shun)   process occurring in the distal convoluted tubule during which substances are added to urine   179

tumor (too'mor)   growth containing cells derived from a single mutated cell that has repeatedly undergone cell division; benign tumors remain at the site of origin and malignant tumors metastasize   403

Turner syndrome (tur'ner sin'drōm)   a condition caused by the inheritance of a single X chromosome   353

twitch (twich)   a brief muscular contraction followed by relaxation   224

tympanic membrane (tim-pan'ik mem'brān)   a membrane located between the external and middle ear; the eardrum   247

U

ulna (ul'nah)   an elongated bone found within the lower arm   219

umbilical arteries and vein (um-bil'i-kal ar'ter-ēz and vān)   fetal blood vessels that travel to and from the placenta   325

umbilical cord (um-bil'i-kal kord)   cord connecting the fetus to the placenta through which blood vessels pass   318, 319

urea (u-re'ah)   primary nitrogenous human waste derived from amino acid breakdown   169

ureter (u-re'ter)   tube from the kidney that takes urine to the bladder   173

urethra (u-re'thrah)   tube that takes urine from bladder to outside   173

uric acid (u'rik as'id)   waste product of nucleotide breakdown   169

urinalysis (u''ri-nal'i-sis)   a medical procedure in which the composition of a patient's urine is determined   181

urinary bladder (u'ri-ner''e blad'der)   an organ where urine is stored before being discharged by way of the urethra   173

uterine cycle (u'ter-in si'kl)   monthly occurring changes in females that are characterized by regularly occurring changes in the uterine lining   284

uterus (u'ter-us)   the female organ in which the fetus develops   283

utricle (u'tre-kl)   an enlarged cavity that makes up part of the membranous labyrinth of the inner ear; contains receptors for static equilibrium   248

V

vaccine (vak'sēn)   antigens prepared in such a way that they can promote active immunity without causing disease   142

vagina (vah-ji'nah)   the copulatory organ and birth canal in females   283

valves (valvz)   membranous extension of a vessel or heart wall that opens and closes, ensuring one-way flow   95

vas deferens (vas def'er-ens)   tube that leads from the epididymis to the urethra of the male reproductive tract   278

vector (vek'tor)   a carrier, e.g., plasmid, for recombinant DNA that introduces a foreign gene into a host cell   388

vein (vān)   a blood vessel that takes blood to the heart   94

vena cava (ve'nah ka'vah)   one of two large veins that returns deoxygenated blood to the right atrium of the heart   97

vena cavae (ve'nah ka'vah)   two large veins, i.e., superior vena cava and inferior vena cava, that return deoxygenated blood to the right atrium of the heart   101

venous duct (ve'nus dukt)   ductus venosus; fetal connection between the umbilical vein and the inferior vena cava   325

ventilation (ven''ti-la'shun)   breathing; the process of moving air into and out of the lungs   155

ventricle (ven'tri-k'l)   a cavity in an organ such as the ventricles of the heart or the ventricles of the brain   96, 201

venule (ven'ūl)   type of blood vessel that takes blood from capillaries to veins   94

vernix caseosa (ver'niks ka-se-o'sah)   cheeselike substance covering the skin of the fetus   324

vertebral column (ver'te-bral kol'um)   the backbone of vertebrates through which the spinal cord passes   218

vestigial structure (ves-tij'e-al struk'tūr)   the remains of a structure that was functional in some ancestor but is no longer functional in the organism in question   419

villi (vil'i)   fingerlike projections that line the small intestine and function in absorption   74

virulence (vir'u-lens)   the ability of a microorganism to cause disease and to invade the tissues of a host   304

vital capacity (vi'tal kah-pas'i-te)   the maximum amount of air a person can exhale after taking the deepest breath possible   157

vitamin (vi'tah-min)   usually coenzymes, needed in small amounts, that the body is no longer capable of synthesizing and therefore must be in the diet   83

vitreous humor (vit're-us hu'mor)   the substance that occupies the space between the lens and retina of the eye   240

vocal cords (vo'kal kordz)   folds of tissue within the larynx that create vocal sounds when they vibrate   152

vulva (vul'vah)   the external genitalia of the female that lie near the opening of the vagina   283

X

X-linked gene (eks lingkt jēn)   an allele located on the X chromosome   370

XYY male (eks-wi-wi māl)   a male that has an extra Y chromosome   353

Y

yellow marrow (yel'o mar'o)   fat storage tissue found in the cavities within certain bones   216

yolk sac (yōk sak)   one of the extraembryonic membranes within which yolk is found; in mammals, the first site of blood-cell formation in the embryo   313

Z

zygote (zi'gōt)   diploid cell formed by the union of two gametes, the product of fertilization   338

Credits

Photographs

© Ray Ellis/Photo Researchers, Inc.; **14.14e,f:** © Bob Coyle; **14.16:** © Dan Heringa/The Image Bank

Chapter 15

Opener: © CNRI/SPL/Photo Researchers, Inc.; **15.1:** Historical Pictures Service, Inc.; **15.2:** © Dr. E. R. Degginger/Color-Pic, Inc.; **15.5a:** © Charles Lightdale/Photo Researchers, Inc.; **15.5b:** © Robert Settineri/Sierra Productions; **15.6a:** © David M. Phillips/Visuals Unlimited; **15.6b:** © David M. Phillips/Visuals Unlimited; **15.6c:** From Dr. R. G. Kessel and Dr. Cy Shih From *Scanning Electron Miscroscopy,* Springer-Verlag, Berlin, Heidelberg, 1976; **15.9:** © John D. Cunningham/Visuals Unlimited; **15.10:** © Biophoto Associates/Photo Researchers, Inc.; **15.11:** Reproduced by permission from Donaldson, David D., *Atlas of External Diseases of the Eye,* Vol. 1, C.V. Mosby Co.; **15.12:** © Janet Barber/Custom Medical Stock Photo; **15.13:** Center for Disease Control, Atlanta, GA; **15.14a,b:** © Carroll Weiss/Camera M.D.; **15.14c:** Center for Disease Control, Atlanta, GA

Chapter 16

Opener: © Comstock/Comstock, Inc.; **16.11 all:** © Claude Edelmann, Petit Format et Guigoz/Photo Researchers, Inc.; **16.12:** © Joe McNally; **16.16:** © Dan Rubin for Mutual of America; **16.18:** © Martin R. Jones/Unicorn Stock Photos

Aids Chapter

Opener: © Alfred Pasieka/SPL/Photo Researchers, Inc.

Chapter 17

Opener: © David M. Phillips/Visuals Unlimited; **17.1:** © Arthur Sirdofsky; **17.3 top:** © Elyse Lewin/The Image Bank; **17.7a,b:** © Eric Grave/Photo Researchers, Inc.; **17.14:** Courtesy of Lennart Nilsson © Boehringer Ingelheim GmbH; *p. 350:* © A. Robinson, National Jewish Hospital and Research Center; **17.18a:** © Jill Cannefax/EKM Nepenthe; **17.20a:** F. A. Davis Company, Philadelphia and RH Kampmeier; **17.20b:** From M. Bartalos and T. A. Barameki, *Medical Cytogenetics,* Williams and Wilkins, Baltimore, MD 1967

Chapter 18

Opener: © Bob Daemmrich/The Image Works; **18.3a,b:** © Bob Coyle; **18.3c,d:** © Tom Ballard/EKM-Nepenthe; **18,3e,f:** © James Shaffer; **18.10c:** © 1989 J. Callahan, M.D./IMS/Univ. of Toronto/Custom Medical Stock Photo; **18.11a:** © Bob Daemmrich/The Image Works, Inc.; **18.12:** Editorial Enterprises; **18.13:** © Robert Burroughs Photography; **18.15b:** © Bill Longcore/Science Source/Photo Researchers, Inc.

Chapter 19

Opener: © Sinclair Stammers/SPL/Custom Medical Stock Photography; **19.3a:** © Jan Halaska/Photo Researchers, Inc.; **19.7b:** Alexander Rich; **19.10a,b,c,d:** Courtesy of Genetech, Inc.; **19.11b:** © E. Hartmann/Magnum, Inc.; **19.12a:** Sanofi Recherche; **19.12b:** Cetus Research Place; **19.12c:** © Phototake, Inc.; **19.13:** General Electric Research and Development Center; **19.14:** © Keith V. Wood/UCSD; **19.15a,b,c,d:** © Runkl Schoenberger/Grant Heilman; **19.17:** From Christian C. Hauden-

schild, News and Comment, Vol. 246, p. 747, *Science,* Nov. 10, 1989, ''Gore Tex Organoids and Genetic Drugs,'' B. Bulliton, © 1989 by AAAS; **19.20:** Hoffmann-LaRoche and Roche Institute

Chapter 20

Opener: © Dr. Brian Eyden/SPL/Photo Researchers, Inc.; **20.1a,b:** Dr. G. Steven Martin; **20.4a:** © Bob Coyle; **20.4b:** © Larry Mulvehill/Science Source/Photo Researchers, Inc.; **20.7:** © Boehringer Ingelheim International GmbH, photo courtesy Lennart Nilsson; **20.8:** © Bob Coyle

Chapter 21

Opener: © N. Pecnik/Visuals Unlimited; **21.1a:** © Dick Young/Unicorn Stock Photos; **21.1b:** © R. W. Jennings/Unicorn Stock Photos; **21.1c:** © Angabe A. Schmidecker/FPG; **21.1d:** © Vega/Photo Researchers, Inc.; **21.2:** Professor John M. Hyes; **21.7:** © Rene Sheret/Marilyn Gartman Agency; **21.8:** © Tom McHugh/National Audubon Society/Photo Researchers, Inc.; **21.11a:** © Martha Reeves/Photo Researchers, Inc.; **21.11b:** © Tom McHugh/Photo Researchers, Inc.; **21.11c:** © George Holton/Photo Researchers, Inc.; **21.11d:** © Tom McHugh/Photo Researchers, Inc.; **21.18:** Courtesy of Dept. of Library Services, American Museum of Natural History

Chapter 22

Opener: © Thomas Kitchin/Tom Stack and Assoc.; **22.1a:** © Peter Kaplan/Photo Researchers, Inc.; **22.1b:** © Frank Miller/Photo Researchers, Inc.; **22.1c:** © Pat O'Hara; **22.3b:** © Gallbridge/Visuals Unlimited; **22.13:** © Kevin Magee/Tom Stack and Assoc.; **22.14:** Jane Windsor, Division of Plant Industry, Florida Dept. of Agriculture, Gainesville, FL; **22.15:** Courtesy of Gary Milburn/Tom Stack and Assoc.

Chapter 23

Opener: © Thomas Kitchin/Tom Stack and Assoc.; **23.5:** © Bob Coyle; **23.6a:** © Sidney Tomson/Animals Animals/Earth Scenes; **23.6b:** © Prance/Visuals Unlimited; **23.6c:** © Nichols/Magnum; **23.7:** © Gary Milburn/Tom Stack and Assoc.; **23.11:** © Don Riepe/Peter Arnold, Inc.; **23.13a,b:** Dr. John Skelley; **23.15a,b:** NASA; **23.17:** © Bob Coyle

Line Art

Chapter 3

Figure 3.2 From John W. Hole, *Human Anatomy and Physiology,* 5th ed. Copyright © 1990 Wm. C. Brown Publishers, Dubuque, Iowa. All Rights Reserved. Reprinted by permission.
Figure 3.14 From Kent M. Van De Graaff and Stuart Ira Fox, *Concepts of Human Anatomy and Physiology,* 2d ed. Copyright © 1989 Wm. C. Brown Publishers, Dubuque, Iowa. All Rights Reserved. Reprinted by permission.
Figure 3.15 From Kent M. Van De Graaff, *Human Anatomy,* 2d ed. Copyright © 1988 Wm. C. Brown Publishers, Dubuque, Iowa. All Rights Reserved. Reprinted by permission.

Chapter 4

Figure 4.3 From Kent M. Van De Graaff, *Human Anatomy,* 2d ed. Copyright © 1988 Wm. C. Brown Publishers, Dubuque, Iowa. All Rights Reserved. Reprinted by permission.

Chapter 5

Figure 5.2 From Kent M. Van De Graaff and Stuart Ira Fox, *Concepts of Human Anatomy and Physiology,* 2d ed. Copyright © 1989 Wm. C. Brown Publishers, Dubuque, Iowa. All Rights Reserved. Reprinted by permission.
Figure 5.6 Crouch, James E.: *Functional Human Anatomy.* Lea & Febiger, Philadelphia, 1985.
Figure 5.13 a and b From Stuart Ira Fox, *Human Physiology,* 3d edition. Copyright © 1990 Wm. C. Brown Publishers, Dubuque, Iowa. All Rights Reserved. Reprinted by permission.

Chapter 6

Illustration, page 118 C. A. Hunt, University of California, San Francisco.

Chapter 8

Figure 8.11 From John W. Hole, *Human Anatomy and Physiology,* 5th ed. © 1990 Wm. C. Brown Publishers, Dubuque, Iowa. All Rights Reserved. Reprinted by permission.

Chapter 9

Figure 9.5 From Kent M. Van De Graaff, *Human Anatomy,* 2d ed. Copyright © 1988 Wm. C. Brown Publishers, Dubuque, Iowa. All Rights Reserved. Reprinted by permission.
Figure 9.9 From Kent M. Van De Graaff and Stuart Ira Fox, *Concepts of Human Anatomy and Physiology,* 2d ed. Copyright © 1989 Wm. C. Brown Publishers, Dubuque, Iowa. All Rights Reserved. Reprinted by permission.
Illustration, page 185 From Kent M. Van De Graaff and Stuart Ira Fox, *Concepts of Human Anatomy and Physiology,* 2d ed. Copyright © 1989 Wm. C. Brown Publishers, Dubuque, Iowa. All Rights Reserved. Reprinted by permission.

Chapter 10

Figure 10.1b From Kent M. Van De Graaff, *Human Anatomy,* 2d ed. Copyright © 1988 Wm. C. Brown Publishers, Dubuque, Iowa. All Rights Reserved. Reprinted by permission.
Figure 10.14 From Kent M. Van De Graaff, *Human Anatomy,* 2d ed. Copyright © 1988 Wm. C. Brown Publishers, Dubuque, Iowa. All Rights Reserved. Reprinted by permission.
Figure 10.17 Montreal Neurological Institute. Used with permission.

Chapter 11

Figure 11.4 From Kent M. Van De Graaff, *Human Anatomy,* 2d ed. Copyright © 1988 Wm. C. Brown Publishers, Dubuque, Iowa. All Rights Reserved. Reprinted by permission.
Figure 11.11 From Kent M. Van De Graaff and Stuart Ira Fox, *Concepts of Human Anatomy and Physiology,* 2d ed. Copyright © 1989 Wm. C. Brown Publishers, Dubuque, Iowa. All Rights Reserved. Reprinted by permission.
Figure 11.16 b and c From John W. Hole, Jr., *Human Anatomy and Physiology,* 5th ed. Copyright © 1990 Wm. C. Brown Publishers, Dubuque, Iowa. All Rights Reserved. Reprinted by permission.

Chapter 14

Figure 14.2 From John W. Hole, Jr., *Human Anatomy and Physiology,* 5th ed. Copyright © 1990 Wm. C. Brown Publishers, Dubuque, Iowa. All Rights Reserved. Reprinted by permission.

Figure 14.3c From Kent M. Van De Graaff and Stuart Ira Fox, *Concepts of Human Anatomy and Physiology,* 2d ed. Copyright © 1989 Wm. C. Brown Publishers, Dubuque, Iowa. All Rights Reserved. Reprinted by permission.

Figure 14.3d From John W. Hole, Jr., *Human Anatomy and Physiology,* 5th ed. Copyright © 1990 Wm. C. Brown Publishers, Dubuque, Iowa. All Rights Reserved. Reprinted by permission.

Figure 14.4 From Kent M. Van De Graaff and Stuart Ira Fox, *Concepts of Human Anatomy and Physiology,* 2d ed. Copyright © 1989 Wm. C. Brown Publishers, Dubuque, Iowa. All Rights Reserved. Reprinted by permission.

Figure 14.7 From John W. Hole, Jr., *Human Anatomy and Physiology,* 5th ed. Copyright © 1990 Wm. C. Brown Publishers, Dubuque, Iowa. All Rights Reserved. Reprinted by permission.

Figure 14.8a From John W. Hole, Jr., *Human Anatomy and Physiology,* 5th ed. Copyright © 1990 Wm. C. Brown Publishers, Dubuque, Iowa. All Rights Reserved. Reprinted by permission.

Figure 14.11 From John W. Hole, Jr., *Human Anatomy and Physiology,* 5th ed. Copyright © 1990 Wm. C. Brown Publishers, Dubuque, Iowa. All Rights Reserved. Reprinted by permission.

Figure 14.12 From Kent M. Van De Graaff and Stuart Ira Fox, *Concepts of Human Anatomy and Physiology,* 2d ed. Copyright © 1989 Wm. C. Brown Publishers, Dubuque, Iowa. All Rights Reserved. Reprinted by permission.

Illustration, page 295 From John W. Hole, Jr., *Human Anatomy and Physiology,* 5th ed. Copyright © 1990 Wm. C. Brown Publishers, Dubuque, Iowa. All Rights Reserved. Reprinted by permission.

Chapter 16

Figure 16.5 From John W. Hole, Jr., *Human Anatomy and Physiology,* 5th ed. Copyright © 1990 Wm. C. Brown Publishers, Dubuque, Iowa. All Rights Reserved. Reprinted by permission.

Figure 16.14 From Kent M. Van De Graaff and Stuart Ira Fox, *Concepts of Human Anatomy and Physiology,* 2d ed. Copyright © 1989 Wm. C. Brown Publishers, Dubuque, Iowa. All Rights Reserved. Reprinted by permission.

Chapter 18

Figure 18.11b From E. Peter Volpe, *Biology and Human Concerns,* 3d ed. Copyright © 1983 Wm. C. Brown Publishers, Dubuque, Iowa. All Rights Reserved. Reprinted by permission.

Figure 18.18 From E. Peter Volpe, *Biology and Human Concerns,* 3d ed. Copyright © 1983 Wm. C. Brown Publishers, Dubuque, Iowa. All Rights Reserved. Reprinted by permission.

Illustrators

Chris Creek

4.1, 7.3, 8.3, 9.2, 9.9, 13.2, 19.1; text art, pages 91, 166, and 185

Fineline Illustrations

1.17a-c, 16.10.

Anne Green

9.6, 12.5, 16.4; text art, page 252

Kathleen Hagelston

1.2, 1.4, 1.5, 2.10b, 3.5a-c, 4.9, 4.10, 5.12, 7.8c, 10.3a, 15.15, 17.17, 19.8a-b, 21.6a-d; text art, page 384; reading art, page 279

Kathleen Hagelston/Marjorie C. Leggitt

I.2, 4.15, 9.12, 17.18b; reading art, page 7

Kathleen Hagelston/Laurie O'Keefe

7.5.

Hans & Cassady, Inc.

18.10a.

Anthony Hunt

Reading art, page 118.

Carlyn Iverson

I.3, 1.3a, 1.7, 1.11, 1.12, 2.11a, 3.13, 3.16, 3.17, 3.18, 4.2, 4.4, 4.5a, 4.7a, 4.11, 5.1a-d, 5.11, 5.15, 6.1, 6.4, 7.6, 8.4, 8.5, 8.13, 9.8a, 9.11b, 9.13a-c, 10.14, 10.18, 11.3a-b, 11.9, 11.10a-b, 11.14a, 11.14c-d, 12.4, 12.12, 12.13, 12.14a-d, 12.15a-b, 14.9, 16.3, 16.6b, 17.2, 19.6, 19.7a, 22.7, 23.12; Chapter 5 Opener, page 93; reading art, pages 293 and 393

Carlyn Iverson/Rolin Graphics

15.7, 17.4, 17.5, 17.9, 17.11, 17.12, 17.13.

Ruth Krabach

16.9, left.

Mark Lefkowitz

The following figures are © 1989 Mark Lefkowitz. All Rights Reserved.
I.6, 10.10, 10.11, 10.13, 10.16, 21.14, 21.15, 21.17, 22.10, and 22.11
The following figures are © 1990 Mark Lefkowitz. All Rights Reserved.
5.3a and b, 5.4a and b, 9.7, and 10.8; text art, page 111

Marjorie C. Leggitt

21.10.

Ron McLean

14.8a.

Robert Margulies

16.5.

Robert Margulies/Tom Waldrop

11.11.

Steve Moon

3.14, 5.2, 14.12.

Diane Nelson

8.11, 11.4a–b, 16.9, right; reading art, page 109, figure b.

Laurie O'Keefe

2.8, 3.10, 6.5, 19.11a, 19.18, 20.2, 20.3.

Precision Graphics

4.18a-c, 7.7b, 8.9b, 8.12a-b, 10.20, 13.15, 18.11b, 18.17, 18.18, 19.19, 22.9; text art, page 330

Mildred Rinehart

19.2.

Rolin Graphics

I.5, I.7, 1.6, 1.8, 1.10, 1.13a, 1.14a-b, 1.15, 1.16a-b, 1.18a-b, 1.19, 1.21, 1.22, 2.1, 2.2a-c, 2.3, 2.5a-c, 2.7b-d, 2.9b, 2.14, 3.11a-2, 3.11b-2, 3.11c-2, 4.13, 5.5, 5.7a-b, 5.8, 5.9, 5.14, 6.2, 6.3b-c, 6.6, 6.7, 6.9a, 6.13a-d, 7.9a, 7.9b, 7.10, 7.14a-e, 8.7, 8.8, 8.10, 9.10, 9.14, 10.5a-d, 10.6, 10.17, 11.2, 11.5, 11.6, 11.7, 11.8, 11.13, 12.2, 12.3, 12.6a-c, 13.3, 13.4, 13.5, 13.6, 13.9, 13.14a-c, 13.18, 13.19, 14.3a, 14.3c-d, 14.6, 14.10, 14.13, 14.15, 15.4, 15.8, 16.1, 16.6a, 16.17, 17.10, 18.1, 18.8, 18.9, 19.4, 19.16, 20.5, 21.3, 21.9a-d, 21.13, 22.2, 22.3a, 22.4a-b, 22.5, 22.6, 22.16, 23.2, 23.4, 23.14a-b; text art, pages 14, 19, 39, 46, 150, 159, 161, 230, and 350; reading art, page 323, 449

Mike Schenk

4.3, 9.5, 10.1b, 16.8, 16.14a-d.

Tom Waldrop

7.2, 10.9a, 10.12, 11.16b, 14.2, 14.4, 14.7, 14.11; text art, page 295; reading art, page 269

John Walter & Associates

5.13a-b.

Index*

A

Abdominal cavity, 59, *62*
ABO blood groups, 127
Accommodation, visual, 242–**43**
Acetylcholine (ACh), **194**, 201, 206, 231, *232*
Acetylcholinesterase (AChE), **194**
Acid chyme, 73, 75
Acid deposition, **463**–65
Acid rain, 16, 462, 463
Acids, **17–18**
Acromegaly, 257 (table), **260**, *261*
Acrosome, **276**, 277
Actin, 19, *40*, 225, **230**
   filaments, 56, 230, 231, *233*
Action potential of nerve impulse, **192–94**
Active immunity, 142, *143*
Active site of enzymes, **41**, *42*
Active transport, 34 (table), **35**
Adaptation of species, **421**, *422*
Addison disease, 265, *266*
Adenosine diphosphate (ADP), 26–27
Adenosine triphosphate. *See* ATP
ADH (antidiuretic hormone), **180**–81, 256–57 (table), **258**, 270
Adipose tissue, 22, *51*, 54, *58*, 330
Adolescence, 328–29
ADP (adenosine diphosphate), 26–27
Adrenal glands, 181, 256–57 (table), 264–65
   disorders of, 265–66
Adrenalin, 201, 256–57 (table), 264
Adrenocorticotropic hormone (ACTH), 256–57 (table), 260, **261**, 265
Adulthood, 329–32
Advanced glycosylation end products (AGEs), 330
Aerobic cellular respiration, **39**, 42–44, **150**, 230, 441
Afferent neuron, 189, *190*, 198, 199
Afterbirth, 328. *See also* Placenta
Agglutination, **127**
Aging, **329–32**
   and replacement reproduction, 456, *457*
Agranular leukocytes, *114*, **124**, 125. *See also* B lymphocytes; T lymphocytes
   formation of, *119*, 132–33
   immune defense function of, 135–41
Agriculture, modern and organic, 445–46, 456, *457*
AIDS (acquired immunodeficiency syndrome), 124, **299**, 322
   blood transfusions and, 118, 371

origin and prevalence of, A-2–A-3
   prevention and future of, A-6–A-7
   symptoms of, 299, A-4–A-6
   transmission of, 299, 301, A-3–A-4
   treatments for, 143, 301, A-6, 387, 391 (table), 392
AIDS related complex (ARC), 299, A-4–A-5
Air pollution, 16, 463–66
Air pressure in lungs, 159, 161
Albinism, 364, 378, *380*
Albumin, **115**, 116 (table)
Alcohol, 410
   abuse of, 207–8, 330
   birth defects and, 322
   cirrhosis and, 78, 207
   formation through fermentation, 44
Aldosterone, *25*, **181**, *182*, 256–57 (table), **265**
Alexeyev, Vasily, *214*
Alkaptonuria, 378, *380*
Allantois, *315*, **318**
Alleles, dominant and recessive, **357**. *See also* Genetics
   autosomal disorders and, 364–66
   detecting defective, 397–98
   multiple, 366–69
Allergen, 144, 146
Allergies, 144, 146
All-or-none law
   in muscle response, **224**
   of neurons, 195
Alpha globulin, **115**, 116 (table)
Alvarez, Walter, 418
Alveoli of breasts, 287
Alveoli of lungs, 152, **154**
   in respiratory ventilation, 156–57
Alzheimer's disease, 206, 332
American Cancer Society, 164, 410
American Heart Association, 107
American Medical Association, 291
Amino acids, **20**, 21
   deamination of, 76–77
   essential, **79**
   evolution of, 422, *423*
   sequences in protein synthesis, 379, *380*–85, *386*
   synthesis of, 78, 79, 266
   transport of, 120, *121*
Ammonia, **169**, 171 (table), *178*, 179, 180, 422
Ammonium (NH_4^+), 443, 462
Amniocentesis, 322, *323*
Amnion, 313, *314*, *318*
Ampulla of ear, **248**, *249*

Amylase, salivary and pancreatic, **78**, 79 (table)
Anabolic steroids, 225, 268, 281
Anaerobic cellular respiration, 43, 45, 230, 424
Analogous structures, **418**
Anaphase
   of meiosis, **344**, *345*
   of mitosis, 340, *342*
Anaplasia, 403
Androgens, 256–57 (table), 267, 268, 278, 324
   regulation of levels of, 280–81, 290
Anemia, **120**, 128, 390
Angina pectoris, 107
Angiotensin I and II, 181, *182*, 265
Animals, 32
   carbon reservoirs in, 441–42
   in ecological food chains, 438, *439*
   evolution of, 415, 418, 419–20, 424
   genetically engineered, 390, 395–96
Anorexia nervosa, 88
Antagonistic muscle pairs, 222
Anthropoids, **425**–27. *See also* Humans
Antibiotics, 298, 304, **305**
Antibodies, **135**
   in ABO and Rh blood types, 127–28
   in active and passive immunity, 142–43
   function and structure of, 125, 137
   and immunological diseases, 144–46
   monoclonal, 144, *145*, *146*, 409
Antibody-mediated immunity, 137, 141 (table)
Antibody titer, 142, *143*
Anticodon, 384, *386*
Antidiuretic hormone (ADH), **180**–81, 256–57 (table), **258**, 270
Antigen-presenting cell (APC), 140, *141*
Antigens, **125**, 141 (table), 368
   action of B cells against, 137–38
   action of T cells against, 138–40
   H-Y, 350
   used in vaccines, 142
Anus, *70*, **75**
Anvil (incus) of ear, **247**, 250
Aorta, 97, *99*, **101**, *102*, *103*, 105, *326*
Aortic bodies, 155
Apes, 2, *6–7*, 425–27, 428
Appendicitis, 75
Appendicular skeleton, *215*, **218–20**

Appendix, *70*, **75**
Aqueous humor, eye, **240**, 241
Arrector pili muscle, *58*, 59, 63
Arteries. *See also names of specific arteries*
   in aging process, 330–31
   blood pressure and flow in, 104–5
   diseases of, 107–10
   of fetal circulation, 319, 325, *326*
   of pulmonary and systemic circulation, *98*, 101, *102–3*
   structure and functions of, 94–95
Arterioles, 94–95, 101, 104
   of nephron, 176, *182*
Arthritis, 222, 265, 332
   rheumatoid, 146, 227
Artificial body components
   blood, 116, 118
   heart, 110
   kidney, 181, 183
Artificial insemination, **292**, 368
Artificial selection, 421
Aspirin, 109, 322
Asters of centrioles, **341**
Astigmatism, *246*
Atherosclerosis, 107, 108
Athletes, *214*
   anabolic steroids and, 225, 268
   maleness and femaleness of, 350
Atmosphere
   greenhouse effect of, 442–43
   primitive, 422–24
Atoms, **13**
   reactions between, 15, 17
   subatomic particles and isotopes of, 13–14
ATP, **26**
   breakdown and production of, 35, 38–39, 436, 440
   in cellular respiration, 42–43, 150
   in muscle contraction, 230, 231, *233*
   structure and cycle of, 26–27
ATPase, 230
Atria, **96**, 97, 98, *99*, 101, *326*
Atrial natriuretic factor, 390, 391 (table)
Atrial natriuretic hormone, 265, 269
Atrioventricular (AV) node, **98**
Atrioventricular (AV) valves, **96**, 97, 98, *99*, *100*, 101
Auditory canal, **247**, 248, *249*, 250
Auditory nerve, **250**
Australopithecines, 428–29
Autoimmune diseases, 146
Autonomic nervous system, 101, **199**–201
Autosomes, **337–38**
   abnormal, 349–52
   dominant disorders of, 364, 366
   recessive disorders of, 364–66